T0181848

Lecture Notes in Computer Science 12269

More information about this series at http://www.springer.com/series/7407

Thomas Bäck · Mike Preuss ·
André Deutz · Hao Wang ·
Carola Doerr · Michael Emmerich ·
Heike Trautmann (Eds.)

Parallel Problem Solving from Nature – PPSN XVI

16th International Conference, PPSN 2020
Leiden, The Netherlands, September 5–9, 2020
Proceedings, Part I

Editors
Thomas Bäck
Leiden University
Leiden, The Netherlands

André Deutz
Leiden University
Leiden, The Netherlands

Carola Doerr
Sorbonne University
Paris, France

Heike Trautmann
University of Münster
Münster, Germany

Mike Preuss
Leiden University
Leiden, The Netherlands

Hao Wang
Sorbonne University
Paris, France

Michael Emmerich
Leiden University
Leiden, The Netherlands

ISSN 0302-9743 ISSN 1611-3349 (electronic)
Lecture Notes in Computer Science
ISBN 978-3-030-58111-4 ISBN 978-3-030-58112-1 (eBook)
https://doi.org/10.1007/978-3-030-58112-1

LNCS Sublibrary: SL1 – Theoretical Computer Science and General Issues

This Springer imprint is published by the registered company Springer Nature Switzerland AG
The registered company address is: Gewerbestrasse 11, 6330 Cham, Switzerland

Preface

Welcome to the two volumes of the proceedings of the Conference on Parallel Problem Solving from Nature, PPSN XVI, September 5–9, 2020, Leiden, The Netherlands! When we applied to host PPSN XVI in Leiden, we were not able to imagine anything like the COVID-19 pandemic. Then the new reality hit us, and we were forced to make decisions under uncertain, dynamically changing conditions and constraints, and certainly with multiple, conflicting objectives. Scientific expertise in evolutionary computation was only partially helpful for this. At the time of writing this preface, June 2020, we believed that a hybrid conference format would be the best approach for dealing with the situation: For those who were not able to travel to Leiden, we decided to run PPSN on-site, with printed posters, workshops, tutorials, keynotes, food, and drinks. For those who could not travel to Leiden, we offered it online, with keynote live streams, poster and tutorial videos, and poster discussion rooms in which attendees could discuss with the poster presenters. The virtual part of the conference also allowed participants to meet other attendees online and start a conversation. The challenging and exciting experiment combining the on-site and online world gave attendees the best of both worlds and the flexibility needed in these difficult times – hopefully giving attendees the best of both worlds and the flexibility needed in these difficult times. Not every detail of our hybrid plan turned out as expected, but we are quite sure that some of the changes to conference organization we have tried will remain, and with the help of applied AI and the digitalization of communication, conference experiences in future will not only change but also improve.

PPSN 2020 was also quite a special event since it was the 30th anniversary of the PPSN conference! In particular for Hans-Paul Schwefel, the founder of PPSN, this is a wonderful confirmation of a successful concept – so our congratulations go to you in particular, Hans-Paul. For Thomas Bäck, who was a first-year PhD student of Hans-Paul in 1990, at PPSN I, it is an honor to be involved in this as a general co-chair, and both Mike Preuss and he share the great experience of having been supervised in their PhD studies by Hans-Paul. Although, as Thomas admits, 1990 was easier since the final conference responsibility was with Hans-Paul. We are particularly proud to have Hans-Paul and Grzegorz Rozenberg, the founder and magician of Natural Computing in Leiden, as our honorary chairs for PPSN 2020.

PPSN 2020 received a total of 268 paper submissions written by 690 authors from 44 different countries. Our Program Committee (PC) comprised 271 members from 39 countries. Together, and despite the individual challenges that the coronavirus crisis imposed on each one of us, the PC members wrote 812 review reports in total, which corresponds to an average 3 reviews per paper. Each review was read and evaluated by one of the PC chairs. Where reviewers disagreed in their assessment, a discussion among PC members was started. In some cases, authors were contacted to provide additional clarification about a technical aspect of their work. In other cases, additional reviews were solicited. The review process resulted in a total number of 99 accepted papers, which corresponds to an acceptance rate of 36.9%. All accepted papers can be found in these LNCS proceedings of PPSN. In addition to the main conference program, an attractive selection of 14 tutorials, 6 workshops, and 3 competitions was offered to participants.

The topics covered classical subjects such as Genetic and Evolutionary Algorithms, Combinatorial Optimization, Multi-objective Optimization, and Real-World Applications of Nature-Inspired Optimization Heuristics. The conference also included quite a number of papers dealing with broader aspects of Artificial Intelligence, reflecting the fact that search and optimization algorithms indeed form an important pillar of modern AI.

As always, PPSN is an interactive forum for inspiring discussions and exchanges, stimulated by on-site and online poster presentations. Three distinguished invited speakers give keynotes at the conference: Carme Torras on assistive and collaborative robotics, Eric Postma on machine learning in image recognition and cognitive modeling, and Christian Stöcker on the direction of AI in general and its effects on society. We are grateful that they accepted our invitation to present their keynotes on-site.

The list of people who made this conference possible is very long, showing the impressive collaborative effort and commitment both of the scientific community that is behind PPSN and of the organizers. This includes all authors, who recognize and acknowledge the scientific quality of this conference series by their submission, and all Program Committee members, who are volunteering although everybody in the community is overloaded with reviewing requests. Our thanks go to the tutorial speakers, workshop organizers, and attendees of the conference and its events.

We are also very grateful for the contributions of the workshop chair, Anna Esparcia-Alcázar, competition chair, Vanessa Volz, and tutorial chair, Ofer Shir. The keynote chair, Aske Plaat, and industrial liaison chair, Bernhard Sendhoff. Our financial chair, Felix Wittleben, who had a difficult time due to the dynamically changing situation. Our publicity chairs, Bas van Stein and Wenjian Luo, who made sure the community heard about PPSN 2020. Our local organization team, Jayshri Murli, Hestia Tamboer, and Roshny Kohabir, who took care of a million things and made the impossible possible. And then, for the conference days, the PhD and master students who helped manage the small but important details. Moreover, all of a sudden, we needed an online conference chair team, for which Bas van Stein, Diederick Vermetten, and Jiawen Kong volunteered to make the online part of the conference happen, and Anna Kononova also joined the team to help with many aspects of the organization. Finally, we would like to express our gratitude to the Leiden Institute of Advanced Computer Science (LIACS), Leiden University for hosting this event, to Leiden University, for its support, particularly to Springer Nature for financing the Best Paper Award, and to the Confederation of Laboratories for Artificial Intelligence

Research in Europe (CLAIRE) and Honda Research Institute Europe GmbH for their invaluable support in countless ways.

Thank you very much to all of you, for making PPSN 2020 possible! We are very proud that we have managed this, under difficult conditions, as a team effort.

July 2020

Thomas Bäck
Mike Preuss
General Chairs

André Deutz
Hao Wang
Proceedings Chairs

Carola Doerr
Michael Emmerich
Heike Trautmann
Program Chairs

Organization

PPSN 2020 was organized and hosted by the Leiden Institute of Advanced Computer Science, Leiden University, The Netherlands. Leiden University was founded in 1575 and is the oldest university of The Netherlands. Sixteen persons associated with Leiden University (either as PhD student or (guest-) researcher) became Nobel prize winners and it was the home of many illustrious individuals such as René Descartes, Rembrandt van Rijn, Christiaan Huygens, Hugo Grotius, Baruch Spinoza, and Baron d'Holbach.

General Chairs

Thomas Bäck	Leiden University, The Netherlands
Mike Preuss	Leiden University, The Netherlands

Honorary Chairs

Hans-Paul Schwefel	TU Dortmund, Germany
Grzegorz Rozenberg	Leiden University, The Netherlands

Program Committee Chairs

Carola Doerr	Sorbonne Université, France
Michael Emmerich	Leiden University, The Netherlands
Heike Trautmann	Westfälische Wilhelms-Universität Münster, Germany

Proceedings Chairs

André Deutz	Leiden University, The Netherlands
Hao Wang	Sorbonne Université, France

Keynote Chair

Aske Plaat	Leiden University, The Netherlands

Workshop Chair

Anna I. Esparcia-Alcázar	SPECIES, Europe

Tutorial Chair

Ofer M. Shir	Tel-Hai College, Israel

Competition Chair

Vanessa Volz — modl.ai, Denmark

Industrial Liaison Chair

Bernhard Sendhoff — Honda Research Institute Europe GmbH, Germany

Financial Chair

Felix Wittleben — Leiden University, The Netherlands

Online Conference Chairs

Bas van Stein — Leiden University, The Netherlands
Diederick Vermetten — Leiden University, The Netherlands
Jiawen Kong — Leiden University, The Netherlands

Publicity Chairs

Bas van Stein — Leiden University, The Netherlands
Wenjian Luo — Harbin Institute of Technology, China

Local Chair

Anna V. Kononova — Leiden University, The Netherlands

Local Organizing Committee

Jayshri Murli — Leiden University, The Netherlands
Roshny Kohabir — Leiden University, The Netherlands
Hestia Tamboer — Leiden University, The Netherlands

Steering Committee

David W. Corne — Heriot-Watt University, UK
Carlos Cotta — Universidad de Málaga, Spain
Kenneth De Jong — George Mason University, USA
Gusz E. Eiben — Vrije Universiteit Amsterdam, The Netherlands
Bogdan Filipič — Jožef Stefan Institute, Slovenia
Emma Hart — Edinburgh Napier University, UK
Juan Julián Merelo Guervós — Universida de Granada, Spain
Günter Rudolph — TU Dortmund, Germany
Thomas P. Runarsson — University of Iceland, Iceland
Robert Schaefer — University of Krakow, Poland
Marc Schoenauer — Inria, France
Xin Yao — University of Birgmingham, UK

Keynote Speakers

Carme Torras	Institut de Robòtica i Informàtica Industrial, Spain
Eric Postma	Tilburg University, The Netherlands
Christian Stöcker	Hochschule für Angewandte Wissenschaften Hamburg, Germany

Program Committee

Michael Affenzeller	Upper Austria University of Applied Sciences, Austria
Hernán Aguirre	Shinshu University, Japan
Youhei Akimoto	University of Tsukuba, Japan
Brad Alexander	The University of Adelaide, Australia
Richard Allmendinger	The University of Manchester, UK
Lucas Almeida	Universidade Federal de Goiás, Brazil
Marie Anastacio	Leiden University, The Netherlands
Denis Antipov	ITMO University, Russia
Dirk Arnold	Dalhousie University, Canada
Dennis Assenmacher	Westfälische Wilhelms-Universität Münster, Germany
Anne Auger	Inria, France
Dogan Aydin	Dumlupinar University, Turkey
Jaume Bacardit	Newcastle University, UK
Samineh Bagheri	TH Köln, Germany
Helio Barbosa	Laboratório Nacional de Computação Científica, Brazil
Thomas Bartz-Beielstein	TH Köln, Germany
Andreas Beham	University of Applied Sciences Upper Austria, Austria
Heder Bernardino	Universidade Federal de Juiz de Fora, Brazil
Hans-Georg Beyer	Vorarlberg University of Applied Sciences, Austria
Mauro Birattari	Université Libre de Bruxelles, Belgium
Aymeric Blot	University College London, UK
Christian Blum	Spanish National Research Council, Spain
Markus Borschbach	FHDW Bergisch Gladbach, Germany
Peter Bosman	Centrum Wiskunde & Informatica, The Netherlands
Jakob Bossek	The University of Adelaide, Australia
Jürgen Branke	University of Warwick, UK
Dimo Brockhoff	Inria, France
Will Browne	Victoria University of Wellington, New Zealand
Alexander Brownlee	University of Stirling, UK
Larry Bull	University of the West of England, UK
Maxim Buzdalov	ITMO University, Russia
Arina Buzdalova	ITMO University, Russia
Stefano Cagnoni	University of Parma, Italy
Fabio Caraffini	De Montfort University, UK
Matthias Carnein	Westfälische Wilhelms-Universität Münster, Germany
Mauro Castelli	Universidade NOVA de Lisboa, Portugal
Josu Ceberio	University of the Basque Country, Spain

Ying-Ping Chen	National Chiao Tung University, Taiwan
Francisco Chicano	Universidad de Málaga, Spain
Miroslav Chlebik	University of Sussex, UK
Sung-Bae Cho	Yonsei University, South Korea
Tinkle Chugh	University of Exeter, UK
Carlos Coello Coello	CINVESTAV-IPN, Mexico
Dogan Corus	The University of Sheffield, UK
Ernesto Costa	University of Coimbra, Portugal
Carlos Cotta	Universidad de Málaga, Spain
Agostinho Da Rosa	ISR-IST, Portugal
Nguyen Dang	St Andrews University, UK
Kenneth A. De Jong	George Mason University, USA
Kalyanmoy Deb	Michigan State University, USA
Antonio Della-Cioppa	University of Salerno, Italy
Bilel Derbel	University of Lille, France
André Deutz	Leiden University, The Netherlands
Benjamin Doerr	École Polytechnique, France
Carola Doerr	Sorbonne Université, France
John Drake	University of Leicester, UK
Rafal Drezewski	AGH University of Science and Technology, Poland
Paul Dufossé	Inria, France
Tome Eftimov	Jožef Stefan Institute, Slovenia
Gusz E. Eiben	Vrije Universiteit Amsterdam, The Netherlands
Mohamed El Yafrani	Aalborg University, Denmark
Talbi El-Ghazali	University of Lille, France
Michael Emmerich	Leiden University, The Netherlands
Anton Eremeev	Sobolev Institute of Mathematics, Russia
Richard Everson	University of Exeter, UK
Pedro Ferreira	Universidade de Lisboa, Portugal
Jonathan Fieldsend	University of Exeter, UK
Bogdan Filipič	Jožef Stefan Institute, Slovenia
Steffen Finck	Vorarlberg University of Applied Sciences, Austria
Andreas Fischbach	TH Köln, Germany
Peter Fleming	The University of Sheffield, UK
Carlos M. Fonseca	University of Coimbra, Portugal
Marcus Gallagher	The University of Queensland, Australia
José García-Nieto	Universidad de Málaga, Spain
António Gaspar-Cunha	University of Minho, Portugal
Mario Giacobini	University of Torino, Italy
Kyriakos Giannakoglou	National Technical University of Athens, Greece
Tobias Glasmachers	Ruhr-Universität Bochum, Germany
Christian Grimme	Westfälische Wilhelms-Universität Münster, Germany
Roderich Gross	The University of Sheffield, UK
Andreia Guerreiro	University of Coimbra, Portugal
Alexander Hagg	Bonn-Rhein-Sieg University of Applied Sciences, Germany

Jussi Hakanen	University of Jyväskylä, Finland
Julia Handl	The University of Manchester, UK
Jin-Kao Hao	University of Angers, France
Emma Hart	Napier University, UK
Verena Heidrich-Meisner	University of Kiel, Germany
Carlos Henggeler Antunes	University of Coimbra, Portugal
Martin Holena	Academy of Sciences of the Czech Republic, Czech Republic
Christian Igel	University of Copenhagen, Denmark
Dani Irawan	TH Köln, Germany
Hisao Ishibuchi	Osaka Prefecture University, Japan
Christian Jacob	University of Calgary, Canada
Domagoj Jakobovic	University of Zagreb, Croatia
Thomas Jansen	Aberystwyth University, UK
Dreo Johann	THALES Research & Technology, France
Laetitia Jourdan	Inria, LIFL CNRS, France
Bryant Julstrom	St. Cloud State University, USA
George Karakostas	McMaster University, Canada
Edward Keedwell	University of Exeter, UK
Pascal Kerschke	Westfälische Wilhelms-Universität Münster, Germany
Marie-Eleonore Kessaci	University of Lille, France
Ahmed Kheiri	Lancaster University, UK
Wolfgang Konen	TH Köln, Germany
Anna Kononova	Leiden University, The Netherlands
Peter Korošec	Jožef Stefan Institute, Slovenia
Lars Kotthoff	University of Wyoming, USA
Oliver Kramer	Universität Oldenburg, Germany
Oswin Krause	University of Copenhagen, Denmark
Krzysztof Krawiec	Poznan University of Technology, Poland
Martin S. Krejca	Hasso-Plattner-Institut, Germany
Timo Kötzing	Hasso-Plattner-Institut, Germany
William La Cava	University of Pennsylvania, USA
Jörg Lässig	University of Applied Sciences Zittau/Görlitz, Germany
William B. Langdon	University College London, UK
Algirdas Lančinskas	Vilnius University, Lithuania
Frederic Lardeux	LERIA, University of Angers, France
Per Kristian Lehre	University of Birmingham, UK
Johannes Lengler	ETH Zurich, Switzerland
Ke Li	University of Exeter, UK
Arnaud Liefooghe	University of Lille, France
Marius Lindauer	Leibniz Universität Hannover, Germany
Giosuè Lo Bosco	Università di Palermo, Italy
Fernando Lobo	University of Algarve, Portugal
Daniele Loiacono	Politecnico di Milano, Italy
Nuno Lourenço	University of Coimbra, Portugal

Jose A. Lozano	University of the Basque Country, Spain
Rodica Ioana Lung	Babeş-Bolyai University, Romania
Chuan Luo	Peking University, China
Gabriel Luque	Universidad de Málaga, Spain
Evelyne Lutton	INRAE, France
Manuel López-Ibáñez	The University of Manchester, UK
Penousal Machado	University of Coimbra, Portugal
Luigi Malagò	Romanian Institute of Science and Technology, Romania
Katherine Malan	University of South Africa, South Africa
Vittorio Maniezzo	University Bologna, Italy
Elena Marchiori	Radboud University, The Netherlands
Luis Marti	Inria, Chile
Asep Maulana	Tilburg University, The Netherlands
Giancarlo Mauri	University of Milano-Bicocca, Italy
Jacek Mańdziuk	Warsaw University of Technology, Poland
James McDermott	National University of Ireland, Ireland
Jörn Mehnen	University of Strathclyde, UK
Alexander Melkozerov	Tomsk State University of Control Systems and Radioelectronics, Russia
Juan J. Merelo	University of Granada, Spain
Marjan Mernik	University of Maribor, Slovenia
Silja Meyer-Nieberg	Bundeswehr Universität München, Germany
Efrén Mezura-Montes	University of Veracruz, Mexico
Krzysztof Michalak	Wroclaw University of Economics, Poland
Kaisa Miettinen	University of Jyväskylä, Finland
Julian Miller	University of York, UK
Edmondo Minisci	University of Strathclyde, UK
Gara Miranda	University of La Laguna, Spain
Mustafa Misir	Istinye University, Turkey
Marco A. Montes De Oca	clypd, Inc., USA
Sanaz Mostaghim	Otto von Guericke Universität Magdeburg, Germany
Mario Andrès Muñoz Acosta	The University of Melbourne, Australia
Boris Naujoks	TH Köln, Germany
Antonio J. Nebro	Universidad de Málaga, Spain
Ferrante Neri	University of Nottingham, UK
Aneta Neumann	The University of Adelaide, Australia
Frank Neumann	The University of Adelaide, Australia
Phan Trung Hai Nguyen	University of Birmingham, UK
Miguel Nicolau	University College Dublin, Ireland
Ellen Norgård-Hansen	NORCE, Norway
Michael O'Neill	University College Dublin, Ireland
Gabriela Ochoa	University of Stirling, UK
Pietro S. Oliveto	The University of Sheffield, UK
Unamay Oreilly	MIT, USA

Shinichi Shirakawa	Yokohama National University, Japan
Moshe Sipper	Ben-Gurion University of the Negev, Israel
Jim Smith	University of the West of England, UK
Christine Solnon	CITI Inria and INSA Lyon, France
Patrick Spettel	Vorarlberg University of Applied Sciences, Germany
Giovanni Squillero	Politecnico di Torino, Italy
Sebastian Urban Stich	École Polytechnique Fédérale de Lausanne, Switzerland
Catalin Stoean	University of Craiova, Romania
Jörg Stork	TH Köln, Germany
Thomas Stützle	Université Libre de Bruxelles, Belgium
Mihai Suciu	Babeş-Bolyai University, Romania
Dirk Sudholt	The University of Sheffield, UK
Andrew Sutton	University of Minnesota, USA
Jerry Swan	University of York, UK
Ricardo H. C. Takahashi	Universidade Federal de Minas Gerais, Brazil
Daniel Tauritz	Auburn University, USA
Olivier Teytaud	Inria, France
Dirk Thierens	Utrecht University, The Netherlands
Sarah Thomson	University of Stirling, UK
Kevin Tierney	Universität Bielefeld, Germany
Renato Tinós	University of São Paulo, Brazil
Julian Togelius	New York University, USA
Marco Tomassini	University of Lausanne, Switzerland
Alberto Tonda	INRA, France
Cheikh Touré	Inria, France
Heike Trautmann	Westfälische Wilhelms-Universität Münster, Germany
Leonardo Trujillo	Instituto Tecnológico de Tijuana, Mexico
Tea Tušar	Jožef Stefan Institute, Slovenia
Ryan J. Urbanowicz	University of Pennsylvania, USA
Koen van der Blom	Leiden University, The Netherlands
Bas van Stein	Leiden University, The Netherlands
Leonardo Vanneschi	Universida de NOVA de Lisboa, Portugal
Sébastien Verel	Université du Littoral Côte d'Opale, France
Diederick Vermetten	Leiden University, The Netherlands
Marco Virgolin	Centrum Wiskunde & Informatica, The Netherlands
Vanessa Volz	modl.ai, Denmark
Markus Wagner	The University of Adelaide, Australia
Stefan Wagner	University of Applied Sciences Upper Austria, Austria
David Walker	University of Plymouth, UK
Hao Wang	Sorbonne Université, France
Hui Wang	Leiden University, The Netherlands
Yali Wang	Leiden University, The Netherlands
Elizabeth Wanner	CEFET, Brazil
Thomas Weise	University of Science and Technology of China, China
Dennis Wilson	ISAE-Supaero, France

Carsten Witt	Technical University of Denmark, Denmark
Man Leung Wong	Lingnan University, China
John Woodward	Queen Mary University of London, UK
Ning Xiong	Mälardalen University, Sweden
Bing Xue	Victoria University of Wellington, New Zealand
Kaifeng Yang	University of Applied Sciences Upper Austria, Austria
Shengxiang Yang	De Montfort University, UK
Furong Ye	Leiden University, The Netherlands
Martin Zaefferer	TH Köln, Germany
Ales Zamuda	University of Maribor, Slovenia
Christine Zarges	Aberystwyth University, UK
Mengjie Zhang	Victoria University of Wellington, New Zealand

Contents – Part I

Automated Algorithm Selection and Configuration

Bayesian- and Surrogate-Assisted Optimization

Combinatorial Optimization

Connection Between Nature-Inspired Optimization and Artificial Intelligence

Contents – Part II

Reinforcement Learning

Theoretical Aspects of Nature-Inspired Optimization

Automated Algorithm Selection and Configuration

Evolving Deep Forest with Automatic Feature Extraction for Image Classification Using Genetic Programming

Ying Bi(✉), Bing Xue, and Mengjie Zhang

School of Engineering and Computer Science, Victoria University of Wellington, Wellington 6140, New Zealand
{Ying.Bi,Bing.Xue,Mengjie.Zhang}@ecs.vuw.ac.nz

Abstract. Deep forest is an alternative to deep neural networks to use multiple layers of random forests without back-propagation for solving various problems. In this study, we propose a genetic programming-based approach to automatically and simultaneously evolving effective structures of deep forest connections and extracting informative features for image classification. First, in the new approach we define two types of modules: forest modules and feature extraction modules. Second, an encoding strategy is developed to integrate forest modules and feature extraction modules into a tree and the search strategy is introduced to search for the best solution. With these designs, the proposed approach can automatically extract image features and find forests with effective structures simultaneously for image classification. The parameters in the forest can be dynamically determined during the learning process of the new approach. The results show that the new approach can achieve better performance on the datasets having a small number of training instances and competitive performance on the datasets having a large number of training instances. The analysis of evolved solutions shows that the proposed approach uses a smaller number of random forests over the deep forest method.

Keywords: Evolutionary deep learning · Genetic programming · Deep forest · Image classification · Feature extraction

1 Introduction

In recent years, deep learning algorithms have achieved a big success in many applications [1]. Image classification is one of the important application areas of deep learning. Many famous deep neural networks (DNNs) have been developed, such as AlexNet, GoogleNet, VGGNet, ResNet, and DenseNet [2–4]. These methods have achieved impressive performance on many large image classification datasets. However, these methods have a number of important limitations. First,

T. Bäck et al. (Eds.): PPSN 2020, LNCS 12269, pp. 3–18, 2020.
https://doi.org/10.1007/978-3-030-58112-1_1

rich domain expertise is needed to design a powerful DNN model/architecture. For example, AlexNet and VGGNet are both convolutional neural networks (CNNs) but have different structures and classification performance. Second, DNNs typically require a large amount of training data to train a model, which often has a huge number of parameters. For example, AlexNet has over 60 million parameters and VGGNet has 138 million parameters, which can only be trained using thousands of training instances. With such parameters, these models cannot be directly applied for solving the tasks with limited training data. In many real-world applications, such as medical applications, the data are difficult or expensive to collect. Learning from limited training data, i.e., also know as few-shot or zero-shot learning [5], for solving a task is important, but the current common DNNs cannot directly solve. Furthermore, interpretability is another problem with them. It is difficult to understand/interpret these models (or the learned features) why they are effective for solving a task. However, in many industrial applications, interpretability is a critical factor for the acceptance and adoption of the model/method.

Motivated by these limitations, many other types of deep models have been developed, such as deep forest (also known as gcForest) [6]. Deep forest builds a deep model based on random forest using a cascade structure and a multi-grained scanning. The cascade structure allows it to have multiple layers of random forests to learn a complex representation of the data for solving a problem. The cascade structure is shown in Fig. 1. The multi-grained scanning scheme uses a sliding window to scan the raw features from sequence data or image data to form a feature vector as the inputs of random forests. Deep forest has been employed to solve many tasks and achieved promising results [6]. However, the deep forest uses fixed structures and parameters of random forests at each layer and the features used are simple without complex transformation. This may limit the performance of deep forest for image classification, especially particular images, e.g., texture images. Although many variants of deep forests have been developed, none of them addressed these two issues.

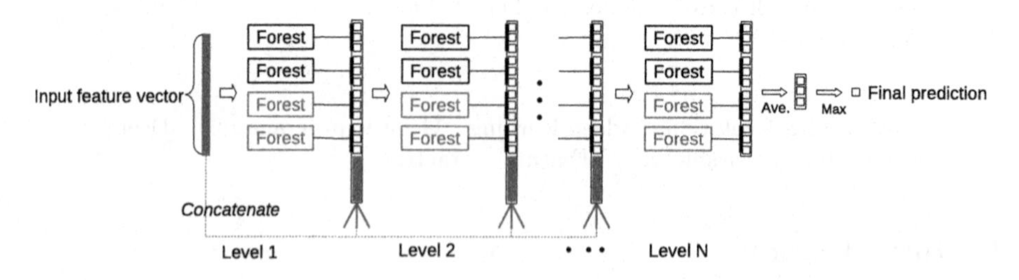

Fig. 1. The cascade forest structure [6].

In this paper, an evolutionary deep forest (EvoDF) approach is developed to automatically extract domain-specific features and find appropriate connections of random forests with effective parameters simultaneously for image classification. Instead of manually designing the structure of forests and choosing its key parameters, EvoDF can automatically find effective structures and parameters of the random forests in deep forest. Feature extraction modules are proposed to perform region selection, descriptor selection and feature description. They allow EvoDF to find effective features, such as local binary patterns (LBP), to solve typical image classification tasks, such as texture classification. Different from deep forest, EvoDF encodes solutions as trees with variable lengths and uses a population to search for the best solution with fewer parameters. In the experiments, EvoDF achieves better results on the datasets of a small number of training instances and competitive results on the datasets of a large number of training instances than deep forest. Further analysis shows that the solution found by EvoDF has a small number of random forests than those in deep forest.

2 Related Work

Evolutionary Algorithms: Evolutionary algorithms can search for the best solutions via a number of generations based on genetic beam search. Evolutionary algorithms are known for their powerful global search ability and non-differentiable requirement [7]. They have been applied to many tasks in machine learning, such as optimizing the architectures of DNNs [8–10] and finding the optimal model of ensemble [11–13]. However, no evolutionary algorithms have been developed to optimize the structure of deep forest. In this study, we use an evolutionary algorithm, i.e., genetic programming (GP) [14], to achieve automatically search for the structure of forests, extract features and select suitable parameters of the forests for image classification.

Feature Extraction: Feature extraction is essential for solving image classification tasks. Traditional methods to feature extraction are local binary patterns (LBP) [15], histogram of oriented gradients (HOG) [16], and scale-invariant feature transform (SIFT) [17]. In recent decades, automatically extracting/learning features from images becomes increasingly popular. Many DNNs can achieve this by learning deep features. More related work can be found in [18]. In deep forest, the features are from images via multi-grained scanning, which may not be effective for classifying particular images, such as texture images. Therefore, we add a feature extraction process into deep forest to achieve automatic feature extraction for image classification.

Ensemble Methods: Deep forest is a typical ensemble method, using multiple decision trees to perform prediction. Ensemble methods train multiple base learners using traditional machine learning algorithms for solving a task [19]. Typically, an ensemble of diverse and accurate classifiers can obtain a better generalization performance than a single classifier [20]. But the diversity is an open issue in ensemble learning [21]. Various ensembles have been developed such as deep super learner [22] and autostacker [11], which make the ensemble model deeper. Recent applications of ensemble methods in machine learning and deep learning can be found in [12,20,23].

Deep Forest Variants: Recently, several variants of deep forests have been developed. Zhou *et al.*[24] extended the deep forest method to generate compact binary code for hashing in image retrieval. Zhu *et al.* [25] proposed an efficient training framework of deep forest on distributed task-parallel platforms. Zhang *et al.* [26] implemented distributed deep forest and tested it on an extra-large dataset of cash-out fraud detection. As a recently proposed method, deep forest is still at early stage and has large research potential. In this study, we aim to improve the performance of deep forest for image classification.

3 The Proposed Approach

In this section, we will describe the proposed approach: evolutionary deep forest (EvoDF). First, we will introduce two main modules, i.e., the forest module and the feature extraction module. Then the encoding and search strategy are presented, followed by the description of the overall algorithm.

3.1 Forest Modules

In deep forest, each layer/level has a predefined number of random forests. To allow EvoDF to search the structures of random forests, we relax this predefined structure of random forests by designing a single module and extending it to deep or wide modules. In EvoDF, the single module is constructed using one random forest, as shown in Fig. 2. The output of the single module is the concatenation of the input and the predictions of the random forest. The predictions of a random forest are the average scores of multiple trees in each class for a classification problem. For example, if the problem has c classes and the dimension of the input feature vector is f (the number of features in the dataset), the dimension of the output is $c + f$.

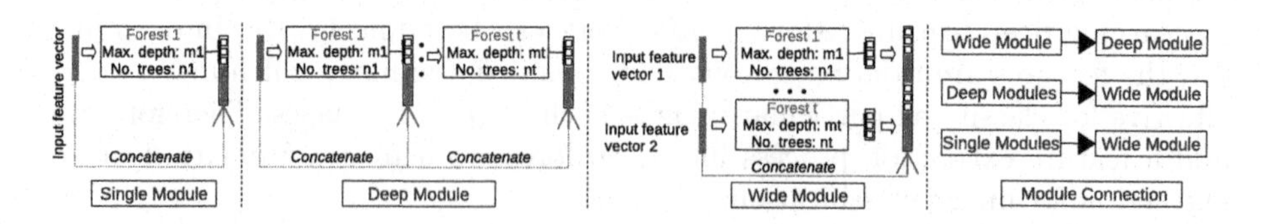

Fig. 2. Forest modules.

The single module can be extended to a deep or wide module, as shown in Fig. 2. The deep module is constructed using multiple single modules sequentially. The output of a single module is the input of the next single module. The final output dimension of a deep module is $f + t \times c$, where t indicates the number of single modules. The wide module is constructed using single modules parallelly. Every single module has various input feature vectors, which can be from different feature extraction modules introduced in the following subsections. The final output dimension of a wide module is $\sum_{i=1}^{t}(f_i + c)$, where f_i represents the input dimension of the ith single module. The deep and wide modules can connect with each other, e.g., using the output feature vector of one module as the input feature vector of another module. This leads to an easy extension of a complex model, which can be constructed by EvoDF.

It is known that the diversity of base learners are important for constructing an ensemble with good generalization ability [19]. The diversity can be enhanced by data sample manipulation, input feature manipulation, learning parameter manipulation, and output representation manipulation [6,19]. In these forest modules, random forest and extremely randomized trees are employed. These two types of random forests can be automatically selected to construct a module. The key parameters, the number of trees and the maximum number of tree depth of each forest, can be automatically determined during the learning process of EvoDF, which will be introduced in the later sections. It is also noted that the input feature vectors for the forest in the wide module are different, which further enhances the diversity.

3.2 Feature Extraction Modules

In deep forest, the inputs of random forests are raw data/images or the concatenation of the predictions and raw data. This may not be effective for solving particular image classification tasks, e.g., texture classification. In the EvoDF method, three new feature extraction modules are developed to extract informative features from images. The three possible feature extraction modules are illustrated in Fig. 3 and Fig. 4.

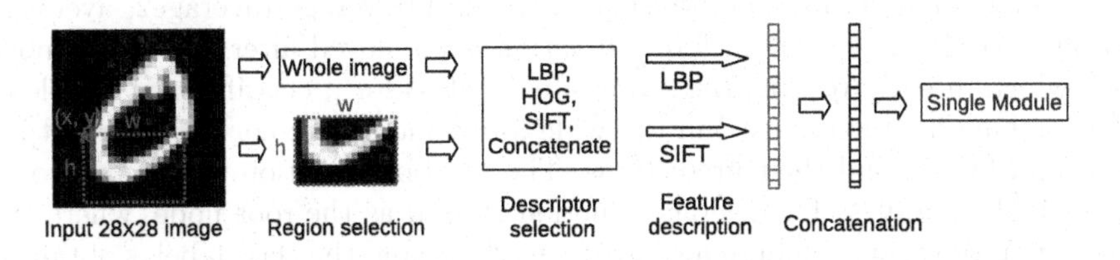

Fig. 3. One possible feature extraction module.

The **first feature extraction module** (module 1 in Fig. 4) consists of descriptor selection and feature description. Four commonly used descriptors, i.e., LBP, HOG, SIFT, and Concatenation [27], are employed to transform an image into features. Descriptor selection is to select one of the four descriptors and feature description is to extract features using the selected descriptor. The output of the feature extraction module is a feature vector. The **second feature extraction module** (module 2 in Fig. 4) has an additional process: region selection, which aims to select a small key region from a large input image. From the selected region, image features can be extracted. To select a region, the position of the top-left point of the region (x, y) and the size of the region (w, h)

are needed. These parameters can be automatically selected during the learning process of EvoDF. The **third feature extraction module** (module 3 in Fig. 4) concatenates the features produced by modules 1 or 2 into a feature vector. It can produce a combination of various features.

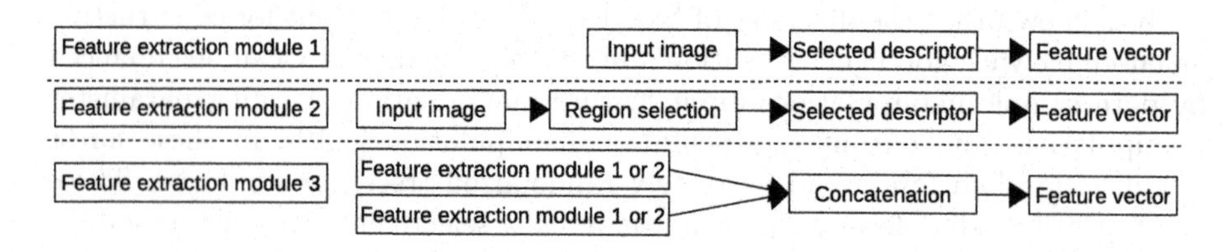

Fig. 4. Feature extraction modules.

3.3 Encoding and Search Strategy

The forest modules and feature extraction modules are connected and represented using a tree structure. A potential advantage is that a tree can be easily extended to be deeper or broader. The encoding of a solution with forest modules and feature extraction modules is based on strongly typed genetic programming [28]. An example solution and its corresponding encoding are shown in Fig. 5.

In this encoding, the internal nodes are functions/models, including the functions in the feature extraction modules and forest modules. These functions are region selection functions, image descriptors (LBP, HOG, SIFT, and Concatenation), single module, random forest, combine, and average (average2, average3 and average 4). The random forest function is employed after the forest modules, which directly returns the score of each class for a classification problem. The combine function is used in the wide forest module to combine the output of several forests and their predictions. The combine function combines two or three single modules. The average function is used as the root node, where the predictions of various random forests are averaged and the class label is obtained according to the maximum score. In EvoDF, the average function can connect with two, three or four random forests. It is noted that the random forest can be replaced by extremely randomized trees during the learning process of EvoDF.

The leaf nodes are parameters and the input image. The parameters are the coordination (x, y) of the top-left point of the selected region, the size (width w and height h) of the selected region, the number of trees (m) in forests, and the maximum tree depth (n). In the EvoDF method, x is in the range of $[0,\ W-10]$, y is in the range of $[0,\ H-10]$, w and h are in the range of $[10,\ 20]$, where W indicates the width of the image and H indicates the height of the image. With these settings, region selection functions can select a region with a size from 10×10 to 20×20. The range of m is $[50,\ 1000]$ with a step of 50. The range of n is $[10,\ 100]$ with a step of 10.

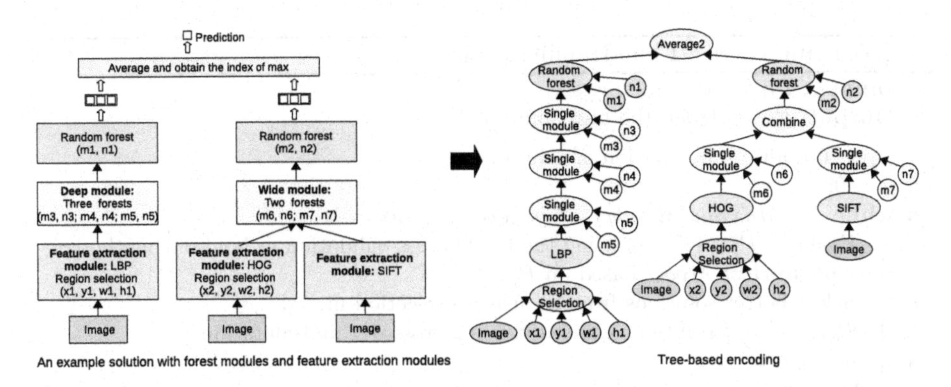

Fig. 5. An example tree/structure with forest modules and feature extraction modules and its encoding in the proposed EvoDF approach.

In EvoDF, a number (population) of trees/solutions are randomly generated by selecting functions to form the internal nodes and selecting parameters from the predefined ranges to form the leaf nodes. Each tree is evaluated on a training set to obtain its classification performance using the k-fold cross-validation or the hold-out method. Then better trees are selected and the mutation and crossover operations are employed to generate new trees from the selected trees. The crossover operation swaps two branches of two trees based on a selected node to generate two new trees. The mutation operation replaces the branch of a tree based on a selected node with a new generated branch. These two operations can change the depths of trees, e.g., from wide to deep. The mutation operation can introduce new branches into the current population. In addition, the functions and parameters of a tree can also be changed by these two operations, leading to a search towards the best tree/solution.

3.4 Algorithm Procedure

Training Process: The training/learning procedure of EvoDF is described in Algorithm 1. In this procedure, a population of solutions are generated and evaluated on the training set $\mathcal{D}_{train}$. In the evaluation procedure, the classification performance of the tree/solution is calculated using the k-fold cross-validation or hold-out methods. Then the tree with the best performance is recorded. A selection method is used to select better trees for crossover or mutation. The better trees may have better forest modules or feature extraction modules, which are inherited to the next generation. Then a new population of trees is generated. The overall process is repeated until the maximum number of generation (G) is reached. Finally, the best solution is returned and tested on a test set.

The parameters for EvoDF are the population size, the maximum number of generations, the mutation rate, the crossover rate, and the selection method. It should be noted that the number of parameters for EvoDF is smaller than that of deep forest or DNNs, as listed in [6]. Furthermore, the parameter settings for EvoDF can follow the commonly used settings of GP [12,27].

Algorithm 1: Algorithm Training Procedure

Input : $\mathcal{D}_{train}$: the training set.
Output: *Best_Model*: the best solution.

```
1  P_0 ← Initialise a population of solutions;
2  g ← 0;
3  while g < Maximal number of generations do
4  |   Evaluate P_g on D_train using the k-fold cross-validation or hold-out methods;
5  |   Update Best_Model based on P_g;
6  |   Select better solutions from P_g using a selection metod;
7  |   P_{g+1} ← New solutions generated using crossover and mutation;
8  |   g ← g + 1;
9  end
10 Return Best_Model.
```

Test Process: In the test process, the random forest in the best solution is trained using $\mathcal{D}_{train}$ and the solution is tested on the test set. The accuracy of the test set is reported. It should be noted that the test set has never been used in the training or learning process.

4 Experiments and Results

4.1 Configuration

In this section, we compare the proposed EvoDF approach with deep forest (gcForest) and several other algorithms. The settings of most comparison algorithms refer to [6]. In EvoDF, the population size is 100 and the maximum number of generation is 20. Note that we use a smaller number of generation due to the high computational cost. The crossover rate is 0.5 and the mutation rate is 0.5. A larger mutation rate than the commonly used one is expected to increase the diversity of the population. We use the same parameter settings for EvoDF on all the datasets for generality. For the datasets having a small number of training instances, 3-fold cross-validation on the training set is used in the learning process of EvoDF (line 4 of Algorithm 1) to improve the generalisation performance. For the datasets having a large number of training instances, the hold-out method is used in the learning process of EvoDF. The training set is split into two subsets, one for training random forests and one for calculating the accuracy of the random forests in the fitness evaluation process. It is noted that evolutionary algorithms often run 30 or 50 times and report the average results. However, many existing image classification methods often report the results of one run due to the high computational cost. Therefore, the experiments of EvoDF run five independent times and the averaged results are reported.

Six different datasets are employed in the experiments. These datasets are ORL [29], Extend Yale B [30], SCENE [31], KTH [32], MNIST [33], and CIFAR-10 [34]. These datasets represent a variety of image classification tasks, i.e., object classification, face recognition, scene classification, and texture classification, aiming to demonstrate the effectiveness of EvoDF on a wide range of image classification tasks with the same parameter settings.

4.2 Classification Results

Face Recognition: ORL [29] is a face recognition dataset, having 400 images in 40 classes. Each class has 10 images. Following the settings in [6], 5/7/9 images are used for training and the remaining images are used for testing, respectively. Table 1 compares the test accuracy of EvoDF and five baseline algorithms. The results of the baseline methods are from [6]. From Table 1, we can find that the proposed EvoDF method achieves better results than any of the baseline methods, including gcForest, using various numbers of training instances. The performance of EvoDF does not degrade significantly when the number of training instances is decreased. Compared with other baseline methods, EvoDF is less affected by reducing the number of training instances.

Table 1. Comparison of test accuracy (%) on **ORL**

	5 images	7 images	9 images
EvoDF	**97.00 (96.30 ± 0.57)**	**97.50 (97.17 ± 0.46)**	**100 (99.0 ± 1.37)**
gcForest	91.00	96.67	97.50
Random Forest	91.00	93.33	95.00
CNN (five layers)	86.50	91.67	95.00
SVM (rbf kernel)	80.50	82.50	85.00
kNN (k=3)	76.00	83.33	92.50

Face Recognition: Extend Yale B [30] contains 2,414 facial images sampled from 38 different people under various illumination conditions. In the experiments, we randomly select 10/20/30 images for training and the remaining

Table 2. Comparison of test accuracy (%) on **Extend Yale B**

	10 images	20 images	30 images
EvoDF	81.36 (80.66 ± 0.86)	**95.37 (94.09 ± 1.33)**	**98.29 (97.93 ± 0.47)**
gcForest (8 forests)	72.21	91.71	96.88
gcForest (4 forests)	72.06	91.05	96.11
Random Forest	75.44	91.47	96.65
CNN (20, 100)	69.96	86.96	94.94
CNN (10, 50)	71.52	92.19	94.31
SVM (linear kernel)	**82.19**	92.55	95.25
kNN (k=1)	39.92	54.57	62.46

images for testing, respectively. Two gcForests, one with four forests and one with eight forests at each level, are used for comparisons. The random forest and the architecture of CNN are the same as that in [6]. The batch size of CNN is set to 10, 20 and the number of epochs is set to 50, 100, respectively. The SVM with a linear kernel achieves better performance than with an RBF kernel. The number of neighbours in k-NN is tuned and the best results obtained when k=1.

The results on the Extend Yale B dataset are listed in Table 2. The EvoDF approach achieves better results than any of the baseline methods using 20 or 30 training images. In the first case, EvoDF achieves worse results than SVM and better results than the remaining algorithms. The results show that EvoDF achieves better results than gcForest with four or eight forests at each level.

Scene and Texture Classification: The **SCENE** [31] dataset has 3,859 natural scene images of 13 classes. The images are sampled under different conditions and have high image variations. The **KTH** [32] dataset is a texture classification task of 10 classes. The total number of images is 810. The images are sampled in nine scales with three poses under four illumination conditions. In the experiments, 100 images per class of the SCENE dataset are used for training and the remaining images are used for testing. For KTH, 40 images per class are used for training and the remaining images are used for testing. kNN uses k=1 on SCENE and k=5 on KTH after tuning. CNN uses 20 batch size and 100 epochs.

The results obtained by EvoDF and the baseline methods on these two datasets are listed in Table 3. The EvoDF method achieves better results than any of the compared methods on these two datasets. Specifically, EvoDF increases the accuracy by 14.42% on SCENE and by 11.95% on KTH. It is also noticeable that EvoDF achieves better results than gcForest. The EvoDF approach can extract LBP and SIFT features, which are typically for texture, shape and appearance description. This may increase the classification performance of EvoDF on the texture and scene datasets.

Table 3. Comparison of test accuracy (%) on **SCENE** and **KTH**

Dataset	SCENE (100 images)	KTH (40 images)
EvoDF	**67.96 (63.07 ± 3.85)**	**87.80 (85.17 ± 2.79)**
CNN	53.54	75.85
gcForest (8 forests)	39.62	64.63
Random Forest	36.19	60.98
SVM (linear kernel)	19.42	41.46
kNN	22.78	36.34

MNIST [33] is the task of handwritten digit classification. It has 60,000 training images and 10,000 testing images. In the learning process of EvoDF, 30,000 images of the training set are used to train forests and the remaining 30,000 images are used to evaluate. In the testing process, the full training set is used to train the forests and the model is tested on the test set. The results of the baseline methods are from the corresponding references on the same test sets. The results are listed in Table 4. It can be found that the accuracy obtained by EvoDF is very close to that by gcForest, i.e., 0.34% lower. The EvoDF approach achieves better results than CascadeForest, deep belief net, deep forest-based hashing, SVM, deep super learner, and random forest.

Table 4. Comparison of test accuracy (%) on **MNIST**

gcForest	99.26 [6]
LeNet-5	99.05 [6]
EvoDF	98.92 (98.83±0.10)
Deep Belief Net	98.75 [35]
SVM (rbf kernel)	98.60 [6]
Deep forest-based hashing	98.50 [24]
Deep super learner	98.42 [22]
CascadeForest	98.02 [6]
Random Forest	96.80 [6]

CIFAR-10 [34] is a complex object classification dataset. CIFAR-10 has 50,000 32×32 color training images and 10,000 colour testing images. In the experiments, we use gray-scale images to reduce computational cost. In the training process of EvoDF, 30,000 images of the training set are used for training and 20,000 images are used for evaluation. Table 5 lists the test accuracy of all the methods. It can be found that EvoDF achieves better results than the random forest, deep forest-based hashing, multilayer perceptron, logistic regression, and SVM. The performance of EvoDF is inferior to state-of-the-art DNNs and a little behind the gcForest (default). The performance of gcForest is further improved by increasing grains and using gradient boosting decision tree (GBDT) [36] as gcForest (gbdt) achieves better results than gcForest (default). The performance of EvoDF can be further improved by using GBDT, or using a large number of population size and generations, or using 3-fold cross-valuation in training. However, due to the limitation of computational resources, we have not tested the performance of EvoDF under that configuration. In addition, it is noteworthy that EvoDF can achieve better results than gcForest or other methods on the image classification datasets using a small number of training instances.

Table 5. Comparison of test accuracy (%) on **CIFAR-10**

ResNet	93.57 [3]
AlexNet	83.00 [2]
gcForest (gbdt)	69.00 [6]
gcForest (5grains)	63.37 [6]
Deep Belief Net	62.20 [37]
gcForest (default)	61.78 [6]
EvoDF	61.27 (60.84 ± 0.33)
Deep forest-based hashing	55.90 [24]
Random Forest	50.17 [6]
Multilayer Perceptron	42.20 [38]
Logistic Regression	37.32 [6]
SVM (linear kernel)	16.32 [6]

5 Further Analysis and Discussions

5.1 Discussions of the Classification Results

From Tables 1, 2, 3, 4 and 5, it can be found that the EvoDF achieves better results than any of the baseline methods on the datasets having a small number of training instances and comparable results on the datasets having a large number of training instances. The results indicate that EvoDF is effective for various types of image classification tasks, especially when the training data is limited. Compared with the original deep forest, the proposed EvoDF approach can automatically search for the structures of deep forests, find appropriate parameters and select effective features as the inputs of the forests. With these designs, the performance of deep forest on image classification has been improved. Although the performance of EvoDF on the large dataset, i.e., CIFAR10, is a little inferior, the comparisons show the potential of EvoDF by automatically searching the structures of forest and extracting features from images. To sum up, EvoDF is suitable and effective for solving tasks with a small number of training instances, such as in the medical, security, or biological domains.

It is also noted that the classification performance of gcForest and EvoDF is worse than deep CNNs, i.e., ResNet and AlexNet, on the large dataset (CIFAR10). These deep CNNs are well-developed algorithms and the area of CNNs for image classification has been developed for over ten years. These methods require a large number of computing resources (GPU) to obtain the current performance on the large datasets. In contrast, deep forest is a newly developed algorithm and its potential has not been comprehensively investigated. The current implementation of deep forest is based on CPU rather than GPU, which may limit its performance. The area of deep forest still has a large research space and needs further investigation in the future.

5.2 Analysis of the Evolved Solution

An example solution found by EvoDF is visualised to show what features are extracted and why it achieves good performance. The best solution is evolved on MNIST, as shown in Fig. 6. It achieves 98.92% test accuracy. It has three branches and the final prediction decision is made from three random forests. Each branch uses a particular feature extraction module to generate features from the input image. The features include SIFT features, LBP features, HOG features, and raw pixels. It is noted that each random forest uses different inputs and has different parameters, which increases the diversity of the classifiers in the ensemble. The solution shows that the connections between the forest modules and the feature extraction modules are very flexible. For example, the output of a single forest module and the output of feature extraction modules can be concatenated to form the input of another random forest.

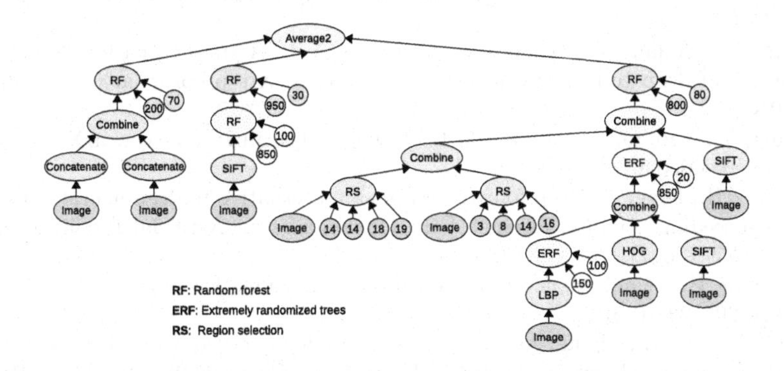

Fig. 6. The solution found by EvoDF on the **MNIST** dataset.

Although EvoDF achieves slightly worse results than gcForest on MNIST, the number of forests in the solution of EvoDF is much smaller than that in gcForest. gcForest uses at least eight random forests at each level (it could use more than 16 random forests), while the solution of EvoDF only uses seven random forests totally. This indicates that EvoDF can improve the utilisation of random forests, i.e., use a small number of random forests to achieve competitive performance by finding an effective structure of forest and feature extraction.

6 Conclusions

In this study, we developed an evolutionary deep forest approach with automatic feature extraction and structure search for image classification. Compared with deep forest, the proposed EvoDF approach found effective connections of forests with various parameters and extracted domain-specific features for image classification. In addition, the new approach used fewer parameters than deep forest and DNNs. The results showed that EvoDF achieved better performance than deep forest and other algorithms on the datasets having a small amount of training data and comparable performance on the datasets having a large number of training data. Further analysis showed that EvoDF can find effective connections of a small number of random forests to achieve competitive performance. In the future, we will further improve the performance of this approach by implementing it distributedly on large datasets.

References

1. LeCun, Y., Bengio, Y., Hinton, G.: Deep learning. Nature **521**(7553), 436 (2015)
2. Krizhevsky, A., Sutskever, I., Hinton, G.E.: ImageNet classification with deep convolutional neural networks. In: Advances in Neural Information Processing Systems, pp. 1097–1105 (2012)
3. He, K., Zhang, X., Ren, S., Sun, J.: Deep residual learning for image recognition. In: Proceedings of IEEE Conference on Computer Vision and Pattern Recognition, pp. 770–778 (2016)
4. Khan, A., Sohail, A., Zahoora, U., Qureshi, A.S.: A survey of the recent architectures of deep convolutional neural networks. arXiv preprint arXiv:1901.06032 (2019)
5. Wang, Y., Yao, Q., Kwok, J., Ni, L.: Few-shot learning: a survey. arXiv preprint arXiv:1904.05046 (2019)
6. Zhou, Z.H., Feng, J.: Deep forest: towards an alternative to deep neural networks. In: Proceedings of International Joint Conferences on Artificial Intelligence, pp. 3553–3559 (2017)
7. Al-Sahaf, H., et al.: A survey on evolutionary machine learning. J. Roy. Soc. NZ **49**(2), 205–228 (2019)
8. Sun, Y., Xue, B., Zhang, M., Yen, G.G.: Evolving deep convolutional neural networks for image classification. IEEE Trans. Evol. Comput. **24**, 394–407 (2019)
9. Bi, Y., Xue, B., Zhang, M.: An evolutionary deep learning approach using genetic programming with convolution operators for image classification. In: Proceedings of IEEE Congress on Evolutionary Computation, pp. 3197–3204 (2019)
10. Baioletti, M., Milani, A., Santucci, V.: Learning Bayesian networks with algebraic differential evolution. In: Auger, A., Fonseca, C.M., Lourenço, N., Machado, P., Paquete, L., Whitley, D. (eds.) PPSN 2018. LNCS, vol. 11102, pp. 436–448. Springer, Cham (2018). https://doi.org/10.1007/978-3-319-99259-4_35
11. Chen, B., Wu, H., Mo, W., Chattopadhyay, I., Lipson, H.: Autostacker: a compositional evolutionary learning system. In: Proceedings of the Genetic and Evolutionary Computation Conference, pp. 402–409 (2018)
12. Bi, Y., Xue, B., Zhang, M.: Genetic programming with a new representation to automatically learn features and evolve ensembles for image classification. IEEE Trans. Cybern., 15 p. (2020). https://doi.org/10.1109/TCYB.2020.2964566
13. Bi, Y., Xue, B., Zhang, M.: An automated ensemble learning framework using genetic programming for image classification. In: Proceedings of the Genetic and Evolutionary Computation Conference, pp. 365–373 (2019)
14. Koza, J.R.: Genetic Programming: On the Programming of Computers by Means of Natural Selection. MIT Press, Cambridge (1992)
15. Ojala, T., Pietikainen, M., Maenpaa, T.: Multiresolution gray-scale and rotation invariant texture classification with local binary patterns. IEEE Trans. Pattern Anal. Mach. Intell. **24**(7), 971–987 (2002)

16. Dalal, N., Triggs, B.: Histograms of oriented gradients for human detection. In: Proceedings of IEEE Conference on Computer Vision and Pattern Recognition, vol. 1, pp. 886–893 (2005)
17. Lowe, D.G.: Distinctive image features from scale-invariant keypoints. Int. J. Comput. Vision **60**(2), 91–110 (2004)
18. Liu, L., et al.: Deep learning for generic object detection: a survey. arXiv preprint arXiv:1809.02165 (2018)
19. Zhou, Z.H.: Ensemble Methods: Foundations and Algorithms. Chapman and Hall/CRC, Boca Raton (2012)
20. Sagi, O., Rokach, L.: Ensemble learning: a survey. Wiley Interdisc. Rev.: Data Min. Knowl. Discov. **8**(4), 1–19 (2018)
21. Dietterich, T.G.: Ensemble methods in machine learning. In: Kittler, J., Roli, F. (eds.) MCS 2000. LNCS, vol. 1857, pp. 1–15. Springer, Heidelberg (2000). https://doi.org/10.1007/3-540-45014-9_1
22. Young, S., Abdou, T., Bener, A.: Deep super learner: a deep ensemble for classification problems. In: Bagheri, E., Cheung, J.C.K. (eds.) Canadian AI 2018. LNCS (LNAI), vol. 10832, pp. 84–95. Springer, Cham (2018). https://doi.org/10.1007/978-3-319-89656-4_7
23. Ding, C., Tao, D.: Trunk-branch ensemble convolutional neural networks for video-based face recognition. IEEE Trans. Pattern Anal. Mach. Intell. **40**(4), 1002–1014 (2017)
24. Zhou, M., Zeng, X., Chen, A.: Deep forest hashing for image retrieval. Pattern Recogn. **95**, 114–127 (2019)
25. Zhu, G., Hu, Q., Gu, R., Yuan, C., Huang, Y.: ForestLayer: efficient training of deep forests on distributed task-parallel platforms. J. Parallel Distrib. Comput. **132**, 113–126 (2019)
26. Zhang, Y.L., et al.: Distributed deep forest and its application to automatic detection of cash-out fraud. ACM Trans. Intell. Syst. Technol. **10**(5), 1–19 (2019)
27. Bi, Y., Xue, B., Zhang, M.: An effective feature learning approach using genetic programming with image descriptors for image classification [research frontier]. IEEE Comput. Intell. Mag. **15**(2), 65–77 (2020)
28. Montana, D.J.: Strongly typed genetic programming. Evol. Comput. **3**(2), 199–230 (1995)
29. Samaria, F.S., Harter, A.C.: Parameterisation of a stochastic model for human face identification. In: Proceedings of 1994 IEEE Workshop on Applications of Computer Vision, pp. 138–142 (1994)
30. Lee, K.C., Ho, J., Kriegman, D.J.: Acquiring linear subspaces for face recognition under variable lighting. IEEE Trans. Pattern Anal. Mach. Intell. **5**, 684–698 (2005)
31. Fei-Fei, L., Perona, P.: A Bayesian hierarchical model for learning natural scene categories. In: Proceedings of IEEE Conference on Computer Vision and Pattern Recognition, vol. 2, pp. 524–531 (2005)
32. Mallikarjuna, P., Targhi, A.T., Fritz, M., Hayman, E., Caputo, B., Eklundh, J.O.: THE KTH-TIPS2 database, pp. 1–10. Computational Vision and Active Perception Laboratory, Stockholm, Sweden (2006)
33. LeCun, Y., Cortes, C., Burges, C.J.: The mnist database (1998). http://yann.lecun.com/exdb/mnist
34. Krizhevsky, A., Nair, V., Hinton, G.: The cifar-10 dataset, no. 55 (2014). http://www.cs.toronto.edu/kriz/cifar.html
35. Hinton, G.E., Osindero, S., Teh, Y.W.: A fast learning algorithm for deep belief nets. Neural Comput. **18**(7), 1527–1554 (2006)

36. Chen, T., Guestrin, C.: XGBoost: a scalable tree boosting system. In: Proceedings of the 22nd ACM SIGKDD International Conference on Knowledge Discovery and Data Mining, pp. 785–794 (2016)
37. Krizhevsky, A., Hinton, G., et al.: Learning multiple layers of features from tiny images. Technical report, Citeseer (2009)
38. Ba, J., Caruana, R.: Do deep nets really need to be deep? In: Advances in Neural Information Processing Systems, pp. 2654–2662 (2014)

Fast Perturbative Algorithm Configurators

George T. Hall, Pietro S. Oliveto(✉), and Dirk Sudholt

The University of Sheffield, Sheffield, UK
{gthall1,p.oliveto,d.sudholt}@sheffield.ac.uk

Abstract. Recent work has shown that the ParamRLS and ParamILS algorithm configurators can tune some simple randomised search heuristics for standard benchmark functions in linear expected time in the size of the parameter space. In this paper we prove a linear lower bound on the expected time to optimise any parameter tuning problem for ParamRLS, ParamILS as well as for larger classes of algorithm configurators. We propose a harmonic mutation operator for perturbative algorithm configurators that provably tunes single-parameter algorithms in polylogarithmic time for unimodal and approximately unimodal (i.e., non-smooth, rugged with an underlying gradient towards the optimum) parameter spaces. It is suitable as a general-purpose operator since even on worst-case (e.g., deceptive) landscapes it is only by at most a logarithmic factor slower than the default ones used by ParamRLS and ParamILS. An experimental analysis confirms the superiority of the approach in practice for a number of configuration scenarios, including ones involving more than one parameter.

Keywords: Parameter tuning · Algorithm configurators · Runtime analysis

1 Introduction

Many algorithms are highly dependent on the values of their parameters, all of which have the potential to affect their performance substantially. It is therefore a challenging but important task to identify parameter values that lead to good performance for a class of problems. This task, called *algorithm configuration* or *parameter tuning*, was traditionally performed by hand: parameter values were updated manually and the performance of each *configuration* assessed, allowing the user to determine which parameter settings performed best. In recent years there has been an increase in popularity of automated *algorithm configurators* [13].

Examples of popular algorithm configurators are *ParamILS*, which uses iterated local search to traverse the *parameter space* (the space of possible configurations) [14]; *irace*, which evaluates a set of configurations in parallel and eliminates those which can be shown statistically to be performing poorly [20];

T. Bäck et al. (Eds.): PPSN 2020, LNCS 12269, pp. 19–32, 2020.
https://doi.org/10.1007/978-3-030-58112-1_2

and *SMAC*, which uses surrogate models to reduce the number of configuration evaluations [15]. Despite their popularity, the foundational understanding of algorithm configurators remains limited. Key questions are still unanswered, such as whether a configurator is able to identify (near) optimal parameter values, and, if so, the amount of time it requires to do so. While analyses of worst-case performance are available, as well as algorithms that provably perform better in worst-case scenarios [18,19,23,24], the above questions are largely unanswered regarding the performance of the popular algorithm configurators used in practice for typical configuration scenarios.

Recently, the performance of ParamRLS and ParamILS was rigorously analysed for tuning simple single-parameter search heuristics for some standard benchmark problems from the literature. It was proved that they can efficiently tune the neighbourhood size k of the randomised local search algorithm (RLS_k) for RIDGE and ONEMAX [10] and the mutation rate of the simple (1+1) EA for RIDGE and LEADINGONES [11]. The analyses, though, also reveal some weaknesses of the search operators used by the two algorithm configurators. The *ℓ-step* mutation operator used by ParamRLS, which changes a parameter value to a neighbouring one at a distance of at most ℓ, may either get stuck on local optima if the neighbourhood size ℓ is too small, or progress too slowly when far away from the optimal configuration. On the other hand, the mutation operator employed by ParamILS, that changes one parameter value uniformly at random, lacks the ability to efficiently fine-tune the current solution by searching locally around the identified parameter values. Indeed both algorithms require linear expected time in the number of parameter values to identify the optimal configurations for the studied unimodal or approximately unimodal parameter spaces induced by the target algorithms and benchmark functions [10,11].

In this paper we propose a more robust mutation operator that samples a step size according to the harmonic distribution [6,7]. The idea is to allow small mutation steps with sufficiently high probability to efficiently fine-tune good parameter values while, at the same time, enabling larger mutations that can help follow the general gradient from a macro perspective, e.g., by tunnelling through local optima. This search operator can be easily used in any perturbative algorithm configurator that maintains a set of best-found configurations and mutates them in search for better ones. Both ParamRLS and ParamILS fall into this large class of configurators.

We first prove that large classes of algorithm configurators, which include ParamRLS and ParamILS with their default mutation operators, require linear expected time in the number of possible configurations to optimise any parameter configuration landscape. Then we provide a rigorous proof that the harmonic search operator can identify the optimal parameter value of single-parameter target algorithms in polylogarithmic time if the parameter landscape is either unimodal or approximately unimodal (i.e., non-smooth, rugged landscapes with an underlying monotonically decreasing gradient towards the optimum). It is also robust as even on deceptive worst-case landscapes it is only by at most a logarithmic factor slower than the default operators of ParamRLS and ParamILS.

Algorithm 1. ParamRLS ($\mathcal{A}, \Theta, \Pi, \kappa, r$). Adapted from [11].

```
1: θ ←initial parameter value chosen uniformly at random
2: while termination condition not satisfied do
3:    θ' ← mutate(θ)
4:    θ ← better(A, θ, θ', π, κ, r) {called eval in [11]}
5: return θ
```

We complement the theory with an experimental analysis showing that both ParamRLS and ParamILS have a statistically significant smaller average optimisation time to identify the optimal configuration in single-parameter unimodal and approximately unimodal landscapes and for a well-studied MAX-SAT configuration scenario where two parameters have to be tuned. The latter result is in line with analyses of Pushak and Hoos that suggests that even in complex configuration scenarios (for instance state-of-the-art SAT, TSP, and MIP solvers), the parameter landscape is often not as complex as one might expect [22].

2 Preliminaries

The ParamRLS Configurator. ParamRLS is a simple theory-driven algorithm configurator defined in Algorithm 1 [11]. The algorithm chooses an initial configuration uniformly at random (u.a.r.) from the parameter space. In each iteration, a new configuration is generated by mutating the current solution. The obtained offspring replaces the parent if it performs better. By default, ParamRLS uses the *ℓ-step* operator which selects a parameter and a step size $d \in \{1, \ldots, \ell\}$ both u.a.r. and then moves to a parameter value at distance[1] $+d$ or $-d$ (if feasible).

The ParamILS Configurator. ParamILS (Algorithm 2) is a more sophisticated iterated local search algorithm configurator [14]. In the initialisation step it selects R configurations uniformly at random and picks the best performing one. In the iterative loop it performs an iterated local search (Algorithm 3) until a local optimum is reached, followed by a perturbation step where up to s random parameters are perturbed u.a.r. A random restart occurs in each iteration with some probability p_{restart}. The default local search operator selects from the neighbourhood uniformly at random without replacement (thus we call this the *random* local search operator). The neighbourhood of a configuration contains all configurations that differ by exactly one parameter value.

The Harmonic-Step Operator. The harmonic-step mutation operator selects a parameter uniformly at random and samples a step size d according to the

[1] Throughout this paper, we consider parameters from an interval of integers for simplicity, where the distance is the absolute difference between two integers. This is not a limitation: if parameters are given as a vector of real values $z_1, z_2, \ldots, z_\phi$, we may simply tune the index, which is an integer from $\{1, \ldots, \phi\}$. Then changing the parameter value means that we change the index of this value.

Algorithm 2. ParamILS pseudocode, recreated from [14].

Require: Initial configuration $\theta_0 \in \Theta$, algorithm parameters r, p_{restart}, and s.
Ensure: Best parameter configuration θ found.
1: **for** $i = 1, \ldots, R$ **do**
2: $\theta \leftarrow$ random $\theta \in \Theta$
3: **if** better(θ, θ_0) **then** $\theta_0 \leftarrow \theta$
4: $\theta_{\text{inc}} \leftarrow \theta_{\text{ils}} \leftarrow$ *IterativeFirstImprovement*(θ_0) {Algorithm 3}
5: **while not** *TerminationCriterion*() **do**
6: $\theta \leftarrow \theta_{\text{ils}}$
7: **for** $i = 1, \ldots, s$ **do** $\theta \leftarrow$ random $\theta' \in Nbh(\theta)$
8: {*Nbh* contains all neighbours of a configuration}
9: $\theta \leftarrow$ *IterativeFirstImprovement*(θ)
10: **if** better$(\theta, \theta_{\text{ils}})$ **then** $\theta_{\text{ils}} \leftarrow \theta$
11: **if** better$(\theta_{\text{ils}}, \theta_{\text{inc}})$ **then** $\theta_{\text{inc}} \leftarrow \theta_{\text{ils}}$
12: **with probability** p_{restart} **do** $\theta_{\text{ils}} \leftarrow$ random $\theta \in \Theta$
13: **return** θ_{inc}

Algorithm 3. IterativeFirstImprovement(θ) procedure, adapted from [14].

1: **repeat**
2: $\theta' \leftarrow \theta$
3: **for all** $\theta'' \in UndiscNbh(\theta')$ in randomised order **do**
4: {*UndiscNbh* contains all undiscovered neighbours of a configuration}
5: **if** better(θ'', θ') **then** $\theta \leftarrow \theta''$; **break**
6: **until** $\theta' = \theta$
7: **return** θ

harmonic distribution. In particular, the probability of selecting a step size d is $1/(d \cdot H_{\phi-1})$, where H_m is the m-th harmonic number (i.e. $H_m = \sum_{k=1}^{m} \frac{1}{k}$) and ϕ is the range of possible parameter values. It returns the best parameter value at distance $\pm d$. This operator was originally designed to perform fast greedy random walks in one-dimensional domains [6] and was shown to perform better than the *1-step* and the random local search (as in ParamILS) operators for optimising the multi-valued OneMax problem [7]. We refer to ParamRLS using the Harmonic-step operator as ParamHS.

3 General Lower Bounds for Default Mutation Operators

To set a baseline for the performance gains obtained by ParamHS, we first show general lower bounds for algorithm configurators, including ParamRLS and ParamILS. Our results apply to a class of configurators described in Algorithm 4. We use a general framework to show that the poor performance of default mutation operators is not limited to particular configurators, and to identify which algorithm design aspects are the cause of poor performance.

We show that mutation operators that only change one parameter by a small amount, such as the ℓ-step operator with constant ℓ, lead to linear expected times in the number of parameter values (sum of all parameter ranges).

Algorithm 4. General scheme for algorithm configurators.

1: Initialise an incumbent configuration uniformly at random
2: **while** optimal configuration not found **do**
3: Pick a mutation operator according to the history of past evaluations.
4: Apply the chosen mutation operator.
5: Apply selection to choose new configuration from the incumbent configuration and the mutated one.

Theorem 1. *Consider a setting with D parameters and ranges $\phi_1, \ldots, \phi_D \geq 2$ such that there is a unique optimal configuration. Let $M = \sum_{i=1}^{D} \phi_i$. Consider an algorithm configurator $\mathcal{A}$ implementing the scheme of Algorithm 4 whose mutation operator only changes a single parameter and does so by at most a constant absolute value (e.g. ParamRLS with local search operator $\pm\{\ell\}$ for constant ℓ). Then $\mathcal{A}$ takes time $\Omega(M)$ in expectation to find the optimal configuration.*

Proof. Consider the L_1 distance of the current configuration $x = (x_1, \ldots, x_D)$ from the optimal one $\text{opt} = (\text{opt}_1, \ldots, \text{opt}_D)$: $\sum_{i=1}^{D} |x_i - \text{opt}_i|$. For every parameter i, the expected distance between the uniform random initial configuration and opt_i is minimised if opt_i is at the centre of the parameter range. Then, for odd ϕ_i, there are two configurations at distances $1, 2, \ldots, (\phi_i - 1)/2$ from opt_i, each being chosen with probability $1/\phi_i$. The expected distance is thus at least $1/\phi_i \cdot \sum_{j=1}^{(\phi_i-1)/2} 2j = (\phi_i - 1)(\phi_i + 1)/(4\phi_i) = (\phi_i - 1/\phi_i)/4 \geq \phi_i/8$. For even ϕ_i, the expectation is at least $\phi_i/4$. By linearity of expectation, the expected initial distance is at least $\sum_{i=1}^{D} \phi_i/8 \geq M/8$. Every mutation can only decrease the distance by $O(1)$, hence the expected time is bounded by $(M/8)/O(1) = \Omega(M)$. □

The same lower bound also applies if the mutation operator chooses a value uniformly at random (with or without replacement), as is done in ParamILS.

Theorem 2. *Consider a setting with D parameters and ranges $\phi_1, \ldots, \phi_D \geq 2$ such that there is a unique optimal configuration. Let $M = \sum_{i=1}^{D} \phi_i$. Consider an algorithm configurator $\mathcal{A}$ implementing the scheme of Algorithm 4 whose mutation operator only changes a single parameter and does so by choosing a new value uniformly at random (possibly excluding values previously evaluated). Then $\mathcal{A}$ takes time $\Omega(M)$ in expectation to find the optimal configuration.*

Proof. Let T_i be the number of times that parameter i is mutated (including the initial step) before it attains its value in the optimal configuration. After $j-1$ steps in which parameter i is mutated, at most j parameter values have been evaluated (including the initial value). The best case is that $\mathcal{A}$ always excludes previous values, which corresponds to a complete enumeration of the ϕ_i possible values in random order. Since every step of this enumeration has a probability of $1/\phi_i$ of finding the optimal value, the expected time spent on parameter i is $\mathrm{E}(T_i) \geq \sum_{j=0}^{\phi_i-1} j/\phi_i = (\phi_i - 1)/2$. The total expected time is at least $\sum_{i=1}^{D} \mathrm{E}(T_i) - D + 1$ as the initial step contributes to all T_i and each following step only contributes to one value T_i. Noting $\sum_{i=1}^{D} \mathrm{E}(T_i) - D + 1 = \sum_{i=1}^{D}(\phi_i - 1)/2 - D/2 + 1 \geq M/4$ (as $\phi_i \geq 2$ for all i) proves the claim. □

ParamILS is not covered directly by Theorem 2 as it uses random sampling during the initialisation that affects all parameters. However, it is easy to show that the same lower bound also applies to ParamILS.

Theorem 3. *Consider a setting with D parameters and ranges $\phi_1, \ldots, \phi_D \geq 2$ such that there is a unique optimal configuration. Let $M = \sum_{i=1}^{D} \phi_i$. Then ParamILS takes time $\Omega(M)$ in expectation to find the optimal configuration.*

Proof. Recall that ParamILS first evaluates R random configurations. If $R \geq M/2$ then the probability of finding the optimum during the first $M/2$ random samples is at most $M/2 \cdot \prod_{i=1}^{D} 1/\phi_i \leq 1/2$ since $M = \sum_{i=1}^{D} \phi_i \leq \prod_{i=1}^{D} \phi_i$. Hence the expected time is at least $1/2 \cdot M/2 = M/4$. If $R < M/2$ then with probability at least 1/2 ParamILS does not find the optimum during the R random steps and starts the IterativeFirstImprovement procedure with a configuration θ_0. This procedure scans the neighbourhood of θ_0, which is all configurations that differ in one parameter; the number of these is $\sum_{i=1}^{D}(\phi_i - 1) = M - D$. If the global optimum is not among these, it is not found in these $M - D$ steps. Otherwise, the neighbourhood is scanned in random order and the expected number of steps is $(M - D - 1)/2$ as in the proof of Theorem 2. In both cases, the expected time is at least $(M - D - 1)/4 \geq M/16$ (as $M \geq 2D$). □

4 Performance of the Harmonic Search Operator

In the setting of Theorem 1, mutation lacks the ability to explore the search space quickly, whereas in the setting of Theorems 2 and 3, mutation lacks the ability to search locally. The harmonic search operator is able to do both. It is able to explore the space, but smaller steps are made with a higher probability, enabling the search to exploit gradients in the parameter landscape.

For simplicity and lack of space we only consider configuring one parameter with a range of ϕ (where the bounds from Theorems 1, 2 and 3 simplify to $\Omega(\phi)$), however the operator improves performance in settings with multiple parameters in the same way. We show that ParamHS is robust in a sense that it performs well on all landscapes (with only a small overhead in the worst case, compared to the lower bounds from Theorem 1, 2 and 3), and it performs extremely well on functions that are unimodal or have an underlying gradient that is close to being unimodal.

To capture the existence of underlying gradients and functions that are unimodal to some degree, we introduce a notion of approximate unimodality.

Definition 1. *Call a function f on $\{1, \ldots, m\}$ (α, β)-approximately unimodal for parameters $\alpha \geq 1$ and $1 \leq \beta \leq m$ if for all positions x with distance $\beta \leq i \leq m$ from the optimum and all positions y with distance $j > \alpha i$ to the optimum we have $f(x) < f(y)$.*

Intuitively, this means that only configurations with distance to the optimal one that is by a factor of α larger than that of the current configuration can be better. This property only needs to hold for configurations with distance to the optimum i with $\beta \leq i \leq m$, to account for landscapes that do not show a clear gradient close to the optimum.

Note that a $(1, 1)$-approximately unimodal function is unimodal and a $(1, \beta)$-approximately unimodal function is unimodal within the states $\{\beta, \ldots, m\}$. Also note that all functions are $(1, m)$-approximately unimodal.

The following performance guarantees for ParamHS show that it is efficient on all functions and very efficient on functions that are close to unimodal.

Theorem 4. *Consider ParamHS configuring an algorithm with a single parameter having ϕ values and a unique global optimum. If the parameter landscape is (α, β)-approximately unimodal then the expected number of calls to* `better()` *before the optimal parameter value is sampled is at most*

$$4\alpha H_{\phi-1} \log(\phi) + 4\alpha\beta H_{\phi-1} = O(\alpha \log^2(\phi) + \alpha\beta \log \phi),$$

where $H_{\phi-1}$ is the $(\phi - 1)$-th harmonic number (i.e. $\sum_{i=1}^{\phi-1} \frac{1}{i}$).

Corollary 1. *In the setting of Theorem 4,*

(a) every unimodal parameter landscape yields a bound of $O(\log^2 \phi)$.
(b) for every parameter landscape, a general upper bound of $O(\phi \log \phi)$ applies.

Hence ParamHS is far more efficient than the $\Omega(\phi)$ lower bound for general classes of tuners (Theorems 1, 2 and 3) on approximately unimodal landscapes and is guaranteed never to be worse than default operators by more than a $\log \phi$ factor.

Proof of Theorem 4. Let $f(i)$ describe the performance of the configuration with the i-th largest parameter value. Then f is (α, β)-approximately unimodal and we are interested in the time required to locate its minimum.

Let d_t denote the current distance to the optimum and note that $d_0 \leq \phi$. Let d_t^* denote the smallest distance to the optimum seen so far, that is, $d_t^* = \min_{t' \leq t} d_{t'}$. Note that d_t^* is non-increasing over time. Since ParamHS does not accept any worsenings, $f(d_t) \leq f(d_t^*)$.

If $d_t^* \geq \beta$ then by the approximate unimodality assumption, for all $j > \alpha d_t^*$, $f(j) > f(d_t^*) \geq f(d_t)$, that is, all points at distance larger than αd_t^* have a worse fitness than the current position and will never be visited.

Now assume that $d_t^* \geq 2\beta$. We estimate the expected time to reach a position with distance at most $\lfloor d_t^*/2 \rfloor$ to the optimum. This includes all points that have distance i to the global optimum, for $0 \leq i \leq \lfloor d_t^*/2 \rfloor$, and distance $d_t - i$ to the current position. The probability of jumping to one of these positions is at least

$$\sum_{i=0}^{\lfloor d_t^*/2 \rfloor} \frac{1}{2(d_t - i)H_{\phi-1}} \geq \sum_{i=0}^{\lfloor d_t^*/2 \rfloor} \frac{1}{2d_t H_{\phi-1}} \geq \frac{d_t^*}{4d_t H_{\phi-1}} \geq \frac{d_t^*}{4\alpha d_t^* H_{\phi-1}} = \frac{1}{4\alpha H_{\phi-1}}.$$

Hence, the expected half time for d_t^* is at most $4\alpha H_{\phi-1}$ and the expected time to reach $d_t^* < 2\beta$ is at most $4\alpha H_{\phi-1} \log \phi$.

Once $d_t^* < 2\beta$, the probability of jumping directly to the optimum is at least $\frac{1}{2d_t H_{\phi-1}} \geq \frac{1}{2\alpha d_t^* H_{\phi-1}} \geq \frac{1}{4\alpha\beta H_{\phi-1}}$ and the expected time to reach the optimum is at most $4\alpha\beta H_{\phi-1}$. Adding the above two times and using the well-known fact that $H_{\phi-1} = O(\log \phi)$ yields the claim. □

5 Experimental Analysis

We have proved that, given some assumptions about the parameter landscape, it is beneficial to use the harmonic-step operator instead of the default operators used in ParamRLS and ParamILS. In this section, we verify experimentally that these theoretical results are meaningful beyond parameter landscapes assumed to be (approximately) unimodal.

We investigated the impact of using the harmonic-step operator on the time taken for ParamRLS and ParamILS to identify the optimal configuration (or in one case a set of near-optimal configurations) in different configuration scenarios. Note that ParamRLS using this operator is equivalent to ParamHS. We analysed the number of configuration comparisons (that is, calls to the `better()` procedure present in both ParamRLS and ParamILS) required for the configurators to identify the optimal mutation rate (the optimal value χ in the mutation rate χ/n) for the (1+1) EA optimising RIDGE and the (1+1) EA optimising LEADINGONES as in [11] and identifying the optimal neighbourhood size k (the number of bits flipped during mutation) for RLS_k optimising ONEMAX as in [10]. Finally, we considered optimising two parameters of the SAT solver SAPS optimising MAX-SAT [17], searching for one of the five best-performing configurations found during an exhaustive search of the parameter space.

In the first two configuration scenarios, with probability $1 - 2^{-\Omega(n^{\varepsilon})}$, the configurator can identify that a neighbouring parameter value is better, hence the landscape is unimodal [11] (see Figs. 1a and 1b). In such landscapes, we expect the harmonic-step operator to perform well. In the third scenario, the parameter landscape is *not* unimodal (see Fig. 1c: $k = 2c + 1$ outperforms $k = 2c$), but it is (2, 1)-approximately unimodal with respect to the expected fitness (as for all k, the parameter value k outperforms all parameter values $k' > 2k$) both empirically (Fig. 1c) and theoretically [8]. In the fourth scenario, the parameter landscape is more complex since we configure two parameters, but it still appears to be approximately unimodal (see Fig. 1d).

5.1 Experimental Setup

In all scenarios we measured the number of calls to the `better()` procedure before the optimal configuration (or a set of near-optimal configurations in the scenario configuring SAPS) is first sampled. We varied the size of the parameter space to investigate how the performance of the mutation operators (i.e. ℓ-step, random, and harmonic-step) depends on the size of the parameter space.

For ParamILS, the BasicILS variant was used. That is, each call to `better()` resulted in the two competing configurations both being run the same, set number of times. For each size of the parameter spaces, the experiment was repeated 200 times and the mean number of calls to `better()` was recorded. For the MAX-SAT scenario 500 repetitions were used to account for the increased complexity of the configuration scenario. The cutoff time κ (the number of iterations for which each configuration is executed for each run in a comparison) varied with the choice of problem class. A fitness-based performance metric was used, as recommended in [10,11], in which the winner of a comparison is the configuration which achieves the highest mean fitness in r runs each lasting κ iterations. In each run, both configurators were initialised uniformly at random. We set $R = 0$ in ParamILS since preliminary experiments indicated that initial random sampling was harmful in the configuration scenarios considered here.

Benchmark Functions. For RIDGE, LEADINGONES and ONEMAX, we used $n = 50$ and 1500 runs per configuration comparison (i.e. $r = 1500$). For RIDGE, we used a cutoff time of $\kappa = 2500$. The value of ℓ in the ℓ-step operator was set to $\ell = 1$. The first parameter space that we considered was $\chi \in \{0.5, 1.0, \ldots, 4.5, 5.0\}$, where χ/n is the mutation rate and $\chi = 1$ is optimal for RIDGE [11]. We increased the size of the parameter space by adding the next five largest configurations (each increasing by 0.5) until the parameter space $\{0.5, \ldots, 25.0\}$ was reached. Following [11], for RIDGE, the (1+1) EA was initialised at the start of the ridge, in order to focus on the search on the ridge (as opposed to the initial approach to the ridge, for which the optimal mutation rate may be different from $1/n$).

When configuring the mutation rate χ/n of the (1+1) EA for LEADINGONES, we initialised the individual u.a.r. and used $\kappa = 2500$ and $\ell = 1$. The size of the parameter space was increased in the same way as in the RIDGE experiments, and the initial parameter space was $\chi \in \{0.6, 1.1, \ldots, 4.6, 5.1\}$ as the optimal value for χ is approximately 1.6 [1,11]. The final parameter space was $\{0.6, \ldots, 25.1\}$.

When configuring the neighbourhood size of RLS_k for ONEMAX, we initialised the individual u.a.r. and set $\kappa = 200$. The initial parameter space was $\{1, 2, \ldots, 9, 10\}$, where $k = 1$ is the optimal parameter [10], and the next five largest integers were added until $\{1, 2, \ldots, 49, 50\}$ was reached. Since this parameter landscape is only approximately unimodal, we set $\ell = 2$ (as recommended in [10]: $\ell = 1$ would fail to reach the optimal value $k = 1$ unless initialised there).

SAPS for MAX-SAT. We considered tuning two parameters of SAPS – α and ρ – for ten instances[2] of the circuit-fuzz problem set (available in AClib [16]). Due to the complexity of the MAX-SAT problem class it was no longer obvious which configurations can be considered optimal. Therefore we conducted an exhaustive search of the parameter space in order to identify configurations that perform well. We did so by running the validation procedure in ParamILS for each configuration with $\alpha \in \{\frac{16}{15}, \frac{17}{15}, \ldots, \frac{44}{15}, \frac{45}{15}\}$ and $\rho \in \{0, \frac{1}{15}, \ldots, \frac{14}{15}, 1\}$. Each configuration was evaluated 2000 times on each of the ten considered circuit-fuzz

[2] Problem instances number 78, 535, 581, 582, 6593, 6965, 8669, 9659, 16905, 16079.

problem instances. In each evaluation, the cutoff time was 10, 000 iterations and the quality of a configuration was the number of satisfied clauses. We selected the set of the five best-performing configurations to be the target.

Since it was not feasible to compute the quality of a configuration each time it was evaluated in a tuner, we instead took the average fitness values generated during the initial evaluation of the parameter landscape to be the fitness of each configuration. As these runs were repeated many times we believe they provide an accurate approximation of the fitness values of the configurations.

In this experiment, we kept the range of values of ρ as the set $\{0, \frac{1}{15}, \ldots, \frac{14}{15}, 1\}$ and the value of the two other parameters of SAPS as $ps = 0.05$ and $wp = 0.01$ (their default values). We then increased the size of the set of possible values of α. The initial range for α was the set $\{\frac{16}{15}, \frac{17}{15}, \frac{18}{15}\}$, which contains all five best-performing configurations. We then generated larger parameter spaces by adding a new value to the set of values for α until the set $\{\frac{16}{15}, \ldots, \frac{45}{15}\}$ was reached.

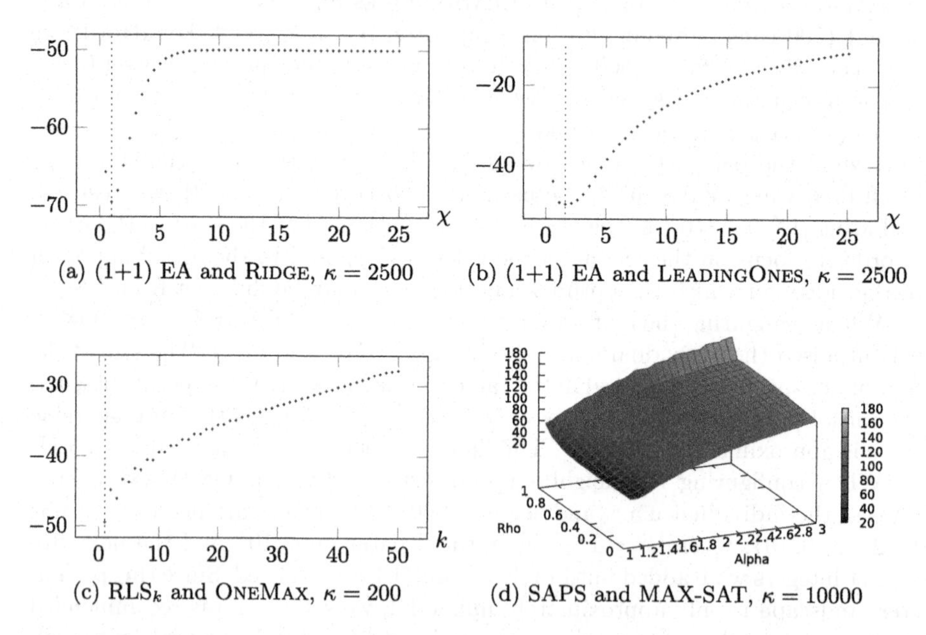

(a) (1+1) EA and Ridge, $\kappa = 2500$ (b) (1+1) EA and LeadingOnes, $\kappa = 2500$

(c) RLS_k and OneMax, $\kappa = 200$ (d) SAPS and MAX-SAT, $\kappa = 10000$

Fig. 1. (a), (b), (c): Mean fitness of the individual in the algorithms with $n = 50$, averaged over 10,000 runs for each parameter value, multiplied by -1 to obtain a minimisation problem. The dotted line indicates the optimal configuration for each scenario. (d): The parameter landscape for SAPS in terms of α and ρ computed for a set of ten SAT instances from the circuit-fuzz dataset. In all figures lower values are better.

5.2 Results

The results from configuring benchmark functions are shown in Figs. 2a, 2b, and 2c. Green lines indicate the random search operator (without replacement),

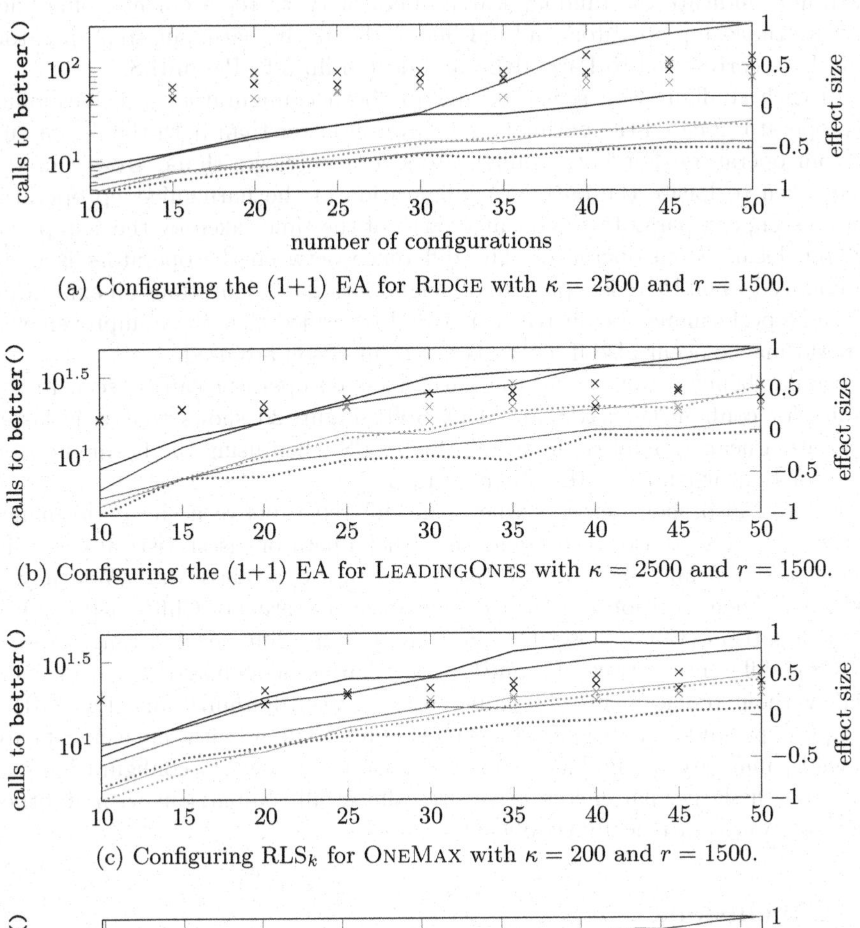

(a) Configuring the (1+1) EA for RIDGE with $\kappa = 2500$ and $r = 1500$.

(b) Configuring the (1+1) EA for LEADINGONES with $\kappa = 2500$ and $r = 1500$.

(c) Configuring RLS_k for ONEMAX with $\kappa = 200$ and $r = 1500$.

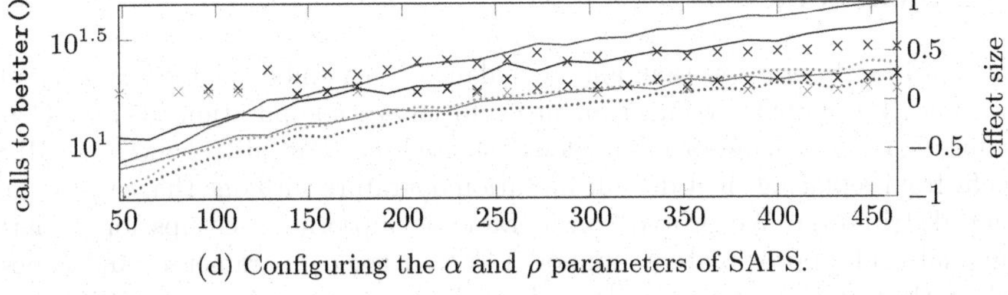

(d) Configuring the α and ρ parameters of SAPS.

Fig. 2. Mean number of calls to `better()` before sampling the optimal configuration. Green lines indicate the random search operator (without replacement), black lines indicate the random search operator (with replacement), blue lines indicate the ℓ-step operator, and red lines indicate the harmonic-step operator. Solid lines correspond to ParamRLS and dotted lines to ParamILS. Crosses show effect size of difference at points where statistically significant for ParamHS versus: ℓ-step (blue); random (without replacement) (green); random (with replacement) (black); and harmonic-step ParamILS vs. default ParamILS (orange). (Color figure online)

black lines indicate the random search operator (with replacement), blue lines indicate the ℓ-step operator, and red lines indicate the harmonic-step operator. Solid lines correspond to ParamRLS and dotted lines to ParamILS.

In each configuration scenario, and for both configurators, the harmonic-step operator located the optimal configuration faster than both the ℓ-step and random operators. For both configurators, the polylogarithmic growth of the time taken to locate the optimal configuration of the harmonic-step operator can be seen, compared to the linear growth of the time taken by the ℓ-step and random local search operators. The difference between the operators is more pronounced when there is a plateau of neighbouring configurations all exhibiting the same performance (as in RIDGE). We also verified that these improvements in performance occur also if few runs per comparison are used.

Similar benefits from using the harmonic-step operator can be seen in the results for configuring SAPS for MAX-SAT. Figure 2d shows that it is faster to locate a near-optimal configuration for SAPS when using the harmonic step operator than when using the other operators.

Figure 2 also shows crosses where the difference between the performance of the harmonic-step operator and the other operators is statistically significant at a significance level of 0.95 (according to a two-tailed Mann-Whitney U test [21]). Their position reflects the effect size (in terms of Cliff's delta [2]) of this comparison (values closer to 1 indicate a larger difference). Orange crosses show the difference between ParamILS using harmonic-step and that using random (without replacement). The differences between ParamHS and ParamRLS using ℓ-step, random (without replacement) and random (with replacement) are shown by blue, green, and black crosses, respectively. In every configuration scenario, for the larger parameter space sizes almost all comparisons with all other operators were statistically significant.

6 Conclusions

Fast mutation operators, that aim to balance the number of large and small mutations, are gaining momentum in evolutionary computation [3–5,9]. Concerning algorithm configuration we demonstrated that ParamRLS and ParamILS benefit from replacing their default mutation operators with one that uses a harmonic distribution. We proved considerable asymptotic speed-ups for smooth unimodal and approximately unimodal (i.e., rugged) parameter landscapes, while in the worst case (e.g., for deceptive landscapes) the proposed modification may only slow down the algorithm by at most logarithmic factor. We verified experimentally that this speed-up occurs in practice for benchmark parameter landscapes that are known to be unimodal and approximately unimodal, as well as for tuning a MAX-SAT solver for a well-studied benchmark set. Indeed other recent experimental work has suggested that the search landscape of algorithm configurations may be simpler than expected, often being unimodal or even convex [12,22]. We believe that this is the first work that has rigorously shown how to provably achieve faster algorithm configurators by exploiting the envisaged parameter landscape, while being only slightly slower if it was to be considerably different. Future theoretical work should estimate the performance of the harmonic mutation operator on larger parameter configuration problem classes, while empirical work should assess the performance of the operator for more sophisticated configurators operating in real-world configuration scenarios.

Acknowledgements. This work was supported by the EPSRC (EP/M004252/1).

References

1. Böttcher, S., Doerr, B., Neumann, F.: Optimal fixed and adaptive mutation rates for the LeadingOnes problem. In: Schaefer, R., Cotta, C., Kołodziej, J., Rudolph, G. (eds.) PPSN 2010. LNCS, vol. 6238, pp. 1–10. Springer, Heidelberg (2010). https://doi.org/10.1007/978-3-642-15844-5_1
2. Cliff, N.: Dominance statistics: ordinal analyses to answer ordinal questions. Psychol. Bull. **114**(3), 494 (1993)
3. Corus, D., Oliveto, P.S., Yazdani, D.: Fast artificial immune systems. In: Auger, A., Fonseca, C.M., Lourenço, N., Machado, P., Paquete, L., Whitley, D. (eds.) PPSN 2018. LNCS, vol. 11102, pp. 67–78. Springer, Cham (2018). https://doi.org/10.1007/978-3-319-99259-4_6
4. Corus, D., Oliveto, P.S., Yazdani, D.: Artificial immune systems can find arbitrarily good approximations for the NP-hard number partitioning problem. Artif. Intell. **247**, 180–196 (2019)
5. Corus, D., Oliveto, P.S., Yazdani, D.: When hypermutations and ageing enable artificial immune systems to outperform evolutionary algorithms. Theor. Comput. Sci. **832**, 166–185 (2020)
6. Dietzfelbinger, M., Rowe, J.E., Wegener, I., Woelfel, P.: Precision, local search and unimodal functions. Algorithmica **59**(3), 301–322 (2011)
7. Doerr, B., Doerr, C., Kötzing, T.: Static and self-adjusting mutation strengths for multi-valued decision variables. Algorithmica **80**, 1732–1768 (2018)
8. Doerr, B., Doerr, C., Yang, J.: Optimal parameter choices via precise black-box analysis. Theor. Comput. Sci. **801**, 1–34 (2020)
9. Doerr, B., Le, H.P., Makhmara, R., Nguyen, T.D.: Fast genetic algorithms. In: Proceedings of the Genetic and Evolutionary Computation Conference, GECCO 2017, pp. 777–784. ACM (2017)
10. Hall, G.T., Oliveto, P.S., Sudholt, D.: On the impact of the cutoff time on the performance of algorithm configurators. In: Proceedings of the Genetic and Evolutionary Computation Conference, GECCO 2019, pp. 907–915. ACM (2019)
11. Hall, G.T., Oliveto, P.S., Sudholt, D.: Analysis of the performance of algorithm configurators for search heuristics with global mutation operators. In: Proceedings of the Genetic and Evolutionary Computation Conference, GECCO 2020. ACM (2020, to appear)
12. Harrison, K.R., Ombuki-Berman, B.M., Engelbrecht, A.P.: The parameter configuration landscape: a case study on particle swarm optimization. In: IEEE Congress on Evolutionary Computation, CEC 2019, pp. 808–814. IEEE (2019)
13. Huang, C., Li, Y., Yao, X.: A survey of automatic parameter tuning methods for metaheuristics. IEEE Trans. Evol. Comput. **24**(2), 201–216 (2020)
14. Hutter, F., Hoos, H.H., Leyton-Brown, K.: ParamILS: an automatic algorithm configuration framework. J. Artif. Intell. Res. **36**(1), 267–306 (2009)

15. Hutter, F., Hoos, H.H., Leyton-Brown, K.: Sequential model-based optimization for general algorithm configuration. In: Coello, C.A.C. (ed.) LION 2011. LNCS, vol. 6683, pp. 507–523. Springer, Heidelberg (2011). https://doi.org/10.1007/978-3-642-25566-3_40
16. Hutter, F., et al.: AClib: a benchmark library for algorithm configuration. In: Pardalos, P.M., Resende, M.G.C., Vogiatzis, C., Walteros, J.L. (eds.) LION 2014. LNCS, vol. 8426, pp. 36–40. Springer, Cham (2014). https://doi.org/10.1007/978-3-319-09584-4_4
17. Hutter, F., Tompkins, D.A.D., Hoos, H.H.: Scaling and probabilistic smoothing: efficient dynamic local search for SAT. In: Van Hentenryck, P. (ed.) CP 2002. LNCS, vol. 2470, pp. 233–248. Springer, Heidelberg (2002). https://doi.org/10.1007/3-540-46135-3_16
18. Kleinberg, R., Leyton-Brown, K., Lucier, B.: Efficiency through procrastination: approximately optimal algorithm configuration with runtime guarantees. In: Proceedings of the Twenty-Sixth International Joint Conference on Artificial Intelligence, IJCAI 2017, pp. 2023–2031. AAAI Press (2017)
19. Kleinberg, R., Leyton-Brown, K., Lucier, B., Graham, D.: Procrastinating with confidence: near-optimal, anytime, adaptive algorithm configuration. In: Advances in Neural Information Processing Systems 32, NeurIPS 2019, pp. 8881–8891. Curran Associates Inc. (2019)
20. López-Ibáñez, M., Dubois-Lacoste, J., Cáceres, L.P., Birattari, M., Stützle, T.: The irace package: iterated racing for automatic algorithm configuration. Oper. Res. Perspect. **3**, 43–58 (2016)
21. Mann, H.B., Whitney, D.R.: On a test of whether one of two random variables is stochastically larger than the other. Ann. Math. Stat. **18**, 50–60 (1947)
22. Pushak, Y., Hoos, H.: Algorithm configuration landscapes: In: Auger, A., Fonseca, C.M., Lourenço, N., Machado, P., Paquete, L., Whitley, D. (eds.) PPSN 2018. LNCS, vol. 11102, pp. 271–283. Springer, Cham (2018). https://doi.org/10.1007/978-3-319-99259-4_22
23. Weisz, G., György, A., Szepesvári, C.: LeapsAndBounds: a method for approximately optimal algorithm configuration. In: Proceedings of the 35th International Conference on Machine Learning, ICML 2018, pp. 5254–5262. PMLR (2018)
24. Weisz, G., György, A., Szepesvár, C.: CapsAndRuns: an improved method for approximately optimal algorithm configuration. In: Proceedings of the 36th International Conference on Machine Learning, ICML 2019, pp. 6707–6715. PMLR (2019)

Dominance, Indicator and Decomposition Based Search for Multi-objective QAP: Landscape Analysis and Automated Algorithm Selection

Arnaud Liefooghe[1(✉)], Sébastien Verel[2], Bilel Derbel[3], Hernan Aguirre[4], and Kiyoshi Tanaka[4]

[1] JFLI – CNRS IRL 3527, University of Tokyo, Tokyo 113-0033, Japan
arnaud.liefooghe@univ-lille.fr
[2] Univ. Littoral Côte d'Opale, LISIC, 62100 Calais, France
[3] Univ. Lille, CNRS, Centrale, Inria, UMR 9189 - CRIStAL, 59000 Lille, France
[4] Shinshu University, Faculty of Engineering, Nagano, Japan

Abstract. We investigate the properties of large-scale multi-objective quadratic assignment problems (mQAP) and how they impact the performance of multi-objective evolutionary algorithms. The landscape of a diversified dataset of bi-, multi-, and many-objective mQAP instances is characterized by means of previously-identified features. These features measure complementary facets of problem difficulty based on a sample of solutions collected along random and adaptive walks over the landscape. The strengths and weaknesses of a dominance-based, an indicator-based, and a decomposition-based search algorithm are then highlighted by relating their expected approximation quality in view of landscape features. We also discriminate between algorithms by revealing the most suitable one for subsets of instances. At last, we investigate the performance of a feature-based automated algorithm selection approach. By relying on low-cost features, we show that our recommendation system performs best in more than 90% of the considered mQAP instances.

1 Introduction

The multi-objective quadratic assignment problem (mQAP) [12,13] appears to be one of the most challenging problem from multi-objective combinatorial optimization. This is probably due to its intrinsic difficulties and the variety of mQAP instances from the literature, having different structures and properties in terms of problem size and data distributions, but also with respect to the number of objectives to be optimized, and their degree of conflict. Evolutionary multi-objective optimization (EMO) algorithms and other population-based

This research was conducted in the scope of the French/Japanese MODŌ international lab, and was partially supported by the French national research agency under Project ANR-16-CE23-0013-01.

T. Bäck et al. (Eds.): PPSN 2020, LNCS 12269, pp. 33–47, 2020.
https://doi.org/10.1007/978-3-030-58112-1_3

multi-objective search heuristics are natural candidates to solve them. They range from dominance-based approaches to indicator- and decomposition-based refinements [6,28,29]. However, they have only been partially investigated for the mQAP, and focused mainly on problems with few (mostly 2) objectives [8,16,19]. There is obviously no single method that is more suitable for all problems, and multi-objective problems are no exception. As such, in the panorama of EMO algorithms, it remains unclear if and how problem characteristics result in differences in the performance of multi-objective selection strategies, and what actually makes an algorithm efficient or not when solving a given problem.

Landscape analysis [17] has emerged as a valuable methodolgy for examining the properties of optimization problems and their effect on search performance. Based on high-level landscape features, it becomes possible to improve our understanding of problems and algorithms, and also to predict algorithm performance, eventually leading to automated algorithm selection [11,22]. There is a large body of literature on single-objective landscape analysis [23], including for the quadratic assignment problem [4,17,21,25]. However, the literature on multi-objective landscapes is more scarce. Interestingly, most papers deal with the mQAP, being about properties from the Pareto set [12,20] or from the solution space [8,9]. However, previous studies were once again mostly devoted to problems with few objectives (mostly 2, sometimes 3), and often require the solution space or the Pareto set to be exhaustively enumerated, making them impractical for prediction. At last, existing multi-objective features were not always related to search performance, and never used for automated algorithm selection.

In a recent paper [15], we revised landscape features for multi-objective combinatorial optimization by building upon those previous studies, and by deriving additional low-cost landscape features that were revealed as highly impactful for EMO algorithms. In this paper, we are interested in analyzing the impact of mQAP instance characteristics on higher-level landscape features, such as ruggedness and multimodality. We also aim at clarifying the impact of mQAP landscape features on problem difficulty and search performance, and at examining if a difference in feature values implies any difference in the performance of EMO algorithms. Our contributions can be summarized as follows:

(1) We characterize the landscape of large-scale mQAP instances with different properties by means of local multi-objective features from [15];
(2) We relate mQAP landscape features with the performance of a dominance-, an indicator-, and a decomposition-based EMO algorithm [6,28,29];
(3) We investigate the performance of feature-based automated algorithm selection by measuring its ability to discriminate between EMO algorithms, and by carefully calibrating the budget allocated to features and search.

The paper is organized as follows. Section 2 gives the necessary background on multi-objective optimization and EMO algorithms. Section 3 presents the mQAP and the instance dataset considered in our analysis. Section 4 introduces multi-objective landscape features, studies how they correlate with one another and with algorithm performance, and highlights their importance to explain search difficulty. Section 5 investigates the prediction accuracy of a feature-based automated algorithm selection system by paying a particular attention to the cost of features. Section 6 concludes the paper and discusses further research.

2 Multi-objective Optimization

2.1 Definitions

Let us consider an objective function vector $f\colon X \mapsto Z$ to be minimized. Each solution from the solution space $x \in X$ maps to a vector in the objective space $z \in Z$, with $Z \subseteq \mathbb{R}^m$, such that $z = f(x)$. In multi-objective combinatorial optimization, the solution space X is a discrete set. Given two objective vectors $z, z' \in Z$, z is dominated by z' iff for all $i \in \{1, \ldots, m\}$ $z'_i \leqslant z_i$, and there is a $j \in \{1, \ldots, m\}$ such that $z'_j < z_j$. Similarly, given two solutions $x, x' \in X$, x is dominated by x' iff $f(x)$ is dominated by $f(x')$. An objective vector $z^\star \in Z$ is non-dominated if there does not exist any $z \in Z$ such that $z^\star$ is dominated by z. A solution $x^\star \in X$ is Pareto optimal (PO), or non-dominated, if $f(x)$ is non-dominated. The set of PO solutions is the Pareto set (PS); its mapping in the objective space is the Pareto front (PF). One of the main challenges in multi-objective optimization is to identify the PS, or a good approximation of it for large-size and complex problems. A number of EMO and other multi-objective heuristics have been designed to this end since the late eighties [3,5].

2.2 EMO Algorithms

We conduct our analysis on three EMO algorithms: NSGA-II, IBEA, MOEA/D. They were selected as representatives of the state-of-the-art in the EMO field, covering dominance-, indicator-, and decomposition-based approaches, respectively. They differ in their selection mechanism, which is described below.

NSGA-II [5] is an elitist *dominance*-based EMO algorithm using Pareto dominance for survival and parent selections. At a given iteration, the current population P_t is merged with its offspring Q_t, and is divided into non-dominated fronts $F = \{F1, F2, \ldots\}$ based on the non-dominated sorting procedure [10]. The front in which a given solution belongs to gives its rank within the population. Crowding distance is also calculated within each front. Selection is based on ranking, and crowding distance is used as a tie breaker. Survival selection consists in filling the new population P_{t+1} with solutions having the best (smallest) ranks. In case a front F_i overfills the population size, the required number of solutions from F_i are chosen based on their crowding distance. Parent selection for reproduction consists of binary tournaments between random individuals, following the lexicographic order induced by ranks first, and crowding distance next.

IBEA [29] introduces a total order between solutions by means of a binary quality *indicator* I. Its selection mechanisms is based on a pairwise comparison of solutions from the current population P_t with respect to I. A fitness value is assigned to each individual $x \in P_t$, measuring the "loss in quality" if x was removed from the current population; i.e. $Fitness(x) = \sum_{x' \in P \setminus \{x\}} (-e^{-I(x',x)/\kappa})$, where $\kappa > 0$ is a user-defined scaling factor. The survival selection mechanism is based on an elitist strategy that combines the current population P_t with its offspring Q_t. It iteratively removes the worst solution until the required population size is reached, and assigns the resulting population into P_{t+1}. Each time a solution is deleted, the fitness values of the remaining individuals are updated. Parent selection for reproduction consists of binary tournaments between randomly chosen individuals. Different indicators can be used within IBEA. We here consider the binary additive ε-indicator ($I_{\varepsilon+}$), as defined by the original authors [29]: $I_{\varepsilon+}(x, x') = \max_{i \in \{1,\ldots,m\}} \{f_i(x) - f_i(x')\}$. It gives the minimum value by which a solution $x \in P_t$ has to, or can be, translated in the objective space in order to weakly dominate another solution $x' \in P_t$.

MOEA/D [28] is a *decomposition*-based EMO algorithm that seek a high-quality solution in multiple regions of the objective space by decomposing the original (multi-objective) problem into a number of scalarizing (single-objective) sub-problems. Let μ be the population size. A set $(\lambda^1, \ldots, \lambda^i, \ldots, \lambda^\mu)$ of uniformly-distributed weighting coefficient vectors defines the scalarizing sub-problems, and a population $P = (x^1, \ldots, x^i, \ldots, x^\mu)$ is maintained such that each individual x^i maps to the sub-problem defined by λ^i. Different scalarizing functions can be used within MOEA/D. We here consider the weighted Chebyshev scalarizing function: $g(x, \lambda) = \max_{i \in \{1,\ldots,m\}} \lambda_i \cdot |z^\star_i - f_i(x)|$, such that x is a solution, λ is a weighting coefficient vector and $z^\star$ is a reference point. In addition, a neighboring relation is defined among sub-problems, based on the assumption that a given sub-problem is likely to benefit from the solution maintained in neighboring sub-problems. The neighborhood $\mathcal{B}(i)$ is defined by considering the T closest weighting coefficient vectors for each sub-problem i. At each iteration, the population evolves with respect to a given sub-problem. Two solutions are selected at random from $\mathcal{B}(i)$ and an offspring is produced by means of variation operators. Then, for each sub-problem $j \in \mathcal{B}(i)$, the offspring is used to replace the current solution x^j if there is an improvement in terms of the scalarizing function. The algorithm iterates over sub-problems until a stopping condition is satisfied.

3 Multi-objective Quadratic Assignment Problem

3.1 Problem Definition

Let us assume a given set of n facilities with e_{ij} being the flow between facilities i and j, and a given set of n locations with d_{ij} being the distance between locations i and j. The Quadratic Assignment Problem (QAP) [24] aims at assigning facilities to locations such that the sum of the products between flows and distances is minimal, and such that each facility is assigned to exactly one location, which is NP-hard [24]. The multi-objective QAP (mQAP) [12,13] considers m flow matrices under the same distance matrix, and can be stated as follows:

$$\min_{x \in X} \sum_{i=1}^{n} \sum_{j=1}^{n} d_{x_i x_j} \, e_{ij}^k \qquad k \in \{1, \dots, m\} \tag{1}$$

where x_i gives the location of facility i in the current solution $x \in X$, and the solution space X is the set of all possible permutations $\{1, \dots, n\}$ (such that $|X| = n!$). By increasing the number flow matrices m, we can define bi-objective ($m = 2$), multi-objective ($m = 3$) and many-objective ($m \geqslant 4$) mQAP instances.

3.2 Instance Dataset

Knowles and Corne [13] provide an instance generator that can produce mQAP instances with different characteristics in terms of the number of variables ($\mathtt{n}$), the number of objectives ($\mathtt{m}$), the correlation among flow matrices (ρ), and the structure of flow matrices (`type`): uniformly random (`uni`) or real-like (`rl`) flow values. Assuming that the dynamics and performance of EMO algorithms are impacted by these parameters, we consider a dataset covering a wide range of problems. In particular, we generate 1 000 mQAP instances following a design of experiments based on random latin hypercube sampling [2]. We consider a problem size in the range $\mathtt{n} \in \{30, \dots, 100\}$, a number of objectives $\mathtt{m} \in \{2, \dots, 5\}$, an objective correlation $\rho \in [-1, 1]$, and two instance types (`uni` and `rl`). We notice that, although the problem size and number of objectives are given, the type and the objective correlation are *unknown* in practice for unseen instances.

3.3 Algorithms Setting and Search Performance

We rely on an out-of-the-box implementation of the considered EMO algorithms with default parameters, as provided in the `jMetal 4.5` framework [7]. In terms of parameters, NSGA-II, IBEA and MOEA/D all use a population of size of 100, an exchange mutation with a rate of 0.2, and a partially-mapped crossover [10] with a rate of 0.95. Preliminary experiments revealed that using the partially-mapped crossover allows the search process to reach better quality in more than 90% of the cases, compared against the 2-point crossover used in a previous setting [15]. All the algorithms stop after 1 000 000 evaluations. We measure algorithm performance in terms of hypervolume ($\mathtt{hv}$) [30], and more particularly in terms of hypervolume relative deviation: $\mathtt{hvrd} = (\mathtt{hv}^\star - \mathtt{hv})/\mathtt{hv}^\star$, where $\mathtt{hv}^\star$ is the best-known hypervolume for the instance under consideration. The hypervolume measures the multi-dimensional area of the objective space covered by an approximation set, and is the only known strictly Pareto-compliant indicator [31]. The hypervolume reference point is set to the upper bound of objective values. For a given instance, each algorithm is executed 20 times, and

Table 1. Considered multi-objective landscape features from [15].

Description	Random walk		Adaptive walk
Correlation among objectives	f_cor_rws		–
average length of walks	–		length_aws
	Average	**Autocorrelation**	**Average**
Prop. dominated neighbors	#inf_avg_rws	#inf_r1_rws	#inf_avg_aws
Prop. dominating neighbors	#sup_avg_rws	#sup_r1_rws	#sup_avg_aws
Prop. incomparable neighbors	#inc_avg_rws	#inc_r1_rws	#inc_avg_aws
Prop. locally non-dominated neighbors	#lnd_avg_rws	#lnd_r1_rws	#lnd_avg_aws
Prop. supported locally non-dom. neighbors	#lsupp_avg_rws	#lsupp_r1_rws	#lsupp_avg_aws
Solution's hypervolume	hv_avg_rws	hv_r1_rws	hv_avg_aws
Solution's hypervolume difference	hvd_avg_rws	hvd_r1_rws	hvd_avg_aws
Neighborhood's hypervolume	nhv_avg_rws	nhv_r1_rws	nhv_avg_aws

the obtained hvrd values are averaged to estimate its expected performance. Significant difference between algorithms is also investigated in terms of statistical test.

4 Feature-Based Landscape Analysis

We start our analysis by characterizing mQAP instances with relevant features from the literature. We rely on the multi-objective landscape features introduced in [15], and particularly on *local* features, based on sampling, that do not require any prior knowledge about the solution space enumeration and/or the Pareto set. We start by recalling their definition. Then, we measure how they relate with each other, and how they individually relate with search performance. At last, we assess their joint effect on performance in an attempt to highlight the main difficulties encountered by EMO algorithms when solving a mQAP instance.

4.1 Multi-objective Landscape Features

The considered multi-objective landscape features are listed in Table 1. When adding the mQAP benchmark parameters (i.e., type, n, m, and ρ), this sums to a total of 30 features. We define the multi-objective landscape for a given mQAP instance as a triplet $(X, f, \mathcal{N})$, such that X is the solution space (i.e., the set of all possible permutations $\{1, \ldots, n\}$), $f \colon X \mapsto Z$ is the objective function vector defined in Eq. (1), and $\mathcal{N} \colon X \mapsto 2^X$ is a neighborhood relation based on the exchange operator, that consists in exchanging the locations of two facilities. The considered features are based on different measures computed on a *sample* of solutions extracted from a walk over the multi-objective landscape [15]. A *walk* is an ordered sequence of solutions $(x_0, x_1, \ldots, x_\ell)$ such that $x_0 \in X$, and $x_t \in \mathcal{N}(x_{t-1})$ for all $t \in \{1, \ldots, \ell\}$. During a *random walk* [27], there is no particular criterion to pick the neighboring solution at each step, a random neighbor is selected. The length of the walk $\ell_{\texttt{rws}}$ is a parameter: the longer the length, the better the features estimation. By contrast, during an *adaptive walk* [26], a dominating neighbor is selected at each step. The length $\ell_{\texttt{aws}}$ corresponds the number of steps performed until no further improvement is possible, and the walk falls into a Pareto local optimal solution (PLO) [18]. Multiple adaptive walks are typically performed to improve the features estimation.

Given an ordered sequence of solutions collected along a walk, we consider the following measures. For each solution, we sample its neighborhood, and we measure the proportion of dominated (`#inf`), dominating (`#sup`), and incomparable (`#inc`) neighbors. We also consider the proportion of non-dominated neighbors (`#lnd`), as well as the proportion of supported solutions therein (`#lsupp`). In addition, we compute the average hypervolume covered by each neighbor (`hv`), the average difference with the hypervolume covered by the current solution (`hvd`), and the hypervolume covered by all neighbors (`nhv`). For samples collected by means of a random walk, we compute both an average over all solutions from the walk and the first autocorrelation coefficient of the measures reported above. We also use solutions from the random walk to estimate the degree of correlation among the objectives (`f_cor_rws`). For adaptive walks, we simply compute average values for each measure, as well as the walk length $\ell_{\texttt{aws}}$ (`length_aws`), which is known to be a good estimator for the number of PLO [26].

Given $\eta_{\texttt{rws}}$ random walks of length $\ell_{\texttt{rws}}$, and a neighborhood sample size $\eta_{\texttt{neig}}$, the computational complexity for random walk features in terms of calls to the objective function is: $\eta_{\texttt{rws}}\left(1 + (1 + \ell_{\texttt{rws}}) \cdot \eta_{\texttt{neig}}\right)$. Similarly, the computational complexity for adaptive walk features is: $\eta_{\texttt{aws}}\left((1+\ell_{\texttt{aws}}) \cdot \eta_{\texttt{neig}} + e_{\texttt{aws}}\right)$, where $\eta_{\texttt{aws}}$ is the number of adaptive walks, $\ell_{\texttt{aws}}$ is the number of steps before the adaptive walk falls into a PLO, and $e_{\texttt{aws}}$ is the total number of evaluations performed for the walk to progress. However, we remark that `length_aws` alone is cheaper to compute, as it does not require any neighborhood exploration. Its complexity is just: $\eta_{\texttt{aws}} \cdot e_{\texttt{aws}}$. Similarly, the complexity of `f_cor_rws` alone is: $\eta_{\texttt{rws}}\,(1 + \ell_{\texttt{rws}})$. We also remark that $\eta_{\texttt{rws}}$, $\ell_{\texttt{rws}}$, $\eta_{\texttt{aws}}$ and $\eta_{\texttt{neig}}$ must be defined by the user for feature estimation. By contrast, the expected value for $\ell_{\texttt{aws}}$ and $e_{\texttt{aws}}$, observed in average over instances from our dataset is 45 and 10 845, respectively.

4.2 Correlation Among Landscape Features

In this section, we consider an expensive budget of $\eta_{\texttt{rws}} = 1$ random walk of length $\ell_{\texttt{rws}} = 1\,000$, and of $\eta_{\texttt{aws}} = 100$ independent adaptive walks, both using a neighborhood sample of $\eta_{\texttt{neig}} = 400$. Figure 1 reports the correlation matrix of all features, as measured on the instance dataset. The correlation is measured in terms of the non-parametric Spearman rank correlation coefficient. The matrix highlights the similarities between features from mQAP, and their association with benchmark parameters. Interestingly, we remark that the number of variables and the instance type are only slightly correlated with landscape features, apart from autocorrelation measures, and the length of adaptive walks for `n`: the larger the search space, the longer the length. By contrast, the number of objectives and their degree of conflict are correlated with average dominance measures, and `m` is also highly positively correlated with average hypervolume measures. Unsurprisingly, there is a high association among average dominance measures, and among average hypervolume measures. This suggests that performing both random and adaptive walks is redundant for those features, and that considering a single walk type might allow us to save computations. At last, we remark in the last column that the correlation between the length of adaptive walks and other features is quite high overall, and we already infer that `length_aws` will be informative for characterizing problem difficulty.

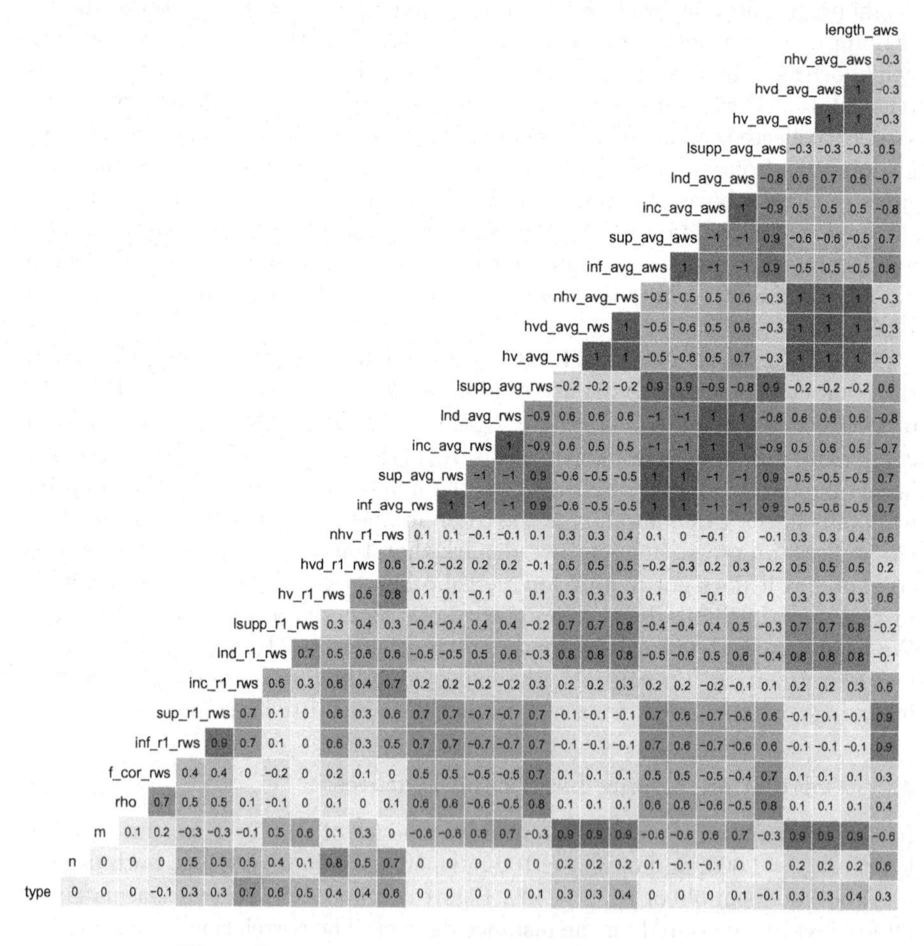

Fig. 1. Pairwise correlation among landscape features.

4.3 Correlation of Landscape Features with Search Performance

We now report in Fig. 2 the Spearman correlation of each feature with the expected performance of the three considered EMO algorithms, measured in terms of hypervolume relative deviation (**hvrd**). The corresponding scatterplots (with locally estimated scatterplot smoothing) for a selected subset of features are shown in Fig. 3 (others are not reported due to space restriction). Firstly, the effect of features on search difficulty has a similar trend for NSGA-II and IBEA, while being quite different for MOEA/D. In particular, the absolute correlation of each feature with the performance of MOEA/D is always below 0.5.

NSGA-II and IBEA are highly impacted by the number of objectives: their relative performance severely decreases with m, whereas MOEA/D performs almost constantly. Similarly, they perform better when average hypervolume measures are small, given that these are correlated with m, as pointed our earlier. IBEA is also impacted by average dominance measures: it performs best when there is not too few (nor too much) locally dominating points. Once again, it does not seem necessary to run both random and adaptive walks to measure average dominance and hypervolume values, given the similar impact of the corresponding features on search difficulty. By contrast, MOEA/D seems more impacted by autocorrelation measures, which quantify the ruggedness of the multi-objective landscape [15]: the rugger the landscape the less efficient MOEA/D. However, we argue that autocorrelation measures alter other algorithms as well; see, e.g., Fig. 3 (middle-left). Unfortunately, this effect is not captured by the correlation coefficients due to the particular trend of features against performance.

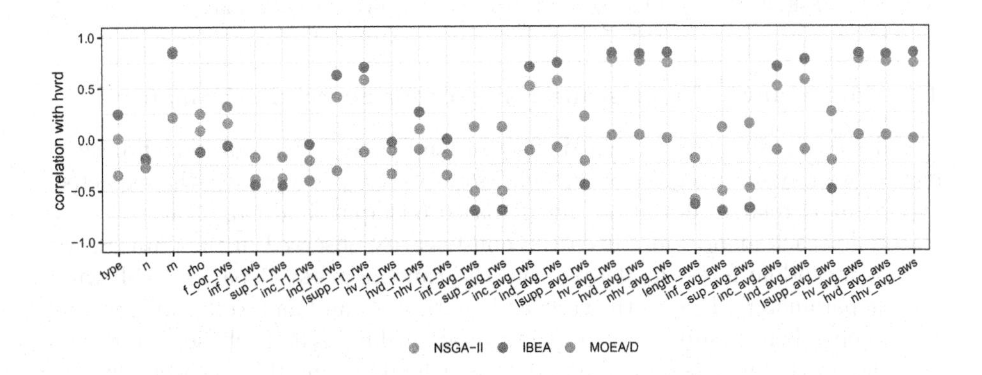

Fig. 2. Correlation between landscape features and algorithm performance.

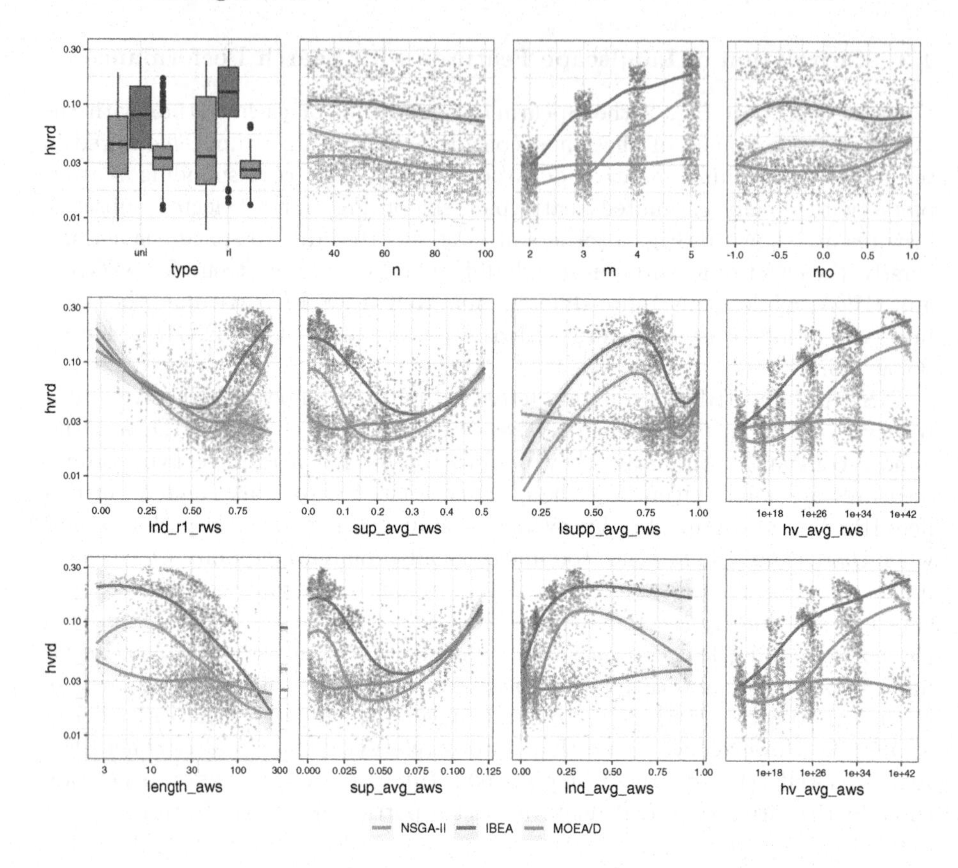

Fig. 3. Selected landscape features vs. algorithm performance.

4.4 Importance of Landscape Features for Search Performance

In order to measure the combined effect of landscape features on search performance, we rely on a regression model. More precisely, we predict the expected hypervolume relative deviation (`hvrd`) based on input landscape features, separately for each algorithm. Given the non-linearity observed in the dataset, we employ a random forest model [1] from the `randomForest` R package [14] with default parameters. Due to the stochastic nature of random forests, we perform 100 independent trainings and report average values. The coefficient of determination of the models on training data is 0.96, 0.98, and 0.78 respectively, for NSGA-II, IBEA, and MOEA/D. This means than more than 75% of the variance in search performance between instances is explained by landscape features.

Beyond prediction accuracy, random forest models have the ability to render the relative importance of each feature for making accurate predictions. In particular, we consider the mean decrease of prediction accuracy after each split on a given predictor [1]: the larger the decrease, the more important the predictor. The importance scores are depicted in Fig. 4. For readability, only the 12 most

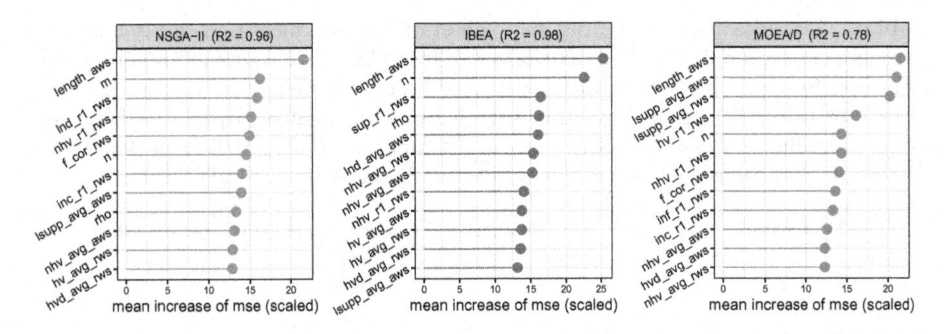

Fig. 4. Importance of landscape features for performance prediction of each algorithm.

important features are depicted for each algorithm, sorted in decreasing order of importance, from top to bottom. As conjectured in Sect. 4.2, `length_aws` turns out to be the most important feature for each algorithm. It relates to the multimodality of the landscape [15,26]: the longer the walk, the fewer the number of Pareto local optima, and the better search performance; see also Fig. 3 (bottom-left). For NSGA-II, the number of objectives `m` is the most important benchmark parameter, whereas it is the number of variables `n` for IBEA and MOEA/D. Some autocorrelation measures also appear for all algorithms, together with the degree of conflict among the objectives (whether ρ or `f_cor_rws`). Interestingly, the proportion of supported non-dominated neighbors is particularly influential for the scalarization-based MOEA/D.

5 Feature-Based Automated Algorithm Selection

5.1 Prediction Accuracy with Expensive Features

Let us examine the ability of landscape features to discriminate between the three algorithms. To do so, we now train a random forest *classification* model to predict whether NSGA-II, IBEA, or MOEA/D performs better, on average, for a given instance. The classification accuracy is reported in Table 2 for models based on different subset of features, corresponding to different costs. A feature-based classification model can predict the algorithm with the best average performance in about 90% of the cases, and an algorithm which is not statistically outperformed by any other in more than 99% of the cases. This is significantly more accurate than a random classifier, a dummy classifier that always predicts the most-frequent best algorithm (here, MOEA/D), and a classifier based on benchmark parameters only. Interestingly, we did not find any significant difference in terms of prediction accuracy between a model using all features and a model using solely features based on random walk plus only the length of adaptive walks. This might actually reduce the computational cost of algorithm selection.

Features importance for algorithm selection is depicted in Fig. 5. The length of adaptive walks is, once again, the most important feature. The subsequent

Table 2. Classification error for different subset of features, measured on random sub-sampling cross-validation (100 repetitions, 80/20% split). Two values are reported: the error rate in predicting the algorithm with the best performance on average, and the error rate in predicting an algorithm that is not statistically outperformed by any other, according to a Mann-Whitney test at a significance level of 0.05 with Bonferroni correction. The dummy classifier always returns the most frequent algorithm.

Subset of features	Classification error	Error predicting statistical best
$\{\texttt{n}, \texttt{m}\}$	.1962	.0332
$\{\texttt{type}, \texttt{n}, \texttt{m}, \rho\}$	.1197	.0072
$\{\star\texttt{\_rws}, \texttt{n}, \texttt{m}\}$	.1114	**.0062**
$\{\star\texttt{\_aws}, \texttt{n}, \texttt{m}\}$	.1125	.0065
$\{\star\texttt{\_rws}, \texttt{length\_aws}, \texttt{n}, \texttt{m}\}$	**.1089**	**.0056**
$\{\star\texttt{\_rws}, \star\texttt{\_aws}, \texttt{n}, \texttt{m}\}$	**.1077**	**.0063**
$\{\star\texttt{\_rws}, \star\texttt{\_aws}, \texttt{type}, \texttt{n}, \texttt{m}, \rho\}$	**.1078**	**.0063**
Random classifier	.6667	.3810
Dummy classifier (MOEA/D)	.4200	.1040

features have a very similar score, and cover complementary landscape characteristics, ranging from autocorrelation coefficients, to average dominance and hypervolume measures and benchmark parameters. Most notably, important adaptive walk features almost always have their random walk counterpart, whether it is for dominance or hypervolume measures.

5.2 Low-Cost Features Subtracted from Search Budget

We conclude our analysis by investigating the performance of a feature-based automated EMO algorithm selection method (AUTO-EMOA for short), while taking the budget allocated to the feature computation into account. Given the results presented above, we focus on a classification model based on features from random walk sampling (`rws`), together with `length_aws` and problem parameters that are given in practical scenarios (dimensions `n` and `m`). In contrast to the previous setting, we now consider a *low-cost* budget for features computation: $\eta_{\texttt{rws}} = 1$ random walk of length $\ell_{\texttt{rws}} = 200$ using a sample of $\eta_{\texttt{neig}} = 100$ neighbors at each step, and $\eta_{\texttt{aws}} = 1$ adaptive walk for estimating `length_aws` only. By measuring the one-to-one correlation between expensive and low-cost features (not reported), we remark that it is always larger than 0.85, apart from locally supported and hypervolume autocorrelation features (between 0.58 and 0.75), that were not detected as important previously.

The total computation of the considered low-cost features sums up to 30 946 evaluations, in average, per instance. Consequently, we deduce 50 000 (>30 946) evaluations from the search process allocated to AUTO-EMOA. In other words, we compare AUTO-EMOA with a search budget of 950 000 evaluations against NSGA-II, IBEA, and MOEA/D with a search budget of 1 000 000 evaluations. Results from 100 repetitions of random sub-sampling cross-validation with a 80/20% split are presented in Fig. 6. The statistical rank of AUTO-EMOA is 0.09 on average, more than three times lower than the best standalone approach

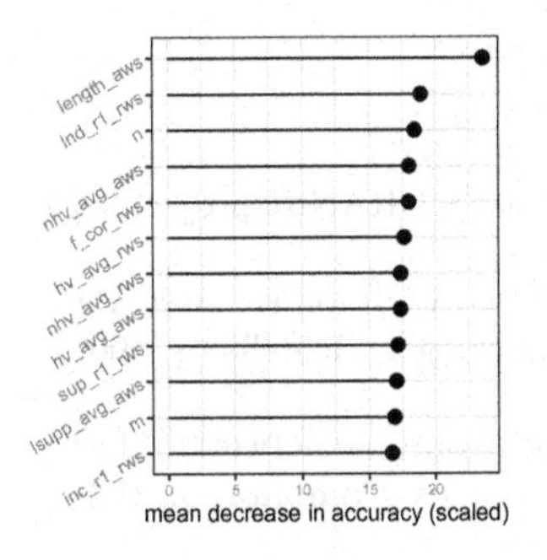

Fig. 5. Importance of features for algorithm selection.

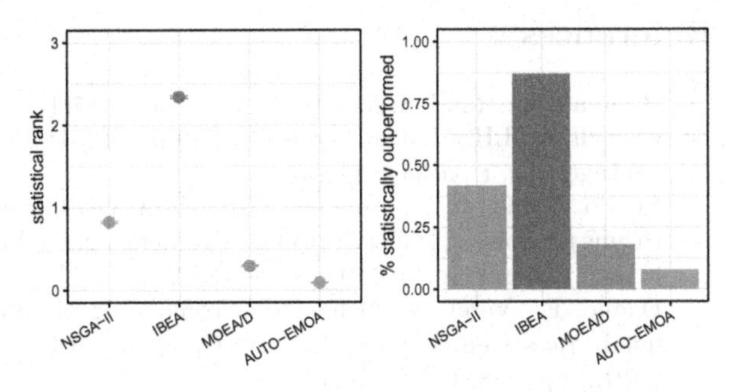

Fig. 6. Performance of AUTO-EMOA compared against other algorithms.

(MOEA/D, with 0.29). Among all instances seen during cross-validation, AUTO-EMOA was *not* significantly outperformed by any other approaches on 92% of the cases (82% for MOEA/D). As such, deducing a small part of the budget allocated to the search process for feature computation appears to be beneficial in order to gain knowledge about the tackled problem, and make better-informed decision on the appropriate multi-objective search strategy to apply for solving it.

6 Conclusions

In this paper, we analyzed the landscape of large-scale bi-, multi- and many-objective mQAP instances, and highlighted the relationship between landscape features and the performance of a dominance-, indicator-, and decomposition-based EMO algorithm. Our study highlights that algorithms are not only impacted by the number of objectives, but that the ruggedness and multimodality of the multi-objective landscape are also crucially important to properly explain search performance. An automated algorithm selection model also revealed the ability of multi-objective landscape features to discriminate between EMO algorithms. By simply allocating less than 5% of the budget to analyze the landscape of a given instance, our recommendation system was shown to perform best on more than 90% of instances under the scenario considered for validation.

Further research includes the investigation of other landscapes, including multi-objective continuous functions that require particular walks for sampling the solution space when computing the features. Additionally, we plan to consider additional EMO algorithms, and in particular highly-configurable frameworks for which we infer that feature-based algorithm configuration is essential.

References

1. Breiman, L.: Random forests. Mach. Learn. **45**(1), 5–32 (2001)
2. Carnell, R.: LHS: Latin Hypercube Samples (2018). https://CRAN.R-project.org/package=lhs, r package version 0.16
3. Coello Coello, C.A., Lamont, G.B., Van Veldhuizen, D.A.: Evolutionary Algorithms for Solving Multi-Objective Problems, 2nd edn. Springer, Heidelberg (2007). https://doi.org/10.1007/978-0-387-36797-2
4. Daolio, F., Verel, S., Ochoa, G., Tomassini, M.: Local optima networks of the quadratic assignment problem. In: Congress on Evolutionary Computation (CEC 2010)., pp. 1–8. IEEE (2010)
5. Deb, K.: Multi-Objective Optimization using Evolutionary Algorithms. Wiley, Hoboken (2001)
6. Deb, K., Pratap, A., Agarwal, S., Meyarivan, T.: A fast and elitist multiobjective genetic algorithm : NSGA-II. IEEE Trans. Evol. Comput. **6**(2), 182–197 (2002)
7. Durillo, J.J., Nebro, A.J.: jMetal: a Java framework for multi-objective optimization. Adv. Eng. Softw. **42**, 760–771 (2011)
8. Garrett, D., Dasgupta, D.: Analyzing the performance of hybrid evolutionary algorithms for the multiobjective quadratic assignment problem. In: IEEE Congress on Evolutionary Computation (CEC 2006), pp. 1710–1717 (2006)
9. Garrett, D., Dasgupta, D.: Multiobjective landscape analysis and the generalized assignment problem. In: Maniezzo, V., Battiti, R., Watson, J.-P. (eds.) LION 2007. LNCS, vol. 5313, pp. 110–124. Springer, Heidelberg (2008). https://doi.org/10.1007/978-3-540-92695-5_9
10. Goldberg, D.E.: Genetic Algorithms in Search, Optimization and Machine Learning. Addison-Wesley, Boston (1989)
11. Kerschke, P., Hoos, H., Neumann, F., Trautmann, H.: Automated algorithm selection: survey and perspectives. Evol. Comput. **27**(1), 3–45 (2019)
12. Knowles, J., Corne, D.: Towards landscape analyses to inform the design of a hybrid local search for the multiobjective quadratic assignment problem. In: Soft Computing Systems: Design, Management and Applications 2002, pp. 271–279 (2002)
13. Knowles, J., Corne, D.: Instance generators and test suites for the multiobjective quadratic assignment problem. In: Fonseca, C.M., Fleming, P.J., Zitzler, E., Thiele, L., Deb, K. (eds.) EMO 2003. LNCS, vol. 2632, pp. 295–310. Springer, Heidelberg (2003). https://doi.org/10.1007/3-540-36970-8_21
14. Liaw, A., Wiener, M.: Classification and regression by randomForest. R News **2**(3), 18–22 (2002)
15. Liefooghe, A., Daolio, F., Verel, S., Derbel, B., Aguirre, H., Tanaka, K.: Landscape-aware performance prediction for evolutionary multi-objective optimization. IEEE Trans. Evol. Comput. 1 (2019, early access)
16. López-Ibáñez, M., Paquete, L., Stützle, T.: Hybrid population-based algorithms for the bi-objective quadratic assignment problem. J. Math. Model. Algorithms **5**(1), 111–137 (2006)
17. Merz, P.: Advanced fitness landscape analysis and the performance of memetic algorithms. Evol. Comput. **12**(3), 303–325 (2004)
18. Paquete, L., Schiavinotto, T., Stützle, T.: On local optima in multiobjective combinatorial optimization problems. Ann. Oper. Res. **156**(1), 83–97 (2007)
19. Paquete, L., Stützle, T.: A study of stochastic local search algorithms for the biobjective QAP with correlated flow matrices. Eur. J. Oper. Res. **169**(3), 943–959 (2006)

20. Paquete, L., Stützle, T.: Clusters of non-dominated solutions in multiobjective combinatorial optimization: an experimental analysis. In: Barichard, V., Ehrgott, M., Gandibleux, X., T'Kindt, V. (eds.) Multiobjective Programming and Goal Programming: Theoretical Results and Practical Applications. Lecture Notes in Economics and Mathematical Systems, vol. 618, pp. 69–77. Springer, Heidelberg (2009). https://doi.org/10.1007/978-3-540-85646-7_7
21. Pitzer, E., Beham, A., Affenzeller, M.: Correlation of problem hardness and fitness landscapes in the quadratic assignment problem. In: Klempous, R., Nikodem, J., Jacak, W., Chaczko, Z. (eds.) Advanced Methods and Applications in Computational Intelligence. Topics in Intelligent Engineering and Informatics, vol. 6, pp. 165–195. Springer, Heidelberg (2014). https://doi.org/10.1007/978-3-319-01436-4_9
22. Rice, J.R.: The algorithm selection problem. Adv. Comput. **15**, 65–118 (1976)
23. Richter, H., Engelbrecht, A. (eds.): Recent Advances in the Theory and Application of Fitness Landscapes, Emergence, Complexity and Computation. Springer, Heidelberg (2014). https://doi.org/10.1007/978-3-642-41888-4
24. Sahni, S., Gonzalez, T.: P-complete approximation problems. J. ACM **23**(3), 555–565 (1976)
25. Smith-Miles, K., Lopes, L.: Measuring instance difficulty for combinatorial optimization problems. Comput. Oper. Res. **39**(5), 875–889 (2012)
26. Verel, S., Liefooghe, A., Jourdan, L., Dhaenens, C.: On the structure of multiobjective combinatorial search space: MNK-landscapes with correlated objectives. Eur. J. Oper. Res. **227**(2), 331–342 (2013)
27. Weinberger, E.D.: Correlated and uncorrelatated fitness landscapes and how to tell the difference. Biol. Cybern. **63**(5), 325–336 (1990)
28. Zhang, Q., Li, H.: MOEA/D: a multiobjective evolutionary algorithm based on decomposition. IEEE Trans. Evol. Comput. **11**(6), 712–731 (2007)
29. Zitzler, E., Künzli, S.: Indicator-based selection in multiobjective search. In: Yao, X., et al. (eds.) PPSN 2004. LNCS, vol. 3242, pp. 832–842. Springer, Heidelberg (2004). https://doi.org/10.1007/978-3-540-30217-9_84
30. Zitzler, E., Thiele, L.: Multiobjective evolutionary algorithms: a comparative case study and the strength pareto approach. IEEE Trans. Evol. Comput. **3**(4), 257–271 (1999)
31. Zitzler, E., Thiele, L., Laumanns, M., Fonseca, C.M., da Fonseca, V.G.: Performance assessment of multiobjective optimizers: an analysis and review. IEEE Trans. Evol. Comput. **7**(2), 117–132 (2003)

Deep Learning as a Competitive Feature-Free Approach for Automated Algorithm Selection on the Traveling Salesperson Problem

Moritz Seiler[1(✉)], Janina Pohl[1], Jakob Bossek[2], Pascal Kerschke[1], and Heike Trautmann[1]

[1] Statistics and Optimization Group, University of Münster, Münster, Germany
{moritz.seiler,janina.pohl,kerschke,trautmann}@uni-muenster.de
[2] Optimisation and Logistics, The University of Adelaide, Adelaide, Australia
jakob.bossek@adelaide.edu.au

Abstract. In this work we focus on the well-known Euclidean Traveling Salesperson Problem (TSP) and two highly competitive inexact heuristic TSP solvers, EAX and LKH, in the context of per-instance algorithm selection (AS). We evolve instances with 1 000 nodes where the solvers show strongly different performance profiles. These instances serve as a basis for an exploratory study on the identification of well-discriminating problem characteristics (features). Our results in a nutshell: we show that even though (1) promising features exist, (2) these are in line with previous results from the literature, and (3) models trained with these features are more accurate than models adopting sophisticated feature selection methods, the advantage is not close to the virtual best solver in terms of penalized average runtime and so is the performance gain over the single best solver. However, we show that a feature-free deep neural network based approach solely based on visual representation of the instances already matches classical AS model results and thus shows huge potential for future studies.

Keywords: Automated algorithm selection · Traveling Salesperson Problem · Feature-based approaches · Deep learning

1 Introduction

The *Traveling Salesperson Problem* (TSP) is a classical $\mathcal{NP}$-hard optimization problem of utmost relevance, e.g., in transportation logistics, bioinformatics or circuit board fabrication. The goal is to route a salesperson through a set of cities such that each city is visited exactly once and the tour is of minimal length. In the past decades tremendous progress has been made in the development of high-performing heuristic TSP solvers. The local search-based *Lin-Kernigham Heuristic* (LKH) [14] and the genetic algorithm *Edge-Assembly-Crossover* (EAX) [35],

T. Bäck et al. (Eds.): PPSN 2020, LNCS 12269, pp. 48–64, 2020.
https://doi.org/10.1007/978-3-030-58112-1_4

along with their respective restart versions introduced in Kotthoff et al. [25], undeniably pose the state-of-the-art in inexact TSP solving.

Automated Algorithm Selection (AS), originally proposed by Rice [39] back in 1976, is a powerful framework to predict the best-performing solver(s) from a portfolio of candidate solvers by means of machine learning. It has been successfully applied to a wide spectrum of challenging optimization problems in both the combinatorial [24,29,30,40,48] and continuous domain [4,21] with partly astonishing performance gains – see the recent survey by Kerschke et al. [19] for a comprehensive overview. In particular, the TSP was subject to several successful AS-studies [20,25,33,34,37] which exploited the complementary performance profiles of simple heuristics on the one hand and the state-of-the-art solvers LKH and EAX on classical TSP benchmark sets on the other hand.

In the classic setting, AS relies on characteristic problem instance properties, termed *(instance) features*. These features are used as predictor variables for classical machine learning algorithms, e.g., random forests or support vector machines. The key idea – and ideal outcome – is that these features can easily be used to automatically derive decision rules that are well-suited to partition the instance space into ideally disjoint sub-spaces of instances, and which then are uniquely solved best by different solvers. However, features have many drawbacks: they are usually hand-crafted in a tedious process [15], partly require time-consuming calculations (which need to be taken into account by the model fitting step) and are problem-tailored (or at least specific to a problem domain, e.g., graph problems). Moreover, we usually prefer light-weight models with few features. Hence, training AS models is frequently combined with automated feature selection methods [11,36,46] or dimensionality reduction techniques [12].

Recently, Alissa et al. [1] took a promising new path after exploring first approaches to avoid manual feature-crafting [42,43]. The authors proposed a deep learning based approach which does not rely on any a-priori calculated feature profiles. Instead, in their study on the 1D-Bin-Packing Problem (BPP) the neural network is given temporal sequence data of the BPP instance as the only input. They were able to achieve drastic improvements. In this paper we adopt and adapt this idea for the TSP. To this end, we evolve a set of instances where LKH and EAX show strongly different behaviour in terms of *Penalized Average Runtime*[1] (PAR10; [2]). We show that with classical AS we can clearly beat the *Single Best Solver* (SBS; the solver with best average performance). However, the gap to the *Virtual Best Solver* (VBS; perfectly predicting oracle) can only be reduced slightly with much room for improvement. This holds true even in the case when we enrich the machine learning pipeline with (1) hand-selected feature subsets (based on exploratory data analysis), (2) different feature selection methods, or (3) a combination of both. After that, we propose a feature-free deep learning approach where the neural networks are trained on the plain image

[1] The PAR10-score is a common measure in AS for combinatorial optimization problems. For a stochastic algorithm A and an instances I it is defined as the average of the running times of A on I where runs which did not reach the optimum within a given time limit T are penalised by a factor of $10 \cdot T$.

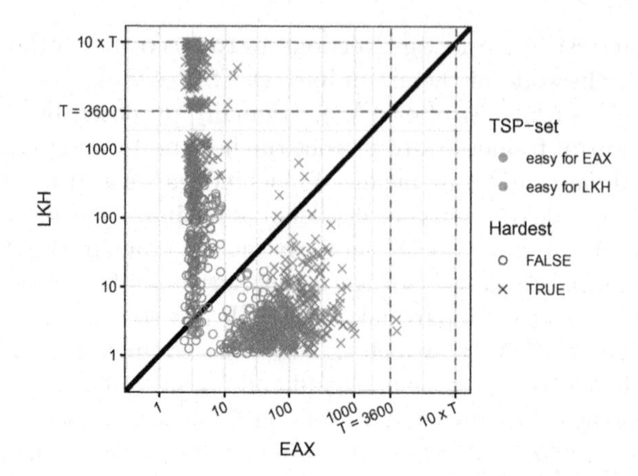

Fig. 1. PAR10 values (log-scaled) of EAX and LKH show complementary performance on the two subsets of instances (easy for EAX or LKH) and thereby suggest huge potential for automated algorithm selection. (Color figure online)

representations of Euclidean TSP instances. This approach achieves competitive performance, but drops the need for manual feature derivation and calculation.

The remainder of this paper is structured as follows. We describe the benchmark set (generation) and pre-selection of feature subsets in Sects. 2 and 3 respectively. In Sect. 4 we present the results that we achieve using classical feature-based AS approaches. Next, in Sect. 5, we detail our feature-free deep learning based approaches, and compare the results with the classical models. We close with a discussion and an outlook in Sect. 6.

2 Evolving TSP Instances

Our benchmark requires a set of Euclidean TSP instances that show strong differences in algorithmic performance. To this end, we adopt an *evolutionary algorithm* (EA) and creative mutation operators recently proposed by Bossek et al. [6]. Their method allows for the tailored generation[2] of TSP instances that (1) have the desired performance difference (in terms of the ratio of PAR10-scores), (2) show multifarious topologies in terms of point arrangement in the Euclidean plane[3], and (3) are well distributed in the space of instances characteristics/features; in particular the latter two properties were not achieved by

[2] Various TSP benchmark libraries exist, e.g., TSPLIB [38] or classical Random Uniform Euclidean (RUE) instances. However, these instances are either inhomogeneous in size or exhibit very little structural difference. For the purpose of algorithm selection though a balanced and homogeneous benchmark set is highly beneficial.

[3] The mutation operators are designed to evolve structures that can be observed in real-world TSP instances, e.g., *Very Large Scale Integration* (VLSI) and are thus closer to the real-world than the often used random uniform problems.

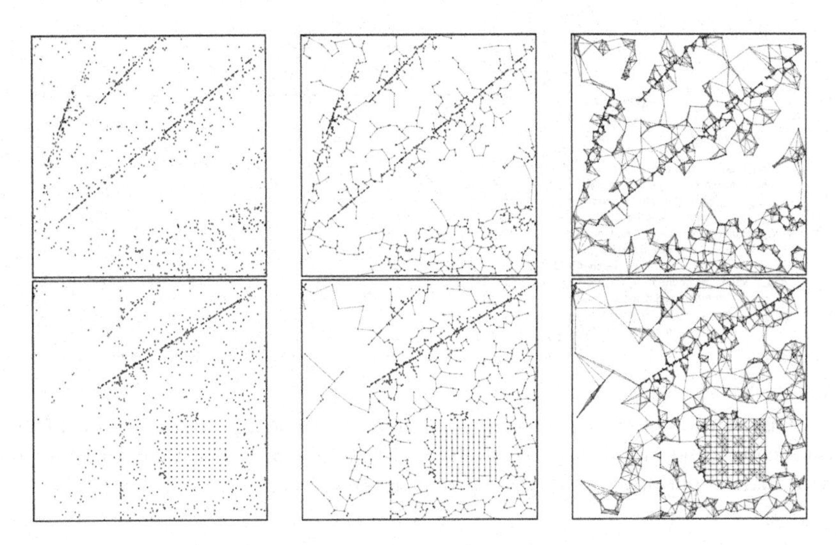

Fig. 2. Exemplary visual representations of two TSP instances in terms of point cloud only (left), a minimum spanning tree (center) and the 5-nearest-neighbor graph (right). The top row shows the instance for which the highest mean PAR10 score was achieved by EAX, the bottom row shows the respective counterpart of LKH.

evolutionary instance generation methods before [33,34]. For sake of brevity we refer the reader to Bossek et al. [6] for more details.

We generated a balanced data set of 1 000 TSP instances with $n = 1\,000$ nodes per instance using the EA parameters from [6]; each 500 being uniquely faster solved to optimality by either EAX or LKH.[4] All generated data is available in a public GitHub repository (https://github.com/mvseiler/PPSN_TSP_DL).

Figure 1 depicts the performance – measured by means of PAR10 – of EAX and LKH on the entire benchmark set. The plot highlights apparent potential for automated algorithm selection due to strong performance differences. On the other hand the data reveals the general superiority of the EAX solver since it is much harder to evolve instances that are hard for EAX (shown as orange points). There are just two instances for which the EAX hits the cutoff time $T = 3\,600\,\text{s}$ (1 h) at least once in all of its ten independent runs on that instance. In contrast, LKH frequently gets stuck in local optima – see the cluster of green points between $T = 3\,600$ and $10 \cdot T$ in the top left corner.

3 Identifying Adequate Subsets of TSP Features

The state-of-the-art TSP-related feature sets [16,33,37] consist of hundreds of hand-crafted features. Features range from statistics (mean, variance etc.) of edge

[4] We work with the restart versions of EAX and LKH which trigger a restart once the internal stopping conditions are met [25] as long as the time limit is not reached.

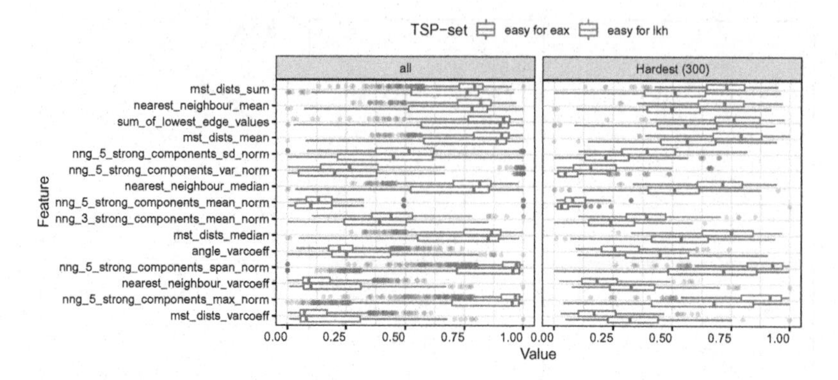

Fig. 3. Distribution of 15 best features according to the significance-test based feature importance method on all instances (left) and the 300 hardest instances with respect to mean PAR10 performance (right). (Color figure online)

lengths, angles of nearest neighbors, to more sophisticated features based on Minimum Spanning Trees (MST) or k-Nearest-Neighbor-Graphs (k-NNG). Figure 2 depicts visual impressions of MSTs and k-NNGs on two evolved instances; these images will be a key ingredient to the neural network in Sect. 5.

Due to the size of feature sets, in the context of algorithm selection, automated feature selection methods have shown their suitability for automatically filtering a (small) subset of discriminating features [20]. Regardless of the sophistication of the feature selection method at hand, feature selection needs to cope with an exponentially sized search space of the underlying subset-selection problem. Hence, in order to assist our feature-based model fitting we conduct a simple univariate exploratory data analysis in order to identify a lucid subset of adequate features *a-priori*. To this end we adopt a simple greedy heuristic. First, all features f are scaled to $[0, 1]$ to allow for a fair comparison across the features. Then, we perform a two-sided non-parametric Wilcoxon-Mann-Whitney test [31] at significance level $\alpha = 0.05$ per feature f, to check the null hypothesis that the distributions of the EAX instances and the LKH instances with respect to f differ by a location shift equal to zero. We extract the 15 most relevant features according to the smallest p-values. This procedure was done once for the entire benchmark set of salesperson features as the most comprehensive feature set (see Sect. 4) and was repeated for subsets of "hardest" instances of decreasing size. Hardest in this context relates to (a) each 300 and 150 hardest instances for each solver with respect to mean PAR10-score, or (b) the ratio of mean PAR10-scores. Both methods pursue to reduce the benchmark set to instances with maximal performance differences.

Figure 3 shows the distribution of selected features both across the whole benchmark set and the subset of each 300 most difficult instances with respect to mean PAR10-score for each solver (marked with crosses in Fig. 1). Noticeably, the 15 most relevant features are identical for both sets. They are mainly composed of summary statistics on strong connected components of the nearest neighbor graph (nng_*) and properties based on minimum spanning trees (mst_*). This is very much in line with crucial features identified in most TSP-related AS-studies [7,8,20,33,34,37] by sophisticated variable importance measurement. These features seem plausible since both MSTs and NNGs capture the global structure, e.g., existence of clusters etc., very well. A close look at Fig. 3 suggests, that instances that are easy for EAX cover a wider range of values which is derived from wider (green) boxes for the features while easy instances for LKH show much more narrow (orange) boxes, with many strong outliers though. The right hand plot, however, shows that the hardest instances seem to be better separable with the features. For instance, for the features in the top 4 rows we observe that 75% of feature values for LKH-friendly instances are higher than 75% of the respective values for EAX-friendly instances.

4 Classical Algorithm Selection

Figure 1 reveals very complementary performances of EAX and LKH, which gives us reason to assume that automated AS might work well in this setting. Further, as outlined in the univariate analysis of the TSP features (see Sect. 3), the features also indicate their potential for distinguishing instances that are beneficial for EAX from instances for which LKH is preferable. As previous works [20,25] already confirmed the effectivity of feature-based AS, we adopted their experimental setup – and only slightly modified it to the scenario at hand. Below, we will outline the considered machine learning algorithms, feature sets, as well as feature selection strategies, which have been used for training our final selectors.

4.1 Experimental Setup

All our candidate AS models are trained using PAR10 [18] as performance measure and assessed with a 10-fold cross-validation (CV). As indicated by Fig. 1, it is much more difficult for LKH to perform well on the instances that were evolved in favor of EAX (green points), rather than vice versa (orange). Therefore, the selectors will likely have a bias towards EAX instances. To adjust for this bias, we additionally tune the classification threshold for all trained models.

For training the potential automated AS models, we considered four different classifiers [13] using the R-package `mlr` [3]: decision trees [45], random forests [27], support vector machines [17] and gradient boosting [9]. Each of them is trained using three different feature sets: the UBC features from Hutter et al. [16], a large set of features by Pihera and Musliu [37], as well as the TSP features from the R-package `salesperson`[5] [5]. The salesperson features provide the up-to-now most comprehensive collection of features; in fact, they are a strict superset of the Pihera and tspmeta features [34]. On the other hand, the UBC and Pihera

[5] The R-package `salesperson` is an efficient and more comprehensive extension of the `tspmeta` feature generator by [34].

Table 1. Overview of all PAR10 results using the classical feature-based AS approach. The best PAR10-scores are highlighted in red. Note that we did not perform exhaustive feature selection on the non-reduced feature sets due to enormous computational costs.

Feature sel. method	ML algor.	All features			Top 15 features				
		Pihera	UBC	Sales.	All	r150	r300	s150	s300
None	rpart	67.25	69.35	67.70	67.32	67.32	67.52	68.12	67.32
	rf	59.09	61.56	62.22	61.21	61.99	60.14	61.51	62.30
	xgboost	65.40	67.00	67.57	65.53	65.53	64.51	65.78	65.53
	ksvm	61.34	61.55	63.09	61.98	61.98	60.54	61.26	61.98
sffs	rpart	66.20	67.92	67.26	66.82	66.82	66.82	66.82	66.82
	rf	325.27	58.42	62.77	60.70	59.74	56.29	56.29	56.29
	xgboost	65.45	66.60	62.61	64.20	64.20	64.20	64.20	64.20
	ksvm	66.42	**56.67**	60.57	62.38	62.38	62.22	62.06	62.38
sfbs	rpart	67.16	66.10	65.61	67.12	67.12	67.19	68.12	67.12
	rf	60.01	61.56	63.24	62.05	61.95	59.62	62.19	62.17
	xgboost	65.47	67.00	67.57	65.88	63.87	64.77	61.87	65.88
	ksvm	61.25	63.07	62.88	60.66	60.62	59.40	62.75	60.47
exh.	rpart				67.19	67.19	66.82	67.10	67.19
	rf				86.70	59.68	56.29	60.57	63.58
	xgboost				61.75	64.28	63.02	64.09	63.02
	ksvm				62.16	62.16	62.16	62.07	62.16

Note: r300 (s300) indicates the set of 300 hardest instances w.r.t. PAR10-ratio (PAR10-score).

features led to the best performing algorithm selectors in previous works [20,25] – which did not consider the salesperson features as the package did not exist back then. Note that there is a large overlap across the three considered feature sets as outlined in [19]. To reduce the noise within and redundancy between the features, we additionally created five small subgroups from the `salesperson` feature set, consisting of 15 features each (see Sect. 3 for details).

In addition to the 32 potential AS models described above (8 feature sets $\times$ 4 learners), the respective feature sets were further reduced using three automated feature selection strategies: sequential floating forward selection (sffs), sequential floating backward selection (sfbs), and – for the reduced feature sets – exhaustive search of the 15 features [19,23]. This resulted in 84 further candidate selectors.

4.2 Findings

Table 1 summarizes the averaged PAR10 performances of all 116 considered AS models, with the best achieved scores highlighted in red. Of course, all shown PAR10-scores already include the costs for the computation of the TSP features. On average those costs account for merely 0.7s at most. According to the listed performances of the best models (61.21 s), the SBS (67.47 s) and the VBS (4.92 s), the best classical AS approaches are able to reduce the SBS-VBS-gap by 10%.

The best found selectors are random forests, which reduced the top 15 features that they were given initially, to the following four features: the sum, arithmetic mean, median and coefficient of variation of the distances of the MST. Moreover, the tuned thresholds varied from 4% to 33% across the ten folds,

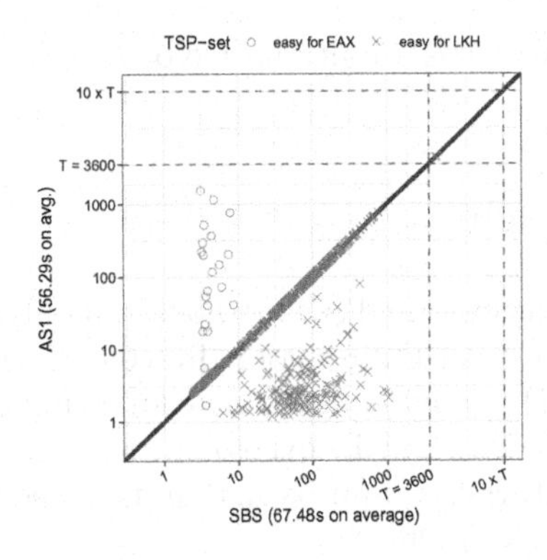

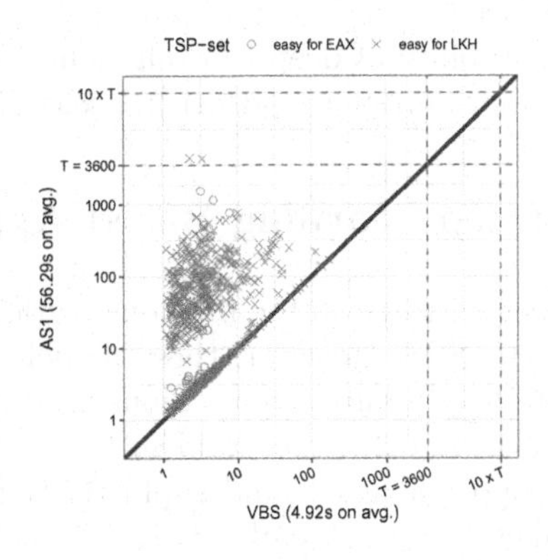

Fig. 4. PAR10-scores (log-scaled) of the best classical AS model reveal the improvement of the best selector over the SBS (left), and the gap towards the VBS (right).

implying that EAX has always been selected once the model predicted EAX with a probability of at least 33%. In consequence, out of all instances, in which LKH was actually the faster solver, it has only been selected 151 times (corresponding to roughly 30%). On the other hand, LKH has only been (wrongly) picked in 4% of the cases, in which EAX would have been the correct choice.

When starting with the full feature sets from Pihera, UBC and salesperson, a support vector machine based on a subset of the UBC features achieved the best performance (printed in bold in Table 1). In fact, the PAR10-score of 56.67 s is only slightly worse than the one of our best selector(s). Noticeably, the SVM also relied on MST features only: the arithmetic mean and standard deviation of the lengths of the edges in the MST, as well as the skewness of its node degrees.

Our findings are also confirmed by the left image of Fig. 4 as only few observations (20) are located above the diagonal. However, looking at the right image, it becomes nearly obvious that the selector is still quite far away from the performance of the VBS, as shown by the many misclassifications (381 out of all 1 000 instances) above the diagonal.

In an attempt to better understand the reason for these rather small thresholds, we investigated the misclassification costs in detail. *Prior* to tuning the threshold, LKH was predicted 154 times when EAX would have been the correct choice – and each of those misclassifications caused (on average) an overhead of 4 423.89 s. In contrast, the 128 cases in which EAX was predicted instead of LKH only came with an average penalty of 95.95 s. *After* tuning the thresholds, each of the 20 wrong predictions of LKH caused only 290.04 s – compared to the average penalty of 126.07 s for the 361 wrong predictions of EAX. Thus, by being rather conservative and only predicting LKH in cases, where the model is highly certain, the selector was able to reduce the misclassification costs significantly.

As our results indicate, the common feature-based approaches are able to improve over the SBS. However, it is also noticeable that the currently available features still have a very hard time in extracting sufficient information from the TSP instance to reliably predict the better solver. Therefore, we will test the suitability of deep learning neural networks as an alternative or supporting means for automated algorithm selection.

5 Deep Learning Based Approach

As demonstrated in the previous section, feature-based AS methods can outperform the SBS. However, these models come with three major drawbacks: they (1) are hand-crafted in a tedious process, (2) partly require time-consuming calculations, and (3) are problem tailored (see Sect. 1). To overcome these issues, we propose a novel, and sophisticated feature-free approach that is based on so-called *Convolutional Neural Networks* (CNN) [26].

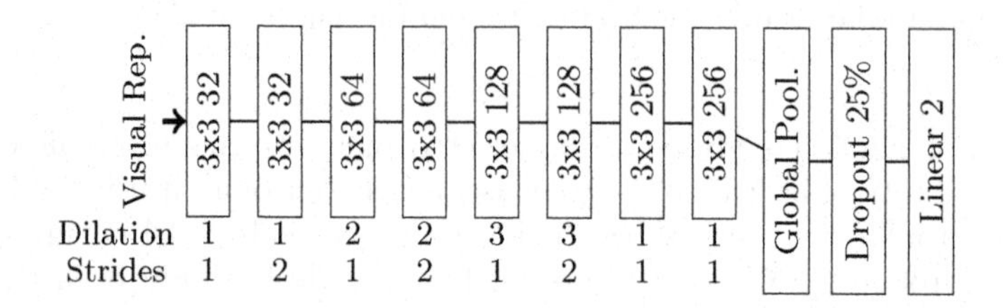

Fig. 5. The chosen neural architecture. All convolutional blocks include a *Group Normalization Layer* and a *Rectified Linear Unit* activation (Conv → GN → ReLU). The strides are used to reduce the feature maps' dimensions and the dilation are used to increase the receptive fields without adding additional parameters.

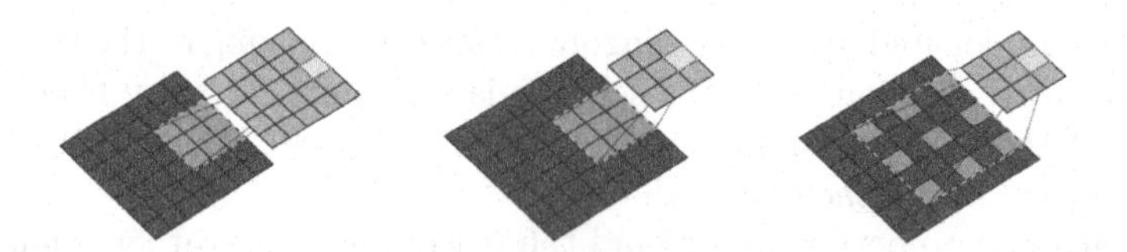

Fig. 6. Exemplary visualization of the operation principles of *Convolutional Neural Networks*. Left: Normal CNN layer with *Dilation* = 1 and *Strides* = 1. Middle: CNN layer with *Strides* = 2 (reducing the output size by half). Right: CNN Layer with *Dilation* = 2 (increasing the receptive field without adding additional weights).

5.1 Experimental Setup

To train CNN based AS models that are independent of the commonly used TSP features, we will produce different visual representations of our TSP instances (see Fig. 2) and use them for training the deep learning networks. Those images are created with a resolution of 512×512 pixels and the coordinates of the instances are scaled to fill out the entire image as exemplarily shown in the two images in the left column of Fig. 2. In addition to these point clouds, images of corresponding *Minimum Spanning Trees* (MST, second column of Fig. 2) and *k-Nearest-Neighbor Graphs* (k-NNG, right column of Fig. 2) with $k = 5$ were generated. We chose MST and 5-NNG as additional visual representations because we found in Sect. 3 that the 15 most important features are based almost exclusively on MST and 5-NNG graphs. In the 5-NNGs, not only mutual (strong) connections but also one-sided (weak) links were considered, in which one city belongs to the nearest neighbor set of another city, but not vice versa.

Admittedly, only networks whose generation was based exclusively on point clouds can be described as *feature-free*. For a better comparison, however, we have additionally evaluated *feature-based* networks that were trained with images of the corresponding MST and 5-NNG. Hence, we considered two different scenarios for our network-based approaches. In the first scenario (S1), the networks were trained based on (a) point clouds of the cities (Points), (b) MST images, and (c) 5-NNG images. In scenario (S2), we combined (a) the scatterplots with the MST images (as two input channels), and (b) the scatterplots with the MST and 5-NNG images (as three input channels). As the costs for generating the images are insignificant, we have not taken their generation time into account when computing the PAR10 scores of the deep learning models. For larger instances, though, these times would have to be taken into account.

To process the visual representations of the instances, we used eight stacked convolutional layers (see Fig. 5). Three of them used $Strides = 2$ to reduce the size of the feature maps and four of them used $Dilation = \{2, 3\}$-Kernels to enlarge the receptive fields and, thus, gain a larger view of the instances (see Fig. 6 to compare the effects of Strides and Dilation). We used *Rectified Linear Unit* (ReLU) [10] as activation function for all layers except for the last linear layer, for which we used a *Softmax* activation. To improve the training speed, the outputs of all convolutional layers are normalized by using *Group Normalization* (GN) [47] with $G = 8$. The GN layers are in-between the convolutional layers and the ReLU activation. For transition from the three-dimensional convolutional ($Width \times Height \times Channels$) layers to the one-dimensional linear layer, a *Global Average Pooling Layer* (GPL) [28] is used. After the GPL, a Dropout (DP) [44] layer with 25% dropout is added to improve regularization. The final layer is a single, linear layer with two output neurons – one for EAX and one for LKH (see Fig. 5). Last, we used 10-fold cross-validation to evaluate the performance of the neural networks. The folds were the same as for the classical feature-based approach. All networks were trained using mini-batches of eight, Adam [22] as optimizer and *Cross-Entropy* [32] as loss function. Note that neural networks are most commonly trained using *Stochastic Gradient Descent* [41], which strongly differs from the training methods used in Sect. 4.

5.2 Findings

The best performing classical approach achieved a mean PAR10-score of 56.29 (see Table 1). Our feature-free networks, which were trained exclusively using the points, achieved a mean PAR10-score of 56.31 after tuning the threshold and thus a similar performance (see Table 2) as the feature-based, classical approaches.

As stated before, we additionally investigated whether adding additional features to the networks could improve the models' performances. Therefore, we also trained feature-based networks using MST and 5-NNG images. As shown in Table 2, both variants perform noticeably better. Besides, we found that while the thresholds between the points and the MST models are rather similar, the thresholds of the NNG models are on average 10% higher. Next, the thresholds of the Points and MST models range from 13% to 34%, and 16% to 35%, respectively, while the thresholds of the NNG models range from 3% to 73%. Thus, networks trained on the NNG images appear to be less stable.

Moreover, the networks based on the Points correctly predicted EAX in 91.8% (and thus 459 times) of the cases, in which EAX was the better solver, compared to only 22.6% (113 cases) for the LKH-friendly cases. This behavior likely results from the fact that a misclassification of an instance, which is favorable for LKH, is cheaper than a misclassification of an instance that is easier for EAX. In contrast, the MST-based networks predict EAX in 91.6% (458) and LKH in 27.2% (136) cases, correctly. Thus, compared to the feature-free networks, which were exclusively based on Points, the MST networks benefit from correctly identifying LKH-friendly instances – without losing accuracy on the EAX-friendly instances. Noticeably, in case of the 5-NNG networks, only 84.6% (423) of the EAX-easy instances are classified correctly, compared to 34.8% (174) among the LKH-easy instances. Thus, despite the improvements among the instances that are favorable for LKH, the PAR10 score of the NNG-based networks is inferior to the MST-based selector, as misclassifying EAX-easy instances is more expensive.

Table 2. PAR10-scores of our deep neural networks for the different scenarios and across all ten CV-folds. In addition, we list the values of the tuned thresholds (TH).

	Scenario S1						Scenario S2			
	Points		MST		NNG		Points + MST		Points + MST + NNG	
Fold	PAR10	TH	PAR10	TH	PAR10	TH	PAR10	TH	PAR10	TH
1	34.40	0.19	30.91	0.27	31.35	0.64	**27.13**	0.21	32.95	0.14
2	78.88	0.15	57.95	0.17	76.99	0.18	**54.65**	0.14	58.06	0.22
3	67.43	0.14	64.14	0.19	**57.94**	0.32	62.40	0.25	62.15	0.15
4	50.50	0.13	**45.89**	0.19	59.35	0.03	54.92	0.19	56.58	0.13
5	**52.41**	0.29	57.76	0.16	59.40	0.43	58.50	0.24	61.43	0.25
6	58.07	0.15	**45.28**	0.21	56.09	0.35	64.59	0.17	54.95	0.19
7	50.32	0.40	58.91	0.29	**42.49**	0.05	49.47	0.25	54.76	0.22
8	36.89	0.23	35.35	0.20	**26.08**	0.73	63.47	0.29	32.07	0.27
9	**93.07**	0.31	95.44	0.23	95.41	0.41	96.89	0.16	93.57	0.27
10	**41.10**	0.34	48.79	0.35	44.52	0.13	52.09	0.42	48.66	0.14
∅	56.31	0.23	54.04	0.23	54.96	0.33	58.41	0.23	55.52	0.20

To investigate whether the combination of the three different input variants would lead to networks that achieve better performances when predicting EAX- and LKH-easy instances, we combined Points and MST, as well as Points, MST

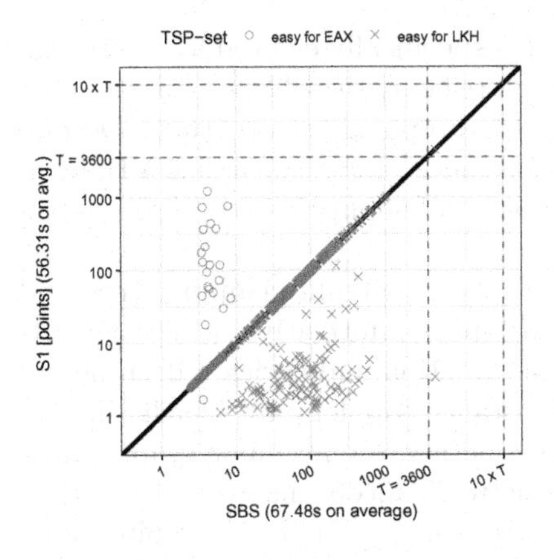

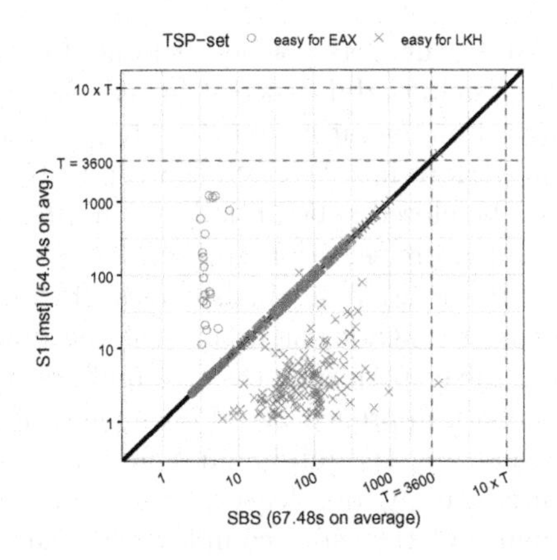

Fig. 7. PAR10-scores (log-scaled) of the points networks of scenario S1 (on the left) and the MST networks of scenario S1 versus the Single-Best-Solver (EAX).

and NNG into two and three input channels (Scenario 2), respectively. However, as shown in Table 2, combining the visual representations does not improve the networks' overall performances. Also, while the network based on Points + MST classifies 31% (155) of the LKH-easy instances correctly, the selector based on Points + MST + NNG only succeeds in 17% (85) of the respective cases. We further observed that the threshold values are quite similar to the ranges of the Points and MST models from scenario S1.

Table 3. Comparison of baseline (VBS and SBS) with our best models.

Measure	Baseline		Classical AS		Deep learning AS		
	VBS	SBS (EAX)	RF	SVM	Points	MST	Points + MST
PAR10	4.92	67.47	56.29	56.67	56.31	54.04	55.52
Accuracy	1.00	0.49	0.62	0.60	0.57	0.59	0.61
F1-Score	1.00	0.65	0.71	0.70	0.68	0.69	0.70

As visualized in Fig. 7, the Points (left) and MST networks (right) prefer EAX over LKH – as there are far more observations below the diagonal. This is also confirmed by the low thresholds (see Table 2). Interestingly, the networks solely based on visual representations of the instance, even perform slightly better than the classical feature-based AS models – but are still clearly inferior to the VBS.

6 Conclusions and Outlook

The conducted experiments shed light on still existing shortcomings of classical feature-based per-instance algorithm selection on the TSP. While previous studies clearly reported successful approaches, the informative character of existing

TSP feature sets reaches its limits for instances specifically evolved for maximum performance difference of the two state-of-the art heuristic solvers. Sophisticated mutation operators here lead to so far unobserved topological structures. Despite outperforming the SBS, the gap to the performance of the oracle-like VBS cannot be closed substantially, even after utilization of sophisticated preprocessing and feature selection approaches.

However, it again becomes obvious that the minimum spanning tree and nearest neighbor structures of the points are most informative in discriminating solver performances. We build on this information and enrich a deep neural network approach based on images of the instance topology by specific images visualizing the minimum spanning tree and the nearest neighbor graph. Most interestingly, our feature-free deep neural network nicely matches the performance of the quite complex classical AS approach (see Table 3), despite being solely based on an image of the instance's points.

This proof-of-concept study thus shows the huge potential of deep learning, feature-free approaches in this domain which we will exploit in future studies by more sophisticated networks and altered loss functions specifically adapted to the designated performance indicators. Moreover, additional image channels will be added, e.g., in terms of heatmaps. Specific investigations have to be conducted with regard to the scaling behaviour of the approach, as image resolutions most probably will have to be carefully adapted to increasing instance sizes.

On the other hand, the observed limitations of classical TSP features show the necessity of enriching the library of TSP features by alternative sets which capture other kinds of—obviously important—instance structures. We will apply our approach to classical feature sets such as RUE or TSPLib as well for a comparison. However, it is specifically noteworthy that, in principle, the deep learning approach nicely generalizes to other graph-based optimization problems while instance features are almost exclusively tailored to the focused domain.

Acknowledgments. The authors acknowledge support by the *European Research Center for Information Systems (ERCIS).*

References

1. Alissa, M., Sim, K., Hart, E.: Algorithm selection using deep learning without feature extraction. In: Proceedings of the Genetic and Evolutionary Computation Conference GECCO 2019, pp. 198–206. Association for Computing Machinery, New York (2019). https://doi.org/10.1145/3321707.3321845
2. Bischl, B., et al.: ASlib: a benchmark library for algorithm selection. Artif. Intell. **237**, 41–58 (2016). https://doi.org/10.1016/j.artint.2016.04.003
3. Bischl, B., et al.: mlr: machine learning in R. J. Mach. Learn. Res. (JMLR) **17**(170), 1–5 (2016). http://jmlr.org/papers/v17/15-066.html
4. Bischl, B., Mersmann, O., Trautmann, H., Preuss, M.: Algorithm selection based on exploratory landscape analysis and cost-sensitive learning. In: Proceedings of the 14th Annual Conference on Genetic and Evolutionary Computation (GECCO), pp. 313–320. ACM, July 2012. https://doi.org/10.1145/2330163.2330209. http://dl.acm.org/citation.cfm?doid=2330163.2330209

5. Bossek, J.: Salesperson: computation of instance features and R interface to the state-of-the-art exact and inexact solvers for the traveling salesperson problem (2017). https://github.com/jakobbossek/salesperson. R package version 1.0.0
6. Bossek, J., Kerschke, P., Neumann, A., Wagner, M., Neumann, F., Trautmann, H.: Evolving diverse TSP instances by means of novel and creative mutation operators. In: Friedrich, T., Doerr, C., Arnold, D. (eds.) Proceedings of the 15th ACM/SIGEVO Workshop on Foundations of Genetic Algorithms (FOGA XV), pp. 58–71. ACM, Potsdam (2019)
7. Bossek, J., Trautmann, H.: Evolving instances for maximizing performance differences of state-of-the-art inexact TSP solvers. In: Festa, P., Sellmann, M., Vanschoren, J. (eds.) LION 2016. LNCS, vol. 10079, pp. 48–59. Springer, Cham (2016). https://doi.org/10.1007/978-3-319-50349-3_4
8. Bossek, J., Trautmann, H.: Understanding characteristics of evolved instances for state-of-the-art inexact TSP solvers with maximum performance difference. In: Adorni, G., Cagnoni, S., Gori, M., Maratea, M. (eds.) AI*IA 2016. LNCS (LNAI), vol. 10037, pp. 3–12. Springer, Cham (2016). https://doi.org/10.1007/978-3-319-49130-1_1
9. Chen, T., et al.: XGBoost: extreme gradient boosting (2019). https://CRAN.R-project.org/package=xgboost. R package version 0.90.0.2
10. Glorot, X., Bordes, A., Bengio, Y.: Deep sparse rectifier neural networks. In: Gordon, G.J., Dunson, D.B., Dudík, M. (eds.) Proceedings of the Fourteenth International Conference on Artificial Intelligence and Statistics, AISTATS 2011, Fort Lauderdale, USA, 11–13 April 2011. JMLR Proceedings, vol. 15, pp. 315–323. JMLR.org (2011). http://proceedings.mlr.press/v15/glorot11a/glorot11a.pdf
11. Guyon, I., Elisseeff, A.: An introduction to feature extraction. In: Guyon, I., Nikravesh, M., Gunn, S., Zadeh, L.A. (eds.) Feature Extraction. STUDFUZZ, vol. 207, pp. 1–25. Springer, Heidelberg (2006). https://doi.org/10.1007/978-3-540-35488-8_1
12. Härdle, W.K., Simar, L.: Applied Multivariate Statistical Analysis, 4th edn. Springer, Heidelberg (2015). https://doi.org/10.1007/978-3-662-45171-7
13. Hastie, T., Tibshirani, R., Friedman, J.: The Elements of Statistical Learning: Data Mining, Inference, and Prediction. Springer, Heidelberg (2009). http://www.springer.com/de/book/9780387848570
14. Helsgaun, K.: An effective implementation of the lin-kernighan traveling salesman heuristic. Eur. J. Oper. Res. **126**(1), 106–130 (2000)
15. Hutter, F., Xu, L., Hoos, H.H., Leyton-Brown, K.: Algorithm runtime prediction: methods & evaluation. Artif. Intell. **206**, 79–111 (2014). https://doi.org/10.1016/j.artint.2013.10.003
16. Hutter, F., Xu, L., Hoos, H.H., Leyton-Brown, K.: Algorithm runtime prediction: methods & evaluation. Artif. Intell. J. (AIJ) **206**, 79–111 (2014). http://www.sciencedirect.com/science/article/pii/S0004370213001082
17. Karatzoglou, A., Smola, A., Hornik, K., Zeileis, A.: kernlab - An S4 package for kernel methods in R. J. Stat. Softw. (JSS) **11**(9), 1–20 (2004). http://www.jstatsoft.org/v11/i09/
18. Kerschke, P., Bossek, J., Trautmann, H.: Parameterization of state-of-the-art performance indicators: a robustness study based on inexact TSP solvers. In: Proceedings of the 20th Genetic and Evolutionary Computation Conference (GECCO) Companion, pp. 1737–1744. ACM, Kyoto (2018). https://doi.org/10.1145/3205651.3208233. http://doi.acm.org/10.1145/3205651.3208233

19. Kerschke, P., Hoos, H.H., Neumann, F., Trautmann, H.: Automated algorithm selection: survey and perspectives. Evol. Comput. (ECJ) **27**(1), 3–45 (2019)
20. Kerschke, P., Kotthoff, L., Bossek, J., Hoos, H.H., Trautmann, H.: Leveraging TSP solver complementarity through machine learning. Evol. Comput. (ECJ) **26**(4), 597–620 (2018)
21. Kerschke, P., Trautmann, H.: Automated algorithm selection on continuous black-box problems by combining exploratory landscape analysis and machine learning. Evol. Comput. **27**(1), 99–127 (2019). https://doi.org/10.1162/evco_a_00236. pMID: 30365386
22. Kingma, D.P., Ba, J.: Adam: a method for stochastic optimization. In: Bengio, Y., LeCun, Y. (eds.) 3rd International Conference on Learning Representations, ICLR 2015, San Diego, CA, USA, 7–9 May 2015, Conference Track Proceedings (2015). http://arxiv.org/abs/1412.6980
23. Kohavi, R., John, G.H., et al.: Wrappers for feature subset selection. Artif. Intell. **97**(1–2), 273–324 (1997)
24. Kotthoff, L.: Algorithm selection for combinatorial search problems: a survey. AI Mag. **35**(3), 48–60 (2014). https://doi.org/10.1609/aimag.v35i3.2460. https://aaai.org/ojs/index.php/aimagazine/article/view/2460
25. Kotthoff, L., Kerschke, P., Hoos, H., Trautmann, H.: Improving the state of the art in inexact TSP solving using per-instance algorithm selection. In: Dhaenens, C., Jourdan, L., Marmion, M.-E. (eds.) LION 2015. LNCS, vol. 8994, pp. 202–217. Springer, Cham (2015). https://doi.org/10.1007/978-3-319-19084-6_18
26. LeCun, Y., Bengio, Y., et al.: Convolutional networks for images, speech, and time series. In: The Handbook of Brain Theory and Neural Networks, vol. 3361, no. 10, p. 1995 (1995)
27. Liaw, A., Wiener, M.: Classification and regression by randomForest. R News **2**(3), 18–22 (2002). https://cran.r-project.org/doc/Rnews/Rnews2002-3.pdf
28. Lin, M., Chen, Q., Yan, S.: Network in network. In: Bengio, Y., LeCun, Y. (eds.) 2nd International Conference on Learning Representations, ICLR 2014, Banff, AB, Canada, 14–16 April 2014, Conference Track Proceedings (2014). http://arxiv.org/abs/1312.4400
29. Lindauer, T.M., Hoos, H.H., Hutter, F., Schaub, T.: AutoFolio: an automatically configured algorithm selector (extended abstract). In: Proceedings of the International Joint Conference on Artificial Intelligence (IJCAI), pp. 5025–5029, August 2017. https://doi.org/10.24963/ijcai.2017/715. https://www.ijcai.org/proceedings/2017/715
30. Malitsky, Y., Sabharwal, A., Samulowitz, H., Sellmann, M.: Algorithm portfolios based on cost-sensitive hierarchical clustering. In: Rossi, F. (ed.) Proceedings of the Twenty-Third International Joint Conference on Artificial Intelligence (IJCAI), vol. 13, pp. 608–614. Association for the Advancement of Artificial Intelligence (AAAI), August 2013. https://www.aaai.org/ocs/index.php/IJCAI/IJCAI13/paper/view/6946
31. Mann, H.B., Whitney, D.R.: On a test of whether one of two random variables is stochastically larger than the other. Ann. Math. Stat. **18**(1), 50–60 (1947). https://doi.org/10.1214/aoms/1177730491

32. Mannor, S., Peleg, D., Rubinstein, R.Y.: The cross entropy method for classification. In: Raedt, L.D., Wrobel, S. (eds.) Machine Learning, Proceedings of the Twenty-Second International Conference (ICML 2005), Bonn, Germany, 7–11 August 2005. ACM International Conference Proceeding Series, vol. 119, pp. 561–568. ACM (2005). https://doi.org/10.1145/1102351.1102422
33. Mersmann, O., Bischl, B., Bossek, J., Trautmann, H., Wagner, M., Neumann, F.: Local search and the traveling salesman problem: a feature-based characterization of problem hardness. In: Hamadi, Y., Schoenauer, M. (eds.) LION 2012. LNCS, pp. 115–129. Springer, Heidelberg (2012). https://doi.org/10.1007/978-3-642-34413-8_9
34. Mersmann, O., Bischl, B., Trautmann, H., Wagner, M., Bossek, J., Neumann, F.: A novel feature-based approach to characterize algorithm performance for the traveling salesperson problem. Ann. Math. Artif. Intell. **69**(2), 151–182 (2013). https://doi.org/10.1007/s10472-013-9341-2. https://link.springer.com/article/10.1007/s10472-013-9341-2
35. Nagata, Y., Kobayashi, S.: A powerful genetic algorithm using edge assembly crossover for the traveling salesman problem. INFORMS J. Comput. **25**(2), 346–363 (2013)
36. Peng, H., Long, F., Ding, C.: Feature selection based on mutual information criteria of max-dependency, max-relevance, and min-redundancy. IEEE Trans. Pattern Anal. Mach. Intell. (TPAMI) **27**(8), 1226–1238 (2005). https://doi.org/10.1109/TPAMI.2005.159. https://ieeexplore.ieee.org/abstract/document/1453511
37. Pihera, J., Musliu, N.: Application of machine learning to algorithm selection for TSP. In: 26th IEEE International Conference on Tools with Artificial Intelligence, ICTAI 2014, Limassol, Cyprus, 10–12 November 2014, pp. 47–54. IEEE Computer Society (2014)
38. Reinelt, G.: TSPLIB-a traveling salesman problem library. ORSA J. Comput. **3**(4), 376–384 (1991)
39. Rice, J.R.: The algorithm selection problem. Adv. Comput. **15**, 65–118 (1976). http://www.sciencedirect.com/science/article/pii/S0065245808605203
40. Rizzini, M., Fawcett, C., Vallati, M., Gerevini, A.E., Hoos, H.H.: Static and dynamic portfolio methods for optimal planning: an empirical analysis. Int. J. Artif. Intell. Tools **26**(01), 1–27 (2017). https://doi.org/10.1142/S0218213017600065. https://www.worldscientific.com/doi/abs/10.1142/S0218213017600065
41. Robbins, H., Monro, S.: A stochastic approximation method. Ann. Math. Stat. **22**, 400–407 (1951)
42. Ross, P., Schulenburg, S., Marín-Bläzquez, J.G., Hart, E.: Hyper-heuristics: learning to combine simple heuristics in bin-packing problems. In: Proceedings of the 4th Annual Conference on Genetic and Evolutionary Computation GECCO 2002, pp. 942–948. Morgan Kaufmann Publishers Inc., San Francisco (2002)
43. Sim, K., Hart, E., Paechter, B.: A hyper-heuristic classifier for one dimensional bin packing problems: improving classification accuracy by attribute evolution. In: Coello, C.A.C., Cutello, V., Deb, K., Forrest, S., Nicosia, G., Pavone, M. (eds.) PPSN 2012. LNCS, vol. 7492, pp. 348–357. Springer, Heidelberg (2012). https://doi.org/10.1007/978-3-642-32964-7_35
44. Srivastava, N., Hinton, G., Krizhevsky, A., Sutskever, I., Salakhutdinov, R.: Dropout: a simple way to prevent neural networks from overfitting. J. Mach. Learn. Res. **15**(1), 1929–1958 (2014)

45. Therneau, T., Atkinson, B.: rpart: recursive partitioning and regression trees (2019). https://CRAN.R-project.org/package=rpart. R package version 4.1-15
46. Urbanowicz, R.J., Meeker, M., La Cava, W., Olson, R.S., Moore, J.H.: Relief-based feature selection: introduction and review. J. Biomed. Inform. **85**, 189–203 (2018). https://doi.org/10.1016/j.jbi.2018.07.014. https://www.sciencedirect.com/science/article/pii/S1532046418301400
47. Wu, Y., He, K.: Group normalization. Int. J. Comput. Vis. **128**(3), 742–755 (2020). https://doi.org/10.1007/s11263-019-01198-w
48. Xu, L., Hutter, F., Hoos, H., Leyton-Brown, K.: Evaluating component solver contributions to portfolio-based algorithm selectors. In: Cimatti, A., Sebastiani, R. (eds.) SAT 2012. LNCS, vol. 7317, pp. 228–241. Springer, Heidelberg (2012). https://doi.org/10.1007/978-3-642-31612-8_18

Automatic Configuration of a Multi-objective Local Search for Imbalanced Classification

Sara Tari[1(✉)], Holger Hoos[2], Julie Jacques[1,3], Marie-Eléonore Kessaci[1], and Laetitia Jourdan[1]

[1] University of Lille, CNRS, UMR 9189 CRIStAL, 59000 Villeneuve d'Ascq, France
{sara.tari,marie-eleonore.kessaci,laetitia.jourdan}@univ-lille.fr
[2] LIACS, Leiden University, Leiden, The Netherlands
hh@liacs.nl
[3] Faculté de Gestion, Economie et Sciences, Lille Catholic University, Lille, France
julie.jacques@univ-catholille.fr

Abstract. MOCA-I is a multi-objective local search algorithm, based on the *Pittsburgh* representation, that has been formerly designed to solve partial classification problems with imbalanced data. Recently, multi-objective automatic algorithm configuration (MO-AAC) has proven effective in boosting the performance of multi-objective local search algorithms for combinatorial optimization problems. Here, for the first time, we apply MO-ACC to multi-objective local search for rule-based classification problems. Specifically, we present the Automatic Configuration of MOCA-I (AC-MOCA-I). AC-MOCA-I uses a methodology based on k-fold cross-validation to automatically configure an extended and improved version of MOCA-I. In a series of experiments on well-known datasets from the literature, we consider 183 456 unique configurations for MOCA-I and demonstrate that AC-MOCA-I leads to substantial improvements in performance. Moreover, we investigate the impact of the running time allotted to AC-MOCA-I on performance and the role of specific parameters and components.

Keywords: Supervised classification · Multi-objective optimization · Automatic algorithm configuration

1 Introduction

Supervised classification aims at predicting the class of unknown observations by using a model built from known observations. Partial (binary) classification focuses on predicting only one class – for instance, only positive observations. In some binary datasets, there are large differences between the number of positive and negative observations for each class. In the medical context (as considered in

Funded by the Pathacov Project of the Interreg France-Wallonie-Vlaanderen program, with the support of the European Regional Development Fund.

T. Bäck et al. (Eds.): PPSN 2020, LNCS 12269, pp. 65–77, 2020.
https://doi.org/10.1007/978-3-030-58112-1_5

the European Pathacov Project), frequent diseases affect only a small percentage of the population. Such datasets are said to be *imbalanced* and raise additional challenges for classification. Various techniques have been proposed to perform supervised classification for imbalanced data. When predicting the onset of disease, the classification must be understandable to the physician; this implies that white-box approaches, such as rule-based classifiers or tree classifiers, are preferred. Jacques *et al.* designed a highly parametric multi-objective local search algorithm to solve imbalanced classification using a rule-based approach [10]. They implemented this algorithm, dubbed MOCA-I (Multi-Objective Classification Algorithm for Imbalanced data), and manually configured its parameters with high experimental effort. However, automatic algorithm configuration (AAC) can be used to reduce this effort and to configure the algorithm automatically. Recently, a multi-objective automatic algorithm configuration (MO-AAC) approach has been proposed to automatically configure multi-objective algorithms [4], such as MOCA-I.

Therefore, we propose AC-MOCA-I (Automatic Configuration of MOCA-I), which uses an adapted MO-AAC methodology to configure an extended version of MOCA-I, which includes additional heuristic mechanisms and exposes additional design choices via new components. The aim of AC-MOCA-I is to produce better solutions than MOCA-I with reduced experimental effort.

We tested AC-MOCA-I on 9 datasets from the literature and compared the results with those of exhaustive runs using the original configuration space of MOCA-I. Our objective in these experiments was to improve the quality of the resulting solutions and to determine the configurations best suited for this purpose.

The contributions we make in this work are (1) a methodology based on MO-AAC for solving classification problems on imbalanced data; (2) the consideration of new components for MOCA-I; and (3) the investigation of the impact of running times on the quality of solutions provided by AC-MOCA-I and its best-performing parameter settings.

The remainder of this paper is organized as follows. Section 2 presents the classification problem we consider, multi-objective local search (MOLS), the MOCA-I algorithm, and the new components and parameters we have added to it. Section 3 introduces multi-objective automatic algorithm configuration and describes how we applied MO-ParamILS to AC-MOCA-I. Section 4 presents our experimental protocol and analysis of the results we obtained using it. Finally, Sect. 5 provides conclusions and perspectives on future work.

2 The MOCA-I Algorithm for the Supervised Imbalanced Classification Problem

2.1 The Supervised Imbalanced Classification Problem

The objective in a binary supervised classification task is to predict the *class* (i.e., the category) to which a given observation belongs. Observations are described by a number of *attributes* (information) that can be of diverse forms, depending on the problem under consideration. In supervised classification, a *classifier* is constructed based on training data, consisting of observations and their known classes, with the goal of predicting, with maximum accuracy, the class of new observations based on their attributes.

In the context of disease prediction, the pathology generally occurs in a minority of the population, leading to *imbalanced datasets*, where ill individuals are less common than healthy individuals. This gives rise to additional challenges that lead most classification algorithms to fail [9]. In addition, the objective in this medical context is the prediction of only ill individuals; this leads to a partial supervised classification problem, where the objective is to predict only a subset of the data (here individuals positive to the disease). Moreover, health professional must be able to challenge the decision of the classifier. So, we will focus on a white-box approach where classifiers give interpretable results.

Finding the best classifier represents a combinatorial problem, which calls for use of optimization techniques. We also want health professionals to be able to adjust the classifier according to their needs – high sensitivity to screen ill patients or high precision for better diagnostic precision. Multi-objective approaches, which produce a set of solutions characterizing possible trade-offs, are well-suited to this goal.

2.2 Multi-objective Imbalanced Classification with MOCA-I

In multi-objective optimization, solutions are evaluated based on at least two criteria or objective functions, and the concept of *Pareto dominance* is often used to compare their qualities. Given solution s_1 and s_2, s_1 *dominates* s_2 iff it is better or equal according to all objectives and at least strictly better w.r.t. one of them, and otherwise, s_1 and s_2 are *non-dominated* by each other. A set S of solutions in which no elements dominates any other is called a *Pareto set* or an *archive*.

MOCA-I [10] is a multi-objective local search specifically designed for imbalanced classification. It uses a *Pittsburgh representation*, where each solution is a ruleset. A ruleset is composed of rules, which are conjunctions of attribute tests (*terms*). A term is described by an attribute, an operator $(<, >, =)$ and a value. The encoding length of a ruleset is variable. For a given ruleset, an observation must trigger at least one rule to be classified as positive.

Three objectives are used simultaneously in MOCA-I: (1) to maximize the *sensitivity* $\frac{TP}{TP+FN} \in [0,1]$, (2) to maximize the *precision* $\frac{TP}{TP+FP} \in [0,1]$ and (3) to minimize the number of terms in the ruleset, where (TP) represents *true positives*, (TN) *true negatives*, (FP) *false positives (FP)* and (FN) *false negatives (FN)* of the confusion matrix.

The sensitivity corresponds to the proportion of samples detected by the ruleset that are positive to the class under investigation. In contrast, the precision corresponds to the probability that an observation detected as positive by the ruleset is a true positive. Minimizing the number of terms helps to reduce complexity and avoid the *bloat* effect that can cause rulesets to contain over-specific rules without any improvement of their quality.

In MOCA-I, the neighborhood of a solution s corresponds to the set of all rulesets that differ in one term that is added to, modified in or removed from s.

2.3 Parameters and Components

A multi-objective local search procedure evolves solutions of an archive using a neighborhood relation [3] by cycling through three phases. The *selection* phase selects, where at least one solution from the archive is selected; the *exploration* of the neighborhood of selected solutions; and the *archiving* of new solutions. Additionally, a `perturbation` phase can be applied to promote diversification during the search. Below we describe the existing strategies and parameters of MOCA-I (underlined) and the ones we newly introduce for AC-MOCA-I (not underlined).

Selection. The selection strategy determines the solutions of the archive to explore. The mechanism `all` selects all unvisited solutions in the archive, whereas `rand`, `oldest` and `newest` select a fixed number of solutions (*selection-size*) uniformly at random, from the oldest and from the newst ones, respectively.

Exploration. The exploration strategy determines how the neighborhood of the selected solutions is explored, and which visited neighbors are kept. During this phase, the visited neighbors are compared to the currently selected solution (`sol`), or to the current archive of solutions with the value (`arch`).

For each selected solution, `all` and `all-imp` explore the entire neighborhood and keep the non-dominated neighbors and dominating solutions, respectively. The mechanism `imp` and `ndom` evaluate the neighborhood until a fixed number (*exploration-size*) of dominating or non-dominated solutions, respectively, are found; an additional mechanism, `imp-ndom`, also include non-dominated neighbors.

Archiving. This phase adds the solutions kept during the exploration phase into the archive and removes all dominated solutions according to the Pareto rule. Here, the archive is bounded, meaning that new solutions are discarded when the maximal size of the archive is reached.

Perturbation. The perturbation phase modifies the archive of solutions using different mechanisms: `restart` restarts the search from a new initial population; `kick` randomly selects a fixed number *kick-size* of solutions that are iteratively replaced by one of their neighbors; and `kick-all` modifies all solutions of the archive. For the two last strategies, the number of random moves applied to each solution is called *perturbation strength*.

In the context of MOCA-I, we consider an additional parameter that specifies the maximum number of rules within any ruleset. Moreover, the initial population is obtained by generating rulesets from existing observations such that each rules matches at least one observation [10].

As MOCA-I aims to produce a classifier, it returns a single ruleset from the final archive [10]. This ruleset is chosen such that it has the best *F-Measure* (F_1), i.e., the highest harmonic mean of sensitivity and precision, computed as follows:

$$F_1 = \frac{2 \cdot Precision \cdot Sensitivity}{Precision + Sensitivity}$$

3 Multi-objective Automatic Algorithm Configuration

The goal of automatic algorithm configuration (AAC) is to automatically determine a configuration (set of parameters) that results in optimal performance of a given algorithm (*target algorithm*) on a given class of problem instances.

In this context, performance can be evaluated using one or more metrics. AAC is typically cast as a single-objective optimization problem, in which the objective is to optimize running time or solution quality. However, for multi-objective optimization problems, several performance indicators are generally used to assess the quality of Pareto sets [12]. These indicators often focus on three main properties: the *accuracy*, the *diversity*, and/or the *cardinality* of a given Pareto set. A direct approach to using AAC for configuring a multi-objective algorithm is to optimize only one indicator [13]. Another approach is to consider at least two indicators, which leads to a multi-objective optimization problem that is known as multi-objective automated algorithm configuration (MO-AAC). MO-AAC typically involves three phases [4]: the training phase, the validation phase, and the testing phase.

For a given configuration space Θ and a given set of *training* instances, the training phase optimizes the configuration of the target algorithm according to the performance metrics under consideration. It returns a Pareto set of configurations $\{\theta^*\}$. Since this phase corresponds to a stochastic process, several runs are performed, resulting in several sets of configurations $\{\{\theta^*\}\}$. The validation phase, usually conducted on training instances, performs runs using the configurations of $\{\{\theta^*\}\}$ and returns a single Pareto set of optimal configurations ($\{\theta^*\}$). Finally, the testing phase assesses the configurations of $\{\theta^*\}$ on a distinct set of *testing* instances.

Multi-objective AAC has been successfully applied to classical multi-objective optimization problems, such as the permutation flowshop scheduling problem and the traveling salesman problem [4]. Usually, on such problems, instances of similar characteristics to the ones to be ultimately solved are used for training and validation. However, in the case of real-world problems such as data classification, having datasets sharing precisely the same characteristics is almost impossible. An additional challenge, in the context of data classification, arises from the need for disjoint datasets for training and assessing classifiers; this is typically resolved by using k-fold cross-validation.

We now present our methodology for automatically configuring multi-objective algorithms for the classification problem using MO-ParamILS. We split our collection of datasets into three groups (κ_1, κ_2, κ_3), such that each group contains the same number of datasets. Two of these groups are then used during the training and validation phases, while the remaining one is used for testing (corresponding to the classical classification protocol). This leads to three combinations of groups (repartitions), as depicted in Fig. 1. In this context, we use Opt_{sets} to denote the datasets used during the MO-AAC training and validation phases of the configurator. From Opt_{sets}, datasets are divided into training sets (Opt_{tra}) and testing sets (Opt_{tst}), following a 5-fold cross-validation protocol. We use CL_{sets} to denote the disjoint set of datasets used in the MO-AAC testing phase. Within this set, we denote the datasets used for the training of the classifier as CL_{tra}, and the datasets used for validation as CL_{tst}.

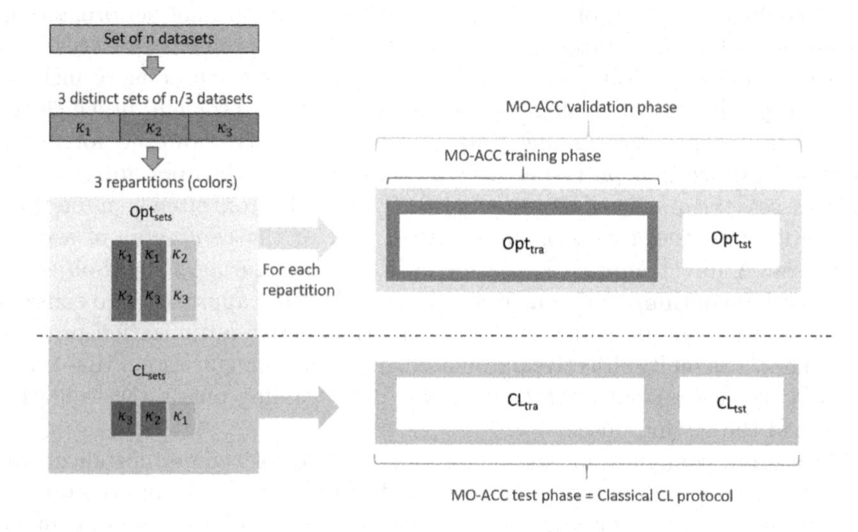

Fig. 1. Repartitioning of datasets for the AAC protocol.

Figure 2 outlines our MO-AAC methodology. In the training phase, the configurator optimizes the target algorithm configurations using the normalized *hypervolume* indicator (HV) [16] and the normalized spread indicator Δ' on the Pareto set produced for the Opt_{tra} datasets. HV assesses the accuracy and the diversity of Pareto sets by measuring the volume between a Pareto set and the reference point $(1, 1)$, and has thus to be maximized. The Δ' spread indicator [4], to be minimized, is a variant of the Δ spread indicator [5] and measures the diversity of solutions in a given Pareto sets. Here, for clarity, we consider 1-HV, so that both objectives are to be minimized. We note that only sensitivity and precision are considered for computing these indicators, since the third objective (the number of terms), mainly serves to guide the search towards small rulesets. The configurations $\{\{\theta^*\}\}$ returned by the configurator after the training phase are assessed during the validation phase, using the same two indicators on the datasets in Opt_{tra} to eliminate any dominated solutions from the set. Then, the classifiers with the resulting configurations $\{\theta^*\}$ are tested on the datasets in Opt_{tst}, and the configuration θ^* with the best average F-measure is

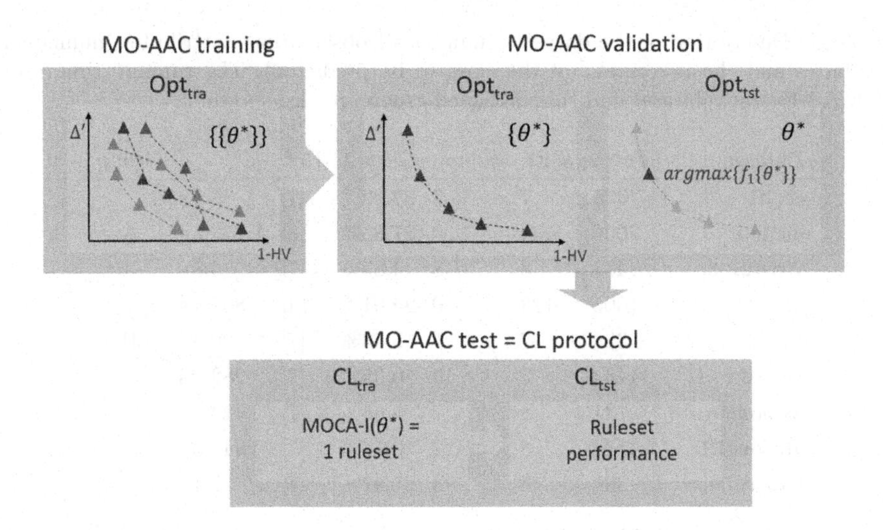

Fig. 2. Illustration of our MO-AAC methodology, as applied to supervised, rule-based classification.

selected as the final result of the configuration process. Finally, AC-MOCA-I, configured with θ^*, builds rulesets on the datasets in CL_{tra}, and the one with the best F-Measure is selected to be evaluated on the datasets in CL_{test}.

4 Experiments

In this section, we present the experiments we have conducted to assess the efficiency of AC-MOCA-I. In addition, we report results from an analysis of the running time of AC-MOCA-I.

4.1 Datasets

Most datasets available from the literature are not binary and imbalanced. While MOCA-I is efficient for this type of classification [10], we transformed datasets into binary ones following the method by Fernández *et al.* [6]. Moreover, real-valued attributes were discretized, using the 10-bin discretization method from the KEEL software [1]. The resulting datasets are listed in Table 1, which also shows the percentage of the class to be predicted for each dataset. These datasets are those previously tackled using MOCA-I [11]. For each dataset, we generated 5 training sets and 5 testing sets following a 5-fold cross-validation approach.

4.2 Experimental Setup

Table 2 shows the components/parameters of our algorithms and their respective values: those from prior work [10] are shown in black, while the new ones introduced here are written in purple. These design choices give rise to 183 456 unique

Table 1. Description of the datasets (number of observations, attributes, numerical attributes and the percentage of the class to be predicted). The running time δ of MOCA-I for each dataset and their assigned group are also shown.

Dataset	#obs.	#att.	#num.	Class 1	Ref.	δ	Group
ecoli1	336	7	7	22.92%	[7]	35.5 s	
lucap0	2000	144	0	27.85%	[8]	1 753.8 s	A
yeast2vs8	482	8	8	4.85%	[7]	4.8 s	
a1a	1605	123	0	24.61%	[14]	864.9 s	
ecoli2	336	7	7	15.48%	[7]	20.3 s	B
yeast3	1484	8	8	10.35%	[7]	282.5 s	
abalone9vs18	731	8	7	5.65%	[7]	423.1 s	
abalone19	4174	8	7	0.77%	[7]	941.3 s	C
haberman	306	3	3	27.42%	[7]	27.1 s	

configurations of AC-MOCA-I. Since the original parameters of MOCA-I only constitute 96 different configurations, we conducted exhaustive runs of MOCA-I to obtain a baseline for our evaluation of AC-MOCA-I.

The nine datasets have different characteristics; thus, we determined the running time allocated to MOCA-I for building a classifier for each of them individually. This was done based on preliminary experiments under the same experimental conditions (not presented here), using a basic configuration to determine the running time for each dataset consisting of the parameter settings underlined in Table 2. We note that this corresponds to a rather time-consuming configuration, since strategy `all` for selection and exploration evaluates the entire neighborhood of each solution in a given archive. The running time for each dataset was then chosen such that this configuration converged (i.e. the Pareto set cannot be improved without restart).

As a MO-AAC configurator, we selected MO-ParamILS [2], in light of its performance and availability.

Table 3 shows the parameters used for MO-ParamILS. Since MO-ParamILS is a randomized procedure, we performed 10 independent runs per scenario. Moreover, in order to prevent overfitting [10] and to assess potential impact on the performance of MOCA-I, we considered 4 different scenarios for AC-MOCA-I, with running times of $\{0.5 \cdot \delta, \delta, 3 \cdot \delta, 5 \cdot \delta\}$, respectively.

4.3 Results

Experiments have been conducted in parallel on two (24 × 3.0 GHz, 64 GB RAM) Intel XEON E5-2687w machines. First, we will analyze the results from an optimization point of view, and then, from a machine learning perspective. Figure 3 (top) presents the Pareto sets of configurations for the exhaustive runs of MOCA-I (on the small configuration space) and each AC-MOCA-I scenario at the end of the MO-AAC validation phase. Clearly, Pareto sets obtained using

Table 2. Configuration space of MOCA-I and AC-MOCA-I. MOCA-I components are reported in black and the additional components for AC-MOCA-I in purple. Parameters used to determine the running time are underlined.

Parameter	Values
Initial population	{50, 100, 200}
Maximum archive size	{100, 300, 500}
Maximum number of rules	{5, 10, 20}
Selection-strategy	{`rand`, `all`, `newest`, `oldest`}
Selection-size	{1, 3, 5, 10}
Exploration-strategy	{`imp-ndom`, `imp`, `ndom`, `all-imp`, `all`}
Exploration-size	{1, 3, 5, 10}
Exploration-reference	{`sol`, `arch`}
Perturbation-strategy	{`restart`, `kick`, `kick-all`}
Perturbation-size	{1, 3, 5, 10}
Perturbation-strength	{3, 5, 7, 10}

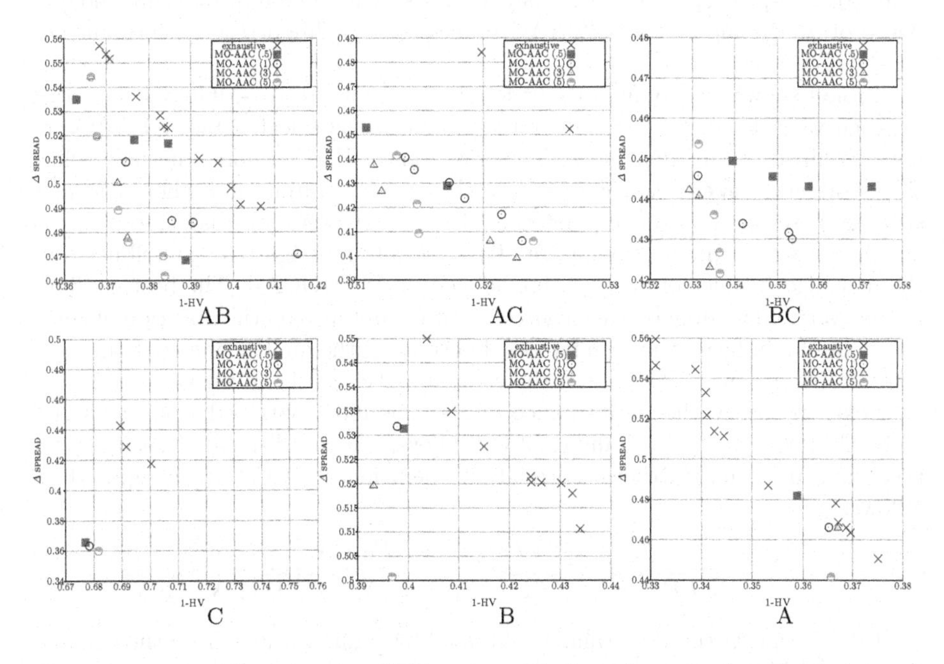

Fig. 3. Pareto configurations on OPT_{tra} in validation phase (top) and solutions on CL_{tra} (bottom).

MO-ParamILS dominate the Pareto set from the exhaustive runs. This clearly indicates that the additional AC-MOCA-I components are beneficial. Moreover, regarding the MO-AAC scenarios, it seems that better performance is obtained for higher running times. Figure 3 (bottom) presents for each selected config-

Table 3. AAC experimental setup.

Phase	Parameters
Training	No default configuration
	1 random configuration
	10 MO-ParamILS runs
	100 MOCA-I run budget
	max 10 MOCA-I runs per configuration
Validation	1 run per instance
Test	30 runs per dataset (6/training set)

uration (from exhaustive runs or AC-MOCA-I) the performance indicators of the Pareto rulesets obtained at the end of the classification training phase. For group C, AC-MOCA-I provides better rulesets than those given by the exhaustive runs of MOCA-I. For group B, AC-MOCA-I achieves better hypervolume, while for group A, AC-MOCA-I yields better spread. These results show no evidence of overfitting of the MO-AAC process and highlight the performance of AC-MOCA-I.

Table 4 shows the configurations θ^* obtained for the AC-MOCA-I scenarios for each validation group. 8 out of 12 configurations selected by MO-AAC are not contained in the smaller configuration space. Furthermore, the new selection strategies `newest` and `oldest` are chosen quite often, and when a former strategy is selected as the `imp-ndom` exploration strategy, the associated size is among the new configuration space. These results highlight the contribution of the new algorithmic components introduced in AC-MOCA-I. The best configurations vary, depending on the group of datasets but also on the duration of runs. These results indicate that the factors determining the quality of a configuration of AC-MOCA-I likely depend on the characteristics of the given dataset.

Next, we study the performance of the classifiers obtained from MOCA-I and AC-MOCA-I, using a number of well-known metrics: F-measure, sensitivity, precision, and the Matthews correlation coefficient (MCC) [15] computed as follows:

$$MCC = \frac{TP \cdot TN - FP \cdot FN}{\sqrt{(TP + FP) \cdot (TP + FN) \cdot (TN + FP) \cdot (TN + FN)}}$$

Table 5 reports the performance obtained by configurations from the exhaustive runs of MOCA-I according to those metrics – we selected the configuration that achieved the best F-Measure on the training sets of each group of datasets – and from the AC-MOCA-I scenarios on CL_{tst}.

To analyze the statistical significance of the observed performance differences, we used Friedman and Wilcoxon paired tests (with $\alpha = 0.05$). The best performance for each metric and each dataset are shown in boldface. Overall, the results obtained in terms of F-measure and MCC are similar for each dataset. On most datasets, the configurations obtained from AC-MOCA-I are able to find

Table 4. Best configuration θ^* for different groups and run durations. ρ refers to the maximum number of rules produced by AC-MOCA-I.

Group	Dur.	Archive size		ρ	Ref.	Exploration		Selection		Perturbation			Score		
		max	init			strat	size	strat	size	strat	size	str	1-HV	Δ'	f1
AB	$0.5 \cdot \delta$	500	50	10	sol	all	–	newest	10	kick-all	–	10	0.363	0.535	0.697
	$1 \cdot \delta$	500	50	10	sol	all	–	all	–	restart	–	–	0.390	0.484	0.712
	$3 \cdot \delta$	500	100	5	sol	imp-ndom	3	oldest	3	kick	10	7	0.368	0.520	0.694
	$5 \cdot \delta$	500	50	10	sol	imp-ndom	10	newest	3	restart	–	–	0.384	0.462	0.707
AC	$0.5 \cdot \delta$	500	50	10	sol	all	–	newest	10	kick	5	10	0.511	0.453	0.485
	$1 \cdot \delta$	500	100	10	arch	all	–	newest	3	restart	–	–	0.514	0.441	0.489
	$3 \cdot \delta$	500	100	20	arch	all	–	newest	3	restart	–	–	0.511	0.437	0.481
	$5 \cdot \delta$	500	100	10	sol	all	–	rand	1	restart	–	–	0.524	0.406	0.480
BC	$0.5 \cdot \delta$	500	200	10	sol	imp-ndom	5	rand	3	kick-all	–	5	0.540	0.449	0.453
	$1 \cdot \delta$	500	100	10	sol	all	–	rand	1	restart	–	–	0.542	0.434	0.449
	$3 \cdot \delta$	500	100	10	sol	all	–	rand	1	restart	–	–	0.534	0.423	0.456
	$5 \cdot \delta$	500	50	20	sol	imp-ndom	10	newest	5	restart	–	–	0.531	0.454	0.468

Table 5. Average F-measure, MCC, sensitivity and precision on each dataset from CL_{tst}. For each metric, the best configurations (accounting for statistical ties) are shown in boldface.

	F-Measure					**MCC**				
dataset	exhau.	MO-AAC				exhau.	MO-AAC			
		$0.5 \cdot \delta$	$1 \cdot \delta$	$3 \cdot \delta$	$5 \cdot \delta$		$0.5 \cdot \delta$	$1 \cdot \delta$	$3 \cdot \delta$	$5 \cdot \delta$
lucap	0.940	**0.948**	**0.948**	**0.950**	0.948	0.779	**0.810**	**0.810**	**0.818**	**0.810**
ecoli1	**0.721**	0.666	0.680	0.686	0.672	**0.643**	0.573	0.594	0.603	0.581
yeast2vs8	**0.560**	**0.522**	**0.522**	**0.520**	0.501	**0.570**	**0.528**	**0.536**	**0.531**	0.497
ala	0.617	0.611	0.617	**0.625**	**0.635**	0.483	0.475	0.482	**0.493**	**0.504**
yeast3	0.666	0.692	0.697	**0.713**	**0.715**	0.630	0.659	0.667	**0.682**	**0.684**
ecoli2	**0.656**	**0.647**	**0.659**	**0.651**	**0.671**	**0.599**	**0.594**	**0.611**	**0.597**	**0.628**
abalone19	**0.007**	**0.000**	**0.017**	**0.004**	**0.000**	**0.000**	-0.006	**0.012**	**-0.002**	**-0.006**
abalone9-18	0.245	**0.34**	**0.302**	0.224	**0.345**	0.255	**0.327**	**0.285**	0.232	**0.331**
haberman	**0.373**	0.323	0.327	**0.362**	0.321	**0.132**	0.081	0.086	**0.113**	0.085

	Sensitivity					**Precision**				
dataset	exhau.	MO-AAC				exhau.	MO-AAC			
		$0.5 \cdot \delta$	$1 \cdot \delta$	$3 \cdot \delta$	$5 \cdot \delta$		$0.5 \cdot \delta$	$1 \cdot \delta$	$3 \cdot \delta$	$5 \cdot \delta$
lucap0	**0.955**	**0.956**	**0.956**	**0.957**	**0.957**	0.927	**0.941**	**0.941**	**0.943**	**0.940**
ecoli1	**0.835**	0.722	0.744	0.772	0.729	**0.650**	**0.643**	**0.655**	**0.648**	**0.650**
yeast2vs8	**0.483**	**0.467**	**0.458**	**0.450**	**0.450**	**0.739**	**0.669**	**0.709**	**0.701**	0.606
ala	0.671	0.684	0.693	0.695	**0.723**	**0.576**	**0.561**	**0.562**	**0.572**	**0.568**
yeast3	**0.718**	0.684	0.686	**0.708**	**0.719**	0.641	**0.715**	**0.727**	**0.730**	**0.722**
ecoli2	**0.694**	**0.652**	**0.644**	**0.633**	**0.627**	0.642	0.670	**0.707**	0.690	**0.745**
abalone19	**0.006**	**0.000**	**0.016**	**0.005**	**0.000**	**0.008**	**0.000**	**0.022**	**0.004**	**0.000**
abalone9-18	0.171	**0.271**	**0.245**	0.155	**0.274**	**0.485**	**0.476**	**0.422**	**0.443**	**0.486**
haberman	**0.442**	0.345	0.350	**0.432**	0.334	**0.337**	**0.321**	**0.324**	**0.326**	**0.324**

rulesets that are either equivalent or better than ones produced by exhaustive configuration of MOCA-I. The single exception is for *ecoli1*, where the exhaustive configuration of MOCA-I is always statistically the best.

For the datasets on which the best configurations of MOCA-I and AC-MOCA-I are statistically tied, the best running time of the latter varies.

For example, on *yeast2vs8* and *lucap* shorter running times produce better results, while on *a1a* and *yeast3* longer running times yield better rulesets. $3 \cdot \delta$ seems to lead toward the best results in terms of MCC and F-Measure, while $5 \cdot \delta$ results in a better sensitivity, and $1 \cdot \delta$ in a better precision. Regarding sensitivity and precision, the performance of AC-MOCA-I is at least as good as that obtained from the exhaustive runs of MOCA-I, except for sensitivity on *ecoli1*.

5 Conclusions and Future Work

In this work, we have demonstrated the efficiency of AC-MOCA-I, which uses MO-AAC to configure an extended version of MOCA-I, a multi-objective local search procedure for rule-based imbalanced classification. We have conducted experiments with several MO-AAC scenarios, for different running times of AC-MOCA-I, on a large configuration space using nine imbalanced datasets from the literature. Our results demonstrate improved performance over the earlier version of MOCA-I.

In future work, we plan to use AC-MOCA-I with more datasets, grouped by characteristics, to find robust configurations of our algorithm and produce even better rulesets. An important next step to further improve the quality of rulesets will be to study the effect of different combinations of performance metrics during the training and validation phases of MO-ParamILS – for example, metrics commonly used in machine learning, such as F-measure – for AC-MOCA-I. An important objective will be to determine a robust configuration in the context of real-world medical data on biomarkers, e.g., for lung cancer detection in the context of the European project 'Pathacov'.

References

1. Alcalá-Fdez, J., et al.: Keel: a software tool to assess evolutionary algorithms for data mining problems. Soft Comput. **13**(3), 307–318 (2009)
2. Blot, A., Hoos, H.H., Jourdan, L., Kessaci-Marmion, M.É., Trautmann, H.: MO-ParamILS: a multi-objective automatic algorithm configuration framework. In: Festa, P., Sellmann, M., Vanschoren, J. (eds.) LION 2016. LNCS, vol. 10079, pp. 32–47. Springer, Cham (2016). https://doi.org/10.1007/978-3-319-50349-3_3
3. Blot, A., Kessaci, M.É., Jourdan, L.: Survey and unification of local search techniques in metaheuristics for multi-objective combinatorial optimisation. J. Heuristics **24**(6), 853–877 (2018). https://doi.org/10.1007/s10732-018-9381-1
4. Blot, A., Marmion, M., Jourdan, L., Hoos, H.H.: Automatic configuration of multi-objective local search algorithms for permutation problems. Evol. Comput. **27**(1), 147–171 (2019)

5. Deb, K., Pratap, A., Agarwal, S., Meyarivan, T.: A fast and elitist multiobjective genetic algorithm: NSGA-II. IEEE Trans. Evol. Comput. **6**(2), 182–197 (2002)
6. Fernández, A., García, S., Luengo, J., Bernadó-Mansilla, E., Herrera, F.: Genetics-based machine learning for rule induction: state of the art, taxonomy, and comparative study. IEEE Trans. Evol. Comput. **14**(6), 913–941 (2010)
7. Frank, A.: UCI machine learning repository (2010). http://archive.ics.uci.edu/ml
8. Guyon, I., et al.: Design and analysis of the causation and prediction challenge. In: Causation and Prediction Challenge, pp. 1–33 (2008)
9. He, H., Garcia, E.A.: Learning from imbalanced data. IEEE Trans. Knowl. Data Eng. **21**(9), 1263–1284 (2009)
10. Jacques, J., Taillard, J., Delerue, D., Dhaenens, C., Jourdan, L.: Conception of a dominance-based multi-objective local search in the context of classification rule mining in large and imbalanced data sets. Appl. Soft Comput. **34**, 705–720 (2015)
11. Jacques, J., Taillard, J., Delerue, D., Jourdan, L., Dhaenens, C.: The benefits of using multi-objectivization for mining Pittsburgh partial classification rules in imbalanced and discrete data. In: GECCO 15th, pp. 543–550. ACM (2013)
12. Knowles, J., Corne, D.: On metrics for comparing nondominated sets. In: CEC 2002. vol. 1, pp. 711–716. IEEE (2002)
13. López-Ibáñez, M., Stützle, T.: The automatic design of multiobjective ant colony optimization algorithms. IEEE Trans. Evol. Comput. **16**(6), 861–875 (2012)
14. Schölkopf, B., Burges, C.J., Smola, A.J., et al.: Advances in Kernel Methods: Support Vector Learning. MIT Press, Cambridge (1999)
15. Weiss, G.M., Provost, F.: Learning when training data are costly: the effect of class distribution on tree induction. J. Artif. Intell. Res. **19**, 315–354 (2003)
16. Zitzler, E., Thiele, L.: Multiobjective evolutionary algorithms: a comparative case study and the strength pareto approach. IEEE Trans. Evol. Comput. **3**(4), 257–271 (1999)

Bayesian- and Surrogate-Assisted Optimization

Multi-fidelity Optimization Approach Under Prior and Posterior Constraints and Its Application to Compliance Minimization

Youhei Akimoto[1,2(✉)], Naoki Sakamoto[1,2,3], and Makoto Ohtani[4]

[1] University of Tsukuba, Tsukuba, Japan
akimoto@cs.tsukuba.ac.jp
[2] RIKEN Center for Advanced Intelligence Project, Tokyo, Japan
[3] JSPS Research Fellow (DC1), Tokyo, Japan
[4] Honda R&D, Haga-gun, Japan

Abstract. In this paper, we consider a multi-fidelity optimization under two types of constraints: prior constraints and posterior constraints. The prior constraints are prerequisite to execution of the simulation that computes the objective function value and the posterior constraint violation values, and are evaluated independently from the simulation with significantly lower computational time than the simulation. We have several simulators that approximately simulate the objective and constraint violation values with different trade-offs between accuracy and computational time. We propose an approach to solve the described constrained optimization problem with as little computational time as possible by utilizing multiple simulators. Based on a covariance matrix adaptation evolution strategy, we combines three algorithmic components: prior constraint handling technique, posterior constraint handling technique, and adaptive simulator selection technique for multi-fidelity optimization. We apply the proposed approach to a compliance minimization problem and show a promising convergence behavior.

Keywords: Prior constraints · Posterior constraints · Multi-fidelity optimization · DD-CMA-ES · Compliance minimization

1 Introduction

We consider the following simulation based constrained optimization problem on n dimensional search space with m_{pri} prior and m_{pos} posterior constraints

$$\begin{aligned} &\text{minimize: } f(\mathcal{S}(x)) \text{ for } x \in \mathbb{R}^n \\ &\text{subject to: } g_i(x) \leqslant 0 \text{ for } i = 1, \dots, m_{\text{pri}} \\ &\qquad\qquad h_j(\mathcal{S}(x)) \leqslant 0 \text{ for } j = 1, \dots, m_{\text{pos}}, \end{aligned} \tag{1}$$

where $\mathcal{S}$ is a simulator that takes x as an input and computes the intermediate information, g_i are the prior constraints (also called QUAK constraints in [7])

T. Bäck et al. (Eds.): PPSN 2020, LNCS 12269, pp. 81–94, 2020.
https://doi.org/10.1007/978-3-030-58112-1_6

that the simulator requires x to satisfy to execute the simulation, and f and h_j are the objective and posterior constraint functions (also called QRSK constraints in [7]) that are computed based on the output of the simulation. The combination of prior constraints and posterior constraints naturally and often appears in engineering optimization problems.

We consider the multi-fidelity situation where we have m_{sim} simulators $\mathcal{S}_k$ ($k = 1, \ldots, m_{\text{sim}}$) whose accuracy and computational time are both higher for a greater k. That is, $\mathcal{S}_1$ is the lowest accuracy simulator with the lowest cost, and $\mathcal{S}_{m_{\text{sim}}}$ is the highest accuracy simulator with the highest cost. We assume that the computational time for g_i are all cheaper than those of $\mathcal{S}_k$, as the prerequisites to the simulator are usually formulated by a simple mathematical expression such as linear functions of x. Our goal is to solve problem (1) with the highest accuracy simulator $\mathcal{S}_{m_{\text{sim}}}$, while we would like to minimize the computational time for the optimization by utilizing $\mathcal{S}_k$ for $k = 1, \ldots, m_{\text{sim}} - 1$ as well.

We consider covariance matrix adaptation evolution strategies [10,11] as the baseline solver as they are recognized as one of the state-of-the-art search strategies for simulation-based continuous optimization. Since they are mainly developed for unconstrained optimization, constraint handling techniques and strategies for multi-fidelity situation are required. There are existing algorithmic components that address each characteristic of the above mentioned problem. Prior constraints are considered in e.g. [4,5,14,16], where a repair operator is applied either in a Lamarckian or Darwinian manner with or without a penalty to the fitness of the solution. Posterior constraints, also called simulation-based constraints, are treated in e.g. [6,8,12,13], where f and h_j values or their rankings are aggregated to virtually transform the problem into an unconstrained one and rank candidate solutions. Efficient use of multiple simulators in evolutionary computation for multi-fidelity optimization has been less addressed in the literature. Adaptive simulator selection mechanisms among more than one possible simulators have been proposed in [2,3] for unconstrained optimization. However, these algorithmic components are mostly designed and applied to problems having solely one of these problem characteristic. It is not trivial how to combine these existing components to address the above mentioned difficulties simultaneously.

The contributions of this paper are summarized as follows.

Firstly, to treat both prior and posterior constraints, we combine a recent prior constraint handling technique and posterior constraint handling technique. As far as the authors know, there are few researches that addresses these two types of constraints in a systematic way, i.e., problem-independent way. We employ ARCH [14] as a prior constraint handling, and MCR-mod [8] as a posterior constraint handling. Each of them considers solely each type of constraints. We combine these two techniques so that the combination does not disturb the adaptation mechanism in these techniques. The resulting technique is designed to be invariant to any strictly increasing transformation of the objective and constraint violation functions: $(f, \{g_i\}_{i=1}^{m_{\text{pri}}}, \{h_j\}_{j=1}^{m_{\text{pos}}}) \mapsto (e^f \circ f, \{e_i^g \circ g_i\}_{i=1}^{m_{\text{pri}}}, \{e_j^h \circ h_j\}_{j=1}^{m_{\text{pos}}})$, where e^f, e_i^g, e_j^h are arbitrary strictly increasing functions satisfying

Algorithm 1. An Iteration of DD-CMA-ES

1: $\{x_1, \ldots, x_\lambda\} = \text{SAMPLE}_{\text{cma}}(m, \Sigma)$ ▷ generate candidate solutions
2: $\{r_\ell\}_\ell = \text{EVALUATE}(x_1, \ldots, x_\lambda)$ ▷ evaluate candidate solutions
3: $\theta \leftarrow \text{UPDATE}_{\text{cma}}(\theta, \{(x_\ell, r_\ell)\}_\ell)$ ▷ update the parameters including m and Σ

Algorithm 2. EVALUATE$(x_1, \ldots, x_\lambda)$ for Unconstrained Optimization

1: **for all** $\ell = 1, \ldots, \lambda$ **do** ▷ possibly in parallel
2: execute $s_\ell = \mathcal{S}(x_\ell)$ and evaluate $f_\ell = f(s_\ell)$
3: **end for**
4: $r_\ell = -\frac{1}{2} + \sum_{\ell'=1}^{\lambda} \mathbf{1}\{f_{\ell'} < f_\ell\} + \frac{1}{2} \sum_{\ell'=1}^{\lambda} \mathbf{1}\{f_{\ell'} = f_\ell\}$ for $\ell = 1, \ldots, \lambda$
5: **return** $\{r_\ell\}_\ell$

$e_i^g(0) = e_j^h(0) = 0$, as well as to any affine transformation of search space coordinates.

Secondly, to exploit multiple simulators in the above stated multi-fidelity constrained optimization scenario, we generalize the adaptive simulator selection mechanism [2], which is proposed for unconstrained optimization, to constrained optimization. Then we combine the generalized simulator selection mechanism with the proposed constraint handling technique. As far as the authors know, this is the first paper that addresses automatic selection of simulators during search process under constraints. The proposed approach is again invariant to any strictly increasing transformation of the objective and constraint violation functions. We need not care about the balance between the objective value and the constraint violations when defining them. Practically, this is desired especially when there are many constraints, where finding a good balance between f, g_i and h_j is a tedious task and is often impossible. Moreover, it is invariant to any affine transformation of search space coordinates so as not to disturb the adaptation mechanism of the baseline unconstrained optimizer, i.e., CMA-ES.

Finally, the proposed technique is fully modularized. We can plug the proposed mechanism in place of the ranking mechanism of candidate solutions in ranking-based evolutionary approaches. To show its high flexibility, we combine the proposed technique to the latest variant of covariance matrix adaptation evolution strategies, namely DD-CMA-ES [1]. We then apply it to a compliance minimization problem [3] as a proof of concept.

2 Existing Algorithmic Components

DD-CMA-ES. DD-CMA-ES [1] is the latest variant of covariance matrix adaptation evolution strategies (CMA-ES) that accelerates the covariance matrix adaptation by decomposing the covariance matrix into the variance matrix and the correlation matrix and updating them separately. It is designed for unconstrained optimization

$$\text{minimize: } f(\mathcal{S}(x)) \text{ for } x \in \mathbb{R}^n. \tag{2}$$

Algorithm 3. EVALUATE($x_1, \ldots, x_\lambda$) with ARCH

1: **for all** $\ell = 1, \ldots, \lambda$ **do** ▷ possibly in parallel
2: $x_\ell^{\mathrm{rep}} = \mathrm{REPAIR_{pri}}(x_\ell)$ ▷ repaired candidate solution
3: $p_\ell = \|x_\ell^{\mathrm{rep}} - x_\ell\|_{\Sigma^{-1}}$ ▷ penalty
4: execute $s_\ell = \mathcal{S}(x_\ell^{\mathrm{rep}})$ and evaluate $f_\ell = f(s_\ell)$
5: **end for**
6: $\alpha \leftarrow \mathrm{UPDATE_{pri}}(\theta)$ ▷ update of the penalty coefficient
7: $\{r_\ell\}_\ell = \mathrm{RANKING_{pri}}(\alpha, \{(f_\ell, p_\ell)\}_\ell)$ ▷ eqs. (4) to (6)
8: **return** $\{r_\ell\}_\ell$

A single iteration of the DD-CMA-ES is displayed in Algorithm 1 in an abstract manner. DD-CMA-ES generates candidate solutions from the n-variate normal distribution $\mathcal{N}(m, \Sigma)$, where m is the mean vector and Σ is the covariance matrix of the distribution. The covariance matrix Σ is decomposed as $\Sigma = \sigma^2 D \cdot C \cdot D$ and they are updated separately, where $\sigma > 0$ is the step-size, D is a diagonal matrix, and C is a positive definite symmetric matrix. Generated candidate solutions are evaluated on the objective function (Algorithm 2). Their rankings are computed based on the objective function values. Then, the distribution parameters and other internal parameters, denoted by θ, are update by using the candidate solutions and their rankings. These steps are repeated until a termination condition is satisfied.

In the update step, the recombination weight w_ℓ is assigned to each candidate solution x_ℓ based on its ranking. Let $\bar{w}_1, \ldots, \bar{w}_\lambda$ be the pre-defined weights. If there is no tie in f-values, the ranking values are $0, 1, \ldots, \lambda - 1$. Then, the recombination weight w_ℓ assigned to x_ℓ is simply $\bar{w}_{1+r_\ell}$. If there are ties, a tie candidate solution x_ℓ receives the same recombination weights that are the average of the pre-defined weights $\bar{w}_{1+\sum_{\ell'=1}^{\lambda} \mathbb{1}\{r_{\ell'} < r_\ell\}}, \ldots, \bar{w}_{\sum_{\ell'=1}^{\lambda} \mathbb{1}\{r_{\ell'} \leqslant r_\ell\}}$.

ARCH. ARCH [14] is a ranking-based constrained handling technique for optimization under prior constraints

$$\begin{aligned} &\text{minimize: } f(\mathcal{S}(x)) \text{ for } x \in \mathbb{R}^n \\ &\text{subject to: } g_i(x) \leqslant 0 \text{ for } i = 1, \ldots, m_{\mathrm{pri}}, \end{aligned} \tag{3}$$

where the input to $\mathcal{S}$ must satisfy the constraints. ARCH is implemented as an EVALUATE function in Algorithm 1. We replace Algorithm 2 with Algorithm 3.

Given a candidate solution x, ARCH repairs x to a boundary point of the feasible domain, which is either the nearest feasible solution to x w.r.t. the square Mahalanobis distance $\|x^{\mathrm{rep}} - x\|_{\Sigma^{-1}}^2 = (x^{\mathrm{rep}} - x)^{\mathrm{T}} \Sigma^{-1} (x^{\mathrm{rep}} - x)$ or the nearest feasible solution under the constraints $g_i(x^{\mathrm{rep}}) = 0$ for constraints g_i that are violated by x. Let $x_\ell^{\mathrm{rep}} = \mathrm{REPAIR_{pri}}(x_\ell)$ be the repaired candidate solution given x_ℓ. Then, the objective function value is computed at x_ℓ^{rep}, denoted by $f_\ell = f(x_\ell^{\mathrm{rep}})$. The penalty for the constraint violation is measured by the Mahalanobis distance between the original and repaired points, $p_\ell = \|x_\ell^{\mathrm{rep}} - x_\ell\|_{\Sigma^{-1}}$. The rankings of the objective value and the penalty value are defined as follows

Algorithm 4. EVALUATE($x_1, \dots, x_\lambda$) with MCR-mod

1: **for all** $\ell = 1, \dots, \lambda$ **do** ▷ possibly in parallel
2: execute $s_\ell = \mathcal{S}(x_\ell)$ and evaluate $f_\ell = f(s_\ell)$
3: $v_\ell^j = \max(h_j(s_\ell), 0)$ for $j = 1, \dots, m_{\mathrm{pos}}$ ▷ constraint violations
4: $N_\ell^v = |\{j : v_\ell^j > 0\}|$ ▷ number of violated constraints
5: **end for**
6: $\{r_\ell\}_\ell = \mathrm{RANKING}_{\mathrm{pos}}(\{(f_\ell, \{v_\ell^j\}_{j=1}^{m_{\mathrm{pos}}}, N_\ell^v)\}_\ell)$ ▷ eqs. (8) and (9)
7: **return** $\{r_\ell\}_\ell$

$$r_\ell^f = -\tfrac{1}{2} + \textstyle\sum_{\ell'=1}^{\lambda} \mathbf{1}_{\{f_{\ell'} < f_\ell\}} + \tfrac{1}{2} \sum_{\ell'=1}^{\lambda} \mathbf{1}_{\{f_{\ell'} = f_\ell\}} \tag{4}$$

$$r_\ell^p = -\tfrac{1}{2} + \textstyle\sum_{\ell'=1}^{\lambda} \mathbf{1}_{\{p_{\ell'} < p_\ell\}} + \tfrac{1}{2} \sum_{\ell'=1}^{\lambda} \mathbf{1}_{\{p_{\ell'} = p_\ell\}} . \tag{5}$$

The final ranking of x_ℓ is then computed as

$$r_\ell = r_\ell^f + \alpha \cdot r_\ell^p , \tag{6}$$

where α is the penalty coefficient that is adapted based on the penalty of the center m of the search distribution.

MCR-mod. MCR-mod [8] is a ranking-based constrained handling technique for optimization under posterior constraints

$$\begin{aligned} &\text{minimize: } f(\mathcal{S}(x)) \text{ for } x \in \mathbb{R}^n \\ &\text{subject to: } h_j(\mathcal{S}(x)) \leqslant 0 \text{ for } j = 1, \dots, m_{\mathrm{pos}}, \end{aligned} \tag{7}$$

where $\mathcal{S}$ accepts any value in $\mathbb{R}^n$ as an input. MCR-mod can be implemented as an EVALUATE function. We replace Algorithm 2 with Algorithm 4.

For each input x_ℓ, the objective function value $f_\ell = f(\mathcal{S}(x_\ell))$ and the constraint violation $v_\ell^j = \max(h_j(\mathcal{S}(x_\ell)), 0)$ are computed. Then calculate the number N_ℓ^v of the violated constraints $N_\ell^v = |\{j : v_\ell^j > 0\}|$. MCR-mod defines the rankings of the objective values, constraint violations, and the number of violated constraints as follows

$$r_\ell^f = \textstyle\sum_{\ell'=1}^{\lambda} \mathbf{1}_{\{f_{\ell'} < f_\ell\}}, \quad r_\ell^{v_j} = \sum_{\ell'=1}^{\lambda} \mathbf{1}_{\{v_{\ell'}^j < v_\ell^j\}}, \quad r_\ell^{N^v} = \sum_{\ell'=1}^{\lambda} \mathbf{1}_{\{N_{\ell'}^v < N_\ell^v\}} . \tag{8}$$

Let $\beta = |\{\ell' : N_{\ell'}^v = 0\}| / \lambda$ be the ratio of the feasible candidate solutions among λ current candidate solutions. Then, the final ranking is computed as follows

$$r_\ell = \beta \cdot r_\ell^f + (1 - \beta) \left(r_\ell^{N^v} + \tfrac{1}{m_{\mathrm{pos}}} \textstyle\sum_{j=1}^{m_{\mathrm{pos}}} r_\ell^{v_j} \right) . \tag{9}$$

Adaptive Simulator Selection. The adaptive objective selection mechanism [2] selects the simulator $\mathcal{S}_k$ among m_{sim} available ones, $\mathcal{S}_1, \dots, \mathcal{S}_{m_{\mathrm{sim}}}$, for an unconstrained multi-fidelity optimization problem

$$\text{minimize: } f(\mathcal{S}_{m_{\mathrm{sim}}}(x)) \text{ for } x \in \mathbb{R}^n . \tag{10}$$

Algorithm 5. EVALUATE$(x_1, \ldots, x_\lambda)$ with Adaptive Simulator Selection

1: $\{r_\ell\}_\ell = \text{EVALUATE}_k(x_1, \ldots, x_\lambda)$
2: $T \leftarrow T + t_k$, where t_k is the total time for $\mathcal{S}_k$ call above
3: **if** $k < m_{\text{sim}}$ and $T > \gamma \cdot T_+$ **then** ▷ check for fidelity index increment
4: $\quad \{r_\ell^+\}_\ell = \text{EVALUATE}_{k+1}(x_1, \ldots, x_\lambda)$
5: $\quad T_+ \leftarrow T_+ + t_{k+1}$, where t_{k+1} is the total time for $\mathcal{S}_{k+1}$ call above
6: $\quad$ compute Kendall's rank coefficient τ for $\{(r_\ell, r_\ell^+)\}_\ell$
7: $\quad$ update $\langle\tau\rangle_+ \leftarrow \langle\tau\rangle_+ + c_\tau^+(\tau - \langle\tau\rangle_+)$
8: **else if** $1 < k$ and $T > \gamma \cdot T_-$ **then** ▷ check for fidelity index decrement
9: $\quad \{r_\ell^-\}_\ell = \text{EVALUATE}_{k-1}(x_1, \ldots, x_\lambda)$
10: $\quad T_- \leftarrow T_- + t_{k-1}$, where t_{k-1} is the total time for $\mathcal{S}_{k-1}$ call above
11: $\quad$ compute Kendall's rank coefficient τ for $\{(r_\ell, r_\ell^-)\}_\ell$
12: $\quad$ update $\langle\tau\rangle_- \leftarrow \langle\tau\rangle_- + c_\tau^-(\tau - \langle\tau\rangle_-)$
13: **end if**
14: **if** $\langle\tau\rangle_+ < \tau_{\text{thresh}}$ **then**
15: $\quad k \leftarrow k + 1$ and reset $T = T_+ = T_- = 0$ and $\langle\tau\rangle_+ = \langle\tau\rangle_- = 0$
16: **else if** $\langle\tau\rangle_- > \tau_{\text{thresh}} + \tau_{\text{margin}}$ **then**
17: $\quad k \leftarrow k - 1$ and reset $T = T_+ = T_- = 0$ and $\langle\tau\rangle_+ = \langle\tau\rangle_- = 0$
18: **end if**
19: **return** $\{r_\ell\}_\ell$

As discussed in the introduction, we assume that $\mathcal{S}_k$ with a lower fidelity index k is less accurate and computationally cheaper.

The adaptive objective selection mechanism works as a wrapper of EVALUATE in Algorithm 2 with $\mathcal{S}_1, \ldots, \mathcal{S}_{m_{\text{sim}}}$, summarized in Algorithm 5, where EVALUATE$_k$ function has the same functionality as EVALUATE in Algorithm 2 except that $\mathcal{S}$ is replaced with $\mathcal{S}_k$. Let k be the current fidelity index. The main idea for the adaptation of the fidelity index is to check if the objective function value sequences evaluated with $\mathcal{S}_k$ and $\mathcal{S}_{k+1}$ (or $\mathcal{S}_{k-1}$) have a low rank correlation (or a high rank correlation). If there is a high correlation between $\mathcal{S}_k$ and $\mathcal{S}_{k+1}$ (or $\mathcal{S}_{k-1}$), there is no need to use the one with a higher fidelity index as they results in similar rankings of candidate solutions and a computationally cheaper simulator is preferred. To avoid spending too much time to check the rank correlation, the execution time for each simulator is recorded and we compute the rank correlation only if the execution time for the current simulator $\mathcal{S}_k$ is γ times more than the execution time for $\mathcal{S}_{k+1}$ or $\mathcal{S}_{k-1}$.

3 Proposed Approach

Combination of ARCH and MCR-mod. As a first step, we combine DD-CMA-ES with ARCH and MCR-mod to tackle optimization problems with prior and posterior constraints (1) with a single simulator $\mathcal{S}$. We combine ARCH (Algorithm 3) and MCR-mod (Algorithm 4) in a way described in Algorithm 6. First, it repairs candidate solutions for prior constraints and computes the penalty values. The repaired candidate solutions are given to the simulator and their objective function values and posterior constraint violations are computed. Then, the final rankings are computed.

Algorithm 6. EVALUATE($x_1, \ldots, x_\lambda$) with ARCH and MCR-mod

1: **for all** $\ell = 1, \ldots, \lambda$ **do** ▷ possibly in parallel
2: repair $x_\ell^{\text{rep}} = \text{REPAIR}_{\text{pri}}(x_\ell)$
3: compute $p_\ell = \|x_\ell^{\text{rep}} - x_\ell\|_{\Sigma^{-1}}$
4: execute $s_\ell = \mathcal{S}(x_\ell^{\text{rep}})$ and evaluate $f_\ell = f(s_\ell)$
5: evaluate $v_\ell^j = \max(h_j(s_\ell), 0)$ for all $j = 1, \ldots, m_{\text{pos}}$
6: calculate $N_\ell^v = |\{j : v_\ell^j > 0\}|$
7: **end for**
8: update $\alpha \leftarrow \text{UPDATE}_{\text{pri}}(\theta)$
9: compute $\{r_\ell^h\}_\ell = \text{RANKING}_{\text{pos}}(\{(f_\ell, v_\ell^1, \ldots, v_\ell^{m_{\text{pos}}}, N_\ell^v)\}_\ell)$ ▷ eqs. (8) and (9)
10: compute $\{r_\ell\}_\ell = \text{RANKING}_{\text{pri}}(\alpha, \{(r_\ell^h, p_\ell)\}_\ell)$ ▷ eqs. (4) to (6)
11: **return** $\{r_\ell\}_\ell$

How to assign the ranking to each point is the only part that requires a careful decision to make for the combination of ARCH and MCR-mod to work well. A straight-forward way to construct the final ranking may simply add all the rankings that appear in Eqs. (6) and (9). However, since the adaptation of α in ARCH is designed to balance r^f and r^p, the additional rankings r^{N^v} and r^{v_j} will disturb the adaptation mechanism. To avoid it, we first compute the rankings by MCR-mod (9), then we pass them to ARCH as if the rankings computed by MCR-mod are the objective function values.

Adaptive Simulator Selection for Constrained Optimization. We now combine Algorithm 5 with Algorithm 6 for multi-fidelity constrained optimization. Since Algorithm 5 is based only on the rankings of the candidate solutions, we simply replace the EVALUATE$_k$ function in Algorithm 5 with the EVALUATE function in Algorithm 6 using $\mathcal{S}_k$. Arguably, this is the most natural combination of these components. The adaptive simulator selection tries to increase or decrease the fidelity index if the resulting rankings of the candidate solutions are different for different simulators, which results in affecting the behavior of the baseline algorithm, here DD-CMA-ES. Even if the rankings of the objective or the rankings of the constraint violations are significantly different for different simulators, the behavior of the algorithm may not change drastically if α in Eq. (6) is very large. Then, a computationally cheaper simulator is preferred. Our design choice is to reflect this demand.[1]

[1] We remark that if a simulator $\mathcal{S}_k$ is computationally very cheap to evaluate, the computation time for EVALUATE$_k$ may be dominated by the time for REPAIR$_{\text{pri}}$, which is computationally demanding compared to the other parts as it internally solves an minimization problem. In such a case, it may be more reasonable to record the computational time for EVALUATE$_k$ to t_k, rather than the total time of $\mathcal{S}_k$ in EVALUATE$_k$. This is also expected to affect the performance of the proposed strategy when it is implemented on a parallel machine, especially when the computational time for Line 2 to 6 in Algorithm 6 differs substantially among $\ell = 1, \ldots, \lambda$. It is highly involved with parallel computation and is out of scope of this paper. Further studies on this line is left for future work.

Discussion on the Choice of Algorithmic Components. For posterior constraint handling, there are other existing approaches such as augmented Lagrangian (AL) approaches [6]. Replacing MCR-mod with AL (in an appropriate way) needs a careful decision since AL aggregates f and h_j with adaptive coefficients and its adaptation mechanism can be broken due to the existence of g_i and the repair operator for them. Moreover, since AL is not invariant to strictly increasing transformation of f and h_j, we observe an undesired behavior that the adaptation of AL fails if the ranges of f and h_j values are hugely different. On the other hand, if the original MCR [13] is preferred to MCR-mod, or once their improvements are proposed, one can easily replace them.[2]

For prior constraint handling, ARCH can treat both linear and non-linear constraints. Limiting the focus on box constraint, there are several approaches such as mirroring, death penalty, resampling, and adaptive penalty method [9]. Even for box constraints, ARCH is reported to have a better performance over other box constraint handling techniques. Moreover, ARCH itself is reported as invariant to strictly increasing transformation of f and g_i, resulting in the invariance of the proposed approach mentioned in the introduction, and invariant to affine transformation of the objective function.

For adaptive simulator selection, one can use an approach proposed in [17]. Though it is proposed for PSO, the idea can be translated to other algorithms such as CMA-ES. One reason for our choice of [2] is that the method in [17] requires one execution of the highest fidelity simulator to check if the fidelity level should be increased or not. It is undesired when the highest accuracy simulator is unnecessarily accurate and computationally expensive, which happens in practice as we often do not know how accurate the simulator should be in advance. Another reason is that it is non-trivial to generalize the fidelity check mechanism to the constrained optimization scenario.

4 Compliance Minimization Problem

A compliance minimization problem is an example problem that can be considered as the multi-fidelity optimization of form (1). A compliance minimization is a classical topology optimization. Given a design space, the material distribution over the design space is parameterized by x. The objective is to find x achieving the optimal material distribution in terms of compliance with different constraints. The objective is typically computed through a finite-element

[2] Both the original MCR and MCR-mod rank the candidate solutions based only on the constraint violations when the candidates are all infeasible. On one hand it is nice as the search distribution is forced to sample feasible solutions with high probability. On the other hand, we empirically observe that it is trapped by a sub-optimal feasible point since it tends to satisfy the constraints at the beginning of the search without considering the objective function landscape. A possible solution is to *include the objective function as a posterior constraint as* $h(\mathcal{S}(x)) = f(\mathcal{S}(x)) - f_{\text{thre}}$, where f_{thre} is the worst allowable objective value. In engineering optimization, it is often the case that the objective is to find a solution that satisfies all demands. In such a situation, f_{thre} is available.

method, and its granularity is a user-parameter. A finer granularity leads to a more accurate but computational more expensive simulation. Hence, it forms a multi-fidelity optimization. Here, we describe the compliance minimization problem formulated in [3] with a Normalized Gaussian network (NGnet) based design representation [15].

NGnet based Material Distribution Representation. The design space is a two dimensional rectangular area $\mathcal{D} = [0, L_h] \times [0, L_v]$. The design space is discretized by square finite elements with length δ. The (i_h, i_v)-th element takes a value in $\{0, 1\}$, representing that there is a material (if 1) or not (if 0).

NGnet based representation places n Gaussian basis functions on the design space and the value at each element is represented by the sum of the basis function values evaluated at each place. Let $(\mu_h^d, \mu_v^d) \in \mathcal{D}$ and $(\sigma_h^d, \sigma_v^d) \in \mathbb{R}_{>0}^2$ be the center and the coordinate-wise standard deviation of the dth basis function, where the basis functions are pre-located by a user. Let $x = ([x]_1, \dots, [x]_n) \in [-1, 1]^n$ be the vector consisting of the height parameters of the basis functions, which is the parameter vector to be optimized. The output of the NGnet is then computed as follows

$$\phi(h, v; x) = \frac{\sum_{d=1}^{n} [x]_d \exp\left(-\frac{(h-\mu_h^d)^2}{2(\sigma_h^d)^2} - \frac{(v-\mu_v^d)^2}{2(\sigma_v^d)^2}\right)}{\sum_{d=1}^{n} \exp\left(-\frac{(h-\mu_h^d)^2}{2(\sigma_h^d)^2} - \frac{(v-\mu_v^d)^2}{2(\sigma_v^d)^2}\right)}. \tag{11}$$

It takes value in $[-1, 1]$. The value of the (i_h, i_v)-th element is $\mathbf{1}\{\phi(h, v; x) \geqslant 0\}$ evaluated at the center of the element, $h = \delta \cdot (i_h - 0.5)$ and $v = \delta \cdot (i_v - 0.5)$.

In practice the distribution of the bases should reflect some prior knowledge so that one can represent a more complex structure in one place than the others. In this paper, we place them on a grid pattern. Let n_h and $n_v = n/n_h$ be the numbers of the basis functions in horizontal and vertical axes, respectively. Let $\Delta_h = L_h/n_h$ and $\Delta_v = L_v/n_v$. The center μ^d and the standard deviation σ^d of the basis function of index $d = d_h + (d_v - 1) \cdot n_h$ is $\mu_h^d = \Delta_h \cdot (d_h - 0.5)$, $\mu_v^d = \Delta_v \cdot (d_v - 0.5)$, $\sigma_h^d = r \cdot \Delta_h$ and $\sigma_v^d = r \cdot \Delta_v$, where $r > 0$ is the scaling parameter.

Compliance Minimization. Let $\Phi_\delta(x)$ be the $L_v/\delta \times L_h/\delta$ dimensional matrix whose (i_h, i_v)-th element is $\mathbf{1}\{\phi(h, v; x) \geqslant 0\}$ evaluated at the center of the element, $h = \delta \cdot (i_h - 0.5)$ and $v = \delta \cdot (i_v - 0.5)$. This corresponds to the material distribution. A finite-element method is performed with $\Phi_\delta(x)$ as an input, and the compliance is calculated based on the simulation results. The compliance of the design $\Phi_\delta(x)$ is the objective function value $f(\mathcal{S}(x))$. The design variable x must live in $[-1, 1]^n$, hence we have $m_{\text{pri}} = 2n$ prior constraints $g_i(x) = x - 1$ (for upper bound) and $g_{n+i}(x) = -x - 1$ (for lower bound) for $i = 1, \dots, n$. A posterior constraint is the volume fraction constraint: the number of the elements of $\Phi_\delta(x)$ being 1 is no greater than $\text{volfrac}_{\max} \cdot (L_v/\delta) \cdot (L_h/\delta)$. The posterior constraint violation is defined as the volume fraction minus the maximum allowed volume fraction $\text{volfrac}_{\max} \in (0, 1]$.

Both the objective value and the posterior constraint violation depend on δ. A smaller δ (a finer granularity) leads to a better accuracy at the risk of longer computational time. Since δ is a user parameter, we can prepare multiple simulators with different δ, leading to multi-fidelity constrained optimization (1).

5 Experiments

We applied the proposed approach to the above mentioned compliance minimization problem to see how the performance-per-cost is improved by the proposed multi-fidelity approach. The baselines are the performance graph of the proposed algorithm without adaptive simulator selection solving the constrained optimization using $\mathcal{S}_k$ (for $k = 1, \dots, m_{\text{sim}}$).

We set the problem parameters as follows: numbers of basis functions in horizontal axis $n_h = 24$ and vertical axis $n_v = 8$, number of elements in horizontal axis $(L_h/\delta) = 12k$ and vertical axis $(L_v/\delta) = 4k$ for simulator $\mathcal{S}_k$, the volume fraction is $\text{volfrac}_{\max} = 0.4$ and the scaling parameter for NGnet is $r = 1$. The number of design variables is $n = n_h \cdot n_v = 192$. All the hyper-parameters for the proposed algorithm are set by following the original references [1,2,14], except that we set $c_\tau^+ = 0.1, 0.3, 0.5.0.7, 0.9$, instead of $c_\tau^+ = 1$ to make the fidelity adaptation more stable. We set $m_{\text{sim}} = 20$.

The experiments were implemented in PYTHON. For each setting, 10 independent runs with different random number generator seeds were conducted in parallel on Intel(R) Core(TM) i9-7900X CPU (10 cores) with 4 8192MB DDR4 RAMs (2133MT/s), and each run was limited to 24 h (86400 s) on a single thread. The code to reproduce the results of this paper is available at https://github.com/akimotolab/multi-fidelity.

Performance Assessment. The performance assessment must be done on $\mathcal{S}_{m_{\text{sim}}}$ as our objective is to locate the optimal solution of (1) with $\mathcal{S}_{m_{\text{sim}}}$. Ideally we can re-evaluate all the candidate solutions on $\mathcal{S}_{m_{\text{sim}}}$ and show the graph of the best-so-far objective value among the feasible solutions. However, it is computationally very expensive, and is not possible in practice. To have a fair comparison, at each iteration t we pick the best ranked repaired candidate solution, $x_{t,\text{rec}}^{\text{rep}}$, and regard it as the recommendation point at iteration t. We re-evaluate the recommendation point on $\mathcal{S}_{m_{\text{sim}}}$, check if the recommendation point is feasible, and record its objective function value $f_{t,\text{rec}} = f(\mathcal{S}_{m_{\text{sim}}}(x_{t,\text{rec}}^{\text{rep}}))$ if it is feasible. Note that g_i are independent but h_j depends on k, hence it is not guaranteed that a feasible solution on $\mathcal{S}_k$ is also feasible on $\mathcal{S}_{m_{\text{sim}}}$. Figure 1 shows percentiles of the best-so-far $f_{t,\text{rec}}$ values, plotted against the elapsed CPU time excluding the time for re-evaluation.

Results. We first summarize the results obtained by optimizing the fixed fidelity problems with $k = 4, 7, 11, 14, 17, 20$. A lower fidelity simulator resulted in a quicker drop of the objective value and a higher (worse) final function value as it is trapped at a local optimum defined on $\mathcal{S}_k$, which is not necessarily a local optimal solution on $\mathcal{S}_{m_{\text{sim}}}$. A higher fidelity simulator tended to result in

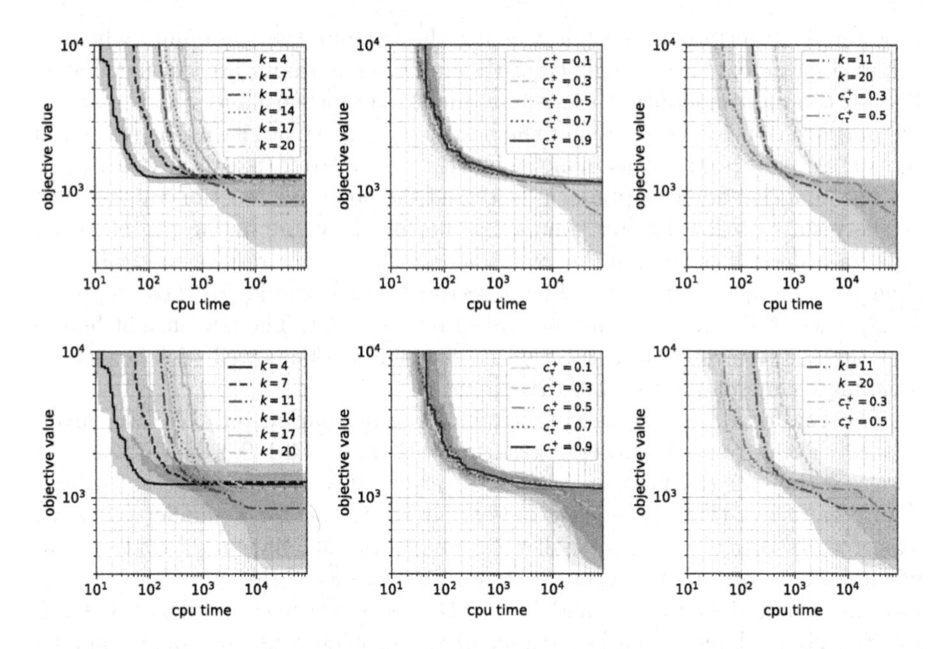

Fig. 1. Best-so-far objective function values of feasible recommendation points evaluated on $\mathcal{S}_{m_{\text{sim}}}$. Median (solid line) and percentile range (transparent area, 25%–75%ile (top figures) and 10%–90%ile (bottom figures)) of 10 independent runs are shown for each cpu time [second]. Left: fixed fidelity approaches. Center: adaptive fidelity approaches. Right: comparison.

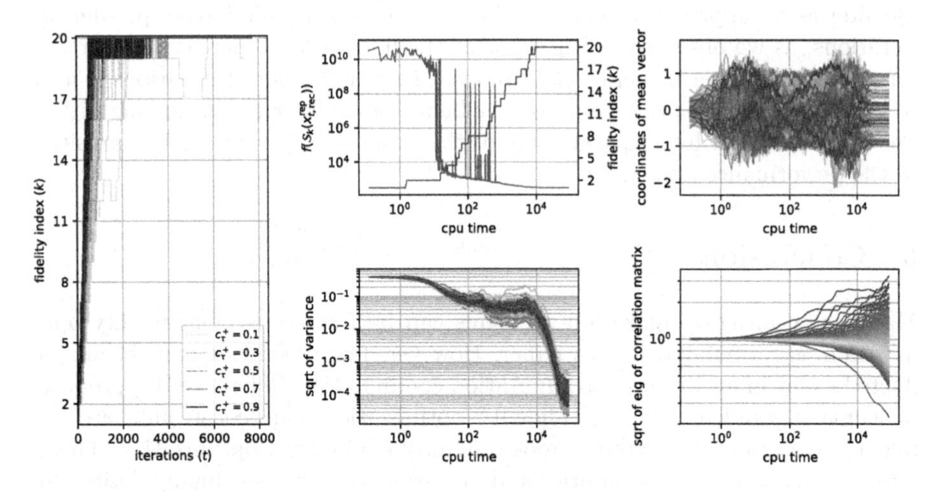

Fig. 2. (a) Fidelity index k during the optimization. Results of 10 independent runs are shown for each setting. (b) Typical run result of the proposed approach. Shown are $f(\mathcal{S}_k(x_{t,\text{rec}}^{\text{rep}}))$ (blue), k (red), the square roots of the variances ($\sigma^2 D^2$), the square roots of the eigenvalues of the correlation matrix (C), and each coordinate of the mean vector m. Different colors mean different coordinates. (Color figure online)

a better final performance while wasting the time at the beginning, where f-values are comparatively high ($\gg 10^4$) and h_j-constraints are often violated. Due to the multimodality, the median and 25%ile performances at the end of the optimization process were not the best for $k = 20$ ($k = 11$ was the best), but the 10%ile performance was monotonic with respect to k (equal for $k \geqslant 14$).

The adaptive fidelity approach decreased the objective function value as fast as the results on low fidelity simulators did and kept improving the objective function value till the end of the experiments, whereas the results on low fidelity simulators stopped improving. The observed trend is similar to those reported in [2], where the constraints are not taken into account. The new insight here is that the adaptive simulator selection mechanism works as well with constraint handling technique proposed in this paper.

The sensitivity of the newly-introduced hyper-parameter, c_τ^+, was investigated. The proposed strategy was run with $c_\tau^+ = 0.1, 0.3, 0.5, 0.7, 0.9$. We observed that all the median graphs are inside 10%–90%ile ranges of others. Figure 2a shows that the fidelity index increased later for a smaller c_τ^+ as we expected. In other words, a smaller c_τ^+ results in spending more time on low fidelity simulators. It is advantageous for adaptation of the parameters of the baseline search algorithm (m and $\Sigma = \sigma^2 D \cdot C \cdot D$ for the case of DD-CMA-ES, see Fig. 2b) as long as the landscapes of the problems on low fidelity simulators translate to those on high fidelity simulators. However, if the landscape is multimodal and an absorbing local optimum on a low fidelity simulator is not a good local optimum on a high fidelity simulator, spending more time on low fidelity simulators might make it more likely for the search to be trapped by a poor local optimum. Such a situation was observed in Fig. 1 for $c_\tau^+ = 0.1$. A simple alternative approach would be to increase the fidelity level every pre-defined iterations, as we observed in Fig. 2a that the fidelity level increased linearly in terms of the number of iterations. However, such a number of iterations is hard to know in advance as it depends on the behavior of the optimization algorithm. We claim that the hyper-parameters of the proposed approach is less sensitive to the performance and easier to set.

6 Conclusion

Many engineering optimization problems can be framed as multi-fidelity optimization as formulated in this paper. However, few researches have been conducted so far in evolutionary computation communities. Reference [17] proposes benchmark problems for multi-fidelity optimization, but constraints are not taken into account. Constraints appear in most engineering optimization. Therefore, a benchmark suite with prior and posterior constraints is highly desired for evaluation and development of multi-fidelity search strategies.

At the same time, the theoretical foundation for the multi-fidelity optimization problem and its approaches should be investigated to guarantee the performance or to reveal the limitation. Some theoretical investigation has been done in [2], revealing when the approach fails to adapt the fidelity level and how to avoid it. In the multi-fidelity optimization, it is assumed that a low fidelity simulator and a high fidelity simulator are somewhat similar, without formally defining the similarity. The assumption on the similarity between simulators can be different between approaches, but they should be reasonable in practice. The development of the benchmark suite should be done in parallel with the theoretical foundation to have a maximal coverage of the difficulties of multi-fidelity optimization with a minimal set of test problems.

This paper proposes an approach to deal with multi-fidelity optimization with prior and posterior constraints. As a proof of concept, the proposed approach has been applied to a multi-fidelity compliance minimization problem, showing its efficacy over the fixed fidelity formulation. Since the compliance minimization shares characteristics with other topology optimization problems, we expect that the proposed approach is applied to other topology optimization problems as well. However, as a generic framework for multi-fidelity constrained optimization, the evaluation of the proposed approach is definitely missing. The above mentioned benchmark suite and theoretical foundation are desired for this purpose.

Acknowledgement. This work is partially supported by JSPS KAKENHI Grant Number 19H04179 and 19J21892. We thank anonymous reviewers for their valuable comments.

References

1. Akimoto, Y., Hansen, N.: Diagonal acceleration for covariance matrix adaptation evolution strategies. Evol. Comput. (2019). https://doi.org/10.1162/evco_a_00260, to appear
2. Akimoto, Y., Shimizu, T., Yamaguchi, T.: Adaptive objective selection for multi-fidelity optimization. In: Proceedings of the Genetic and Evolutionary Computation Conference, GECCO 2019, pp. 880–888. ACM, New York (2019). https://doi.org/10.1145/3321707.3321709
3. Andreassen, E., Clausen, A., Schevenels, M., Lazarov, B.S., Sigmund, O.: Efficient topology optimization in MATLAB using 88 lines of code. Struct. Multi. Optim. **43**(1), 1–16 (2011). https://doi.org/10.1007/s00158-010-0594-7
4. Arnold, D.V.: An active-set evolution strategy for optimization with known constraints. In: Handl, J., Hart, E., Lewis, P.R., López-Ibáñez, M., Ochoa, G., Paechter, B. (eds.) PPSN 2016. LNCS, vol. 9921, pp. 192–202. Springer, Cham (2016). https://doi.org/10.1007/978-3-319-45823-6_18
5. Arnold, D.V.: Reconsidering constraint release for active-set evolution strategies. In: Proceedings of the Genetic and Evolutionary Computation Conference, GECCO 2017, pp. 665–672. Association for Computing Machinery, New York (2017). https://doi.org/10.1145/3071178.3071294
6. Atamna, A., Auger, A., Hansen, N.: Augmented Lagrangian constraint handling for CMA-ES—case of a single linear constraint. In: Handl, J., Hart, E., Lewis, P.R., López-Ibáñez, M., Ochoa, G., Paechter, B. (eds.) PPSN 2016. LNCS, vol. 9921, pp. 181–191. Springer, Cham (2016). https://doi.org/10.1007/978-3-319-45823-6_17
7. Digabel, S.L., Wild, S.M.: A taxonomy of constraints in simulation-based optimization. arXiv:1505.07881 (2015)

8. Dwianto, Y.B., Fukumoto, H., Oyama, A.: On improving the constraint-handling performance with modified multiple constraint ranking (MCR-mod) for engineering design optimization problems solved by evolutionary algorithms. In: Proceedings of the Genetic and Evolutionary Computation Conference, GECCO 2019, pp. 762–770. ACM, New York (2019). https://doi.org/10.1145/3321707.3321808
9. Hansen, N., Niederberger, A.S.P., Guzzella, L., Koumoutsakos, P.: A method for handling uncertainty in evolutionary optimization with an application to feedback control of combustion. IEEE Trans. Evol. Comput. **13**(1), 180–197 (2009). https://doi.org/10.1109/TEVC.2008.924423
10. Hansen, N., Müller, S.D., Koumoutsakos, P.: Reducing the time complexity of the derandomized evolution strategy with covariance matrix adaptation (cma-es). Evol. Comput. **11**(1), 1–18 (2003). https://doi.org/10.1162/106365603321828970
11. Hansen, N., Ostermeier, A.: Completely derandomized self-adaptation in evolution strategies. Evol. Comput. **9**(2), 159–195 (2001). https://doi.org/10.1162/106365601750190398
12. Hellwig, M., Beyer, H.: A matrix adaptation evolution strategy for constrained real-parameter optimization. In: 2018 IEEE Congress on Evolutionary Computation (CEC), pp. 1–8 (2018). https://doi.org/10.1109/CEC.2018.8477950
13. de Paula Garcia, R., de Lima, B.S.L.P., de Castro Lemonge, A.C., Jacob, B.P.: A rank-based constraint handling technique for engineering design optimization problems solved by genetic algorithms. Comput. Struct. **187**, 77–87 (2017). https://doi.org/https://doi.org/10.1016/j.compstruc.2017.03.023
14. Sakamoto, N., Akimoto, Y.: Adaptive ranking based constraint handling for explicitly constrained black-box optimization. In: Proceedings of the Genetic and Evolutionary Computation Conference, GECCO 2019, pp. 700–708. ACM, New York (2019). https://doi.org/10.1145/3321707.3321717
15. Sato, T., Watanabe, K., Igarashi, H.: Multimaterial topology optimization of electric machines based on normalized gaussian network. IEEE Trans. Mag. **51**(3), 1–4 (2015). https://doi.org/10.1109/TMAG.2014.2359972
16. Spettel, P., Beyer, H., Hellwig, M.: A covariance matrix self-adaptation evolution strategy for optimization under linear constraints. IEEE Trans. Evol. Comput. **23**(3), 514–524 (2019). https://doi.org/10.1109/TEVC.2018.2871944
17. Wang, H., Jin, Y., Doherty, J.: A generic test suite for evolutionary multifidelity optimization. IEEE Trans. Evol. Comput. **22**(6), 836–850 (2018). https://doi.org/10.1109/TEVC.2017.2758360

Model-Based Algorithm Configuration with Default-Guided Probabilistic Sampling

Marie Anastacio[1(✉)] and Holger Hoos[1,2]

[1] Leiden University, Leiden, The Netherlands
m.i.a.anastacio@liacs.leidenuniv.nl, hh@liacs.nl
[2] University of British Columbia, Vancouver, Canada

Abstract. In recent years, general-purpose automated algorithm configuration procedures have enabled impressive improvements in the state of the art in solving a wide range of challenging problems from AI, operations research and other areas. To search vast combinatorial spaces of parameter settings for a given algorithm as efficiently as possible, the most successful configurators combine techniques such as racing, estimation of distribution algorithms, Bayesian optimisation and model-free stochastic search. Two of the most widely used general-purpose algorithm configurators, SMAC and irace, can be seen as combinations of Bayesian optimisation and racing, and of racing and an estimation of distribution algorithm, respectively. Here, we propose a first approach that combines all three of these techniques into one single configurator, while exploiting prior knowledge contained in expert-chosen default parameter values. We demonstrate significant performance improvements over irace and SMAC on a broad range of running time optimisation scenarios from AClib.

1 Introduction

For many combinatorial problems, the performance of state-of-the-art algorithms critically depends on parameter settings [24]. Traditionally, algorithm developers have manually tuned these parameters to find a configuration that performs well across a range of benchmarks. However, this manual configuration process tends to be tedious and inefficient. For the past decade, automated methods for configuring those algorithms have been broadly established as an effective alternative to this manual approach. Prominent examples of general-purpose automated algorithm configurators include irace [31], ParamILS [24], GGA/GGA++ [3,4] and SMAC [23]. These fully automated configuration procedures are based on advanced machine learning and optimisation methods. Many examples of prominent and challenging computational problems on which automated configurators are strikingly effective can be found in the literature; these include propositional satisfiability (SAT) (see, *e.g.*, [25]), mixed integer programming (MIP) (see, *e.g.*, [22]), automated planning (see, *e.g.*, [12]) and supervised machine learning (ML) (see, *e.g.*, [27,36]).

T. Bäck et al. (Eds.): PPSN 2020, LNCS 12269, pp. 95–110, 2020.
https://doi.org/10.1007/978-3-030-58112-1_7

The availability of those highly effective automated algorithm configurators encourages algorithm designers to expose design choices as parameters. This has lead to a design paradigm known as programming by optimisation (PbO) [21]. PbO is based on the idea that by avoiding premature choices in the design of algorithms, we can let automated methods, such as general-purpose algorithm configurators, make design choices specifically adapted to a use case and thus broadly achieve better performance. However, the size of the combinatorial configuration spaces encountered in this context grows exponentially with the number of exposed parameters. To search these spaces effectively, state-of-the-art algorithm configuration procedures combine sophisticated methods, such as racing, estimation of distribution algorithms, Bayesian optimisation and model-free stochastic local search.

In particular, SMAC and irace can be seen as combinations of Bayesian optimisation and racing, and of racing and an estimation of distribution algorithm, respectively. We propose a first approach that combines all three of these techniques into a single general-purpose automated algorithm configuration procedure we call SMAC+PS. Based on work exploring the exploitation of prior knowledge contained in expert-chosen default parameter values [2], we bias the sampling of new configurations towards these values using truncated normal distributions centred on them. To test our new configuration approach, we applied it to a broad range of widely used running time minimisation scenarios from AClib [26]. As we will report later, we obtained significant performance improvements over irace and SMAC, the two state-of-the-art configurators SMAC+PS is based on. Our sampling method improves over SMAC in more than two thirds, and over irace in three quarters of the benchmark scenarios we considered.

The remainder of this paper is structured as follows: After a brief introduction to the algorithm configuration problem and several current state-of-the-art configurators (Sect. 2), we introduce our approach and the underlying motivation (Sect. 3). Next, we describe the experimental protocol we used to evaluate SMAC+PS (Sect. 4), followed by our results (Sect. 5), some general conclusions and a brief outlook on future work (Sect. 6).

2 Automated Algorithm Configuration

The algorithm configuration problem (see, *e.g.*, [20]) can be defined as follows: Given a target algorithm A with parameters $p_1, p_2, ..., p_k$, a domain D_j of possible values and a default value $d_j \in D_j$ for each parameter p_j; a configuration space C, containing all valid combinations of parameter values of A; a set of problem instances I; and a performance metric m that measures the performance of a configuration $c \in C$ of target algorithm A on I; find $c^* \in C$ that optimises the performance of A on instance set I, according to metric m.

Configuration spaces may contain different types of parameters. Categorical parameters have an unordered, finite set of values; they are often used to select between several heuristic components or mechanisms. As a special case,

Boolean parameters typically activate or deactivate optional algorithm components. Numerical, integer- or real-valued parameters are often used to quantitatively control performance-relevant aspects of a given target algorithm or heuristic, such as the strength of a diversification mechanism. Some parameters may conditionally depend on others, in that they are only active for certain values of other (often categorical or Boolean) parameters.

In recent years, much work has been done on automatic algorithm configuration, resulting in several high-performance, general-purpose automatic algorithm configurators, based on different approaches. The most prominent configurators include irace [8,31], GGA++ [3] and SMAC [23].

Among the main challenges of automated algorithm configuration are the time required to evaluate the performance of a configuration and the size of the configuration space. To minimise the number of time-consuming evaluations, SMAC and GGA++ are using an empirical performance model to predict how well a configuration will perform without running the target algorithm. irace leverages racing methods based on a statistical test to discard configurations when they are unlikely to perform well. Approaches to terminate less promising runs before they finish have shown great potential (see *e.g.*, [11,32]), but can only be applied to any-time algorithms, thus excluding many prominent algorithms for decision problems, such as SAT.

To handle the combinatorial space of possible configurations, a key component of any algorithm configurator is the way it samples the parameter values to be evaluated. irace uses a truncated normal distribution around the best known values to sample promising configurations. GGA++ uses its surrogate model to genetically engineer part of the offspring when creating a new generation of configurations. SMAC generates two sets of random configurations; one of these is used for diversification and the other is improved by local search, using an empirical performance model and an expected improvement criterion.

Each of the state-of-the-art configurators we mentioned (and indeed, any such configurator we are aware of) combines several fundamental mechanisms into a sophisticated search procedure.

3 SMAC+PS

Our recent study [2] has demonstrated that algorithm developers tend to provide default parameter configurations that have been chosen to perform reasonably well across a broad range of problem instances.

We propose a new method to exploit the prior knowledge contained in such expert-chosen default parameter values. Our method is based on combining the widely used and freely available, state-of-the-art general-purpose automated algorithm configurator, SMAC, with the sampling approach of irace. We call our new configurator SMAC+PS, as it combines the sequential model-based approach of SMAC with probabilistic sampling. Before explaining SMAC+PS, we briefly review SMAC.

3.1 The SMAC Configurator

SMAC is based on a sequential model-based optimisation approach (also known as Bayesian optimisation). It creates a random forest model to predict the performance of the given target algorithm for arbitrary parameter configurations, based on performance data collected from specific target algorithm runs. In each iteration, SMAC selects a candidate configuration and runs it on some problem instances from the given training set. The random forest model is then updated with the performance observed in this run. This process continues until a given time budget (usually specified in terms of wall-clock time) has been exhausted.

The step that is of interest in the context of this work is the way in which SMAC selects new configurations. This is done using two different mechanisms. On the one hand, to exploit current knowledge from its random forest model, SMAC performs local search from many randomly generated configurations, attempting to maximise expected improvement over the best configuration encountered so far. On the other hand, to diversify, explore unknown parts of the search space and avoid being misguided by its model, SMAC selects new configurations uniformly at random from the entire (usually vast) configuration space. SMAC alternates between these two mechanisms, so that overall, each of them contributes half of the new target algorithm configurations that are evaluated.

While running random configurations helps SMAC to balance exploration and exploitation, those runs can also waste substantial time and resources. Especially when dealing with a set of challenging training instances, randomly chosen configurations can easily time out and thus provide little useful information.

3.2 Extension with Probabilistic Sampling

As shown in previous work, simply reducing the ranges of numerical parameters around the given default values can lead to significant performance improvements for SMAC [2]. This was inspired by two other recent findings; one from a study of the dependence of algorithm performance on the setting of numerical parameters, which showed that the configuration landscapes tend to be surprisingly benign [35] and suggests that limited manual tuning may lead to useful information about promising parameter values; and one indicating that SMAC can be sped up considerably by starting it from a performance model learned on another set of benchmark instances [29], thus suggesting that knowledge regarding high-quality configurations can be transferred between sets of problem instances. However, by pruning the search space, our previous approach completely excluded regions that in specific cases could be relevant to explore, and we did so in a configurator-agnostic, yet somewhat ad-hoc manner.

Building on those findings, our intuition was that sampling near good configurations will very likely lead to other good configurations, while sampling uniformly, as done in SMAC, will likely waste time on some very bad parameter settings. Here, we propose a principled approach for exploiting the prior knowledge contained in the default value of parameters, by including probabilistic sampling in the context of sequential model-based optimisation. As we will

Algorithm 1. SMAC+PS
C: configuration space. c_d: the default configuration.
C_{rand}, C_{prom} and C_{new}: sets of configurations.
$\mathcal{R}$: target algorithm runs performed. $\mathcal{M}$: performance model.

1: $inc \leftarrow c_d$
2: $\mathcal{R} \leftarrow run\,(c_d)$
3: **while** Budget not exhausted **do**
4: $\quad \mathcal{M} \leftarrow update\,(\mathcal{M}, \mathcal{R})$
5: $\quad C_{prom} \leftarrow uniform\_sample\_configurations\,(C)$
6: $\quad C_{prom} \leftarrow local\_search\,(\mathcal{M}, C_{prom})$
7: $\quad C_{rand} \leftarrow normal\_sample\_configurations\,(C)$
8: $\quad C_{new} \leftarrow interleave\,(C_{rand}, C_{prom})$
9: $\quad inc \leftarrow intensify\,(C_{new}, inc)$
10: **end while**
11: **Return** *inc*

demonstrate, this can yield significant improvement for general-purpose algorithm configuration. To implement our approach, we extended SMAC with a simple mechanism for sampling new values according to a truncated normal distribution centred around the given default values, which replaces the uniform random sampling used in the original version of SMAC.

Algorithm 1 outlines our new configuration method, which we refer to as SMAC+PS, at a high level. SMAC+PS differs from SMAC only in the sampling distribution used in line 7; further details of SMAC can be found in the original paper [23]. New configurations are sampled in two places: as starting points for the local search process (line 5) and for non-model-based diversification (line 7). The two sets of configurations thus obtained are then interleaved and raced against the current incumbent (line 9). We change the sampling distribution used for non-model-based diversification (line 7). Each parameter is sampled independently from a distribution that depends on the parameter type and domain. For each numerical parameter p_n, we first normalise the range to $[0, 1]$ (if p_n is specified as "log scale", after applying a log base ten transformation). We then sample a value from a truncated normal distribution with mean equal to the default value of p_n and variance 0.05; this value was chosen based on preliminary experiments on configuration scenarios different from the ones used in our evaluation. For a categorical parameter p_c with k values, we sample the default value with probability 0.5, and each other value with probability $0.5/(k-1)$.

In cases where the default configuration is far away from the most promising areas of a given configuration space, this approach might be counterproductive. We note, however, that model-based local search starting from uniformly sampled configurations has the potential of counterbalancing this effect, and preliminary experiments in the appendix provide some evidence that this is sufficient for preventing major performance degradation compared to SMAC.

Our probabilistic sampling approach is inspired by estimation of distribution algorithms, which are known to provide an effective way for leveraging prior knowledge when solving complex optimisation problems [19]. It is also conceptually related to the approach taken by irace, which – unlike SMAC – does not make use of an empirical performance model mapping parameter configurations to performance values, but uses the best known configurations as a basis for estimating where promising parameters may be located. Indeed, irace uses an intensification mechanism based on sampling new values for numerical parameters from truncated normal distributions, in combination with a racing mechanism for updating the incumbent configuration and the location of the sampling distributions. We note that the probabilistic sampling process we use in SMAC is simpler, since it keeps the sampling distributions fixed throughout the configuration process. We decided on this design in order to evaluate to which extent a simple probabilistic sampling mechanism solely focused on exploiting information from expert-chosen default values would already enable improvements over state-of-the-art algorithm configurators, such as SMAC and irace.

4 Experimental Setup

To evaluate SMAC+PS (available at `ada.liacs.nl/projects/smacps`), we selected 16 configuration scenarios for running time minimisation from the general algorithm configuration library, AClib [26], covering three prominent and widely studied combinatorial problems: propositional satisfiability (SAT), AI planning and mixed integer programming (MIP). The algorithms and benchmarks included in AClib are well known and widely used in the automatic algorithm configuration literature, and running time minimisation is an important and widely studied special case of automated algorithm configuration. This section describes the benchmark instance sets and target algorithms, as well as our experimental protocol and execution environment.

4.1 Benchmark Instance Sets

The AClib scenarios we considered are based on 10 sets of randomly generated and real-world SAT, MIP and AI planning instances [26]. We used the training and testing sets provided by AClib, as shown in Table 1.

Our three SAT benchmarks originate from the configurable solver SAT challenge [25]: a set of instances generated by a CNF fuzzing tool (CF) [9], a set of low auto-correlation binary sequence problems converted into SAT (LABS) [33], and a set of 5-SAT problems generated uniformly at random from which only unsatisfiable instances have been kept (UNSAT).

Our three automated planning benchmarks originate from the third international planning competition [30]; a set that combines transportation and blocks problem (Depots), a set about the control and observation scheduling of satellites (Satellite), and a set of route planning problem (Zenotravel) [34].

Table 1. Benchmark instance sets from AClib considered in our experiments.

Problem	Benchmark	Instances		Reference
		Train	Test	
SAT	CF	298	301	[9,25]
	LABS	350	350	[25,33]
	UNSAT	299	249	[25]
Planning	Depots	2000	2000	[30]
	Satellite	2000	2000	[30]
	Zenotravel	2000	2000	[30,34]
MIP	CLS	50	50	[5,22]
	COR-LAT	1000	1000	[18,22]
	RCW2	495	495	[1,38]
	REG200	999	999	[22,28]

Three of our four MIP benchmarks originate from a study on MIP solver configuration [22]: a set of capacitated lot-sizing benchmark (CLS) [5], a set of MIP problem instances generated with the combinatorial auction test suite [28], and a set of real-life data for wildlife corridors for grizzly bears in the Northern Rockies [18]. The fourth benchmark stems from work on combining algorithm configuration and selection [38]: a set of MIP-encoded habitat preservation data for the endangered red-cockaded woodpecker [1].

Table 2. Target algorithms used in our experiments (as provided by AClib).

Problem	Solver	Number of parameters			
		Total	Real	Integers	Conditionals
SAT	Clasp	75	7	30	55
	Lingeling	322	0	186	0
	SpToRiss	222	16	36	176
Planning	LPG	67	14	5	22
MIP	CPLEX	74	7	16	4

4.2 Target Algorithms

In our experiments, we used five prominent solvers for SAT, AI planning and MIP (see Table 2). Our SAT solvers were selected based on their performance in the Configurable SAT Solver Challenge (CSSC) 2014 [25]: Lingeling [7] ranked first on the *industrial SAT+UNSAT* track and second on the *crafted SAT+UNSAT* track, Clasp [14] first on the *crafted SAT+UNSAT* and *Random SAT+UNSAT* tracks and SparrowToRiss (SpToRiss) [6] second on the *Random SAT* track. For automated planning, we selected LPG [15–17], as it has been successfully configured previously [13,37] and is also available through AClib. For MIP, we chose IBM's CPLEX solver, as it is widely used in practice and has shown great potential for performance improvement through automated configuration [22].

4.3 Configurators

We compare our approach to SMAC [23] (SMAC3 v0.10.0), the state-of-the-art general-purpose algorithm configurator on which SMAC+PS is based. As a second baseline, we chose irace with capping [10,31] (v3.3.2238:2239), another widely used, state-of-the-art configurator, which provided some inspiration for our sampling method. We did not compare to GGA++ [3], since it is not publicly available and, unlike the other configurators we considered, appears to critically require parallel execution.

In configuration scenarios where the budget is total CPU time, irace evaluates the running time of the target algorithm to estimate the length of each iteration and the number of possible iterations given the time budget specified by the user. In cases where the given cutoff time is significantly larger than the typical running time of the target algorithm, irace has difficulty estimating how many runs of the target-algorithm can be performed within the budget and may either refuse to run, due to insufficient budget, or exceed the budget. This affected 4 of the 16 scenarios we studied, namely the AClib MIP scenarios, but even if qualitatively different results were obtained when modifying those scenarios, this would not affect the overall conclusions drawn from our experiments (Sect. 5). We decided to terminate irace after 150% of the allowed overall time budget – that is, in our case, 3 instead of 2 days of configuration time; as a result, some results for irace are based on larger time budgets than those of SMAC. In two scenarios, too few of the configuration runs finished within this time budget, and we do not report results for irace in those cases (see Sect. 4.5).

4.4 Configuration Scenarios

All configuration scenarios were run according to the setup defined in AClib, including default configurations, and making sure that the configurators used those defaults (see Table 3). As a configuration objective, we used minimisation of PAR10 (average running time of the target algorithm, with timed-out runs counted as ten times the cutoff time).

4.5 Evaluation Protocol

We ran each configurator independently 24 times on each scenario and evaluated the 24 resulting parameter configurations on our training and testing sets.

Algorithm configurators are randomised, and their performance is known to vary substantially between multiple independent runs on the same scenario.

Table 3. Configuration scenarios from AClib used in our experiments.

Problem	Benchmark	Algorithm	Cutoff [s]	Budget [s]
SAT	CF	Clasp	300	172800
	LABS			
	UNSAT			
	CF	Lingeling		
	LABS			
	UNSAT			
	CF	SpToRiss		
	LABS			
	UNSAT			
Planning	Depots	LPG		
	Satellite			
	Zenotravel			
MIP	CLS	CPLEX	10000	
	COR-LAT			
	RCW2			
	REG200			

To leverage this, it is common practice to perform multiple independent runs of a configurator on a given scenario (usually in parallel), and to report the best configuration (evaluated on the training instances) as the final result of the overall configuration process.

To capture the statistical variability of this standard protocol, we repeatedly sampled 8 runs uniformly at random and identified the best of these according to performance on the training set. We used 10 000 such samples to estimate the probability distribution of the quality of the result produced by each configurator on each configuration scenario. We then compared the medians of these empirical distributions, using a one-sided Mann-Whitney U-test ($\alpha = 0.05$) to assess the statistical significance of observed performance differences.

For irace, when the estimated running time of the algorithm was too long and it refused to run within the given time budget, we decided to apply the same protocol to the successfully completed runs. This happened in particular for the CPLEX scenarios, where AClib prescribes a cutoff time of 10 000 s. In those cases, when the random seed (ranging from 1 to 24) leads to the selection of a hard instance for evaluating the running time of the default configuration, irace determines that the running time is too high for the given configuration budget. The decision to apply the standard protocol to successful runs of irace leads to positive bias on the irace results for CPLEX on CLS, where 19 runs finished successfully. However, in cases such as the scenarios for CPLEX on RCW2 and on REG200, where only 7 and 9 runs, respectively, terminated successfully, we omitted the results from our analysis, as application of the standard protocol would lead to extreme distortions from realistically achievable performance.

All experiments were performed on a computing cluster with CentOS on Dual 16-core 2.10 GHz Intel Xeon E5-2683 CPUs with 40 MB cache and 94 GB RAM.

5 Results

We now present the results of the experiments described in Sect. 4.

5.1 Comparison Based on Median PAR10 Scores

We compare the performance obtained for the configuration scenarios we studied, following the protocol described in Sect. 4.5, which produces statistics over the way state-of-the-art configurators are commonly used in practice. Table 4 shows the median PAR10 values we obtained; the missing results for CPLEX on RCW2 and REG200 are due to the fact that irace refused to start more than half of the 24 runs, since it considered the configuration budget to be insufficient.

Table 4. Results for SMAC, SMAC+PS and irace; median PAR10 (in CPU sec); best results are underlined, while boldface indicates results that are statistically tied to the best, according to a one-sided Mann-Whitney test ($\alpha = 0.05$). Right columns highlight the result of the pairwise comparison between SMAC+PS and the two baselines; ✓ if it is better, ✗ if not; parentheses indicate that the difference is not statistically significant.

Solver	Benchmark	Default	SMAC	irace	SMAC+PS	$\succ$ SMAC	$\succ$ irace
Clasp	CF	193.87	193.00	**174.00**	**173.61**	✓	(✓)
	LABS	745.74	**837.93**	847.10	**837.61**	(✓)	✓
	UNSAT	0.885	**0.362**	**0.359**	**0.359**	(✓)	(✓)
Lingeling	CF	327.00	**261.06**	328.214	300.40	✗	✓
	LABS	873.67	866.99	959.35	**863.73**	✓	✓
	UNSAT	2.41	**1.59**	2.36	1.65	✗	✓
SpToRiss	CF	472.78	**226.51**	236.61	253.72	✗	✗
	LABS	911.25	857.67	846.40	**805.46**	✓	✓
	UNSAT	222.26	1.56	1.56	**1.55**	✓	✓
LPG	Depots	34.68	1.14	**1.08**	1.25	(✗)	✗
	Satellite	22.40	5.43	8.04	**5.19**	✓	✓
	Zenotravel	29.64	2.61	2.93	**2.56**	✓	✓
CPLEX	CLS	4.06	3.36	4.06	**2.95**	✓	✓
	COR-LAT	24.81	21.84	**10.04**	21.26	✓	✗
	RCW2	82.51	**71.98**	–	78.10	✗	✓
	REG200	13.08	5.56	–	**5.09**	✓	✓

Comparing the results for SMAC and irace, we note that in most cases, SMAC achieves better performance than irace. SMAC reached a statistically significantly lower median PAR10 score for 10 out of the 16 configuration scenarios we studied, and irace outperformed SMAC for the remaining 6 scenarios. This indicates complementary strengths of our two baseline configurators. While SMAC has an edge on most of the scenarios, there is at least one case where irace finds substantially better configurations (CPLEX on COR-LAT).

Comparing SMAC+PS against SMAC, which is not only the stronger of our two baselines, but also served as the starting point for our new configuration procedure, we notice that for 11 of the 16 scenarios, SMAC+PS reaches a lower median PAR10 score. In all but two of those cases (Clasp on LABS and UNSAT), the performance differences are statistically significant.

Finally, compared to irace, SMAC+PS achieves better performance on 13 of our 16 scenarios. In all but two cases (Clasp on CF and UNSAT), the differences are statistically significant. SpToRiss on LABS shows a case where SMAC could not achieve better result than irace, while SMAC+PS outperforms irace.

Overall, these results indicate clearly that SMAC+PS represents a significant improvement over both baselines, and hence an advance in the state of the art in automated algorithm configuration for running time minimisation.

As SMAC+PS relies on the assumption that the provided default values are well-chosen, we performed some preliminary experiments to study its robustness to misleading default values. The results, available in the appendix, show no significant loss in performance compared to SMAC.

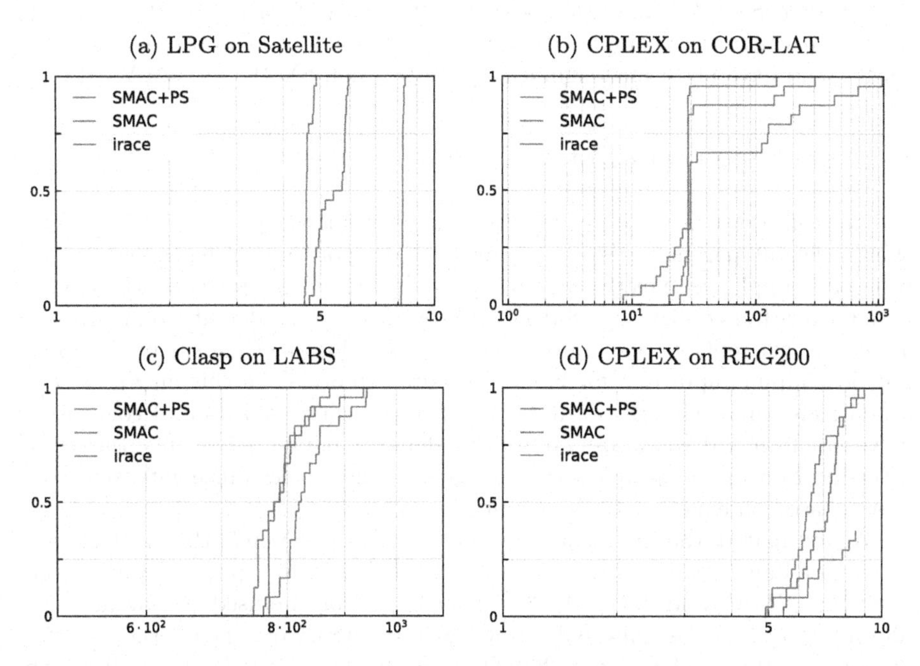

Fig. 1. Cumulative distribution functions for PAR10 scores (in CPU seconds, x-axis) over independent configurator runs on the testing set.

5.2 Distributions of PAR10 Scores over Multiple Configurator Runs

To examine the performance of SMAC, irace and SMAC+PS in more detail, we studied the empirical cumulative distribution functions over individual configurator runs (without applying the standard protocol) – see Fig. 1. As we minimise PAR10 scores, better performance is indicated by CDFs closer to the top left corner of the plots. We show results for one scenario on which SMAC+PS performs better than the two other configurators (Fig. 1a), one scenario on which irace performs better than SMAC+PS and SMAC (Fig. 1b, one in which SMAC and SMAC+PS are tied (Fig. 1c), as well as one on which a large number of irace did not terminate successfully (Fig. 1d).

LPG on Satellite (Fig. 1a) is a case in which the difference between the CDFs for SMAC and SMAC+PS is particularly pronounced. In this scenario, SMAC+PS clearly dominates the two other configurators.

CPLEX on COR-LAT (Fig. 1b) is a case in which irace performs better than the two other configurators. SMAC and SMAC+PS give rise to similar performance distributions. irace has an edge, reinforced by the standard protocol, which leverages the left tail of the performance distributions.

Clasp on LABS (Fig. 1c), a scenario on which SMAC and SMAC+PS are statistically tied, looks qualitatively similar, except that the difference in the left tail between SMAC and SMAC+PS is too small to be reliably exploitable using the standard protocol. irace, on the other side, is probabilistically dominated on this scenario.

Figure 1d shows a scenarios on which irace terminated prematurely (as described in Sect. 4). However, looking at the partial CDFs for those configurator runs that were completed successfully, there is no reason to expect that irace would have performed significantly better than SMAC and SMAC+PS.

6 Conclusion and Future Work

Building on previous work on exploiting the prior knowledge included in expert-chosen default parameter values in automated algorithm configuration [2], we propose a simple, yet effective way to probabilistically bias the sequential model-based algorithm configurator SMAC [26] towards a given default configuration. To do so, we replaced its uniform random sampling mechanism by a probabilistic sampling approach for numerical parameters. While this may seem like a relatively minor change, we note that it represents a substantial conceptual departure from SMAC, whose uniform random sampling mechanism ensures the diversification of the search process, while our new mechanism intensifies the search based on expert-defined default parameter values.

We evaluated the resulting procedure, dubbed SMAC+PS, against two widely used and freely available state-of-the-art general-purpose algorithm configurators, SMAC and irace [31]. For this comparison, we used 16 running time optimisation scenarios from AClib, a widely used library of benchmarks for automated algorithm configuration [26]. The scenarios we selected cover well-known algorithms and benchmark instance sets from three widely studied combinatorial problems – propositional satisfiability (SAT), mixed integer programming (MIP) and AI planning.

We found that SMAC+PS performs better than SMAC on 11 of those 16 scenarios, and better than irace on 12 of them, and thus represents a significant improvement over the state of the art in automated algorithm configuration for running time minimisation. Whether similar results can be obtained for different performance metrics is an open question.

Our results are consistent with recent work showing that the configuration landscapes (i.e., the functions relating parameter values to target algorithm performance) are far more benign than one might have expected [35], which suggests that sampling around known good parameters values provides an efficient way towards finding new good values, an assumption also leveraged by irace.

In future work, we plan to extend our approach with a mechanism for adapting the median and variance of the distributions used for sampling values for each parameter, such that we overcome more effectively the prior knowledge from poorly chosen defaults, while good defaults are still exploited from the beginning of the configuration process. Such a mechanism also offers several avenues for better exploiting prior knowledge from the default values of categorical parameters. Finally, our work on SMAC+PS opens an interesting path towards richer mechanisms for allowing algorithm designers to express prior knowledge about parameter values.

Appendix: Robustness to Misleading Default Values

To obtain better insights into the robustness of our approach to misleading default values, we generated random configurations until we found one that performed worse than the default, but produced time-outs on fewer than a third of the training instances from our three SAT benchmarks. Then, we repeated a few configuration experiments from Sect. 5.1, using these new, misleading default configurations (using the same protocol as described earlier). The results of this experiment are shown in Table 5.

Table 5. Results for SMAC and SMAC+PS when given a default generated to be misleading. The numbers shown are median PAR10 scores in CPU seconds; best results are underlined, while boldface indicates results that are statistically tied to the best, according to a one-sided Mann-Whitney test ($\alpha = 0.05$).

Solver	Benchmark	Misleading default	SMAC	SMAC+PS
SpToRiss	CF	650.27	246.94	**<u>227.08</u>**
	UNSAT	165.80	**<u>1.49</u>**	1.51
	LABS	966.75	822.42	**<u>814.75</u>**

References

1. Ahmadizadeh, K., Dilkina, B., Gomes, C.P., Sabharwal, A.: An empirical study of optimization for maximizing diffusion in networks. In: Cohen, D. (ed.) CP 2010. LNCS, vol. 6308, pp. 514–521. Springer, Heidelberg (2010). https://doi.org/10.1007/978-3-642-15396-9_41
2. Anastacio, M., Luo, C., Hoos, H.: Exploitation of default parameter values in automated algorithm configuration. In: Workshop Data Science Meets Optimisation (DSO), IJCAI 2019, August 2019
3. Ansótegui, C., Malitsky, Y., Samulowitz, H., Sellmann, M., Tierney, K.: Model-based genetic algorithms for algorithm configuration. In: Proceedings of the IJCAI 2015, pp. 733–739 (2015)
4. Ansótegui, C., Sellmann, M., Tierney, K.: A gender-based genetic algorithm for the automatic configuration of algorithms. In: Gent, I.P. (ed.) CP 2009. LNCS, vol. 5732, pp. 142–157. Springer, Heidelberg (2009). https://doi.org/10.1007/978-3-642-04244-7_14
5. Atamtürk, A., Muñoz, J.C.: A study of the lot-sizing polytope. Math. Program. **99**(3), 443–465 (2004). https://doi.org/10.1007/s10107-003-0465-8
6. Balint, A., Manthey, N.: SparrowToRiss. In: Proceedings of the SAT Competition 2014, pp. 77–78 (2014)
7. Biere, A.: Yet another local search solver and lingeling and friends entering the SAT competition 2014 (2014)
8. Birattari, M., Yuan, Z., Balaprakash, P., Stützle, T.: F-Race and iterated F-Race: an overview. In: Bartz-Beielstein, T., Chiarandini, M., Paquete, L., Preuss, M. (eds.) Experimental Methods for the Analysis of Optimization Algorithms, pp. 311–336. Springer, Heidelberg (2010). https://doi.org/10.1007/978-3-642-02538-9_13
9. Brummayer, R., Lonsing, F., Biere, A.: Automated testing and debugging of SAT and QBF solvers. In: Strichman, O., Szeider, S. (eds.) SAT 2010. LNCS, vol. 6175, pp. 44–57. Springer, Heidelberg (2010). https://doi.org/10.1007/978-3-642-14186-7_6
10. Cáceres, L.P., López-Ibáñez, M., Hoos, H., Stützle, T.: An experimental study of adaptive capping in irace. In: Battiti, R., Kvasov, D.E., Sergeyev, Y.D. (eds.) LION 2017. LNCS, vol. 10556, pp. 235–250. Springer, Cham (2017). https://doi.org/10.1007/978-3-319-69404-7_17
11. Domhan, T., Springenberg, J.T., Hutter, F.: Speeding up automatic hyperparameter optimization of deep neural networks by extrapolation of learning curves. In: Yang, Q., Wooldridge, M.J. (eds.) IJCAI 2015, pp. 3460–3468. AAAI Press (2015)
12. Fawcett, C., Helmert, M., Hoos, H., Karpas, E., Röger, G., Seipp, J.: FD-Autotune: domain-specific configuration using fast downward. In: Proceedings of the ICAPS Workshop, PAL 2011, pp. 13–20 (2011)
13. Fawcett, C., Hoos, H.H.: Analysing differences between algorithm configurations through ablation. J. Heuristics **22**(4), 431–458 (2015). https://doi.org/10.1007/s10732-014-9275-9
14. Gebser, M., Kaufmann, B., Schaub, T.: Conflict-driven answer set solving: from theory to practice. Artif. Intell. **187–188**, 52–89 (2012)
15. Gerevini, A., Saetti, A., Serina, I.: Planning through stochastic local search and temporal action graphs in LPG. J. Artif. Intell. Res. **20**, 239–290 (2003)
16. Gerevini, A., Saetti, A., Serina, I.: An approach to efficient planning with numerical fluents and multi-criteria plan quality. Artif. Intell. **172**, 899–944 (2008)

17. Gerevini, A., Saetti, A., Serina, I.: An empirical analysis of some heuristic features for planning through local search and action graphs. Fundamenta Informaticae **107**(2–3), 167–197 (2011)
18. Gomes, C.P., van Hoeve, W.-J., Sabharwal, A.: Connections in networks: a hybrid approach. In: Perron, L., Trick, M.A. (eds.) CPAIOR 2008. LNCS, vol. 5015, pp. 303–307. Springer, Heidelberg (2008). https://doi.org/10.1007/978-3-540-68155-7_27
19. Hauschild, M., Pelikan, M.: An introduction and survey of estimation of distribution algorithms. Swarm Evol. Comput. **1**(3), 111–128 (2011)
20. Hoos, H.H.: Automated algorithm configuration and parameter tuning. In: Hamadi, Y., Monfroy, E., Saubion, F. (eds.) Autonomous Search, pp. 37–71. Springer, Heidelberg (2011). https://doi.org/10.1007/978-3-642-21434-9_3
21. Hoos, H.H.: Programming by optimization. Commun. ACM **55**(2), 70–80 (2012)
22. Hutter, F., Hoos, H.H., Leyton-Brown, K.: Automated configuration of mixed integer programming solvers. In: Lodi, A., Milano, M., Toth, P. (eds.) CPAIOR 2010. LNCS, vol. 6140, pp. 186–202. Springer, Heidelberg (2010). https://doi.org/10.1007/978-3-642-13520-0_23
23. Hutter, F., Hoos, H.H., Leyton-Brown, K.: Sequential model-based optimization for general algorithm configuration. In: Coello, C.A.C. (ed.) LION 2011. LNCS, vol. 6683, pp. 507–523. Springer, Heidelberg (2011). https://doi.org/10.1007/978-3-642-25566-3_40
24. Hutter, F., Hoos, H.H., Leyton-Brown, K., Stützle, T.: ParamILS: an automatic algorithm configuration framework. J. Artif. Intell. Res. **36**, 267–306 (2009)
25. Hutter, F., Lindauer, M., Balint, A., Bayless, S., Hoos, H.H., Leyton-Brown, K.: The configurable SAT solver challenge (CSSC). Artif. Intell. **243**, 1–25 (2017)
26. Hutter, F., et al.: AClib: a benchmark library for algorithm configuration. In: Pardalos, P.M., Resende, M.G.C., Vogiatzis, C., Walteros, J.L. (eds.) LION 2014. LNCS, vol. 8426, pp. 36–40. Springer, Cham (2014). https://doi.org/10.1007/978-3-319-09584-4_4
27. Kotthoff, L., Thornton, C., Hoos, H.H., Hutter, F., Leyton-Brown, K.: Auto-WEKA 2.0: automatic model selection and hyperparameter optimization in WEKA. J. Mach. Learn. Res. **18**, 25:1–25:5 (2017)
28. Leyton-Brown, K., Pearson, M., Shoham, Y.: Towards a universal test suite for combinatorial auction algorithms. In: Jhingran, A., Mackie-Mason, J., Tygar, D.J. (eds.) Proceedings of the 2nd ACM Conference on Electronic Commerce (EC-00), Minneapolis, MN, USA, 17–20 October 2000, pp. 66–76. ACM (2000)
29. Lindauer, M., Hutter, F.: Warmstarting of model-based algorithm configuration. In: Proceedings of the AAAI-18, IAAI-18, and EAAI-18, pp. 1355–1362 (2018)
30. Long, D., Fox, M.: The 3rd international planning competition: results and analysis. J. Artif. Intell. Res. **20**, 1–59 (2003)
31. López-Ibáñez, M., Dubois-Lacoste, J., Pérez Cáceres, L., Stützle, T., Birattari, M.: The irace package: iterated racing for automatic algorithm configuration. Oper. Res. Perspect. **3**, 43–58 (2016)
32. Luo, C., Hoos, H.H., Cai, S., Lin, Q., Zhang, H., Zhang, D.: Local search with efficient automatic configuration for minimum vertex cover. In: IJCAI-19, pp. 1297–1304. International Joint Conferences on Artificial Intelligence Organization (2019)
33. Mugrauer, F., Balint, A.: Sat encoded low autocorrelation binary sequence (labs) benchmark description. In: Proceedings of the SAT Competition 2013, pp. 117–118 (2013)
34. Penberthy, J.S., Weld, D.S.: Temporal planning with continuous change. In: Proceedings of the AAAI-94, vol. 2, pp. 1010–1015 (1994)

35. Pushak, Y., Hoos, H.: Algorithm configuration landscapes: more benign than expected? In: Auger, A., Fonseca, C.M., Lourenço, N., Machado, P., Paquete, L., Whitley, D. (eds.) PPSN 2018. LNCS, vol. 11102, pp. 271–283. Springer, Cham (2018). https://doi.org/10.1007/978-3-319-99259-4_22
36. Thornton, C., Hutter, F., Hoos, H.H., Leyton-Brown, K.: Auto-WEKA: combined selection and hyperparameter optimization of classification algorithms. In: The 19th ACM SIGKDD, KDD 2013, pp. 847–855 (2013)
37. Vallati, M., Fawcett, C., Gerevini, A., Hoos, H.H., Saetti, A.: Automatic generation of efficient domain-optimized planners from generic parametrized planners. In: Proceedings of the RCRA 2011, pp. 111–123 (2011)
38. Xu, L., Hutter, F., Hoos, H.H., Leyton-Brown, K.: Hydra-MIP: automated algorithm configuration and selection for mixed integer programming. In: Proceedings of the RCRA 2011, pp. 16–30 (2011)

Evolving Sampling Strategies for One-Shot Optimization Tasks

Jakob Bossek[1(✉)], Carola Doerr[2], Pascal Kerschke[3], Aneta Neumann[1], and Frank Neumann[1]

[1] The University of Adelaide, Adelaide, Australia
jakob.bossek@adelaide.edu.au
[2] Sorbonne Université, CNRS, LIP6, Paris, France
[3] University of Münster, Münster, Germany

Abstract. One-shot optimization tasks require to determine the set of solution candidates prior to their evaluation, i.e., without possibility for adaptive sampling. We consider two variants, classic one-shot optimization (where our aim is to find at least one solution of high quality) and one-shot regression (where the goal is to fit a model that resembles the true problem as well as possible). For both tasks it seems intuitive that well-distributed samples should perform better than uniform or grid-based samples, since they show a better coverage of the decision space. In practice, quasi-random designs such as Latin Hypercube Samples and low-discrepancy point sets are indeed very commonly used designs for one-shot optimization tasks.

We study in this work how well low star discrepancy correlates with performance in one-shot optimization. Our results confirm an advantage of low-discrepancy designs, but also indicate the correlation between discrepancy values and overall performance is rather weak. We then demonstrate that commonly used designs may be far from optimal. More precisely, we evolve 24 very specific designs that each achieve good performance on one of our benchmark problems. Interestingly, we find that these specifically designed samples yield surprisingly good performance across the whole benchmark set. Our results therefore give strong indication that significant performance gains over state-of-the-art one-shot sampling techniques are possible, and that evolutionary algorithms can be an efficient means to evolve these.

Keywords: One-shot optimization · Regression · Fully parallel search · Surrogate-assisted optimization · Continuous optimization

1 Introduction

When dealing with costly to evaluate problems under high time pressure, a decision maker is often left with the only option of evaluating a few possible decisions in parallel, in the hope that one of them proves to be a reasonable alternative. The problem of designing strategies that guarantee a fair chance of

T. Bäck et al. (Eds.): PPSN 2020, LNCS 12269, pp. 111–124, 2020.
https://doi.org/10.1007/978-3-030-58112-1_8

finding a good solution is studied under the term *one-shot optimization*. One-shot optimization is studied in numerous variants and contexts, including classic Operations Research [12] and numerical analysis [26,27]. Most recently, one-shot optimization has gained momentum in the context of Machine Learning applications, including hyper-parameter optimization for deep neural networks and for heuristic optimization techniques [2,4,9].

We study in this work two variants of one-shot optimization tasks, classic one-shot optimization and one-shot regression. In **classic one-shot optimization,** n solution candidates are evaluated in parallel. We only care about the best one of them, x^{best} and measure its simple regret $f(x^{\text{best}}) - \inf f$. In **one-shot regression**, in contrast, we use all n evaluated samples to build an approximation $\hat{f}$ of the actual, unknown function f. The objective is to determine a surrogate $\hat{f}$ which resembles f as well as possible. The quality of $\hat{f}$ is measured, for example, by the mean squared error (MSE) $\sum_{x \in X} (\hat{f}(x) - f(x))^2/|X|$. One-shot regression is also studied under the term *global surrogate modeling* [12].

Several works, in particular the one of Bousquet et al. [4] and the more recent work by Cauwet et al. [9], show that quasi-random designs of low discrepancy are more suitable for the classic one-shot optimization task than i.i.d. uniform samples or grid search. The overall recommendation propagated in [4] are randomly scrambled Hammersley point sets with random shifts. Other low-discrepancy point sets also perform well in the experiments reported there. Also for one-shot regression quasi-random constructions such as Latin Hypercube Samples (LHS [30]) and again low-discrepancy point sets [12,28] are quite common, leaving us with *the question if there is a correlation between the discrepancy of a point set and its performance in one-shot optimization.* If such a correlation existed, one could hope to find even better one-shot designs by searching for point sets of small discrepancy – a problem that is much easier (yet very hard [14]) to address than the original one-shot optimization problem. Interestingly, no such direct comparison has been attempted in the literature, although several works have investigated the suitability of various sampling designs for one-shot regression, see [12,28] for examples and although such a correlation is well known to hold in the context of numerical integration, via the Koksma-Hlawka inequality [21,24].

We compare five different experimental designs, three generalized Halton point sets, one LHS construction, and i.i.d. uniform sampling, see Sect. 2 for more details. Our test bed are the 24 noiseless BBOB functions [19,20], a standard benchmark set for numerical black-box optimization, which covers a wide range of different problems encountered in real-world optimization. We focus on *star discrepancy* [14] as diversity measure for the point sets, since this is the one that also appears in the mentioned Koksma-Hlawka inequality. For the regression task, we compare four standard regression techniques, support vector machines (SVMs) [11], decision trees [7], random forests [6], and Kriging [10], see Sect. 3.

1.1 Summary of Results

Results for Standard Sampling Designs. In the context of *classic one-shot optimization,* our experiments confirm the superiority of low-discrepancy point sets over random sampling. However, no clear correlation could be identified between the star discrepancy value of a point set and its performance as one-shot optimizer, somewhat refuting our hope that point sets with optimized discrepancy values could substantially boost performance in one-shot optimization.

For the *one-shot regression* task, we observe that there is no clear winning design, nor any obvious correlation between discrepancy and performance, indicating that we cannot rely on simple recommendations suggesting to use a specific design and/or surrogate model. Rather, we observe that competence maps, which provide recommendations based on some high-level features of the problem, can be crucial to achieve peak performance in one-shot optimization.

Constructing High-Performing Designs with Evolutionary Algorithms. In the absence of theoretical bounds, we investigate in Sect. 6 how the performances obtained by the tested (design, surrogate) combinations in the one-shot regression task compare against sampling strategies that are explicitly designed for minimizing the MSE individually for each of the 24 benchmark problems[1]. To this end, we apply an off-the-shelf evolutionary algorithm and evolve designs of low MSE for each BBOB function. To our surprise, we find that some of these designs perform very well not only on the problem that they have been designed for, but across all 24 functions, indicating that substantial performance gains over the state-of-the-art one-shot optimization strategies might exist.

Discussion. While our results might appear negative with respect to the original question about the correlation between the discrepancy of a sampling strategy and its performance in one-shot optimization, they reveal a clear need and may pave a way for identifying other diversity measures showing a better correlation with the performance results. The evolved designs clearly indicate that such investigations could significantly improve the state of the art. We note that previous attempts to construct low discrepancy samples [15] or well-performing LHS designs [22,28] can be found in the literature. A wider application of such constructions, however, seems to be lagging behind its potential. We therefore believe more research is needed to test these methods in various applications, and to make them easily applicable and accessible.

Finally, while we only focus on one-shot optimization in this work, we note that good one-shot optimization designs are likely to be useful in the context of *sequential model-based optimization (SMBO)* [23]. SMBO is also studied under the notion of *global optimization* or *surrogate-based optimization*, sand entails iterative methods for the optimization of black-box functions that are computationally expensive to evaluate. In SMBO, one uses the evaluated samples of an

[1] Note here that for the classic one-shot optimization task, this question is not meaningful, as the design $\{x\}$ with $x = \arg\min f$ is optimal with zero regret.

initial design to build a model of the true objective function, which is computationally fast or at least much faster to evaluate than the true objective function. In a sequential process this initial design is augmented by injecting further design points in order to improve the function approximation. So-called infill criteria or acquisition functions which usually balance exploitation of the current model and exploration of areas with high model uncertainty are used to decide which point(s) seem(s) adequate to evaluate next with the true objective function. Classic model-based approaches, such as the efficient global optimization algorithm (EGO) by Jones et al. [23], typically use well-distributed, space-filling point sets to initialize the search (see, e.g., [17]).

Reproducibility. We can only show a small set of results here in this extended abstract. Detailed data for both one-shot optimization tasks, for all 29 designs, the 4 surrogate models, 5 sample sizes, and each of the 24 tested BBOB functions is available on our public GitHub repository [3].

2 Low-Discrepancy Designs

The discrepancy of a point set measures how far it deviates from a perfectly distributed set. Various discrepancy measures exist, providing different performance guarantees in quasi-Monte Carlo integration and other applications [1,25,29]. The arguably most common discrepancy metric is the *star discrepancy*, which measures the largest absolute difference between the volume V_y of any origin-anchored box $[0,y] := \prod_{i=1}^{d}[0,y_i]$ and the fraction of points contained in this box. Hence, the star discrepancy of a point set $\{x^1,\ldots,x^n\} \in [0,1]^d$ is defined as $D^*(X) := \sup_{y\in[0,1]^d} \left|V_y - \left|[0,y]\cap X\right|/n\right|$.

Low-discrepancy designs provide a proven guarantee on their asymptotic discrepancy value. They are well-studied objects in numerical analysis, because of the good error guarantees that they provide for numerical integration. The interested reader is referred to the survey [14], which covers in particular the computational aspects of star discrepancies relevant to our work. In our experiments we consider four different designs of low discrepancy, and we compare them to uniform sampling. More precisely, we study Latin Hypercube Samples (LHS [30]; we use the maximin LHS implementation of the R package `lhs` [8]) and three variants of so-called Halton sequences: the original one suggested by Halton [18], an improved version introduced by Braaten and Wellter [5], and a third design which we obtain from a full enumeration and evaluation of all generalized Halton sequences (for the sample sizes $n \in \{125, 1\,000\}$). For our four-dimensional setting, these are 34 560 different designs each. Those are evaluated using the algorithm by Dobkin et al. [13], which has running time $n^{1+d/2}$. This exact approach becomes infeasible for larger sample sizes, and we use the best generator for $n = 1\,000$ instead. The so-minimized Halton designs are referred to as "Best" in the remainder of this work. The discrepancy values of the different designs used in this paper are summarized in Table 1.

Table 1. Discrepancy value of the best design and the relative overhead of the other designs. Values for LHS, UNIFORM, and EVOLVED designs are averaged.

n	Best	BW	Halton	LHS	UNIFORM	EVOLVED
125	0.056	12%	48%	49%	185%	156%
1000	0.013	20%	35%	109%	316%	343%
2500	0.008	0%	6%	295%	371%	–
5000	0.005	2%	6%	376%	413%	–

3 Experimental Setup and Availability of Data

For our experiments we have chosen the 24 noiseless problems from the *black-box optimization benchmark (BBOB)* by Hansen et al. [20]. For computational reasons, we limit our attention to the first instance of each problem. The BBOB functions assume $[-5, 5]^d$ as search space. We therefore scale our designs, which are initially constructed in $[0, 1]^d$, accordingly.

Our study summarizes the results from a total of 124 080 scenarios. We considered designs of three (deterministic) Halton sequences, as well as LHS and random uniform samples. As the latter two are stochastic, we generated ten samples each to account for their stochasticity. Moreover, each design was generated for the five sample sizes $n \in \{125, 1\,000, 2\,500, 5\,000, 10\,000\}$. For each design, we then computed surrogates using the following four machine learning algorithms: support vector machines (SVMs) [11], decision trees [7], random forests [6], and Kriging [10]. Note that the latter could not be computed on designs of size 10 000 due to memory issues. Also, as (except for the decision trees) the considered algorithms are stochastic – or at least contain stochastic elements within their R implementations – we replicated all experiments ten times. In addition to these 104 880 scenarios we further evaluated a total number of 1 920 "evolved" designs, which will be introduced in detail in Sect. 5 (one-shot regression).

4 Classic One-Shot Optimization

In the classic one-shot optimization scenario we are asked to provide a point set $\{x^1, \ldots, x^n\}$ for which the quality $f(x^{\text{best}})$ of the best point $x^{\text{best}} := \arg\min_{x^i \in \{x^1,\ldots,x^n\}} f(x^i)$ is as good as possible.

In line with the machine learning literature, where the one-shot problem originates from, we consider simple regret $f(x^{\text{best}}) - f^*$ as performance measure, where $f^* = \inf_x f(x)$ denotes the best function value. In optimization, this measure is referred to as the *target precision* of the best design point. Of course, this performance criterion requires that f^* is known. This is usually not the case for real-world applications, but for the BBOB benchmarks these values are available [20], so that the regret can be computed straightforwardly. Minimizing simple regret is also the standard objective in other related domains, including evolutionary computation [19]. Since this performance depends on a single point,

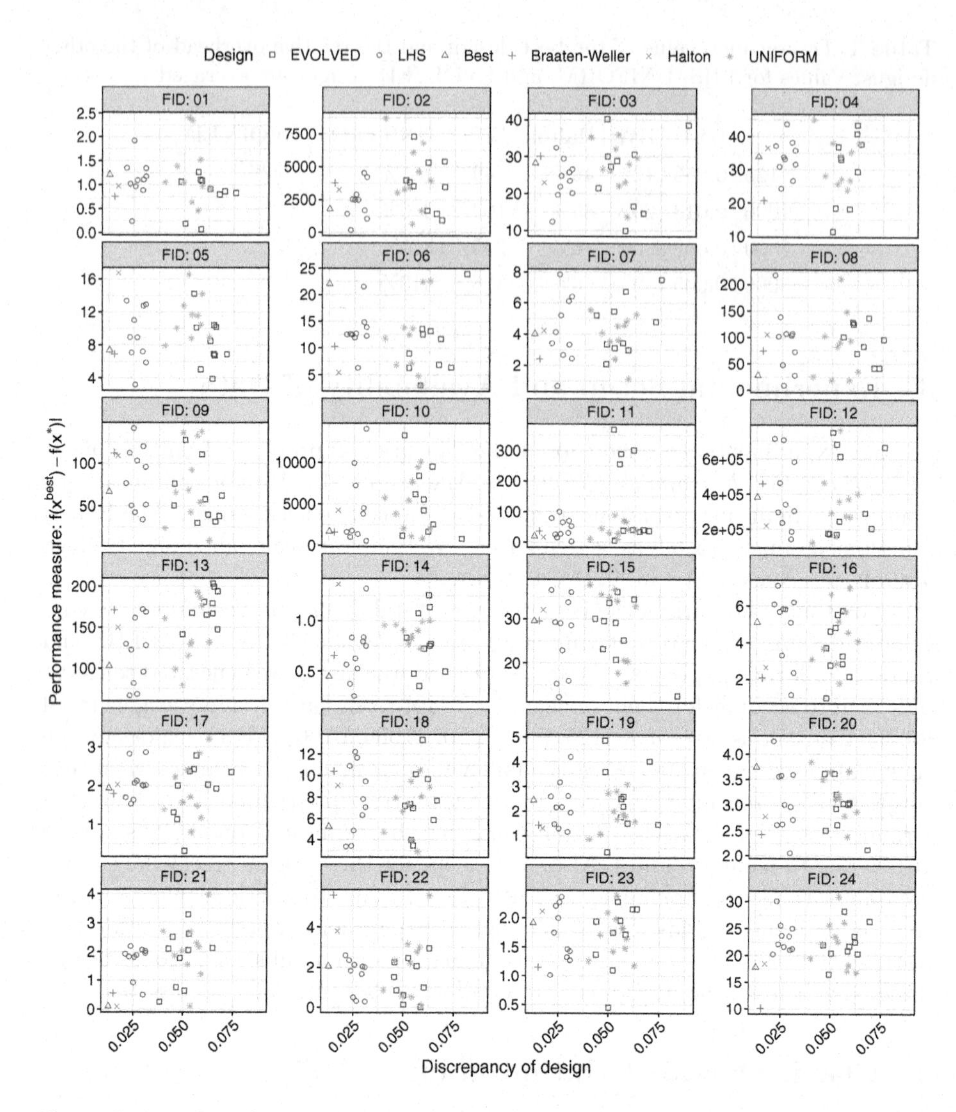

Fig. 1. Scatterplots showing the relationship between the discrepancy of designs of size $n = 1\,000$ and the one-shot performance $f(x^{\text{best}}) - f^*$ for all 24 BBOB problems. The EVOLVED instances were designed for Kriging surrogates.

the variance of the results can be tremendous, and it is therefore interesting to compare different designs over different sets of problems (and to perform several independent runs in case of the stochastic designs LHS and uniform sampling).

Figure 1 compares the average regret for each pair of function and design, and plots the respective performance (y-axis) in dependence of the design's discrepancy (x-axis). Due to different scales of the problems, absolute performances should not be directly compared across functions (we will use relative perfor-

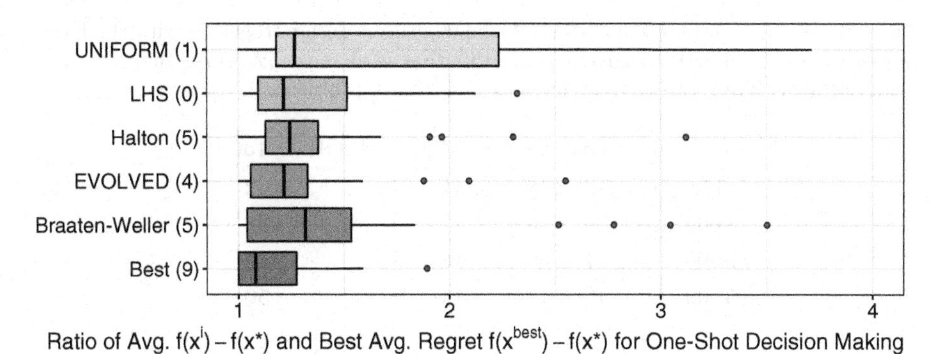

Fig. 2. Boxplots for factors by which the average one-shot result is worse than that of the best design (one data point per BBOB function). The x-axis is capped at 4 (outliers not shown in this plot: UNIFORM at 4.5 and 4.7, and Best at 7.2). Numbers in brackets indicate on how many functions the design achieved the best (average) result. All numbers are for $n = 1\,000$ points and use Kriging surrogates.

mances instead). As already mentioned before, the results for LHS, UNIFORM and EVOLVED sampling are based on ten independent designs. Note that the concept of the EVOLVED designs will be discussed in more detail later in this work, but we are already showing its results for completeness.

The plots in Fig. 1 indicate that the correlation between discrepancy and one-shot-performance is rather weak, as we do not see any obvious trend. However, one has to keep in mind that these performances depend on a single point only – similar to a *lucky punch* in sports. Therefore, we additionally analyze the aggregated performances in Fig. 2. The boxplots display the distribution of the factor by which each design is worse than the best design for the respective function. According to this aggregated view, the "Best" design – whose discrepancy is the smallest among all sets (recall Table 1) – is also the one achieving the smallest mean and median result. The Braaten-Weller-design had the best performance in 5 out of the 24 benchmarks. Although LHS showed good (average) performance as well, it did not achieve the best average result on any of the benchmark functions. Interestingly, uniform sampling achieves a good median score. In fact, we can see in Fig. 1 that the best uniform design often outperforms all other designs, but at the same time there is (with few exceptions) always at least one of the uniform samples which is worse than all other designs. Of course, our benchmark set is small compared to broad range of numerical problems encountered in practice. An extension to more use-cases, possibly grouped by type of application, forms an important direction for future work. In particular, we suggest to not only consider more instances of the BBOB functions, but to extend our approach to other problems, such as those provided by Nevergrad [32,33], the problems from the black-box optimization competition BBComp (https://bbcomp.ini.rub.de), and to hyper-parameter tuning.

Table 2. Number of functions for which the respective design, together with the Kriging surrogates, achieved (on average) a MSE that is at most 5% worse than the best achieved MSE. We recall that we have 24 benchmark problems in total.

n	Best	BW	Halton	LHS	UNIFORM	Total
125	6	13	8	11	4	**42**
1 000	9	9	6	11	3	**38**
2 500	11	7	8	16	3	**45**
5 000	12	10	11	14	3	**50**

5 One-Shot Regression

We now turn our attention to the *one-shot regression* problem, in which we aim to build a regression model $\hat{f}$ that predicts the function values of the true function f as accurately as possible. The accuracy of the *one-shot regression models* is measured by the mean squared error (MSE), for which we evaluate both f and our proxy $\hat{f}$ in t i.i.d. points $y^1, \ldots, y^t$, which are selected from the domain $[-5, 5]^d$ uniformly at random. The MSE is then computed as

$$\mathrm{MSE}(\hat{f}) := \frac{1}{t} \sum_{j=1}^{t} \left(f(y^j) - \hat{f}(y^j) \right)^2 .$$

In our evaluation, we use $t = 100\,000$ i.i.d. samples. For LHS and UNIFORM designs, we compute the MSE for each of the ten random designs, and average the results.

In Table 2 we compare the five designs for different sample sizes. For each sample size, we count the number of functions for which the design achieved an MSE that is at most 5% worse than the best one for the respective sample size. The displayed results are based on Kriging but results for the other surrogate models are similar. Uniform samples seem to enable less accurate regression models than the Halton and LHS designs. However, there are three cases in which the uniform design yields the best MSE: for function F16 (Weierstrass) with $n = 125$ points, F22 (Gallagher's Gaussian 21-hi Peaks) with $n = 1\,000$, and F3 (Rastrigin) with $n = 2\,500$. In the latter case no other design achieves an MSE within the 5% margin, whereas for the first two combinations the other designs achieve just slightly worse MSEs.

Figure 3 provides a more detailed impression of the regression quality for the different (design, function)-pairs. This chart includes the EVOLVED designs, which we introduce and discuss in the next section. The results in Fig. 3 are for Kriging, but those for the other models look alike. We observe clear patterns: uniform designs, in general, produce surrogate models with high mean MSE and high variance and hence a poor global approximation of the target function f on average. An exception is F21, Gallagher's Gaussian 101-me peaks function, for which uniform samples obtain the best median results with far reaching whiskers

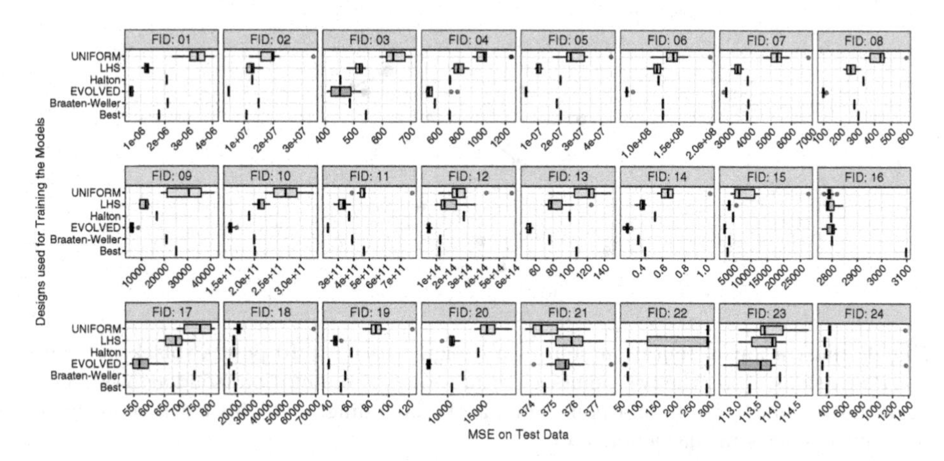

Fig. 3. MSE of the Kriging models, individually trained for each of the six designs with $n = 1\,000$ points and each of the 24 BBOB problems. Each model was assessed on a set of 100 000 i.i.d. uniform samples. Boxplots show the distribution of the 10 independent constructions.

though. We attribute this to lucky sampling. In contrast, models fitted to LHS and low-discrepancy designs tend to be much more accurate approximations of the true function. However, there is no obvious winner among the five designs, indicating that the correlation between discrepancy and performance is more complex than one might have hoped for.

6 Evolving Designs for One-Shot Regression

Given a target function f and a surrogate model we do not know what quality (w.r.t. the MSE on test data) one can achieve in the best-case with an optimal design of n points in d dimensions – a baseline is missing. In order to get an impression for the absolute quality of our tested designs, as well as for the potential of further improvement, we have approximated optimal n-point designs by means of an evolutionary algorithm (EA) [16]. That is, we evolve sampling plans in a heuristically guided stochastic manner.

Our algorithm starts with an initial LHS design x of n points in $[-5, 5]^d$. In each iteration, a new candidate design y is created from the current-best design x by applying Gaussian perturbations to a subset of $\lfloor n/10 \rfloor$ points. Points falling off the $[-5, 5]^d$ boundaries are repaired by projecting the violating components to the boundary, see Fig. 4 for an illustration. If y is no worse w.r.t. the fitness function, replace x by y, otherwise discard y. The process is repeated for a fixed number of 2 000 iterations. The fitness function fits a surrogate model based on the given design in a first step. Next, the quality of the surrogate is assessed by means of the MSE for ten random uniform designs with 10 000 points each. The fitness value is the average of these MSE values and is meant to be minimized.

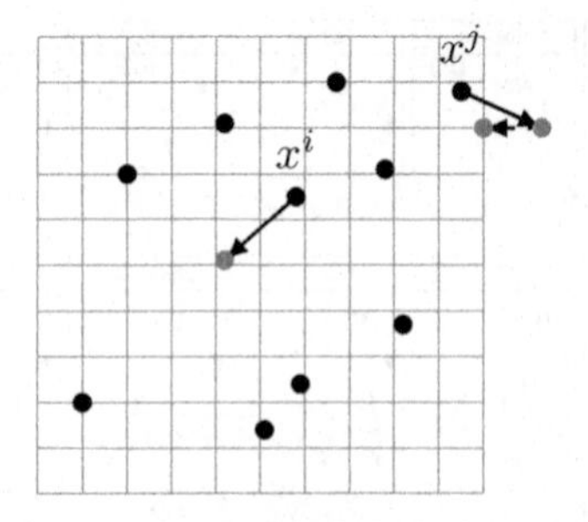

Fig. 4. Illustration of mutation step of a design with $n = 10$ points (black dots) in two dimensions. Here, the two points x^i, x^j are subject to mutation (solid arrows). The perturbation of x^j results in a point outside the bounding box. This is where a repair mechanism comes in (dashed arrow).

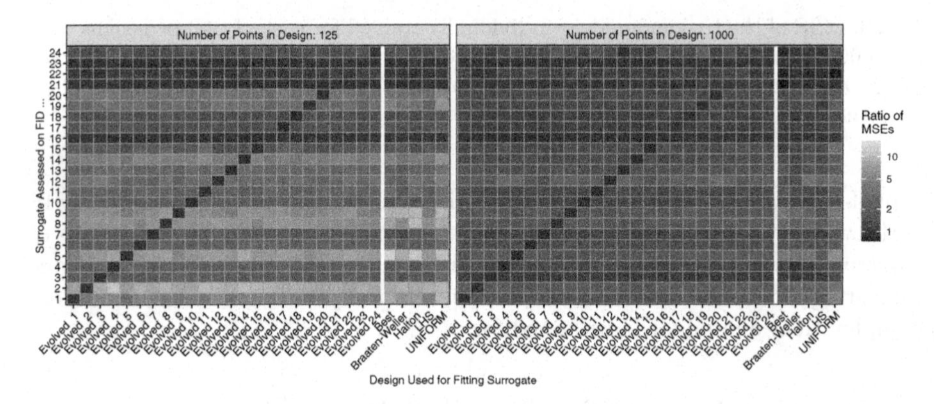

Fig. 5. Illustration of MSEs across the 24 functions from the BBOB suite. The first 24 columns correspond to the evolved designs (the i-th evolved design has been optimized for the i-th BBOB function), and the remaining five columns show the results for the five one-shot designs (Best, Braaten-Weller, Halton, LHS and UNIFORM). For each of the 29 designs, a Kriging model has been fitted to the BBOB function of the respective row and assessed by means of the MSE. The cell colors illustrate the ratio of the respective model's MSE and the MSE of the corresponding problem-tailored (i.e., evolved) design.

Note that each run of the EA produces a large set of interim solutions, but we only keep the final design for further evaluation.

We evolved ten designs (to account for randomness of the EA approach) for each combination of surrogate modelling approach, BBOB function, and size $n \in \{125, 1\,000\}$ of sampling plan resulting in 1 920 EVOLVED designs. We neglected larger sampling sizes to keep computational costs reasonable (each fitness evaluation requires fitting a surrogate on n points, which becomes computationally expensive for increasing n).

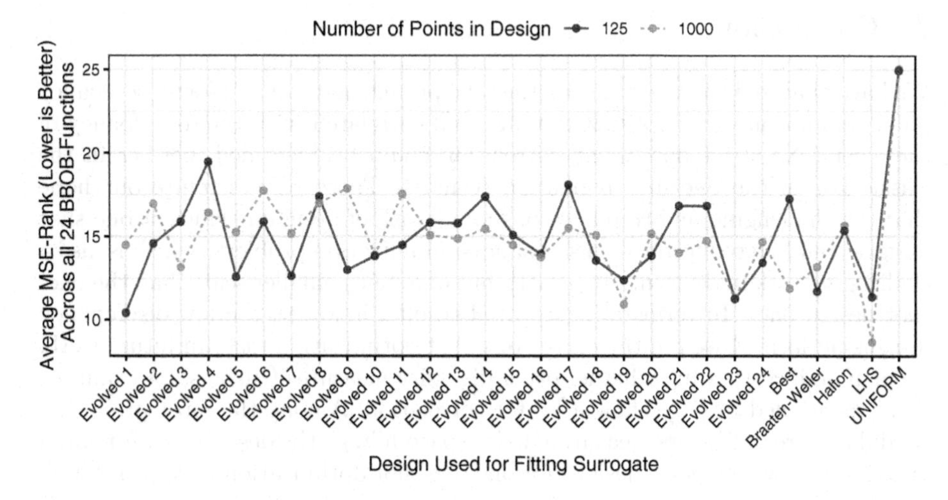

Fig. 6. Visualization of average MSE-ranks (lower is better) for all 29 designs. This figure aggregates the detailed MSE values in Fig. 5 across all 24 BBOB functions.

Returning to Fig. 3 we observe that the evolved designs lead to drastic improvements w.r.t. the MSE (and low variance) for the majority of BBOB functions; in particular for FIDs 1–14 (first three BBOB groups with mainly unimodal functions with global structure). Contrary, for FIDs 16 and 21–23—i.e., functions which are characterized by a highly rugged landscape with many local optima and weak global structure – the evolving process is far less successful w.r.t. MSE improvement.

Recall that we evolved designs for specific combinations of target function, surrogate-modelling approach, and sample size. However, as depicted in Fig. 5, the problem-specifically evolved designs are not necessarily inferior to any of the established sampling strategies. While the designs resulted indeed in significantly superior performances on the problems they have been evolved on – as can be seen by the diagonal of dark blue cells – their MSE ratios are usually comparable, if not even better, than the respective ratios of Best, Braaten-Weller, etc. When comparing the average ranks obtained by the 29 designs, see Fig. 6, the design evolved for F1 (Sphere) achieves the best average score (10.4) of all 29 tested designs for $n = 125$, closely followed by LHS (11.4) and Braaten-Weller (11.7). For $n = 1\,000$ LHS has an average rank of 8.7, while the runner-ups are "Evolved 23" (11.3) and Best (11.9). Several other evolved designs obtain fine average ranks. Noticeably, the uniform design is clearly the worst, with an average rank of 25.0 for both $n = 125$ and $n = 1\,000$. That is, the MSE of uniform sampling is on average more than 5 ranks worse than any of the other 28 designs.

Figure 5 also reveals quite noticeable differences across the functions on which the trained surrogates are assessed (rows). Non-surprisingly, we observed a decrease in the MSE ratios for an increase in sample size.

7 Conclusion

We have analyzed the question whether the promising results of low-discrepancy point sets for one-shot optimization are well correlated with the discrepancy of these sets. No strict one-to-one correlation could be identified, neither in the classic nor in the one-shot regression scenario. These results refute our hope that the challenging and resource-consuming task of designing efficient one-shot designs could be reduced to a discrepancy-minimization problem (which is also a challenging task in its own, see [15,31], but of a much smaller scale than the one-shot design one). In terms of aggregated results, however, the low-discrepancy designs performed well in the classic one-shot optimization task. In future work, we plan on investigating whether other diversity measures (such as, for example, those mentioned in [12]) show a better correlation. Among the most promising candidates are indicators measuring how "space-filling" the designs are. A related question is how well good designs for one one-shot optimization task perform on other tasks.

The decent performance of the problem-specific designs obtained through our evolutionary approach was a big surprise. Not only did they improve quite considerably over the standard designs for one-shot regression for the problem and learner they were evolved for, but some of them even rank in the top places when evaluated across the whole benchmark set. A cross-validation of the evolutionary approach on other benchmarks and an extension to other dimensions forms another line of research that seems very promising in the context of one-shot optimization.

Acknowledgments. We thank François-Michel de Rainville for help with his implementation of the generalized Halton sequences. We also thank the reviewers for providing useful comments and references. This work was financially supported by the Paris Ile-de-France Region, by ANR-11-LABX-0056-LMH, by the Australian Research Council (ARC) through grant DP190103894, and by the South Australian Government through the Research Consortium "Unlocking Complex Resources through Lean Processing". Moreover, P. Kerschke acknowledges support by the *European Research Center for Information Systems (ERCIS).*

References

1. Beck, J.: Irregularities of Distribution. Cambridge University Press, Cambridge (1987)
2. Bergstra, J., Bengio, Y.: Random search for hyper-parameter optimization. J. Mach. Learn. Res. (JMLR) **13**, 281–305 (2012). http://dl.acm.org/citation.cfm?id=2188395
3. Bossek, J., Doerr, C., Kerschke, P., Neumann, A., Neumann, F.: Github repository with project data (2020). https://github.com/jakobbossek/PPSN2020-oneshot/
4. Bousquet, O., Gelly, S., Kurach, K., Teytaud, O., Vincent, D.: Critical hyper-parameters: no random, no cry. arXiv preprint arXiv:1706.03200 (2017)
5. Braaten, E., Weller, G.: An improved low-discrepancy sequence for multidimensional quasi-Monte Carlo integration. J. Comput. Phys. **33**(2), 249–258 (1979)

6. Breiman, L.: Random forests. Mach. Learn. **45**(1), 5–32 (2001). https://doi.org/10.1023/A:1010933404324
7. Breiman, L., Friedman, J.H., Stone, C.J., Olshen, R.A.: Classification and Regression Trees. Wadsworth & Brooks/Cole Advanced Books & Software, Monterey (1984). https://doi.org/10.1201/9781315139470
8. Carnell, R.: lhs: Latin Hypercube Samples, r package version 1.0.2 (2020). https://CRAN.R-project.org/package=lhs
9. Cauwet, M., et al.: Fully parallel hyperparameter search: reshaped space-filling. arXiv preprint arXiv:1910.08406 (2019)
10. Chilès, J.-P., Desassis, N.: Fifty years of Kriging. In: Daya Sagar, B.S., Cheng, Q., Agterberg, F. (eds.) Handbook of Mathematical Geosciences, pp. 589–612. Springer, Cham (2018). https://doi.org/10.1007/978-3-319-78999-6_29
11. Cortes, C., Vapnik, V.: Support-vector networks. Mach. Learn. **20**(3), 273–297 (1995). https://doi.org/10.1007/BF00994018
12. Crombecq, K., Laermans, E., Dhaene, T.: Efficient space-filling and non-collapsing sequential design strategies for simulation-based modeling. Eur. J. Oper. Res. **214**(3), 683–696 (2011). https://doi.org/10.1016/j.ejor.2011.05.032
13. Dobkin, D.P., Eppstein, D., Mitchell, D.P.: Computing the discrepancy with applications to supersampling patterns. ACM Trans. Graph. **15**, 354–376 (1996)
14. Doerr, C., Gnewuch, M., Wahlström, M.: Calculation of discrepancy measures and applications. In: Chen, W., Srivastav, A., Travaglini, G. (eds.) A Panorama of Discrepancy Theory. LNM, vol. 2107, pp. 621–678. Springer, Cham (2014). https://doi.org/10.1007/978-3-319-04696-9_10
15. Doerr, C., Rainville, F.D.: Constructing low star discrepancy point sets with genetic algorithms. In: Proceedings of Genetic and Evolutionary Computation Conference (GECCO), pp. 789–796. ACM (2013). https://doi.org/10.1145/2463372.2463469
16. Eiben, A.E., Smith, J.E.: Introduction to Evolutionary Computing. NCS. Springer, Heidelberg (2015). https://doi.org/10.1007/978-3-662-44874-8
17. Forrester, A.I.J., Sobester, A., Keane, A.J.: Engineering Design via Surrogate Modelling - A Practical Guide. Wiley, Chichester (2008)
18. Halton, J.H.: Algorithm 247: radical-inverse quasi-random point sequence. Commun. ACM **7**(12), 701–702 (1964). https://doi.org/10.1145/355588.365104
19. Hansen, N., Auger, A., Mersmann, O., Tušar, T., Brockhoff, D.: COCO: a platform for comparing continuous optimizers in a black-box setting. arXiv e-prints arXiv:1603.08785 (2016)
20. Hansen, N., Finck, S., Ros, R., Auger, A.: Real-parameter black-box optimization benchmarking 2009: noiseless functions definitions. Technical report RR-6829, Inria (2009). https://hal.inria.fr/inria-00362633/document
21. Hlawka, E.: Funktionen von beschränkter variation in der theorie der gleichverteilung. Ann. Mat. Pura Appl. **54**, 325–333 (1961). https://doi.org/10.1007/BF02415361
22. Jin, R., Chen, W., Sudjianto, A.: An efficient algorithm for constructing optimal design of computer experiments. J. Stat. Plan. Infer. **134**(1), 268–287 (2005). https://doi.org/10.1016/j.jspi.2004.02.014
23. Jones, D.R., Schonlau, M., Welch, W.J.: Efficient global optimization of expensive black-box functions. J. Global Optim. **13**, 455–492 (1998). https://doi.org/10.1023/A:1008306431147
24. Koksma, J.F.: Een algemeene stelling uit de theorie der gelijkmatige verdeeling modulo 1. Mathematica B (Zutphen) **11**, 7–11 (1942/3)
25. Kuipers, L., Niederreiter, H.: Uniform Distribution of Sequences. Wiley, New York (1974)

26. Lemieux, C.: Monte Carlo and Quasi-Monte Carlo Sampling. Springer, New York (2009). https://doi.org/10.1007/978-0-387-78165-5
27. Leobacher, G., Pillichshammer, F.: Introduction to Quasi-Monte Carlo Integration and Applications. CTM. Springer, Cham (2014). https://doi.org/10.1007/978-3-319-03425-6
28. Liu, L.: Could enough samples be more important than better designs for computer experiments? In: Proceedings of Annual Symposium on Simulation (ANSS 2005), pp. 107–115. IEEE (2005). https://doi.org/10.1109/ANSS.2005.17
29. Matoušek, J.: Geometric Discrepancy, 2nd edn. Springer, Berlin (2009). https://doi.org/10.1007/978-3-642-03942-3
30. McKay, M.D., Beckman, R.J., Conover, W.J.: A comparison of three methods for selecting values of input variables in the analysis of output from a computer code. Technometrics **21**, 239–245 (1979). http://www.jstor.org/stable/1268522
31. Rainville, F.D., Gagné, C., Teytaud, O., Laurendeau, D.: Evolutionary optimization of low-discrepancy sequences. ACM Trans. Model. Comput. Simul. **22**, 9:1–9:25 (2012). https://doi.org/10.1145/2133390.2133393
32. Rapin, J., Gallagher, M., Kerschke, P., Preuss, M., Teytaud, O.: Exploring the MLDA benchmark on the nevergrad platform. In: Proceedings of the 21st Annual Conference on Genetic and Evolutionary Computation (GECCO 2019) Companion, pp. 1888–1896. ACM (2019). https://doi.org/10.1145/3319619.3326830
33. Rapin, J., Teytaud, O.: Nevergrad - A Gradient-Free Optimization Platform (2018). https://GitHub.com/FacebookResearch/Nevergrad

A Surrogate-Assisted Evolutionary Algorithm with Random Feature Selection for Large-Scale Expensive Problems

Guoxia Fu[1], Chaoli Sun[1(✉)], Ying Tan[1], Guochen Zhang[1], and Yaochu Jin[2]

[1] Department of Computer Science and Technology, Taiyuan University of Science and Technology, Taiyuan 030024, China
s20180536@stu.tyust.edu.cn, {chaoli.sun,tanying}@tyust.edu.cn, imzgc@hotmail.com

[2] Department of Computer Science, University of Surrey, Guildford GU2 7XH, UK
yaochu.jin@surrey.ac.uk

Abstract. When optimizing large-scale problems an evolutionary algorithm typically requires a substantial number of fitness evaluations to discover a good approximation to the global optimum. This is an issue when the problem is also computationally expensive. Surrogate-assisted evolutionary algorithms have shown better performance on high-dimensional problems which are no larger than 200 dimensions. However, it is very difficult to train sufficiently accurate surrogate models for a large-scale optimization problem due to the lack of training data. In this paper, a random feature selection technique is utilized to select decision variables from the original large-scale optimization problem to form a number of sub-problems, whose dimension may differ to each other, at each generation. The population employed to optimize the original large-scale optimization problem is updated by sequentially optimizing each sub-problem assisted by a surrogate constructed for this sub-problem. A new candidate solution of the original problem is generated by replacing the decision variables of the best solution found so far with those of the sub-problem that has achieved the best approximated fitness among all sub-problems. This new solution is then evaluated using the original expensive problem and used to update the best solution. In order to evaluate the performance of the proposed method, we conduct the experiments on 15 CEC'2013 benchmark problems and compare to some state-of-the-art algorithms. The experimental results show that the proposed method is more effective than the state-of-the-art algorithms, especially on problems that are partially separable or non-separable.

Keywords: Large-scale optimization problems · Surrogate models · Expensive problems · Random feature selection

T. Bäck et al. (Eds.): PPSN 2020, LNCS 12269, pp. 125–139, 2020.
https://doi.org/10.1007/978-3-030-58112-1_9

1 Introduction

Meta-heuristic algorithms, including evolutionary algorithms and swarm intelligent methods, have been proposed and shown efficient for solving large-scale optimization problems (LSOPs) that normally have hundreds, thousands, millions, or even billions of decision variables [4,5]. LSOPs can be classified into two categories: decomposition based and non-decomposition based [8]. In the decomposition based category, the problem will be divided into two or more smaller sub-problems, and the solutions of the sub-problems are combined to form a solution to the original problem, which is called the divide-and-conquer strategy. The cooperative coevolutionary (CC) framework [19] is a popular or well-known divide-and-conquer method [15], and different decomposition based strategies have been proposed, such as random grouping [17,32], differential grouping (DG) [16,18,34], and recursive differential grouping [23,24]. Different to the decomposition based strategy, in the non-decomposition based category, the large-scale problem will not be decomposed and will be optimized by an enhanced optimization strategy, for example, the competitive swarm optimization (CSO) [2], the social learning particle swarm optimization (SL-PSO) [3], and the level-based learning swarm optimizer (LLSO) [31]. However, most of these approaches are impractical to be applied to solve computationally expensive large-scale optimization problems because a great number of fitness evaluations are required to obtain a good approximation to the optimal solution. A wide range of surrogate models, such as polynomial regression (PR) [26], Gaussian processes (GP, also named Kriging) [27], radial basis function networks (RBFNs) [22], artificial neural networks (ANNs) [10,12], and support vector regression (SVR) [11], have been proposed to assist meta-heuristic algorithms to solve computationally expensive problems. Different surrogate-assisted meta-heuristic algorithms have been presented in [1,22,27,28,33] for solving high-dimensional expensive problems. However, few of them are suited for the optimization of large-scale expensive problems. Not many methods have been proposed for computationally expensive large-scale problems. Falco et al. [6,7] proposed a surrogate-assisted cooperative coevolution (SACC) algorithm for large-scale optimization, in which RBF, GP, SVR and the quadratic polynomial approximation (QPA) were used as the surrogate to model the low-dimensional sub-problems resulted from the problem decomposition, and the adaptive DE algorithm (JADE) proposed by Zhang and Sanderson [35] was applied to optimize each sub-problem. Ren et al. [20] proposed an RBF-SHADE-SACC algorithm, in which the RBF and the success-history based adaptive differential evolution (SHADE) [25] were integrated into the surrogate model assisted cooperative coevolution (SACC) framework. Sun et al. [21] proposed a fitness approximation assisted competitive swarm optimizer (CSO) for large-scale optimization problems, in which, instead of a surrogate model, a fitness estimation strategy based on the positional relationship between individuals at the same iteration was used to approximate the fitness value of an individual. However, its performance heavily relies on the search capability of the CSO algorithm itself.

Different to existing algorithms for optimization of computationally expensive large-scale problems, the method proposed in this paper optimizes a large-scale expensive problem through sequential evolutionary optimization of its sub-problems. Each sub-problem contains a randomly selected subset of the decision variables of the original problem, and the fitness function of each sub-problem is approximated by a surrogate. Consequently, evolutionary optimization of each sub-problem is assisted by the surrogate constructed for each sub-problem.

The remainder of the paper is organized as follows. Section 2 briefly introduces the background the work. In Sect. 3, a detailed description of the proposed method is given. Studies comparing the proposed method with SACC [7] and two non-decomposition based algorithms without surrogates are presented in Sect. 4. Finally, Sect. 5 concludes the paper with a summary and some ideas for future work.

2 Background

2.1 Radial Basis Functions Network

The radial basis function (RBF) network is a neural network that is composed of only three layers: an input, a hidden, and an output layer. Given a D-dimensional input $\mathbf{x}$, the output of the RBF network can be given as follows:

$$y = \sum_{i=1}^{nc} \omega_i \varphi(\mathbf{x}, \mathbf{c}_i) + \omega_0 \tag{1}$$

In Eq. (1), ω_i is the i-th weight, ω_0 is a bias term, and nc denotes the number of neurons in the hidden layer. Each neuron has an activate function $\varphi(\mathbf{x}, \mathbf{c}_i) = \phi(\|\mathbf{x}-\mathbf{c}_i\|)$, where $\phi(\|\cdot\|)$ is called the radial function that provides the non-linear feature of the model, and $\mathbf{c}_i$ is the i-th RBF center. Different functions, such as linear, cubic, thin plate spline, multiquadric and Gaussian, have been used as the radial basic function. In this work, the cubic function is adopted because it is simple and there are no extra parameters that need to be regulated.

2.2 Random Feature Selection

Feature selection, also known as variable selection, is often used in machine learning to reduce the number of the variables when constructing a model. Generally, there are two main types of feature selection algorithms: filter methods and wrapper methods [13]. The filter feature selection methods apply a statistical measure to assign a score to each feature, which will be used as the basis to choose the variables to be used in the model. The wrapper feature selection methods compare many models with different subsets of variables and select those features that result in the best performance.

Recently, random feature selection (RFS) becomes popular in the multi-classifier-systems (MCS) to produce higher classification accuracies [29,30] because it can improve the diversity by sampling different features for each model. The random feature selection can further be classified into uniform RFS and non-uniform RFS. This work employs the uniform random feature selection method.

3 Proposed Algorithm

As is well known, the larger the dimension of a problem is, the more data for training the surrogate is required, making it impractical for assisting optimization of large-scale expensive problems. Therefore, in this paper, we try to optimize LSOPs by sequently optimizing its sub-problems, which is expected to reduce the search space and make it easier to build high-quality surrogates using limited amount of training data. Each sub-problem is composed of a randomly selected subset of the original decision variables. By decomposing the original large-scale problem into a number of smaller sub-problems, we are able to reduce the search space and perform more efficient localized search. A temporary population will be initialized by copying all solutions from the current population of the original problem before we start optimizing all sub-problems. Then, a surrogate model will be trained for each sub-problem and used to assist evolutionary optimization of the sub-problems. At the end of the optimization of each sub-problem, the decision variables of each individual will substitute those corresponding decision variables of the temporary population individual by individual. Note that the population for optimizing each sub-problem is initialized by copying the corresponding decision variables of the temporary population individual by individual. Once the optimization of the last sub-problem is completed and the temporary population is updated using the corresponding individuals in the generation, the temporary population will be the population for the original problem of the next generation. The solution with the best estimated fitness value among all sub-problems will be kept together with its surrogate model, denoted as M_{best}. And this solution will be compared to the solution in the final updated temporary population of the current generation, which contains the optimal solution of the last sub-problem, in terms of the values approximated by the surrogate model M_{best} on its related dimensions. After that, the best solution for the original problem will be updated.

The pseudocode of our proposed surrogate-assisted evolutionary algorithm with random feature selection, denoted as SAEA-RFS, is given in Algorithm 1. Before the optimization starts, a number of solutions will be generated using the Latin hypercube sampling technique, evaluated using the real expensive fitness function and saved in an archive Arc. Then NP individuals will be initialized for a population pop. The current population pop will then be updated by a temporary population pop_t, which has been updated by replacing the corresponding decision variables of each individual in pop_t with the decision variables of the individuals in the population for evolving the sub-problem. (lines 6–13 in Algorithm 1). The solution with the best estimated fitness among all sub-problems will be kept together with its surrogate model. Suppose sub-problem

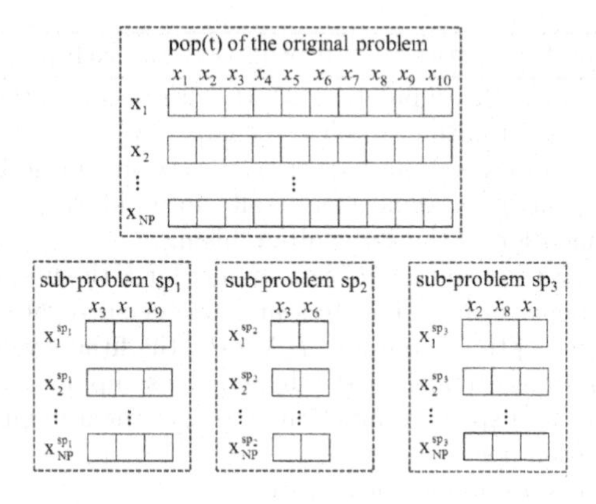

Fig. 1. The formation of sub-problems.

$k \in \{1, 2, \dots, K\}$, K is the number of sub-problems, has the best estimated fitness. In the next, the optimal solution of sub-problem k ($\mathbf{x}_b^{sp_k}$) will be compared to the solution in the final temporary population pop_t of the current generation (i.e., the solution contains the optimal solution of the K-th sub-problem $\mathbf{x}_b^{sp_K}$) with respect to the fitness values approximated by M_k of sub-problem k. The one with better estimated fitness of sp_k will be used to replace the value of the decision variables of the k-th sub-problem in the best solution found so far for the original large-scale problem to generate a new candidate solution. The new solution will then be evaluated using the expensive objective function of the original problem (lines 15–16 in Algorithm 1), and used to update the best solution found so far (line 17 in Algorithm 1). From Algorithm 1, we can see that there is one and only one solution that will be evaluated using the exact expensive function and will be saved in the archive *Arc* for training the surrogate models. In the following, we will give a detailed description of the sequential optimization of the sub-problems and the update of the global best solution.

3.1 Sequential Optimization of the Sub-problems

Sub-problem Formation. In Algorithm 1, a number of sub-problems, each of which is formed by randomly selecting a subset of the decision variables from the original problem (line 7 in Algorithm 1), will be optimized sequentially. Figure 1 gives an example to show how a sub-problem will be formed using the random feature selection technique. In Fig. 1, the dimension of the original problem is 10. Suppose the next population for the original problem is obtained after sequentially optimizing three sub-problems, which are sub-problem sp_1, sub-problem sp_2, and sub-problem sp_3. From Fig. 1, we can see the decision variables x_3, x_1 and x_9 in the original problem will be selected to be decision variables of sub-problem sp_1, i.e., $(x_1^{sp_1}, x_2^{sp_1}, x_3^{sp_1}) = (x_3, x_1, x_9)$, decision variables x_3 and x_6 in the original problem will be selected to be the decision variables of sub-problem sp_2, i.e., $(x_1^{sp_2}, x_2^{sp_2}) = (x_3, x_6)$, and so on. It can be clearly seen

Algorithm 1. The framework of the proposed algorithm.

Input: NP: the size of the population; D: the dimension of the problem;
ts_inum: the size of initial sampling; ts_num: the size of the training set;
K: the number of the sub-problems; sp_k: the k-th sub-problem;
$MaxD_{sp}$: the maximum dimension of all sub-problems;
D_{sp_k}: the dimension of the k-th sub-problem;
ND_{sp_k}: the data to train a surrogate model for k-th sub-problem;
$nIter$: the number of iterations to search on the surrogate model.
Output: $Xbest$: the optimal solution; $f(Xbest)$: the fitness value of $Xbest$.

1 Generate ts_inum data using the Latin hypercube sampling, evaluate their fitness using the exact expensive function, and save them in an archive Arc;
2 Initialize a population pop;
3 **while** the stopping criterion is not met **do**
4 $k = 1$;
5 Initialize a temporary population pop_t by copying all solutions from pop;
6 **while** $k \leq K$ **do**
7 Generate an integer number within $[1, MaxD_{sp}]$ and denote it as D_{sp_k};
8 Randomly select D_{sp_k} decision variables from the original D variables, and form a sub-problem sp_k. The decision space of sp_k is denoted as $\mathbf{x}^{sp_k} = (x_1^{sp_k}, x_2^{sp_k}, \ldots, x_{D_{sp}}^{sp_k})$;
9 Randomly select ND_{sp_k} samples from Arc and the data on the corresponding variables of the sub-problem sp_k will be used to train a surrogate model, denoted as M_k;
10 Find the optimal solution of the sub-problem sp_k, i.e., $\mathbf{x}_b^{sp_k}$ using an evolutionary algorithm assisted by the surrogate model M_k. Record $\mathbf{x}_b^{sp_k}$, sp_k and M_k;
11 Replace the population pop_t on the corresponding dimension of the sub-problem sp_k using the current population pop^{sp_k} for the sub-problem sp_k optimization;
12 $k = k + 1$;
13 **end while**
14 Update the population pop of the original problem using the temporary population pop_t;
15 Generate a new candidate solution based on the decision variables of the sub-problem that achieves the best approximated fitness and the best solution found so far;
16 Evaluate the solution using the exact fitness function and save it in the archive Arc;
17 Update the optimal solution of the original problem $Xbest$ and its fitness value $f(Xbest)$ as well;
18 **end while**
19 **return** $(Xbest, f(Xbest))$

from Fig. 1 that a decision variable may be optimized in different sub-problems, but it will not be selected more than once for the same sub-problem.

Population Updating. A temporary population pop_t will be initialized by copying all solutions from the current population $pop(t)$ of the original problem before the optimization of all sub-problems starts, and will be updated immediately by using the population of the sub-problem after the optimization of this sub-problem is completed. Suppose the decision variables of sub-problem sp_k are determined, ND_{sp_k} samples in the archive Arc will be randomly drawn and used to train a surrogate model M_k on the corresponding variables selected for the sub-problem. The sub-problem will be optimized by an evolutionary algorithm assisted by surrogate model M_k, and its selected decision variables, the optimal solution $\mathbf{x}_b^{sp_k}$ and the model M_k will be stored. After the temporary population pop_t is updated by using the population of the sub-problem sp_K when the optimization is completed, the temporary population pop_t will become the next population of the original problem, i.e., $pop(t+1) = pop_t$.

An example is given in Fig. 2 showing how to update the population. A temporary population pop_t is generated by copying the current population $pop(t)$ of the original problem, i.e., $pop_t = pop(t)$. From Fig. 2, we can see that the population pop_t will be updated by using the optimized populations of the three sub-problems, i.e., sub-problems sp_1, sp_2 and sp_3, sequentially. The initial population for optimization of the sub-problem sp_1 are copied from the corresponding decision variables, i.e., x_3, x_1 and x_9, individual by individual, from pop_t. Once the optimization of sub-problem sp_1 is finished, the individuals in the final generation will replace those in population pop_t on the corresponding decision variables, i.e., x_3, x_1 and x_9, as is shown in Fig. 2. Then the initial population for the optimization of sub-problem sp_2 will be copied from pop_t on the corresponding decision variables, i.e., x_3 and x_6, again individual by individual. The process will be repeated until the optimization of all sub-problems has completed and the temporary population pop_t will become the population of the original problem for the next generation $pop(t+1)$. From Fig. 2, we can see that it can happen that only a subset of the decision variables are not updated at the current generation, such as x_4, x_5, x_7, and x_{10}, while some other decision variables may be updated for more than once, such as x_1 and x_3.

Note that the surrogate models, the decision variables selected to form each sub-problem and the best solution of each sub-problem at each generation are kept for updating the best solution found for the original large-scale problem.

The Update of the Global Best Position. Algorithm 2 gives the pseudocode to update the best solution of the original large-scale problem. All optimal solutions of sub-problems will be compared to each other on their approximated fitness values. Suppose the optimal solution of sub-problem sp_k has the best approximated fitness value among all optimal solutions, i.e., $\mathbf{x}_b^{sp_k} = argmin\{\hat{f}^{M_1}(\mathbf{x}_b^{sp_1}), \hat{f}^{M_2}(\mathbf{x}_b^{sp_2}), \ldots, \hat{f}^{M_K}(\mathbf{x}_b^{sp_K})\}$, then its approximated fitness $\hat{f}^{M_k}(\mathbf{x}_b^{sp_k})$, its index k among all sub-problems, its surrogate model M_k, and its sub-problem sp_k selected from original large-scale problem will be saved (line 1

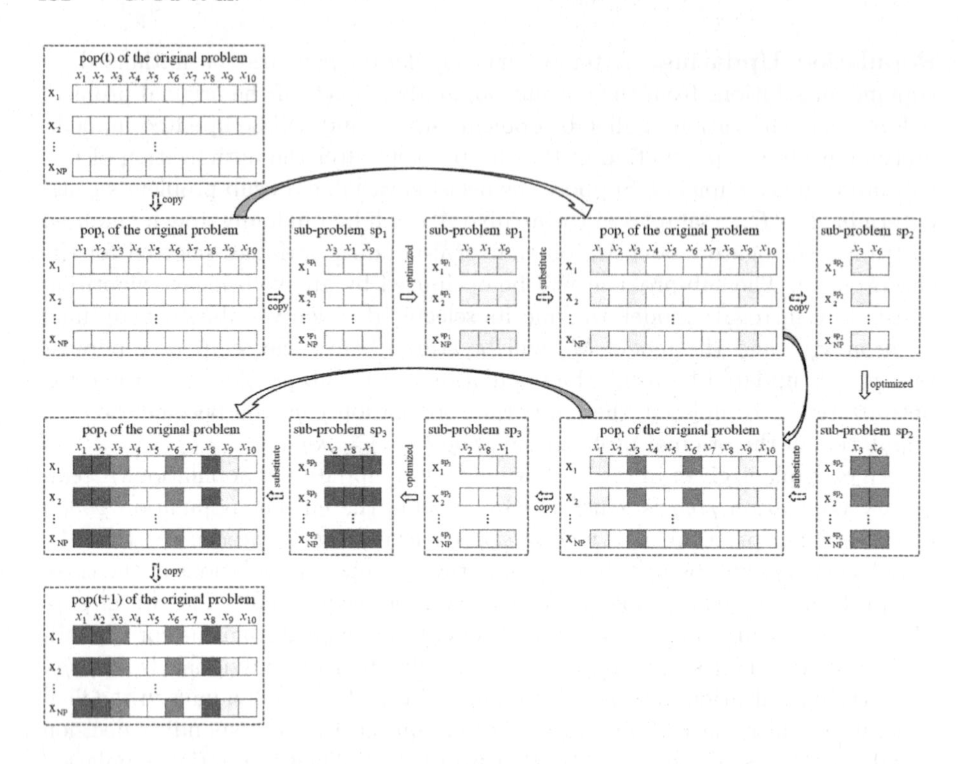

Fig. 2. Population update by sequentially optimizing a number of sub-problems.

in Algorithm 2). Note that the decision variables in the sub-problem sp_k may be updated in the optimization of the following sub-problems. Therefore, in this work, the fitness of the individual, which includes the optimal solution of the last sub-problem sp_K, i.e., $\mathbf{x}_b^{sp_K}$, in the pop_t will be approximated on the sub-problem sp_k using the surrogate model M_k, and compared to $\hat{f}^{M_k}(\mathbf{x}_b^{sp_k})$. The solution of the sub-problem sp_k with better estimated fitness will be used to replace the optimal solution found so far for the original problem on the corresponding decision variables, which will be a new solution $Xbest_t$ (lines 3–9 in Algorithm 2). Then, this new solution will be evaluated using the exact expensive fitness function and used to update the best solution found so far. Note that the solution that has been evaluated using the expensive fitness function will be saved in the archive Arc as the training data.

An example of the procedure for updating the best position of the original large-scale problem found so far is given in Fig. 3. In Fig. 3, $Xbest(t)$ represents the best solution for the problem at t generation. Suppose the approximated fitness of the optimal solution of the sub-problem sp_1 is best among all sub-problems, denoted as $\mathbf{x}_b^{sp_1} = argmin\{\hat{f}^{M_1}(\mathbf{x}_b^{sp_1}), \hat{f}^{M_2}(\mathbf{x}_b^{sp_2}), \hat{f}^{M_3}(\mathbf{x}_b^{sp_3})\}$. Suppose the optimal solution of the last sub-problem sp_3 ($K = 3$ in this example) is included in the solution $\mathbf{x}_{b_3}$ in the final temporary population pop_t. The solution of $\mathbf{x}_{b_3}$ of sub-problem sp_1 is denoted as $\mathbf{x}_{b_3}^{sp_1}$. The fitness value of solution

$\mathbf{x}_{b_3}^{sp_1}$ will be approximated by the surrogate model M_1 using its related decision variables x_3, x_1 and x_9. Now we compare the fitness values of $\mathbf{x}_b^{sp_1}$ and $\mathbf{x}_{b_3}^{sp_1}$ approximated using the surrogate model M_1. If $\hat{f}^{M_1}(\mathbf{x}_b^{sp_1})$ is better than $\hat{f}^{M_1}(\mathbf{x}_{b_3}^{sp_1})$, then a new candidate solution $Xbest_t$ will be generated by combining the solution of $\mathbf{x}_b^{sp_1}$ on x_3, x_1 and x_9 and the best solution found so far of the population on the other dimensions. Otherwise, the solution of $\mathbf{x}_{b_3}^{sp_1}$ on x_3, x_1 and x_9 will be used in $Xbest_t$. After that, $Xbest_t$ will be evaluated using the real computationally expensive problem, which is used to update the optimal solution of the population found so far.

Algorithm 2. Update of the best solution of the original problem.

Input:
pop_t: the temporary updated population after the optimization of all sub-problems is completed;
K: the number of the sub-problems;
($\mathbf{x}_b^{sp_k}$, $\hat{f}^{M_k}(\mathbf{x}_b^{sp_k})$): the optimal solution of each sub-problem sp_k ($\mathbf{x}_b^{sp_k}$) and its fitness value approximated by its surrogate model M_k ($\hat{f}^{M_k}(\mathbf{x}_b^{sp_k})$);
$\mathbf{x}_{b_K}$: the solution in pop_t with the best approximated fitness on sub-problem K;
$(Xbest(t), f(Xbest(t)))$: the optimal solution of the original large-scale problem found so far ($Xbest(t)$) and its fitness value ($f(Xbest(t))$);
Output: $(Xbest(t+1), f(Xbest(t+1)))$

1 $\mathbf{x}_b^{sp_k} = argmin\{\hat{f}^{M_1}(\mathbf{x}_b^{sp_1}), \hat{f}^{M_2}(\mathbf{x}_b^{sp_2}), \ldots, \hat{f}^{M_K}(\mathbf{x}_b^{sp_K})\}$, record its approximated fitness $\hat{f}^{M_k}(\mathbf{x}_b^{sp_k})$, its index k among all sub-problems, its surrogate model M_k, and its sub-problem sp_k selected from the original large-scale problem;
2 Approximate the fitness of solution $\mathbf{x}_{b_K}$ in pop_t on sub-problem sp_k using the surrogate model M_k, denoted as $\hat{f}^{M_k}(\mathbf{x}_{b_K}^{sp_k})$;
3 $Xbest_t = Xbest(t)$;
4 **if** $\hat{f}^{M_k}(\mathbf{x}_b^{sp_k}) < \hat{f}^{M_k}(\mathbf{x}_{b_K}^{sp_k})$ **then**
5 | Replace $Xbest_t$ on the sub-problem sp_k using $\mathbf{x}_b^{sp_k}$;
6 **else**
7 | Replace $Xbest_t$ on the sub-problem sp_k using $\mathbf{x}_{b_K}^{sp_{sp_k}}$;
8 **end if**
9 Evaluate the fitness of $Xbest_t$ using the exact expensive fitness function;
10 **if** $f(Xbest_t) < f(Xbest(t))$ **then**
11 | $Xbest(t+1) = Xbest_t$; $f(Xbest(t+1)) = f(Xbest_t)$;
12 **else**
13 | $Xbest(t+1) = Xbest(t)$; $f(Xbest(t+1)) = f(Xbest(t))$;
14 **end if**
15 **return** $(Xbest(t+1), f(Xbest(t+1)))$

4 Experimental Studies

In order to evaluate the performance of the proposed SAEA-RFS, we conducted an empirical study on the CEC'2013 special session and competition on large-scale global optimization with 1000 dimensions (F1–F12, F15) and 905 dimensions (F13–F14), respectively [14]. The characteristics of benchmark problems can be found in [14], which can be classified into four categories: fully-separable functions (F1–F3), partially additively separable functions with a separable subcomponent(F4–F7), partially additively separable functions with no separable subcomponents (F8–F11), overlapping functions (F12–F14) and non-separable functions (F15).

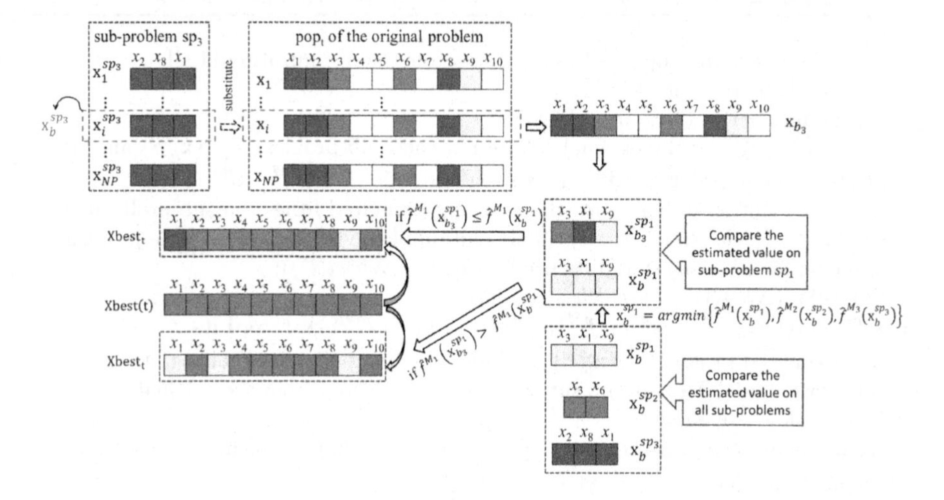

Fig. 3. Update of the best position found so far.

All compared algorithms are implemented in Matlab R2016a and run on a computer with Intel(R) Core(TM) i5-3210M, CPU @ 2.50 GHz and 4.00 GB RAM. The surrogate-assisted CC (SACC) optimizer [7], which uniformly divides the large-scale optimization problem into a number of sub-problems and optimized assisted by a surrogate model (RBF, SVR, GP and QPA utilized in their paper), will be adopted to compare the performance of our proposed SAEA-RFS. We did not adopt RBF-SHADE-SACC algorithm proposed by Ren et al. [20] because the precondition to use this algorithm is that an ideal decomposition of problem is known, which is highly unlikely in the optimization of a real-world application. We also compare the proposed algorithm to two PSO variants (SL-PSO [3] and CSO [2]) without surrogate models. All parameters of the compared algorithms are set as recommended in the original papers.

In our method, the RBF surrogate model is adopted because it appears to have better performance than other surrogate models on the high-dimensional expensive problems [33]. The maximum dimension of each sub-problem $MaxD_{sp}$

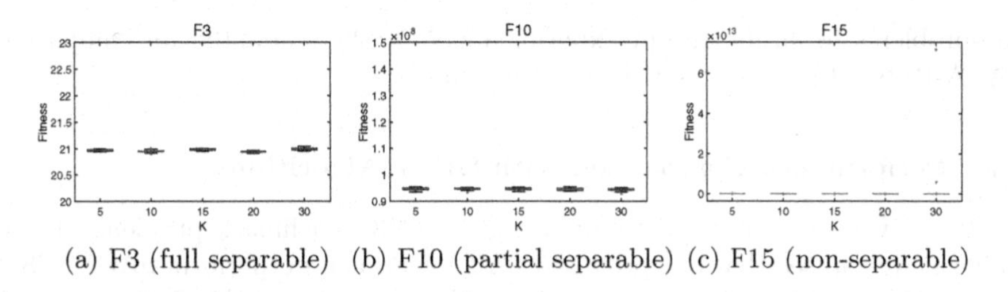

(a) F3 (full separable) (b) F10 (partial separable) (c) F15 (non-separable)

Fig. 4. The box plots on the fitness values obtained by 25 independently runs of SAEA-RFS with different number of sub-problems.

is set to 100, i.e., the dimension of a sub-problem will be a random integer in the range $[1, 100]$. The minimum number of data for training an RBF model is set to be $D+1$ [9], so in our experiments, the number of training data is set to twice of the sub-problem dimension, i.e., $ts\_num = 2 \times D_{sp_k}$. Therefore, the sample size is set to 200 to ensure that the number of data is sufficient to train a surrogate model for a sub-problem with 100 decision variables. The population size NP is set 10, $K = 20$ sub-problems will be formed at each generation and the maximum iterations of sub-problem optimization is set to 5. The differential evolution (DE) is used as the optimizer for each sub-problem, in which DE/best/1 and binomial crossover is used as mutation and crossover strategy, respectively. The crossover rate CR is set to 1 and the mutation scaling factor F is set to 0.8.

In all experiments, the stopping criterion is that the maximum number of fitness evaluations $11 * D$ in all experiments is reached, where D is the search dimension of the problem. Each algorithm will be run independently for 25 times on each problem, and the Wilcoxon rank-sum test with Bonferroni correction at a significance level of 0.05 is applied to assess whether the performance of a solution obtained by one of the two compared algorithms is expected to be better than the other. Here, $+$, $\approx$, and $-$ represent that SAEA-RFS is significantly better, equivalent to, and worse than the compared algorithms, respectively, according to the Wilcoxon's rank sum test on the median fitness values.

4.1 Sensitive Analysis on the Number of the Sub-problems

The number of sub-problems (K) is a parameter introduced in the proposed algorithm. Therefore, we conduct experiments on F3 (fully-separable function), F10 (partially separable function) and F15 (non-separable function) in CEC'2013 large-scale benchmark problems to investigate the effect of the number of sub-problems on the final performance. Figure 4 shows the boxplots of the optimal solutions obtained in 25 independently runs of SAEA-RFS when the number of sub-problems equals 5, 10, 15, 20, and 30, respectively. From Fig. 4, we can see that the results obtained by SAEA-RFS are relatively insensitive to the changes in the number of sub-problems K. However, the performance starts to become less robust when $K = 30$. Therefore, we recommend to set the number of the sub-problems to [5, 20]. In our experiments, $K = 20$ is adopted for comparing SAEA-RFS with the state-of-the-art algorithms.

4.2 Performance Comparison with Other Algorithms

Table 1 gives the statistical results on CEC'2013 benchmark problems. From Table 1, we can see that our proposed SAEA-RFS can outperform SACC, CSO and SL-PSO on 9 CEC'2013 test problems. To be specific, SAEA-RFS performed better than both non-decomposition CSO and SL-PSO on 13/15 problems. Compared to the SACC framework, our proposed method obtained 10/15 better results than SACC-RBFN and 11/15 better results than all others. Specially, we can find from Table 1 that the performance of our proposed SAEA-RFS algorithm becomes better when the problem is less separable.

Table 1. Median and median absolute deviation of optimal fitness values obtained by different algorithms on CEC'2013 benchmark problems. The best median result in each row is shown in bold.

Func		SAEA-RFS	CSO	SL-PSO	SACC-RBFN	SACC-SVR	SACC-GP	SACC-QPA
F1	Median	7.1040E+09	7.7588E+10 (+)	6.2529E+10 (+)	2.0809E+09 (−)	1.3459E+09 (−)	**9.5219E+08 (−)**	3.2809E+09 (−)
	Mad	5.5284E+08	1.6197E+09 (+)	2.3644E+09 (+)	1.1371E+08 (−)	9.4539E+07 (−)	**4.1661E+07 (−)**	2.9115E+08 (−)
F2	Median	1.9991E+04	3.9968E+04 (+)	3.8062E+04 (+)	9.5890E+03 (−)	**7.7214E+03 (−)**	9.1251E+03 (−)	1.1407E+04 (−)
	Mad	4.7842E+02	4.8994E+02 (+)	5.8755E+02 (+)	1.7731E+02 (−)	**9.1652E+01 (−)**	1.5203E+02 (−)	1.7638E+02 (−)
F3	Median	2.0923E+01	2.1627E+01 (+)	2.1632E+01 (+)	2.0591E+01 (−)	2.0580E+01 (−)	2.0576E+01 (−)	**2.0532E+01 (−)**
	Mad	1.1895E-02	4.3042E-03 (+)	5.5112E-03 (+)	1.1257E-02 (−)	9.9047E-03 (−)	1.2985E-02 (−)	**1.0999E-02 (−)**
F4	Median	**1.1983E+12**	3.2850E+12 (+)	3.8134E+12 (+)	4.8469E+12 (+)	6.7659E+12 (+)	4.8422E+12 (+)	7.9752E+12 (+)
	Mad	**2.9443E+11**	2.9211E+11 (+)	2.8020E+11 (+)	1.3494E+12 (+)	2.2941E+12 (+)	9.5760E+11 (+)	2.5130E+12 (+)
F5	Median	2.3937E+07	1.6367E+07 (−)	**1.4772E+07 (−)**	2.4079E+07 (≈)	2.7173E+07 (+)	2.6763E+07 (+)	2.7873E+07 (+)
	Mad	2.3474E+06	4.6182E+05 (−)	**3.4225E+05 (−)**	2.0450E+06 (≈)	2.7448E+06 (+)	3.4908E+06 (+)	1.8535E+06 (+)
F6	Median	**1.0606E+06**	1.0662E+06 (+)	1.0675E+06 (+)	1.0666E+06 (+)	1.0659E+06 (+)	1.0665E+06 (+)	1.0690E+06 (+)
	Mad	**1.7974E+03**	1.0801E+03 (+)	6.2046E+02 (+)	1.3233E+03 (+)	2.1647E+03 (+)	2.4745E+03 (+)	1.6541E+03 (+)
F7	Median	**6.9321E+09**	1.7028E+12 (+)	1.0419E+12 (+)	2.8180E+10 (+)	2.7217E+10 (+)	3.4172E+10 (+)	2.9452E+11 (+)
	Mad	**2.1430E+09**	3.0629E+11 (+)	3.3487E+11 (+)	9.1957E+09 (+)	5.8699E+09 (+)	1.7415E+10 (+)	1.4778E+11 (+)
F8	Median	**1.9640E+16**	9.8246E+16 (+)	1.1896E+17 (+)	3.1381E+17 (+)	2.7842E+17 (+)	3.0519E+17 (+)	4.4840E+17 (+)
	Mad	**6.4764E+15**	8.7887E+15 (+)	2.6829E+16 (+)	1.1836E+17 (+)	8.5000E+16 (+)	1.0684E+17 (+)	1.6490E+17 (+)
F9	Median	1.6062E+09	1.2756E+09 (−)	**1.1363E+09 (−)**	1.9539E+09 (+)	2.0153E+09 (+)	1.9795E+09 (+)	2.1442E+09 (+)
	Mad	2.3199E+08	1.5676E+07 (−)	**2.8179E+07 (−)**	1.7351E+08 (+)	2.2247E+08 (+)	2.3673E+08 (+)	2.8067E+08 (+)
F10	Median	**9.4551E+07**	9.5200E+07 (+)	9.5407E+07 (+)	9.5867E+07 (+)	9.5968E+07 (+)	9.5985E+07 (+)	9.6228E+07 (+)
	Mad	**4.2776E+05**	1.0668E+05 (+)	7.7644E+04 (+)	4.2800E+05 (+)	2.4781E+05 (+)	1.6207E+05 (+)	3.5806E+05 (+)
F11	Median	**8.8613E+11**	2.6977E+14 (+)	2.1319E+14 (+)	3.8546E+12 (+)	3.2749E+12 (+)	4.1133E+12 (+)	2.3197E+13 (+)
	Mad	**3.0049E+11**	4.6461E+13 (+)	3.2508E+13 (+)	8.1992E+11 (+)	1.4255E+12 (+)	1.6889E+12 (+)	1.0380E+13 (+)
F12	Median	4.8910E+11	2.0427E+12 (+)	1.6398E+12 (+)	2.5921E+10 (−)	**8.1716E+09 (−)**	2.8418E+10 (−)	5.9437E+10 (−)
	Mad	2.8418E+10	2.4209E+10 (+)	3.9238E+10 (+)	1.9425E+09 (−)	**7.6608E+08 (−)**	2.1324E+09 (−)	2.4218E+09 (−)
F13	Median	**1.0892E+11**	1.0776E+12 (+)	1.6420E+14 (+)	8.2671E+11 (+)	4.3888E+11 (+)	7.1633E+11 (+)	2.3902E+13 (+)
	Mad	**1.3510E+10**	1.3174E+11 (+)	3.1747E+13 (+)	1.6500E+11 (+)	1.0526E+11 (+)	1.8392E+11 (+)	1.3336E+13 (+)
F14	Median	**1.1205E+12**	7.4289E+12 (+)	3.1356E+14 (+)	4.4563E+12 (+)	2.9075E+12 (+)	4.2607E+12 (+)	2.2505E+13 (+)
	Mad	**1.1850E+11**	1.0571E+12 (+)	6.6102E+13 (+)	5.8627E+11 (+)	5.7475E+11 (+)	1.1633E+12 (+)	1.0025E+13 (+)
F15	Median	**3.1404E+08**	5.8871E+14 (+)	3.8143E+14 (+)	4.0026E+09 (+)	3.4749E+09 (+)	4.0142E+09 (+)	1.1288E+10 (+)
	Mad	**8.1643E+07**	5.7063E+13 (+)	4.3945E+13 (+)	1.6703E+09 (+)	1.4752E+09 (+)	1.5701E+09 (+)	3.8612E+09 (+)
+/≈/-			13/0/2	13/0/2	10/1/4	11/0/4	11/0/4	11/0/4

5 Conclusion

A surrogate-assisted evolutionary algorithm with random feature selection technique was proposed in this paper. In the proposed method, the large-scale optimization problems was optimized by sequentially optimizing a number of sub-problems which are formed using the random feature selection technique. The experimental results on CEC'2013 F1–F12 and F15 test problems with 1000 dimensions and F13–F14 problems with 905 dimensions show that our proposed method has better performance than the algorithms compared in this work, especially on the less separable problems. This work, which is still preliminary, indicates that it is a promising approach to optimize a large-scale expensive problem by sequentially optimizing a number of sub-problems using the surrogate-assisted methods. In the future, different problem decomposition methods will be investigated and both sequential and parallel optimization of the sub-problems will be examined. In addition, application of the proposed algorithm to real-world large-scale optimization problems will be considered.

Acknowledgements. The authors would like to thank Professor Jonathan E. Fieldsend in University of Exeter for his work on improving the quality of the paper. This work was supported in part by National Natural Science Foundation of China (Grant No. 61876123), Natural Science Foundation of Shanxi Province (201801D121131, 201901D111264, 201901D111262), Shanxi Science and Technology Innovation project for Excellent Talents (201805D211028), the Doctoral Scientific Research Foundation of Taiyuan University of Science and Technology (20162029), and the China Scholarship Council (CSC).

References

1. Cai, X., Gao, L., Li, X., Qiu, H.: Surrogate-guided differential evolution algorithm for high dimensional expensive problems. Swarm Evol. Comput. **48**, 288–311 (2019)
2. Cheng, R., Jin, Y.: A competitive swarm optimizer for large scale optimization. IEEE Trans. Cybern. **45**(2), 191–204 (2015)
3. Cheng, R., Jin, Y.: A social learning particle swarm optimization algorithm for scalable optimization. Inf. Sci. **291**, 43–60 (2015)
4. Deb, K., Myburgh, C.: Breaking the billion-variable barrier in real-world optimization using a customized evolutionary algorithm. In: Proceedings of the Genetic and Evolutionary Computation Conference, pp. 653–660 (2016)
5. Deb, K., Reddy, A.R., Singh, G.: Optimal scheduling of casting sequence using genetic algorithms. Mater. Manuf. Process. **18**(3), 409–432 (2003)
6. Falco, I.D., Cioppa, A.D., Trunfio, G.A.: Large scale optimization of computationally expensive functions. In: Proceedings of the Genetic and Evolutionary Computation Conference, pp. 1788–1795. ACM Press (2017)
7. Falco, I.D., Cioppa, A.D., Trunfio, G.A.: Investigating surrogate-assisted cooperative coevolution for large-scale global optimization. Inf. Sci. **482**, 1–26 (2019)
8. Ge, Y.F., et al.: Distributed differential evolution based on adaptive mergence and split for large-scale optimization. IEEE Trans. Cybern. **48**(7), 2166–2180 (2017)
9. Gutmann, H.M.: A radial basis function method for global optimization. J. Glob. Optim. **19**(3), 201–227 (2001)
10. Hamody, S.F., Adra, A.I.: A hybrid multi-objective evolutionary algorithm using an inverse neural network for aircraft control system design. In: Proceedings of the IEEE Congress on Evolutionary Computation (2005)
11. Jin, Y.: A comprehensive survey of fitness approximation in evolutionary computation. Soft Comput. **9**(1), 3–12 (2003)
12. Jin, Y., Olhofer, M., Sendhoff, B.: A framework for evolutionary optimization with approximate fitness functions. IEEE Trans. Evol. Comput. **6**(5), 481–494 (2002)

13. Kuhn, M., Johnson, K.: Applied Predictive Modeling, vol. 26. Springer, Heidelberg (2013)
14. Li, X., Tang, K., Omidvar, M.N., Yang, Z., Qin, K.: Benchmark functions for the CEC 2013 special session and competition on large-scale global optimization. Evolutionary Computation and Machine Learning Group, RMIT University, Australia, Technical report (2013)
15. Liu, Y., Yao, X., Zhao, Q., Higuchi, T.: Scaling up fast evolutionary programming with cooperative coevolution. In: Proceedings of the 2001 IEEE Congress on Evolutionary Computation (2002)
16. Omidvar, M., Li, X., Mei, Y., Yao, X.: Cooperative co-evolution with differential grouping for large scale optimization. IEEE Trans. Evol. Comput. **18**(3), 378–393 (2014)
17. Omidvar, M.N., Li, X., Yang, Z., Yao, X.: Cooperative co-evolution for large scale optimization through more frequent random grouping. In: Proceedings of 2010 IEEE Congress on Evolutionary Computation, pp. 1–8. IEEE (2010)
18. Omidvar, M.N., Yang, M., Mei, Y., Li, X., Yao, X.: DG2: a faster and more accurate differential grouping for large-scale black-box optimization. IEEE Trans. Evol. Comput. **21**(6), 929–942 (2017)
19. Potter, M.A., Jong, K.A.D.: A cooperative coevolutionary approach to function optimization. Third Parallel Probl. Sol. Form Nat. **866**, 249–257 (1994)
20. Ren, Z., et al.: Surrogate model assisted cooperative coevolution for large scale optimization. Appl. Intell. **49**(2), 513–531 (2019)
21. Sun, C., Ding, J., Zeng, J., Jin, Y.: A fitness approximation assisted competitive swarm optimizer for large scale expensive optimization problems. Memetic Comput. **10**(2), 123–134 (2016)
22. Sun, C., Jin, Y., Cheng, R., Ding, J., Zeng, J.: Surrogate-assisted cooperative swarm optimization of high-dimensional expensive problems. IEEE Trans. Evol. Comput. **21**(4), 644–660 (2017)
23. Sun, Y., Kirley, M., Halgamuge, S.K.: A recursive decomposition method for large scale continuous optimization. IEEE Trans. Evol. Comput. **22**(5), 647–661 (2018)
24. Sun, Y., Omidvar, M.N., Kirley, M., Li, X.: Adaptive threshold parameter estimation with recursive differential grouping for problem decomposition. In: Proceedings of the Genetic and Evolutionary Computation Conference, pp. 889–896. ACM Press (2018)
25. Tanabe, R., Fukunaga, A.: Success-history based parameter adaptation for differential evolution. In: Proceedings of the 2013 IEEE Congress on Evolutionary Computation, pp. 71–78. IEEE (2013)
26. Tang, Y., Chen, J., Wei, J.: A surrogate-based particle swarm optimization algorithm for solving optimization problems with expensive black box functions. Eng. Optim. **45**(5), 557–576 (2013)
27. Tian, J., Tan, Y., Zeng, J., Sun, C., Jin, Y.: Multiobjective infill criterion driven Gaussian process-assisted particle swarm optimization of high-dimensional expensive problems. IEEE Trans. Evol. Comput. **23**(3), 459–472 (2018)
28. Wang, H., Jin, Y., Sun, C., Doherty, J.: Offline data-driven evolutionary optimization using selective surrogate ensembles. IEEE Trans. Evol. Comput. **23**(2), 203–216 (2018)
29. Waske, B., van der Linden, S., Benediktsson, J.A., Rabe, A., Hostert, P.: Sensitivity of support vector machines to random feature selection in classification of hyperspectral data. IEEE Trans. Geosci. Rem. Sens. **48**(7), 2880–2889 (2010)

30. Yang, J.M., Kuo, B.C., Yu, P.T., Chuang, C.H.: A dynamic subspace method for hyperspectral image classification. IEEE Trans. Geosci. Rem. Sens. **48**(7), 2840–2853 (2010)
31. Yang, Q., Chen, W.N., Da Deng, J., Li, Y., Gu, T., Zhang, J.: A level-based learning swarm optimizer for large-scale optimization. IEEE Trans. Evol. Comput. **22**(4), 578–594 (2017)
32. Yang, Z., Tang, K., Yao, X.: Large scale evolutionary optimization using cooperative coevolution. Inf. Sci. **178**(15), 2985–2999 (2008)
33. Yu, H., Tan, Y., Zeng, J., Sun, C., Jin, Y.: Surrogate-assisted hierarchical particle swarm optimization. Inf. Sci. **454**, 59–72 (2018)
34. Yuan, S., Kirley, M., Halgamuge, S.K.: Extended differential grouping for large scale global optimization with direct and indirect variable interactions. In: Proceedings of the Genetic and Evolutionary Computation Conference (2015)
35. Zhang, J., Sanderson, A.C.: JADE: adaptive differential evolution with optional external archive. IEEE Trans. Evol. Comput. **13**(5), 945–958 (2009)

Designing Air Flow with Surrogate-Assisted Phenotypic Niching

Alexander Hagg[1,3(✉)], Dominik Wilde[1,2], Alexander Asteroth[1], and Thomas Bäck[3]

[1] Bonn-Rhein-Sieg University of Applied Sciences, Sankt Augustin, Germany
{alexander.hagg,dominik.wilde,alexander.asteroth}@h-brs.de
[2] Chair of Fluid Mechanics, University of Siegen, Siegen, Germany
[3] Leiden Institute of Advanced Computer Science, Leiden University, Leiden, The Netherlands
t.h.w.baeck@liacs.leidenuniv.nl

Abstract. In complex, expensive optimization domains we often narrowly focus on finding high performing solutions, instead of expanding our understanding of the domain itself. But what if we could quickly understand the complex behaviors that can emerge in said domains instead? We introduce surrogate-assisted phenotypic niching, a quality diversity algorithm which allows to discover a large, diverse set of behaviors by using computationally expensive phenotypic features. In this work we discover the types of air flow in a 2D fluid dynamics optimization problem. A fast GPU-based fluid dynamics solver is used in conjunction with surrogate models to accurately predict fluid characteristics from the shapes that produce the air flow. We show that these features can be modeled in a data-driven way while sampling to improve performance, rather than explicitly sampling to improve feature models. Our method can reduce the need to run an infeasibly large set of simulations while still being able to design a large diversity of air flows and the shapes that cause them. Discovering diversity of behaviors helps engineers to better understand expensive domains and their solutions.

Keywords: Evolutionary computation · Quality diversity · Phenotypic niching · Computational fluid dynamics · Surrogate models · Bayesian optimization.

1 Introduction

We design objects with the expectation that they will exhibit a certain behavior. In fluid dynamics optimization, we want an airplane wing to experience low drag forces, but also have a particular lift profile, depending on angle of attack and air speed. We want to understand how the design of our public transportation hubs, dealing with large influxes of travelers, can cause congestion at maximal flow rates. We want our buildings to cause as little wind nuisance as possible and understand how their shape and the wind turbulence they cause are linked.

T. Bäck et al. (Eds.): PPSN 2020, LNCS 12269, pp. 140–153, 2020.
https://doi.org/10.1007/978-3-030-58112-1_10

In all these cases, it is not easy to design without prior experience and we often require long iterative design processes or trial-and-error methods.

What if we could quickly understand the possible types of behavior in expensive engineering problems and get an early intuition about how shape and behavior are related? In this work, we try to answer these questions, and in particular, whether we can discover different high performing behaviors of shapes, designing air flow simultaneously to the shapes that causes it. An overview of related work is given in Sect. 2, where we explain quality diversity (QD) algorithms and the use of surrogate assistance. In Sect. 3 we introduce a new QD algorithm that performs surrogate-assisted phenotypic niching. Two problem domains are used (Sect. 4): one inexpensive domain that optimizes the symmetry of polygons, allowing us to perform an in depth evaluation of various QD methods, and an expensive air flow domain (Sect. 5).

2 Quality Diversity

QD algorithms combine performance based search with "blind" novelty search, which searches for novel solutions without taking into account performance [14]. QD finds a diverse set of high performing optimizers [3,15] by only allowing solutions to compete in local niches. Niches are based on features that describe phenotypic aspects, like shape, structure or behavior. It keeps track of an archive of niches and solutions are added if their phenotype fills an empty niche or their quality is higher than that of the solution that was previously placed inside.

QD became applicable to expensive optimization problems after the introduction of surrogate-assisted illumination (SAIL) [7]. In this Bayesian interpretation of QD, a multi-dimensional archive of phenotypic elites (MAP-Elites) [3] is created based on *upper confidence bound* (UCB) [1] sampling, which takes an optimistic view at surrogate-assisted optimization. A Gaussian Process (GP) regression [18] model predicts the performance of new solutions based on the distance to previous examples, which is modeled using a covariance function. A commonly used covariance function is the squared exponential, which has two hyperparameters: the length scale (or sphere of influence) and the signal variance, which are found by minimizing the negative loglikelihood of the process. For any location, the GP model predicts a mean value μ and confidence intervals σ of the prediction. σ is added to μ with the idea that a location where the model has low confidence also has the promise of holding a better performing solution: $UCB(x) = \mu(x) + \kappa \cdot \sigma(x)$. The parameter κ allows us to tune the UCB function between exploitation ($\kappa = 0$) and exploration ($\kappa >> 0$).

In SAIL, after MAP-Elites fills the acquisition map which contains "optimistic" solution candidates, a random selection of those candidates is analyzed in the expensive evaluation function to form additional training samples for the GP model. This loop continues until the evaluation budget is exhausted. Then κ is set to 0 in a final MAP-Elites run to create a feature map that now contains a diverse set of solutions that is predicted to be high-performing. SAIL needs a budget orders of magnitudes smaller than MAP-Elites because it can

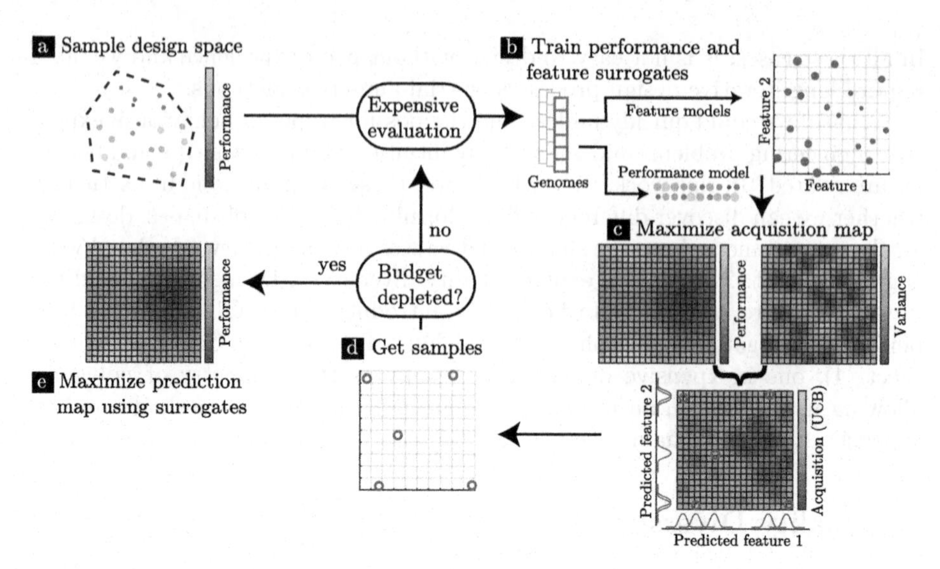

Fig. 1. Surrogate-assisted Phenotypic Niching. An initial sample set is evaluated (a), then models are trained to predict performance and feature coordinates (b), MAP-Elites is used to produce an acquisition map, balancing exploitation and exploration with the UCB of the performance model. Feature models predict the niche of new individuals (c). New samples are selected from the acquisition map (d). After the evaluation budget is depleted, the surrogate models are used to generate the prediction map with MAP-Elites, ignoring model confidence (e).

exploit the surrogate model without "wasting" samples. SAIL, however, is constrained to features that are cheap to calculate, like shape features [7] that can be determined without running the expensive evaluation.

With SAIL it became possible to use performance functions of expensive optimization domains. But the strength of QD, to perform niching based on behavior, cannot be applied when determining those behaviors is expensive. In this work we evaluate whether we can include surrogate models for such features.

3 Surrogate-Assisted Phenotypic Niching

To be able to handle expensive features, we introduce surrogate-assisted phenotypic niching (SPHEN) (Fig. 1 and Algorithm 1). By building on the insight that replacing the performance function with a surrogate model decreases the necessary evaluation budget, we replace the exact feature descriptors as well.

The initial training sample set, used to form the first seeds of the acquisition map, is produced by a space-filling, quasi-random low-discrepancy Sobol [21] sequence in the design space (Fig. 1a). Due to the lack of prior knowledge in black-box optimization, using space-filling sequences has become a standard method to ensure a good coverage of the search domain. The initial set is evaluated, for example in a computational fluid dynamics simulator. Performance

Algorithm 1. Surrogate-assisted Phenotypic Niching

Set $budget$, $maxGens$, $numInitSamples$ ▷ Configure
$\mathcal{X}' \leftarrow \mathcal{X} \leftarrow Sobol(numInitSamples)$ ▷ Initial samples
while $|\mathcal{X}| < budget$ **do**
  $(\mathbf{f}', \mathbf{p}') \leftarrow Sim(\mathcal{X}')$ ▷ Precisely evaluate performance and features
  $(\mathcal{X}, \mathbf{f}, \mathbf{p}) \leftarrow (\mathcal{X} \cup \mathcal{X}', \mathbf{f} \cup \mathbf{f}', \mathbf{p} \cup \mathbf{p}')$
  $(\mathbf{M}_p, \mathbf{M}_f) \leftarrow Train(\mathcal{X}, \mathbf{f}, \mathbf{p})$ ▷ Train surrogate models
  $\mathbf{A}_{map} \leftarrow$ MAP-ELITES$(20, \mathcal{X}, Predict(\mathcal{X}, \mathbf{M}_f), Predict(\mathcal{X}, \mathbf{M}_p), \mathbf{M}_p, \mathbf{M}_f)$ ▷ Produce acquisition map based on predicted sample performance and features
  $\mathcal{X}' \leftarrow Sobol(\mathbf{A}_{map})$ ▷ Select new (optimized) samples from acquisition map
end while
$\mathbf{P}_{map} \leftarrow$ MAP-ELITES$(acq(), 0, feat(), \mathcal{X}, \mathbf{f}, \mathbf{p}, \mathbf{M}_p, \mathbf{M}_f)$ ▷ Produce prediction map

procedure MAP-ELITES$(\sigma_{ucb}, \mathcal{X}, \mathbf{f}, \mathbf{p}, \mathbf{M}_p, \mathbf{M}_f)$
  $\mathbf{I}_{map} \leftarrow (\mathcal{X}, \mathbf{f}, \mathbf{p})$ ▷ Create initial map
  while $gens < maxGens$ **do**
    $\mathbf{P} \leftarrow Sobol(\mathbf{I}_{map})$ ▷ Evenly, pseudo-randomly select parents from map
    $\mathbf{C} \leftarrow Perturb(\mathbf{P})$ ▷ Perturb parents to get children
    $\mathbf{f} \leftarrow Predict(\mathbf{C}, \mathbf{M}_f)$ ▷ Predict features
    $\mathbf{p} \leftarrow UCB(\mathbf{C}, \sigma_{ucb}, \mathbf{M}_p)$ ▷ Predict performance (Upper Confidence Bound)
    $\mathbf{I}_{map} \leftarrow Replace(\mathbf{I}_{map}, \mathbf{C}, \mathbf{f}, \mathbf{p})$ ▷ Replace bins if empty or better
  end while
end procedure

and phenotypic features of those samples are derived from the results, or, in the case of simpler non-behavioral features, from the solutions' expression or shape themselves. The key issue here is to check the range of the initial set's features. Since we do not know what part of the phenotypic feature space will be discovered in the process, the initial set's feature coordinates only give us a hint of the reachable feature space. Just because we used a space-filling sampling technique in the design space, does not mean the samples are space-filling in feature space.

After collecting performance and feature values, the surrogate models are trained (Fig. 1b). We use GP models, which limit the number of samples to around 1,000, as the training and prediction becomes quite expensive. A squared exponential covariance function is used and the (constant) mean function is set to the training samples' mean value. The covariance function's hyperparameters, length scale and signal variance, are deduced using the GP toolbox GPML's [19] conjugate gradients based minimization method for 1,000 iterations.

MAP-Elites then creates the acquisition map by optimizing the UCB of the performance model (with a large exploration factor $\kappa = 20$), using feature models to assign niches for the samples and new solutions (Fig. 1c). Notably, we do not take into account the confidence of those feature models. Surrogate assisted QD works, because, although the search takes place in a high-dimensional space, QD only has to find the *elite hypervolume* [23], or *prototypes* [9], the small volumes consisting of high-performing solutions. Only the performance function can guide the search towards the hypervolume. Taking into account the feature

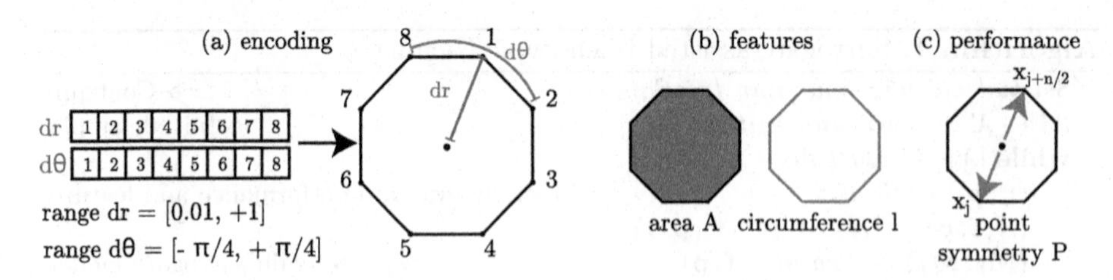

Fig. 2. The genome (a), consisting of 16 parameters that define axial and radial deformations, shape features (b) and performance (c) of polygons in the domain.

models' confidence intervals adds unnecessary complexity to the modeling problem. SPHEN's goal is to be able to only predict features for high-performing solutions, so we let feature learning "piggyback" on this search. We use a Sobol sequence on the bins of the acquisition map to select new solutions (Fig. 1d) that are then evaluated to continue training the surrogate models. This process iterates as long as the evaluation budget is not depleted. Finally, MAP-Elites is used to create a prediction map, ignoring the models' confidence (Fig. 1e), which is filled with diverse, high-performing solutions.

4 Domains

Phenotypic features describe phenomena that can be related to complex domains, like behavioral robotics, mechanical systems, or computational fluid dynamics (CFD). Before we apply SPHEN to an expensive CFD domain, we compare its performance to MAP-Elites and SAIL in a simpler, inexpensive domain.

4.1 Polygons

To be able to calculate all performance and feature values, we optimize free form deformed, eight-sided polygons. The polygons are encoded by 16 parameters controlling the polar coordinate deviation of the control points (Fig. 2a). The first half of the genome determines the corner points' radial deviation ($dr \in [0.01, 1]$). The second half of the genome determines their angular deviation ($d\theta \in [-\pi/4, \pi/4]$). The phenotypic features are the area of the polygon A and its circumference l (Fig. 2b). These values are normalized between 0 and 1 by using predetermined ranges ($A \in [0.01, 0.6]$ and $l \in [1, 4]$). The performance function (Fig. 2c) is defined as the point symmetry P. The polygon is sampled at $n = 100$ equidistant locations on the polygon's circumference, after which the symmetry metric is calculated (Eq. 1), based on the symmetry error E_s, the sum of Euclidean distances of all $n/2$ opposing sampling locations to the center:

$$f_P(x_i) = \frac{1}{1 + E_s(x_i)}, E_s(x) = \sum_{j=1}^{n/2} \left\| \boldsymbol{x}_j - \boldsymbol{x}_{j+n/2} \right\| \tag{1}$$

4.2 Air Flow

The air flow domain is inspired by the problem of wind nuisance in the built environment. Wind nuisance is defined in building norms [10,16] and uses the wind amplification factor measured in standardized environments, with respect to the hourly mean wind speed. In a simplified 2D setup, we translate this problem to that of minimizing maximum air flow speed (u_{Max}) based on a fixed flow input speed. The performance is determined as the inverse over the normalized maximum velocity: $p(x) = \frac{2}{(1+u_{Max}(x))} - 1$. However, we only need to keep u_{Max} within a *nuisance threshold*, which we set to $u_{Max} \leq 0.12$.

The encoding from the polygon domain is used to produce 2D shapes that are then placed into a CFD simulation. To put emphasis on the architectural nature of the domain, we use two features, area and air flow turbulence. The chaotic behavior of turbulence provokes oscillations around a mean flow velocity, which influences the maximum flow velocity. Both features are not optimization goals. Rather, we want to analyze, under the condition of keeping the flow velocity low, how the size of the area and turbulence are related to each other. We want to produce polygons that are combinations between their appearance (small to large) and their effect on the flow (low to high turbulence). Concretely, at the lowest and highest values of area and turbulence, regular intuitive shapes should be generated by the algorithm such as slim arrow-like shapes for low turbulence and area, or regular polygons for high turbulence and area. However, for area/turbulence combinations in between, the design of the shape is not unique and will possibly differ from intuition.

Lattice Boltzmann Method. The Lattice Boltzmann method (LBM) is an established tool for the simulation of CFD [13]. Instead of directly solving the Navier-Stokes equations, the method operates a stream and collide algorithm of particle distributions derived from the Boltzmann equation. In this contribution, LBM is used on a 2D grid with the usual lattice of nine discrete particle velocities. At the inlets and outlets, the distribution function values are set to equilibrium according to the flow velocity. The full bounce-back boundary condition is used at the solid grid points corresponding to the polygon. Although there are more sophisticated approaches for the boundaries, this configuration is stable throughout all simulations. In addition, the bounce-back boundary condition is flexible, as the boundary algorithm is purely local with respect to the grid points. As an extension of the Bhatnagar-Gross-Krook (BGK) collision model [2], a Smagorinsky subgrid model [6] is used to account for the under-resolved flow in the present configuration. For a more detailed description of the underlying mechanisms, we refer to [13]. Note that the results of the 2D domain do not entirely coincide with results that will be found in 3D, caused by the difference in turbulent energy transport [22].

The simulation domain consists of $300 \cdot 200$ grid points. A bitmap representation of the polygon is placed into this domain, occupying up to $64 \cdot 64$ grid points. As the Lattice Boltzmann method is a solver of weakly compressible flows, it is necessary to specify a Mach number (0.075), a compromise between

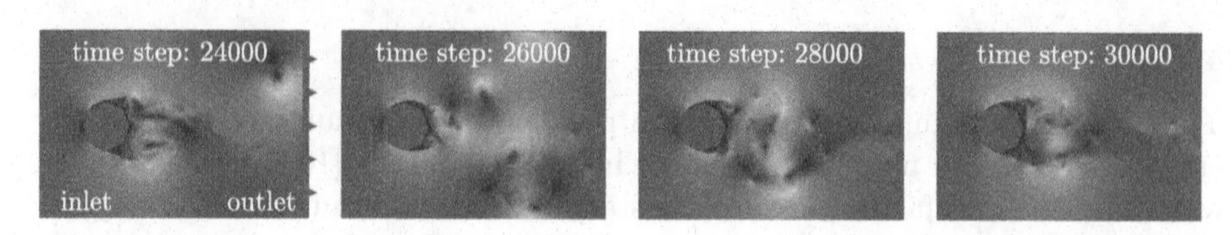

Fig. 3. Air flow around a circular polygon shape at four different time steps. (Color figure online)

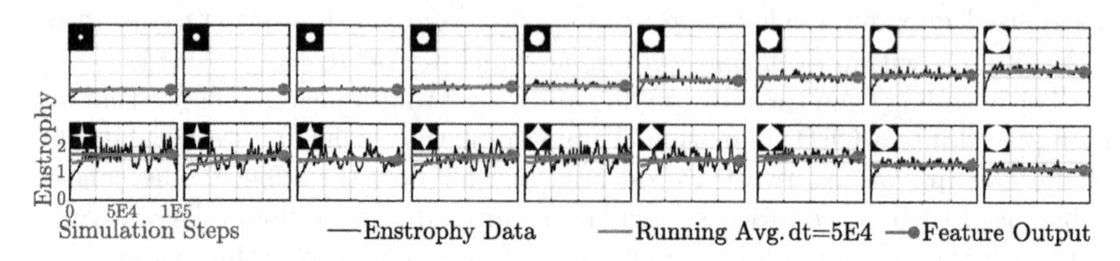

Fig. 4. Enstrophy values during simulation of circles and stars. The running average of the last 50,000 time steps converges to the final feature output.

computation time and accuracy. The Reynolds number is $Re = 10,000$ with respect to the largest possible extent of the polygon. For the actual computation, the software package *Lettuce* is used [11], which is based on the PyTorch framework [17], allowing easy access to GPU functionality. The fluid dynamics experiment was run on a cluster with four GPU nodes, each simulation taking ten minutes. Figure 3 shows the air flow around a circular polygon at four different, consecutive time steps. Brighter colors represent higher magnitudes of air flow velocity. Throughout the 100,000 time steps of the simulation, maximum velocity and enstrophy are measured. The enstrophy, a measure for the turbulent energy dissipation in the system with respect to the resolved flow quantities [8,12], increases as turbulence intensity increases in the regarded volume.

Validation and Prediction of Flow Features. The maximum velocity u_{max} and enstrophy E are measured every 50 steps. We employ a running average over the last 50,000 time steps. To test whether we indeed converge to a stable value, we run simulations with different shapes (nine varied-size circles and nine deformed star shapes) and calculate the moving average of the enstrophy values, which is plotted in Fig. 4. The value converges to the final feature value (red).

To validate the two measures, we simulate two small shape sets of circles and stars. Increasing the radius of the circles set should lead to higher u_{max} and E, as more air is displaced by the larger shapes. The stars set is expected to have larger u_{max} and E for the more irregular shapes. This is confirmed in Fig. 5.

Next, we investigate whether we can predict the simple shapes' flow feature values correctly. Although GP models are often called "parameter free", this is not entirely accurate. The initial guess for the hyperparameter's values, before minimization of the negative log likelihood of the model takes place, can have large effects on the accuracy of the model. The log likelihood landscape can

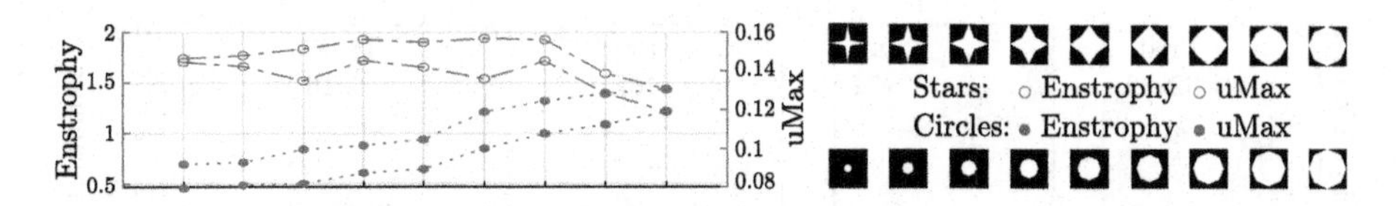

Fig. 5. Enstrophy and maximum velocity of circles and stars.

Table 1. Parameter settings for MAP-Elites, SAIL[A], restricted SAIL[B] and SPHEN.

Parameter	MAP-Elites	SAIL[A]	SAIL[B]	SPHEN
Generations	4,096	1,024	63	1,024
Descendants	16	32	16	32
Budget (per iteration)	-	1,024 (16)	1,024 (16)	1,024 (16)
Resolution (acquisition)	-	16×16	16×16	32×32

[A] Due to the number of feature evaluations, MAP-Elites uses $4{,}096 \cdot 16 = 65{,}536$ and SAIL uses $16 + 1{,}024 \cdot 32 \cdot (\frac{1{,}024}{16}) + 1{,}024 = 2{,}098{,}192$ evaluations.
[B] Here, SAIL is restricted to the number of evaluations that was used in MAP-Elites. Number of generations ($\frac{4{,}096 \cdot 16 - 1{,}024 - 16}{1{,}024} \approx 63$).

contain local optima. We perform a grid search on the initial guesses for length scale and signal variance. Using leave-one-out cross validation, GP models are trained on all but one shape, after which we measure the accuracy using the mean absolute percentage error (MAPE), giving a good idea about the magnitude of the prediction error. The process is repeated until all examples were part of the test set once. The MAPE on u_{Max} was 2.4% for both sets. The enstrophy was harder to model, at 4.9% and 10.3% for the respective sets, but still giving us confidence that these two small hypervolumes can be modeled.

5 Evaluation

We evaluate how well SPHEN performs in comparison to SAIL and MAP-Elites when we include the cost of calculating the features, how accurate the feature models are when trained with a performance based acquisition function, and whether we can apply SPHEN to an expensive domain.

5.1 Quality Diversity Comparison

We run QD optimization without (MAP-Elites) and with surrogate model(s) (SAIL, SPHEN) on the polygon domain (Sect. 4.1). This allows us to check all ground truth performance and feature values in a feasible amount of time. The shape features should be easier to learn than the flow features of the air flow domain. The working hypothesis is that we expect SPHEN to perform somewhere between SAIL and MAP-Elites, as it has the advantage of using a surrogate model but also has to learn two phenotypic features. But in the end,

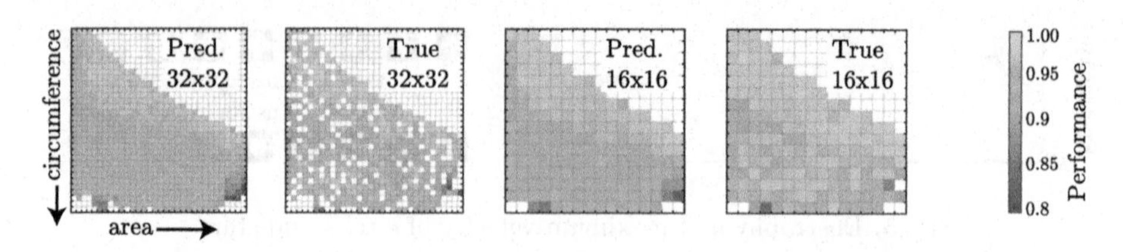

Fig. 6. Predicted and true SPHEN maps on symmetry domain, trained in 32×32 resolution (left), then reduced to 16×16 resolution to remove holes (right).

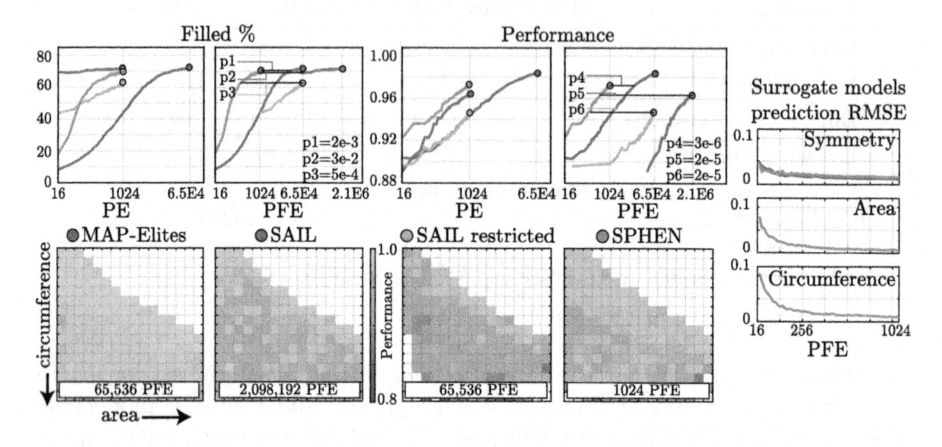

Fig. 7. Comparison of MAP-Elites, SAIL and SPHEN based on performance evaluations (PE) and performance/feature evaluations (PFE). Experiments were repeated five times to produce the mean percentage of map filled and mean performance values. Prediction errors are included on the right and example prediction maps at the bottom. The experiments include a version of SAIL that is restricted to the number of PFE used in MAP-Elites.

since the ultimate goal is to be able to use QD on expensive features, SPHEN will be our only choice. The parameterization of all algorithms is listed in Table 1. The initial sample set of 16 examples as well as the selection of new samples (16 in every iteration) is handled by a pseudo-random Sobol sequence. The mutation operator adds a value drawn from a Gaussian distribution with $\sigma = 10\%$.

Due to the expected inaccuracy of the feature models, misclassifications will decrease the accuracy of the maps. Figure 6 shows a prediction map at a resolution of 32×32 and the true performance and feature map. Holes appear due to misclassifications, which is why we train SPHEN on a higher resolution map and then reduce the prediction map to a resolution of 16×16. Most bins are now filled. In this experiment all prediction maps have a resolution of 16×16 solutions.

The mean amount of filled map bins and performance values for five replicates are shown in Fig. 7. SAIL and SPHEN find about the same number of solutions using the same number of performance evaluations (PE). Notably, the mean performance of SPHEN's solutions is higher than that of SAIL. However, in domains with expensive feature evaluations we need to take into account the performance or feature evaluations (PFE). SAIL now needs more than two million PFE to perform almost as well as SPHEN, which only needs 1,024, which is over three orders of magnitude less and still more than an order of magnitude less than MAP-Elites. Since in expensive real world optimization problems we cannot expect to run more than about 1,000 function evaluations, due to the infeasibly large computational investment, the efficiency gain of SPHEN is substantial. If we lower the number of PFE of SAIL to the same budget as MAP-Elites and give it more time to search the iteratively improving surrogate model before running out of the budget of 65,536 PFE (see Table 1), SAIL still takes a big hit, not being able to balance out quality and diversity. The example prediction maps are labeled with the number of PFE necessary to achieve those maps. Although we do not sample new training examples to improve the feature models specifically, their root mean square error (RMSE) ended up at 0.012 and 0.016 respectively. Finally, we test SPHEN to the three alternative algorithms on the null hypothesis that they need the same number of PFE to reach an equally filled map or equal performance. Significance levels, calculated using a two-sample t-test, are shown in Fig. 7. In all cases, the null hypothesis is improbable ($p < 0.05$), although for the comparison of filled levels to SAIL it is rejected with less certainty.

We conclude that we do not need to adjust the acquisition function. SPHEN and SAIL search for the same elite hypervolume, which is only determined by the performance function.

5.2 Designing Air Flow

After showing that SPHEN can learn both performance as well as feature models, we now run SPHEN in the air flow domain (Sect. 4.2). The objective is to find a diverse set of air flows using a behavioral feature, turbulence, and one shape feature, the surface area of the polygon. We want to find out how the size of the area and turbulence are related to each other and which shapes do not pass the wind nuisance threshold. We use the same parameters for SPHEN as were listed in Table 1, but allow 4,096 generations in the prediction phase. The enstrophy and velocity are normalized between 0 and 1 using a predetermined value range of $E \in [0.15, 1.1]$ and $u_{Max} \in [0.05, 0.20]$.

The resulting map of solutions in Fig. 8 shows that turbulence and surface area tend to increase the mean maximum air flow velocity, as expected. A small selection of air flows is shown in detail. RMSE of the models is 0.06, 0.01 and 0.10 respectively and Kendall's tau rank correlation to the ground truth amounts to 0.78, 1.00 and 0.73 (1.00 for A, B, C and D).

Due to the chaotic evolution of turbulent and transient flows, a static snapshot of the velocity field provides only limited information about the flow structures. Therefore, dynamic mode decomposition (DMD) is used to extract and visualize coherent structures and patterns over time from the flow field [4,20].

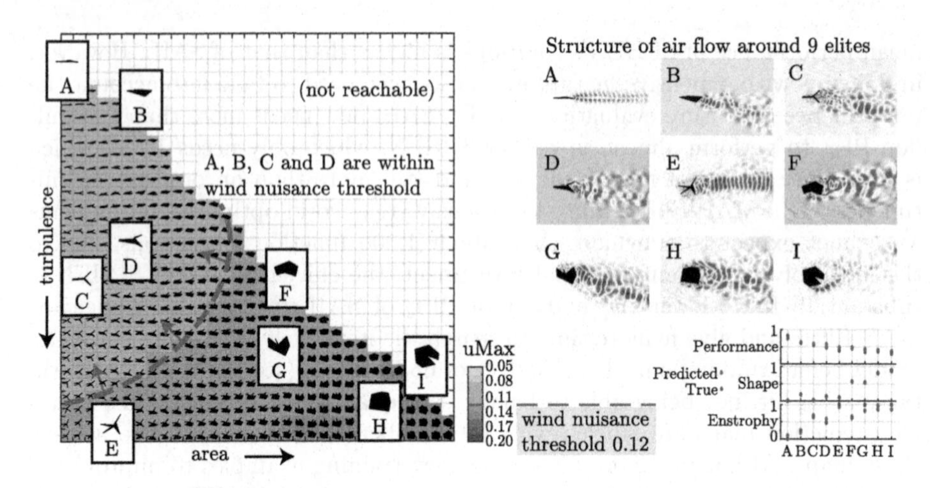

Fig. 8. A diversity of shapes and air flows that shows which designs conform to the wind nuisance threshold. The dominant DMD mode shows the structure of the air flow around nine selected shapes. A, B, C and D are within the wind nuisance threshold.

Especially those shapes at the extrema of area and turbulence align with the aerodynamic expectations as detailed in Sect. 4.2. At low turbulence intensity, the shapes tend to be slim and long with respect to the flow direction (shapes A and B). High turbulence levels at small shape areas are achieved if the shapes are oriented perpendicularly to the flow (shape E). Pentagons or hexagons evoke high turbulence levels at large areas (shapes H and I). However, impressively, there is an enormous variety of nuances in between these extrema with non-intuitive shapes, enabling the designer to determine a shape for given flow parameters down to a lower turbulence bound for each area value. Furthermore, the algorithm also suggests known tricks to manipulate the flow. Side arms are an appropriate measure to vary the turbulence intensity in the wake (shapes C, D, E, and G). Indentations or curved shapes redirect the flow and extract kinetic energy similar to turbine blades [5], which can be observed in shape D. Conclusively, for the highest and lowest area and turbulence values, SPHEN matches the expectations while for the shapes in between SPHEN exceeds expectations by introducing unusual shape nuances, which encourage further investigation.

5.3 Discussion

In the polygon domain, both surrogate-assisted algorithms are able to find a large variety of solutions. When features do not have to be modeled, they show similar performance, although SAIL converges much sooner. However, when taking into account the number of feature evaluations, SPHEN clearly outperforms SAIL as well as MAP-Elites. Modeling features does not lower the performance of a prediction map. In terms of solution performance, both surrogate-assisted algorithms are outperformed by MAP-Elites in the simple domain, but SPHEN clearly beats MAP-Elites by requiring less evaluations. The feature models become more accurate even when sampling only to improve the performance model.

When designing diverse air flows, one SPHEN run took 23 h, producing 494 different air flow profiles. With SAIL, obtaining the same result would have taken over five years. Although MAP-Elites outperformed SAIL in the simple polygon domain, and might have outperformed it in the air flow domain as well, it still would have taken two months to calculate with uncertain result. Figure 8 shows that we can find structure in the air flows that can appear in this problem domain. We can efficiently combine variations (area) of the object we want to design as well as their effect on the environment (turbulence). Even when only using two phenotypic features, the nuances between the variations give us an idea which shapes do not pass the wind nuisance threshold and which ones do, and could continue the design process based on our new intuition.

6 Conclusion

In this work we showed that expensive phenotypic features can be learned along with an expensive performance function, allowing SPHEN, an evolutionary QD algorithm, to find a large diversity of air flows. In an inexpensive domain we showed that, when we take into account the number of feature evaluations, SPHEN clearly outperforms state of the art algorithms like MAP-Elites and SAIL. The result clears the way for QD to find diverse phenotypes as well as behaviors in engineering domains without the need for an infeasible number of expensive simulations. This is made possible because only the elite hypervolume needs to be modeled. Fluid dynamics domains count as some of the most complicated. Although often solved in ingenious ways by engineers relying on experience, QD can add *automated intuition* to the design process. Variations of the object we want to optimize as well as variations in the effects on the object's environment can be seen "at a glance", which is what intuition is all about.

The most urgent future work is to study whether we can make adjustments to the acquisition function, taking into account feature models' confidence intervals to improve SPHEN. Furthermore, the solution diversity should be analyzed in higher-dimensional feature spaces and applied to 3D shapes.

We showed what expected and unexpected behavioral patterns can emerge in complicated problem domains using surrogate-assisted phenotypic niching. Our main contribution, automatic discovery of a broad intuition of the interaction between shape and behavior, allows engineers to think more out-of-the-box.

Acknowledgments. This work was funded by the Ministry for Culture and Science of the state of Northrhine-Westphalia (grant agreement no. 13FH156IN6) and the German Research Foundation (DFG) project FO 674/17-1. The authors thank Andreas Krämer for the discussions about the Lettuce solver.

References

1. Auer, P.: Using confidence bounds for exploitation-exploration trade-offs. J. Mach. Learn. Res. **3**(Nov), 397–422 (2002)
2. Bhatnagar, P.L., Gross, E.P., Krook, M.: A model for collision processes in gases. i. Small amplitude processes in charged and neutral one-component systems. Phys. Rev. **94**(3), 511 (1954)
3. Cully, A., Clune, J., Tarapore, D., Mouret, J.B.: Robots that can adapt like animals. Nature **521**(7553), 503–507 (2015)
4. Demo, N., Tezzele, M., Rozza, G.: PyDMD: Python dynamic mode decomposition. J. Open Sour. Softw. **3**(22), 530 (2018)
5. Dorschner, B., Chikatamarla, S.S., Karlin, I.V.: Transitional flows with the entropic lattice Boltzmann method. J. Fluid Mech. **824**, 388–412 (2017)
6. Gaedtke, M., Wachter, S., Rädle, M., Nirschl, H., Krause, M.J.: Application of a lattice Boltzmann method combined with a smagorinsky turbulence model to spatially resolved heat flux inside a refrigerated vehicle. Comput. Math. Appl. **76**(10), 2315–2329 (2018)
7. Gaier, A., Asteroth, A., Mouret, J.B.: Data-Efficient Exploration, Optimization, and Modeling of Diverse Designs through Surrogate-Assisted Illumination (2017)
8. Gassner, G.J., Beck, A.D.: On the accuracy of high-order discretizations for under-resolved turbulence simulations. Theor. Comput. Fluid Dyn. **27**(3–4), 221–237 (2013)
9. Hagg, A., Asteroth, A., Bäck, T.: Prototype discovery using quality-diversity. In: Auger, A., Fonseca, C.M., Lourenço, N., Machado, P., Paquete, L., Whitley, D. (eds.) PPSN 2018. LNCS, vol. 11101, pp. 500–511. Springer, Cham (2018). https://doi.org/10.1007/978-3-319-99253-2_40
10. Janssen, W., Blocken, B., van Hooff, T.: Pedestrian wind comfort around buildings: comparison of wind comfort criteria based on whole-flow field data for a complex case study. Build. Environ. **59**, 547–562 (2013)
11. Krämer, A., Wilde, D., Bedrunka, M.: Lettuce: PyTorch-based lattice Boltzmann solver (2020)
12. Krämer, A., Wilde, D., Küllmer, K., Reith, D., Foysi, H.: Pseudoentropic derivation of the regularized lattice Boltzmann method. Phys. Rev. E **100**(2), 1–16 (2019)
13. Krüger, T., Kusumaatmaja, H., Kuzmin, A., Shardt, O., Silva, G., Viggen, E.M.: The Lattice Boltzmann Method: Principles and Practice (2017)
14. Lehman, J., Stanley, K.O.: Abandoning objectives: evolution through the search for novelty alone. Evol. Comput. **19**(2), 189–223 (2011)
15. Lehman, J., Stanley, K.O.: Evolving a diversity of virtual creatures through novelty search and local competition. In: Proceedings of the 13th Annual Conference on Genetic and Evolutionary Computation, pp. 211–218 (2011)
16. Wind comfort and wind danger in the built environment (in Dutch). Norm NEN 8100 (2006)
17. Paszke, A., et al.: PyTorch: an imperative style, high-performance deep learning library. In: Wallach, H., Larochelle, H., Beygelzimer, A., D'Alché-Buc, F., Fox, E., Garnett, R. (eds.) Advances in Neural Information Processing Systems, vol. 32, pp. 8024–8035. Curran Associates, Inc. (2019)
18. Rasmussen, C.E.: Evaluation of Gaussian processes and other methods for non-linear regression. Ph.D. thesis, University of Toronto Toronto, Canada (1997)
19. Rasmussen, C.E., Nickisch, H.: Gaussian processes for machine learning (GPML) toolbox. J. Mach. Learn. Res. **11**, 3011–3015 (2010)

20. Schmid, P.J.: Dynamic mode decomposition of numerical and experimental data. J. Fluid Mech. **656**, 5–28 (2010)
21. Sobol', I.M.: On the distribution of points in a cube and the approximate evaluation of integrals. Zhurnal Vychislitel'noi Matematiki i Matematicheskoi Fiziki **7**(4), 784–802 (1967)
22. Tennekes, H.: Turbulent flow in two and three dimensions. Bull. Am. Meteorol. Soc. **59**(1), 22–28 (1978)
23. Vassiliades, V., Mouret, J.B.: Discovering the elite hypervolume by leveraging interspecies correlation. In: Proceedings of the Genetic and Evolutionary Computation Conference, pp. 149–156 (2018)

Variance Reduction for Better Sampling in Continuous Domains

Laurent Meunier[1,2(✉)], Carola Doerr[3], Jeremy Rapin[1], and Olivier Teytaud[1]

[1] Facebook Artificial Intelligence Research (FAIR), Paris, France
laurentmeunier@fb.com
[2] PSL, Université Paris-Dauphine, Miles Team, Paris, France
[3] Sorbonne Université, CNRS, LIP6, Paris, France

Abstract. Design of experiments, random search, initialization of population-based methods, or sampling inside an epoch of an evolutionary algorithm uses a sample drawn according to some probability distribution for approximating the location of an optimum. Recent papers have shown that the optimal *search* distribution, used for the sampling, might be more peaked around the center of the distribution than the *prior* distribution modelling our uncertainty about the location of the optimum.We confirm this statement, provide explicit values for this reshaping of the search distribution depending on the population size λ and the dimension d, and validate our results experimentally.

1 Introduction

We consider the setting in which one aims to locate an optimal solution $x^* \in \mathbb{R}^d$ for a given black-box problem $f : \mathbb{R}^d \to \mathbb{R}$ through a parallel evaluation of λ solution candidates. A simple, yet effective strategy for this *one-shot optimization* setting is to choose the λ candidates from a normal distribution $\mathcal{N}(\mu, \sigma^2)$, typically centered around an *a priori* estimate μ of the optimum and using a variance σ^2 that is calibrated according to the uncertainty with respect to the optimum. Random independent sampling is – despite its simplicity – still a very commonly used and performing good technique in one-shot optimization settings. There also exist more sophisticated sampling strategies like Latin Hypercube Sampling (LHS [19]), or quasi-random constructions such as Sobol, Halton, Hammersley sequences [7,18] – see [2,6] for examples. However, no general superiority of these strategies over random sampling can be observed when the benchmark set is sufficiently diverse [4]. It is therefore not surprising that in several one-shot settings – for example, the design of experiments [1,13,19,21] or the initialization (and sometimes also further iterations) of evolution strategies – the solution candidates are frequently sampled from random independent distributions (though sometimes improved by mirrored sampling [27]). A surprising finding was recently communicated in [6], where the authors consider the setting in which the optimum x^* is known to be distributed according to

Arxiv version [20] of the present document includes bigger plots and the appendices.

T. Bäck et al. (Eds.): PPSN 2020, LNCS 12269, pp. 154–168, 2020.
https://doi.org/10.1007/978-3-030-58112-1_11

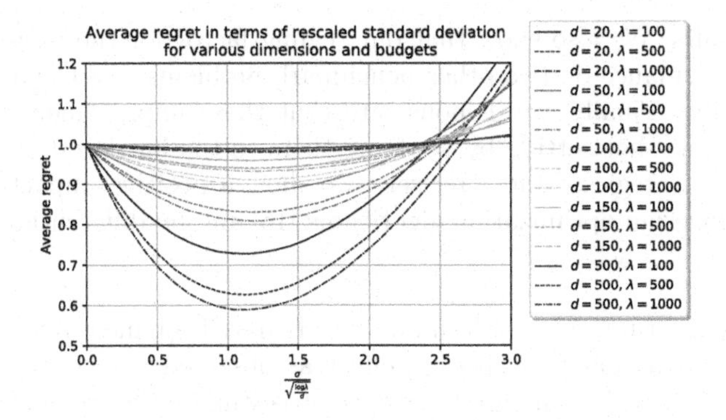

d	λ	σ^*	$\sigma = 1$
20	100	**0.73**	0.88
	500	**0.63**	0.72
	1000	**0.59**	0.66
50	100	**0.89**	1.23
	500	**0.83**	1.10
	1000	**0.81**	1.05
100	100	**0.94**	1.44
	500	**0.91**	1.33
	1000	**0.90**	1.29
150	100	**0.96**	1.53
	500	**0.94**	1.44
	1000	**0.93**	1.41
500	100	**0.99**	1.74
	500	**0.98**	1.68
	1000	**0.98**	1.66

Fig. 1. Average regret, normalized by d, on the sphere function for various dimensions and budgets in terms of rescaled standard deviation. Each mean has been estimated from 100,000 samples. Table on the right: Average regret for $\sigma^* = \sqrt{\log(\lambda)/d}$ and $\sigma = 1$.

a standard normal distribution $\mathcal{N}(0, I_d)$, and the goal is to minimize the distance of the best of the λ samples to this optimum. In the context of evolution strategies, one would formulate this problem as minimizing the sphere function with a normally distributed optimum. Intuitively, one might guess that sampling the λ candidates from the same prior distribution, $\mathcal{N}(0, I_d)$, should be optimal. This intuition, however, was disproved in [6], where it is shown that – unless the sample size λ grows exponentially fast in the dimension d – the median quality of sampling from $\mathcal{N}(0, I_d)$ is worse than that of sampling a single point, namely the center point 0. A similar observation was previously made in [22], without mathematically proven guarantees.

Our Theoretical Result. It was left open in [6] how to optimally scale the variance σ^2 when sampling the λ solution candidates from a normal distribution $\mathcal{N}(0, \sigma^2 I_d)$. While the result from [6] suggests to use $\sigma = 0$, we show in this work that a more effective strategy exists. More precisely, we show that setting $\sigma^2 = \min\{1, \Theta(\log(\lambda)/d)\}$ is asymptotically optimal, as long as λ is subexponential, but growing in d. Our variance scaling factor reduces the median approximation error by a $1-\varepsilon$ factor, with $\varepsilon = \Theta(\log(\lambda)/d)$. We also prove that no constant variance nor any other variance scaling as $\omega(\log(\lambda)/d)$ can achieve such an approximation error. Note that several optimization algorithms operate with rescaled sampling. Our theoretical results therefore set the mathematical foundation for empirical rules of thumb such as, for example, used in e.g. [6,8–10,17,22,28].

Our Empirical Results. We complement our theoretical analyses by an empirical investigation of the rescaled sampling strategy. Experiments on the sphere function confirm the results. We also show that our scaling factor for the variance yields excellent performance on two other benchmark problems, the Cigar and the Rastrigin function. Finally, we demonstrate that these improvements are not restricted to the one-shot setting by applying them to the initialization of iterative optimization strategies. More precisely, we show a positive impact on the initialization of Bayesian optimization algorithms [15] and on differential evolution [25].

Related Work. While the most relevant works for our study have been mentioned above, we briefly note that a similar surprising effect as observed here is the "Stein phenomenon" [14,24]. Although an intuitive way to estimate the mean of a standard gaussian distribution is to compute the empirical mean, Stein showed that this strategy is sub-optimal w.r.t. mean squared error and that the empirical mean needs to be rescaled by some factor to be optimal.

2 Problem Statement and Related Work

The context of our theoretical analysis is *one-shot optimization.* In one-shot optimization, we are allowed to select λ points $x_1, \ldots, x_\lambda \in \mathbb{R}^d$. The quality $f(x_i)$ of these points is evaluated, and we measure the performance of our samples in terms of simple regret [5] $\min_{i=1,\ldots,\lambda} f(x_i) - \inf_{x\in\mathbb{R}^d} f(x)$.[1] That is, we aim to minimize the distance – measured in *quality space* – of the best of our points to the optimum. This formulation, however, also covers the case in which we aim to minimize the distance to the optimum in the *search space*: we simply take as f the root of the sphere function $f_{x^*} : \mathbb{R}^d \to \mathbb{R}, x \mapsto \|x - x^*\|^2$, where here and in the following $\|.\|$ denotes the Euclidean norm.

Rescaled Random Sampling for Randomly Placed Optimum. In the setting studied in Sect. 3 we assume that the optimum x^* is sampled from the standard multivariate Gaussian distribution $\mathcal{N}(0, I_d)$, and that we aim to minimize the regret $\min_{i=1,\ldots,\lambda}\|x_i - x^*\|^2$ through i.i.d. samples $x_i \sim \mathcal{N}(0, \sigma^2 I_d)$. That is, in contrast to the classical *design of experiments* (DoE) setting, we are only allowed to choose the scaling factor σ, whereas in DoE more sophisticated (often quasi-random and space-filling designs – which are typically not i.i.d. samples) are admissible. Intuitively, one might be tempted to guess that $\sigma = 1$ should be a good choice, as in this case the λ points are chosen from the same distribution as the optimum x^*. This intuition, however, was refuted in [6, Theorem 1], where is was shown that the middle point sampling strategy, which uses $\sigma = 0$ (i.e., all λ points collapse to $(0, \ldots, 0)$) yields smaller regret than sampling from $\mathcal{N}(0, I_d)$ unless λ grows exponentially in d. More precisely, it is shown in [6] that, for this regime of λ and d, the median of $\|x^*\|^2$ is smaller than the median of $\|x_i - x^*\|^2$

[1] This requires knowledge of $\inf_x f(x)$, which may not be available in real-world applications. In this case, without loss of generality (this is just for the sake of plotting regret values), the infimum can be replaced by an empirical minimum. In all applications considered in this work the value of $\inf_x f(x)$ is known.

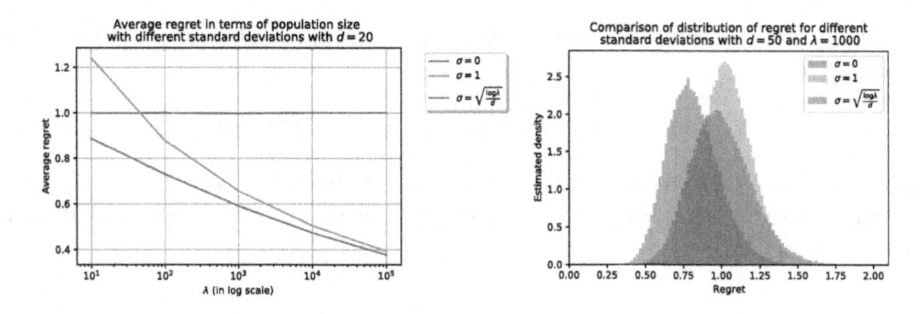

Fig. 2. Comparison of methods: without rescaling ($\sigma = 1$), middle point sampling ($\sigma = 0$), and our rescaling method ($\sigma = \sqrt{\frac{\log \lambda}{d}}$). Each mean has been estimated from 10^5 samples. (On left) Average regret, normalized by d, on the sphere function for diverse population sizes λ at fixed dimension $d = 20$. The gain of rescaling decreases as λ increases. (On right) Distribution of the regret for the strategies on the $50d$-sphere function for $\lambda = 1000$.

for i.i.d. $x_i \in \mathcal{N}(0, I_d)$. This shows that sampling a single point can be better than sampling λ points with the wrong scaling factor, unless the budget λ is very large.

Our goal is to improve upon the middle point strategy, by deriving a scaling factor σ such that the λ i.i.d. samples yield smaller regret with a decent probability. More precisely, we aim at identifying σ such that

$$\mathbb{P}\left[\min_{1\leq i\leq\lambda} \|x_i - x^*\|^2 \leq (1-\varepsilon)\|x^*\|^2\right] \geq \delta, \qquad (1)$$

for some $\delta \geq 1/2$ and $\varepsilon > 0$ as large as possible. Here, in line with [6], we have switched to regret, for convenience of notation. [6] proposed, without proof, such a scaling factor: our proposal is dramatically better in some regimes.

3 Theoretical Results

We derive sufficient and necessary conditions on the scaling factor σ such that Eq. (1) can be satisfied. More precisely, we prove that Eq. (1) holds with approximation gain $\varepsilon \approx \log(\lambda)/d$ when the variance σ^2 is chosen proportionally to $\log \lambda/d$ (and λ does not grow too rapidly in d). We then show that Eq. (1) cannot be satisfied for $\sigma^2 = \omega(\log(\lambda)/d)$. Moreover, we prove that $\varepsilon = O(\log(\lambda)/d)$, which, together with the first result, shows that our scaling factor is asymptotically optimal. The precise statements are summarized in Theorems 1, 2, and 3, respectively. Proof sketches are available in Sect. 3. Proofs are left in the full version available on the ArXiv version [20].

Theorem 1 *(Sufficient condition on rescaling).* *Let $\delta \in [\frac{1}{2}, 1)$. Let $\lambda = \lambda_d$, satisfying*

$$\lambda_d \to \infty \text{ as } d \to \infty \text{ and } \log(\lambda_d) \in o(d). \qquad (2)$$

Then there exist two positive constants c_1, c_2, *and* d_0, *such that for all* $d \geq d_0$ *it holds that*

$$\mathbb{P}\left[\min_{i=1,\ldots,\lambda} \|x^* - x_i\|^2 \leq (1-\varepsilon)\, \|x^*\|^2\right] \geq \delta \tag{3}$$

when x^* *is sampled from the standard Gaussian distribution* $\mathcal{N}(0, I_d)$, $x_1, \ldots, x_\lambda$ *are independently sampled from* $\mathcal{N}(0, \sigma^2 I_d)$ *with* $\sigma^2 = \sigma_d^2 = c_2 \log(\lambda)/d$ *and* $\varepsilon = \varepsilon_d = c_1 \log(\lambda)/d$.

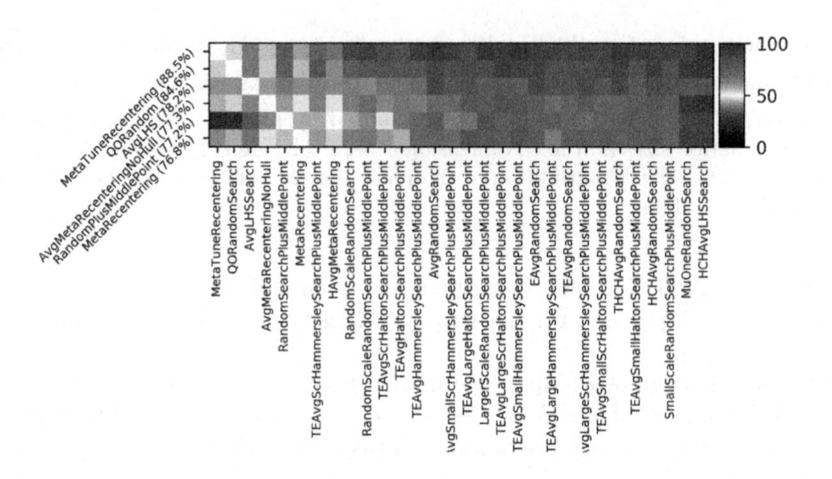

Fig. 3. Comparison of various one-shot optimization methods from the point of view of the simple regret. Reading guide in Sect. 4.2. Results are averaged over objective functions Cigar, Rastrigin, Sphere in dimension 20, 200, 2000, and budget 30, 100, 3000, 10000, 30000, 100000. `MetaTuneRecentering` performs best overall. Only the 30 best performing methods are displayed as columns, and the 6 best as rows. Red means superior performance of row vs col. Rows and cols ranked by performance. (Color figure online)

Theorem 1 shows that i.i.d. Gaussian sampling can outperform the middle point strategy derived in [6] (i.e., the strategy using $\sigma^2 = 0$) if the scaling factor σ is chosen appropriately. Our next theorem summarizes our findings for the conditions that are *necessary* for the scaling factor σ^2 to outperform this middle point strategy. This result, in particular, illustrates why neither the natural choice $\sigma = 1$, nor any other constant scaling factor can be optimal.

Theorem 2 *(Necessary condition on rescaling).* *Consider* $\lambda = \lambda_d$ *satisfying assumptions* (2). *There exists an absolute constant* $C > 0$ *such that for all* $\delta \in [\frac{1}{2}, 1)$, *there exists* $d_0 > 0$ *such that, for all* $d > d_0$ *and for all* σ *the property*

$$\exists \varepsilon > 0, \mathbb{P}\left[\min_{i=1,\ldots,\lambda} \|x^* - x_i\|^2 \leq (1-\varepsilon)\, \|x^*\|^2\right] \geq \delta \tag{4}$$

for $x^* \sim \mathcal{N}(0, I_d)$ *and* $x_1, \ldots, x_\lambda$ *independently sampled from* $\mathcal{N}(0, \sigma^2 I_d)$, *implies that* $\sigma^2 \leq C \log(\lambda)/d$.

While Theorem 2 induces a necessary condition on the scaling factor σ to improve over the middle point strategy, it does not bound the gain that one can achieve through a proper scaling. Our next theorem shows that the factor derived in Theorem 1 is asymptotically optimal.

Theorem 3 *(Upper bound for the approximation factor). Consider $\lambda = \lambda_d$ satisfying assumptions* (2)*. There exists an absolute constant $C' > 0$ such that for all $\delta \in [\frac{1}{2}, 1)$, there exists $d_0 > 0$ such that, for all $d > d_0$ and for all $\varepsilon, \sigma > 0$, it holds that if $\mathbb{P}\left[\min_{i=1,\ldots,\lambda} \|x^* - x_i\|^2 \leq (1-\varepsilon)\|x^*\|^2\right] \geq \delta$ for $x^* \sim \mathcal{N}(0, I_d)$ and $x_1, \ldots, x_\lambda$ independently sampled from $\mathcal{N}(0, \sigma^2 I_d)$, then $\varepsilon \leq C' \log(\lambda)/d$.*

Proof Sketches. We first notice that as x^* is sampled from a standard normal distribution $\mathcal{N}(0, I_d)$, its norm satisfies $\|x^*\|^2 = d + o(d)$ as $d \to \infty$. We then use that, conditionally to x^*, it holds that

$$\mathbb{P}\left[\min_{i\in[\lambda]} \|x^* - x_i\|^2 \leq (1-\varepsilon)\|x^*\|^2 \middle| x^*\right] = 1 - \left(1 - \mathbb{P}\left[\|x - x^*\|^2 \leq (1-\varepsilon)\|x^*\|^2 \middle| x^*\right]\right)^\lambda$$

We therefore investigate when the condition

$$\mathbb{P}\left[\|x - x^*\|^2 \leq (1-\varepsilon)\|x^*\|^2 \middle| x^*\right] > 1 - (1-\delta)^{\frac{1}{\lambda}} \tag{5}$$

is satisfied. To this end, we make use of the fact that the squared distance $\|x^*\|^2$ of x^* to the middle point 0 follows the central $\chi^2(d)$ distribution, whereas, for a given point $x^* \in \mathbb{R}^d$, the distribution of the squared distance $\|x - x^*\|^2/\sigma^2$ for $x \sim \mathcal{N}(0, \sigma^2 I_d)$ follows the non-central $\chi^2(d, \mu)$ distribution with non-centrality parameter $\mu := \|x^*\|^2/\sigma^2$. Using the concentration inequalities provided in [29, Theorem 7] for non-central χ^2 distributions, we then derive sufficient and necessary conditions for condition (5) to hold. With this, and using assumptions (2), we are able to derive the results from Theorems 1, 2, and 3.

4 Experimental Performance Comparisons

The theoretical results presented above are in asymptotic terms, and do not specify the constants. We therefore complement our mathematical investigation with an empirical analysis of the rescaling factor. Whereas results for the setting studied in Sect. 3 are presented in Sect. 4.1, we show in Sect. 4.2 that the advantage of our rescaling factor is not limited to minimizing the distance in search space. More precisely, we show that the rescaled sampling achieves good results also in a classical DoE task, in which we aim for minimizing the regret for the Cigar and for the Rastrigin functions. Finally, we investigate in Sect. 4.3 the impact of initializing two common optimization heuristics, Bayesian Optimization (BO) and differential evolution (DE), by a population sampled from the Gaussian distribution $\mathcal{N}(0, \sigma^2 I_d)$ using our rescaling factor $\sigma = \sqrt{\log(\lambda)/d}$.

4.1 Validation of Our Theoretical Results on the Sphere Function

Figure 1 displays the normalized average regret $\frac{1}{d}\mathbb{E}\left[\min_{i=1,\ldots,\lambda}\|x^* - x_i\|^2\right]$ in terms of $\sigma/\sqrt{\log(\lambda)/d}$ for different dimensions and budgets. We observe that the best parametrization of σ is around $\sqrt{\log(\lambda)/d}$ in all displayed cases. Moreover, we also see that – as expected – the gain of the rescaled sampling over the middle point sampling ($\sigma = 0$) goes to 0 as $d \to \infty$ (i.e. we get a result closer to the case $\sigma = 0$ as dimension goes to infinity). We also see that, for the regimes plotted in Fig. 1, the advantage of the rescaled variance grows with the budget λ. Figure 2 (on left) displays the average regret (average over multiple samplings and multiple positions of the optimum) as a function of increasing values of λ for the different rescaling methods ($\sigma \in \{0, \sqrt{\log \lambda/d}, 1\}$). We remark, unsurprisingly, that the gain of rescaling is diminishing as $\lambda \to \infty$. Finally, Fig. 2 (on right) shows the distribution of regrets for the different rescaling methods. The improvement of the expected regret is not at the expense of a higher dispersion of the regret.

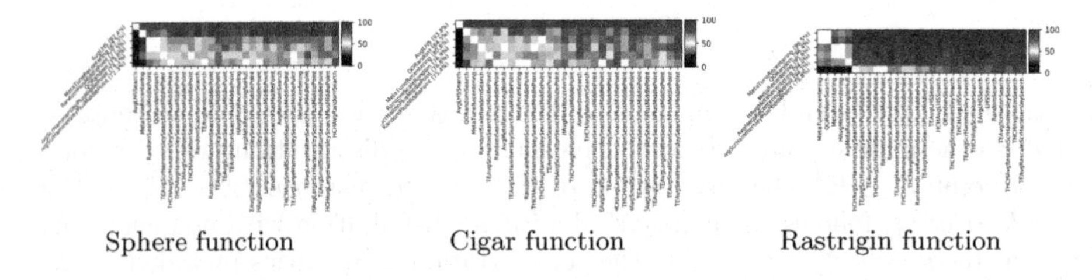

Fig. 4. Same experiment as Fig. 3, but separately over each objective function. Results are still averaged over 6 distinct budgets (30, 100, 3000, 10000, 30000, 100000) and 3 distinct dimensionalities (20, 200, 2000). `MetaTuneRecentering` performs well in each case, and is not limited to the sphere function for which it was derived. Variants of LHS are sometimes excellent and sometimes not visible at all (only the 30 best performing methods are shown).

4.2 Comparison with the DoEs Available in Nevergrad

Motivated by the significant improvements presented above, we now investigate whether the advantage of our rescaling factor translates to other optimization tasks. To this end, we first analyze a DoE setting, in which an underlying (and typically not explicitly given) function f is to be minimized through a parallel evaluation of λ solution candidates $x_1, \ldots, x_\lambda$, and regret is measured in terms of $\min_i f(x_i) - \inf_x f(x)$. In the broader machine learning literature, and in particular in the context of hyper-parameter optimization, this setting is often referred to as *one-shot optimization* [2,6].

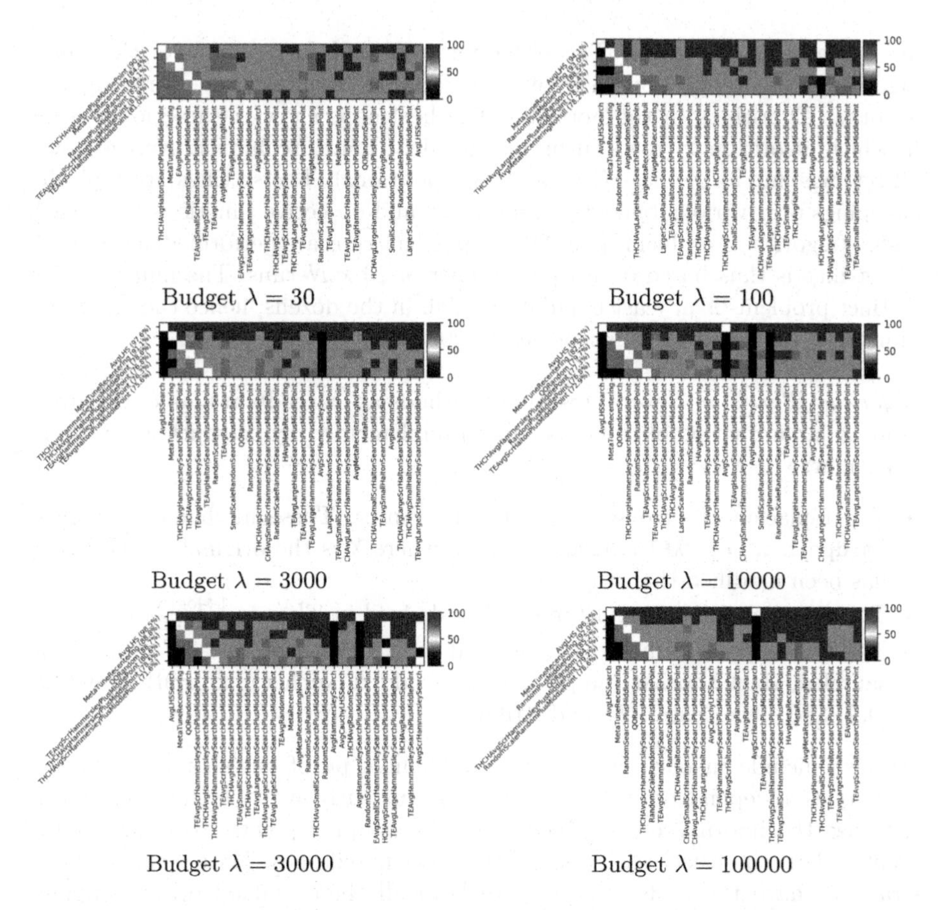

Fig. 5. Methods ranked by performance on the sphere function, per budget. Results averaged over dimension 20, 200, 2000. `MetaTuneRecentering` performs among the best in all cases. LHS is excellent on this very simple setting, namely the sphere function.

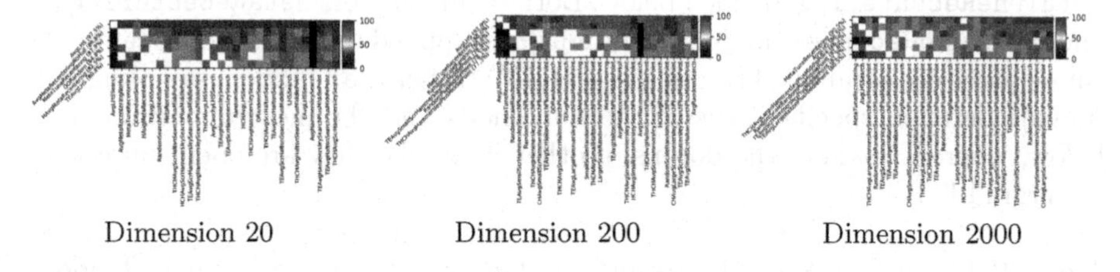

Fig. 6. Results on the sphere function, per dimensionality. Results are averaged over 6 values of the budget: 30, 100, 3000, 10000, 30000, 100000. Our method becomes better and better as the dimension increases.

Experimental Setup. All our experiments are implemented and freely available in the Nevergrad platform [23]. Results are presented as shown in Fig. 3. Typically, the six best methods are displayed as rows. The 30 best performing methods are presented as columns. The order for rows and for columns is the same: algorithms are ranked by their average winning frequency, measured against all other algorithms in the portfolio. The heatmaps show the fraction of runs in which algorithm x (row) outperformed algorithm y (column), averaged over all settings and all replicas (i.e. random repetitions). The settings are typically sweepings over various budgets, dimensions, and objective functions.[2] For each tested (algorithm, problem) pair, 20 independent runs are performed: a case with N settings is thus based on a total number of $20 \times N$ runs. The number N of distinct problems is at least 6 and often high in the dozens, hence the minimum number of independent runs is at least 120.

Algorithm Portfolio. Several rescaling methods are already available on Nevergrad. A large fraction of these have been implemented by the authors of [6]; in particular:

- The replacement of one sample by the center. These methods are named "`midpointX`" or "`XPlusMiddlePoint`", where `X` is the original method that has been modified that way.
- The rescaling factor `MetaRecentering` derived in [6]: $\sigma = \frac{1+\log(\lambda)}{4\log(d)}$.
- The quasi-opposite methods suggested in [22], with prefix "`QO`": when x is sampled, then another sample $c - rx$ is added, with r uniformly drawn in $[0,1]$ and c the center of the distribution.

We also include in our comparison a different type of one-shot optimization techniques, independent of the present work, currently available in the platform: they use the information obtained from the sampled points to recommend a point x that is not necessarily one of the λ evaluated ones. These *"one-shot+1"* strategies have the prefix "`Avg`". We keep all these and all other sampling strategies available in Nevergrad for our experiments. We add to this existing Nevergrad portfolio our own rescaling strategy, which uses the scaling factor derived in Sect. 3; i.e., $\sigma = \sqrt{\log(\lambda)/d}$. We refer to this sampling strategy as `MetaTuneRecentering`, defined below. Both scaling factors `MetaRecentering` [6] and `MetaTuneRecentering` (our equations) are applied to quasirandom sampling (more precisely, scrambled Hammersley [1,13]) rather than random sampling. We provide detailed specifications of these methods and the most important ones below, whereas we skip the dozens of other methods: they are open sourced in Nevergrad [23].

From $[0,1]^d$ to Gaussian Quasi-random, Random or LHS Sampling: Random sampling, quasi-random sampling, Latin Hypercube Sampling (or others) have a well known definition in $[0,1]^d$ (for quasi-random, see Halton [12] or Hammersley [13], possibly boosted by scrambling [1]; for LHS, see [19]). To extend to multidimensional Gaussian sampling, we use that if U is a uniform random variable on $[0,1]$ and Φ the standard Gaussian CDF, then $\Phi^{-1}(U)$ simulates

[2] Detailed results for individual settings are available at http://dl.fbaipublicfiles.com/nevergrad/allxps/list.html.

a $\mathcal{N}(0,1)$ distribution. We do so on each dimension: this provides a Gaussian quasi-random, random or LHS sampling.

Then, one can rescale the Gaussian quasi-random sampling with the corresponding factor σ for `MetaRecentering` ($\sigma = \frac{1+\log(\lambda)}{4\log(d)}$ [6]) and `MetaTuneRecentering` ($\sigma = \sqrt{\log(\lambda)/d}$): for $i \leq \lambda$ and $j \leq d$, $x_{i,j} = \sigma\phi^{-1}(h_{i,j})$ where $h_{i,j}$ is the jth coordinate of a ith Scrambled-Hammersley point.

Results for the Full DoE Testbed in Nevergrad. Figure 3 displays aggregated results for the Sphere, the Cigar, and the Rastrigin functions, for three different dimensions and six different budgets. We observe that our `MetaTuneRecentering` strategy performs best, with a winning frequency of 80%. It positively compares against all other strategies from the portfolio, with the notable exception of `AvgLHS`, which, in fact, compares favorably against every single other strategy, but with a lower average winning frequency of 73.6%. Note here that `AvgLHS` is one of the **"oneshot+1"** strategies, i.e., it has not only one more sample, but it is also allowed to sample its recommendation adaptively, in contrast to our fully parallel `MetaTuneRecentering` strategy. It performs poorly in some cases (Rastrigin) and does not make sense as an initialization (Sect. 4.3).

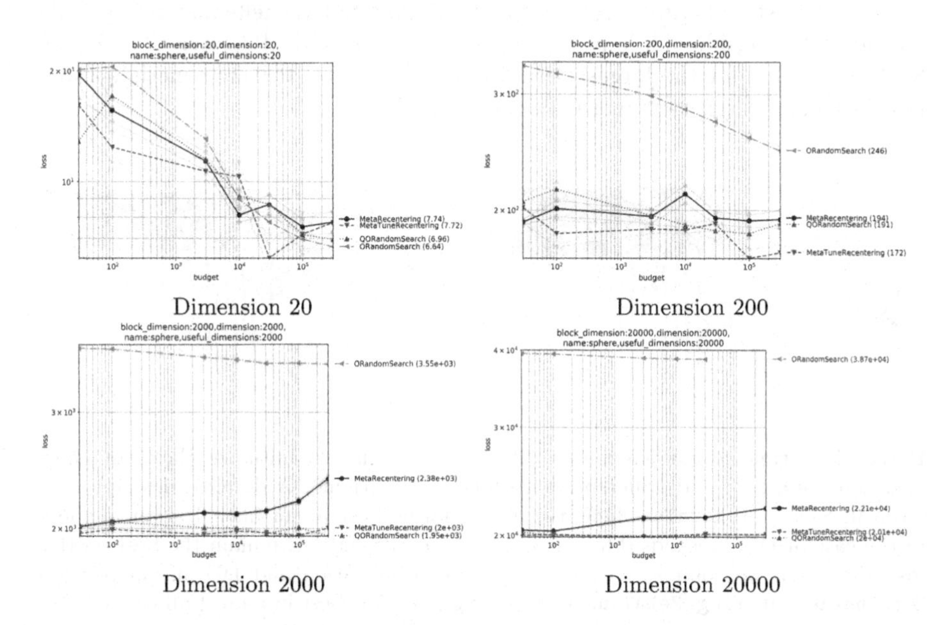

Fig. 7. Same context as Fig. 6, with x-axis = budget and y-axis = average simple regret. We see the failure of `MetaRecentering` in the worsening performance as budget goes to infinity: the budget has an impact on σ which becomes worse, hence worse overall performance. We note that quasi-opposite sampling can perform decently in a wide range of values. Opposite Sampling is not much better than random search in high-dimension. Our `MetaTuneRecentering` shows decent performance: in particular, simple regret decreases as $\lambda \to \infty$.

Selected DoE Tasks. Figure 4 breaks down the aggregated results from Fig. 3 to the three different functions. We see that `MetaTuneRecentering` scores second on sphere (where `AvgLHS` is winning), third on Cigar (after `AvgLHS` and `QORandom`), and first on Rastrigin. This fine performance is remarkable, given that the portfolio contains quite sophisticated and highly tuned methods. In addition, the `AvgLHS` methods, sometimes performing better on the sphere, besides using more capabilities than we do (as it is a "oneshot+1" method), had poor results for Rastrigin (not even in the 30 best methods). On sphere, the difference to the third and following strategies is significant (87.3% winning rate against 77.5% for the next runner-up). On Cigar, the differences between the first four strategies are greater than 4% points each, whereas on Rastrigin the average winning frequencies of the first five strategies is comparable, but significantly larger than that of the sixth one (which scores 78.8% against >94.2% for the first five DoEs). Figure 5 zooms into the results for the sphere function, and breaks them further down by available budget λ (note that the results are still averaged over the three tested dimensions). `MetaTuneRecentering` scores second in all six cases. A breakdown of the results for sphere by dimension (and aggregated over the six available budgets) is provided in Fig. 6 and Fig. 7. For dimension 20, we see that `MetaTuneRecentering` ranks third, but, interestingly, the two first methods are "oneshot+1" style (`Avg` prefix). In dimension 200, `MetaTuneRecentering` ranks second, with considerable advantage over the third-ranked strategy (88.0% vs. 80.8%). Finally, for the largest tested dimension, $d = 2000$, our method ranks first, with an average winning frequency of 90.5%.

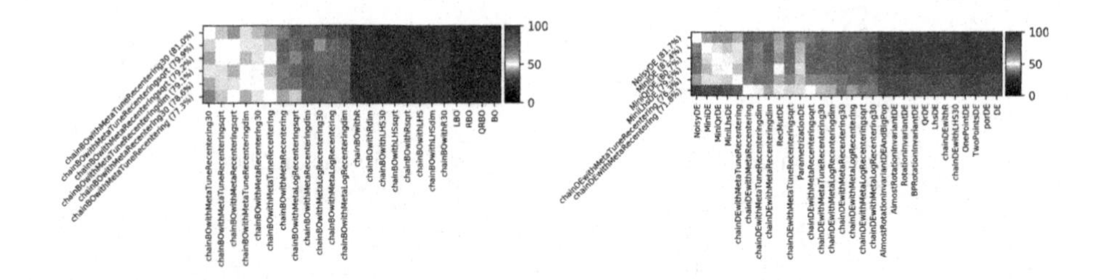

Fig. 8. Performance comparison of different strategies to initialize Bayesian Optimization (BO, left) and Differential Evolution (DE, right). A detailed description is given in Sect. 4.3. `MetaTuneRecentering` performs best as an initialization method. In the case of DE, methods different from the traditional DE remain the best on this testcase: when we compare DE with a given initialization and DE initialized with `MetaTuneRecentering`, `MetaTuneRecentering` performs best in almost all cases.

4.3 Application to Iterative Optimization Heuristics

We now move from the one-shot settings considered thus far to *iterative optimization*, and show that our scaling factor can also be beneficial in this context.

More precisely, we analyze the impact of initializing efficient global optimization (EGO [15], a special case of Bayesian optimization) and differential evolution (DE [25]) by a population that is sampled from a distribution that uses our variance scaling scheme. It is well known that a proper initialization can be very critical for the performance of these solvers; see [3,11,16,22,26] for discussions. Figure 8 summarizes the results of our experiments. As in the previous setups, we compare against existing methods from the Nevergrad platform, to which we have just added our rescaling factor termed `MetaTuneRecentering`. For each initialization scheme, four different initial population sizes are considered: denoting by d the dimension, by w the parallelism (i.e., the number of workers), and by b the total budget that the algorithms can spend on optimizing the given optimization task, the initial population λ is set as $\lambda = \sqrt{b}$ for `Sqrt`, as $\lambda = d$ for `Dim`, $\lambda = w$ for no suffix, and as $\lambda = 30$ when the suffix is `30`. As in Sect. 4.2 we superpose our scaling scheme on top of the quasi-random Scrambled Hammersley sequence suggested in [6], but we also consider random initialization rather than quasi-random (indicated by the suffix "`R`") and Latin Hypercube Sampling [19] (suffix "`LHS`"). The left chart in Fig. 8 is for the Bayesian optimization case. It aggregates results for 48 settings, which stem from Nevergrad's "parahdbo4d" suite. It comprises the four benchmark problems Sphere, Cigar, Ellipsoid and Hm. Results are averaged over the total budgets $b \in \{25, 31, 37, 43, 50, 60\}$, dimension $d \in \{20, 2000\}$, and parallelism $w = \max(d, \lfloor b/6 \rfloor)$. We observe that a BO version using our `MetaTuneRecentering` performs best, and that several other variants using this scaling appear among the top-performing configurations. The chart on the right of Fig. 8 summarizes results for Differential Evolution. Since DE can handle larger budgets, we consider here a total number of 100 settings, which correspond to the testcase named "paraalldes" in Nevergrad. In this suite, results are averaged over budgets $b \in \{10, 100, 1000, 10000, 100000\}$, dimensions $d \in \{5, 20, 100, 500, 2500\}$, parallelism $w = \max(d, \lfloor b/6 \rfloor)$, and again the objective functions Sphere, Cigar, Ellipsoid, and Hm. Specialized versions of DE perform best for this testcase, but we see that DE initialized with our `MetaTuneRecentering` strategy ranks fifth (outperformed only by ad hoc variants of DE), with an overall winning frequency that is not much smaller than that of the top-ranked `NoisyDE` strategy (76.3% for `ChainDEwithMetaTuneRecentering` vs. 81.7% for `NoisyDE`) - and almost always outperforms the rescaling used in the original Nevergrad.

5 Conclusions and Future Work

We have investigated the scaling of the variance of random sampling in order to minimize the expected regret. While previous work [6] had already shown that, in the context of the sphere function, the optimal scaling factor is not identical to that of the prior distribution from which the optimum is sampled (unless the sample size is exponentially large in the dimension), it did not answer the question how to scale the variance optimally. We have proven that a standard deviation scaled as $\sigma = \sqrt{\log(\lambda)/d}$ gives, with probability at least 1/2, a sample that is significantly closer to the optimum than the previous known strategies. We have also proven that the gain achieved by our scaling strategy is asymptotically optimal and that any decent scaling factor is asymptotically at most as large as our suggestion.

The empirical assessment of our rescaled sampling strategy confirmed decent performance not only on the sphere function, but also on other classical benchmark problems. We have furthermore given indications that the sampling might help improve state-of-the-art numerical heuristics based on differential evolution or using Bayesian surrogate models. Our proposed one-shot method performs best in many cases, sometimes outperformed by e.g. `AvgLHS`, but is stable on a wide range of problems and meaningful also as an initialization method (as opposed to `AvgLHS`). Whereas our theoretical results can be extended to quadratic forms (by conservation of barycenters through linear transformations), an extension to wider families of functions (e.g., families of functions with order 2 Taylor expansion) is not straightforward. Apart from extending our results to broader function classes, another direction for future work comprises extensions to the multi-epoch case. Our empirical results on DE and BO gives a first indication that a properly scaled variance can also be beneficial in iterative sampling. Note, however, that in the latter case, we only adjusted the initialization, not the later sampling steps. This forms another promising direction for future work.

Acknowledgements. This work was initiated at Dagstuhl seminar 19431 on Theory of Randomized Optimization Heuristics.

References

1. Atanassov, E.I.: On the discrepancy of the Halton sequences. Math. Balkanica (NS) **18**(1–2), 15–32 (2004)
2. Bergstra, J., Bengio, Y.: Random search for hyper-parameter optimization. J. Mach. Learn. Res. **13**, 281–305 (2012)
3. Bossek, J., Doerr, C., Kerschke, P.: Initial design strategies and their effects on sequential model-based optimization. In: Proceeding of the Genetic and Evolutionary Computation Conference (GECCO 2020). ACM (2020). https://arxiv.org/abs/2003.13826
4. Bossek, J., Kerschke, P., Neumann, A., Neumann, F., Doerr, C.: One-shot decision-making with and without surrogates. CoRR abs/1912.08956 (2019). http://arxiv.org/abs/1912.08956
5. Bubeck, S., Munos, R., Stoltz, G.: Pure exploration in multi-armed bandits problems. In: Gavaldà, R., Lugosi, G., Zeugmann, T., Zilles, S. (eds.) ALT 2009. LNCS (LNAI), vol. 5809, pp. 23–37. Springer, Heidelberg (2009). https://doi.org/10.1007/978-3-642-04414-4_7
6. Cauwet, M.L., et al.: Fully parallel hyperparameter search: reshaped space-filling. arXiv preprint arXiv:1910.08406 (2019)
7. Dick, J., Pillichshammer, F.: Digital Nets and Sequences. Cambridge University Press, Cambridge (2010)
8. Ergezer, M., Sikder, I.: Survey of oppositional algorithms. In: 14th International Conference on Computer and Information Technology (ICCIT 2011), pp. 623–628 (2011)

9. Esmailzadeh, A., Rahnamayan, S.: Enhanced differential evolution using center-based sampling. In: 2011 IEEE Congress of Evolutionary Computation (CEC), pp. 2641–2648 (2011)
10. Esmailzadeh, A., Rahnamayan, S.: Center-point-based simulated annealing. In: 2012 25th IEEE Canadian Conference on Electrical and Computer Engineering (CCECE), pp. 1–4 (2012)
11. Feurer, M., Springenberg, J.T., Hutter, F.: Initializing Bayesian hyperparameter optimization via meta-learning. In: AAAI (2015)
12. Halton, J.: On the efficiency of certain quasi-random sequences of points in evaluating multi-dimensional integrals. Numer. Math. **2**, 84–90 (1960). http://eudml.org/doc/131448
13. Hammersley, J.M.: Monte-Carlo methods for solving multivariate problems. Ann. N. Y. Acad. Sci. **86**(3), 844–874 (1960)
14. James, W., Stein, C.: Estimation with quadratic loss. In: Proceedings of the Fourth Berkeley Symposium on Mathematical Statistics and Probability, Contributions to the Theory of Statistics, vol. 1, pp. 361–379. University of California Press (1961). https://projecteuclid.org/euclid.bsmsp/1200512173
15. Jones, D.R., Schonlau, M., Welch, W.J.: Efficient global optimization of expensive black-box functions. J. Glob. Optim. **13**(4), 455–492 (1998)
16. Maaranen, H., Miettinen, K., Mäkelä, M.: Quasi-random initial population for genetic algorithms. Comput. Math. Appl. **47**(12), 1885–1895 (2004)
17. Mahdavi, S., Rahnamayan, S., Deb, K.: Center-based initialization of cooperative co-evolutionary algorithm for large-scale optimization. In: 2016 IEEE Congress on Evolutionary Computation (CEC), pp. 3557–3565 (2016)
18. Matoušek, J.: Geometric Discrepancy, 2nd edn. Springer, Berlin (2010)
19. McKay, M.D., Beckman, R.J., Conover, W.J.: A comparison of three methods for selecting values of input variables in the analysis of output from a computer code. Technometrics **21**(2), 239–245 (1979)
20. Meunier, L., Doerr, C., Rapin, J., Teytaud, O.: Variance reduction for better sampling in continuous domains (2020)
21. Niederreiter, H.: Random Number Generation and Quasi-Monte Carlo Methods. Society for Industrial and Applied Mathematics, Philadelphia (1992)
22. Rahnamayan, S., Wang, G.G.: Center-based sampling for population-based algorithms. In: 2009 IEEE Congress on Evolutionary Computation, pp. 933–938, May 2009. https://doi.org/10.1109/CEC.2009.4983045
23. Rapin, J., Teytaud, O.: Nevergrad - a gradient-free optimization platform (2018). https://GitHub.com/FacebookResearch/Nevergrad
24. Stein, C.: Inadmissibility of the usual estimator for the mean of a multivariate normal distribution. In: Proceeding of the Third Berkeley Symposium on Mathematical Statistics and Probability, Contributions to the Theory of Statistics, vol. 1, pp. 197–206. University of California Press (1956). https://projecteuclid.org/euclid.bsmsp/1200501656
25. Storn, R., Price, K.: Differential evolution - a simple and efficient heuristic for global optimization over continuous spaces. J. Glob. Optim. **11**(4), 341–359 (1997)
26. Surry, P.D., Radcliffe, N.J.: Inoculation to initialise evolutionary search. In: Fogarty, T.C. (ed.) AISB EC 1996. LNCS, vol. 1143, pp. 269–285. Springer, Heidelberg (1996). https://doi.org/10.1007/BFb0032789

27. Teytaud, O., Gelly, S., Mary, J.: On the ultimate convergence rates for isotropic algorithms and the best choices among various forms of isotropy. In: Runarsson, T.P., Beyer, H.-G., Burke, E., Merelo-Guervós, J.J., Whitley, L.D., Yao, X. (eds.) PPSN 2006. LNCS, vol. 4193, pp. 32–41. Springer, Heidelberg (2006). https://doi.org/10.1007/11844297_4
28. Yang, X., Cao, J., Li, K., Li, P.: Improved opposition-based biogeography optimization. In: The Fourth International Workshop on Advanced Computational Intelligence, pp. 642–647 (2011)
29. Zhang, A., Zhou, Y.: On the non-asymptotic and sharp lower tail bounds of random variables (2018)

High Dimensional Bayesian Optimization Assisted by Principal Component Analysis

Elena Raponi[1(✉)], Hao Wang[2], Mariusz Bujny[3], Simonetta Boria[1], and Carola Doerr[2]

[1] University of Camerino, Via Madonna delle Carceri 9, 62032 Camerino, Italy
elena.raponi@unicam.it
[2] Sorbonne Université, CNRS, LIP6, Paris, France
[3] Honda Research Institute Europe GmbH, 63073 Offenbach am Main, Germany

Abstract. Bayesian Optimization (BO) is a surrogate-assisted global optimization technique that has been successfully applied in various fields, e.g., automated machine learning and design optimization. Built upon a so-called infill-criterion and Gaussian Process regression (GPR), the BO technique suffers from a substantial computational complexity and hampered convergence rate as the dimension of the search spaces increases. Scaling up BO for high-dimensional optimization problems remains a challenging task.

In this paper, we propose to tackle the scalability of BO by hybridizing it with a Principal Component Analysis (PCA), resulting in a novel PCA-assisted BO (PCA-BO) algorithm. Specifically, the PCA procedure learns a linear transformation from all the evaluated points during the run and selects dimensions in the transformed space according to the variability of evaluated points. We then construct the GPR model, and the infill-criterion in the space spanned by the selected dimensions.

We assess the performance of our PCA-BO in terms of the empirical convergence rate and CPU time on multi-modal problems from the COCO benchmark framework. The experimental results show that PCA-BO can effectively reduce the CPU time incurred on high-dimensional problems, and maintains the convergence rate on problems with an adequate global structure. PCA-BO therefore provides a satisfactory trade-off between the convergence rate and computational efficiency opening new ways to benefit from the strength of BO approaches in high dimensional numerical optimization.

Keywords: Bayesian optimization · Black-box optimization · Principal Component Analysis · Dimensionality reduction

1 Introduction

Over the last few years, Gaussian Process Regression (GPR) [20] has been proven to be a very flexible and competitive tool in the modeling of functions that are

T. Bäck et al. (Eds.): PPSN 2020, LNCS 12269, pp. 169–183, 2020.
https://doi.org/10.1007/978-3-030-58112-1_12

expensive to evaluate or characterized by strong nonlinearities and measurement noises. A Gaussian Process (GP) is a stochastic process, i.e., a family of random variables, such that any finite collection of them have joint Gaussian distributions. GP is used in many application fields (engineering, geology, etc.) in order to evaluate datasets, predict unknown function values, and perform surrogate model-based optimization [7]. Based on a certain number of evaluations of the objective function (training data), surrogate models allow for the construction of computationally cheap-to-evaluate approximations and for replacing the direct optimization of the real objective with the model. As such, many more evaluations can be performed on the approximate model. GP uses a measure of the similarity between points – the kernel function – to predict the value for an unseen point from training data. GP is often chosen because it also provides a theoretical uncertainty quantification of the prediction error, among a variety of kernel-based methods, e.g., Radial Basis Functions [4] and Support Vector Regression [9,36]. Indeed, GP not only provides the estimate of the value of a function (the mean value), but also the variance at each domain point. This variance defines the uncertainty of the model while performing a prediction of the function value on an untested point of the search space. In particular, when GP is used in surrogate model-based optimization, at each iteration of the optimization algorithm, the estimate of the error is a crucial information in the update phase of the approximation model.

A GP can be used as a prior probability distribution over functions in Bayesian Optimization (BO) [24,25], also known as Efficient Global Optimization (EGO) [16]. BO is a sequential design strategy targeting black-box global optimization problems. It generates an initial set of training points in the search space where the objective function is evaluated. Based on this training data, a GPR model is built to approximate the real objective function. Afterwards, the optimization procedure starts by iteratively choosing new candidate points for which the objective function is evaluated. These points are obtained through the optimization of an acquisition function. However, it has to be noted that the optimization of the acquisition function and the model construction become quite complicated and time-consuming as the dimension of the problem increases, due to the well-known *curse of dimensionality.* In particular, this is the case for most real-world optimization problems, where the number of design variables can range from a few (for simple parameter optimization problems) up to millions (for complex shape [12] or topology optimization [1] problems). In spite of this difficulty, BO techniques have been successfully applied in many fields of engineering, including materials science [17,35], aerospace engineering [6,15,18,21], turbomachinery design [2,14], or even fluid and structural topology optimization [22,27–29,41], thanks to a reduction of the problem dimensionality on the representation level. Nevertheless, since the efficiency of these methods decreases strongly with the rising number of design variables, the development of BO approaches able to address high-dimensional problems is crucial for their future applications in industry [26,37].

Different dimensionality reduction techniques using other tools than PCA have been introduced in the literature for the optimization of expensive problems through GPs. Huang et al. [13] proposed a scalable GP for regression by training a neural network with a stacked denoising auto-encoder and performing BO afterwards on the top layer of the trained network. In [40], Wang et al. proposed to embed a lower-dimensional subspace into the original search space, through the so-called random embedding technique. Moreover, Blanchet-Scalliet et al. [3] presented four algorithms to substitute classical kernels, which provide poor predictions of the response in high-dimensional problems. In [8], a kernel-based approach is devised to perform the dimensionality reduction in a feature space for the shape optimization problem in Computer Aided Design (CAD) systems. To the authors' knowledge, the only ones who used PCA to discover hidden features in GP regression are Vivarelli and Williams [38]. However, their strategy mainly allowed for estimating which are the relevant directions in a prescribed feature space, rather than for pursuing the intrinsic optimization process. Finally, Kapsoulis et al. [19] embedded Kernel and Linear PCA in optimization strategies using Evolutionary Algorithms (EAs). Nevertheless, they identify the principal component directions and map the design space to a new feature space only when a prescribed iteration is reached, while in the first iterations PCA is not applied. In this paper, we propose to tackle this problem by using the well-known Principal Component Analysis (PCA) procedure to identify a proper subspace, on which we deploy the BO algorithm. This treatment leads to a novel hybrid surrogate modeling method - the PCA-assisted Bayesian Optimization (PCA-BO), which starts with a Design of Experiments (DoE) [7] and adaptively learns a linear map to reduce the dimensionality by a weighted PCA procedure. Here, the weighting scheme is meant for taking the objective values into account. We assessed the empirical performance of PCA-BO by testing it on the well-known BBOB problem set [11]. Among these problems, we focus on the multi-modal ones, which are most representative of real-world optimization challenges.

2 Bayesian Optimization

In this paper, we consider the minimization of a real-valued objective function $f : \mathrm{S} \subseteq \mathbb{R}^D \rightarrow \mathbb{R}$ with simple box constraints, i.e., $\mathrm{S} = [\underline{\mathbf{x}}, \overline{\mathbf{x}}]$. Bayesian Optimization [16,25] (BO) starts with sampling an initial DoE of size n_0: $\mathbf{X} = [\mathbf{x}_1, \mathbf{x}_2, \ldots, \mathbf{x}_{n_0}]^\top \subseteq \mathrm{S}^{n_0}$. The DoE can be uniform random samples, or obtained through more sophisticated sampling methods [31]. We choose the so-called Optimal Latin Hypercube Sampling (OLHS) [5] in this paper. The corresponding objective function values are denoted as $\mathbf{y} = (f(\mathbf{x}_1), f(\mathbf{x}_2), \ldots, f(\mathbf{x}_{n_0}))^\top$. Conventionally, a centered Gaussian process prior is assumed on the objective function: $f \sim gp(0, k(\cdot,\cdot))$, where $k\colon \mathrm{S} \times \mathrm{S} \rightarrow \mathbb{R}$ is a positive definite function (a.k.a. *kernel*) that computes the autocovariance of the process. Often, a Gaussian likelihood is taken, leading to a conjugate posterior process [30], i.e., $f \mid \mathbf{y} \sim gp(\hat{f}(\cdot), k'(\cdot,\cdot))$, where $\hat{f}$ and k' are the posterior mean and covariance function, respectively. On an unknown point $\mathbf{x}$, $\hat{f}(\mathbf{x})$ yields the maximum a posteriori (MAP) estimate of $f(\mathbf{x})$ whereas $\hat{s}^2(\mathbf{x}) := k'(\mathbf{x}, \mathbf{x})$ quantifies the uncertainty of this estimation. We shall refer to the posterior process as the Gaussian process regression (GPR) model. Based on the posterior process, we usually identify promising points via the so-called *infill-criterion* which balances $\hat{f}$ with $\hat{s}^2$ (exploitation vs. exploration). A variety of infill-criteria has been proposed in the literature, e.g., Probability of Improvement [7,25], Expected Improvement [7], and the Moment-Generating Function of Improvement [39]. In this work, we adopt the Penalized Expected Improvement (PEI) criterion [28] to handle the box constraints,

$$\mathrm{PEI}(\mathbf{x}) = \begin{cases} E\{\max\{0, \min \mathbf{y} - f(\mathbf{x})\} \mid \mathbf{y}\} & \text{if } \mathbf{x} \text{ is feasible,} \\ -P(\mathbf{x}) & \text{if } \mathbf{x} \text{ is infeasible,} \end{cases} \tag{1}$$

which essentially calculates the expected improvement when $\mathbf{x}$ is feasible, and returns a negative penalty value otherwise. The penalty function $P(\mathbf{x})$ is proportional to the degree of infeasibility of the point $\mathbf{x}$. This will be further described in Sect. 3. BO then proceeds to select the next point by maximizing PEI, namely, $\mathbf{x}^* = \arg\min_{\mathbf{x} \in \mathcal{S}} \mathrm{PEI}(\mathbf{x})$. After evaluating $\mathbf{x}^*$, BO augments[1] the data set, $\mathbf{X} \leftarrow \mathbf{X} \cup \{\mathbf{x}^*\}, \mathbf{y} \leftarrow \mathbf{y} \cup \{f(\mathbf{x}^*)\}$, on which the GPR model is retrained.

3 PCA-Assisted Bayesian Optimization

In this section, we introduce the coupling of the principal component analysis procedure and Bayesian optimization method, called **PCA-BO**. Loosely speaking, this approach learns a linear transformation (from the initial design points of BO) that would, by design, identify directions (a.k.a. principal components) in $\mathbb{R}^D$ along which the objective value changes rapidly. Also, this approach adapts this transformation through the optimization process. Intuitively, we could further drop the components to which objective function values are less sensitive, resulting in a lower-dimensional search space $\mathbb{R}^r$ $(r < D)$. Particularly for BO, the surrogate model shall be trained in this lower-dimensional search space, thus benefiting from the dimensionality reduction. Also, the infill-criterion optimization, which is typically an expensive sub-procedure in BO, becomes less costly because it now operates in the lower-dimensional space as well.

The PCA-BO Algorithm. We first present a high-level overview of the proposed PCA-BO algorithm. A graphical representation is provided in Fig. 1, whereas a pseudo-code description is presented in Algorithm 1. PCA-BO works as follows:

1. We perform an initial DoE in the original search space, generating as an outcome a set of evenly distributed points.
2. We design a weighting scheme to embed the information from objective function into the DoE points, where smaller weights are assigned to the points with worse function values.

[1] With an abuse of terminology, the operation $\mathbf{X} \cup \{\mathbf{x}^*\}$ is understood as appending $\mathbf{x}^*$ at the bottom row of $\mathbf{X}$ throughout this paper. $\mathbf{y} \cup \{f(\mathbf{x}^*)\}$ is defined similarly.

3. We apply a PCA procedure to obtain a linear map from the original search space to a lower-dimensional space, using the weighted DoE points.
4. In the lower-dimensional space, we train a GPR model and maximize an infill-criterion to find a promising candidate point.
5. We map the candidate point back to the original search space, and evaluate it with the objective function.
6. We augment the data set with the candidate point and its objective value, and then proceed to step 2 for the next iteration.

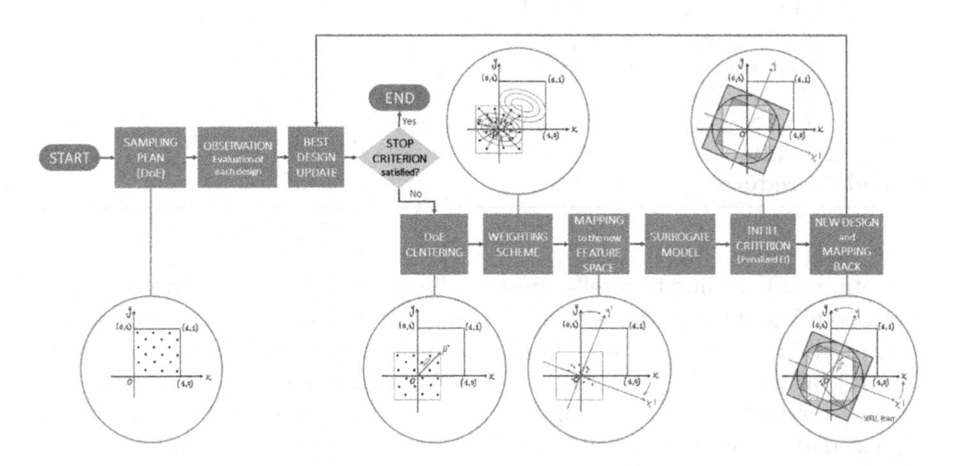

Fig. 1. Flowchart of the PCA-BO optimization algorithm in which PCA-related operations are depicted in circles.

Rescaling Data Points for PCA. Because the goal of applying PCA, as stated above, is to spot directions in $\mathbb{R}^D$ to which the function value is sensitive, it is then necessary to take into account the information on the objective function. We implement this consideration by rescaling the data points according to a weighting scheme, which depends on the corresponding function values. The weighting scheme aims to adjust the "importance" of data points when applying the PCA procedure, such that a point associated with a better (here lower) function value is assigned with a larger weight. We propose to use a rank-based weighting scheme, in which the weight assigned to a point is solely determined by the rank thereof. Let $\{r_i\}_{i=1}^n$ be the ranking of the points in $\mathbf{X}$ according to their function values $\mathbf{y}$. We calculate the rank-based (pre-)weights as follows:

$$\tilde{w}_i = \ln n - \ln r_i, \tag{2}$$

where n stands for the current number of data points. Afterwards, we normalize the pre-weights: $w_i = \tilde{w}_i / \sum \tilde{w}_i$. Notably, this weighting scheme will gradually decrease the weights of worse points when a better point is found and added to

Algorithm 1. PCA-assisted Bayesian Optimization

1: **procedure** PCA-BO($f, \mathscr{A}$, S) ▷ f: objective function, $\mathscr{A}$: infill-criterion, S: search space, S′: lower-dimensional search space
2: Create $\mathbf{X} = [\mathbf{x}_1, \mathbf{x}_2, \ldots, \mathbf{x}_{n_0}]^\top \subset \mathrm{S}^{n_0}$ with optimal Latin hypercube sampling
3: $\mathbf{y} \leftarrow (f(\mathbf{x}_1), f(\mathbf{x}_2), \ldots, f(\mathbf{x}_{n_0}))^\top$, $n \leftarrow n_0$
4: **while** the stop criteria are not fulfilled **do**
5: Centering: $\bar{\mathbf{X}} \leftarrow \mathbf{X} - \mathbf{1}_n \boldsymbol{\mu}$
6: Rescaling: $\mathbf{X}' \leftarrow \bar{\mathbf{X}}\mathbf{W}$
7: $\boldsymbol{\mu}', \mathbf{P}_r \leftarrow$ PCA($\mathbf{X}'$)
8: Mapping to $\mathbb{R}^r$: $\mathbf{Z}_r \leftarrow \mathbf{P}_r(\bar{\mathbf{X}} - \mathbf{1}_n\boldsymbol{\mu}')^\top$
9: GPR training: $\hat{f}, \hat{s}^2 \leftarrow$ GPR($\mathbf{Z}_r, \mathbf{y}$)
10: $\mathbf{z}' \leftarrow \arg\min_{\mathbf{z} \in \mathrm{S}'} \mathscr{A}(\mathbf{z}; \hat{f}, \hat{s}^2)$
11: Mapping to $\mathbb{R}^D$: $\mathbf{x}' \leftarrow \mathbf{P}_r^\top \mathbf{z} + \boldsymbol{\mu}' + \boldsymbol{\mu}$
12: $y' \leftarrow f(\mathbf{x}')$
13: $\mathbf{X} \leftarrow \mathbf{X} \cup \{\mathbf{x}'\}$, $\mathbf{y} \leftarrow (\mathbf{y}^\top, y')^\top$, $n \leftarrow n + 1$
14: **end while**
15: **end procedure**

the data, hence leading to a self-adjusting discount factor on the data set.[2] It is worth pointing out that this weighting scheme resembles the weight calculation in the well-known Covariance Matrix Adaptation Evolution Strategy (CMA-ES) [10], although they are motivated differently. For ease of discussion, we will use a weight matrix $\mathbf{W} = \mathrm{diag}(w_1, w_2, \ldots, w_n)$ henceforth. We rescale the original data $\mathbf{X}$ by removing its sample mean, i.e., $\bar{\mathbf{X}} = \mathbf{X} - \mathbf{1}_n\boldsymbol{\mu}$, where $\boldsymbol{\mu} = \left(\frac{1}{n}\sum_{i=1}^n X_{i1}, \ldots, \frac{1}{n}\sum_{i=1}^n X_{id}\right)$ and $\mathbf{1}_n = (1, \ldots, 1)^\top$ (n 1's), and then adjust each point with the corresponding weight, namely, $\mathbf{X}' = \bar{\mathbf{X}}\mathbf{W}$.

Dimensionality Reduction. The PCA procedure starts with computing the sample mean of data matrix $\mathbf{X}'$, i.e., $\boldsymbol{\mu}' = \left(\frac{1}{n}\sum_{i=1}^n X'_{i1}, \ldots, \frac{1}{n}\sum_{i=1}^n X'_{id}\right)$, after which it calculates the unbiased sample covariance matrix $\mathbf{C} = \frac{1}{n-1}\mathbf{X}''^\top\mathbf{X}'', \mathbf{X}'' = \mathbf{X}' - \mathbf{1}_n\boldsymbol{\mu}'$. The principle components (PCs) are computed via the eigendecomposition $\mathbf{C} = \mathbf{P}\mathbf{D}\mathbf{P}^{-1}$, where column vectors (a.k.a. eigenvectors) of $\mathbf{P}$ identify the PCs, and the diagonal matrix $\mathbf{D} = \mathrm{diag}(\sigma_1^2, \ldots, \sigma_d^2)$ contains the variance of $\mathbf{X}''$ along each principal component, i.e., $\sigma_i^2 = \mathrm{var}\{\mathbf{X}''\mathbf{p}_i\}$ ($\mathbf{p}_i$ is the ith column of $\mathbf{P}$). For clarity of discussion, we consider the orthonormal matrix $\mathbf{P}$ that defines a rotation operation, as a change of the standard basis in $\mathbb{R}^D$ such that PCs are taken as the basis after the transformation. After changing the basis, we proceed to reduce the dimensionality of $\mathbb{R}^D$ by sorting the PCs according to the decreasing order of their variances and keeping the first r components. Here r is chosen as the smallest integer such that the top-r variances sum up to at least

[2] Alternatively, the weight can also be computed directly from the function value, e.g., through a parameterized hyperbolic function. However, we do not prefer this approach since it introduces extra parameters that require tuning, and does not possess the discount effect of the rank-based scheme since the weights remain static throughout the optimization.

α percent of the total variability of the data (we use $\alpha = 95\%$ in our experiments). By denoting the ordering of variances and PCs as $\sigma^2_{1:d}, \sigma^2_{2:d} \dots, \sigma^2_{d:d}$ and $\mathbf{p}_{1:d}, \mathbf{p}_{2:d}, \dots, \mathbf{p}_{d:d}$, respectively, we formulate the selection of PCs as follows:

$$\mathbf{P}_r := [\mathbf{p}_{1:d}, \mathbf{p}_{2:d}, \dots, \mathbf{p}_{r:d}]^\top \in \mathbb{R}^{r \times d},\ r = \inf \left\{ k \in [1..d] \colon \sum_{i=1}^{k} \sigma^2_{i:d} \geq \alpha \sum_{i=1}^{d} \sigma^2_i \right\}.$$

Now, we could transform the data set $\mathbf{X}$ into a lower-dimensional space, i.e., $\mathbf{Z}_r = \mathbf{P}_r(\bar{\mathbf{X}} - \mathbf{1}_n \boldsymbol{\mu}')^\top$. Note that, 1) in this transformation, the centered data set $\bar{\mathbf{X}}$ is not scaled by the weights because we only intend to incorporate the information on the objective values when determining $\mathbf{P}_r$. Using the scaled matrix $\bar{\mathbf{X}}\mathbf{W}$ will make it cumbersome to define the inverse mapping (see below). 2) It is crucial to substract $\boldsymbol{\mu}'$ from $\bar{\mathbf{X}}$ since it estimates the center of ellipsoidal contours when f resembles a quadratic function globally. In this way, the principal axis of contours will be parallel to the PCs after applying $\mathbf{P}_r$. 3) Here, $\mathbf{P}_r \colon \mathbb{R}^D \to \mathbb{R}^r$ is not injective $(\dim(\ker \mathbf{P}_r) = D - r)$. Therefore when transforming the centered data set to the lower-dimensional space, it seems that we might loose some data points in $\bar{\mathbf{X}}$. However, we argue that such an event $\{\mathbf{x} - \mathbf{x}' \in \ker \mathbf{P}_r \colon \mathbf{x}, \mathbf{x}' \in \mathbf{X}\}$ is a null set with respect to any non-singular measure on $\mathbb{R}^D$, and the probability distribution of $\mathbf{X}$, no matter which forms it takes, should not be singular (otherwise BO will only search in a subspace of $\mathbb{R}^D$). Thus, it is safe to ignore the possibility of losing data points. Now we are ready to train a GPR model on $(\mathbf{Z}_r, \mathbf{y})$, and propose candidate points by optimizing an infill-criterion in $\mathbb{R}^r$. Importantly, to evaluate a candidate point $\mathbf{z} \in \mathbb{R}^r$, we map it back to $\mathbb{R}^D$ on which the objective function takes its domain, using the following linear map L:

$$L \colon \mathbb{R}^r \to \mathbb{R}^D, \quad \mathbf{z} \mapsto \mathbf{P}_r^\top \mathbf{z} + \boldsymbol{\mu}' + \boldsymbol{\mu}.$$

Note that, 1) the linear map L is an injection; 2) it is necessary to include $\boldsymbol{\mu}$ and $\boldsymbol{\mu}'$ in this transformation since we subtract them from the design matrix $\mathbf{X}$ when mapping it to the lower-dimensional space; 3) After evaluating point $\mathbf{z}$, i.e., $y = f(L(\mathbf{z}))$, we augment the data set in $\mathbb{R}^D$: $\mathbf{X} \cup \{L(\mathbf{z})\}$ and $\mathbf{y} \cup \{y\}$, from which the linear map $\mathbf{P}_r$ is re-computed using the procedure described above; 4) After re-computing $\mathbf{P}_r$, we have to re-calculate $\mathbf{Z}_r$ by mapping the augmented $\mathbf{X}$ into $\mathbb{R}^r$: $\mathbf{Z}_r = \mathbf{P}_r(\bar{\mathbf{X}} - \mathbf{1}_n \boldsymbol{\mu}')^\top$, and retrain the GPR model on $(\mathbf{Z}_r, \mathbf{y})$.

Penalized Infill-Criterion. Once the regression model is fitted, we use it to construct and maximize the acquisition function that determines the next infill point. As previously stated, in this work we use a penalized version of the well-known EI, referred to as PEI. By definition, we search for the maximum of PEI through DE over a prescribed cubic domain C, which is orientated according to the axes of the transformed coordinate system and must contain the projection of the PCA-transform of the old domain S onto the lower-dimensional feature space. We hence define and penalize a bounding cube C in order to prevent 1) that the optimum is found in a region of C that falls outside S when mapped back, and 2) that we neglect some good regions for new query points, i.e., the regions that belong S but not to C. To achieve this, if ρ is the radius of the sphere tangent to S, we define $C = [C_1 - \rho, C_1 + \rho] \times \cdots \times [C_r - \rho, C_r + \rho]$, where C_i, $i = 1, \ldots, r$ is the ith component of the center of the new feature space. Since, for simplicity reasons, C is wider than necessary, we introduce a penalty to be assigned to the points that would not fall into S when mapped back through the inverse PCA linear transformation – the *infeasible points*. In order to avoid a stagnation of the DE search, the penalty P assigned to infeasible points is defined as the additive inverse of the distance of their images through the PCA inverse transformation from the boundary of S. As a result, the infeasible points are automatically discarded in the maximization of PEI (Eq. (1)).

4 Experiments

Experimental Setup. We assess the performance of PCA-BO on ten multi-modal functions taken from the BBOB problem set [11], and compare the experimental result to a standard BO. This choice of test functions aims to verify the applicability of PCA-BO to more difficult problems, which potentially resemble the feature of real-world applications. We test PCA-BO on three different dimensions $D \in \{10, 20, 40\}$ with the following budget of $10D+50$ function evaluations, of which the initial DoE takes up to 20%. To allow for a statistically meaningful comparison, we performed 30 independent runs for each combination of algorithm, function, and dimension, on an Intel(R) Xeon(R) CPU E5620 @ 2.40GHz machine in single-threaded mode. Also, we recorded the best-so-far function value and measured CPU time taken by those algorithms, which are considered as performance metrics in this paper. We select a squared-exponential kernel function for the GPR model that underpins both PCA-BO and BO. A Differential Evolution (DE) [34] algorithm[3] is adopted to maximize the PEI criterion in every iteration. We implement both algorithms in `Python`, where the `PyDoE`, `PyOpt`, and `Scikit-learn` open-source packages are used to handle the DoE, modeling, and PCA-transformation phases, respectively.

Results. In this section, we illustrate the results we obtained in the optimization of multi-modal benchmark functions from the BBOB suite by using PCA-BO and the state-of-the-art BO. In particular, each algorithm is tested on functions **F15–F24**, which can be categorized in two main groups: 1) multi-modal functions with adequate global structure: **F15–19**, and 2) multi-modal functions with weak global structure: **F20–24**. By keeping 95% of the variability of data points, our PCA-assisted approach, on average, reduces about 40% dimensions from the original search space across all problems and

[3] We take the `Scipy` implementation of DE (https://docs.scipy.org/doc/scipy/reference/generated/scipy.optimize.differential_evolution.html) with a population size of $20r$ and the "best1bin" strategy, which uses the binary crossover and calculates the differential vector based on the current best point. Here, we set the evaluation budget to $20020r^2$ to optimize the infill-criterion.

dimensions. Also, we compare the two algorithms in terms of the empirical convergence and CPU time. In Fig. 2, we report the mean target precision achieved on each test function, which is the difference between the best-so-far function value and the optimal function value averaged over 30 runs with the same initial DoE points. Also, we plot the mean target precision against the iteration number that counts the number of function evaluations after the DoE phase and illustrate the 95% confidence interval of the mean target precision.[4] On each dimension, the plots clearly illustrate that PCA-BO is either superior or at least comparable to the standard BO when the test function exhibits an adequate global structure (**F15–19**). For each function in this group, PCA-BO reveals a steep convergence at the beginning of the optimization procedure. Notably, such an early and steep convergence is extremely valuable when we apply it in real-world scenarios, e.g., industrial optimizations, where we could only afford a very limited evaluation budget for expensive optimization problems. In contrast, on multi-modal functions with weak global structure (**F20–24**), PCA-BO has difficulties in maintaining the convergence speed, particularly on functions F21 and F22 whose global contour lines are composed of multiple local ellipsoids that have different orientations (see contour plots in [11]). We speculate that the poor performance of PCA-BO on those two functions is because we apply the PCA procedure on a global scale, which could be potentially confused by the landscape of F21 and F22. However, we argue that this issue could be alleviated effectively with a simple extension of applying the PCA procedure locally, which can be realized by either searching for the optimum with the trust region approach [42], or through clustering techniques [33]. We shall investigate this issue in future work. Moreover, we performed the well-known Wilcoxon rank-sum test on the target precision values reached by each algorithm upon termination in each dimension. Under the null hypothesis of this test, the target precision values has an equal distribution in BO and PCA-BO [23]. Based on the data, we reject the null hypothesis at the 5% significance level on all functions and dimensions except F23, which belongs to the category of multi-modal functions with weak global structure. Therefore, we argue that PCA-BO converges at least as fast as the standard BO algorithm except on problems **F20–F24**, where the problem landscape exhibits high local irregularities and lacks a global structure. This scenario will be addressed by the authors in future works. More importantly, the PCA-BO algorithm achieves this convergence speed with a significantly less computational power, which is supported by measured CPU times in Fig. 3. Here, we observe that the CPU time of PCA-BO, on average, is about 1.2209 times that of BO on 10D. Although this ratio does not imply any benefits from PCA-BO on $10D$, it, however, decreases to 0.7519 for 20D, and further to 0.5910 for 40D. Therefore, PCA-BO can effectively reduce the CPU time incurred on higher-dimensional problems, leading to a 25% and a 41% reduction in the 20- and 40-variables tests, respectively.

[4] On the 10-dimensional **F20** problem, we observed that the standard deviation of BO over 30 runs gradually shrinks to zero after 50 iterations, making the confidence interval disappear in the corresponding subplot.

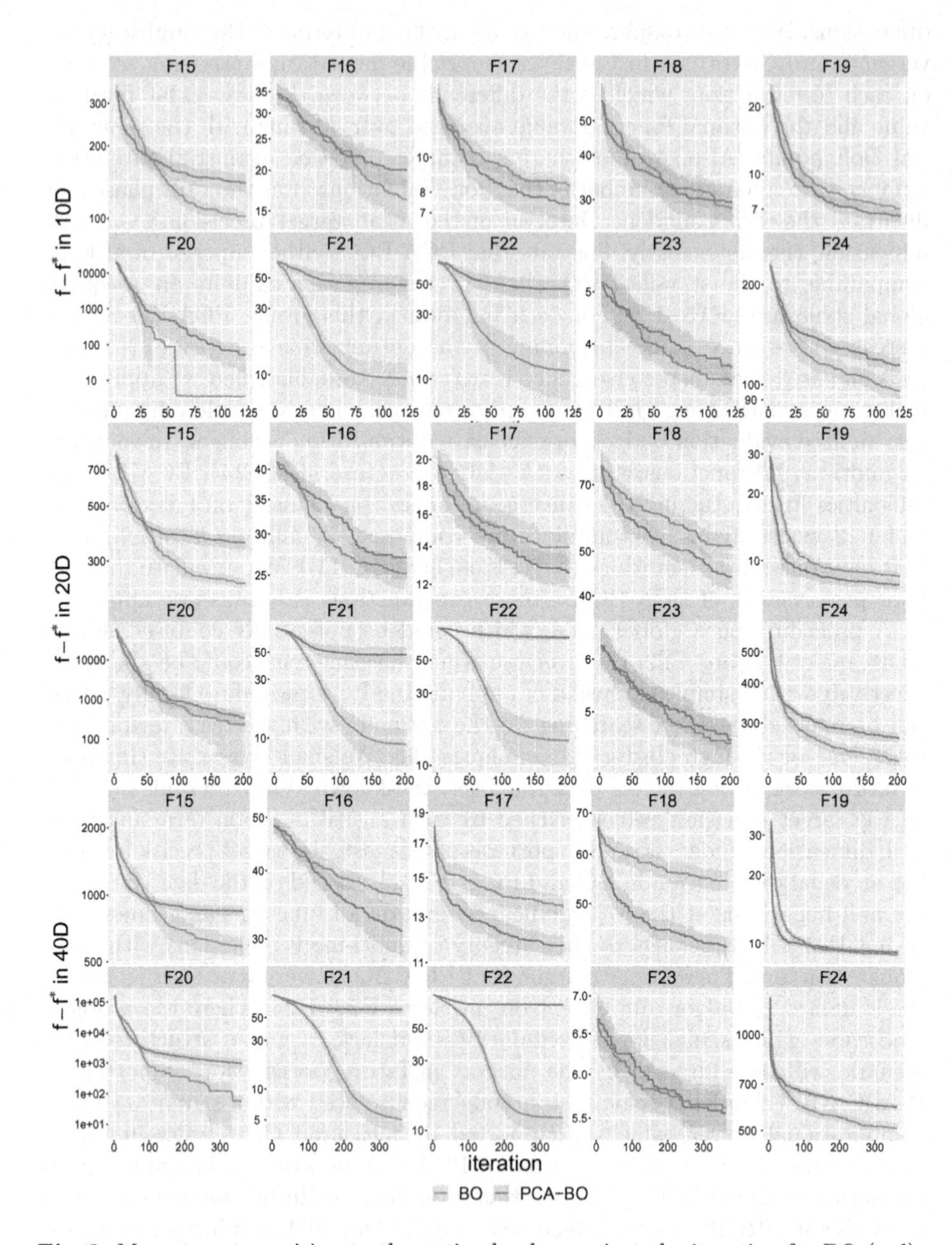

Fig. 2. Mean target precision to the optimal value against the iteration for BO (red) and PCA-BO (blue), on multi-modal functions F15–24 from the BBOB problem set, in three dimensionalities, 10D (top), 20D (middle), and 40D (bottom). 30 independent runs are conducted and the shade area indicates the 95% confidence interval of the mean target precision. The iteration counts the number of function evaluations after evaluating the initial design points. (Color figure online)

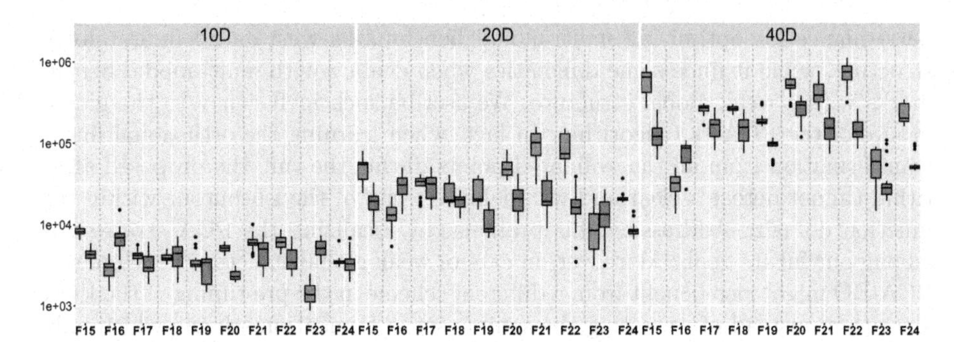

Fig. 3. CPU time (in seconds) taken by PCA-BO and BO to complete the evaluation budget for all functions and three dimensions. For each pair of problems and dimensions, we also applied the well-known Mann–Whitney U test on the CPU time with the Holm–Bonferroni correction of p-values and a significance level of 0.01. The red-colored test problems indicate that the CPU time of PCA-BO is significantly smaller than BO based on the test. (Color figure online)

5 Conclusions and Future Research

In this paper, we presented the novel PCA-assisted Bayesian Optimization (PCA-BO) method, which aims to efficiently address high-dimensional black-box optimization problems by means of surrogate modeling techniques. In fact, despite BO has been successfully applied in various fields, it demonstrated to be restricted to problems of moderate dimension, typically up to about 15. Therefore, since many real-word applications require the solution of very high-dimensional optimization problems, it becomes necessary to scale BO to more complex parameter spaces, e.g., through dimensionality reduction techniques. Essentially, PCA-BO allows for achieving this by performing iteratively an orthogonal linear transformation of the original data-set, in order to map it to a new feature space where the problem becomes *more separable* and features are selected in order to inherit the maximum possible variance from the original sample set. Once the transformation is applied, the model construction and the optimization of the acquisition function – i.e., the costly steps characterizing the BO sequential strategy – are performed in the lower-dimensional space so that we can find a promising point by saving computational resources. Afterwards, the selected point is mapped back to the original feature space, it is evaluated, added to the sample set to increase the accuracy of the model, and compared to the previous points to eventually update the optimum.

We empirically evaluated PCA-BO by testing it on the well-known COCO/BBOB benchmark suite. In particular, we used the presented method to optimize two sets of multi-modal benchmarks, characterized by either an adequate or weak global structure. We also ran the standard BO for comparison purposes. It is clear from the results that PCA-BO achieves state-of-the-art BO performance for optimizing multi-modal benchmarks with an adequate global structure, while it finds some difficulties when dealing with multi-modal benchmarks with a weak global structure. We ascribe such inefficiency to the global nature of the PCA-BO algorithm. In fact, when learning the orthogonal linear transformation from all the evaluated points during the run, the proposed algorithm cannot detect a clear and well-definite trend of the isocontour within the function domain. Addressing this problem, by localizing the PCA strategy, is among our future goals. Moreover, in case of weak global structure benchmarks, PCA-BO might also benefit from a different scheme in the pre-tuning of the PCA algorithm: instead of weighting the sample set, a truncation scheme which only selects a prescribed percentage of samples – ordered according to their function values – to train the linear transformation parameters can be adopted. Within this study, we have assessed the performance of PCA-BO not only in terms of the empirical convergence rate, but also tracking the CPU time elapsed to complete the runs. Results showed that, except for the 10-variables test cases that can be efficiently addressed with standard BO, the average CPU time throughout the different tested functions can be consistently reduced by approaching the optimization problem with PCA-BO. In fact, training the GP model and optimizing the acquisition function is a much cheaper procedure in a search space of reduced dimensionality than in the full-dimensional space.

In the future, we will extend our study by applying a non-linear transformation when tuning the parameters for the PCA-mapping to the lower-dimensional feature space. One option is represented by the Kernel Principal Component Analysis (KPCA) [32], which is a non-linear dimensionality reduction through the use of kernels, more capable to capture the isocontour of the cost functions when variables are not linearly correlated. Moreover, we will extend PCA-BO to make it capable of addressing real-world constrained applications, such as the optimization of mechanical structures subjected to static/dynamic loads. These problems are usually characterized by a great number of design variables and by costly evaluations, requiring optimization techniques of affordable computational cost.

Acknowledgments. Our work was supported by the Paris Ile-de-France Region and by COST Action CA15140 "Improving Applicability of Nature-Inspired Optimisation by Joining Theory and Practice (ImAppNIO)".

References

1. Aage, N., Andreassen, E., Lazarov, B.S.: Topology optimization using PETSc: an easy-to-use, fully parallel, open source topology optimization framework. Struct. Multidisciplinary Optim. **51**(3), 565–572 (2014). https://doi.org/10.1007/s00158-014-1157-0
2. Arsenyev, I., Duddeck, F., Fischersworring-Bunk, A.: Adaptive surrogate-based multi-disciplinary optimization for vane clusters. In: ASME Turbo Expo 2015: Turbine Technical Conference and Exposition. American Society of Mechanical Engineers Digital Collection (2015). https://doi.org/10.1115/GT2015-42164

3. Blanchet-Scalliet, C., Helbert, C., Ribaud, M., Vial, C.: Four algorithms to construct a sparse kriging kernel for dimensionality reduction. Comput. Stat. **34**(4), 1889–1909 (2019). https://doi.org/10.1007/s00180-019-00874-2
4. Broomhead, D.S.: Radial Basis Functions, Multi-variable Functional Interpolation and Adaptive Networks. Royals Signals and Radar Establishment (1988)
5. Fang, K.T., Li, R., Sudjianto, A.: Design and Modeling for Computer Experiments. CRC Press, Boca Raton (2005)
6. Forrester, A.I., Bressloff, N.W., Keane, A.J.: Optimization using surrogate models and partially converged computational fluid dynamics simulations. Proc. R. Soc. A: Math. Phys. Eng. Sci. **462**(2071), 2177–2204 (2006). https://doi.org/10.1098/rspa.2006.1679
7. Forrester, A.I.J., Sóbester, A., Keane, A.J.: Engineering Design via Surrogate Modelling - A Practical Guide. Wiley, Hoboken (2008). https://doi.org/10.1002/9780470770801
8. Gaudrie, D., Le Riche, R., Picheny, V., Enaux, B., Herbert, V.: From CAD to eigenshapes for surrogate-based optimization. In: 13th World Congress of Structural and Multidisciplinary Optimization. Beijing. (2019). https://hal.archives-ouvertes.fr/hal-02142492
9. Gunn, S.R.: Support Vector Machines for Classification and Regression, Technical report (1998)
10. Hansen, N.: The CMA evolution strategy: a comparing review. In: Lozano, J.A., Larrañaga, P., Inza, I., Bengoetxea, E. (eds.) Towards a New Evolutionary Computation. Studies in Fuzziness and Soft Computing, vol. 192., pp. 75–102. Springer, Berlin, Heidelberg (2006). https://doi.org/10.1007/3-540-32494-1_4
11. Hansen, N., et al.: COmparing continuous optimizers: numbbo/COCO on Github (2019). https://doi.org/10.5281/zenodo.2594848
12. Hojjat, M., Stavropoulou, E., Bletzinger, K.U.: The vertex morphing method for node-based shape optimization. Comput. Methods Appl. Mech. Eng. **268**, 494–513 (2014). https://doi.org/10.1016/j.cma.2013.10.015
13. Huang, W., Zhao, D., Sun, F., Liu, H., Chang, E.: Scalable Gaussian process regression using deep neural networks. In: Proceedings of the 24th International Conference on Artificial Intelligence, IJCAI 2015, pp. 3576–3582. AAAI Press, Buenos Aires (2015)
14. Huang, Z., Wang, C., Chen, J., Tian, H.: Optimal design of aeroengine turbine disc based on kriging surrogate models. Comput. Struct. **89**(1–2), 27–37 (2011). https://doi.org/10.1016/j.compstruc.2010.07.010
15. Jeong, S., Murayama, M., Yamamoto, K.: Efficient optimization design method using kriging model. J. Aircraft **42**(2), 413–420 (2005). https://doi.org/10.1023/A:1008306431147
16. Jones, D.R., Schonlau, M., Welch, W.J.: Efficient global optimization of expensive black-box functions. J. Global Optim. **13**(4), 455–492 (1998). https://doi.org/10.1023/A:1008306431147
17. Ju, S., Shiga, T., Feng, L., Hou, Z., Tsuda, K., Shiomi, J.: Designing nanostructures for phonon transport via Bayesian optimization. Phys. Rev. X **7**(2), 021024 (2017). https://doi.org/10.1103/PhysRevX.7.021024
18. Kanazaki, M., Takagi, H., Makino, Y.: Mixed-fidelity efficient global optimization applied to design of supersonic wing. Procedia Eng. **67**(1), 85–99 (2013). https://doi.org/10.1016/j.proeng.2013.12.008

19. Kapsoulis, D., Tsiakas, K., Asouti, V., Giannakoglou, K.: The use of Kernel PCA in evolutionary optimization for computationally demanding engineering applications. In: 2016 IEEE Symposium Series on Computational Intelligence (SSCI), pp. 1–8 (2016). https://doi.org/10.1109/SSCI.2016.7850203
20. Kleijnen, J.P.C.: Kriging metamodeling in simulation: a review. Eur. J. Oper. Res. **192**(3), 707–716 (2009). https://doi.org/10.1016/j.ejor.2007.10.013
21. Lam, R., Poloczek, M., Frazier, P., Willcox, K.E.: Advances in Bayesian optimization with applications in aerospace engineering. In: 2018 AIAA Non-Deterministic Approaches Conference, p. 1656 (2018). https://doi.org/10.2514/6.2018-1656
22. Liu, K., Detwiler, D., Tovar, A.: Metamodel-based global optimization of vehicle structures for crashworthiness supported by clustering methods. In: Schumacher, A., Vietor, T., Fiebig, S., Bletzinger, K.-U., Maute, K. (eds.) WCSMO 2017, pp. 1545–1557. Springer, Cham (2018). https://doi.org/10.1007/978-3-319-67988-4_116
23. MATLAB: version 9.5.0.944444 (R2018b). The MathWorks Inc., Natick, Massachusetts (2018)
24. Močkus, J.: On Bayesian methods for seeking the extremum. In: Marchuk, G.I. (ed.) Optimization Techniques 1974. LNCS, vol. 27, pp. 400–404. Springer, Heidelberg (1975). https://doi.org/10.1007/3-540-07165-2_55
25. Mockus, J.: Bayesian Approach to Global Optimization: Theory and Applications, vol. 37. Springer, Dordrecht (2012). https://doi.org/10.1007/978-94-009-0909-0
26. Palar, P.S., Liem, R.P., Zuhal, L.R., Shimoyama, K.: On the use of surrogate models in engineering design optimization and exploration: the key issues. In: Proceedings of the Genetic and Evolutionary Computation Conference Companion, pp. 1592–1602 (2019). https://doi.org/10.1145/3319619.3326813
27. Raponi, E., Bujny, M., Olhofer, M., Aulig, N., Boria, S., Duddeck, F.: Kriging-guided level set method for crash topology optimization. In: 7th GACM Colloquium on Computational Mechanics for Young Scientists from Academia and Industry, GACM, Stuttgart (2017)
28. Raponi, E., Bujny, M., Olhofer, M., Aulig, N., Boria, S., Duddeck, F.: Kriging-assisted topology optimization of crash structures. Comput. Methods Appl. Mech. Eng. **348**, 730–752 (2019). https://doi.org/10.1016/j.cma.2019.02.002
29. Raponi, E., Bujny, M., Olhofer, M., Boria, S., Duddeck, F.: Hybrid Kriging-assisted level set method for structural topology optimization. In: 11th International Conference on Evolutionary Computation Theory and Applications, Vienna (2019). https://doi.org/10.5220/0008067800700081
30. Rasmussen, C., Williams, C.: Gaussian Processes for Machine Learning. Adaptative Computation and Machine Learning Series. University Press Group Limited (2006)
31. Santner, T.J., Williams, B.J., Notz, W.I.: The Design and Analysis of Computer Experiments. Springer Series in Statistics. Springer, Dordrecht (2003). https://doi.org/10.1007/978-1-4757-3799-8
32. Schölkopf, B., Smola, A., Müller, K.R.: Nonlinear component analysis as a kernel eigenvalue problem. Neural Comput. **10**(5), 1299–1319 (1998). https://doi.org/10.1162/089976698300017467
33. van Stein, B., Wang, H., Kowalczyk, W., Emmerich, M., Bäck, T.: Cluster-based Kriging approximation algorithms for complexity reduction. Appl. Intell. **50**(3), 778–791 (2019). https://doi.org/10.1007/s10489-019-01549-7
34. Storn, R., Price, K.: Differential evolution - a simple and efficient heuristic for global optimization over continuous spaces. J. Global Optim. **11**(4), 341–359 (1997). https://doi.org/10.1023/A:1008202821328

35. Ueno, T., Rhone, T.D., Hou, Z., Mizoguchi, T., Tsuda, K.: COMBO: an efficient Bayesian optimization library for materials science. Mater. Discov. **4**, 18–21 (2016). https://doi.org/10.1016/j.md.2016.04.001
36. Vapnik, V.N.: The Nature of Statistical Learning Theory. Springer, New York (1995). https://doi.org/10.1007/978-1-4757-3264-1
37. Viana, F.A., Simpson, T.W., Balabanov, V., Toropov, V.: Special section on multidisciplinary design optimization: metamodeling in multidisciplinary design optimization: how far have we really come? AIAA J. **52**(4), 670–690 (2014). https://doi.org/10.2514/1.J052375
38. Vivarelli, F., Williams, C.K.I.: Discovering hidden features with Gaussian processes regression. In: Kearns, M.J., Solla, S.A., Cohn, D.A. (eds.) Advances in Neural Information Processing Systems, vol. 11, pp. 613–619. MIT Press (1999)
39. Wang, H., van Stein, B., Emmerich, M., Bäck, T.: A new acquisition function for Bayesian optimization based on the moment-generating function. In: 2017 IEEE International Conference on Systems, Man, and Cybernetics, SMC 2017, Banff, AB, Canada, 5–8 October 2017, pp. 507–512. IEEE (2017). https://doi.org/10.1109/SMC.2017.8122656
40. Wang, Z., Hutter, F., Zoghi, M., Matheson, D., De Freitas, N.: Bayesian optimization in a billion dimensions via random embeddings (2016)
41. Yoshimura, M., Shimoyama, K., Misaka, T., Obayashi, S.: Topology optimization of fluid problems using genetic algorithm assisted by the Kriging model. Int. J. Numer. Method Eng. **109**(4), 514–532 (2016). https://doi.org/10.1002/nme.5295
42. Yuan, Y.X.: A Review of Trust Region Algorithms for Optimization

Simple Surrogate Model Assisted Optimization with Covariance Matrix Adaptation

Lauchlan Toal and Dirk V. Arnold(✉)

Faculty of Computer Science, Dalhousie University,
Halifax, NS B3H 4R2, Canada
{lc790935,dirk}@dal.ca

Abstract. We aim to observe differences between surrogate model assisted covariance matrix adaptation evolution strategies applied to simple test problems. We propose a simple Gaussian process assisted strategy as a baseline. The performance of the algorithm is compared with those of several related strategies using families of parameterized, unimodal test problems. The impact of algorithm design choices on the observed differences is discussed.

1 Introduction

Evolution strategies [5] are black-box optimization algorithms that sample candidate solutions probabilistically, with no bias. In most cases, the majority of the sampled candidate solutions will be relatively poor and thus discarded. If objective function evaluations are expensive, then efficiency can be gained by using surrogate modelling techniques. Surrogate models are built based on information gained from evaluations of candidate solutions made in prior iterations. Common types of surrogate models include low-order polynomials fitted through regression, ranking support vector machines, and Gaussian processes. Surrogate models can be used to either replace costly evaluations of the objective function, or to determine promising candidate solutions to sample. Evolutionary algorithms that employ surrogate models need to carefully balance the computational savings from avoiding evaluations of the objective function with the potentially reduced quality of steps resulting from inaccurate or biased models. Overviews of surrogate model assisted evolutionary optimization have been compiled by Jin [9] and Loshchilov [13].

Several surrogate model assisted variants of covariance matrix adaptation evolution strategies (CMA-ES) [6,7] have been proposed [1,4,11,16]. The algorithms use different types of surrogate models and employ those models in different ways. Comparisons of algorithms are most often based on the COCO benchmark (coco.gforge.inria.fr), which includes problems ranging from the very simple to the highly difficult. We contend that additional insights regarding the impact of algorithm design choices can be obtained by performing more fine

T. Bäck et al. (Eds.): PPSN 2020, LNCS 12269, pp. 184–197, 2020.
https://doi.org/10.1007/978-3-030-58112-1_13

grained experiments on families of parameterized, simple test problems. While algorithm designers likely perform such experiments in the process of making design choices, those results most often remain unpublished.

The contributions of this paper are twofold. First, we combine the surrogate model assisted evolution strategy of Yang and Arnold [19] with the approach to covariance matrix adaptation implemented in CMA-ES. The algorithm uses Gaussian process surrogate models and is referred to as GP-CMA-ES. We consider it the simplest surrogate model assisted CMA-ES and use it as a baseline to compare other algorithms against. And second, we employ families of parameterized, unimodal test problems to compare GP-CMA-ES with several other algorithms, including the surrogate model assisted evolution strategies of Loshchilov et al. [16] and Hansen [4] as well as a quasi-Newton method that obtains gradient estimates through finite differencing. The purpose of the experiments is to observe scaling behaviour with regard to dimension, conditioning and deviation from being quadratic.

2 Related Work

Kern et al. [11] were among the first to propose a surrogate model assisted variant of CMA-ES, termed lmm-CMA-ES ("lmm" standing for local meta-model). They employ locally weighted regression to fit quadratic models and use the strategy's covariance matrix in the computation of distances when weighting the impact of points evaluated in prior iterations. lmm-CMA-ES in each iteration generate a batch of candidate solutions, evaluate them using the surrogate model, and then evaluate a fraction of those candidate solutions deemed best by the surrogate model using the true objective function. That fraction is adapted based on the consistency of the ranking established by the surrogate model with the results observed when using the true objective function. Bouzarkouna et al. [2] propose a variant of the algorithm that is more effective when batch sizes are large.

Loshchilov et al. [15] propose ^{s*}ACM-ES, a surrogate model assisted CMA-ES variant that preserves the CMA-ES's invariance with regard to strictly monotonic transformations of the objective function by employing ranking support vector machines as surrogate models. The importance of preserving invariance properties is further discussed by Loshchilov et al. [14]. ^{s*}ACM-ES alternate the optimization of the objective function using a surrogate model assisted CMA-ES with the adaptation of the surrogate model's parameters using another CMA-ES. Loshchilov et al. [16] introduce a mechanism for more intensively exploiting the surrogate models. They find that while beneficial on unimodal problems, more intensive exploitation of the surrogate model may increase the propensity to fail on multimodal functions.

Bajer et al. [1] consider Gaussian process surrogate model assisted variants of CMA-ES and conduct a comparison of algorithm variants with several other evolutionary black-box optimization algorithms using the COCO benchmark. They find that convergence speeds of published Gaussian process surrogate model assisted CMA-ES variants are often lower than those of lmm-CMA-ES and ^{s*}ACM-ES, which rely on quadratic regression and ranking support vector machines as surrogate models instead. They introduce Gaussian process assisted DTS-CMA-ES ("DTS" standing for Doubly Trained Surrogate) and find that they perform well particularly on some classes of multimodal test problems. ^{s*}ACM-ES are found to be the most competitive evolutionary algorithms on the unimodal problems, where they perform similarly to DTS-CMA-ES in five dimensions, but enjoy a significant performance advantage in twenty dimensions. DTS-CMA-ES are observed to perform similarly to lmm-CMA-ES on many COCO benchmark functions.

Hansen [4] presents lq-CMA-ES ("lq" standing for linear/quadratic), a further evolution of lmm-CMA-ES that employs global rather than local surrogate models. He uses the COCO benchmark to conduct a thorough comparison between several surrogate model assisted CMA-ES variants, including lmm-CMA-ES, ^{s*}ACM-ES, and DTS-CMA-ES. He find that despite their relative simplicity, lq-CMA-ES perform similarly to the other surrogate model assisted algorithms and appear to have an edge for function evaluation budgets up to ten times the number of dimensions of the problems being solved. Hansen also presents results from more fine grained experiments on four parameterized test function families. Those experiments involve CMA-ES without surrogate model assistance, lq-CMA-ES, and an approach based on sequential quadratic programming, but no further surrogate model assisted CMA-ES variants.

Kayhani and Arnold [10] consider a simple $(1+1)$-ES that uses surrogate models in order to determine whether to evaluate a candidate solution using the true objective function, or to reject it outright. They propose a step size adaptation mechanism based on the implementation of the 1/5th rule [18] by Kern et al. [12]: decrease the step size if either the candidate solution is rejected outright or it is rejected after evaluation using the true objective function; increase it if the offspring is successful. Yang and Arnold [19] generalize that algorithm by allowing for more intensive exploitation of the surrogate models by means of preselection. A batch of trial points is generated and evaluated using the surrogate model. A fraction of the seemingly best trial points are averaged to form a candidate solution that is then assessed by the surrogate model in order to determine whether it should be evaluated using the true objective function or rejected outright. They find that more intensive exploitation of the surrogate model necessitates fundamentally different rates of change for the step size: while the algorithm examined by Kayhani and Arnold [10] achieves optimal performance by rejecting most candidate solutions outright, that considered by Yang and Arnold [19] relies on the effects of preselection to ensure that objective function evaluations are mostly reserved for successful offspring. As Loshchilov et al. [16], they find that more intensive exploitation of surrogate models is most often beneficial on unimodal problems.

Our approach incorporates covariance matrix adaptation in the algorithms of Kayhani and Arnold [10] and Yang and Arnold [19]. It shares the use of the algorithm's covariance matrix in the computation of distances with the lmm-CMA-ES by Kern et al. [11]. As Bajer et al. [1] we use Gaussian process surrogate models. In contrast to much of the work presented above, focus here is on performance on unimodal problems, and we thus do not consider restart algorithms or other techniques aimed at non-local optimization.

Required: candidate solution $\mathbf{x} \in \mathbb{R}^n$, step size parameter $\sigma \in \mathbb{R}_{>0}$, covariance matrix $\mathbf{C} \in \mathbb{R}^{n \times n}$, search path $\mathbf{s} \in \mathbb{R}^n$, archive $\mathcal{A} = \{(\mathbf{x}_k, f(\mathbf{x}_k)) \mid k = 1, 2 \dots, m\}$

1: Build a surrogate model from archive $\mathcal{A}$.
2: Compute $\mathbf{A} = \mathbf{C}^{1/2}$.
3: Generate trial step vectors $\mathbf{z}_i \sim \mathcal{N}(\mathbf{0}, \mathbf{I}_{n \times n})$, $i = 1, 2, \dots, \lambda$.
4: Evaluate $\mathbf{y}_i = \mathbf{x} + \sigma \mathbf{A} \mathbf{z}_i$ using the surrogate model, yielding $f_\epsilon(\mathbf{y}_i)$, $i = 1, 2, \dots, \lambda$.
5: Let $\mathbf{z} = \sum_{j=1}^{\lambda} w_j \mathbf{z}_{j;\lambda}$, where $j; \lambda$ is the index of the jth smallest of the $f_\epsilon(\mathbf{y}_i)$.
6: Evaluate $\mathbf{y} = \mathbf{x} + \sigma \mathbf{A} \mathbf{z}$ using the surrogate model, yielding $f_\epsilon(\mathbf{y})$.
7: **if** $f_\epsilon(\mathbf{y}) > f(\mathbf{x})$ **then**
8: Let $\sigma \leftarrow \sigma \, \mathrm{e}^{-d_1/D}$.
9: **else**
10: Evaluate $\mathbf{y}$ using the objective function, yielding $f(\mathbf{y})$.
11: Add $(\mathbf{y}, f(\mathbf{y}))$ to $\mathcal{A}$.
12: **if** $f(\mathbf{y}) > f(\mathbf{x})$ **then**
13: Let $\sigma \leftarrow \sigma \, \mathrm{e}^{-d_2/D}$.
14: **else**
15: Let $\mathbf{x} \leftarrow \mathbf{y}$ and $\sigma \leftarrow \sigma \, \mathrm{e}^{d_3/D}$.
16: Update $\mathbf{s}$ and $\mathbf{C}$.
17: **end if**
18: **end if**

Fig. 1. Single iteration of the GP-CMA-ES.

3 Algorithm

We consider the task of minimizing an objective function $f : \mathbb{R}^n \to \mathbb{R}$. Figure 1 presents a single iteration of the GP-CMA-ES. The state of the algorithm consists of candidate solution $\mathbf{x} \in \mathbb{R}^n$, step size parameter $\sigma \in \mathbb{R}_{>0}$, positive definite $n \times n$ matrix $\mathbf{C}$ that is referred to as the covariance matrix, vector $\mathbf{s} \in \mathbb{R}^n$ that is referred to as the search path, and an archive of m candidate solutions that have been evaluated in prior iterations along with their objective function values. In Line 1, a Gaussian process surrogate model is constructed from the archive, resulting in a function $f_\epsilon : \mathbb{R}^n \to \mathbb{R}$ that approximates the objective function in the vicinity of previously evaluated points, but is assumed to be much cheaper to evaluate. Details of this step are described below. The algorithm then proceeds to compute positive definite matrix $\mathbf{A}$ as the principal square root of $\mathbf{C}$. Line 3 generates $\lambda \geq 1$ trial step vectors that are independently drawn from a multivariate Gaussian distribution with zero mean and unit covariance matrix. Line 4 uses the surrogate model to evaluate the corresponding trial points $\mathbf{y}_i = \mathbf{x} + \sigma \mathbf{A} \mathbf{z}_i$, and Line 5 computes $\mathbf{z} \in \mathbb{R}^n$ as a weighted sum of the trial vectors, using rank based weights. Altogether, Lines 3 through 5 implement a form of preselection. Line 6 generates candidate solution $\mathbf{y} = \mathbf{x} + \sigma \mathbf{A}\mathbf{z}$ that is then evaluated using the surrogate model. If the surrogate model suggests that $\mathbf{y}$ is inferior to parental candidate solution $\mathbf{x}$, then the step size is reduced and the iteration is complete. Otherwise, $\mathbf{y}$ is evaluated using the true objective function and added to the archive. If $\mathbf{y}$ is inferior to the parental candidate solution, then the step size is reduced and the iteration is complete. Otherwise, the offspring candidate solution replaces the parental one, the step size is increased, and the search path $\mathbf{s}$ and covariance matrix $\mathbf{C}$ are updated as described below. Notice that at most one evaluation of the objective function is performed in each iteration of the algorithm. Also notice that without the update of the covariance matrix and for a particular choice of weights in Line 5, the algorithm is identical to that described by Yang and Arnold [19].

Parameter Settings. Parameter λ determines to what degree the surrogate model is exploited as discussed by Yang and Arnold [19]. The impact of the value used for that parameter is explored experimentally in Sect. 4. The remaining parameters of the algorithm are set dependent on whether $\lambda = 1$ or $\lambda > 1$ as follows:

- If $\lambda = 1$, then the sole weight used in Line 5 of the algorithm is $w_1 = 1$. The parameters that determine the relative rates of change of the step size parameter in Lines 8, 13, and 15 are set to $d_1 = 0.05$, $d_2 = 0.2$, and $d_3 = 0.6$ according to the recommendation by Kayhani and Arnold [10].
- If $\lambda > 1$, then the rank based weights used in Line 5 are set as proposed by Hansen [3]. That is, weights w_1 through $w_{\lfloor \lambda/2 \rfloor}$ form a strictly monotonically decreasing sequence of positive values that sum to one; the remaining weights are zero. The parameters that determine the relative rates of change of the step size parameter are set to $d_1 = 0.2$, $d_2 = 1.0$, and $d_3 = 1.0$ according to the recommendation by Yang and Arnold [19].

In accordance with prior work [10,19], parameter D, which scales the rates of the step size parameter changes, is set to $\sqrt{1+n}$ in both cases.

Covariance Matrix Adaptation. For $\lambda > 1$, the covariance matrix update performed in Line 16 of the algorithm is that proposed by Hansen et al. [6,7], with some modifications as described by Hansen [3]. Search path $\mathbf{s}$ is an exponentially fading record of past steps that is updated according to

$$\mathbf{s} \leftarrow (1-c)\mathbf{s} + \sqrt{\mu_{\text{eff}} c(2-c)}\mathbf{A}\mathbf{z},$$

where μ_{eff} is computed from the rank based weights as described by Hansen [3]. After updating the search path, the covariance matrix is updated according to

$$\mathbf{C} \leftarrow \mathbf{C} + c_1 \mathbf{s}\mathbf{s}^{\mathrm{T}} + c_\mu \mathbf{A} \left(\sum_{j=1}^{\lambda} w_j^{\circ} \mathbf{z}_{j;\lambda} \mathbf{z}_{j;\lambda}^{\mathrm{T}} \right) \mathbf{A}^{\mathrm{T}}.$$

Settings for parameters c, c_1 and c_μ as well as the weights w_j° are as described by Hansen [3] and, by virtue of normalization, preserve the positive definiteness of $\mathbf{C}$ even as some of the w_j° are negative. If $\lambda = 1$, then the covariance matrix update described by Igel et al. [8] with the parameter settings suggested there is used instead.

Surrogate Model. The use of Gaussian processes for regression is described in detail by Rasmussen and Williams [17]. In order to build a surrogate model from past observations $\{(\mathbf{x}_k, f(\mathbf{x}_k)) \mid k = 1, 2, \ldots, m\}$, in Line 1 of the algorithm in Fig. 1 we generate $m \times m$ matrix $\mathbf{K}$ with entries $k_{ij} = k(\mathbf{x}_i, \mathbf{x}_j)$, where kernel function k is defined as

$$k(\mathbf{x}, \mathbf{y}) = \exp\left(-\frac{(\mathbf{x}-\mathbf{y})^{\mathrm{T}}\mathbf{C}^{-1}(\mathbf{x}-\mathbf{y})}{2h^2\sigma^2}\right)$$

with the length scale parameter set to $h = 8n$. Notice that this is the commonly employed squared exponential kernel with the Mahalanobis distance using matrix $\sigma^2\mathbf{C}$ replacing the Euclidean distance. Its use here is akin to that by Kern et al. [11] who employ the same distance measure for quadratic regression. We have experimented with maximum likelihood estimation to adapt the length scale parameter, but have not been able to consistently improve on the performance of the constant setting.

In order to evaluate the surrogate model at a point $\mathbf{y} \in \mathbb{R}^n$, in Lines 4 and 6 of the algorithm in Fig. 1 we compute $m \times 1$ vector $\mathbf{k}$ with entries $k_i = k(\mathbf{x}_i, \mathbf{y})$ and define

$$f_\epsilon(\mathbf{y}) = f(\mathbf{x}) + \mathbf{k}^{\mathrm{T}}\mathbf{K}^{-1}\mathbf{f},$$

where $m \times 1$ vector $\mathbf{f}$ has entries $f_i = f(\mathbf{x}_i) - f(\mathbf{x})$. That is, the parental objective function value $f(\mathbf{x})$ is used as the Gaussian process's prior mean function. In contrast to Bajer et al. [1], we do not use the uncertainty prediction of the Gaussian process in our algorithm.

In order to avoid increasing computational costs resulting from the need to invert matrix $\mathbf{K}$ as the size of the archive grows, we use at most the most recent $m = (n+2)^2$ points from the archive in order to construct the surrogate model. The use of a quadratic number of points is motivated by attempting to capture an amount information that scales linearly with the number of variables in the covariance matrix.

Initialization and Start-Up. The search path $\mathbf{s}$ is initialized to the zero vector; covariance matrix $\mathbf{C}$ to the identity matrix. The initialization of $\mathbf{x}$ and σ usually is problem specific. We avoid constructing models based on insufficient data by not using surrogate models in the first $2n$ iterations of the algorithm. This can be accomplished by setting $\lambda = 1$ and defaulting $f_\epsilon(\cdot)$ to $-\infty$ as long as the archive contains fewer than $2n$ points. For the duration of this start-up phase, the algorithm is thus a model-free $(1+1)$-CMA-ES and we use parameter settings $d_2 = 0.25$ and $d_3 = 1.0$ for step size adaptation.

4 Evaluation

This section experimentally evaluates the performance of GP-CMA-ES relative to that of several related algorithms. Section 4.1 briefly outlines the comparator algorithms, Sect. 4.2 describes the testing environment, and Sect. 4.3 presents experimental data along with a discussion of the findings.

4.1 Comparator Algorithms

We compare GP-CMA-ES with $\lambda \in \{1, 10, 20, 40\}$ with four other algorithms:

- CMA-ES without surrogate model assistance; we use Version 3.61.beta of the *Matlab* implementation provided by N. Hansen at `cma.gforge.inria.fr`.
- ^{s*}ACM-ES; we use Version 2 of the *Matlab* implementation provided by I. Loshchilov at `loshchilov.com`.
- lq-CMA-ES; we use Version 3.0.2 of the *Python* implementation provided by N. Hansen at `cma.gforge.inria.fr`.
- `fminunc` as provided by *Mathworks* in the Optimization Toolbox (Version 8.1). The *Matlab* function provides an implementation of a quasi-Newton method that obtains gradient estimates through finite differencing.

All algorithm specific parameters are set to their default values. All of the algorithms are variable-metric methods. All evolution strategies but lq-CMA-ES, which employ quadratic models with diagonal scaling before switching to fully quadratic models once sufficiently many data points are available, are invariant to rotations of the coordinate system. Both CMA-ES and ^{s*}ACM-ES are invariant to strictly monotonic transformations of objective function values; GP-CMA-ES, lq-CMA-ES, and the quasi-Newton method are not. We do not include lmm-CMA-ES in the experiments as we consider the algorithm superseded by lq-CMA-ES, and we omit DTS-CMA-ES as their strengths play out predominantly on multimodal problems.

4.2 Test Environments

In order to compare the relative merits of the five approaches we consider the following three parameterized families of unimodal test problems:

- Spherically symmetric functions $f(\mathbf{x}) = (\mathbf{x}^{\mathrm{T}}\mathbf{x})^{\alpha/2}$. We refer to this family as sphere functions. For $\alpha = 2$ this family includes the quadratic sphere.
- Convex quadratic functions $f(\mathbf{x}) = \mathbf{x}^{\mathrm{T}}\mathbf{B}\mathbf{x}$ where symmetric $n \times n$ matrix $\mathbf{B}$ has eigenvalues $b_{ii} = \beta^{(i-1)/(n-1)}$, $i = 1, \ldots, n$, with condition number $\beta \geq 1$. The quadratic sphere is included for $\beta = 1$. We refer to this family as ellipsoid functions.
- Quartic functions $f(\mathbf{x}) = \sum_{i=1}^{n-1} \left[\gamma(x_{i+1} - x_i^2)^2 + (1 - x_i)^2\right]$. For $\gamma = 100$ this family includes the generalized Rosenbrock function, and we refer to its members as Rosenbrock functions.

All of those are considered in dimensions $n \in \{2, 4, 8, 16\}$. The optimal function value for all problems is zero. Sphere functions are perfectly well conditioned and do not require the learning of axis scales, but for $\alpha \neq 2$ may pose difficulties for algorithms that internally build quadratic or near-quadratic models. The primary difficulty inherent in ellipsoid functions is their conditioning and thus the need to learn appropriate axis scales. Rosenbrock functions for the range of parameter γ considered here are only moderately ill-conditioned, but in contrast to the other function families considered require a constant relearning of axis scales as their local Hessian matrix changes throughout the runs of the algorithms. Rosenbrock functions for $n \geq 4$ possess a local minimizer different from the global one and, depending on initialization, a minority of the runs of all of the algorithms converge to that merely local optimizer. As global optimization ability is outside of the focus of this paper, we discard such runs and repeat them until convergence to the global optimizer is observed.

All runs of all algorithms are initialized by sampling starting points uniformly at random in $[-4, 4]^n$. The step size parameter is initially set to $\sigma = 2$ for all of the evolution strategies. Runs are terminated when a candidate solution with an objective function value no larger than 10^{-8} is generated and evaluated. Except as noted for `fminunc` below and those runs on the Rosenbrock functions that were discarded for converging to the merely local minimizer, all runs of all algorithms located the globally optimal solutions with no restarts required.

4.3 Results

We have performed fifteen runs of each of the algorithms for each test problem instance considered. Figures 2, 3 and 4 illustrate the numbers of objective function evaluations required by the different algorithms to locate the optimal solutions to within the required accuracy. Lines connect the median values and error bars illustrate the ranges of values observed (i.e., they span the range from the smallest to the largest values observed across the fifteen runs). The different batch sizes used in the case of GP-CMA-ES are distinguished by their line styles (solid for $\lambda = 10$, dash-dotted for $\lambda = 20$, and dotted for $\lambda = 40$). Results are discussed in detail in the following paragraphs.

Sphere Functions: Figure 2 plots the number of objective function evaluations divided by the dimension for sphere functions with parameter $\alpha \in [1, 4]$. CMA-ES without surrogate model assistance require the largest number of function evaluations throughout. ^{s*}ACM-ES achieve a nearly constant speed-up with the gap widening with increasing dimension to a factor of about four for $n = 16$, reflecting the invariance of the algorithm to strictly monotonic transformations of function values. The remaining algorithms do not possess this invariance property and achieve their highest speed-ups for $\alpha = 2$, where lq-CMA-ES and `fminunc` excel as the quadratic models that they either build or implicitly use perfectly match the objective function. Both of those algorithms are less well suited for the optimization of sphere functions with values of α near the edges of

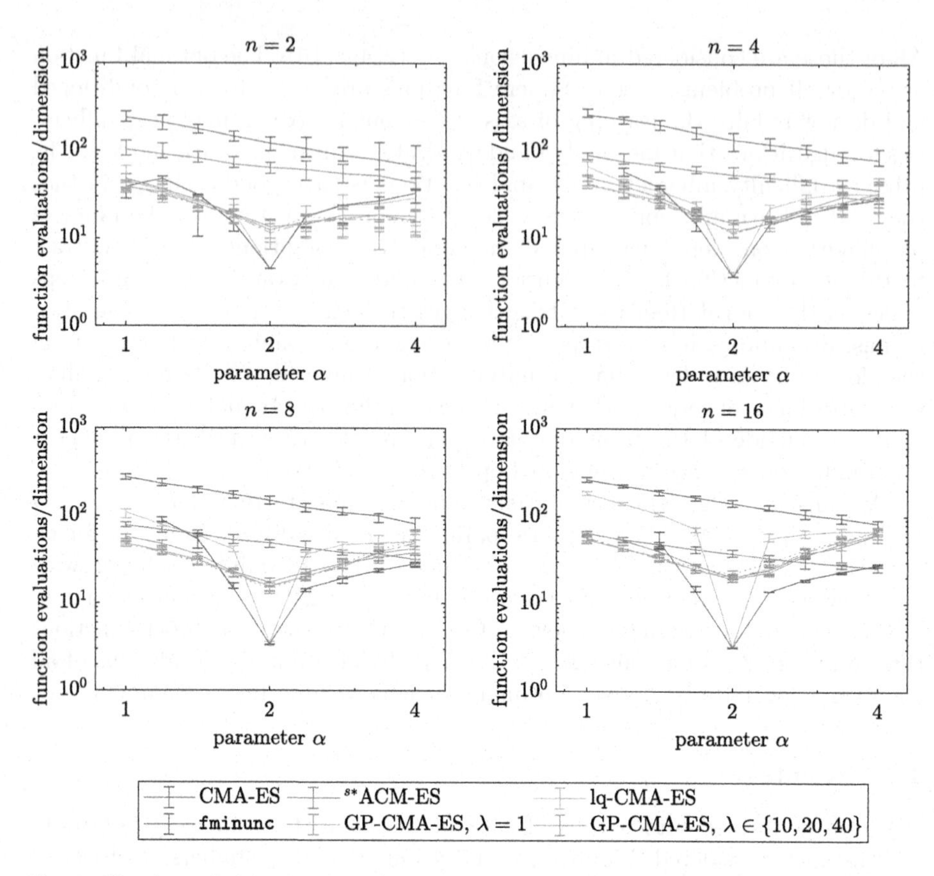

Fig. 2. Number of objective function evaluations per dimension required to optimize sphere functions with parameter $\alpha \in [1, 4]$. The lines connect median values; the error bars reflect the full range of values observed for the respective algorithms.

the domain, where for $n \geq 8$ they are outperformed by ^{s*}ACM-ES. For values of α near 1, `fminunc` in some runs fails to attain termination accuracy as the gradient vector approximations obtained through finite differencing are effectively zero and the corresponding data are missing from the plots.

Speed-ups of GP-CMA-ES for $\alpha = 2$ fall in between those of lq-CMA-ES and `fminunc` on the one hand and ^{s*}ACM-ES on the other. We hypothesize that this is both because the Gaussian process surrogate models that they employ fail to model quadratic objective functions with the same accuracy as quadratic models do, and because they do not proceed based on surrogate model predictions only. For smaller values of α, GP-CMA-ES prove to be more robust than `fminunc` and more efficient than lq-CMA-ES, but with increasing n and decreasing α lose their performance advantage over ^{s*}ACM-ES. For values of α in excess of two, the quality of the Gaussian process models deteriorates and GP-CMA-ES are eventually outperformed by both ^{s*}ACM-ES and `fminunc`, providing smaller and smaller speed-ups compared to CMA-ES that do not make use of surrogate

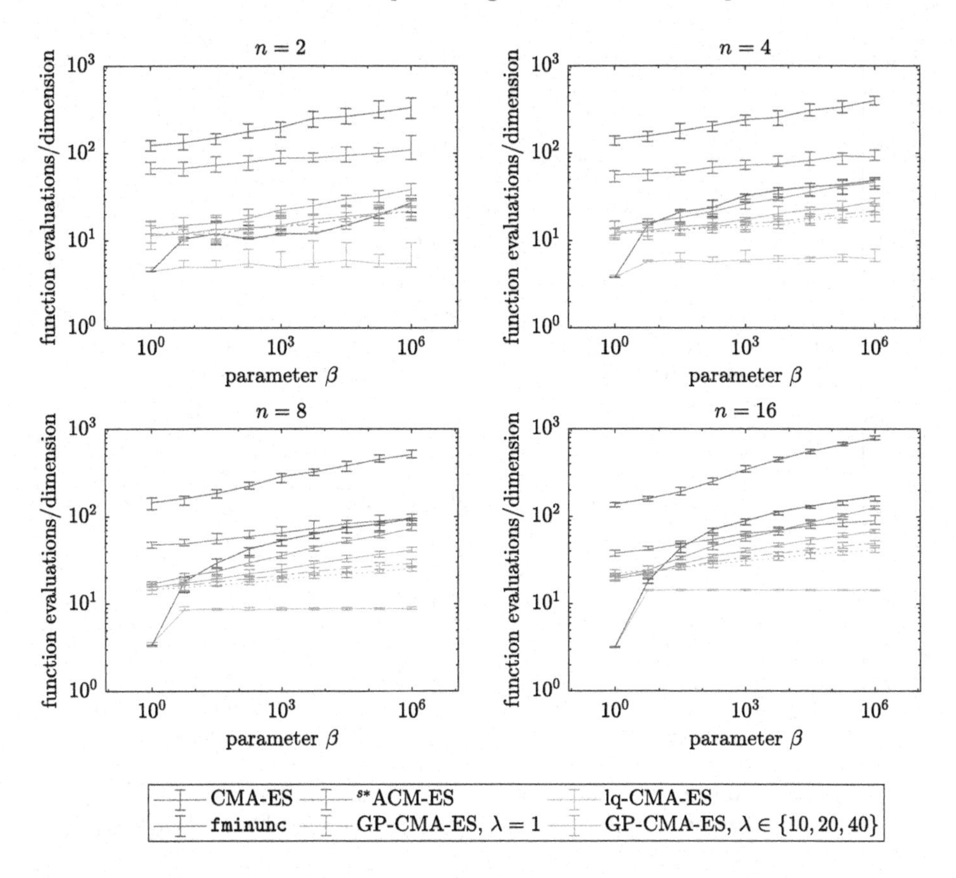

Fig. 3. Number of objective function evaluations per dimension required to optimize ellipsoid functions with parameter $\beta \in [10^0, 10^6]$. The lines connect median values; the error bars reflect the full range of values observed for the respective algorithms.

model assistance. The impact of batch size parameter λ on the performance of the algorithm overall appears rather minor, with more intensive exploitation of the surrogate models in most cases yielding a small improvement.

Ellipsoid Functions: Figure 3 plots the number of objective function evaluations divided by the dimension for ellipsoid functions with parameter $\beta \in [10^0, 10^6]$. The quadratic sphere function is at the left hand edge of the subplots, and the degree of ill-conditioning in each subplot increases from left to right. It can be seen that the performance advantage of ^{s*}ACM-ES over CMA-ES without surrogate model assistance widens with increasing β. Even though the problem is quadratic and thus optimally suited for a quasi-Newton approach, ^{s*}ACM-ES outperform `fminunc` for the higher condition numbers at $n = 16$.

GP-CMA-ES can be seen to generally outperform ^{s*}ACM-ES. Using larger values of λ and thus more intensively exploiting the surrogate models is never

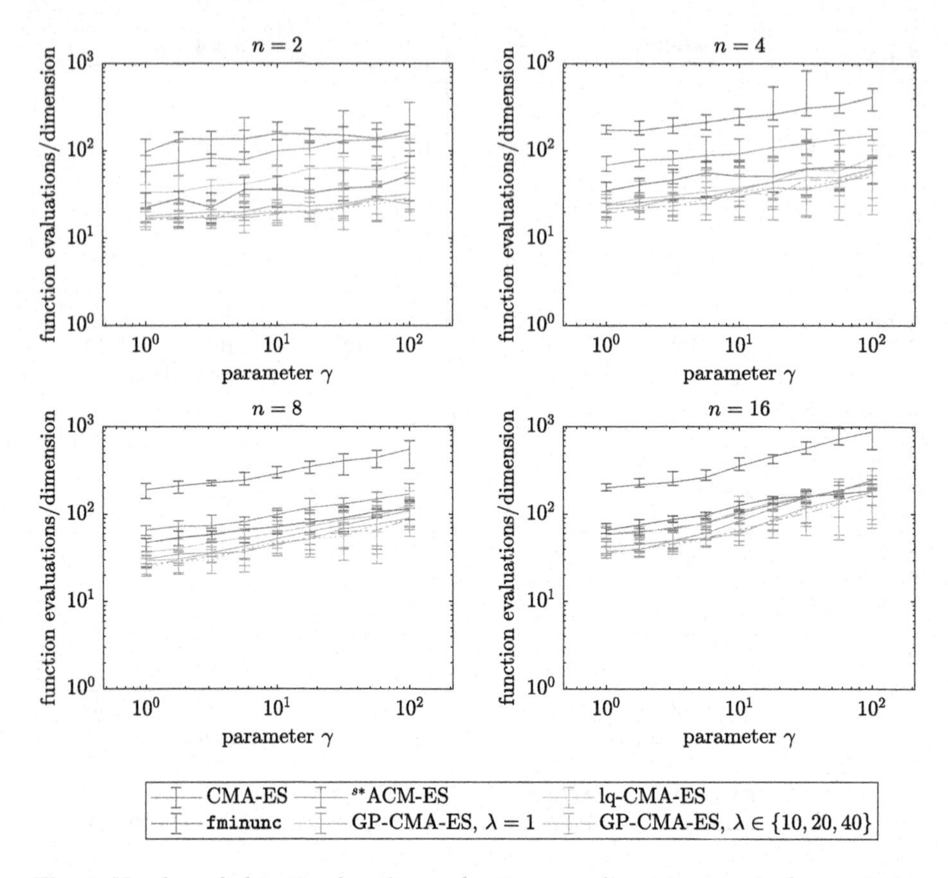

Fig. 4. Number of objective function evaluations per dimension required to optimize Rosenbrock functions with parameter $\gamma \in [10^0, 10^2]$. The lines connect median values; the error bars reflect the full range of values observed for the respective algorithms.

observed to be detrimental and except for very low dimensions and condition numbers provides a significant speed-up over the algorithm with $\lambda = 1$. The speed-up compared to ^{s*}ACM-ES for $\lambda = 40$ appears to be nearly independent of the degree of ill-conditioning, but decreases with increasing dimension. While speed-ups in the vicinity of five are observed for $n = 2$, corresponding values for $n = 16$ are in the vicinity of two.

lq-CMA-ES match the performance of `fminunc` for the sphere function, and they significantly outperform all of the other algorithms for $\beta > 1$. Moreover, the performance of lq-CMA-ES for all ellipsoids considered other than the sphere is insensitive to the conditioning parameter. Once the algorithm has performed sufficiently many objective function evaluations to build an accurate model, it is able to locate the optimizer of the objective based on that model alone.

Rosenbrock Functions: Figure 4 plots the number of objective function evaluations divided by the dimension for Rosenbrock functions with parameter $\gamma \in [10^0, 10^2]$. The original generalized Rosenbrock function is located at the right hand edge of the subplots. As the error bars show, variations from the median are much larger for this family of functions than for the other families considered. With the potential exception of $n = 2$, ^{s*}ACM-ES provide a sizable speed-up compared to CMA-ES that do not make use of surrogate model assistance. Their performance approaches or even exceeds that of the quasi-Newton algorithm for $n = 16$. lq-CMA-ES exhibit a performance advantage over ^{s*}ACM-ES that decreases with increasing dimension and disappears for $n = 16$.

Considering median values, GP-CMA-ES require fewer objective function evaluations to optimize Rosenbrock functions than any of the other algorithms considered. The algorithm achieves a speed-up over ^{s*}ACM-ES by a factor of about four for $n = 2$. However, that gap narrows for higher dimensions, and for $n = 16$ median values of all surrogate model assisted algorithms as well as of `fminunc` are within a factor of two of each other.

5 Conclusions

To conclude, we have introduced GP-CMA-ES, a Gaussian process assisted evolution strategy with covariance matrix adaptation. The algorithm uses the evolution strategy's covariance matrix in the kernel function of the Gaussian process. The only parameter to be set by the user is the number of trial points generated per iteration. That parameter controls the intensity of surrogate model exploitation, and its setting has been found to be uncritical for the test problems considered, with larger values affording a moderate speed-up on some problems. We do not expect the algorithm to match the performance of the more sophisticated surrogate model assisted evolution strategies when applied to more difficult optimization problems, such as those that include noise or multiple local optimizers. Instead, we consider the algorithm notable for its relative simplicity and a baseline for other algorithms to compare against.

A number of interesting observations were made in the comparison that involved CMA-ES without surrogate model assistance as well as several surrogate model assisted CMA-ES variants and a quasi-Newton algorithm. lq-CMA-ES excel predominantly on the quadratic test problems, where they are able to locate optimal solutions with almost no objective function evaluations once accurate models of the objective have been built. For the ill-conditioned ellipsoids, they very significantly outperform all other algorithms considered. For problems that deviate from being quadratic, the invariance with regard to order preserving transformations of function values of ^{s*}ACM-ES is attractive. That algorithm does not appear to be as well tuned as the other surrogate model assisted CMA-ES variants when the problem dimension is very small, but it becomes increasingly competitive with higher dimensions. GP-CMA-ES do not employ quadratic models and are thus less efficient than lq-CMA-ES when applied to quadratic problems. They do not share the invariance properties of ^{s*}ACM-ES, but they exhibit at least the second best performance of all of the surrogate model assisted CMA-ES variants across all problems considered.

Acknowledgements. This research was supported by the Natural Sciences and Engineering Research Council of Canada (NSERC).

References

1. Bajer, L., Pitra, Z., Repický, J., Holeňa, M.: Gaussian process surrogate models for the CMA evolution strategy. Evol. Comput. **27**(4), 665–697 (2019)
2. Bouzarkouna, Z., Auger, A., Ding, D.Y.: Investigating the local-meta-model CMA-ES for large population sizes. In: Di Chio, C., et al. (eds.) EvoApplications 2010. LNCS, vol. 6024, pp. 402–411. Springer, Heidelberg (2010). https://doi.org/10.1007/978-3-642-12239-2_42
3. Hansen, N.: The CMA evolution strategy: a tutorial. arxiv:1604.00772 (2016)
4. Hansen, N.: A global surrogate assisted CMA-ES. In: Genetic and Evolutionary Computation Conference – GECCO 2019, pp. 664–672. ACM Press (2019)
5. Hansen, N., Arnold, D.V., Auger, A.: Evolution strategies. In: Kacprzyk, J., Pedrycz, W. (eds.) Springer Handbook of Computational Intelligence, pp. 871–898. Springer, Heidelberg (2015). https://doi.org/10.1007/978-3-662-43505-2_44
6. Hansen, N., Müller, S.D., Koumoutsakos, P.: Reducing the time complexity of the derandomized evolution strategy with covariance matrix adaptation (CMA-ES). Evol. Comput. **11**(1), 1–18 (2003)
7. Hansen, N., Ostermeier, A.: Completely derandomized self-adaptation in evolution strategies. Evol. Comput. **9**(2), 159–195 (2001)
8. Igel, C., Suttorp, T., Hansen, N.: A computational efficient covariance matrix update and a (1+1)-CMA for evolution strategies. In: Genetic and Evolutionary Computation Conference – GECCO 2006, pp. 453–460. ACM Press (2006)
9. Jin, Y.: Surrogate-assisted evolutionary computation: recent advances and future challenges. Swarm Evol. Comput. **1**(2), 61–70 (2011)
10. Kayhani, A., Arnold, D.V.: Design of a surrogate model assisted (1 + 1)-ES. In: Auger, A., Fonseca, C.M., Lourenço, N., Machado, P., Paquete, L., Whitley, D. (eds.) PPSN 2018. LNCS, vol. 11101, pp. 16–28. Springer, Cham (2018). https://doi.org/10.1007/978-3-319-99253-2_2
11. Kern, S., Hansen, N., Koumoutsakos, P.: Local meta-models for optimization using evolution strategies. In: Runarsson, T.P., Beyer, H.-G., Burke, E., Merelo-Guervós, J.J., Whitley, L.D., Yao, X. (eds.) PPSN 2006. LNCS, vol. 4193, pp. 939–948. Springer, Heidelberg (2006). https://doi.org/10.1007/11844297_95
12. Kern, S., Müller, S.D., Hansen, N., Büche, D., Ocenasek, J., Koumoutsakos, P.: Learning probability distributions in continuous evolutionary algorithms – a comparative review. Nat. Comput. **3**(1), 77–112 (2004)
13. Loshchilov, I.: Surrogate-assisted evolutionary algorithms. Ph.D. thesis, Université Paris Sud - Paris XI (2013)
14. Loshchilov, I., Schoenauer, M., Sebag, M.: Comparison-based optimizers need comparison-based surrogates. In: Schaefer, R., Cotta, C., Kołodziej, J., Rudolph, G. (eds.) PPSN 2010. LNCS, vol. 6238, pp. 364–373. Springer, Heidelberg (2010). https://doi.org/10.1007/978-3-642-15844-5_37
15. Loshchilov, I., Schoenauer, M., Sebag, M.: Self-adaptive surrogate-assisted covariance matrix adaptation evolution strategy. In: Genetic and Evolutionary Computation Conference – GECCO 2012, pp. 321–328. ACM Press (2012)

16. Loshchilov, I., Schoenauer, M., Sebag, M.: Intensive surrogate model exploitation in self-adaptive surrogate-assisted CMA-ES. In: Genetic and Evolutionary Computation Conference – GECCO 2013, pp. 439–446. ACM Press (2013)
17. Rasmussen, C.E., Williams, C.K.I.: Gaussian Processes for Machine Learning. MIT Press, Cambridge (2006)
18. Rechenberg, I.: Evolutionsstrategie – Optimierung technischer Systeme nach Prinzipien der biologischen Evolution. Friedrich Frommann Verlag (1973)
19. Yang, J., Arnold, D.V.: A surrogate model assisted (1 + 1)-ES with increased exploitation of the model. In: Genetic and Evolutionary Computation Conference – GECCO 2019, pp. 727–735. ACM Press (2019)

Benchmarking and Performance Measures

Proposal of a Realistic Many-Objective Test Suite

Weiyu Chen, Hisao Ishibuchi(✉), and Ke Shang

Shenzhen Key Laboratory of Computational Intelligence,
University Key Laboratory of Evolving Intelligent Systems of Guangdong Province,
Department of Computer Science and Engineering,
Southern University of Science and Technology (SUSTech), Shenzhen, China
11711904@mail.sustech.edu.cn, hisao@sustech.edu.cn, kshang@foxmail.com

Abstract. Real-world many-objective optimization problems are not always available as easy-to-use test problems. For example, their true Pareto fronts are usually unknown, and they are not scalable with respect to the number of objectives or decision variables. Thus, the performance of existing multi-objective evolutionary algorithms is always evaluated using artificial test problems. However, there exist some differences between frequently-used many-objective test problems and real-world problems. In this paper, first we clearly point out one difference with respect to the effect of changing the value of each decision variable on the location of the objective vector in the objective space. Next, to make artificial test problems more realistic, we introduce a coefficient matrix to their problem structure. Then, we demonstrate that a different coefficient matrix leads to a different difficulty of a test problem. Finally, based on these observations, we propose a realistic many-objective test suite using various coefficient matrices.

Keywords: Many-objective optimization · Multi-objective evolutionary algorithm · Many-objective test suite · Coefficient matrix

1 Introduction

A multi-objective optimization problem (MOP) aims to optimize m conflicting objectives simultaneously. When the number of objectives of an MOP is larger than three (i.e., $m > 3$), it is referred to as a many-objective optimization problem (MaOP). In recent years, MaOPs have received increasing attention in the evolutionary multi-objective optimization field since they are widespread in real-world problems [3,12,13] and pose new challenges to optimization algorithms. An MOP or MaOP can be written as follows:

$$Minimize\ \boldsymbol{f}(\boldsymbol{x}) = (f_1(\boldsymbol{x}), f_2(\boldsymbol{x}), ..., f_m(\boldsymbol{x})),\ subject\ to\ \boldsymbol{x} \in S, \tag{1}$$

where $S \subseteq \mathbb{R}^n$ is the decision space and $\boldsymbol{x} = (x_1, x_2, \ldots, x_n)$ is the decision vector.

T. Bäck et al. (Eds.): PPSN 2020, LNCS 12269, pp. 201–214, 2020.
https://doi.org/10.1007/978-3-030-58112-1_14

Multi-objective evolutionary algorithms (MOEAs) (e.g., [5], [19], and [21]) have shown promising performance on MOPs. However, when the number of objectives increases, some of these algorithms do not perform well due to the severe decrease in selection pressure. To tackle this issue, some many-objective evolutionary algorithms (MaOEAs) (e.g., [4], [1], and [14]) which are specifically designed for MaOPs have been proposed in recent years.

Many scalable multi-objective test suites (e.g., DTLZ [6], WFG [9], LSMOP [2], and DBMOPP [8]) have been used to evaluate the performance of existing MOEAs. These test suites provide a wide variety of properties. Among these test suites, DTLZ [6] and WFG [9] are the most frequently used test suites[1]. These two test suites use the bottom-up approach [6] to design various problems. In the problems designed by this approach, decision variables are divided into two groups: position variables $\boldsymbol{x^p}$ and distance variables $\boldsymbol{x^d}$. Each objective function involves a position function $h(\boldsymbol{x^p})$ and a distance function $g(\boldsymbol{x^d})$. The inputs of the distance function $h(\boldsymbol{x^p})$ are distance variables while the inputs of the position function $g(\boldsymbol{x^d})$ are position variables.

The main difficulty of DTLZ and WFG as test suites is that the same distance function is used in all objectives in each test problem. As a result, a change of the value of a single distance variable leads to changes of all objective values in the same direction. This property is not very realistic and makes DTLZ and WFG relatively easy to solve [10]. To avoid this special property, a modular approach was proposed in [18]. In the modular approach, each distance function has different distance variables, and each distance variable is related to only a single objective. As a result, a change of the value of a single distance variable leads to a change of only a single objective value. This is also a rare occurrence.

In this paper, we examine the relation between the decision variables and the objectives in DTLZ and WFG and some real-world problems. Our analysis clearly shows a difference between the artificial test problems and the examined real-world problems with respect to the effect of changing the value of each decision variable on the values of the objectives. Based on this analysis, we propose a realistic many-objective test suite by introducing a coefficient matrix. We show that some existing test problems can be characterized by coefficient matrices. For example, all elements of the coefficient matrices of DTLZ are one. The modular DTLZ can be characterized by a diagonal coefficient matrix (i.e., all non-diagonal elements are zero). The main characteristic feature of the proposed test suite is that the coefficient matrix of each test problem is defined by real number elements in $[-1, 1]$ instead of zero-one elements. Thus, a change of the value of a single distance variable leads to different changes of objective values, which is consistent with the real-world problems examined in this paper.

The rest of the paper is organized as follows. The widely used artificial test problems and some real-world problems are compared in Sect. 2. In Sect. 3, we show the relation between the coefficient matrix and the difficulty of MaOPs. In

[1] The problem formulations of DTLZ7 and WFG7-9 are slightly different from the general structures mentioned in this paper. Due to the page limitation, the discussions in this paper do not include DTLZ7 and WFG7-9.

Sect. 4, we proposed a new test suite using various coefficient matrices. In Sect. 5, we report performance comparison results of some state-of-the-art algorithms on the proposed test suite. Finally, we conclude this paper in Sect. 6.

2 Analysis of Real-World and Artificial Test Problems

In the field of evolutionary many-objective optimization, the DTLZ [6] and WFG [9] test suites are the two most widely-used test suites. They are designed using similar methods and have similar problem structures. Generally, a DTLZ or WFG problem with m objective functions can be formulated as:

$$f_i(\boldsymbol{x}) = h_i(\boldsymbol{x^p}) + \alpha h_i(\boldsymbol{x^p}) g_i(\boldsymbol{x^d}) + \beta g_i(\boldsymbol{x^d}),\ i = 1, 2, \ldots, m, \tag{2}$$

where $\boldsymbol{x} = (\boldsymbol{x^p}, \boldsymbol{x^d})$ is the decision vector, $\boldsymbol{x^p} = (x_1, \ldots, x_{m-1})$ is the position vector, and $\boldsymbol{x^d} = (x_m, ..., x_n)$ is the distance vector. In Eq. (2), when $\alpha = 1$ and $\beta = 0$, it is DTLZ. When $\alpha = 0$ and $\beta = 1$, it is WFG.

For a problem in the DTLZ and WFG test suites, the distance functions $g_i(\boldsymbol{x^d})$, $i = 1, \ldots, m$ in Eq. (2) are the same (i.e., $g_1(\boldsymbol{x^d}) = g_2(\boldsymbol{x^d}) = ... = g_m(\boldsymbol{x^d})$). This characteristic can hardly be found in real-world problems. Therefore, a modular approach [18] was proposed to avoid this special characteristic. In modular DTLZ problems [18], each distance function is correlated with different parts of the distance variables. Formally, the distance function of the *i-th* objective is $g_i(\boldsymbol{x^d} | J_i)$ where $J_i = \{j | mod(j - m + 1 - i, m) = 0, j = m, m + 1, ..., n\}$. In this way, each distance function is different. However, this type of problems also have an unrealistic characteristic.

Figure 1 demonstrates the effect of changing the value of a randomly selected distance variable on the location of the objective vector for the three-objective DTLZ1 and modular DTLZ1 problems. Each solution (open circle) is generated from a single red solution in each figure by iterating the following two-step procedure. First, a single distance variable is randomly selected. Then, its value is randomly changed. A single distance variable is selected in each iteration since the mutation probability is often specified as $1/n$ where n is the number of decision variables (i.e., since the expected number of mutated variables is one). The results on the other DTLZ and WFG problems are similar to Fig. 1 (a), and the results on the other modular DTLZ problems are similar to Fig. 1 (b).

For problems with the same distance function for all objectives (e.g., DTLZ and WFG problems), all objective functions are increased or decreased simultaneously by changing the value of a single distance variable. The generated solutions are on the same line as shown in Fig. 1(a). Thus, each of the generated solutions is dominated by or dominates the original red solution (i.e., each solution is not non-dominated with the original red solution). In the modular DTLZ, the value of a single objective is changed by changing the value of a single distance variable. The generated solutions are on the lines parallel to the axes of the objective space as shown in Fig. 1 (b). Thus each of the generated solutions is dominated by or dominates the original red solution. These observations are inconsistent with the general case that almost all solutions are

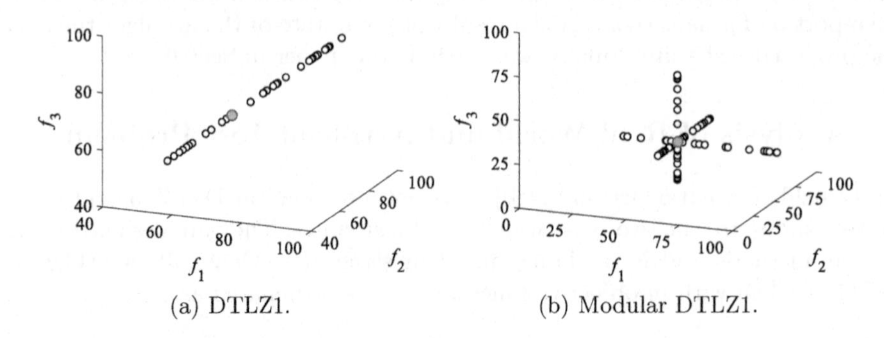

Fig. 1. Solutions generated by randomly changing the value of a single distance variable of the red solution. A different distance variable is randomly selected to generate each solution. (Color figure online)

non-dominated with each other in many-objective optimization. Since the number of non-dominated solutions generated by mutation is much smaller than the general case, these two types of problems are relatively easy for MOEA.

We use the same method as in Fig. 1 to analyze real-world MOPs. As real-world MOPs, we use RE problems [15] which are reformulations of some well-known real-world problems in the literature. Figure 2 shows the solutions generated by randomly changing one decision variable in the vehicle crashworthiness design problem [11] and the rocket injector design problem [16]. As we can see from Fig. 2, the generated solutions are not always dominated by or dominate the original solution. This is different from DTLZ and WFG in Fig. 1.

To further address the effect of changing the value of a decision variable on the changes of the objective values, we classify the decision variables into the following four types.

Type I: By the change of a decision variable of this type, all objective values increase or decrease simultaneously. Typical examples of Type I decision variables are distance variables in DTLZ and WFG test suites.
Type II: By the change of a decision variable of this type, some objective values increase or decrease simultaneously, and all the others remain unchanged.
Type III: By the change of a decision variable of this type, some objective values increase, some others decrease, and all the others remain unchanged.
Type IV: By the change of a decision variable of this type, some objective values increase, and the others decrease.

Table 1 summarizes the number of each type of decision variables in some artificial test problems and real-world problems. In some real-world problems, some objectives do not decrease after being less than zero. In this table, we only analyze the case when all objective values are larger than zero. In Table 1, there is a clear difference between frequently-used test suites and the examined real-world problems. Type I and Type II variables can hardly be found in real-world

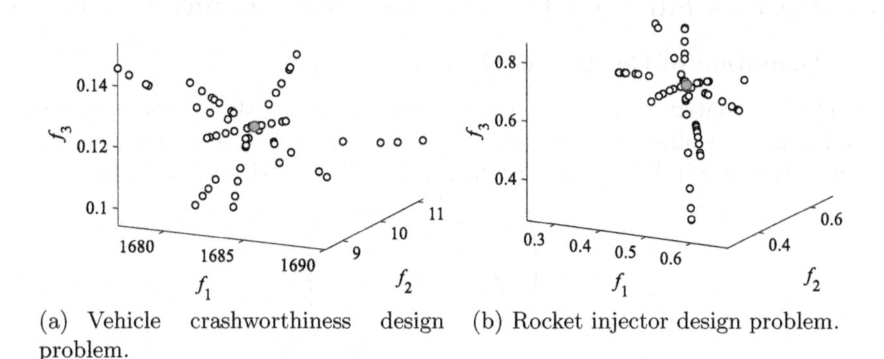

(a) Vehicle crashworthiness design problem.

(b) Rocket injector design problem.

Fig. 2. Solutions generated by randomly changing the value of a single decision variable of the red solution for the two real-world three-objective problems.

Table 1. The number of four types of decision variables in each test problem (m: the number of objectives, n: the total number of decision variables).

Multi-objective problems	m	n	Type I	II	III	IV
DTLZ1	5	14	10	0	0	4
WFG4	5	14	10	0	0	4
Modular DTLZ1	5	14	0	10	0	4
Vehicle crashworthiness design problem	3	5	0	0	0	5
Rocket injector design problem	3	4	0	0	0	4
Car side impact design problem	4	7	0	0	5	2
Conceptual marine design problem	4	6	0	0	0	6
Water resource planning problem	6	3	0	0	0	3

problems, especially when the number of objectives is large. However, they widely exist in artificial test problems. As shown in Fig. 1, generated solutions by the mutation of a Type I or Type II decision variable are dominated by or dominate the original solution. However, generated solutions by the mutation of a Type III or Type IV decision variable are likely to be non-dominated with the original solution as shown in Fig. 2. This makes artificial test problems easier than real-world problems, especially when the number of objectives is large. From these discussions, we can see the necessity of constructing realistic test problems with no Type I or Type II decision variables to evaluate the performance of MOEAs.

3 Distance Function Formulation with Coefficient Matrix

3.1 Definition of Coefficient Matrix

A coefficient matrix is used to design a test problem with a different distance function in each objective. For the m-objective optimization problem with the general structure in Eq. (2), the coefficient matrix $\boldsymbol{C}$ is defined as follows:

$$\boldsymbol{C} = \begin{pmatrix} c_{1,1} & c_{1,2} & \cdots & c_{1,m} \\ c_{2,1} & c_{2,2} & \cdots & c_{2,m} \\ \vdots & \vdots & \ddots & \vdots \\ c_{m,1} & c_{m,2} & \cdots & c_{m,m} \end{pmatrix}, \tag{3}$$

where $c_{i,j} \in \mathbb{R}$. The distance variables are divided into m groups ($\boldsymbol{x^d} = \boldsymbol{x^d_1} \cup \boldsymbol{x^d_2} \cup ... \cup \boldsymbol{x^d_m}$ and $\boldsymbol{x^d_i} \cap \boldsymbol{x^d_j} = \emptyset$ for $i \neq j$) where $\boldsymbol{x^d_j}$ is the j-th group of the distance variables. Then the distance function $g_i(\boldsymbol{x^d})$ is defined as

$$g_i(\boldsymbol{x^d}) = max\{0, \sum_{j=1}^{m} l(c_{i,j} \cdot \boldsymbol{x^d_j})\}, \tag{4}$$

where $\times$ is the standard arithmetic product operator and $l(c_{i,j} \times \boldsymbol{x^d_j})$ is the basic function that defines the basic fitness landscape.

Whereas we use the same basic function $l(\cdot)$ for all objectives, the coefficient for each variable is different. Thus the distance function is different for each objective. Note that the coefficient matrix in this paper is different from the correlation matrix in [2]. The correlation matrix shows the correlation between the decision variables and the objective functions. The elements in the correlation matrix can be only zero or one and they are multiplied with the basic function rather than decision variables.

3.2 Representation of Distance Functions in Existing Test Suites

We can represent some widely-used multi-objective test suites using coefficient matrices $\boldsymbol{C_1}$ or $\boldsymbol{C_2}$.

$$\boldsymbol{C_1} = \begin{pmatrix} 1 & 1 & \cdots & 1 \\ 1 & 1 & \cdots & 1 \\ \vdots & \vdots & \ddots & \vdots \\ 1 & 1 & \cdots & 1 \end{pmatrix}, \quad \boldsymbol{C_2} = \begin{pmatrix} 1 & 0 & \cdots & 0 \\ 0 & 1 & \cdots & 0 \\ \vdots & \vdots & \ddots & \vdots \\ 0 & 0 & \cdots & 1 \end{pmatrix}. \tag{5}$$

The coefficient matrix in the DTLZ test suites is $\boldsymbol{C_1}$. In a test problem with this coefficient matrix, each distance variable is involved in all distance functions (i.e., the distance functions are the same for all objectives). In the modular DTLZ problems, the coefficient matrix is $\boldsymbol{C_2}$. Each distance variable is involved in only a single distance function. In test problems with these two coefficient matrices, all distance variables are Type I or Type II variables. Therefore, as we discussed in Sect. 2, test problems based on these two coefficient matrices are not very realistic.

3.3 Effects of Real Number Coefficients in [0, 1]

We construct four 10-objective test problems with the following coefficient matrices to examine the effect of different coefficients.

Table 2. Average GD values of each algorithm over 31 runs on the problems with coefficient matrices $\boldsymbol{M_1}$ - $\boldsymbol{M_4}$.

	MOEA/D-PBI	RVEA	NSGA-III
$\boldsymbol{M_1}$	0.0011	0.0011	0.0011
$\boldsymbol{M_2}$	1.6016	3.6681	0.0010
$\boldsymbol{M_3}$	9.0694	3.3553	0.0010
$\boldsymbol{M_4}$	4.8037	15.639	0.0010

$$\boldsymbol{M_1} = \begin{pmatrix} 1 & 1 & \cdots & 1 \\ 1 & 1 & \cdots & 1 \\ \vdots & \vdots & \ddots & \vdots \\ 1 & 1 & \cdots & 1 \end{pmatrix}, \quad \boldsymbol{M_2} = \begin{pmatrix} 1 & 0.8 & \cdots & 0.8 \\ 0.8 & 1 & \cdots & 0.8 \\ \vdots & \vdots & \ddots & \vdots \\ 0.8 & 0.8 & \cdots & 1 \end{pmatrix},$$
$$\boldsymbol{M_3} = \begin{pmatrix} 1 & 0.6 & \cdots & 0.6 \\ 0.6 & 1 & \cdots & 0.6 \\ \vdots & \vdots & \ddots & \vdots \\ 0.6 & 0.6 & \cdots & 1 \end{pmatrix}, \quad \boldsymbol{M_4} = \begin{pmatrix} 1 & 0.4 & \cdots & 0.4 \\ 0.4 & 1 & \cdots & 0.4 \\ \vdots & \vdots & \ddots & \vdots \\ 0.4 & 0.4 & \cdots & 1 \end{pmatrix}. \tag{6}$$

We set α and β in Eq. (2) to 0 and 1, respectively. We use the position function of the DTLZ1 problem for all the four test problems. As the basic fitness landscape function $l(\boldsymbol{x})$, we use Rastrigin function [18] for all the four test problems:

$$l(\boldsymbol{x}) = 100(|\boldsymbol{x}| + \sum_{i=1}^{|\boldsymbol{x}|}(x_i^2 - cos(20\pi x_i)) \quad x_i \in [-0.5, 0.5], \tag{7}$$

where $|\boldsymbol{x}|$ means the number of dimensions of $\boldsymbol{x}$. When each element of the coefficient matrix is 0 or 1, each distance variable has the same effect (by the element value 1) or no effect (by the element value 0) on the distance function in each objective. Each decision variable is Type I or Type II. However, when each element of the coefficient matrix has a different non-zero value as in $\boldsymbol{M_2}$, $\boldsymbol{M_3}$ and $\boldsymbol{M_4}$, each distance variable has a different effect on the distance function in each objective. As a result, there are no Type I or Type II decision variables in the three test problems with $\boldsymbol{M_2}$, $\boldsymbol{M_3}$ and $\boldsymbol{M_4}$. These test problems may be more similar to real-world problems. Table 2 shows the average generational distance (GD) [17] values of the solutions obtained by MOEA/D-PBI [19], RVEA [1] and NSGA-III [4] on the four 10-objective test problems. The population size is set to 275 and the number of solution evaluations is set to 300,000. Each algorithm is executed 31 times for each test problem.

As shown in Table 2, all of the three algorithms converge well on the first test problem with the coefficient matrix $\boldsymbol{M_1}$. However, when the coefficient matrices include different non-zero elements (i.e., the other three test problems with $\boldsymbol{M_2}$, $\boldsymbol{M_3}$ and $\boldsymbol{M_4}$), MOEA/D-PBI and RVEA cannot achieve good convergence while the performance of NSGA-III does not change too much. The reason is that newly generated solutions by the mutation of a single distance variable are not always on a line (i.e., different from Fig. 1). This causes difficulties to some MOEAs which have a small improved region of a current solution in their fitness evaluation mechanisms.

3.4 Effects of Negative Coefficients

Negative coefficients can also be used in the coefficient matrix. In this section, we design four 10-objective test problems with the following four coefficient matrices to show the effect of negative coefficients.

$$\boldsymbol{M_5} = \begin{pmatrix} 1 & 0 & 0 & \cdots & 0 \\ 0 & 1 & 0 & \cdots & 0 \\ 0 & 0 & 1 & \cdots & 0 \\ \vdots & \vdots & \vdots & \ddots & \vdots \\ 0 & 0 & 0 & \cdots & 1 \end{pmatrix}, \boldsymbol{M_6} = \begin{pmatrix} 1 & 0 & -0.1 & \cdots & -0.1 \\ 0 & 1 & -0.1 & \cdots & -0.1 \\ 0 & 0 & 1 & \cdots & -0.1 \\ \vdots & \vdots & \vdots & \ddots & \vdots \\ 0 & 0 & -0.1 & \cdots & 1 \end{pmatrix},$$
$$\boldsymbol{M_7} = \begin{pmatrix} 1 & -0.1 & -0.1 & \cdots & -0.1 \\ 0 & 1 & -0.1 & \cdots & -0.1 \\ 0 & -0.1 & 1 & \cdots & -0.1 \\ \vdots & \vdots & \vdots & \ddots & \vdots \\ 0 & -0.1 & -0.1 & \cdots & 1 \end{pmatrix}, \boldsymbol{M_8} = \begin{pmatrix} 1 & -0.1 & -0.1 & \cdots & -0.1 \\ -0.1 & 1 & -0.1 & \cdots & -0.1 \\ -0.1 & -0.1 & 1 & \cdots & -0.1 \\ \vdots & \vdots & \vdots & \ddots & \vdots \\ -0.1 & -0.1 & -0.1 & \cdots & 1 \end{pmatrix}. \quad (8)$$

The problem structure and the position function are the same as in Sect. 3.3. We use a much easier basic landscape function $l(\boldsymbol{x}) = 100\sum_{i=1}^{|\boldsymbol{x}|} x_i$ where $x_i \in [0, 0.5]$. The experimental settings are the same as in Sect. 3.3. The average GD values of the solutions obtained by MOEA/D-PBI, RVEA and NSGA-III on the four test problems with the coefficient matrix $\boldsymbol{M_5}$, $\boldsymbol{M_6}$, $\boldsymbol{M_7}$ and $\boldsymbol{M_8}$ are shown in Table 3.

Table 3. Average GD values of each algorithm over 31 runs on the problems with coefficient matrices $\boldsymbol{M_5}$ - $\boldsymbol{M_8}$.

	MOEA/D-PBI	RVEA	NSGA-III
$\boldsymbol{M_5}$	0.00128	0.00125	0.21898
$\boldsymbol{M_6}$	0.01858	0.26805	0.42065
$\boldsymbol{M_7}$	0.66308	0.62469	0.73549
$\boldsymbol{M_8}$	0.97179	0.57320	0.68193

From Table 3, we can see that the problems defined by the coefficient matrices with negative elements are difficult for all the three algorithms, especially the

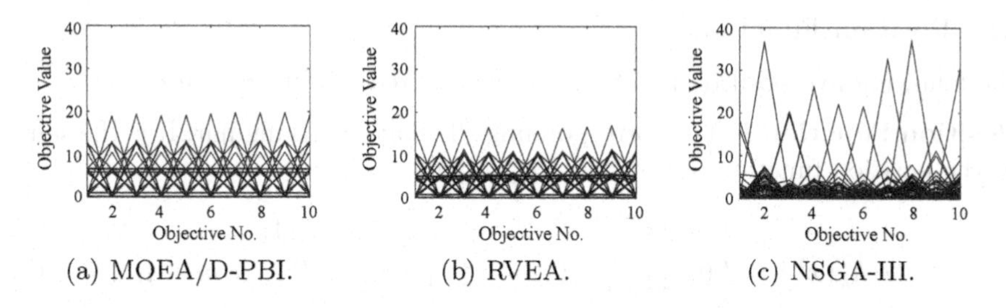

(a) MOEA/D-PBI. (b) RVEA. (c) NSGA-III.

Fig. 3. Solutions obtained by each algorithm on the problem with the coefficient matrix $\boldsymbol{M_8}$. A single run with the median GD value over 31 runs is selected for each algorithm.

two problems with the coefficient matrices $\boldsymbol{M_7}$ and $\boldsymbol{M_8}$. To further explain the reason, we show the solutions obtained by MOEA/D-PBI, RVEA and NSGA-III on the problem with the coefficient matrix $\boldsymbol{M_8}$ in Fig. 3. The obtained solutions do not converge well in Fig. 3. The reason is that if we change a single distance variable of the problem with the coefficient matrix $\boldsymbol{M_8}$, only one objective decreases while all the other objectives increase. Therefore, new solutions are always non-dominated with the original solution. This makes it very difficult for mutation operators to produce good solutions.

4 The Proposed Many-Objective Test Suite

4.1 Problem Characteristics

Coefficient Matrix. As we have already explained, the use of different non-zero elements in the coefficient matrix leads to a realistic problem formulation by decreasing the number of Type I and Type II distance variables. In the proposed test suite, we use different non-zero elements (including negative values) in the coefficient matrix in order to make each test problem realistic and challenging.

PF shape. The PF shape is considered as an essential characteristic when designing MOPs [9]. MOPs can have various PF shapes such as convex, linear, concave, disconnected, and degenerate. Besides, as pointed out by Ishibuchi et al. [10], the inverted PF shape has a strong effect on the performance of some MOEAs. We also include several problems with inverted PF shapes.

Problem Structure. If we set $\alpha = 1$ and $\beta = 0$ in Eq. (2), it is the structure of DTLZ. The DTLZ structure is not used in our test suite because it is not suitable for problems with different distance functions. In this paper, we use the WFG structure (i.e., $\alpha = 0$ and $\beta = 1$ in (2)) and the mixed structure (i.e., $\alpha = 1$ and $\beta = 1$ in (2)).

Due to the page limitation, we do not include some other characteristics such as bias, modality, parameter dependencies and constraints in this paper. Users can easily extend the proposed test suite to incorporate these characteristics.

4.2 Position Functions

The following five position functions are considered in our test suite.

Position Function I. The corresponding PF shape is triangular. It is the same as the position function of the DTLZ1 problem.

$$h_k^I(\boldsymbol{x^p}) = \begin{cases} 0.5\prod_{i=1}^{m-1} x_i^p, & k = 1, \\ 0.5\prod_{i=1}^{m-k} x_i^p(1 - x_{m-k+1}^p), & k = 2, ..., m-1, \\ 0.5(1 - x_1^p), & k = m. \end{cases} \tag{9}$$

Position Function II. The corresponding PF shape is concave. It is the same as the position function of the DTLZ2 problem.

$$h_k^{II}(\boldsymbol{x^p}) = \begin{cases} \prod_{i=1}^{m-1} cos(0.5\pi x_i^p), & k = 1, \\ (\prod_{i=1}^{m-k} cos(0.5\pi x_i^p))(sin(0.5\pi x_{m-k+1}^p)), & k = 2, ..., m-1, \\ sin(0.5\pi x_1^p), & k = m. \end{cases} \tag{10}$$

Position Function III. The corresponding PF shape is discontinuous. It is the same as the position function of the WFG2 problem but we

$$h_k^{III}(\boldsymbol{x^p}) = \begin{cases} h_k^{II}, & k = 1, ..., m-1, \\ 1 - x_i^p cos^2(5\pi x_1^p), & k = m. \end{cases} \tag{11}$$

Position Function IV. The corresponding PF shape is inverted triangular.

$$h_k^{IV}(\boldsymbol{x^p}) = 1 - h_k^I(\boldsymbol{x^p}). \tag{12}$$

Position Function V. The corresponding PF shape is inverted spherical.

$$h_k^V(\boldsymbol{x^p}) = 1 - h_k^{II}(\boldsymbol{x^p}). \tag{13}$$

4.3 Basic Landscape Functions

The following three basic landscape functions are considered in our test suite.

Basic Landscape Function I. Rastrigin function [6].

$$l(\boldsymbol{x}) = 100(|\boldsymbol{x}| + \sum_{i=1}^{|\boldsymbol{x}|}(x_i^2 - cos(20\pi x_i)), \quad x_i \in [-0.5, 0.5]. \tag{14}$$

Basic Landscape Function II. Linear function.

$$l(\boldsymbol{x}) = 100\sum_{i=1}^{|\boldsymbol{x}|} x_i, \quad x_i \in [0, 0.5]. \tag{15}$$

Basic Landscape Function III.

$$l(\boldsymbol{x}) = 100\sum_{i=1}^{|\boldsymbol{x}|}(x_i + sin(x_i)), \quad x_i \in [0, 0.5]. \tag{16}$$

4.4 Problem Definitions

The general formulation of the proposed problems is shown in Eqs. (1)–(3). The position functions, basic landscape functions and coefficient matrices of the proposed test problems are listed in Table 4. The true PFs of the proposed problems depend on the position functions and are the same as DTLZ or WFG problems with the same position functions.

5 Experimental Results

We examine the performance of some widely used MOEAs on DTLZ1-4 and the problems proposed in this paper. The examined algorithms include two classic algorithms (NSGA-II [5], MOEA/D-PBI [19]) and three algorithms that are specifically designed for MaOPs (NSGA-III [4], RVEA [1], and MaOEA-IGD [14]). The proposed problems are scalable to any number of decision variables and any number of objectives. In our experiments, we set the number of objectives to 10. The number of position variables is 9 and the number of distance variables is 10. The maximum number of solution evaluations is 300,000. The polynomial mutation and the simulated binary crossover (SBX) [7] are used in all algorithms. The SBX probability is set to 1 and the polynomial mutation probability is set to $1/n$. The distribution index is specified as 20.The inverted generational distance (IGD) indicator [20] is used to evaluate the performance of each algorithm, which is executed 31 times for each test problem. Experimental results are analyzed by the Wilcoxon rank-sum test with a significance level of 0.05. The "+", "−" and "=" indicate the performance of one algorithm is "significantly better than", "significantly worse than" and "statistically similar to" the baseline algorithm, respectively.

Table 4. Definitions of the proposed problems.

Problem	α	β	Coefficient matrix	Position function	Basic landscape function
CMP1	1	1	$\boldsymbol{M_2}$	IV	I
CMP2	1	1	$\boldsymbol{M_3}$	II	I
CMP3	0	1	$\boldsymbol{M_3}$	III	I
CMP4	1	1	$\boldsymbol{M_4}$	V	I
CMP5	0	1	$\boldsymbol{M_7}$	I	II
CMP6	1	1	$\boldsymbol{M_7}$	IV	II
CMP7	0	1	$\boldsymbol{M_8}$	II	II
CMP8	1	1	$\boldsymbol{M_8}$	IV	II
CMP9	1	1	$\boldsymbol{M_8}$	V	III

Table 5 shows the average IGD values of solutions obtained by the five algorithms on the 10-objective DTLZ problems and the proposed test suite. As we can observe from the table, performance comparison results on the DTLZ

Table 5. Average IGD values by each algorithm on the 10-objective DTLZ problems and the proposed test problems.

Problem	NSGA-II	MOEA/D-PBI	NSGA-III	RVEA	MaOEA-IGD
DTLZ1	41.568	0.1095 +	0.1308 +	**0.1083 +**	2.5304 +
DTLZ2	2.2026	0.4220 +	0.4292 +	**0.4212 +**	0.4384 +
DTLZ3	1390.6	0.4712 +	0.5215 +	**0.4201 +**	5.5938 +
DTLZ4	1.9357	0.5491 +	0.4429 +	**0.4204 +**	0.4294 +
+/−/=	baseline	4/0/0	4/0/0	4/0/0	4/0/0
CMP1	**0.1342**	257.47 −	0.1436 −	11.847 −	34.834 −
CMP2	0.5234	102.01 −	**0.4780 +**	5.8131 −	37.645 −
CMP3	0.1968	95.209 −	**0.0996 +**	0.1099 +	47.212 −
CMP4	**0.5656**	128.79 −	0.5992 −	37.331 −	29.424 −
CMP5	21.866	14.884 +	4.8238 +	**4.6258 +**	16.361 +
CMP6	43.399	20.407 +	6.1768 +	**6.1532 +**	24.105 +
CMP7	16.928	10.107 +	**4.2698 +**	5.8289 +	16.276 =
CMP8	24.580	14.104 +	**6.5961 +**	7.9268 +	18.581 +
CMP9	35.747	50.658 =	**16.344 +**	19.058 +	48.024 =
+/−/=	baseline	4/4/1	7/2/0	6/3/0	3/4/2

problems are different from those on the proposed test suite. On DTLZ1 to DTLZ4, RVEA always gets the best results, and NSGA-II is outperformed by all the other algorithms. However, on the proposed test problems, these new algorithms do not always perform better than NSGA-II. Besides, none of the examined algorithms can always achieve a good IGD value on all the proposed test problems. Especially on CMP9, the IGD values by all the algorithms are larger than 10, which means that the obtained solutions are inferior. That is, the proposed test suite is more difficult than DTLZ.

6 Conclusion and Future Work

In this paper, first we analyzed the difference between frequently-used many-objective test problems and some real-world problems. Next, to construct more realistic test problems, a coefficient matrix was introduced. We showed that the choice of a coefficient matrix has a large effect on the difficulty of a test problem. Then, we proposed a new many-objective test suite with various features. Our test problems in the proposed test suite have no Type I variables, and only two problems with $\boldsymbol{M_7}$ have a single group of Type II variables (since the first column of $\boldsymbol{M_7}$ has only zero and one elements). This implies that our test problems are more similar to real-world problems than the frequently-used many-objective test problems with Type I and Type II distance variables. The experimental results showed that the proposed problems are more difficult than DTLZ, which poses new challenges to existing MOEAs.

The problem formulation based on the coefficient matrix is simple. By defining other coefficient matrices, we can formulate test problems with a wide range of difficulty. Examination of other coefficient matrices is an interesting future research topic. It is also an interesting future research direction to include many other characteristic features in the proposed test suite to make it more realistic and more diverse.

Acknowledgements. This work was supported by National Natural Science Foundation of China (Grant No. 61876075), the Program for Guangdong Introducing Innovative and Enterpreneurial Teams (Grant No. 2017ZT07X386), Shenzhen Science and Technology Program (Grant No. KQTD2016112514355531), the Science and Technology Innovation Committee Foundation of Shenzhen (Grant No. ZDSYS201703031748284), the Program for University Key Laboratory of Guangdong Province (Grant No. 2017KSYS008).

References

1. Cheng, R., Jin, Y., Olhofer, M., Sendhoff, B.: A reference vector guided evolutionary algorithm for many-objective optimization. IEEE Trans. Evol. Comput. **20**(5), 773–791 (2016)
2. Cheng, R., Jin, Y., Olhofer, M., Sendhoff, B.: Test problems for large-scale multiobjective and many-objective optimization. IEEE Trans. Cybern. **47**(12), 4108–4121 (2017)
3. Chikumbo, O., Goodman, E., Deb, K.: Approximating a multi-dimensional Pareto front for a land use management problem: a modified MOEA with an epigenetic silencing metaphor. In: 2012 IEEE Congress on Evolutionary Computation, pp. 1–9 (2012)
4. Deb, K., Jain, H.: An evolutionary many-objective optimization algorithm using reference-point-based nondominated sorting approach, part I: Solving problems with box constraints. IEEE Trans. Evol. Comput. **18**(4), 577–601 (2014)
5. Deb, K., Pratap, A., Agarwal, S., Meyarivan, T.: A fast and elitist multiobjective genetic algorithm: NSGA-II. IEEE Trans. Evol. Comput. **6**(2), 182–197 (2002)
6. Deb, K., Thiele, L., Laumanns, M., Zitzler, E.: Scalable multi-objective optimization test problems. In: Proceedings of the 2002 Congress on Evolutionary Computation CEC2002 (Cat. No.02TH8600), vol. 1, pp. 825–830 (2002)
7. Deb, K., Tiwari, S.: Omni-optimizer: a generic evolutionary algorithm for single and multi-objective optimization. Eur. J. Oper. Res. **185**(3), 1062–1087 (2008)
8. Fieldsend, J.E., Chugh, T., Allmendinger, R., Miettinen, K.: A feature rich distance-based many-objective visualisable test problem generator. In: Proceedings of the Genetic and Evolutionary Computation Conference GECCO 2019, pp. 541–549. Association for Computing Machinery, New York (2019). https://doi.org/10.1145/3321707.3321727
9. Huband, S., Hingston, P., Barone, L., While, L.: A review of multiobjective test problems and a scalable test problem toolkit. IEEE Trans. Evol. Comput. **10**(5), 477–506 (2006)
10. Ishibuchi, H., Setoguchi, Y., Masuda, H., Nojima, Y.: Performance of decomposition-based many-objective algorithms strongly depends on Pareto front shapes. IEEE Trans. Evol. Comput. **21**(2), 169–190 (2017)

11. Liao, X., Li, Q., Yang, X., Zhang, W., Li, W.: Multiobjective optimization for crash safety design of vehicles using stepwise regression model. Struct. Multi. Optim. **35**, 561–569 (2008). https://doi.org/10.1007/s00158-007-0163-x10.1007/s00158-007-0163-x
12. Lygoe, R.J., Cary, M., Fleming, P.J.: A real-world application of a many-objective optimisation complexity reduction process. In: Purshouse, R.C., Fleming, P.J., Fonseca, C.M., Greco, S., Shaw, J. (eds.) EMO 2013. LNCS, vol. 7811, pp. 641–655. Springer, Heidelberg (2013). https://doi.org/10.1007/978-3-642-37140-0_48
13. Narukawa, K., Rodemann, T.: Examining the performance of evolutionary many-objective optimization algorithms on a real-world application. In: 2012 Sixth International Conference on Genetic and Evolutionary Computing, pp. 316–319 (2012)
14. Sun, Y., Yen, G.G., Yi, Z.: IGD indicator-based evolutionary algorithm for many-objective optimization problems. IEEE Trans. Evol. Comput. **23**(2), 173–187 (2019)
15. Tanabe, R., Ishibuchi, H.: An easy-to-use real-world multi-objective optimization problem suite. Appl. Soft Comput. **89**, 106078 (2020)
16. Vaidyanathan, R., Tucker, P.K., Papila, N., Shyy, W.: Computational-fluid-dynamics-based design optimization for single-element rocket injector. J. Propul. Power **20**(4), 705–717 (2004)
17. Van Veldhuizen, D.A.: Multiobjective evolutionary algorithms: classifications, analyses, and new innovations. Ph.D. thesis, USA (1999)
18. Wang, Z., Ong, Y., Ishibuchi, H.: On scalable multiobjective test problems with hardly dominated boundaries. IEEE Trans. Evol. Comput. **23**(2), 217–231 (2019)
19. Zhang, Q., Li, H.: MOEA/D: a multiobjective evolutionary algorithm based on decomposition. IEEE Trans. Evol. Comput. **11**(6), 712–731 (2007)
20. Zhang, Q., Zhou, A., Jin, Y.: RM-MEDA: a regularity model-based multiobjective estimation of distribution algorithm. IEEE Trans. Evol. Comput. **12**(1), 41–63 (2008)
21. Zitzler, E., Künzli, S.: Indicator-based selection in multiobjective search. In: Yao, X., Burke, E.K., Lozano, J.A., Smith, J., Merelo-Guervós, J.J., Bullinaria, J.A., Rowe, J.E., Tiňo, P., Kabán, A., Schwefel, H.-P. (eds.) PPSN 2004. LNCS, vol. 3242, pp. 832–842. Springer, Heidelberg (2004). https://doi.org/10.1007/978-3-540-30217-9_84

Approximate Hypervolume Calculation with Guaranteed or Confidence Bounds

A. Jaszkiewicz(✉), R. Susmaga, and P. Zielniewicz

Institute of Computing Science, Poznan University of Technology, Piotrowo 2, 60-965 Poznań, Poland
andrzej.jaszkiewicz@cs.put.poznan.pl

Abstract. We present a new version of the Quick Hypervolume algorithm allowing calculation of guaranteed lower and upper bounds for the value of hypervolume, which is one of the most often used and recommended quality indicators in multiobjective optimization. To ensure fast convergence of these bounds, we use a priority queue of subproblems instead of the depth-first search applied in the original recursive Quick Hypervolume algorithm. We also combine this new algorithm with the Monte Carlo sampling approach, which allows obtaining better confidence intervals than the standard Monte Carlo sampling. The performance of the two proposed methods is compared with that of a straightforward adaptation of recursive Quick Hypervolume algorithm and the standard Monte Carlo sampling in a comprehensive computational experiment.

Keywords: Multiobjective optimization · Hypervolume indicator · Hypervolume calculation · Approximate hypervolume

1 Introduction

Hypervolume is one of the most often used and recommended quality indicators in multiobjective optimization [22]. It is used for experimental evaluation of multiobjective algorithms, tuning their parameters, and controlling so-called indicator-based multiobjective evolutionary algorithms [2,4,12,23], which constitute one of the main classes of evolutionary multiobjective optimization (EMO) algorithms [14]. Although a significant progress in exact calculation of hypervolume has been obtained in recent years [3,10,17,18,21], its exact calculation may remain impractical for larger sets of points and/or larger number of objectives, owing to exponential time complexity of exact algorithms [2,3]. Although efficient algorithms exist for the number of objectives $d = 3$ [3] and $d = 4$ [8], the best known algorithm for the general case has $\mathcal{O}(n^{d/3} \operatorname{polylog} n)$ time complexity [6]. Because of this fact, hypervolume estimation algorithms, e.g. ones based on Monte Carlo sampling, are often used [2,7,19,20]. A disadvantage of such methods is, however, that they can provide only an estimated value and, in some cases, a confidence interval with no guaranteed bounds.

T. Bäck et al. (Eds.): PPSN 2020, LNCS 12269, pp. 215–228, 2020.
https://doi.org/10.1007/978-3-030-58112-1_15

In this paper, we modify the Quick Hypervolume (QHV) algorithm [10,17, 18] to allow calculation of guaranteed lower and upper bounds for the value of hypervolume with fast convergence of these bounds. Furthermore, we show that by combining the proposed modification of QHV with Monte Carlo sampling, we can obtain better confidence intervals than using the standard Monte Carlo approach given the same computing time.

The rest of the paper is organized as follows. Sections 2 and 3 state the problem and present existing methods, respectively. Section 4 describes the methods proposed in this paper, while Sect. 5 presents the experiments. The paper ends with conclusions and directions for further research.

2 Problem Statement

A point $s^1 \in \mathbb{R}^d$, where $\mathbb{R}^d$ is a space of maximized objectives, *dominates* a point $s^2 \in \mathbb{R}^d$ if, and only if $s^1_j \geq s^2_j \; \forall j \in \{1,\ldots,d\} \wedge \exists j \in \{1,\ldots,d\} : s^1_j > s^2_j$. We denote this relation by $s^1 \succ s^2$. A *hypercuboid* in $\mathbb{R}^d$ parallel to the axes is defined by two extreme points $r_* \in \mathbb{R}^d$ and $r^* \in \mathbb{R}^d$, $r^* \succ r_*$, such that $H(r^*, r_*) = \{s \in \mathbb{R}^d \;\; | \forall j \in \{1,\ldots,d\} \; r_{*j} \leq s_j \leq r^*_j\}$.

Assume that a finite set of points in $\mathbb{R}^d$ is given $S \subset H(r^*, r_*)$. This set may be, for example, an outcome of an EMO algorithm (i.e. a set of points in the objective space corresponding to the generated solutions) or its current population. The hypervolume of the space dominated by S within hypercuboid $H(r^*, r_*)$, denoted by $\mathcal{H}(S, H(r^*, r_*))$, is the Lebesgue measure of the set $\bigcup_{s \in S} H(s, r_*)$. Simultaneously, $\mathcal{H}(Z)$ denotes the Lebesgue measure of the set Z (in particular: a hypercuboid).

3 Related Works

Because of the excessive computational complexity of the exact hypervolume calculation, most of the research in this field is devoted to speeding up this process. Various approaches have been proposed, some focused on improving theoretical and practical efficiency of exact algorithms [3,10,17,18,21], while others on quicker algorithms providing approximated values of hypervolume.

Probably the first hypervolume estimation algorithm is presented in [2]. The evolutionary algorithm presented in this paper, known as HypE, is controlled by hypervolume, but uses Monte Carlo sampling for faster hypervolume calculation. Points generated randomly in the underlying hypercuboid are classified as either dominated or non-dominated using a non-dominance test. Because the additional points are generated uniformly within the hypercuboid, the relative number of dominated points becomes an estimation of the relative hypervolume of the dominated region, and the quality of this estimation increases with the number of the generated points. In other words, the process is equivalent to estimating the probability of success in a binomial distribution.

In [1] Monte Carlo sampling was adapted to faster calculation of the contribution of a single point to the hypervolume indicator. Another Monte Carlo-based approach for incremental hypervolume estimation is presented in [7]. In this method the number of domination comparisons is reduced by exploiting incremental properties of the approximated Pareto front. By using information from previous time steps, the estimation of hypervolume monotonically improves over time. The problems considered in these papers are, however, different from the one tackled in this paper because they concern estimation of contributions of individual points [1] or dynamic estimation of hypervolume [7], while we deal with static approximation of hypervolume for a known set of solutions.

The algorithm proposed in [20] is also based on the Monte Carlo sampling method, but it uses a Partial Precision and Partial Approximation (PPPA) decomposition strategy to decrease the number of points to be sampled for the hypervolume calculation. This strategy is based on a particular way of splitting the hypercuboids, which is similar but different from that used in QHV [17] and QHV-II [10], and on choosing the number of sample points, which depends on the size of hypervolume. The algorithm has been compared to the standard Monte Carlo sampling on data sets with up to 300 points. Since, however, this method does not provide a confidence interval, the difference between the exact and estimated values was used as an quality indicator.

A different approach was developed in a series of papers starting with [9], which introduces what is called the 'R2 indicator' of hypervolume. That value, based on computing distances instead of volumes, is computed much quicker that the actual value. This indicator was later developed in [15] and [19]. Note that methods based on R2 indicator do not provide a confidence interval but only an estimation of the hypervolume. Further, experiments reported in the two earlier publications generally concern multi-objective optimization, of which calculating the hypervolume constitutes only a small part. On the other hand, [19] is directly concerned with calculating R2 (which approximates the hypervolume), but uses only relatively easy data sets with up to 5 objectives and does not compare its results with that of Monte Carlo sampling.

We are not aware of any algorithm aimed at providing guaranteed lower and upper bounds for hypervolume.

4 Proposed Methods

4.1 QHV with Bounds Calculation Using Priority Queue

Quick hypervolume (QHV) algorithm is one of the fastest exact algorithms for hypervolume calculation [10,17,18]. It is based on the following observations [10]:

1. $\forall_{s^p \in S}\ \mathcal{H}(S, H(r^*, r_*)) = \mathcal{H}(H(s^p, r_*)) + \mathcal{H}\Big(\big(\bigcup_{s \in S \setminus \{s^p\}} H(s, r_*)\big) \setminus H(s^p, r_*)\Big)$,
   i.e. hypervolume of the space dominated by S is equal to the hypervolume of the hypercuboid defined by a single point $s^p \in S$ and r_*, i.e. $\mathcal{H}(H(s^p, r_*))$, plus the hypervolume of the region dominated by remaining points, i.e. $S \setminus \{s^p\}$ excluding the hypervolume of hypercuboid $H(s^p, r_*)$.

2. The remaining region $H(r^*, r_*) \setminus H(s^p, r_*)$ can be expressed as a union of non-overlapping hypercuboids $\{H_1, ..., H_L\}$, where $H_l = H(r_l^*, r_{l*}), l = 1, ..., L$.
3. For a point $s \notin H_l \wedge s \succ r_{l*}$, the hypervolume of the space dominated by s within H_l is equal to the hypervolume of the space dominated by a projection of s onto H_l. Projection means that the coordinates of the point to be projected which are larger than the corresponding coordinates of r_l^* are replaced by the corresponding coordinates of r_l^* (for non-dominated sets S these projections are only needed when $d \geq 2$).

These observations allow calculation of hypervolume in a divide-and-conquer manner. QHV algorithm selects a pivot point, calculates hypervolume of the region dominated by the pivot point, and then splits the remaining problem (corresponding to the remaining region) into a number of sub-problems corresponding to hypercuboids $\{H_1, ..., H_L\}$. If the number of points is sufficiently small, it uses simple geometric properties to calculate the hypervolume.

Different versions of QHV differ in the way they split the remaining region into hypercuboids. In this paper, we use the splitting scheme proposed in Improved Quick Hypervolume algorithm (QHV-II) [10], which splits the remaining region $H(r^*, r_*) \setminus H(s^p, r_*)$ into d smaller hypercuboids in the following way:

- H_1 is defined by the condition $s_1 > s_1^p$,
- $\forall_{j \in \{1,...,d\}}$: H_j is defined by the conditions $s_l \leq s_l^p \, \forall \, l = 1, \ldots, j-1 \wedge s_j > s_j^p$

Note, however, that the algorithm proposed in this paper is general and could also be used with the splitting scheme proposed in the original QHV algorithm [17], where $L = 2^d - 2$. As the pivot point s^p the point $s \in S$ with maximum $\mathcal{H}(H(s, r_*))$ is selected.

The algorithm for calculation of guaranteed hypervolume bounds proposed in this paper is based on the observation that whenever a pivot point is selected and the remaining region is split into smaller hypercuboids, the lower and upper bounds may be updated. The starting lower bound and upper bounds are 0 and $\mathcal{H}(H(r^*, r_*))$, respectively. In each iteration, the hypervolume of the region dominated by the pivot point is added to the lower bound, and the hypervolume of all empty (i.e. containing no points in S or projections of points in S) remaining hypercuboids is subtracted from the upper bound.

The original QHV algorithm works in a recursive manner. Of course, it is possible to use the same recursive scheme and update lower and upper bounds in each step. Such a bound generating, straightforward adaptation of QHV will be referred to as QHV-BR (QHV with Bounds calculation using Recursion).

The recursive algorithm used in QHV (and QHV-BR) corresponds, however, to the depth-first search of the tree of subproblems (and the associated hypercuboids). In result, some very small subproblems (i.e. with small hypervolume of the hypercuboids) are processed before other much larger subproblems. This leads to a poor any-time behaviour, i.e. to slow convergence of the lower and upper bounds, because processing small subproblems may reduce the gap between these bounds only slightly, while other larger subproblems, that could yield much larger reduction of this gap, remain unprocessed.

Thus, we propose the following modification of QHV. New subproblems are placed in a priority queue that is sorted according to the hypervolumes of the corresponding hypercuboids. At each iteration, a subproblem with the largest (or relatively large) hypervolume is selected for processing. Such approach ensures a good any-time behaviour of the proposed method, i.e. the gap between lower and upper bound is reduced faster at the beginning, because larger subproblems are processed first. We call this algorithm QHV-BQ (QHV with Bounds calculation using priority Queue). Note that although other exact algorithms for hypervolume calculation could probably be also adapted for calculation of lower and/or upper bounds, the advantage of QHV is that it is possible to organize the process such that the largest hypercuboids are processed first.

While the above idea is relatively simple, it raises a number of technical issues. The number of subproblems in the priority queue may grow very fast. The worst case size of the queue may be derived in a way similar to the worst case time complexity for QHV-II algorithm reported in [10], i.e. assuming that all points except of the pivot point are assigned to each sub-problem. In such case $|Q(n)| = d|Q(n-1)|$, where $|Q(n)|$ is the size of the queue needed to process n points. Solving the above recurrence $|Q(n)| = cd^{n-1}$, where c is an arbitrary constant. Thus, $|Q(n)| = \mathcal{O}(d^{n-1})$. In the presented computational experiment the size of the queues exceeds 200 million (see Fig. 2). Even if contemporary computers may store this amount of data in the memory, the processes of memory management and of updating the priority queue become bottlenecks of the algorithm. Thus in our C++ implementation applied in the reported experiments we decided to use a number of specific design decisions:

- Instead of the standard C++ `priority_queue` class we use our own data structure that uses bucket sort. The queue is divided into a number of buckets (10000 in our case). Each bucket corresponds to a range of hypervolume decreasing in a logarithmic way. In each bucket an unsorted list of subproblems is kept. The next subproblem to be processed is taken from the first non-empty (current) bucket containing largest unprocessed subproblems. To ensure that the subproblem with the largest hypervolume is selected, the current bucket could be sorted, but we did not observe any practical advantage of this additional sorting, so we keep each bucket unsorted. This 'internal' sorting is not necessary because all subproblems placed in one bucket have very similar hypervolume, so the order in which they are processed has negligible influence on the performance of the algorithm. Furthermore, the set of used (non-empty) bucket indices is stored in the `set` class. Whenever the current bucket is emptied, the next non-empty bucked is chosen from this set (which takes advantage of the internal sorting of the `set` class to speed up this selection). Note that new subproblems cannot be larger than the current one, so they will never be placed in a preceding bucket.
- To avoid costly memory allocation/deallocation of subproblems we implemented our own pool of subproblems with `getNew` and `free` operations. Internally, the pool keeps a list of subproblems and a priority queue (`priority_queue` class) of indices of free positions in the list. `getNew` operation returns

the first free position (and extends the list if necessary), while `free` operation adds new free index to the priority queue of free indices.
- As suggested in [10,17,18], the projected points are not constructed explicitly, but their coordinates are calculated on demand to save memory requirements. The indices of (unprojected) points belonging to particular subproblems are stored in a single list in continuous regions and for each subproblem only the starting and ending indices of this region are kept. In result, for each subproblem the same amount of memory is required, which makes it possible to apply the allocation/deallocation approach described above.

The proposed QHV-BQ method is summarized in Algorithm 1. Note, that it is general enough to use either the original QHV splitting scheme with $L = 2^d - 2$ [17], or the QHV-II splitting scheme with $L = d$ [10].

4.2 QHV-BQ Combined with Monte Carlo Sampling

The advantage of the proposed QHV-BQ algorithm over hypervolume estimation algorithms, and over Monte Carlo (MC) sampling in particular, is that it provides guaranteed bounds for hypervolume, while in the case of MC only confidence bounds could be calculated. Such guaranteed bounds, may be, for example, used for conclusive hypervolume-based comparison of two sets of points. If the lower bound for one of them is larger than the upper bound for the other set, the former is guaranteed to have a better hypervolume value. At the same time, estimation algorithms allow probabilistic comparison only.

In some cases, however, confidence intervals may be sufficient in a given context. As we show in the computational experiment described below, the confidence intervals yielded by Monte Carlo sampling may be smaller than the gaps between the guaranteed bounds yielded by QHV-BQ in the same time. Thus, we ask the question, if better confidence intervals could be obtained by combining QHV-BQ with Monte Carlo sampling.

The idea of the second method proposed in this paper is to start with a relatively short run of QHV-BQ and then switch to the Monte Carlo sampling. After stopping the QHV-BQ algorithm, a number of unprocessed (remaining) subproblems and corresponding hypercuboids are located in the queue. All regions outside these remaining hypercuboids are either dominated or not-dominated by points in S and the hypervolumes of these two regions (dominated and not-dominated) correspond to the lower bound and the difference between the hypervolume of the initial hypercuboid and the upper bound. Thus, to estimate the total hypervolume, only the remaining hypercuboids need to be sampled by the Monte Carlo method. Since the sum of hypervolumes of these remaining hypercuboids will be equal to the difference between the upper and lower bound and will usually be much smaller than the hypervolume of the initial hypercuboid, such an application of the Monte Carlo sampling could provide a much better confidence interval with the same number of samples.

To sample uniformly points in a set of hypercuboids, in each iteration at first a hypercuboid is drawn with the probability proportional to its hypervolume,

Algorithm 1. QHV-BQ algorithm

Input: r^*, r_* and $S \subset H(r^*, r_*)$
Output: *LowerBound* and *UpperBound* of $\mathcal{H}(S, H(r^*, r_*))$

Initialize the priority queue Q with subproblem defined by $H(r^*, r_*)$ and S
$LowerBound \leftarrow 0$
$UpperBound \leftarrow \mathcal{H}(H(r^*, r_*))$
repeat
  Take the next subproblem defined by S' and $H(r'^*, r'_*)$ from Q
  if S' contains one or two points **then**
    Calculate $\mathcal{H}(S', H(r'^*, r'_*))$ using simple geometric properties
    $LowerBound \leftarrow LowerBound + \mathcal{H}(S', H(r'^*, r'_*))$
    $UpperBound \leftarrow UpperBound - (\mathcal{H}(H(r'^*, r'_*)) - \mathcal{H}(S', H(r'^*, r'_*)))$
  else
    Select the pivot point $s^p \in S'$
    $LowerBound \leftarrow LowerBound + \mathcal{H}(H(s^p, r'_*))$
    Split $H(r'^*, r'_*) \setminus H(s^p, r'_*)$ into non-overlapping hypercuboids $\{H_1, \ldots, H_L\}$
    for all $H_l \in \{H_1, \ldots, H_L\}$ **do**
      Construct set S_l of points dominating r_{l*}, if necessary projected onto H_l
      if $S_l = \emptyset$ **then**
        $UpperBound \leftarrow UpperBound - \mathcal{H}(H_l)$
      else
        Add subproblem defined by S_l and H_l to Q
until stopping conditions are met or Q is empty
return *LowerBound* and *UpperBound*

and then a random point in this hypercuboid is drawn. Again, some optimization techniques are used to perform this process efficiently. Consider a set of remaining hypercuboids $\{H_1, \ldots, H_K\}$ with total hypervolume $\mathcal{H}_T$. For each hypercuboid $H_l, l = 1, \ldots, K$, its relative hypervolume $\mathcal{H}_R(H_l) = \mathcal{H}(H_l)/\mathcal{H}_T$ and the sum of relative hypervolumes of preceding hypercuboids $\mathcal{SH}_R(H_l) = \sum_{k=1}^{l-1} \mathcal{H}_R(H_k)$ is calculated. To draw a hypercuboid, a random number $r \in [0, 1)$ is drawn with uniform distribution. Then the hypercuboid H_p such that $\mathcal{SH}_R(H_p) \leq r < \mathcal{SH}_R(H_{p+1})$ is found by the bisection search. To test efficiently if a sampled point is dominated by any point in S the ND-Tree data structure is used [11]. We call this algorithm QHV-BQ+MC (QHV-BQ combined with Monte Carlo sampling).

5 Computational Experiment

To verify the proposed methods we performed a computational experiment using the data sets proposed in [13] (concave, convex, linear and hard instances) and data sets described in [17] (uniformly spherical), generated by ourselves. With the exception of the hard ones, we use 10 data instances of each type with 7 through 10 objectives and exactly 1000 points (in the case of the hard type only single instances with 6, 8 and 10 objectives and the maximum provided numbers of points are available)[1]. We do not use instances with fewer objectives and/or fewer points, as approximate hypervolume calculation is of interest for data sets requiring relatively long time for obtaining the exact value. For each data instance, the values of objectives were normalized to be in the range $[0, 1]$. All the computations were performed on a PC computer with Core i7-8700 CPU and 64GB RAM.

The goal of this experiment is to verify the quality and convergence of the results generated by the proposed two methods (QHV-BQ and QHV-BQ+MC) by comparing it with the straightforward adaptation of QHV (i.e. QHV-BR) and the standard Monte Carlo sampling approach (MC)[2]. As the main quality measure we use the gap between lower and upper bound.

Note, however, that there is an important difference in the nature of the bounds generated by the four methods, which should be stressed here. The bounds generated by QHV-BR and QHV-BQ are guaranteed lower and upper bounds for the hypervolume, i.e. the actual hypervolume is always contained in the resulting interval. In the case of MC and QHV-BQ+MC the bounds are just bounds for the confidence interval calculated for the assumed confidence level (MC: from the very beginning, QHV-BQ+MC: only after the switch takes place). Thus, the actual hypervolume is merely expected (not guaranteed) to be contained in the resulting interval with probability $1 - \alpha$. In result, the gaps between confidence bounds and the gaps between guaranteed bounds are not directly comparable and guaranteed bounds may still be preferred over confidence bounds even if the gap is lower in the latter case. Anyway, since we use relatively high confidence level in the experiment ($\alpha = 0.01$), we decided to use these gaps for a rough comparison of the methods.

Throughout the experiment the following is assumed:

- For each instance, the exact QHV-II method is run first, and then the computing time of each of the four verified approximate methods is limited to the computing time taken by QHV-II. Methods QHV-BR and QHV-BQ after some time converge to the exact value of the hypervolume. However, because they need additional operations related to bounds calculation and because of the technical issues discussed above, they usually need more (up to several times) CPU time than QHV-II to reach the final convergence. In practice, however, one would be interested in using such methods only when the available CPU time is shorter than the time needed for the exact method. That is why it is the convergence of the methods in time shorter than needed for the exact method that is of interest here.
- In QHV-BQ+MC method the switch to the Monte Carlo sampling depends on the number of sub-problems corresponding to the split hypercuboids. With the concave, convex, linear and uniformly spherical types of data sets, in the case of the 7-objective instances QHV-BQ+MC calculates the guaranteed bounds until 4000 hypercuboids are placed in the queue and then switches to

[1] All data instances used in this experiment, source code and the detailed results are available at https://chmura.put.poznan.pl/s/c6dClctKDDo3vSc.

[2] In all cases we use our own C++ implementations of the methods.

the Monte Carlo sampling. In the case of the 8-, 9- and 10-objective instances this switch parameter is chosen to be 20000, 100000 and 500000 hypercuboids, respectively. With the hard types of data sets, the switch always occurs after 10000 hypercuboids are placed in the queue. Those values ensure that in each case the switch takes place before 5% of the available running time has elapsed. We do not claim, however, that this switching rule is optimal or even close to optimal. Our goal is merely to show that better confidence bounds than those provided by the standard Monte Carlo sampling may be obtained by combining QHV-BQ with MC even with an ad hoc switching rule.
- The calculation of the confidence interval in the Monte Carlo sampling is performed with the use of the Wilson score interval with continuity correction for binomial distribution [16] with $\alpha = 0.01$.

Figure 1 presents a set of 15 (3 data types $\times$ 4 numbers of objectives + 1 data type $\times$ 3 numbers of objectives) charts that illustrate gaps (i.e. differences) between the upper and the lower bounds for the hypervolume for concave, convex, linear, and hard data sets[3]. Note that a logarithmic scale is used for the gaps. The running time is given in seconds. In the first top rows each individual chart contains 40 (4 methods $\times$ 10 instances) series of points: red for QHV-BR, orange for QHV-BQ, blue for MC and green for QHV-BQ+MC. The sooner a series winds down, the better (this demonstrates high speed of convergence of the bounds). In the bottom row each individual chart contains 4 (4 methods $\times$ 1 instance) such series. Note that because of excessive memory demands, we were not able to complete the runs of QHV-BQ for the instances of type concave and d = 10 even on a computer with 64 GB memory. This is why these series terminate earlier in the corresponding chart.

Simultaneously, Fig. 2 presents a set of 4 charts depicting queue sizes used by QHV-BQ in the assumed computing time.

As can be observed:

- in all cases, QHV-BR (a straightforward adaption of QHV for bounds calculation) is clearly worse than all the other methods, and QHV-BQ+MC (after the switch to the Monte Carlo sampling), is clearly better than MC,
- the convergence pattern of QHV-BQ and MC is much different, bounds generated by MC (and QHV-BQ+MC after switch to MC) converge very fast at the beginning but further improvement is very slow, QH-BQ converges relatively constantly during the whole run,
- in result, up to a certain time (approximately half of the whole time limit), the convergence of QHV-BQ is slower than that of MC and QHV-BQ+MC, but relative performance of QHV-BQ improves with growing running time,
- QHV-BQ performs especially well for linear instances.

[3] To save space, Fig. 1 omits charts for the uniformly spherical data sets (which are basically indistinguishable from those for the convex data sets).

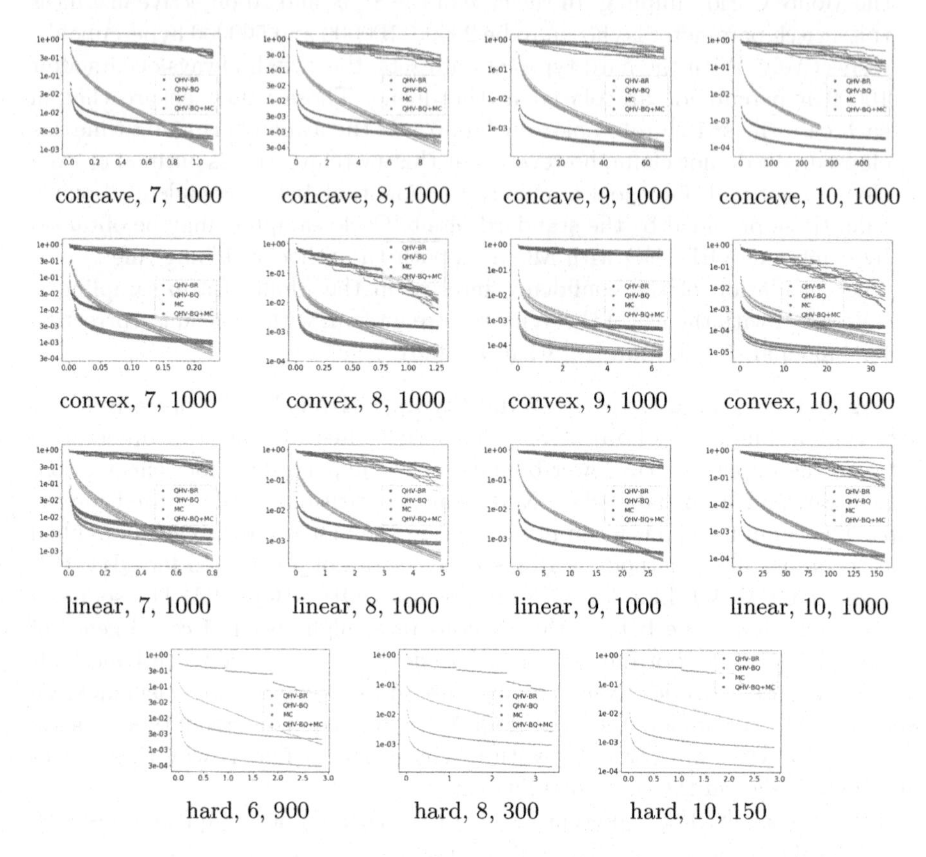

Fig. 1. Gaps between lower and upper bounds generated by the four considered methods. Top three rows: sets of 10 instances. Bottom row: single instances. Horizontal axis (linear): time [s], vertical axis (logarithmic, base 10): gap [unitless]. Chart description: 'instance type, objective count, point count'. (Color figure online)

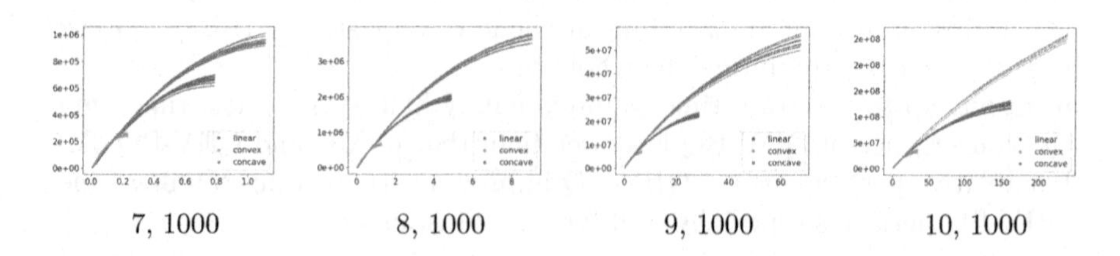

Fig. 2. Queue sizes in QHV-BQ for the sets of 10 instances of three instance types. Horizontal axis: time [s], vertical axis: size [elements]. Chart description: 'objective count, point count'.

Although the ordering of the four considered methods may be easily established from Fig. 1, Table 1 additionally presents the statistical comparison of these methods based on the Friedman post-hoc test with Bergmann and Hommel's correction at the significance level $\alpha = 0.05$ after 5%, 10%, 25%, 50% and 100% of the computing time taken by the exact method for each case. The numbers in parentheses indicate the average ranks of particular methods calculated for 163 data instances[4] (4 data types $\times$ 4 numbers of objectives $\times$ 10 non-hard instances + 3 hard instances). The notation $A \succ B$ indicates that method A is significantly better than method B, while the notation $A \sim B$ shows that there is no significant difference between these methods.

The resulting ranking of the methods (consistently with what is evident from the charts) is: QHV-BQ+MC $\succ$ MC $\succ$ QHV-BQ $\succ$ QHV-BR, with the exception of the two longest running times. For the running time equal to 50% of the available time QHV-BQ performs similarly to MC and for 100% of the available time equal to 100% QHV-BQ performs similarly to QHV-BQ+MC.

Furthermore, to ascertain the quality of the confidence bounds produced by MC and QHV-BQ+MC, we additionally verify the number and the extent of bounds violations (i.e. situations, in which the actual value of hypervolume is not included in the interval defined by the bounds). Notice that the assumed $\alpha = 0.01$ determines the expected percentage of such cases to be about $\alpha \cdot 100\% = 1\%$. As it turns out, the observed number of violations amounts to 0.23% and 1.35% for MC and QHV-BQ+MC, respectively. At the same time, the mean extent of the violations for MC equals $2.5 \pm 1.7\%$ (the maximum being 7.4%) of the gap, while for QHV-BQ+MC it equals $7.1 \pm 5.4\%$ (the maximum being 24.4%) of the gap. We are not sure what is the source of the difference in the numbers and extents of violations for MC and QHV-BQ+MC. One possibility is that it is caused by differences in the typical values of success probabilities being estimated by Monte Carlo sampling. Anyway, the number of violations for QHV-BQ+MC is still quite close to the expected value of 1%. This makes the confidence bounds generated by the methods quite reliable.

Table 1. Statistical results

% of time limit	Result
5	QHV-BQ+MC (1.00) $\succ$ MC (2.00) $\succ$ QHV-BQ (3.00) $\succ$ QHV-BR (4.00)
10	QHV-BQ+MC (1.00) $\succ$ MC (2.00) $\succ$ QHV-BQ (3.00) $\succ$ QHV-BR (4.00)
25	QHV-BQ+MC (1.00) $\succ$ MC (2.01) $\succ$ QHV-BQ (2.99) $\succ$ QHV-BR (4.00)
50	QHV-BQ+MC (1.00) $\succ$ MC (2.49) $\sim$ QHV-BQ (2.51) $\succ$ QHV-BR (4.00)
100	QHV-BQ (1.46) $\sim$ QHV-BQ+MC (1.56) $\succ$ MC (2.99) $\succ$ QHV-BR (4.00)

[4] As already mentioned, because of the 10 prematurely terminated runs of QHV-BQ for the instances of type concave and d = 10 only 153 data instances were used in the test performed after 100% of the computing time.

6 Conclusions and Directions for Further Research

We have presented a new QHV-BQ algorithm for calculation of guaranteed hypervolume bounds with a good any-time behaviour, which provides much better guaranteed bounds than a straightforward adaptation of QHV given the same CPU time, and its combination with Monte Carlo sampling (QHV-BQ+MC), which provides better confidence intervals than the standard Monte Carlo sampling (MC). The effectiveness of the proposed methods has been shown in a computational experiment on standard benchmarks.

QHV-BQ+MC method starts from running QHV-BQ and then switches to Monte Carlo sampling. Our goal was to show that with such switching better confidence intervals could be obtained. We do not claim, however, that we used the optimal switching rule in the presented experiment. In fact, the convergence patterns observed in the computational experiment suggest that the best approach would be to run QHV-BQ for a vast majority of the available time and then perform a relatively short run of MC. The choice of the best switching rule remains an open problem and an interesting direction for further research.

Another interesting question is, if it is possible to combine QHV-BQ with Monte Carlo sampling in some other way, without a single switching moment, to ensure best confidence intervals at any moment. Note that a method combining recursive split of hypercuboids with Monte Carlo sampling has been recently proposed by Tang et al. [20]. This method differs from our approach though, in particular, it does not use guaranteed bounds, nor provides confidence bounds.

Another potential direction for further research could be trying to reduce high memory requirements of QHV-BQ. We envision some possibilities like ignoring sufficiently small subproblems or delaying insertion of subproblems to the queue until they may be needed for processing (which would, however, require storing the parent subproblem until all its child subproblems are processed).

In this paper, the proposed methods were compared with Monte Carlo sampling. Another direction for future research is to compare it with other hypervolume estimation methods, like that from [20], or the methods based on R2 indicator [9,15,19]. Note, however, that since these methods provide neither guaranteed nor confidence bounds, a different evaluation approach would be required.

Further directions for future research could be to analyze the interplay of space requirements and computation time in QHV-BQ, as well as the possibility to express the time/space complexity in terms of the gap between the bounds.

Finally, it would be certainly interesting to design hypervolume-related methods taking advantage of the guaranteed bounds. We expect, for example, that these bounds could be used to efficiently find points with the minimum or maximum contribution to the hypervolume [5,12], which is an important step in some indicator-based EMO algorithms.

Acknowledgement. We are grateful to our Reviewers for their insightful suggestions. Our paper has been supported by local Statutory Funds (SBAD-2020) of the affiliated institute.

References

1. Bader, J., Deb, K., Zitzler, E.: Faster hypervolume-based search using Monte Carlo sampling. In: Ehrgott, M., Naujoks, B., Stewart, T.J., Wallenius, J. (eds.) Multiple Criteria Decision Making for Sustainable Energy and Transportation Systems, pp. 313–326. Springer, Berlin. https://doi.org/10.1007/978-3-642-04045-0_27
2. Bader, J., Zitzler, E.: HypE: an algorithm for fast hypervolume-based many-objective optimization. Evol. Comput. **19**(1), 45–76 (2011)
3. Beume, N., Fonseca, C.M., Lopez-Ibanez, M., Paquete, L., Vahrenhold, J.: On the complexity of computing the hypervolume indicator. IEEE Trans. Evol. Comput. **13**(5), 1075–1082 (2009)
4. Beume, N., Naujoks, B., Emmerich, M.: SMS-EMOA: multiobjective selection based on dominated hypervolume. Euro. J. Oper. Res. **181**(3), 1653–1669 (2007)
5. Bringmann, K., Friedrich, T.: Approximating the least hypervolume contributor: NP-hard in general, but fast in practice. Theor. Comput. Sci. **425**, 104–116 (2012)
6. Chan, T.M.: Klee's measure problem made easy. In: 2013 IEEE 54th Annual Symposium on Foundations of Computer Science, pp. 410–419 (2013)
7. Fieldsend, J.E.: Efficient real-time hypervolume estimation with monotonically reducing error. In: Proceedings of the Genetic and Evolutionary Computation Conference, GECCO 2019, New York, USA, pp. 532–540. Association for Computing Machinery (2019)
8. Guerreiro, A.P., Fonseca, C.M.: Computing and updating hypervolume contributions in up to four dimensions. IEEE Trans. Evol. Comput. **22**(3), 449–463 (2018)
9. Ishibuchi, H., Tsukamoto, N., Sakane, Y., Nojima, Y.: Hypervolume approximation using achievement scalarizing functions for evolutionary many-objective optimization. In: 2009 IEEE Congress on Evolutionary Computation, pp. 530–537. IEEE (2009)
10. Jaszkiewicz, A.: Improved quick hypervolume algorithm. Comput. Oper. Res. **90**, 72–83 (2018)
11. Jaszkiewicz, A., Lust, T.: ND-tree-based update: a fast algorithm for the dynamic nondominance problem. IEEE Trans. Evol. Comput. **22**(5), 778–791 (2018)
12. Jiang, S., Zhang, J., Ong, Y.S., Zhang, A.N., Tan, P.S.: A simple and fast hypervolume indicator-based multiobjective evolutionary algorithm. IEEE Trans. Cybern. **45**(10), 2202–2213 (2015)
13. Lacour, R., Klamroth, K., Fonseca, C.M.: A box decomposition algorithm to compute the hypervolume indicator. Comput. Oper. Res. **79**, 347–360 (2017)
14. Li, B., Li, J., Tang, K., Yao, X.: Many-objective evolutionary algorithms: a survey. ACM Comput. Surv. **48**(1), 1–35 (2015)
15. Ma, X., Zhang, Q., Tian, G., Yang, J., Zhu, Z.: On Tchebycheff decomposition approaches for multiobjective evolutionary optimization. IEEE Trans. Evol. Comput. **22**, 226–244 (2018)
16. Newcombe, R.G.: Two-sided confidence intervals for the single proportion: comparison of seven methods. Stat. Med. **17**(8), 857–872 (1998)
17. Russo, L.M.S., Francisco, A.P.: Quick hypervolume. IEEE Trans. Evol. Comput. **18**(4), 481–502 (2014)
18. Russo, L.M.S., Francisco, A.P.: Extending quick hypervolume. J. Heuristics **22**(3), 245–271 (2016)
19. Shang, K., Ishibuchi, H., Zhang, M.L., Liu, Y.: A new R2 indicator for better hypervolume approximation. In: Proceedings of the Genetic and Evolutionary Computation Conference, GECCO 2018, New York, USA, pp. 745–752. Association for Computing Machinery (2018)

20. Tang, W., Liu, H., Chen, L.: A fast approximate hypervolume calculation method by a novel decomposition strategy. In: Huang, D.-S., Bevilacqua, V., Premaratne, P., Gupta, P. (eds.) ICIC 2017. LNCS, vol. 10361, pp. 14–25. Springer, Cham (2017). https://doi.org/10.1007/978-3-319-63309-1_2
21. While, L., Bradstreet, L.: Applying the WFG algorithm to calculate incremental hypervolumes. In: 2012 IEEE Congress on Evolutionary Computation, New York, USA. IEEE International Conference on Fuzzy Systems/International Joint Conference on Neural Networks/IEEE Congress on Evolutionary Computation/IEEE World Congress on Computational Intelligence, Brisbane, Australia, 10–15 June 2012
22. Zitzler, E., Thiele, L., Laumanns, M., Fonseca, C.M., da Fonseca, V.G.: Performance assessment of multiobjective optimizers: an analysis and review. IEEE Trans. Evol. Comput. **7**(2), 117–132 (2003)
23. Zitzler, E., Künzli, S.: Indicator-based selection in multiobjective search. In: Yao, X., et al. (eds.) PPSN 2004. LNCS, vol. 3242, pp. 832–842. Springer, Heidelberg (2004). https://doi.org/10.1007/978-3-540-30217-9_84

Can Compact Optimisation Algorithms Be Structurally Biased?

Anna V. Kononova[1], Fabio Caraffini[2(✉)], Hao Wang[3], and Thomas Bäck[1]

[1] Leiden Institute of Advanced Computer Science (LIACS), Leiden University, Leiden, The Netherlands
{a.kononova,t.h.w.baeck}@liacs.leidenuniv.nl

[2] Institute of Artificial Intelligence, De Montfort University, Leicester, UK
fabio.caraffini@dmu.ac.uk

[3] LIP6, Sorbonne Université Paris, Paris, France
hao.wang@lip6.fr

Abstract. In the field of stochastic optimisation, the so-called *structural bias* constitutes an undesired behaviour of an algorithm that is unable to explore the search space to a uniform extent. In this paper, we investigate whether algorithms from a subclass of *estimation of distribution algorithms*, the *compact algorithms*, exhibit structural bias. Our approach, justified in our earlier publications, is based on conducting experiments on a test function whose values are uniformly distributed in its domain. For the experiment, 81 combinations of compact algorithms and strategies of dealing with infeasible solutions have been selected as test cases. We have applied two approaches for determining the presence and severity of structural bias, namely an (existing) visual and an (updated) statistical (Anderson-Darling) test. Our results suggest that compact algorithms are more immune to structural bias than their counterparts maintaining explicit populations. Both tests indicate that strong structural bias is found only in the `cBFO` algorithm, regardless of the choice of strategy of dealing with infeasible solutions, and `cPSO` with `mirror` strategy. For other test cases, statistical and visual tests *disagree* on some cases classified as having mild or strong structural bias: the former one tends to make harsher decisions, thus needing further investigation.

Keywords: Structural bias · Compact algorithm · Continuous optimisation · Estimation of distribution algorithm · Infeasible solution

1 Introduction

Evolutionary algorithms (EAs) [1,9] are based on a *biological metaphor* which creates an *ontological link* between a set of solutions of the optimisation problem, which iteratively approximate its optimum, and a population of biological individuals, which adapt to their environment through evolution. An essential part of this metaphor is an *individual*, an atomic part of the *population*, that has been

T. Bäck et al. (Eds.): PPSN 2020, LNCS 12269, pp. 229–242, 2020.
https://doi.org/10.1007/978-3-030-58112-1_16

created by some combination of one or more of its parent individuals in an attempt to build upon previously successful approximations of the optimum. Both biological and (most) computational populations typically do not explicitly 'record' their history, thus, potentially loosing the already exploited information regarding the 'successes' in the past generations. Following the biological metaphor, a 'success' in some generation is directly translated into the individual's reproductive advantage and, therefore, an opportunity to pass on its 'achievements'.

Striving to exploit the historical information contained in the sequential populations of an evolutionary algorithm, a special class of algorithms has been been proposed in the 1990s [18,19] which attempts to build *explicit probabilistic models of promising solutions* as the optimisation process progresses and steer the subsequent simulated evolutionary progress towards such solutions. These new algorithms, just like other heuristics [7,15,20], are probabilistic, iterative, and thus can suffer from undesirable algorithmic behaviours such as premature convergence, stagnation and presence of *structural bias* (SB) [7,15]. The latter is the *focus of this paper.*

The aforementioned class of algorithms, referred to as *estimation of distribution algorithms* (EDAs) [10], do not maintain explicit populations but rather have *virtual sampling populations.* They work through updating their models incrementally, starting from some uninformed prior and, ideally, leading up to the model producing only the optimum solution. Clearly, the problem of constructing such a model in itself is by far not trivial and can only be solved with some simplifications. It is the scope and extent of such *simplifications* that define the sub-classes of EDAs.

This paper addresses the question of whether a *subclass* of algorithms with virtual populations exhibit such algorithmic deficiency as structural bias – the tendency of an algorithm to 'prefer' some parts of the domain irrespective of the objective function. This paper continues the effort of the authors to investigate a wide range of heuristic optimisation algorithms for possible structural bias deficiencies [5,7,14,15]. The paper is organised as follows: Sect. 2 discusses compact algorithms in general and the particular instances investigated in this study, Sect. 3 describes the experimental methodology and methods for assessing SB, Sect. 4 discusses results concerning SB in compact algorithms, and Sect. 5 provides the conclusions.

2 Compact Algorithms

The term 'compact algorithm' refers to a subclass of EDAs that mimic the behaviour of established population-based algorithms [11] through a *'memory-saving'* probabilistic model where design variables are assumed to be *fully uncorrelated.* This minimalist model is fully described with a $2\times n$ matrix (n is problem dimensionality) that defines the generating distribution[1] $\mathcal{D}_\theta$, where $\boldsymbol{\theta} = [\boldsymbol{\mu}, \boldsymbol{\sigma}]$,

[1] It is called 'probability vector' in the original publications [11]; a terminology which we find somewhat misleading in case of a continuous search space and a Gaussian generating distribution.

$\boldsymbol{\mu} \in \mathbb{R}^n, \boldsymbol{\sigma} \in (\mathbb{R}^+)^n$ are the vectors containing the chosen mean and the standard deviation values for a truncated Gaussian distribution (the optimisation process takes places in the *re-normalised domain* $[-1,1]^n$).

All 'elitist' real-valued compact algorithms share the structure outlined in Algorithm 1 and only differ by the logic used to generate a new solution $\mathbf{x}$.

Algorithm 1. Skeleton of a *generic* elitist compact algorithm

given: objective function f, generating distribution $\mathcal{D}_\theta$ with parameters $\theta = [\boldsymbol{\mu}, \boldsymbol{\sigma}]$
initialise $\boldsymbol{\mu}, \boldsymbol{\sigma}$ with $\mu_i = 0$ and $\sigma_i \gg 1$ ▷ e.g. $\sigma_i = 10$ as in [11]
draw initial solution $\boldsymbol{x}_{\text{elite}}$ from $\mathcal{D}_\theta$ and evaluate its fitness $f_{\text{elite}} = f(\boldsymbol{x}_{\text{elite}})$
while budget condition is not met **do**
  draw i.i.d. samples $\mathcal{P} = \{\mathbf{x}_1, \mathbf{x}_2, \ldots\}$ from $\mathcal{D}_\theta$ ▷ $|\mathcal{P}|$ depends on the specific
  generate a new candidate solution $\mathbf{x}$ from $\mathcal{P}$ operator (Section 2.1)
  evaluate $f(\mathbf{x})$;
  if $f(\boldsymbol{x}) \bowtie f_{\text{elite}}$ **then** ▷ $\bowtie \in \{\leq, \geq\}$ for minimisation/maximisation
    $\mathbf{l} \leftarrow \mathbf{x}_{\text{elite}}$; $\mathbf{w} \leftarrow \mathbf{x}$; $\mathbf{x}_{\text{elite}} \leftarrow \mathbf{x}$; ▷ $\mathbf{w}$ is the winner, $\mathbf{l}$ loser
  else
    $\mathbf{l} \leftarrow \mathbf{x}$; $\mathbf{w} \leftarrow \mathbf{x}_{\text{elite}}$;
  end if
  $\boldsymbol{\mu}_{\text{old}} \leftarrow \boldsymbol{\mu}$
  $\boldsymbol{\mu} \leftarrow \boldsymbol{\mu} + \frac{1}{V_{\text{ps}}}(\mathbf{w} - \mathbf{l})$ ▷ user defined virtual population size V_{ps} [11]
  $\boldsymbol{\sigma} \leftarrow \sqrt{\boldsymbol{\sigma} \circ \boldsymbol{\sigma} + \boldsymbol{\mu}_{\text{old}} \circ \boldsymbol{\mu}_{\text{old}} - \boldsymbol{\mu} \circ \boldsymbol{\mu} + \frac{1}{V_{\text{ps}}}(\mathbf{w} \circ \mathbf{w} - \mathbf{l} \circ \mathbf{l})}$
end while ▷ $\circ$ is the Hadamard product
Output: $\boldsymbol{x}_{\text{elite}}$

2.1 Compact Algorithms Employed in This Study

All compact algorithms employed in this study follow the logic described in Algorithm 1. Details on these algorithms, including their suggested and adopted parameters setting, are available in [11]. A brief description of each algorithm is given below. These algorithms are equipped with various strategies of dealing with infeasible solutions (SDIS) generated, see Sect. 3.4.

Configurable compact differential evolution (`cDE/x/y/z`): similar to non-compact variants of differential evolution (DE) [21], a variety of compact configurations can be obtained with the combinations `x/y/z`, where `z` is either the binary `bin` or the exponential `exp` crossover [6,21], while the `x/y` component is taken from these options[2]: (i) `rand/1` (ii) `rand/2` (iii) `best/1` (iv) `best/2` (v) `current-to-best/1` (vi) `rand-to-best/2` (vii) `current-to-rand/1` (does not require a crossover). It must be highlighted that in a DE algorithm, the `x/y/z` operators require a number of randomly selected individuals from the population to produce $\mathbf{x}$. Due to the absence of a stored population, these individuals are drawn from the generating distribution $\mathcal{D}_\theta$ in the compact representation. This implies that logically `current-to-best/1` $\equiv$ `rand-to-best/1`.

[2] This list clearly does not exhaust all possibilities.

Compact differential evolution light: `cDE-Light` is a DE-inspired compact algorithm that requires a smaller number of computationally expensive operations with respect to its predecessor algorithm `cDE`, thus being faster and lighter in terms of memory consumption. This algorithm employs a specific mutation referred to as `mutation-light`, which mimics the behaviour of the `rand/1` mutation, and specific crossover operator referred to as `crossover-light`, which emulates the `exp` crossover without the need of looping through the solutions to exchange their variables.

Compact particle swarm optimisation (`cPSO`): generates novel candidate solutions $\mathbf{x}$ through the simple `PSO` perturbation logic based on a weighted sum of the currently available solution and the so-called 'velocity' vector $\mathbf{v}$, i.e. $\mathbf{x} \leftarrow \gamma_1\mathbf{x} + \gamma_2\mathbf{v}$. Before perturbing the position of $\mathbf{x}$ in the search space with the previous formula, $\mathbf{v}$ must be updated through the standard method $\mathbf{v} \leftarrow \phi_1\mathbf{v} + \phi_2\mathbf{u_1} \circ (\mathbf{x_{lb}} - \mathbf{x}) + \phi_3\mathbf{u_2} \circ (\mathbf{x_{gb}} - \mathbf{x})$, in which $\mathbf{u_1}$ and $\mathbf{u_2}$ are two n-dimensional vectors containing uniformly drawn random numbers; $\mathbf{x_{lb}}$ is the 'local best' solution, which is not present in the compact representation and therefore has to be drawn from $\mathcal{D}_\theta$ and evaluated; $\mathbf{x_{gb}}$ is the 'global best' solution, i.e., $\mathbf{x_{gb}} \leftarrow \mathbf{x_{elite}}$. It must be pointed out that $\mathcal{D}_\theta$ is updated with $\mathbf{w}$ and $\mathbf{l}$ obtained by comparing the objective function values $\mathbf{x_{lb}}$ and $\mathbf{x}$ while the $\mathbf{x_{elite}}$ solution is subsequently updated.

Compact bacterial foraging optimisation (`cBFO`) reproduces the same search logic of the original BFO algorithm [8] with the difference that, at each iteration, a candidate solution $\mathbf{x}$ is drawn from $\mathcal{D}_\theta$ rather than being taken from a population. Such solution undergoes a series of perturbations to perform the so called 'chemotaxis', 'tumble' and 'swim' moves in the search space by means of the operator $\mathbf{x} \leftarrow \mathbf{x} + \frac{\mathbf{c} \circ \boldsymbol{\Delta}}{\sqrt{\boldsymbol{\Delta}^T\boldsymbol{\Delta}}}$, where $\mathbf{c}$ is an n-dimensional vector whose components are the so-called 'run-length' unit parameters [8], which control the step-size, and $\boldsymbol{\Delta}$ is an n-dimensional vector whose components are uniformly sampled in the interval $[-1, 1]$ as indicated in [8] for each one of the three moves.

Compact genetic algorithm: the real-valued compact genetic algorithm `rcGA` [11], or `cGA` here, is the simplest example of compact algorithm as it only draws a new solution from $\mathcal{D}_\theta$ (i.e. $\mathcal{P} = \{\mathbf{x}\}$) to produce a new candidate solution.

3 Methodology

3.1 Structural Bias

The field of EAs is saturated with a multitude of nature inspired algorithms [2,4]. For practical reasons, these algorithms need to be compared and characterised. Amongst *dimensions* over which the quality of an optimisation algorithm can be measured are: (i) values of the best or average improvement of the objective function attained over a series of independent runs on some function or class of functions; (ii) best or average ranking of the algorithm among other algorithms on some function or class of functions; (iii) the distance from the found solution to the known optima; (iv) whether the algorithm has stagnated or converged prematurely; (v) typical or peak memory consumption required by the algorithm to solve the problem; (vi) scalability of the algorithm; (viii) proportion of the previously-non-visited solutions; etc.

In the EA/EC community, variations of the first two of the aforementioned dimensions are traditionally used. However, most performance measures come with a difficulty: dependence on the objective function [23]. Moreover, in practice, classes of objective functions are typically hard to be defined exhaustively and extensively and benchmarking over a set of diverse functions strongly depends on the choice of such functions.

In an attempt to characterise the performance of optimisation algorithms *from a different angle*, an additional fitness-free comparison 'dimension' has been suggested in [15]: the so-called *structural bias* (SB) has been defined as an *intrinsic deficiency* of a probabilistic iterative algorithm dictated solely by its structure. An algorithm is said to possess SB when it is unable to explore all areas of the search space to a *uniform* extent, *irrespective* of the objective function.

In other words, characterising the algorithm in terms of SB allows one to judge how much *general-purpose* the algorithm is, since a fully general-purpose optimisation algorithm is expected to be able to locate the optima regardless of where they are located in the search space. It has been established [15] that for a *general* objective function, the movement of solutions in the populations evolving over time is dictated by the superposition of two forces: the gradient formed by the values of objective function in the current population and the force originating from the structure of algorithm. These *two forces are not necessarily in agreement in terms of direction and strength.* The problem with the existence of the second force is that it can potentially pull the search away from some areas of the domain, thus limiting the algorithm's ability to find the optima therein.

It must be remarked that due to the *stochastic nature* of the utilised test function f_0 (see Sect. 3.2), there is no sense in tracking objective function improvements over time. The goal of tests on f_0 is *only* to establish deficiencies in movements of the populations during the optimisation process and *not to rank* the methods according to their 'objective-function-improvement' on f_0.

3.2 Structural Bias via Visual Tests

The procedure for testing for presence of SB is based on a *theoretical result* [15] that *true minima/maxima of*

$$f_0 : [0,1]^n \to [0,1] \mid \forall \mathbf{x}\ \mathbf{f_0}(\mathbf{x}) \sim \mathcal{U}(0,1) \tag{1}$$

are distributed uniformly in its domain (where $\mathcal{U}(0,1)$ denotes a scalar random value sampled independently from the uniform distribution on $[0,1]$). Thus, through examination of the distribution of locations of the optima of f_0 identified by the algorithm and its subsequent comparison to the true uniform distribution across the domain, one can establish whether the algorithm exhibits any SB [15]. To date, such comparison has been done *visually* due to the lack of a good *'all-encompassing'* measure, see Sect. 3.3 for more discussion. Plotting locations of final best solutions in a series of independent runs in parallel coordinates [12] is an established technique that facilitates the analysis.

3.3 Structural Bias via Statistical Tests

To identify SB, we build on the previous studies [14,15] where Kolmogorov-Smirnov test has been used for hypothesis testing. Here, we propose a different statistical approach which *tests the uniformity* of final points per dimension *via a non-parametric goodness-of-fit* test – the Anderson-Darling (AD) test is chosen given its high statistical power [22]. The motivation behind this approach is two-fold: first, testing the multivariate uniformity is known to be a *challenging* task [13]; second, it is methodologically *erroneous to merge samples* from all dimensions to perform one univariate good-of-fit test as the design variables could be correlated and not identically distributed, thus resulting in a potential loss of information on each dimension.

Hence, for each dimension $i \in [1..n]$ the AD test is applied to the ith component of final points $\{x_i^{(1)}, \ldots, x_i^{(N_r)}\}$ obtained over N_r independent runs ($N_r = 50$ here). When testing the uniformity of the sample distribution along each dimension, the AD test-statistic is formulated as: $A^2 = \int_0^1 (\widehat{F}_{N_r}(t) - t)^2 / t(1-t)\mathrm{d}t$, where $\widehat{F}_{N_r}(t) = \sum_{k=1}^{N_r} \mathbb{1}(x_i^{(k)} \leq t)$ is the empirical cumulative distribution function (ECDF) of the ith component. Intuitively, A^2 quantifies the proximity between the ECDF and the theoretical distribution function of the uniform distribution. We shall denote the resulting test statistics and p-values as $\{A_i^2\}_{i=1}^n$ and $\{p_i\}_{i=1}^n$ respectively. The significance level $\alpha = 0.01$ is used to reject the null hypotheses H_0. Whenever H_0 is rejected we conclude that the ECDF differs from the uniform distribution by an amount of A^2, with an error rate of α. The SB 'degree' is then determined by counting the *rejected* dimensions.

Moreover, we propose an *aggregated measure of SB* over results from all dimensions, defined as the sum of A_i^2 test statistics that are associated with a statistical significance over all dimensions: $\mathrm{SB} = \frac{1}{n}\sum_{i=1}^n A_i^2 \mathbb{1}(p_i \leq \alpha)$, where $\mathbb{1}$ stands for the indicator function. We shall contrast this new measure of SB with the visual test shown in Sect. 4.2. Note that methods to combine p-values (e.g., Kost's method [16]), which performs a test on the results from several tests, is not suitable here since we also interested in combining the statistical effect from several dimensions.

3.4 Strategy of Dealing with Infeasible Solutions as Operator

Practical optimisation problems to be solved via computer simulations are defined in *bounded domains* whose most typical shape is hyperrectangular. Research into the algorithmic design of optimisation methods from the field of computational intelligence [7] has shown that the chosen strategy of dealing with the solutions generated outside such domain – the *infeasible solutions* (ISs) – is an *essential part of the algorithm* that to a large extent decides the success of the optimisation method. Unfortunately, in the majority of papers in the field, the choice of such strategy is overlooked or omitted from the publications, thus limiting the reproducibility of the results and lowering the overall impact of such studies.

To highlight the importance of this algorithmic operator, we employ five different strategies of dealing with ISs (SDIS):

1. Complete One-tailed normal correction strategy (`COTN`) [7] – only in infeasible dimensions, moves an infeasible solution inside the domain to a position resampled from the rescaled one-sided Normal distribution centred on the boundary;
2. `dismiss`[6] – dismisses an infeasible solution and replaces it with one of the parent/generating points);
3. `mirror` [7] – mirrors the position of an infeasible solution in infeasible dimensions only inwards off the closest boundary;
4. `saturation` [5,7] – moves an infeasible solution onto the closets boundary only in infeasible dimensions;
5. `toroidal` [5,7] – reflects an infeasible solution inwards off the opposite boundary.

3.5 Experimental Setup

This experimentation involves 13 `cDE/x/y/z` variants and the 4 other algorithms described in Sect. 3.4. All of them but cGA (which generates only feasible solutions) are considered with 5 SDIS – the total of $16 \times 5 + 1 = 81$ configurations considered.

Results on the SB presented in this paper are based on *experiments*: (i) *minimising* the test function f_0 (see Sect. 3.2) for $n = 30$ (ii) by 81 algorithmic configurations described in Sects. 2.1 and 3.4; (iii) each configuration is run 50 times; (iv) each run has independently seeded *Java random.utils* pseudorandom generator – seed is initialised with the current time since January 1, 1970 in milliseconds via Java's `System.currentTimeMillis`; (v) each run is budgeted in terms of the number of objective function evaluations as $10000n$.

All algorithms refer to their *persistent elitist* variants. All experiments are executed on a standard desktop using the SOS platform [4] implemented in Java (algorithms' source code is available online [3]). It is worth mentioning that the aforementioned pseudorandom generator used here for all experiments is considered *on the better side of the scale for linear congruential generators* [17].

4 Discussion of Results

Using the approaches described in Sects. 3.2 and 3.3, all 81 configurations have been investigated. Results in these figures are shown in *parallel coordinates* [12] and *should be read as follows*: final positions attained in a series of 50 independent runs of each configuration are shown with 50 '+' markers on each of the $n = 30$ parallel vertical 'axes'. Positions of these 'axes' identify dimensions and are

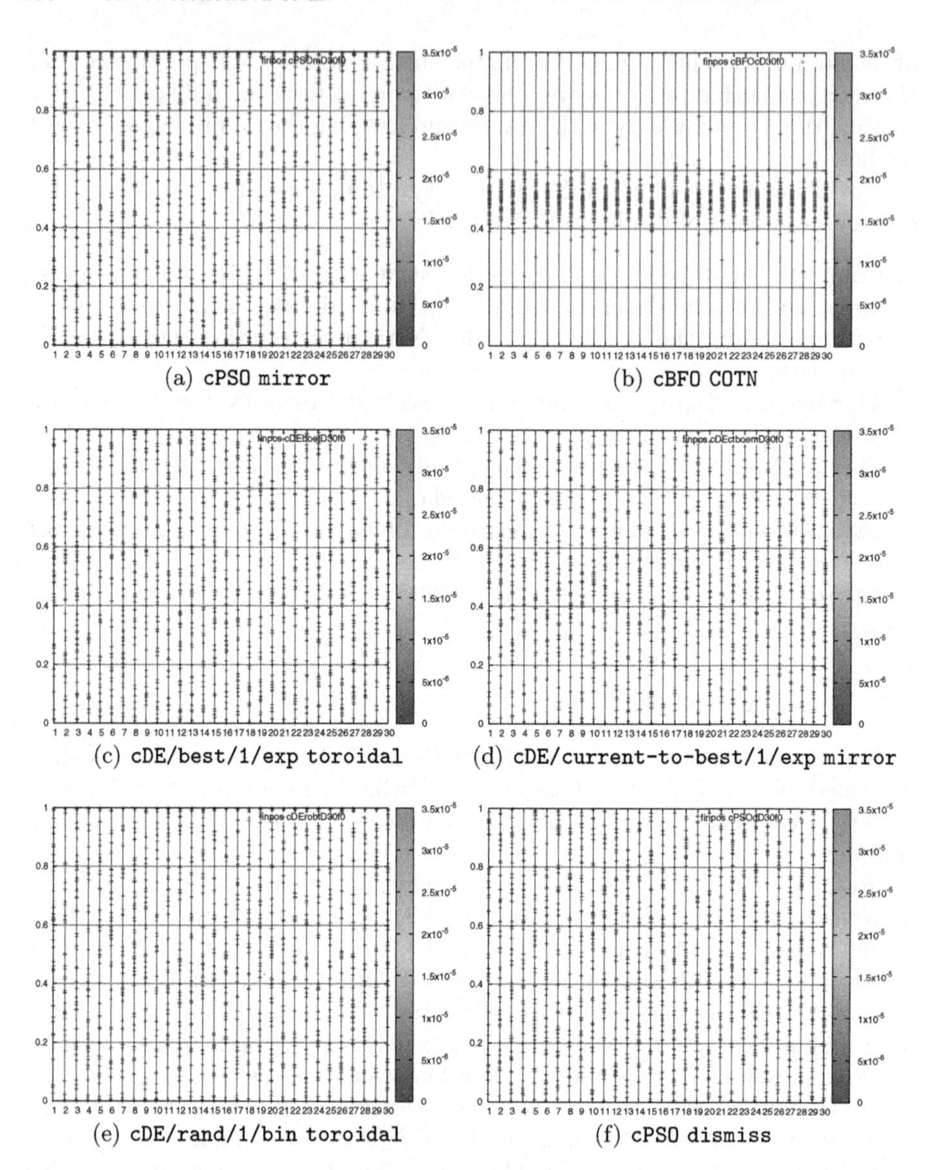

(a) cPSO mirror (b) cBFO COTN

(c) cDE/best/1/exp toroidal (d) cDE/current-to-best/1/exp mirror

(e) cDE/rand/1/bin toroidal (f) cPSO dismiss

Fig. 1. Distribution of locations of final best solutions: *example* configurations that exhibit *strong* SB in (a), (b), *mild* SB with: *local clustering* in (c), *clustering across domain* in (d), *clustering on boundaries domain* in (e) and *large gaps* in (f). See Sect. 4 for explanation on how to read this figure.

shown on the traditional horizontal axis; meanwhile, the traditional vertical axis shows the range of the dimension ($[0, 1]$ here). Values of f_0 attained by the final solutions are shown in colour (a recap: this is a minimisation problem). Due to the page limit in this publication, only a few figures are shown in Fig. 1. All results can be obtained from [6].

4.1 Visual Tests

Following the methodology of visual testing described in Sect. 3.2, out of 81 configurations considered in this paper, only 6 configurations have been found to be strongly structurally biased (e.g. Figs. 1(a), 1(b)), meanwhile 40 configurations exhibit only mild SB. It is worth highlighting that decisions in visual tests on whether mild SB is present are *highly subjective* and should be contrasted with results from statistical testing in Sect. 4.2.

The summary of results discussed in this Section can be found in Table 1 in the columns marked as 'visual SB test' for all basic compact configurations (rows) and all strategies of dealing with IS (smaller columns)[3].

Based on the *visual tests only*, overall, compact configurations appear to be *more 'immune' to the strong SB* than their equivalents maintaining explicit populations [6,7,15]. SB, if at all present, is more *subtle* across all configurations of compact algorithms considered. The resulting distributions of locations of final best solutions differ from the true uniform distribution in clustering of points and not in the span of the domain (with exception of all `cBFO` configurations as discussed below). It means that, on the whole, compact configurations of algorithms considered in this study *should have more exploratory potential* and be more successful in finding optima wherever they are situated in the domain. The latter one, however, is *not guaranteed* without the use of good exploitative operators (such investigation is *out of the scope* of this paper).

One of the exceptions to the above statement is all the `cBFO` configurations that have turned out to be badly biased towards the middle of the domain regardless of the choice of correction strategy, e.g. Fig. 1(b). More precisely, `cBFO` appears to be *unable* to find optima on f_0 outside the region $[0.4, 0.6]^{30}$ (with only a handful of exceptions per configuration).

Another exception to the above statement is the `cPSO mirror` configuration which exhibits strong SB towards all corners of the domain (see Fig. 1(a)) – *interestingly enough, such situation resembles the case of SB found in non-compact* `PSO` *with a small population size* [15].

When talking about *mild SB*, resulting distributions of the locations of final best solutions appear to *marginally* deviate from the uniform distribution in the following non-exclusive aspects:

1. 'higher-than-expected' clustering of points *within* the domain (e.g. Fig. 1(c));
2. 'higher-than-expected' clustering of points *across* the domain (e.g. Fig. 1(d);
3. 'higher-than-expected' clustering of points on the *boundaries*[4] (e.g. Fig. 1(e));
4. *large empty gaps* consistently identified in all 30 dimensions (e.g. Fig. 1(f)).

When analysing results for `cDE/x/y/z` only, out of 30 `bin` and 30 `exp` considered configurations, 16 and 13, respectively, appear to be *mildly biased*. Out

[3] To avoid complicating Table 1 further, results for `cGA` that requires no SDIS are shown as `dismiss` – it is the closest to how `cGA` deals with infeasible solutions.

[4] This is easily explained if `saturation` is used but is not trivial if `toroidal` is used.

Table 1. Comparison of results on the presence and strength of structural bias based on *visual* and *statistical* tests across all 81 configurations (see [6]). For both tests, cells with background in **black** mark configurations exhibiting strong SB, in **grey** - configurations with mild SB and in **white** - configuration with no SB identified based on the corresponding tests (i.e. *colour marks the corresponding decision of the test*). Cells containing '×' mark configurations that are not possible by design. *Symbols mark results of comparing the two tests*: symbol '=' stands for cells where results of the visual and statistical tests coincide (colour of the symbol has no meaning) and '•' - for the differences in results from visual and statistical tests (colour of the symbol has no meaning). Values shown in columns for statistical test are the *corresponding values of the statistic*. Thresholds for decisions based on these values are given in Sect. 4.2.

Kind of SB test:	visual					statistical				
Configuration:	COTN	dismiss	mirror	saturation	toroidal	COTN	dismiss	mirror	saturation	toroidal
cDE/rand/1/bin	=	=	•	•	=	0.00	0.00	0.02	∞	0.13
cDE/rand/1/exp	=	•	=	•	=	0.00	0.00	0.00	∞	0.02
cDE/rand/2/bin	•	=	•	•	•	0.02	0.00	0.02	∞	0.21
cDE/rand/2/exp	=	=	=	•	=	0.00	0.00	0.00	∞	0.06
cDE/current-to-rand/1	=	=	=	•	•	0.00	0.02	0.00	∞	0.00
cDE/best/1/bin	=	•	•	•	=	0.01	0.00	0.00	∞	0.00
cDE/best/1/exp	•	=	=	•	•	0.00	0.00	0.00	∞	0.00
cDE/best/2/bin	=	=	=	•	•	0.01	0.00	0.00	∞	0.01
cDE/best/2/exp	=	=	=	•	•	0.00	0.00	0.00	∞	0.01
cDE/current-to-best/1/bin	•	=	=	•	•	0.00	0.00	0.01	∞	0.02
cDE/current-to-best/1/exp	•	=	=	•	•	0.00	0.00	0.00	1.00	0.01
cDE/rand-to-best/2/bin	•	=	=	•	=	0.00	0.00	0.00	∞	0.01
cDE/rand-to-best/2/exp	•	•	•	•	•	0.02	0.00	0.00	∞	0.02
cDE-Light	•	•	•	•	•	0.00	0.00	0.00	0.01	0.00
cPSO	=	•	=	•	=	0.03	0.00	0.44	∞	0.01
cBFO	=	=	=	=	=	1.00	0.92	1.00	0.96	0.93
cGA (no SDIS, shown as dismiss)	×	•	×	×	×	×	0.01	×	×	×
Found strong SB/total cases:	1/16	1/17	2/16	1/16	1/16	1/16	1/17	2/16	15/16	2/16
Found mild SB/total cases:	8/16	6/17	4/16	13/16	9/16	5/16	2/17	3/16	1/16	10/16
Strong/mild/no SB cases:	6/40/35					21/26/59				
Agreement between	56	65	69	6	44	all cases				
visual and statistical	100	100	100	100	100	cases with strong SB only*				
tests (in %, calculated	38	17	25	0	55	cases with mild SB only				
'post factum'):	71	90	80	0	17	cases with no SB only				

5 `cDE/current-to-rand/1` configurations that require *no crossover*, 3 appear to be *mildly biased*. To some extent, it is fair to say that *simpler* `cDE/x/y/z` *configurations with* y> 1 appear to be *freer of mild SB*.

4.2 Statistical Tests

Here, we present the calculated values of the statistical measure of structural bias (defined in Sect. 3.3) in the 'statistical SB test' column of Table 1 (the meaning of symbols and colour scales are explained in the table caption). We use the 20- (0.00) and 90-quantiles (0.158)[5] of the statistical values over all combinations as *thresholds to determine the level of SB*. More specifically, zero values of statistic shall be classified as having *no* SB; ranges for *mild* and *strong* SB are $(0, 0.158]$ and $(0.158, 1] \cup \{+\infty\}$, respectively.

From results presented in the table, it is obvious that `cBFO` is exceptionally biased regardless of the SDIS. Also, the `saturation` SDIS seems to yield strong SB for all the algorithms except `cDE-Light`. For the remaining combinations, we observe either no or mild SB.

Comparing to the visual test on the same combinations, it seems that cases classified as strongly biased by the visual tests are always indicated as strongly biased as well from the statistical side – see the third line from the bottom in Table 1, marked with a *. However, since there are at most two discoveries of the strong bias from both tests, *the reliability of this agreement is questionable.* In contrast, cases with *mild* SB in the visual test are largely *mis-classified* as possessing no SB in the statistical approach. Also, most of the algorithms with the `saturation` SDIS are indicated as strongly biased by the statistical measure while those cases are considered mildly biased in the visual test. We conjecture the observed mismatches between those two approaches as follows: (i) the SB measure is calculated from a multiple testing procedure, where the p-value is corrected, thus the SB measure can suffer from a reduction of its statistical power (i.e., more false-negative decisions are made). This leads to a scenario that the Anderson-Darling test is rejected on all dimensions for those cases with mild SB in the visual test and hence the statistical measure classifies them as not biased; (ii) the SB measure is not scale-invariant and can be less informative after the performed normalisation. In this light, when no bias is displayed, we shall conclude that some SB degree is exhibited but negligible if compared to the bias shown by the most biased algorithm (i.e., `cBFO`). Such relativity in the statistical approach might be different from that in the visual test, which leads to the observed discrepancy.

5 Conclusions

The extensive experimentation presented in this piece of research has unveiled the presence of mild structural biases for most compact algorithms except `cBFO` and `cPSO` – the former one especially carries a so strong SB that can be categorically detected via the visual inspection of the generated graphs. More precisely, in `cBFO`, regardless of the employed SDISs, only the middle section of the search domain is populated with the found best solutions, while its peripheral areas are

[5] The quantiles are chosen *ad hoc*, based on the distribution of statistical measure over all combinations of algorithms and SDISs.

left completely out. This undesired algorithmic behaviour suggests that cBFO is not suitable for general-purpose optimisation, since it displays design flaws that limit its applicability to problems whose optimum/optima is/are at the centre of the search space. Similarly, also cPSO mirror displays a visible strong bias. However, it is interesting to observe that in this case, the solutions obtained over multiple runs accumulate towards the corners of the search space. This behaviour is in line with the one of the standard PSO algorithm – when employed with a small population size [15].

In a similar way, also the mild SB individuated in the remaining algorithms under study mainly reveals itself in the form of 'higher-than-expected' clustering of final solution located either across the domain or on the boundaries. However, in a few cases, uniformly distributed large empty gaps are also visible on each dimension of the generated graphs. Such gaps clearly flag the presence of SB, but final solutions do not accumulate in specific areas of the search space and thus do not seem to cause deleterious effect in terms of coverage of the whole domain. It is interesting to point out that amongst the cDE/x/y/z variants tested in this study, a mild SB is mainly visible only for those cases equipped with mutation operators using one difference vector – e.g. this is evident for the best/1 mutation, in particular when used in combination with binomial crossover bin. cDE variants equipped with mutation operators using two difference vectors, on the other hands, seem to be freer from SB – e.g. the case of rand/2, in particular when followed by exponential crossover exp.

To summarise, it can be stated that the compact algorithms under investigations appeared to be more 'immune' to the SB than their population-based equivalents according to the proposed visual test. However, it is important to conclude this study by observing that the proposed statistical SB detection method agrees with the visual test on strong SB cases while disagrees on most of the visually detected mild SB cases. We speculate that this discrepancy is caused by the insufficient sample size as well as the conservative nature of this testing procedure and we commit to investigating this aspect further in our future studies. We plan to increase the sample-size in future experimentation and, most importantly, improve upon the sensitivity of the proposed statistical measure with respect to the number of independent runs.

Acknowledgments. The work of Hao Wang was supported by the Paris Ile-de-France Region.

References

1. Bäck, T.: Evolutionary Algorithms in Theory and Practice. Oxford University Press, New York (1996)
2. Campelo, F., Aranha, C.: EC bestiary: a bestiary of evolutionary, swarm and other metaphor-based algorithms, June 2018. https://doi.org/10.5281/zenodo.1293352
3. Caraffini, F.: The stochastic optimisation software (SOS) platform, June 2019. https://doi.org/10.5281/zenodo.3237023

4. Caraffini, F., Iacca, G.: The SOS platform: designing, tuning and statistically benchmarking optimisation algorithms. Mathematics **8**(5), 785 (2020). https://doi.org/10.3390/math8050785
5. Caraffini, F., Kononova, A.V.: Structural bias in differential evolution: a preliminary study. In: 14th International Workshop on Global Optimization, LeGO 2018, vol. 2070, p. 020005. AIP, Leiden (2018)
6. Caraffini, F., Kononova, A.V.: Structural Bias in Optimisation Algorithms: Extended Results (2020). https://doi.org/10.17632/zdh2phb3b4.2. Mendeley Data
7. Caraffini, F., Kononova, A.V., Corne, D.W.: Infeasibility and structural bias in differential evolution. Inf. Sci. **496**, 161–179 (2019). https://doi.org/10.1016/j.ins.2019.05.019
8. Das, S., Biswas, A., Dasgupta, S., Abraham, A.: Bacterial foraging optimization algorithm: theoretical foundations, analysis, and applications. In: Abraham, A., Hassanien, A.E., Siarry, P., Engelbrecht, A. (eds.) Foundations of Computational Intelligence. SCI, vol. 203, pp. 23–55. Springer, Heidelberg (2009). https://doi.org/10.1007/978-3-642-01085-9_2
9. De Jong, K.A.: An analysis of the behavior of a class of genetic adaptive systems. Ph.D. thesis, University of Michigan, USA (1975)
10. Hauschild, M., Pelikan, M.: An introduction and survey of estimation of distribution algorithms. Swarm Evol. Comput. **1**(3), 111–128 (2011). https://doi.org/10.1016/j.swevo.2011.08.003
11. Iacca, G., Caraffini, F.: Compact optimization algorithms with re-sampled inheritance. In: Kaufmann, P., Castillo, P.A. (eds.) EvoApplications 2019. LNCS, vol. 11454, pp. 523–534. Springer, Cham (2019). https://doi.org/10.1007/978-3-030-16692-2_35
12. Inselberg, A.: The plane with parallel coordinates. Vis. Comput. **1**(2), 69–91 (1985). https://doi.org/10.1007/BF01898350
13. Justel, A., Peña, D., Zamar, R.: A multivariate Kolmogorov-Smirnov test of goodness of fit. Stat. Probab. Lett. **35**(3), 251–259 (1997)
14. Kononova, A.V., Caraffini, F., Wang, H., Bäck, T.: Can single solution optimisation methods be structurally biased? MDPI Preprints (2020). https://doi.org/10.20944/preprints202002.0277.v1
15. Kononova, A.V., Corne, D.W., Wilde, P.D., Shneer, V., Caraffini, F.: Structural bias in population-based algorithms. Inf. Sci. **298**, 468–490 (2015). https://doi.org/10.1016/j.ins.2014.11.035
16. Kost, J.T., McDermott, M.P.: Combining dependent p-values. Stat. Probab. Lett. **60**(2), 183–190 (2002)
17. L'Ecuyer, P., Simard, R.: TestU01: a C library for empirical testing of random number generators. ACM Trans. Math. Softw. **33**(4) (2007). https://doi.org/10.1145/1268776.1268777
18. Mühlenbein, H., Paaß, G.: From recombination of genes to the estimation of distributions I. Binary parameters. In: Voigt, H.-M., Ebeling, W., Rechenberg, I., Schwefel, H.-P. (eds.) PPSN 1996. LNCS, vol. 1141, pp. 178–187. Springer, Heidelberg (1996). https://doi.org/10.1007/3-540-61723-X_982
19. Pelikan, M., Goldberg, D., Lobo, F.: A survey of optimization by building and using probabilistic models. In: Proceedings of the 2000 American Control Conference, vol. 5, pp. 3289–3293 (2000)
20. Piotrowski, A.P., Napiorkowski, J.J.: Searching for structural bias in particle swarm optimization and differential evolution algorithms. Swarm Intell. **10**(4), 307–353 (2016). https://doi.org/10.1007/s11721-016-0129-y

21. Price, K.V., Storn, R., Lampinen, J.: Differential Evolution: A Practical Approach to Global Optimization. Springer, Heidelberg (2005). https://doi.org/10.1007/3-540-31306-0
22. Razali, N.M., Wah, Y.B.: Power comparisons of Shapiro-Wilk, Kolmogorov-Smirnov, Lilliefors and Anderson-Darling tests. J. Stat. Model. Anal. **2**(1), 21–33 (2011)
23. Wolpert, D., Macready, W.: No free lunch theorems for optimization. IEEE Trans. Evol. Comput. **1**, 67–82 (1997). https://doi.org/10.1109/4235.585893

Parallelized Bayesian Optimization for Expensive Robot Controller Evolution

Margarita Rebolledo[1(✉)], Frederik Rehbach[1], A. E. Eiben[2], and Thomas Bartz-Beielstein[1]

[1] Institute For Data Science, Engineering, and Analytics, TH Köln, Cologne, Germany
{margarita.rebolledo,frederik.rehbach, thomas.bartz-beielstein}@th-koeln.de
[2] Department of Computer Science, Vrije Universiteit Amsterdam, Amsterdam, Netherlands
a.e.eiben@vu.nl

Abstract. An important class of black-box optimization problems relies on using simulations to assess the quality of a given candidate solution. Solving such problems can be computationally expensive because each simulation is very time-consuming. We present an approach to mitigate this problem by distinguishing two factors of computational cost: the number of trials and the time needed to execute the trials. Our approach tries to keep down the number of trials by using Bayesian optimization (BO) –known to be sample efficient– and reducing wall-clock times by parallel execution of trials. We compare the performance of four parallelization methods and two model-free alternatives. Each method is evaluated on all 24 objective functions of the Black-Box-Optimization-Benchmarking (BBOB) test suite in their five, ten, and 20-dimensional versions. Additionally, their performance is investigated on six test cases in robot learning. The results show that parallelized BO outperforms the state-of-the-art CMA-ES on the BBOB test functions, especially for higher dimensions. On the robot learning tasks, the differences are less clear, but the data do support parallelized BO as the 'best guess', winning on some cases and never losing.

Keywords: Parallelization · Bayesian optimization · CMA-ES · BBOB benchmarking · Robotics

1 Introduction

Many real-world optimization problems are expensive to evaluate and require a considerable amount of computation time for each candidate solution evaluation. In such cases, the total evaluation budget is usually severely limited. Bayesian optimization (BO) [9,22] and parallel computing are two state-of-the-art methods for solving budget limited black-box optimization problems.

T. Bäck et al. (Eds.): PPSN 2020, LNCS 12269, pp. 243–256, 2020.
https://doi.org/10.1007/978-3-030-58112-1_17

In Bayesian optimization, a data-driven surrogate of the expensive function is fitted and an extensive search on the cheap surrogate is feasible. Only the points that are considered promising on the surrogate are evaluated on the expensive function. This makes BO very sample efficient compared to other algorithms.

Parallel computing makes use of the ever-increasing amount of available CPU cores in modern server systems. Running multiple simulations in parallel requires more computational resources and more energy, yet, it does not increase the real-time spent on the computation. Population-based algorithms like evolutionary algorithms (EAs), which can propose multiple candidate solutions per iteration, have an inherent efficiency benefit on parallel computing systems. In this area the covariance matrix adaptation evolution strategy (CMA-ES) [13] is the state-of-the-art evolutionary algorithm (EA) for real valued optimization. Other than CMA-ES, the standard BO approach does only evaluate a single candidate solution per iteration. Yet, several approaches have been proposed through which BO can be adapted to a parallel environment. An overview of these methods is given in [11].

One example of expensive, yet parallelizable, objective functions can be found in the field of evolutionary robotics (ER). ER aims to automatically design robots that are well suited to their environment and can perform a given task [4,7]. This can be achieved by evolving the robot morphologies (bodies) and controllers (brains) through iterated cycles of selection and reproduction. As outlined in the triangle of life (ToL) framework proposed by [6], this implies that newborn robots (with a new body form that is different from the bodies of the parents) must start their lifetime with a learning stage. In this stage, a robot with the given morphology needs to learn to control its body and maximize its task performance.

In this paper we investigate six robot learning test cases, specified by two different tasks, gait learning (moving in any direction) and directed locomotion, and three bio-inspired robot shapes, a snake, a spider, and a gecko. Testing the behavior of these robots requires a computationally expensive simulation. In future stages, learning will be conducted in the real world where 3D-printed robots are automatically configured with the proposed controllers and their behavior is tested in a robot arena. Each of these trials will require a considerable amount of time, ranging from multiple minutes up to an hour. This makes the real world evaluation function even more expensive than the simulator we are currently using. Due to the expensive nature of the problem at hand, extensive tests and algorithm comparisons are less feasible. Therefore, additional to the robots simulations, the well-known set of Black-Box-Optimization-Benchmarking (BBOB) test functions [17] are used for extensively comparing the performance of each algorithm on different problem classes and dimensionalities. Given the results on the artificial functions, comparisons can be drawn about real-world applications. Considering this approach the following research questions arise:

RQ-1. Can parallel variants of BO outperform inherently parallel algorithms like CMA-ES if the evaluation budget is severely constrained?

RQ-2. Which parallel variants of BO show the best performance on which of the tested problem landscapes?
RQ-3. How do BO and CMA-ES compare on the robot application?

The rest of this paper is structured as follows: We will first give an overview of the implemented parallelization methods in Sect. 2. We will introduce the robot application for the real-world scenario in Sect. 3. The setup of our experiments is explained in Sect. 4. The obtained results are presented in Sect. 5 and finally discussed in Sect. 6.

2 Overview of Implemented Optimization Methods

2.1 Bayesian Optimization

Bayesian optimization (BO) is an iterative global optimization framework useful for expensive black-box derivative free problems [9,22]. The main components of BO are a surrogate model and an acquisition function. We use Gaussian process (GP) [19] with radial basis function kernel [8] as the surrogate model. The kernel is defined as $\Sigma^{(i)} = \exp(-\sum_{j=1}^{n} \theta_j |x_j^{(i)} - x'_j|^{p_j})$, where p_j determines the correlation function smoothness and θ_j the extend of a point's x_i influence. Two variants of this kernel will be used in this work: **P2**, where $p_i = 2$, and **FitP**, where p_i is part of the optimization loop.

Three of the most common acquisition functions are considered: expected improvement (EI), lower confidence bound (LCB) and predicted value (PV) [22]. PV is a heavy exploitation approach. It uses the surrogate model's best prediction as the next candidate solution. Under the right conditions this assures quick convergence but might lead to getting stuck in local optima.

LCB is an optimistic acquisition function. At a point x, it underestimates the mean using the uncertainty, $\alpha_{LCB}(x, \beta)$, where $\beta \leq 0$ is the exploration-exploitation trade-off factor. Following [23] we set $\beta = 1$ for every instance of BO using LCB as acquisition function.

EI [8] is a more explorative acquisition function. It calculates the amount of improvement a new point can achieve based on its mean and variance. The point with the highest expected improvement is selected as the next candidate solution.

2.2 Parallelization Approaches

Depending on the selected acquisition function BO can balance model exploitation (to quickly converge to the global optimum) and model exploration (to increase model quality). Most parallelization techniques for BO create multiple candidates per iteration by searching for multiple differently weighted compromises of these two goals. A total of q candidate solutions is generated per iteration, where q defines the number of objective function evaluations that can be run in parallel. In the following, three of these techniques are presented: investment portfolio improvement (IPI), multi-point expected improvement (q-EI), multi-objective infill criteria (MOI), and our implementation, multi-kernel Bayesian optimization (mK-BO). For more information about these methods please refer to the presented bibliography.

Investment Portfolio Improvement. [24] views the suggestion of new candidate solutions as handling an investment portfolio. It tries to balance high- and low-risk investments. On the one hand, candidate solutions which have a very high probability of improvement are safe investments. Yet, they often yield a minimal improvement over the best-known candidate solutions. On the other hand, candidates with a high expected improvement usually incorporate a high uncertainty and thus high risk. [24] proposes a sequential switching criterion that cycles between a high-, medium,- and a low-risk point. This sequential approach can directly be adapted to a parallel application. Instead of just three candidates, a different balance between exploration and exploitation can be defined for q candidate solutions. The candidates are then evaluated in parallel.

Multi-point Expected Improvement. In q-EI, the definition of EI is adapted to a set of points with shared EI. A detailed description of an efficient implementation is given by Ginsbourger et al. [10]. For this study, the q-EI implementation of the 'DiceOptim' R-package [21] is used together with the model implementation of the 'DiceKriging' package [21].

A property of q-EI that is worth mentioning is that it favors solutions that are spread throughout the search space. As sequential EI already leans towards exploration, this effect is fortified with each added parallel candidate. If two points of a set are too close to each other, the expected improvement of one of them will tend to zero.

Multi-objective Infill Criteria. [2] approaches the compromise between exploration and exploitation as a multi-objective optimization problem. The predicted value of the GP model defines the first objective. The models' uncertainty defines the second. An evolutionary algorithm searches for the Pareto-front of the bi-objective acquisition function. Each point on the front is a good compromise between the objectives and could be considered in the parallel evaluation. To narrow down the number of proposed points Bischl et al. consider two distance-based techniques (nearest neighbor and nearest better neighbor) as a third objective on the Pareto-front. For our experiments, we chose the implementation that was considered best in their benchmarks, described in [2] with ID 10. Our implementation is directly taken from the author's R-package 'mlrMBO' [1].

Multi-kernel BO. Additionally, we inspect a rather simple way to parallelize BO. When faced with a black-box optimization problem, often the choice for the right kernel parameter setting or acquisition function for BO is not clear.

While some settings might perform well on specific landscape types or problem dimensionalities, they might fail in others. In a parallel environment, this problem can be circumvented by running different BO configurations in parallel. Instead of choosing a single setup for BO, as many configurations are created as objective function evaluations can be run in parallel. Since our parallelization system is capable of running six robot simulations in parallel, we describe six distinct BO configurations to be run in parallel: Three acquisition functions, EI, LCB, and PV, are combined with the previously mentioned variants of kernel configuration, P2 and FitP.

After sampling an initial design, the algorithm will build the six distinctly configured GP models in parallel, one on each available core. Their respective acquisition functions are optimized, and the candidate solutions are evaluated. The algorithm waits for all instances to complete their evaluations in a synchronization step. After that, the next iteration starts with a new set of models built on all so far evaluated candidate solutions. Thus, the models share knowledge and synchronize after each iteration.

2.3 CMA-ES

CMA-ES is a state-of-the-art EA for optimizing non-linear non-convex black-box functions. In short, CMA-ES samples in each iteration a population, λ, from a multivariate normal distribution $\mathcal{N} \sim (m, \sigma^2\mathbf{C})$, where m is the weighted mean value of selected candidates, σ is the step size and $\mathbf{C}$ is the covariance matrix. The population is evaluated on the objective function and ranked. According to the ranking results parameters m, σ and $\mathbf{C}$ are updated to give more probability to good samples on the next iteration. A small λ accelerates the convergence, while larger values are helpful for noisy functions [13]. The python 'cma' implementation by Hansen et al. [14], was used in our experiments. To fit the general R framework that was used for experimenting, the python code is called via the R to python interface "reticulate" [25].

3 Robotic Application

Following the triangle of life framework proposed by [6], we focus on the infancy phase of a robot's life cycle. After a new robot is generated with a specific morphology and controller, it needs to go through a learning stage in order to adapt the configuration of its controller to complete a given task as efficient as possible. Usually this stage is only part of a longer process in which the goal is to find the best morphology-controller combination through evolution. In a real-time real-space scenario this process can be extremely long and costly.

The robot's dynamics, the controller's structure, and, the task a robot needs to complete are variable or unknown factors and can be seen as a black-box function. Given the time every simulation carries it can also be considered expensive. The aim of our optimization task is to find controller configurations that make it possible for the robots to move faster on a restricted number of simulation tests.

The robots are simulated using the **R**obot **Evolve** (Revolve) [16] toolkit. Revolve works on top of Gazebo[1] and incorporates a set of tools to allow an easy definition of the robots, environments to execute the simulations and objective functions to evaluate a robot's performance.

The robots used in this work are based on the framework RoboGen[2]. The robot's bodies contain three components: A core component housing the robot's microcontroller and battery unit, fixed bricks which allow to attach other components to any of its faces, and lastly, active hinges which are powered by servo motors. Each hinge adds an extra degree of freedom to the robot, thus increasing the input dimensionality in our controller design.

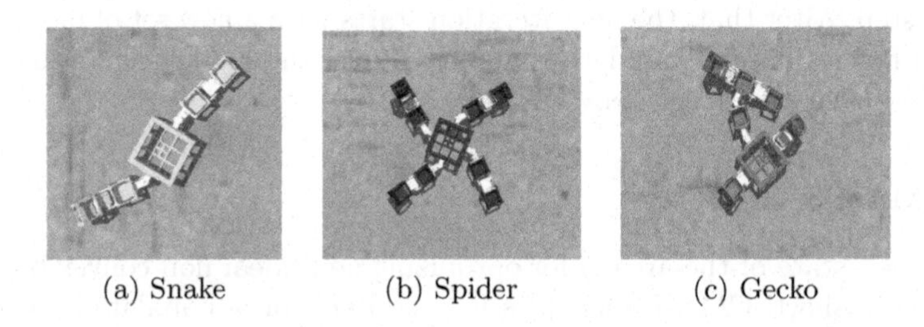

(a) Snake (b) Spider (c) Gecko

Fig. 1. Tested robot morphologies, each simulating an animal structure. The three different body components can be easily distinguished by shape and color. The biggest brick corresponds to the core component, each robot can only have one. Fixed bricks look similar but are smaller in size. Active hinges are illustrated in white. (Color figure online)

Three different robot morphologies are tested, each is built to simulate the structure of a snake, spider and gecko respectively. In Fig. 1 the different robot morphologies and their components are presented. All the servo motors on the robot's body are controlled by an output oscillator. This oscillator depends only on a sinusoid signal determined by three parameters: amplitude, period and phase offset. To reduce the controller's dimensionality the amplitude parameter is fixed to an unit value. Since the tested tasks require the robot to be in constant motion, we assume a fixed amplitude value will not affect the robot's speed as it would in start/stop scenarios.

4 Experiments

The source code of all algorithm configurations, software, and all experiments results presented in this work are freely available for reproducibility at: https://github.com/frehbach/rebo19a/.

[1] https://gazebosim.org/.

[2] http://robogen.org.

Two test scenarios are considered to test the viability and performance of the different optimization algorithms. Firstly, the algorithms are extensively benchmarked on the BBOB test suite to assess their performance on varying landscapes and dimensionalities. We simulate the environment of an expensive function by limiting the algorithm's budget to match the amount of permitted iterations on the robot application. Secondly, the algorithms are applied to the robot controller problem.

Since the problem can be evaluated in parallel, we do not count the individual objective function evaluations, but rather the sets of parallel evaluations (iterations) done by each algorithm. The machine used for the experiments allows to run six robot simulations efficiently in parallel. Thus, each algorithm can evaluate up to six candidate solutions per iteration.

The different algorithms are started with an initial Latin hypercube design of ten points. q-EI and MOI implementations require the amount of initial samples to be larger than the input dimensionality. Therefore, their amount of initial samples was set to the next multiplier of six which is greater than the respective problem dimensionality. To match the number of available processors the population size for CMA-ES is set to 6 and the initial step size is left at the default value 0.5. After the initial design, each algorithm is given a total of 15 parallel iterations, resulting in a maximum budget of 100 function evaluations. Each experiment is repeated 30 times for statistical analysis.

4.1 First Scenario: BBOB Functions

The Black-Box-Optimization-Benchmarking (BBOB) test suite [12] is one of the most well-known benchmark suites in the evolutionary computation community. The 24 BBOB functions are selected to test the performance of algorithms on landscapes with known difficulty. Every function is scalable to varying input dimensionalities. A detailed description of each function, its optima and properties is available in [17].

In the BBOB suite, each function is available in multiple instances. An instance is a rotated or shifted version of the original objective function. In our experiments the algorithms are run on all 15 standard instances in their five, ten and 20 dimensional version. This will account for possible performance variations in the algorithms caused by problem dimensionality. This is an important point given that the robots can change their morphology and thus reduce or increase the controller dimensionality. All described experiments were run with a recent GitHub version of BBOB, v2.3.1 [15].

4.2 Second Scenerio: Robot Controller

Performance on the application case is measured as the maximum speed (m/s) a robot can achieve in a fixed number of function evaluations (simulations). All robots are simulated in a flat world without obstacles. The robot is able to move along the x- and y-axis (ground). A fixed simulation time of 60 s is set for each robot.

Two tasks with different degree of difficulty are used, gait learning and directed locomotion. In gait learning the Euclidian distance between the robots starting and end point is measured and divided by the simulation time. The resulting speed is considered to be the robot's fitness.

For directed locomotion, the fitness function takes into account the direction in which the robot moves. Only distance traveled on the y-axis will be measured and then divided by the simulation time.

The input dimensionality of the robots application varies across robots as a results of the different number of active hinge elements. The snake, gecko and spider have 8, 12 and 16 parameters respectively.

Test Problem Publication. The software required to run the robot controller application does not run on all platforms. To overcome this issue and to create an open and easy process for everyone to access the robot application, a dockerized version was developed as part of this work. Docker is an open source tool [3] and works on any major operating system. The docker container for running the robot application with a brief usage manual is also available at https://github.com/frehbach/rebo19a/.

5 Results and Discussion

All convergence and significance plots for all functions can be found in https://github.com/frehbach/rebo19a/.

We do not assume that the collected results are normally distributed. Therefore, we apply non-parametric tests that make less assumptions about the underlying data, as suggested by Derrac et al. [5]. We accept results to be statistically significant if the corresponding p-values are smaller than $\alpha = 0.05$. The Kruskal-Wallis rank-sum test (base-R package 'stats') is used to determine, whether a significant difference is present in the set of applied algorithms or not. If this test is positive, a posthoc test according to Conover (PMCMR R package [18]) for pairwise multiple comparisons is used to check for differences in each algorithm pair. The pairwise comparisons are further used to rank the algorithms on each problem class as follows: The set of algorithms that is never outperformed with statistical significance is considered rank one and removed from the list. This process is repeated until all algorithms are ranked from 1 to 6.

To answer our first research question the initial focus lies on the BBOB test function experiments. First we compare single-core BO and CMA-ES. The combination of different kernel configurations and acquisition functions explained in Sect. 2 make up the six tested variants of single-core BO.

BO outperforms CMA-ES on most problems and dimensionalities in the single-core scenario. Figure 2 illustrates this behavior on the ten-dimensional version of the Büche-Rastrigin function, where all BO approaches ranked better than CMA-ES and random search. The different kernel configurations and acquisition functions were observed to have statistically significant difference between

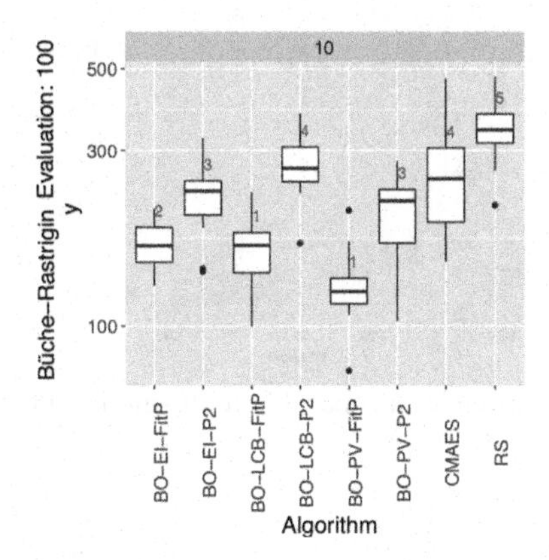

Fig. 2. Single-core BO vs. CMA-ES on the Büche-Rastrigin function. The notation refers to the different combination of acquisition function and kernel configuration. An experiment denoted as BO-PV-P2 then refers to BO using predicted value kernel with parameter $p = 2$. The red number refers to the algorithm rank as determined by the pairwise multiple comparisons test.

each other. Interestingly, greedy acquisition functions, like PV, show better performance than the more popular EI in 16 out of the 24 tested functions. This agrees with results presented in [20]. In the multi-core tests, parallel-BO seems to converge faster and have better performance than CMA-ES on most test problems. This specially holds true in higher dimensions. Convergence plots for functions Rastrigin, Sharp Ridge and Schwefel in Fig. 3 illustrate the performance difference on all tested dimensions. A clear advantage was not always the case for functions with high conditioning or weak global structure. However, only in functions $Weierstrass$ and $Katsura$ did BO not achieved equal or better performance than CMA-ES or random search. Both functions are highly rugged and repetitive.

Our second research question can also be answered by looking at the BBOB results. As shown on Fig. 3, q-EI and IPI frequently perform worse than MOI or mK-BO. Based on the single-core BO results it is not surprising that the multi-point adaptation of EI did not perform well. The focus on exploration is disadvantageous on problems with limited function evaluations. This can also be seen for IPI were exploration may win over exploitation.

MOI authors remark the algorithm favors exploitative behaviors. If the suggested points tend to exploitation, the performance of MOI is in accordance to the results from single-core experiments, where greedy approaches achieved better values.

For our last research question all algorithms are tested on the expensive black-box robot application. In contrast to the last experiments our aim here is to maximize the objective function. This represents the need to have robots that can move faster.

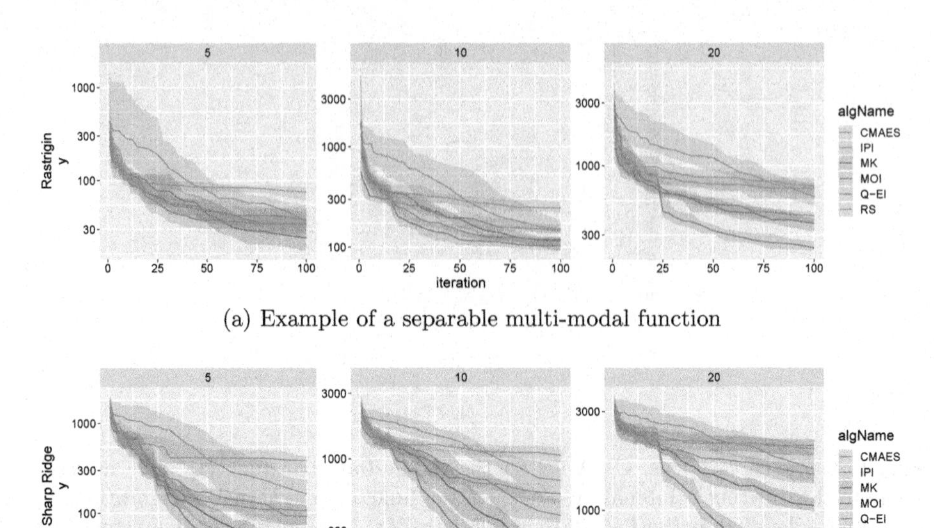

(a) Example of a separable multi-modal function

(b) Example of a unimodal function with high conditioning

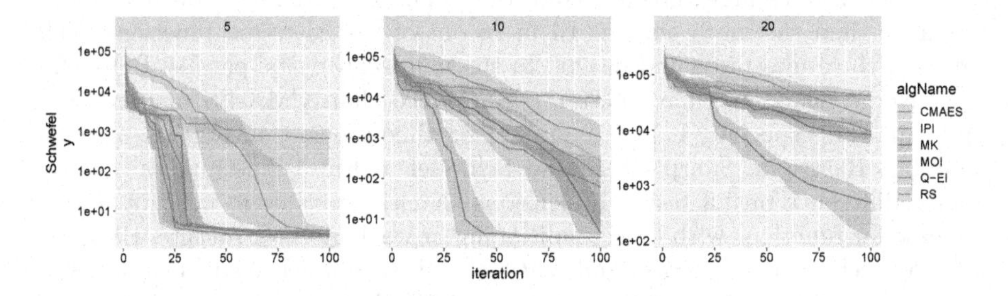

Fig. 3. Parallel algorithms performance on three different BBOB subgroups with different dimensionalities. The convergence plots show the median, upper and lower quartiles. MOI parallel implementation shows a clear advantage in performance for most cases, especially on higher dimensions. Parallel-BO using IPI and q-EI are in many cases outperformed by the simple mK-BO approach.

Interestingly there are no clear visible performance differences on some of the robot problems. Random search often performs similarly to all parallel methods. Figure 4 Illustrates the multi-comparison test results for the last iteration in both robot experiments. BO methods rank better than CMA-ES but not always better than random search. If we refer to the benchmark results, a similar behavior was found for highly multi-modal functions with weak global structures. This is a good indication of the complex controller's landscape and the complex interactions between its parameters.

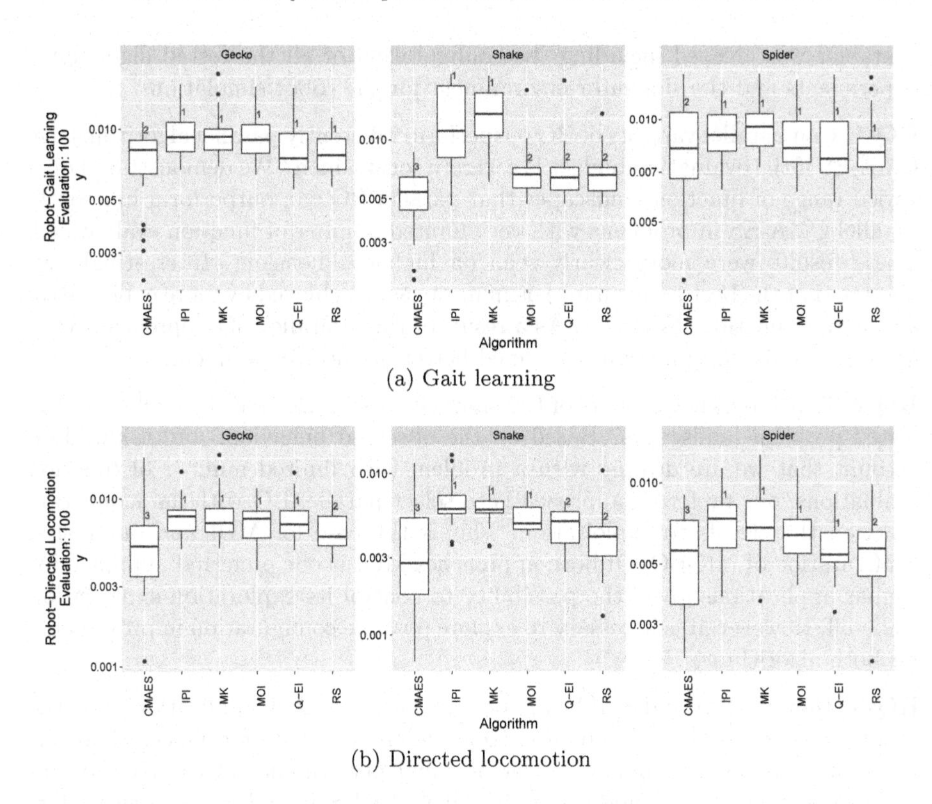

(a) Gait learning

(b) Directed locomotion

Fig. 4. Last iteration ranked multi-comparision test results for the robot application. The red number indicates the algorithm's achieved rank. BO methods maintain a better performance. However similar results by random search suggest BO is not able to efficiently exploit the optimization landscape. (Color figure online)

During the simulations it was possible to visualize the locomotive behaviors of the robots. For the snake it was noticed that it tended to roll in order to achieve greater speeds. This brought an advantage in gait learning but not in directed locomotion. This specific case can be the reason for the performance divergence present for the snake. However, why IPI and mK-BO in particular seem to perform better on this scenario is still open to investigation.

6 Conclusions

To answer our research questions, we compared the performance of sample efficient parallel-BO with the state-of-the-art EA, CMA-ES and random search on expensive black-box problems with a maximum of 100 function evaluations. A first experimental stage was conducted using the BBOB test-functions with the assumption they are expensive to evaluate. As a second stage the algorithms were tested in a real world robot application. To enable easy replicability, a docker container was created including the configuration of all the tested algorithms, experiments and the simulator environment for the robot simulations.

RQ-1. Can parallel variants of BO outperform inherently parallel algorithms like CMA-ES if the evaluation budget is severely constrained? We demonstrated on a varied range of function landscapes that parallel-BO can outperform inherently parallel CMA-ES in problems with very limited number of function evaluations. These results were more clearly seen on higher dimensions. Interestingly we observe that the best performance, is in most cases, achieved by more exploitative acquisition functions (see [20]). As a result of this, Parallel-BO approaches with greedier acquisition function performed better on most experiments.

RQ-2. Which parallel variants of BO show the best performance on which of the tested problem landscapes? Based on the observed behaviour and taking into account that we are dealing with a problem with limited number of function evaluations, our preferred approach is to select parallel-BO with an acquisition function that tends to exploitation. This is the case for MOI and our tested configuration of mK-BO. Of both approaches MOI is our preferred approach for similar applications given the possibility to control its exploration-exploitation trade-off. However, it is necessary to explore possible configuration improvements for both algorithms.

RQ-3. How do BO and CMA-ES compare on the robot application? On the robotic test cases there is no prominent difference in performance. However, the statistical tests support a better or equal performance of the parallel-BO implementations over random search and CMA-ES for all cases. The performance similarity of random search suggests that neither BO nor CMA-ES could take advantage of the objective function landscape. This is an indication that the problem of robot learning implies a highly multi-modal complex objective function.

In summary, the main contributions of this work are: an efficient BO parallelization method, a detailed performance comparison of BO parallelization methods, and a freely available robotics test suite via a docker image for (comparative) challenging problems for optimization methods.

It remains open to further investigate the contribution that each of the different BO configurations had on the final response of mK-BO. Non-stationary kernels for GP can be studied in order to find better configurations for the complex robot task. Furthermore, we are working on developing and testing robot controllers of different structure and re-evaluate the efficiency and efficacy of the optimizers discussed in this paper.

References

1. Bischl, B., Richter, J., Bossek, J., Horn, D., Thomas, J., Lang, M.: mlrMBO: A Modular Framework for Model-Based Optimization of Expensive Black-Box Functions (2017)

2. Bischl, B., Wessing, S., Bauer, N., Friedrichs, K., Weihs, C.: MOI-MBO: multi-objective infill for parallel model-based optimization. In: Pardalos, P.M., Resende, M.G.C., Vogiatzis, C., Walteros, J.L. (eds.) LION 2014. LNCS, vol. 8426, pp. 173–186. Springer, Cham (2014). https://doi.org/10.1007/978-3-319-09584-4_17
3. Boettiger, C.: An introduction to docker for reproducible research. ACM SIGOPS Oper. Syst. Rev. **49**(1), 71–79 (2015)
4. Bongard, J.: Evolutionary robotics. Commun. ACM **56**(8), 74–85 (2013)
5. Derrac, J., García, S., Molina, D., Herrera, F.: A practical tutorial on the use of nonparametric statistical tests as a methodology for comparing evolutionary and swarm intelligence algorithms. Swarm Evol. Comput. **1**(1), 3–18 (2011). https://doi.org/10.1016/j.swevo.2011.02.002
6. Eiben, A., et al.: The triangle of life: evolving robots in real-time and real-space, September 2013. https://doi.org/10.7551/978-0-262-31709-2-ch157
7. Floreano, D., Husbands, P., Nolfi, S.: Evolutionary robotics. In: Siciliano, B., Khatib, O. (eds.) Handbook of Robotics, 1st edn, pp. 1423–1451. Springer, Heidelberg (2008). https://doi.org/10.1007/978-3-540-30301-5_62
8. Forrester, A., Sobester, A., Keane, A.: Engineering Design via Surrogate Modelling: A Practical Guide. Wiley, Hoboken (2008)
9. Frazier, P.I.: A tutorial on Bayesian optimization (2018)
10. Ginsbourger, D., Le Riche, R., Carraro, L.: Kriging is well-suited to parallelize optimization. In: Tenne, Y., Goh, C.-K. (eds.) Computational Intelligence in Expensive Optimization Problems. ALO, vol. 2, pp. 131–162. Springer, Heidelberg (2010). https://doi.org/10.1007/978-3-642-10701-6_6
11. Haftka, R.T., Villanueva, D., Chaudhuri, A.: Parallel surrogate-assisted global optimization with expensive functions-a survey. Struct. Multidiscip. Optim. **54**(1), 3–13 (2016)
12. Hansen, N., Auger, A., Mersmann, O., Tusar, T., Brockhoff, D.: COCO: a platform for comparing continuous optimizers in a black-box setting. arXiv e-prints, August 2016
13. Hansen, N., Ostermeier, A.: Completely derandomized self-adaptation in evolution strategies. Evol. Comput. **9**(2), 159–195 (2001)
14. Hansen, N., Akimoto, Y., Baudis, P.: CMA-ES/pycma on Github. Zenodo, February 2019. https://doi.org/10.5281/zenodo.2559634
15. Hansen, N., et al.: COmparing Continuous Optimizers: numbbo/COCO on Github, March 2019. https://doi.org/10.5281/zenodo.2594848
16. Hupkes, E., Jelisavcic, M., Eiben, A.E.: Revolve: a versatile simulator for online robot evolution. In: Sim, K., Kaufmann, P. (eds.) EvoApplications 2018. LNCS, vol. 10784, pp. 687–702. Springer, Cham (2018). https://doi.org/10.1007/978-3-319-77538-8_46
17. Nikolaus, H., Steffen, F., Raymond, R., Auger, A.: Real-parameter black-box optimization benchmarking 2009: noiseless functions definitions. Research Report INRIA - 00362633v2, INRIA (2009)
18. Pohlert, T.: The pairwise multiple comparison of mean ranks package (PMCMR) (2014). http://CRAN.R-project.org/package=PMCMR. Accessed 12 Jan 2016
19. Rasmussen, C., Williams, C.: Gaussian Processes for Machine Learning. Adaptive Computation and Machine Learning. MIT Press, Cambridge (2006)
20. Rehbach, F., Zaefferer, M., Naujoks, B., Bartz-Beielstein, T.: Expected improvement versus predicted value in surrogate-based optimization (2020)
21. Roustant, O., Ginsbourger, D., Deville, Y.: DiceKriging, DiceOptim: two R packages for the analysis of computer experiments by kriging-based metamodeling and optimization. J. Stat. Softw. **51**(1), 1–55 (2012)

22. Shahriari, B., Swersky, K., Wang, Z., Adams, R.P., de Freitas, N.: Taking the human out of the loop: a review of Bayesian optimization. Proc. IEEE **104**(1), 148–175 (2016). https://doi.org/10.1109/JPROC.2015.2494218
23. Srinivas, N., Krause, A., Kakade, S.M., Seeger, M.W.: Information-theoretic regret bounds for Gaussian process optimization in the bandit setting. IEEE Trans. Inf. Theory **58**(5), 3250–3265 (2012). https://doi.org/10.1109/tit.2011.2182033
24. Ursem, R.K.: From expected improvement to investment portfolio improvement: spreading the risk in kriging-based optimization. In: Bartz-Beielstein, T., Branke, J., Filipič, B., Smith, J. (eds.) PPSN 2014. LNCS, vol. 8672, pp. 362–372. Springer, Cham (2014). https://doi.org/10.1007/978-3-319-10762-2_36
25. Ushey, K., Allaire, J., Tang, Y.: Reticulate: Interface to 'Python' (2019). https://CRAN.R-project.org/package=reticulate, r package version 1.13

Revisiting Population Models in Differential Evolution on a Limited Budget of Evaluations

Ryoji Tanabe(✉)

Yokohama National University, Yokohama, Japan
rt.ryoji.tanabe@gmail.com

Abstract. No previous study has reported that differential evolution (DE) is competitive with state-of-the-art black-box optimizers on a limited budget of evaluations (i.e., the expensive optimization scenario). This is true even for surrogate-assisted DEs. The basic framework of DE should be reconsidered to improve its performance substantially. In this context, this paper revisits population models in DE on a limited budget of evaluations. This paper analyzes the performance of DE with five population models on the BBOB function set. Results demonstrate that the traditional synchronous model is unsuitable for DE in most cases. In contrast, the performance of DE can be significantly improved by using the plus-selection model and the worst improvement model. Results also demonstrate that DE with a suitable population model is competitive with covariance matrix adaptation evolution strategy depending on the number of evaluations and the dimensionality of a problem.

1 Introduction

Single-objective black-box numerical optimization involves finding a solution $\boldsymbol{x} = (x_1, ..., x_n)^\top$ that minimizes a given objective function $f : \mathbb{X} \rightarrow \mathbb{R}$. Here, $\mathbb{X}$ is the n-dimensional solution space. Any explicit knowledge of f is unavailable.

This paper considers black-box optimization on a limited budget of evaluations. Optimization with a small number of function evaluations (e.g., $100 \times n$ evaluations) is generally called *expensive optimization*. In contrast, this paper denotes optimization with a relatively large number of function evaluations (e.g., $10\,000 \times n$ evaluations) as *cheap optimization*. Some real-world problems require a long computation time to evaluate a solution $\boldsymbol{x}$ by expensive computer simulations [3,24] (e.g., CFD [6]). Also, expensive optimization frequently appears in the filed of hyperparameter optimization of machine learning models [10].

In the evolutionary computation community, surrogate-assisted evolutionary algorithms are representative approaches for expensive optimization [3,24]. In general, surrogate-assisted evolutionary algorithms replace an expensive objective function with a cheap surrogate model. Then, the objective value of a new solution is predicted by the surrogate model based on past solutions found during

T. Bäck et al. (Eds.): PPSN 2020, LNCS 12269, pp. 257–272, 2020.
https://doi.org/10.1007/978-3-030-58112-1_18

the search process. It is expected that surrogate-assisted evolutionary algorithms can effectively reduce the number of evaluations by an actual objective function.

Covariance matrix adaptation evolution strategy (CMA-ES) is a variant of evolution strategies (ES) for numerical optimization [13,19]. Surrogate-assisted approaches have been intensively studied in the field of ES, especially CMA-ES [4,14,26,33]. Surrogate-assisted CMA-ES has shown state-of-the-art performance for expensive optimization. Even a non-surrogate-assisted CMA-ES performs well for expensive optimization. Although Bayesian optimizers (e.g., EGO [25]) generally show good performance within a very small number of evaluations, a non-surrogate-assisted CMA-ES performs well after that in most cases [32,36].

Similar to CMA-ES, differential evolution (DE) is a numerical optimizer [42]. The results in the annual IEEE CEC competitions have demonstrated that some DEs are competitive with more complex optimizers for *cheap optimization* (e.g., [46]). However, the number of previous studies on DE algorithms for *expensive optimization* is much smaller than that for *cheap optimization*. In contrast to the ES community, as pointed out in [7], surrogate-assisted approaches have not received much attention in the DE community. Also, a surrogate-assisted DE has not been competitive even with a non-surrogate-assisted CMA-ES. Most previous studies "avoided" to compare DE with CMA-ES for expensive optimization.

Nevertheless, the simplicity of DE is still attractive in practice. The easy-to-use property of DE is valuable for a non-expert user who just uses an evolutionary algorithm as a black-box optimization tool. In this context, we want to substantially improve the performance of DE for expensive optimization. To achieve this goal while keeping the simplicity of DE, this paper revisits population models in DE for expensive optimization. A population model determines how to update the population for each iteration. The original DE [42] uses the synchronous model, which updates all individuals in the population simultaneously by using one-to-one survivor selection. In addition to the synchronous model, this paper analyzes the performance of DE with the following four population models: the asynchronous [8,53,54], $(\mu + \lambda)$ [37,49], worst improvement [1], and subset-to-subset [12] models. Although some previous studies (e.g., [8]) investigated the effect of population models in DE, such an analysis is only for cheap optimization, not for expensive optimization. Thus, little is known about the influence of the population model on the performance of DE for expensive optimization.

Our contributions in this paper are at least threefold:

1. We demonstrate that the performance of DE can be significantly improved by replacing the traditional synchronous model with the $(\mu + \lambda)$ or worst improvement models. This means that a more efficient surrogate-assisted DE for expensive optimization could be designed by using the two models.
2. We compare DE with the two models to CMA-ES and other optimizers. We show that DE with a suitable population model performs better than or similar to CMA-ES depending on the number of function evaluations and the dimensionality of a problem. This is the first study to report such a promising performance of DE for expensive optimization.

Algorithm 1: Synchronous model

1 Initialize $\boldsymbol{P} = \{\boldsymbol{x}_1, ..., \boldsymbol{x}_\mu\}$ randomly;
2 **while** The termination criteria are not met **do**
3   **for** $i \in \{1, ..., \mu\}$ **do**
4         $\boldsymbol{u}_i \leftarrow$ generateTrialVector($\boldsymbol{P}$);
5   **for** $i \in \{1, ..., \mu\}$ **do**
6   **if** $f(\boldsymbol{u}_i) \leq f(\boldsymbol{x}_i)$ **then** $\boldsymbol{x}_i \leftarrow \boldsymbol{u}_i$;

3. DE with the two models can be viewed as a base-line. Any surrogate-assisted DE should outperform the two non-surrogate-assisted DEs. Some surrogate-assisted DEs have been proposed (e.g., [30,34,51,56]). However, benchmarking a surrogate-assisted DE has not been standardized in the DE community. It is also difficult to accurately reproduce experimental results of most existing surrogate-assisted DEs since their source code is not available through the Internet. Consequently, the progress is unclear. Our base-line addresses this issue, facilitating a constructive development of a surrogate-assisted DE.

The rest of this paper is organized as follows. Section 2 explains the five population models in DE. Section 3 describes the setting of our computational experiments. Section 4 shows analysis results. Section 5 concludes this paper.

2 Five Population Models in DE

Here, we do not consider any method that aims to maintain the diversity in the population, such as crowding DE [47], island/distributed models [8,52], and the cellular topology [8,35]. These methods aim to prevent the premature convergence of DE to find a good solution with a large number of function evaluations. Thus, these methods are not suitable for expensive optimization, where DE needs to quickly find a good solution with a small number of function evaluations.

We carefully surveyed the literature on population models that aim to accelerate the convergence speed of DE. As a result, we found the following four population models: the asynchronous model [8,53,54], the $(\mu+\lambda)$ model [37,49], the worst improvement model [1], and the subset-to-subset (STS) model [12]. Algorithms 1–5 show the overall procedure of the basic DE with the traditional synchronous model and the four population models. We extracted only the population models from their corresponding original algorithms. For example, the worst improvement model is derived from DE with generalized differential (DEGD) [1], which consists of multiple components. In this study, we want to focus only on the population model. For this reason, we generalized it so that we can examine its effectiveness in an isolated manner. For the same reason, we do not use any surrogate model and any parameter adaptation method for the scale factor F and the crossover rate C [45]. Below, we explain the five models.

- **Synchronous model (Algorithm 1)**

Table 1. Seven representative mutation strategies for DE.

Strategies	Definitions
rand/1	$\boldsymbol{v}_i := \boldsymbol{x}_{r_1} + F\left(\boldsymbol{x}_{r_2} - \boldsymbol{x}_{r_3}\right)$
rand/2	$\boldsymbol{v}_i := \boldsymbol{x}_{r_1} + F_i\left(\boldsymbol{x}_{r_2} - \boldsymbol{x}_{r_3}\right) + F_i\left(\boldsymbol{x}_{r_4} - \boldsymbol{x}_{r_5}\right)$
best/1	$\boldsymbol{v}_i := \boldsymbol{x}_{\text{best}} + F_i\left(\boldsymbol{x}_{r_1} - \boldsymbol{x}_{r_2}\right)$
best/2	$\boldsymbol{v}_i := \boldsymbol{x}_{\text{best}} + F_i\left(\boldsymbol{x}_{r_1} - \boldsymbol{x}_{r_2}\right) + F_i\left(\boldsymbol{x}_{r_3} - \boldsymbol{x}_{r_4}\right)$
current-to-best/1	$\boldsymbol{v}_i := \boldsymbol{x}_i + F_i\left(\boldsymbol{x}_{\text{best}} - \boldsymbol{x}_i\right) + F_i\left(\boldsymbol{x}_{r_1} - \boldsymbol{x}_{r_2}\right)$
current-to-*p*best/1	$\boldsymbol{v}_i := \boldsymbol{x}_i + F_i\left(\boldsymbol{x}_{p\text{best}} - \boldsymbol{x}_i\right) + F_i\left(\boldsymbol{x}_{r_1} - \tilde{\boldsymbol{x}}_{r_2}\right)$
rand-to-*p*best/1	$\boldsymbol{v}_i := \boldsymbol{x}_{r_1} + F_i\left(\boldsymbol{x}_{p\text{best}} - \boldsymbol{x}_{r_1}\right) + F_i\left(\boldsymbol{x}_{r_2} - \tilde{\boldsymbol{x}}_{r_3}\right)$

The synchronous model is used in the original DE [42] and most recent DE variants, including jDE [5], JADE [57], CoDE [50], and SHADE [43]. Here, we explain the synchronous model and some basic operations in DE.

At the beginning of the search, the population $\boldsymbol{P} = \{\boldsymbol{x}_1, ..., \boldsymbol{x}_\mu\}$ is initialized (line 1 in Algorithm 1), where μ is the population size. For each $i \in \{1, ..., \mu\}$, $\boldsymbol{x}_i$ is the i-th individual in the population $\boldsymbol{P}$. Here, $\boldsymbol{x}_i$ is an n-dimensional solution of a problem. For each $j \in \{1, ..., n\}$, $x_{i,j}$ is the j-th element of $\boldsymbol{x}_i$. According to the DE terminology, we use the terms "individual" and "vector" synonymously.

After the initialization of $\boldsymbol{P}$, the following steps (lines 2–6 in Algorithm 1) are repeatedly performed until a termination condition is satisfied. For each $i \in \{1, ..., \mu\}$, a trial vector (child) $\boldsymbol{u}_i$ is generated (lines 3–4 in Algorithm 1). In this operation, first, a mutant vector $\boldsymbol{v}_i$ is generated by applying a differential mutation to some individuals in $\boldsymbol{P}$. Table 1 shows seven representative mutation strategies in DE. The scale factor F controls the magnitude of the mutation. Parent indices $r_1, r_2, ...$ are randomly selected from $\{1, ..., \mu\} \setminus \{i\}$ such that they differ from each other. In Table 1, $\boldsymbol{x}_{\text{best}}$ is the best individual in $\boldsymbol{P}$. For each $i \in \{1, ..., \mu\}$, $\boldsymbol{x}_{p\text{best}}$ is randomly selected from the top $\max(\lfloor p\,\mu \rfloor, 2)$ individuals in $\boldsymbol{P}$, where $p \in [0, 1]$ controls the greediness of current-to-*p*best/1 [57] and rand-to-*p*best/1 [55]. Also, $\tilde{\boldsymbol{x}}_{r_2}$ and $\tilde{\boldsymbol{x}}_{r_3}$ in current-to-*p*best/1 and rand-to-*p*best/1 are randomly selected from the union of $\boldsymbol{P}$ and an external archive $\boldsymbol{A}$.

After the mutant vector $\boldsymbol{v}_i$ has been generated, a trial vector $\boldsymbol{u}_i$ is generated by applying crossover to $\boldsymbol{x}_i$ and $\boldsymbol{v}_i$. In this study, we use binomial crossover [42]:

$$u_{i,j} := \begin{cases} v_{i,j} & \text{if } q_j \leq C \text{ or } j = j_{\text{rand}} \\ x_{i,j} & \text{otherwise} \end{cases}, \tag{1}$$

where q_j is randomly selected from $[0, 1]$, and j_{rand} is randomly selected from $\{1, ..., n\}$. The crossover rate C in (1) controls the number of inherited elements from the target vector $\boldsymbol{x}_i$ to the trial vector $\boldsymbol{u}_i$.

After the μ trial vectors have been generated, all individuals are updated simultaneously (lines 5–6 in Algorithm 1). If $f(\boldsymbol{u}_i) \leq f(\boldsymbol{x}_i)$ for each $i \in \{1, ..., \mu\}$, $\boldsymbol{x}_i$ is replaced with $\boldsymbol{u}_i$. The individuals that were worse than the trial vectors are

Algorithm 2: Asynchronous model

1 Initialize $\boldsymbol{P} = \{\boldsymbol{x}_1, ..., \boldsymbol{x}_\mu\}$ randomly;
2 **while** The termination criteria are not met **do**
3   **for** $i \in \{1, ..., \mu\}$ **do**
4     $\boldsymbol{u} \leftarrow \text{generateTrialVector}(\boldsymbol{P})$;
5   **if** $f(\boldsymbol{u}) \leq f(\boldsymbol{x}_i)$ **then** $\boldsymbol{x}_i \leftarrow \boldsymbol{u}$;

Algorithm 3: $(\mu + \lambda)$ model

1 Initialize $\boldsymbol{P} = \{\boldsymbol{x}_1, ..., \boldsymbol{x}_\mu\}$ randomly;
2 **while** The termination criteria are not met **do**
3   $\boldsymbol{Q} \leftarrow \emptyset$;
4   **for** $i \in \{1, ..., \lambda\}$ **do**
5     $\boldsymbol{u} \leftarrow \text{generateTrialVector}(\boldsymbol{P})$, $\boldsymbol{Q} \leftarrow \boldsymbol{Q} \cup \{\boldsymbol{u}\}$;
6   $\boldsymbol{P} \leftarrow$ Select the μ best individuals from the union $\boldsymbol{P} \cup \boldsymbol{Q}$;

stored in the external archive $\boldsymbol{A}$. When $|\boldsymbol{A}|$ exceeds a pre-defined size, randomly selected individuals are deleted to keep the archive size constant.

- **Asynchronous model (Algorithm 2)**

In contrast to the synchronous model, individuals in the population are updated in an asynchronous manner. Thus, immediately after the trial vector $\boldsymbol{u}$ has been generated, $\boldsymbol{x}_i$ can be replaced with $\boldsymbol{u}$ (lines 4–5 in Algorithm 2). It is difficult to find out the first study that proposed the asynchronous model in DE since such an idea is general. In fact, some previous studies did not explicitly describe that they dealt with the asynchronous model (e.g., [53]). While DE with the asynchronous model can immediately use a new superior individual for the search, DE with the synchronous model needs to "wait" by the next iteration. For this reason, the asynchronous model is generally faster than the synchronous model in terms of the convergence speed of the population [7,8].

- **$(\mu + \lambda)$ model (Algorithm 3)**

The elitist $(\mu + \lambda)$ model is general in the field of evolutionary algorithms, including genetic algorithm and ES. For each iteration, a set of λ trial vectors $\boldsymbol{Q} = \{\boldsymbol{u}_1, ..., \boldsymbol{u}_\lambda\}$ are generated (lines 3–5 in Algorithm 3). For each $\boldsymbol{u}$, the target vector is randomly selected from the population. Then, the best μ individuals in $\boldsymbol{P} \cup \boldsymbol{Q}$ survive to the next iteration (line 6 in Algorithm 3). Unlike other evolutionary algorithms, the $(\mu + \lambda)$ model has not received much attention in the DE community. Only a few previous studies (e.g., [37,39,49]) considered the $(\mu + \lambda)$ model. As pointed out in [37], the synchronous model may discard a trial vector that performs worse than its parent but performs better than other individuals in the population. The $(\mu + \lambda)$ model addresses such an issue.

Algorithm 4: Worst improvement model

```
1 Initialize P = {x_1, ..., x_μ} randomly;
2 while The termination criteria are not met do
3     K ← Select indices of the λ worst individuals in P;
4     for i ∈ K do
5         u_i ← generateTrialVector(P);
6     for i ∈ K do
7         if f(u_i) ≤ f(x_i) then x_i ← u_i;
```

- **Worst improvement model (Algorithm 4)**

In the worst improvement model [1], only λ worst individuals can generate trial vectors (lines 3–5 in Algorithm 4). Then, the λ individuals and their λ trial vectors are compared as in the synchronous model (lines 6–7 in Algorithm 4).

Ali [1] demonstrated that a better individual is rarely replaced with its trial vector. In other words, the better the individual $\boldsymbol{x}_i$ ($i \in \{1, ..., \mu\}$) is, the more difficult it is to generate $\boldsymbol{u}_i$ such that $f(\boldsymbol{u}_i) \leq f(\boldsymbol{x}_i)$. In contrast, a worse individual is frequently replaced with its trial vector. Based on this observation, Ali proposed the worst improvement model that allows only the λ worst individuals to generate their λ trial vectors. The number of function evaluations could possibly be reduced by not generating the remaining $\mu - \lambda$ trial vectors that are unlikely to outperform their $\mu - \lambda$ parent individuals.

- **Subset-to-subset (STS) model (Algorithm 5)**

The aim of the STS model [12] is the same as that of the $(\mu + \lambda)$ model. The STS model uses the index-based ring topology. In the STS model, μ individuals and μ trial vectors are grouped based on their indices and the subset size $s \geq 2$. After μ trial vectors have been generated (lines 4–5 in Algorithm 5), an index i is randomly selected from $\{1, ..., \mu\}$ (line 6 in Algorithm 5), where i determines the start position on the index-based ring topology. First, l is set to s or the size of the remaining individuals (line 8 in Algorithm 5). Then, a set of l individuals and l trial vectors are stored into $\boldsymbol{X}$ based on the index-based ring topology (line 9 in Algorithm 5). In Algorithm 5, the function "modulo(a, b)" returns the remainder of the division of a by b. Finally, the l best individuals in $\boldsymbol{X}$ survive to the next iteration (lines 10–11 in Algorithm 5).

For example, suppose that $\mu = 5$, $s = 2$, and $i = 3$. In this case, 5+5 individuals are grouped as follows: $\{\boldsymbol{x}_3, \boldsymbol{x}_4, \boldsymbol{u}_3, \boldsymbol{u}_4\}$, $\{\boldsymbol{x}_5, \boldsymbol{x}_1, \boldsymbol{u}_5, \boldsymbol{u}_1\}$, and $\{\boldsymbol{x}_2, \boldsymbol{u}_2\}$. When $f(\boldsymbol{x}_3) = 0.3$, $f(\boldsymbol{x}_4) = 0.1$, $f(\boldsymbol{u}_3) = 0.4$, and $f(\boldsymbol{u}_4) = 0.2$, $\boldsymbol{x}_4$ and $\boldsymbol{u}_4$ survive to the next iteration as follows: $\boldsymbol{x}_3 := \boldsymbol{x}_4$ and $\boldsymbol{x}_4 := \boldsymbol{u}_4$. Notice that $\boldsymbol{u}_4$ cannot survive to the next iteration in the synchronous model since $f(\boldsymbol{u}_4) > f(\boldsymbol{x}_4)$.

Algorithm 5: STS model

```
1  Initialize P = {x_1, ..., x_μ} randomly;
2  h ← μ/s ;                                  // If modulo(μ, s) ≠ 0, h ← μ/s + 1
3  while The termination criteria are not met do
4      for i ∈ {1, ..., μ} do
5          u_i ← generateTrialVector(P);
6      i ← Randomly select an index from {1, ..., μ};
7      for j ∈ {1, ..., h} do
8          l ← min{μ − ((j − 1) × s), s};
9          K ← {modulo(i, μ), ..., modulo(i + l − 1, μ)},
           X ← {x_k | k ∈ K} ∪ {u_k | k ∈ K};
10         for k ∈ K do
11             x_k ← argmin_{x∈X} {f(x)}, X ← X \ {x_k};
12         i ← modulo(i + s, μ);
```

3 Experimental Setup

We performed all experiments using the COCO software [16]. The source code used in our experiments is available at https://github.com/ryojitanabe/de_expensiveopt. We used the 24 BBOB noiseless functions $f_1, ..., f_{24}$ [18], which are grouped into the following five categories: separable functions ($f_1, ..., f_5$), functions with low or moderate conditioning ($f_6, ..., f_9$), functions with high conditioning and unimodal ($f_{10}, ..., f_{14}$), multimodal functions with adequate global structure ($f_{15}, ..., f_{19}$), and multimodal functions with weak global structure ($f_{20}, ..., f_{24}$). The dimensionality n of the functions was set to $2, 3, 5, 10, 20$, and 40. For each function, 15 runs were performed. These settings adhere to the procedure in COCO. According to the expensive optimization scenario in COCO, the maximum number of function evaluations was set to $100 \times n$.

In general, the best parameter setting in EAs depends on the allowed budget of evaluations [9,38,41,44]. Although parameter studies on DE have been well performed for cheap optimization (e.g., [5,11,58]), little is known about a suitable parameter setting for expensive optimization. Thus, it is unclear how to set control parameters in DE. We try to address this issue by "hand-tuning" and "automated algorithm configuration".

We performed "hand-tuning" to select μ, the mutation strategy, and some parameters (i.e., λ and s) in the five population models. We mainly used the Sphere function $f(\boldsymbol{x}) = \sum_{i=1}^{n} x_i^2$ for benchmarking. As a result, we set μ to $\lfloor \alpha_\mu \ln n \rfloor$, and $\alpha_\mu = 13$. The population was initialized using Latin hypercube sampling. We used rand-to-$p$best/1 in Table 1. As in [57], we set the parameters of rand-to-$p$best/1 as follows: $p = 0.05$ and $|\boldsymbol{A}| = \mu$. As in the standard setting in DE for cheap optimization, we set F and C to 0.5 and 0.9, respectively. These settings were suitable for most population models. In the $(\mu + \lambda)$ and worst improvement models, we set λ to 1. We also set s in the STS model to 2.

Table 2. Parameters tuned by SMAC, where $\mu = \max\{\lfloor\alpha_\mu \ln n\rfloor, 6\}$, $|\boldsymbol{A}| = \lfloor\alpha_{\mathrm{arc}}\mu\rfloor$, $\lambda = \max\{\lfloor\alpha_\lambda\mu\rfloor, 1\}$, and $s = \max\{\lfloor\alpha_s\mu\rfloor, 2\}$.

	α_μ	Strategy	p	α_{arc}	F	C	α_λ	α_s
Range	[5, 20]	See Table 1	[0, 1]	[0, 3]	[0, 1]	[0, 1]	[0, 1]	[0, 0.2]
Default	10	rand/1	0.05	1.0	0.5	0.5	0	0
Synchronous	9.27	rand-to-pbest/1	0.17	0.58	0.51	0.52	–	–
Asynchronous	9.09	rand-to-pbest/1	0.29	1.81	0.50	0.62	–	–
$(\mu + \lambda)$	9.50	rand-to-pbest/1	0.34	1.96	0.53	0.65	0.64	–
Worst improvement	7.33	rand-to-pbest/1	0.89	1.62	0.58	0.75	0.22	–
STS	5.44	rand-to-pbest/1	0.36	1.95	0.61	0.61	–	0.18

We performed "automatic algorithm configuration" using SMAC [22], which is a surrogate-model based configurator. We used the latest version of SMAC (version 2.10.03) downloaded from the authors' website. We set the cost function in SMAC (i.e., the estimated performance of a configuration) to the error value $|f(\boldsymbol{x}^{\mathrm{bsf}}) - f(\boldsymbol{x}^*)|$, where $\boldsymbol{x}^{\mathrm{bsf}}$ is the best-so-far solution found by DE, and $\boldsymbol{x}^*$ is the optimal solution of a training problem. We used the 28 CEC2013 functions [28] with $n \in \{2, 5, 10, 20, 40\}$ as the training problems. For each run of DE, the maximum number of function evaluations was set to $100 \times n$. For each run of SMAC, the maximum number of configuration evaluations was set to 5 000. For each population model, five independent SMAC runs were performed. Then, we evaluated the performance of DE with the five configurations found by SMAC on the CEC2013 set with $n \in \{2, 5, 10, 20, 40\}$. Finally, we selected the best one from the five configurations based on their average rankings by the Friedman test. Table 2 shows the range of each parameter, the default configuration, and the best configuration for each model. Interestingly, all configurations include rand-to-*p*best/1. For all configurations, F and C values are relatively similar.

4 Results

This section analyzes the performance of DE with the five population models. We mainly discuss our results based on the anytime performance of optimizers, rather than the end-of-the-run results exactly at $100 \times n$ evaluations. For the sake of simplicity, we refer to "a DE with a population model" as "a population model". We also use the following abbreviations in Figs. 1, 2, and 3: the synchronous (Syn), asynchronous (Asy), $(\mu + \lambda)$ (Plus), worst improvement (WI), and subset-to-subset (STS) models. Section 4.1 demonstrates the performance of the five population models with the hand-tuned parameters. Section 4.2 examines the effect of automatic algorithm configuration in the population models. Section 4.3 compares two population models to CMA-ES and other optimizers.

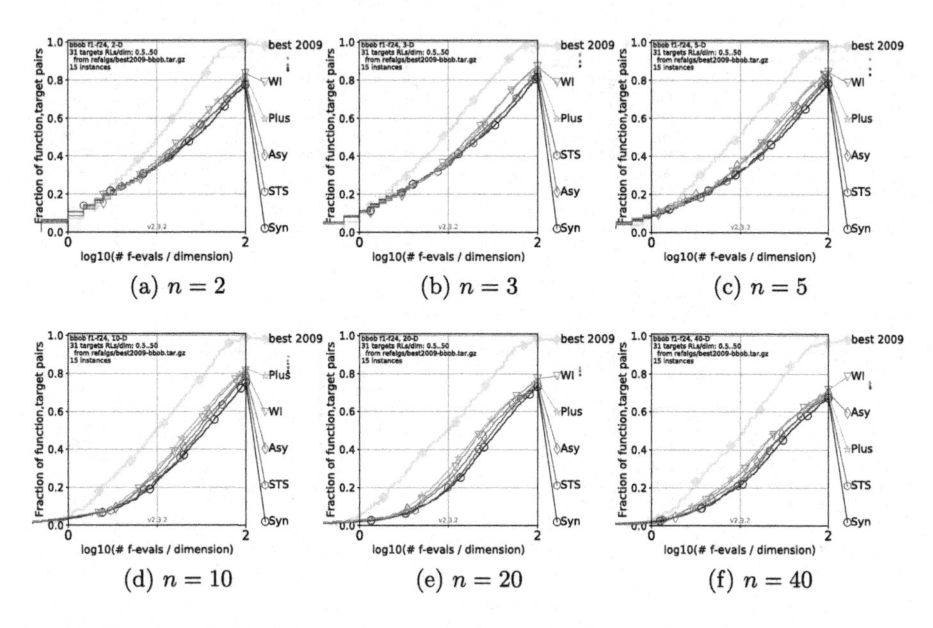

Fig. 1. Results of the five population models with the hand-tuned parameters. For each abbreviation (Syn, Asy, Plus, WI, and STS), see the beginning of Sect. 4.

4.1 Comparison of the Five Population Models

Figure 1 shows results of the five population models with the hand-tuned parameters on all 24 BBOB functions with $n \in \{2, 3, 5, 10, 20, 40\}$. In Fig. 1, "best 2009" is a *virtual* algorithm portfolio that consists of the performance data of 31 algorithms participating in the GECCO BBOB 2009 workshop [17].

Figure 1 shows the bootstrapped empirical cumulative distribution (ECDF) of the number of function evaluations (FEvals) divided by n (FEvals/n) for 31 targets for all 24 BBOB functions. We used the COCO software with the "`--expensive`" option to generate all ECDF figures in this paper. Although the target error values are usually in $\{10^2, ..., 10^{-8}\}$, they are adjusted based on "best 2009". In ECDF figures, the vertical axis indicates the proportion of target error values reached by an optimizer within specified function evaluations. For example, in Fig. 1 (e), the synchronous model reaches about 20% of all 31 target error values within $10 \times n$ evaluations on all 24 BBOB functions with $n = 20$ in all 15 runs. For more details of ECDF, see [15]. In addition to ECDF, we analyzed the performance of the population models based on average run-time with the rank-sum test ($p = 0.05$), but the results are consistent with the ECDF figures in most cases. For this reason, we show only ECDF figures.

Figure 1 shows that the $(\mu + \lambda)$ and worst improvement models show a good performance on any dimensional problems. Although the $(\mu + \lambda)$ model performs slightly worse than the worst improvement model exactly at $100 \times n$ evaluations, it performs better than the worst improvement model at the early stage, especially for $n \in \{20, 40\}$. The asynchronous model performs better than the

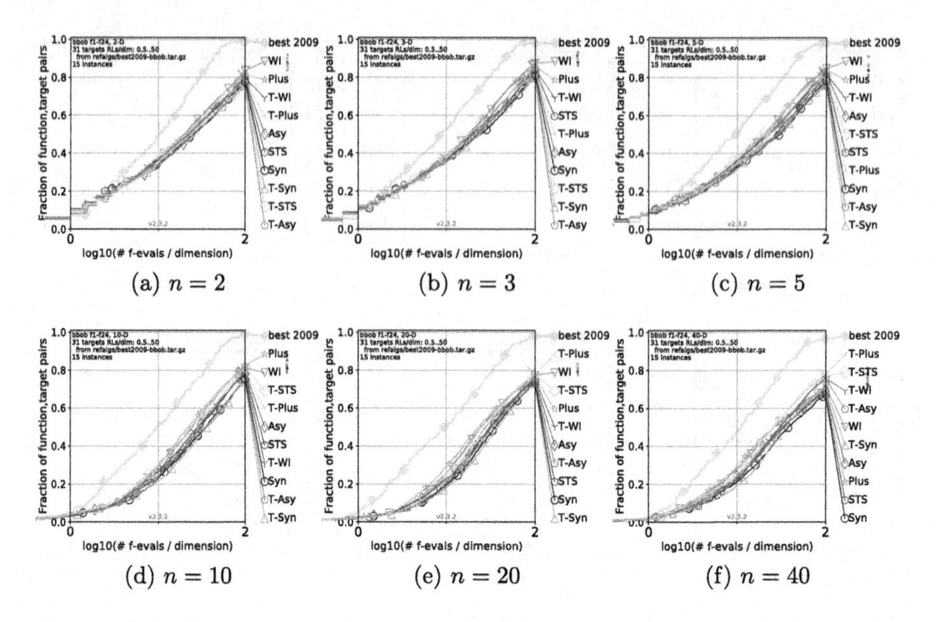

(a) $n = 2$ (b) $n = 3$ (c) $n = 5$

(d) $n = 10$ (e) $n = 20$ (f) $n = 40$

Fig. 2. Results of the five population models with automatically-tuned parameters. The prefix "T-" means that the model uses the automatically-tuned parameters.

synchronous model. This is consistent with the results in previous studies (e.g., [7,8]). The asynchronous model is also competitive with the $(\mu + \lambda)$ model for $n = 40$. Compared to the STS model, the asynchronous model has the advantage that it does not require any parameter such as the subset size s.

Overall, we can say that the choice of a population model significantly influences the performance of DE for expensive optimization. The $(\mu + \lambda)$ and worst improvement models are the best when using the hand-tuned parameters. In contrast, the traditional synchronous model performs the worst in the five population models for all dimensions. Based on these observations, we do not recommend the use of the synchronous model for expensive optimization.

4.2 Effect of Automatic Algorithm Configuration

This section investigates the effect of automatic algorithm configuration in the population models. Figure 2 shows results of the five population models with the automatically-tuned parameters. For details of the parameters, see Table 2.

Except for the STS model, the hand-tuned parameters are more suitable than the automatically-tuned parameters for $n \in \{2, 3, 5, 10\}$. In contrast, all the five models with the automatically-tuned parameters outperform those with the hand-tuned parameters for $n = 40$. The reason is discussed in Sect. 4.3. Although the $(\mu + \lambda)$ model with the automatically-tuned parameters performs the best exactly at $100 \times n$ evaluations for $n \in \{20, 40\}$, that with the hand-tuned parameters shows a good performance at the early stage.

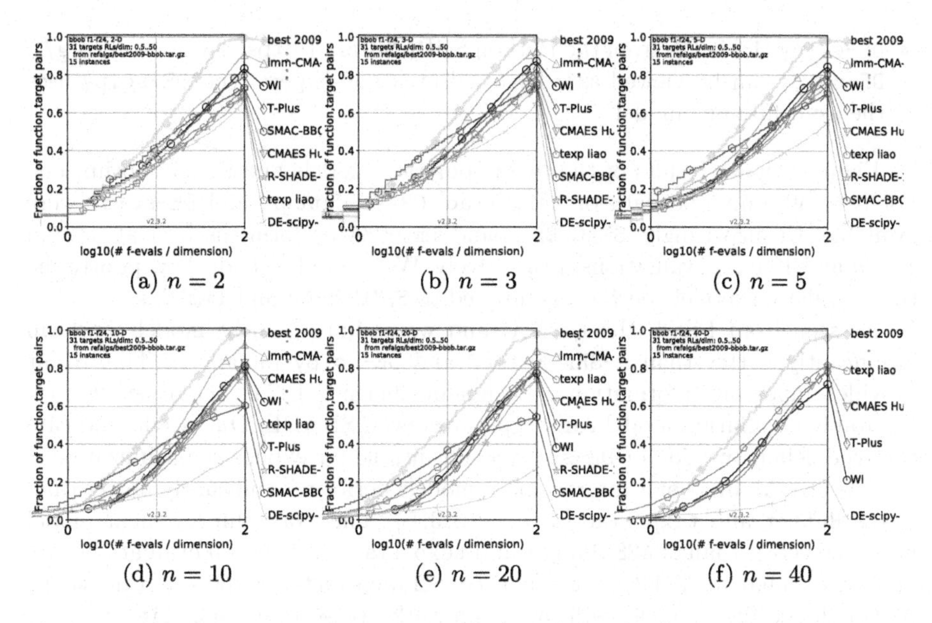

(a) $n = 2$ (b) $n = 3$ (c) $n = 5$

(d) $n = 10$ (e) $n = 20$ (f) $n = 40$

Fig. 3. Results of the two population models (WI and T-Plus) and the six optimizers.

In summary, we can obtain almost the same conclusion in Sect. 4.1 even when using the automatically-tuned parameters. For example, the synchronous model performs poorly even when using the automatically-tuned parameters.

4.3 Comparison to Other Optimizers

The results in Sect. 4.1 and 4.2 show that the worst improvement model with the hand-tuned parameters (WI) and the $(\mu + \lambda)$ model with the automatically-tuned parameters (T-Plus) perform the best in the five population models for $n \leq 10$ and $n \geq 20$, respectively. Here, we compare the two best population models with the following six optimizers. We downloaded the performance data of the six optimizers from the COCO data archive.

- `SMAC-BBOB` [23] is a Bayesian optimizer. Although the original version of SMAC [22] is an algorithm configurator as explained in Sect. 3, this version of SMAC (`SMAC-BBOB`) is a numerical optimizer. `SMAC-BBOB` is similar to EGO [25].
- `lmm-CMA` [2] is a surrogate-assisted CMA-ES with local meta-models [4].
- `CMAES_Hutter` [23] is a CMA-ES with the default parameters.
- `texp_liao` [29] is a CMA-ES with automatically-tuned parameters for expensive optimization as in Sect. 4.2. The irace tool [31] was used for tuning.
- `R-SHADE-10e2` [44] is a SHADE [43] with automatically-tuned parameters for expensive optimization as in Sect. 4.2. SHADE is one of the state-of-the-art adaptive DE algorithms. `R-SHADE-10e2` uses the synchronous model.

- DE-scipy [48] is DE from the Python SciPy library (https://www.scipy.org/). DE-scipy can be viewed as a DE used by a non-expert user. DE-scipy uses the asynchronous model.

Figure 3 shows results of the two population models and the six optimizers. For $n = 40$, only the data of texp_liao, CMAES_Hutter, and DE-scipy were available. Unsurprisingly, SMAC-BBOB and lmm-CMA perform the best at around $10 \times n$ and $100 \times n$ evaluations, respectively. We wanted to know how poorly the two population models perform compared to SMAC-BBOB and lmm-CMA.

As seen from Fig. 3, the $(\mu + \lambda)$ and worst improvement models perform significantly better than R-SHADE-10e2 and DE-scipy for any n. This result provides important information to design an efficient DE, i.e., the basic DE with a suitable population model can outperform even SHADE. Our results indicate that the default version of DE-scipy is unsuitable for expensive optimization.

The $(\mu+\lambda)$ and worst improvement models perform significantly better than CMAES_Hutter and texp_liao for $n \in \{2, 3, 5\}$. The worst improvement model performs better than CMAES_Hutter and texp_liao until $100 \times n$ evaluations and performs similar to CMAES_Hutter and texp_liao exactly at $100 \times n$ evaluations. Although the two models perform significantly worse than texp_liao for $n \in \{20, 40\}$, they have a better performance than CMAES_Hutter at the early stage. The $(\mu+\lambda)$ model is also competitive with CMAES_Hutter at $100 \times n$ evaluations.

The $(\mu+\lambda)$ model in this section and texp_liao use the automatically-tuned parameters for expensive optimization. Interestingly, both optimizers perform well for high-dimensional functions ($n \geq 20$), but they perform poorly for low-dimensional functions ($n \leq 10$). This unintuitive observation may come from the difficulty in finding a parameter set with a good scalability to n. As reported in [40], automatically-tuned parameters are likely to fit for hard-to-solve training problems. Here, parameters in the $(\mu+\lambda)$ model were tuned on the 28 CEC2013 functions with $n \in \{2, 5, 10, 20, 40\}$, and parameters in texp_liao were tuned on the 19 SOCO functions [20] with $n \in \{5, 10, 20, 40\}$. It seems that parameters in the two optimizers fit only for high-dimensional functions. Although addressing this issue is beyond the scope of this paper, an in-depth analysis is needed.

5 Conclusion

We analyzed the performance of the five population models in DE for expensive optimization on the 24 BBOB functions. The traditional synchronous model performs the worst in most cases. In contrast, the worst improvement and $(\mu + \lambda)$ models are suitable for DE on a limited budget of evaluations. The worst improvement model with the hand-tuned parameters and the $(\mu+\lambda)$ model with the automatically-tuned parameters perform significantly better than CMA-ES for $n \in \{2, 3, 5\}$ and are competitive with CMA-ES for $n = 10$. In summary, our results demonstrated that DE with a suitable population model performs better than or similar to CMA-ES depending on the number of evaluations and dimensionality of a problem. This is the first study to report such a promising performance of DE for expensive optimization. DE with the two models should be a base-line for benchmarking new surrogate-assisted DE algorithms.

We believe that the poor performance of DE for $n \in \{20, 40\}$ compared to CMA-ES can be addressed by using an efficient parameter adaptation method for F and C [45]. It is promising to design a surrogate-assisted DE with the worst improvement and $(\mu + \lambda)$ models. It is also promising to design an algorithm portfolio [21,27,36] that consists of DE, CMA-ES, and Bayesian optimizers.

Acknowledgments. This work was supported by Leading Initiative for Excellent Young Researchers, MEXT, Japan.

References

1. Ali, M.M.: Differential evolution with generalized differentials. J. Comput. Appl. Math. **235**(8), 2205–2216 (2011)
2. Auger, A., Brockhoff, D., Hansen, N.: Benchmarking the local metamodel CMA-ES on the noiseless BBOB'2013 test bed. In: GECCO (Companion), pp. 1225–1232 (2013)
3. Bartz-Beielstein, T., Zaefferer, M.: Model-based methods for continuous and discrete global optimization. Appl. Soft Comput. **55**, 154–167 (2017)
4. Bouzarkouna, Z., Auger, A., Ding, D.Y.: Local-meta-model CMA-ES for partially separable functions. In: GECCO, pp. 869–876 (2011)
5. Brest, J., Greiner, S., Bošković, B., Mernik, M., Žumer, V.: Self-adapting control parameters in differential evolution: a comparative study on numerical benchmark problems. IEEE TEVC **10**(6), 646–657 (2006)
6. Daniels, S.J., Rahat, A.A.M., Everson, R.M., Tabor, G.R., Fieldsend, J.E.: A suite of computationally expensive shape optimisation problems using computational fluid dynamics. In: Auger, A., Fonseca, C.M., Lourenço, N., Machado, P., Paquete, L., Whitley, D. (eds.) PPSN 2018. LNCS, vol. 11102, pp. 296–307. Springer, Cham (2018). https://doi.org/10.1007/978-3-319-99259-4_24
7. Das, S., Mullick, S.S., Suganthan, P.N.: Recent advances in differential evolution - an updated survey. Swarm Evol. Comput. **27**, 1–30 (2016)
8. Dorronsoro, B., Bouvry, P.: Improving classical and decentralized differential evolution with new mutation operator and population topologies. IEEE TEVC **15**(1), 67–98 (2011)
9. Dymond, A.S.D., Engelbrecht, A.P., Heyns, P.S.: The sensitivity of single objective optimization algorithm control parameter values under different computational constraints. In: IEEE CEC, pp. 1412–1419. IEEE (2011)
10. Feurer, M., Hutter, F.: Hyperparameter optimization. In: Hutter, F., Kotthoff, L., Vanschoren, J. (eds.) Automated Machine Learning. TSSCML, pp. 3–33. Springer, Cham (2019). https://doi.org/10.1007/978-3-030-05318-5_1
11. Gämperle, R., Müller, S.D., Koumoutsakos, P.: A parameter study for differential evolution. In: International Conference on Advances in Intelligent Systems, Fuzzy Systems, Evolutionary Computation, pp. 293–298 (2002)
12. Guo, J., Li, Z., Yang, S.: Accelerating differential evolution based on a subset-to-subset survivor selection operator. Soft. Comput. **23**(12), 4113–4130 (2019)
13. Hansen, N.: The CMA evolution strategy: a tutorial. CoRR, abs/1604.00772 (2016)
14. Hansen, N.: A global surrogate assisted CMA-ES. In: Auger, A., Stützle, T. (eds.) GECCO, pp. 664–672 (2019)

15. Hansen, N., Auger, A., Brockhoff, D., Tušar, D., Tušar, T.: COCO: performance assessment. CoRR, abs/1605.03560 (2016)
16. Hansen, N., Auger, A., Mersmann, O., Tušar, T., Brockhoff, D.: COCO: a platform for comparing continuous optimizers in a black-box setting. CoRR (2016)
17. Hansen, N., Auger, A., Ros, R., Finck, S., Posík, P.: Comparing results of 31 algorithms from the black-box optimization benchmarking BBOB-2009. In: GECCO (Companion), pp. 1689–1696 (2010)
18. Hansen, N., Finck, S., Ros, R., Auger, A.: Real-parameter black-box optimization benchmarking 2009: noiseless functions definitions. Technical report, INRIA (2009)
19. Hansen, N., Ostermeier, A.: Completely derandomized self-adaptation in evolution strategies. Evol. Comput. **9**(2), 159–195 (2001)
20. Herrera, F., Lozano, M., Molina, D.: Test suite for the spec. iss. of Soft Computing on scalability of evolutionary algorithms and other metaheuristics for large scale continuous optimization problems. Technical report, Univ. of Granada (2010)
21. Hoffman, M.D., Brochu, E., de Freitas, N.: Portfolio allocation for Bayesian optimization. In: UAI, pp. 327–336 (2011)
22. Hutter, F., Hoos, H.H., Leyton-Brown, K.: Sequential model-based optimization for general algorithm configuration. In: Coello, C.A.C. (ed.) LION 2011. LNCS, vol. 6683, pp. 507–523. Springer, Heidelberg (2011). https://doi.org/10.1007/978-3-642-25566-3_40
23. Hutter, F., Hoos, H.H., Leyton-Brown, K.: An evaluation of sequential model-based optimization for expensive blackbox functions. In: GECCO (Companion), pp. 1209–1216 (2013)
24. Jin, Y.: Surrogate-assisted evolutionary computation: recent advances and future challenges. Swarm Evol. Comput. **1**(2), 61–70 (2011)
25. Jones, D.R., Schonlau, M., Welch, W.J.: Efficient global optimization of expensive black-box functions. J. Glob. Optim. **13**(4), 455–492 (1998)
26. Kayhani, A., Arnold, D.V.: Design of a surrogate model assisted (1 + 1)-ES. In: Auger, A., Fonseca, C.M., Lourenço, N., Machado, P., Paquete, L., Whitley, D. (eds.) PPSN 2018. LNCS, vol. 11101, pp. 16–28. Springer, Cham (2018). https://doi.org/10.1007/978-3-319-99253-2_2
27. Kerschke, P., Trautmann, H.: Automated algorithm selection on continuous black-box problems by combining exploratory landscape analysis and machine learning. Evol. Comput. **27**(1), 99–127 (2019)
28. Liang, J.J., Qu, B.Y., Suganthan, P.N., Hernández-Díaz, A.G.: Problem definitions and evaluation criteria for the CEC 2013 special session on real-parameter optimization. Technical report, Nanyang Technological Univ. (2013)
29. Liao, T., Stützle, T.: Expensive optimization scenario: IPOP-CMA-ES with a population bound mechanism for noiseless function testbed. In: GECCO (Companion), pp. 1185–1192 (2013)
30. Liu, B., Zhang, Q., Gielen, G.G.E.: A Gaussian process surrogate model assisted evolutionary algorithm for medium scale expensive optimization problems. IEEE TEVC **18**(2), 180–192 (2014)
31. López-Ibáñez, M., Dubois-Lacoste, J., Cáceres, L.P., Birattari, M., Stützle, T.: The irace package: iterated racing for automatic algorithm configuration. Oper. Res. Perspect. **3**, 43–58 (2016)
32. Loshchilov, I., Hutter, F.: CMA-ES for hyperparameter optimization of deep neural networks. CoRR, abs/1604.07269 (2016)
33. Loshchilov, I., Schoenauer, M., Sebag, M.: Self-adaptive surrogate-assisted covariance matrix adaptation evolution strategy. In: GECCO, pp. 321–328 (2012)

34. Lu, X., Tang, K., Sendhoff, B., Yao, X.: A new self-adaptation scheme for differential evolution. Neurocomputing **146**, 2–16 (2014)
35. Lynn, N., Ali, M.Z., Suganthan, P.N.: Population topologies for particle swarm optimization and differential evolution. Swarm Evol. Comput. **39**, 24–35 (2018)
36. Mohammadi, H., Le Riche, R., Touboul, E.: Making EGO and CMA-ES complementary for global optimization. In: Dhaenens, C., Jourdan, L., Marmion, M.-E. (eds.) LION 2015. LNCS, vol. 8994, pp. 287–292. Springer, Cham (2015). https://doi.org/10.1007/978-3-319-19084-6_29
37. Noman, N., Iba, H.: A new generation alternation model for differential evolution. In: GECCO, pp. 1265–1272 (2006)
38. Piotrowski, A.P.: Review of differential evolution population size. Swarm Evol. Comput. **32**, 1–24 (2017)
39. Rahnamayan, S., Tizhoosh, H.R., Salama, M.M.A.: Opposition-based differential evolution. IEEE TEVC **12**(1), 64–79 (2008)
40. Smit, S.K., Eiben, A.E.: Parameter tuning of evolutionary algorithms: generalist vs. specialist. In: Di Chio, C., et al. (eds.) EvoApplications 2010. LNCS, vol. 6024, pp. 542–551. Springer, Heidelberg (2010). https://doi.org/10.1007/978-3-642-12239-2_56
41. Socha, K.: The influence of run-time limits on choosing ant system parameters. In: GECCO, pp. 49–60 (2003)
42. Storn, R., Price, K.: Differential evolution - a simple and efficient heuristic for global optimization over continuous spaces. J. Glob. Optim. **11**(4), 341–359 (1997)
43. Tanabe, R., Fukunaga, A.: Success-history based parameter adaptation for differential evolution. In: IEEE CEC, pp. 71–78 (2013)
44. Tanabe, R., Fukunaga, A.: Tuning differential evolution for cheap, medium, and expensive computational budgets. In: IEEE CEC, pp. 2018–2025 (2015)
45. Tanabe, R., Fukunaga, A.: Reviewing and benchmarking parameter control methods in differential evolution. IEEE Trans. Cyber. **50**(3), 1170–1184 (2020)
46. Tanabe, R., Fukunaga, A.S.: Improving the search performance of SHADE using linear population size reduction. In: IEEE CEC, pp. 1658–1665 (2014)
47. Thomsen, R.: Multimodal optimization using crowding-based differential evolution. In: IEEE CEC, pp. 1382–1389 (2004)
48. Varelas, K., Dahito, M.: Benchmarking multivariate solvers of SciPy on the noiseless testbed. In: GECCO (Companion), pp. 1946–1954 (2019)
49. Wang, Y., Cai, Z.: Constrained evolutionary optimization by means of $(\mu + \lambda)$-differential evolution and improved adaptive trade-off model. Evol. Comput. **19**(2), 249–285 (2011)
50. Wang, Y., Cai, Z., Zhang, Q.: Differential evolution with composite trial vector generation strategies and control parameters. IEEE TEVC **15**(1), 55–66 (2011)
51. Wang, Y., Yin, D., Yang, S., Sun, G.: Global and local surrogate-assisted differential evolution for expensive constrained optimization problems with inequality constraints. IEEE Trans. Cyber. **49**(5), 1642–1656 (2019)
52. Weber, M., Neri, F., Tirronen, V.: Distributed differential evolution with explorative-exploitative population families. GPEM **10**(4), 343–371 (2009)
53. Wormington, M., Panaccione, C., Matney, K.M., Bowen, D.K.: Characterization of structures from X-ray scattering data using genetic algorithms. Phil. Trans. R. Soc. Lond. A **357**(1761), 2827–2848 (1999)
54. Zhabitsky, M., Zhabitskaya, E.: Asynchronous differential evolution with adaptive correlation matrix. In: GECCO, pp. 455–462 (2013)

55. Zhang, J., Sanderson, A.: Adaptive Differential Evolution: A Robust Approach to Multimodal Problem Optimization, vol. 1. Springer, Heidelberg (2009). https://doi.org/10.1007/978-3-642-01527-4
56. Zhang, J., Sanderson, A.C.: DE-AEC: a differential evolution algorithm based on adaptive evolution control. In: IEEE CEC, pp. 3824–3830. IEEE (2007)
57. Zhang, J., Sanderson, A.C.: JADE: adaptive differential evolution with optional external archive. IEEE TEVC **13**(5), 945–958 (2009)
58. Zielinski, K., Weitkemper, P., Laur, R., Kammeyer, K.D.: Parameter study for differential evolution using a power allocation problem including interference cancellation. In: IEEE CEC, pp. 1857–1864 (2006)

Continuous Optimization Benchmarks by Simulation

Martin Zaefferer[(✉)] and Frederik Rehbach

Institute for Data Science, Engineering, and Analytics, TH Köln, 51643 Gummersbach, Germany
{martin.zaefferer,frederik.rehbach}@th-koeln.de

Abstract. Benchmark experiments are required to test, compare, tune, and understand optimization algorithms. Ideally, benchmark problems closely reflect real-world problem behavior. Yet, real-world problems are not always readily available for benchmarking. For example, evaluation costs may be too high, or resources are unavailable (e.g., software or equipment). As a solution, data from previous evaluations can be used to train surrogate models which are then used for benchmarking. The goal is to generate test functions on which the performance of an algorithm is similar to that on the real-world objective function. However, predictions from data-driven models tend to be smoother than the ground-truth from which the training data is derived. This is especially problematic when the training data becomes sparse. The resulting benchmarks may not reflect the landscape features of the ground-truth, are too easy, and may lead to biased conclusions.

To resolve this, we use simulation of Gaussian processes instead of estimation (or prediction). This retains the covariance properties estimated during model training. While previous research suggested a decomposition-based approach for a small-scale, discrete problem, we show that the spectral simulation method enables simulation for continuous optimization problems. In a set of experiments with an artificial ground-truth, we demonstrate that this yields more accurate benchmarks than simply predicting with the Gaussian process model.

Keywords: Simulation · Benchmarking · Test function · Continuous optimization · Gaussian process regression · Kriging

1 Introduction

For the design and development of optimization algorithms, benchmarks are indispensable. Benchmarks are required to test hypotheses about algorithm behavior, to understand the impact of algorithm parameters, to tune those parameters, or to compare algorithms with each other. Multiple benchmarking frameworks exist, with BBOB/COCO being a prominent example [11,12].

One issue of benchmarks is their relevance to real-world problems. The employed test functions may be of an artificial nature, yet should reflect the

T. Bäck et al. (Eds.): PPSN 2020, LNCS 12269, pp. 273–286, 2020.
https://doi.org/10.1007/978-3-030-58112-1_19

behavior of algorithms on real-world problems. An algorithm's performance on test functions and real-world problems should be similar. Yet, real-world problems may not be available in terms of functions, but only as data (i.e., observations from previous experiments). This can be due to real objective function evaluations being too costly or not accessible (in terms of software or equipment).

In those cases, using a data-driven approach may be a viable alternative: surrogate models can be trained and subsequently used to benchmark algorithms. The intent is not to replace artificial benchmarks such as BBOB (which have their own advantages), but rather to augment them with problems that have a closer connection to real-world problems. This approach has been considered in previous investigations [2,5,7,8,17,18]. Additionally, recent benchmark suites offer access to real-world problems, e.g., the Computational Fluid Dynamics (CFD) test problem suite [6] and the Games Benchmark for Evolutionary Algorithms (GBEA) [20]. Notably, the authors of the GBEA accept data provided by other researchers as a basis for surrogate model-based benchmarking[1].

As pointed out by Zaefferer et al. [23], surrogate model-based benchmarks face a crucial issue: the employed machine learning models may smoothen the training data, especially if the training data is sparse. Hence, these models are prone to produce optimization problems that lack the ruggedness and difficulty of the underlying real-world problems. Thus, algorithm performances may be overrated, and comparisons become biased. Focusing on a discrete optimization problem from the field of computational biology, Zaefferer et al. proposed to address this issue via simulation with Gaussian Process Regression (GPR). In contrast to estimation (or prediction) with GPR, simulation may provide a more realistic assessment of an algorithm's behavior. The response of the simulation retains the covariance properties determined by the model [15].

The decomposition-based simulation approach used by Zaefferer et al. relies on the selection of a set of simulation samples [23]. The simulation is evaluated at these sample locations. The simulation samples are distinct from and less sparse than the observed training samples. They are not restricted by evaluation costs. Still, using a very large number of simulation samples can quickly become computationally infeasible. In small discrete search spaces, all samples in the search space can be simulated. In larger search spaces, the simulation has to be interpolated between the simulation samples. The interpolation step might again introduce undesirable smoothness. Thus, decomposition-based simulation may work well for (small-scale) combinatorial optimization problems. Conversely, it is not suited for continuous benchmarks. Hence, our research questions are:

- Q1. How can simulation with GPR models be used to generate benchmarks for continuous optimization?
- Q2. Do simulation-based benchmarks provide better results than estimation-based benchmarks, for continuous optimization?

For Q1, we investigate the spectral method for GPR-simulation [4]. The required background on GPR, estimation, and simulation is given in Sect. 2. Then, we

[1] See the GBEA website, at http://www.gm.fh-koeln.de/~naujoks/gbea/gamesbench_doc.html\#subdata. Accessed on 2020-08-03.

describe a benchmark experiment to answer Q2 in Sect. 3, and the results in Sect. 4. The employed code is made available.[2] We discuss critical issues of GPR and simulation in Sect. 5. Section 6 concludes the paper with a summary and outlook.

2 Gaussian Processes Model

In the following, we assume that we deal with an objective function $\mathrm{f}(\mathbf{x})$, which is expensive to evaluate or has otherwise limited availability. Here, $\mathbf{x} \in \mathbb{R}^n$ are the variables of the optimization problem. Respectively, we have to learn models that regress data sets with m training samples $\mathbf{X} = \{\mathbf{x}_1, \ldots, \mathbf{x}_m\}$, and the corresponding observations $\mathbf{y} \in \mathbb{R}^m$, with $y_j = \mathrm{f}(\mathbf{x}_j)$, and $j = 1, \ldots, m$.

2.1 Gaussian Process Regression

GPR (also known as Kriging) assumes that the training data $\mathbf{X}$, $\mathbf{y}$ is sampled from a stochastic process of Gaussian distribution. Internally, it interprets data based on their correlations. These correlations are determined by a kernel $\mathrm{k}(\mathbf{x}, \mathbf{x}')$. A frequently chosen kernel is

$$\mathrm{k}(\mathbf{x}, \mathbf{x}') = \exp\left(\sum_{i=1}^{n} -\theta_i |x_i - x'_i|^2\right). \tag{1}$$

Here, $\theta_i \in \mathbb{R}$ is a parameter that is usually determined by Maximum Likelihood Estimation (MLE). In the following, we assume that the model has already been trained via MLE, based on the data $\mathbf{X}$, $\mathbf{y}$. The kernel $\mathrm{k}(\mathbf{x}, \mathbf{x}')$ yields the correlation matrix $\mathbf{K}$, which collects all pairwise correlations of the training samples $\mathbf{X}$. The vector of correlations between each training sample $\mathbf{x}_j$, and a single, new sample $\mathbf{x}$ is denoted by $\mathbf{k}$. Further details on GPR, including model training by MLE, are given by Forrester et al. [9].

In the context of GPR, the term *estimation* denotes the prediction of the model at some unknown, new location. It is performed with the predictor

$$\hat{y}(\mathbf{x}) = \hat{\mu} + \mathbf{k}^T \mathbf{K}^{-1} (\mathbf{y} - \mathbf{1}\hat{\mu}). \tag{2}$$

Here, $\mathbf{1}$ is a vector of ones and the parameter $\hat{\mu}$ is determined by MLE. Estimation intends to give an accurate response value at a single location $\mathbf{x}$.

2.2 Simulation by Decomposition

Conversely to estimation, *simulation* intends to reproduce the covariance structure of a set of samples as accurately as possible [4,14]. Intuitively, this is exactly what we require for the generation of optimization benchmarks: We are interested

[2] Reproducible code and a complete set of the presented figures is provided at https://github.com/martinzaefferer/zaef20b. For easily accessible interfaces and demonstrations see https://github.com/martinzaefferer/COBBS.

in the topology of the landscape (here: captured by the covariance structure), rather than accurate predictions of isolated function values [17].

One approach towards simulation is based on the decomposition of a covariance matrix $\mathbf{C}_s$ [4]. This matrix is computed for a set of n_{sim} simulation samples $\mathbf{X}_s = \{\mathbf{x}_1, ..., \mathbf{x}_{n_{\text{sim}}}\}$, with $\mathbf{x}_t \in \mathbb{R}^n$ and $t = 1, ..., n_{\text{sim}}$. Here, n_{sim} is usually much larger than the number of training samples m. Using Eq. (1), $\mathbf{X}_s$ yields the correlation matrix $\mathbf{K}_s$ of all simulation samples, and the respective covariance matrix is $\mathbf{C}_s = \hat{\sigma}^2 \mathbf{K}_s$. Here, $\hat{\sigma}^2$ is a model parameter (determined by MLE). Decomposition can, e.g., be performed with the Cholesky decomposition $\mathbf{C}_s = \mathbf{L}\mathbf{L}^T$. This yields the *unconditionally* simulated values $\hat{\mathbf{y}}_s = \mathbf{1}\hat{\mu} + \mathbf{L}\boldsymbol{\epsilon}$, where $\boldsymbol{\epsilon}$ is a vector of independent normal-distributed random samples, $\epsilon_i \sim N(0, 1)$. In this context, 'unconditional' means that the simulation reproduces only the covariance structure, but not the observed values $\mathbf{y}$. Additional steps are required for conditioning, so that the observed values are reproduced, too [4].

Obviously, the simulation only produces a discrete number of values $\hat{\mathbf{y}}_s$, at specific locations $\mathbf{X}_s$. Initially, we do not know the locations where our optimization algorithms will attempt to evaluate the test function. Hence, subsequent evaluations at arbitrary locations rely on interpolation. The predictor from Eq. (2) can be used, replacing all values linked to the training data with the respective values from the simulation ($\mathbf{X}_s$, $\hat{\mathbf{y}}_s$, $\mathbf{K}_s$ instead of $\mathbf{X}$, $\mathbf{y}$, $\mathbf{K}$). The model parameters $\hat{\sigma}^2$, $\hat{\mu}$, and θ_i remain unchanged.

Unfortunately, this interpolation step is a critical weakness when applied to continuous optimization problems. The locations $\mathbf{X}_s$ have to be sufficiently dense, to avoid that the interpolation introduces undesirable smoothness. Yet, computational restrictions limit the density of $\mathbf{X}_s$. Even a rather sparse grid of 20 samples in each dimension requires $m = 20^n$ simulated samples. Then, $\mathbf{C}_s$ is of dimension $20^n \times 20^n$, which is prohibitively large in terms of memory consumption for $n \geq 4$. Even a mildly multimodal function may easily require a much denser sample grid. This renders the approach infeasible for continuous optimization problems with anything but the lowest dimensionalities.

2.3 Simulation by the Spectral Method

Following up on [23], we investigate a different simulation approach that is well suited for continuous optimization problems: the spectral method [4]. This approach directly generates a function that can be evaluated at arbitrary locations, without interpolation. It yields a superposition of cosine functions [4],

$$\mathrm{f}_s(\mathbf{x}) = \hat{\sigma}\sqrt{\frac{2}{N}} \sum_{v=1}^{N} \cos(\boldsymbol{\omega}_v \cdot \mathbf{x} + \phi_v),$$

with ϕ_v being an i.i.d. uniform random sample from the interval $[-\pi, \pi]$. The sampling of $\boldsymbol{\omega}_v$ requires the spectral density function of the GPR model's kernel [4,15]. That is, $\boldsymbol{\omega}_v \in \mathbb{R}^n$ are i.i.d. random samples from a distribution with that same density. For the kernel from Eq. (1), the respective distribution for the i-th dimension is the normal distribution with zero mean and variance $2\theta_i$. A simulation conditioned on the training data can be generated with

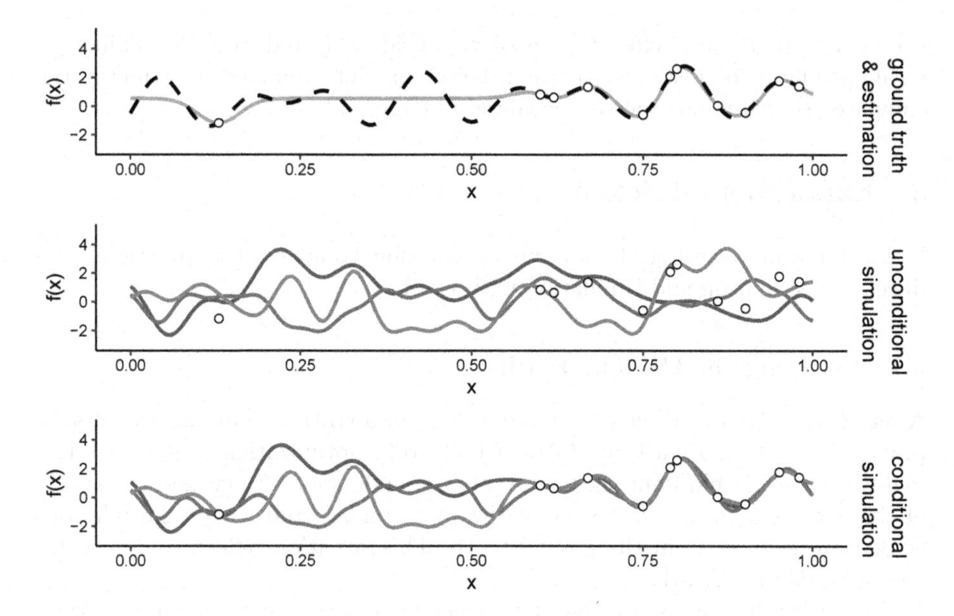

Fig. 1. Top: Ground-truth f(x) (dashed line), training data (circles), and GPR model estimation (gray solid line). Middle: Three instances of an unconditional simulation (same model). Bottom: Three instances of a conditional simulation (same model). The three different instances are generated by re-sampling of $\boldsymbol{\omega}_v$ and ϕ_v.

$\mathrm{f_{sc}}(\mathbf{x}) = \mathrm{f_s}(\mathbf{x}) + \hat{y}^*(\mathbf{x})$ [4], where $\hat{y}^*(\mathbf{x}) = \hat{\mu} + \mathbf{k}^T\mathbf{K}^{-1}(\mathbf{y}_{sc} - \mathbf{1}\hat{\mu})$ is the predictor from Eq. (2) with the training observations $\mathbf{y}$ replaced by $\mathbf{y}_{sc}$, and $\hat{\mu} = 0$. Here, $\mathbf{y}_{sc}$ are the unconditionally simulated values at the training samples, that is, $y_{\mathrm{sc}_j} = \mathrm{f_s}(\mathbf{x}_j)$.

2.4 Simulation for Benchmarking

To employ these simulations in a benchmarking context, we roughly follow the approach by Zaefferer et al. [23]. First, a data set is created by evaluating the true underlying problem (if not already available in the form of historical data). Then, a GPR model is trained with that data. Afterwards, the spectral method is used to generate conditional or unconditional simulations. These simulations are finally used as test functions for optimization algorithms.

Here, the advantage of simulation over estimation is the ability to reproduce the topology of functions, rather than predicting a single, isolated value. As an illustration, let us assume an example for $n = 1$, where the ground-truth is $\mathrm{f}(x) = \sin(33x) + \sin(49x - 0.5) + x$. A GPR model is trained with the samples $\mathbf{X} = \{0.13, 0.6, 0.62, 0.67, 0.75, 0.79, 0.8, 0.86, 0.9, 0.95, 0.98\}$. The resulting estimation, unconditional simulation, and conditional simulation of the GPR model are presented in Fig. 1. This example shows how estimation might be unsuited to reproduce an optimization algorithm's behavior. In the sparsely sampled region, the GPR estimation is close to constant, considerably reducing the number of local optima in the (estimated) search landscape. The number of optima in the simulated search landscapes is considerably larger.

3 Experimental Setup

In the following, we describe an experiment that compares test functions produced by estimation and simulation with GPR.

3.1 Selecting the Ground-Truth

A set of objective functions is required as a ground-truth for our experiments. In practice, the ground-truth would be a real-world optimization problem. Yet, a real-world case would limit the comparability, understandability, and the extent of the experimental investigation. We want to understand where and why our emulation deviates from the ground-truth. This situation reflects the need for model-based benchmarks.

Hence, we chose a well-established artificial benchmark suite for optimization: the single-objective, noiseless BBOB suite from the COCO framework [11,12]. The BBOB suite allows us to compare in-detail how algorithms behave on the actual problem (ground-truth) and how they behave on an estimation or simulation with GPR. Moreover, important landscape features of the BBOB suite are known (e.g., modality, symmetry/periodicity, separability), which enables us to understand and explain where GPR models fail.

The function set that we investigated is described in [13]. This set consists of 24 unimodal and multimodal functions. For each function, 15 randomized instances are usually produced. We followed the same convention. In addition, all test functions are scalable in terms of search space dimensionality n. We performed our experiments with $n = 2, 3, 5, 10, 20$.

3.2 Generating the Training Data

We generated training data by running an optimization algorithm on the original problem. The data observed during the optimization run was used to train the GPR model. This imitates a common scenario that occurs in real-world applications: Some algorithm has already been run on the real-world problem, and the data from that preliminary experiment provides the basis for benchmarking. Moreover, this approach allows us to determine the behavior of the problem on a local and global scale. An optimization algorithm (especially, a population-based evolutionary algorithm) will explore the objective function globally as well as performing smaller, local search steps.

Specifically, we generated our training data as follows: For each BBOB function (1, ..., 24) and each function instance (1, ..., 15), we ran a variant of Differential Evolution (DE) [19] with $50n$ function evaluations, and a population size of $20n$. All evaluations were recorded. We used the implementation from the DEoptim R-package, with default configuration [1]. This choice is arbitrary. Other population-based algorithms would be equally suited for our purposes.

3.3 Generating the Model

We selected the $50n$ data samples provided by the DE runs. Based on that data, we trained a GPR model, using the SPOT R-package [3]. Three non-default parameters of the model were specified with: `useLambda=FALSE` (no nugget effect, see also [9]), `thetaLower=1e$-$6` (lower bound on θ_i), and `thetaUpper=1e12` (upper bound on θ_i). For the spectral simulation, we used $N = 100n$ cosine functions. We only created conditional simulations, to reflect each BBOB instance as closely as possible. Scenarios where an unconditional simulation is preferable have been discussed by Zaefferer et al. [23].

3.4 Testing the Algorithms

We tested three algorithms:

- DE: As a global search strategy, we selected DE. We tested the same DE variant as mentioned in Sect. 3.2, but with a population size of $10n$ and a different random number generator seed for initialization. All other algorithm parameters remained at default values.
- NM: As a classical local search strategy, the Nelder-Mead (NM) simplex algorithm was selected [16]. We employed the implementation from the R-package nloptr [22]. All algorithm parameters remained at default values.
- RS: We also selected a Random Search (RS) algorithm, which evaluates the objective function with i.i.d. samples from a uniform random distribution.

This selection was to some extent arbitrary. The intent was not to investigate these specific algorithms. Rather, we selected these algorithms to observe a range of different convergence behaviors. Essentially, we made a selection that scales from very explorative (RS), to balanced exploration/exploitation (DE), to very exploitative (NM).

All three algorithms receive the same test instances and initial random number generator seeds. For each test instance, each algorithm uses $1000n$ function evaluations. Overall, each algorithm was run on each instance, function, and dimension of the BBOB test suite ($24 \times 15 \times 5 = 1800$ runs, each run with $1000n$ evaluations). Additionally, each of these runs was repeated with an estimation-based test function, and with a simulation-based test function.

4 Results

4.1 Quality Measure

Our aim is to measure how well algorithm performance is reproduced by the test functions. One option would be to measure the error as the difference of observed values along the search path compared to the ground-truth values at those same locations. But this is problematic. Let us assume that the ground truth is $\mathrm{f}(x) = x^2$, and two test functions are $\mathrm{f}_{\mathrm{t1}}(x) = (x-1)^2$ and $\mathrm{f}_{\mathrm{t2}}(x) = 0.5$, with $x \in [0,1]$. Clearly, f_{t1} is a reasonable oracle for most algorithms' performance (e.g., in terms of convergence speed) while f_{t2} is not. Yet, the mean squared error of f_{t1} would usually be larger than the error of f_{t2}. The error on f_{t1} even increases when an algorithm approaches the optimum.

Hence, we measured the error on the performance curves. For each algorithm run, the best observed function values after each objective function evaluation were recorded (on all test instances, including estimation, simulation, and ground-truth). In the following, this will be referred to as the performance of the algorithm. The resulting performance values were scaled to values between zero and one, for each problem instance (i.e., each BBOB function, each BBOB instance, each dimension, and also separately for the ground-truth, estimation, and simulation variants). This yielded what we term the scaled performance. The error of the scaled performance was then calculated as the absolute deviation of the performance on the model-based functions, compared to the performance on the ground-truth problem. For example, let us assume that DE achieved a (scaled) function value on the ground-truth of 0.25 after 200 objective function evaluations. But the same algorithm only achieved 0.34 on the estimation-based test function after 200 evaluations. Then, the error of the estimation-based run is $|0.34 - 0.25| = 0.09$ (after 200 evaluations).

4.2 Observations

In Figs. 2 and 3, we show the resulting errors over run time for a subset of the 24 BBOB functions. Due to space restrictions, we only show the error for the DE and NM algorithms. Similar patterns are visible in the omitted curves for RS. We also omit the curves for $n = 3$, which closely mirror those for $n = 2$.

For the simulation, we mostly observe decreasing errors or constant errors over time. The decrease can be explained by the algorithms' tendency to find the best values for the respective problem instance later on in the run, regardless of the objective function. Earlier, the difference is usually larger. For the estimation-based test functions, the error often increases, and sometimes decreases again during the later stages of a run.

When comparing estimation and simulation, the modality and the dimensionality n are important. For low-dimensional unimodal BBOB functions ($n = 2, 3, 5$, and function IDs: 1, 2, 5–7, 10–14), the simulation yields larger errors than estimation. In most of the multimodal cases, the simulation seems to perform equally well or better. This can be explained: The larger activity of the simulation-based functions may occasionally introduce additional optima (turning unimodal into multimodal problems). The estimation is more likely to reproduce the single valley of the ground-truth. Conversely, the simulation excels for the multimodal cases because it does not remove optima by interpolation. For higher-dimensional cases ($n = 10, 20$), this situation changes: the simulation

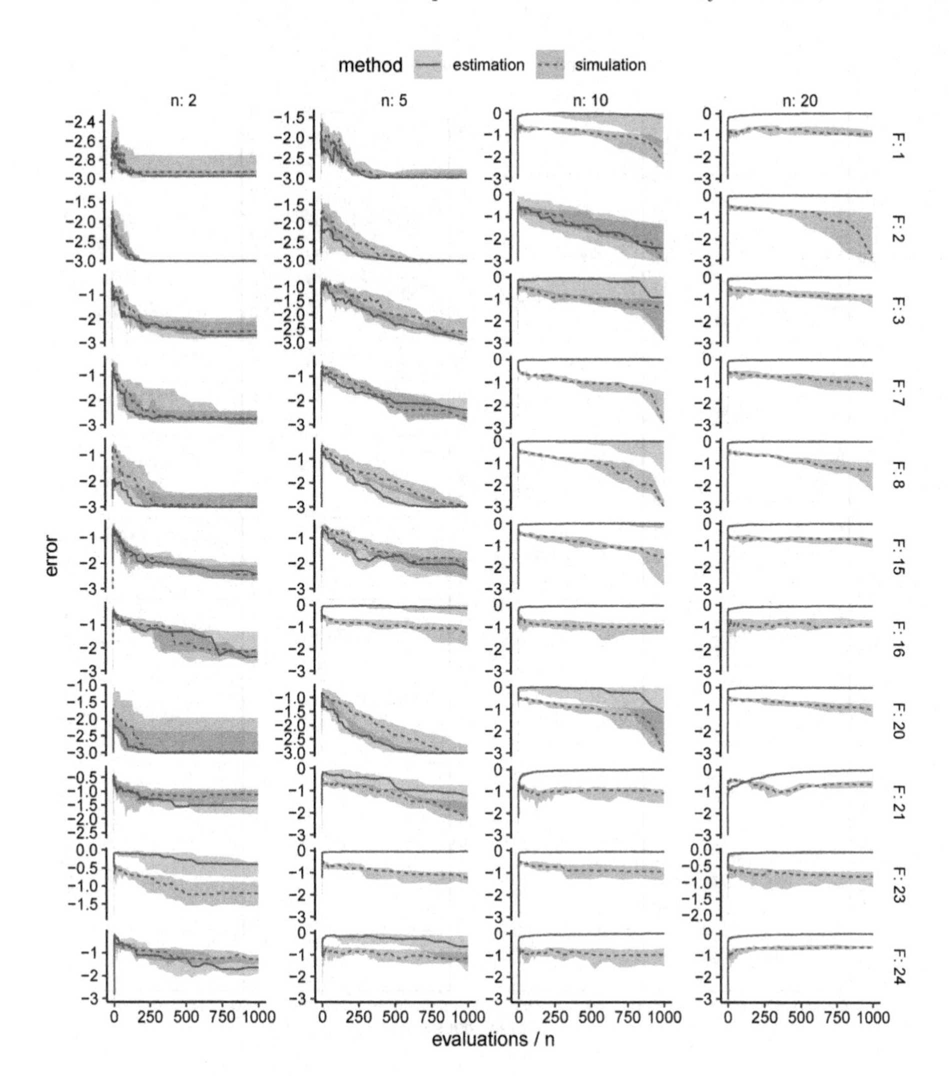

Fig. 2. Error of algorithm performance with simulation- and estimation-based test functions. The curves are based on the performance of a DE run on the model-based test functions, compared against the performance values on the ground-truth. Errors are log-scaled via $\log_{10}(\text{error} + 0.001)$. The labels on the right-hand side specify the respective IDs of the BBOB functions. Top-side labels indicate dimensionality. The lines indicate the median, the colored areas indicate the first and third quartile. These statistics are calculated over the 15 instances for each BBOB function.

produces lower errors, regardless of modality. This is explained by the increasing sparseness of the training data, which in case of estimation will frequently lead to extremely poor search landscapes. The estimation will mostly produce a constant value, with the exception of very small areas close to the training data.

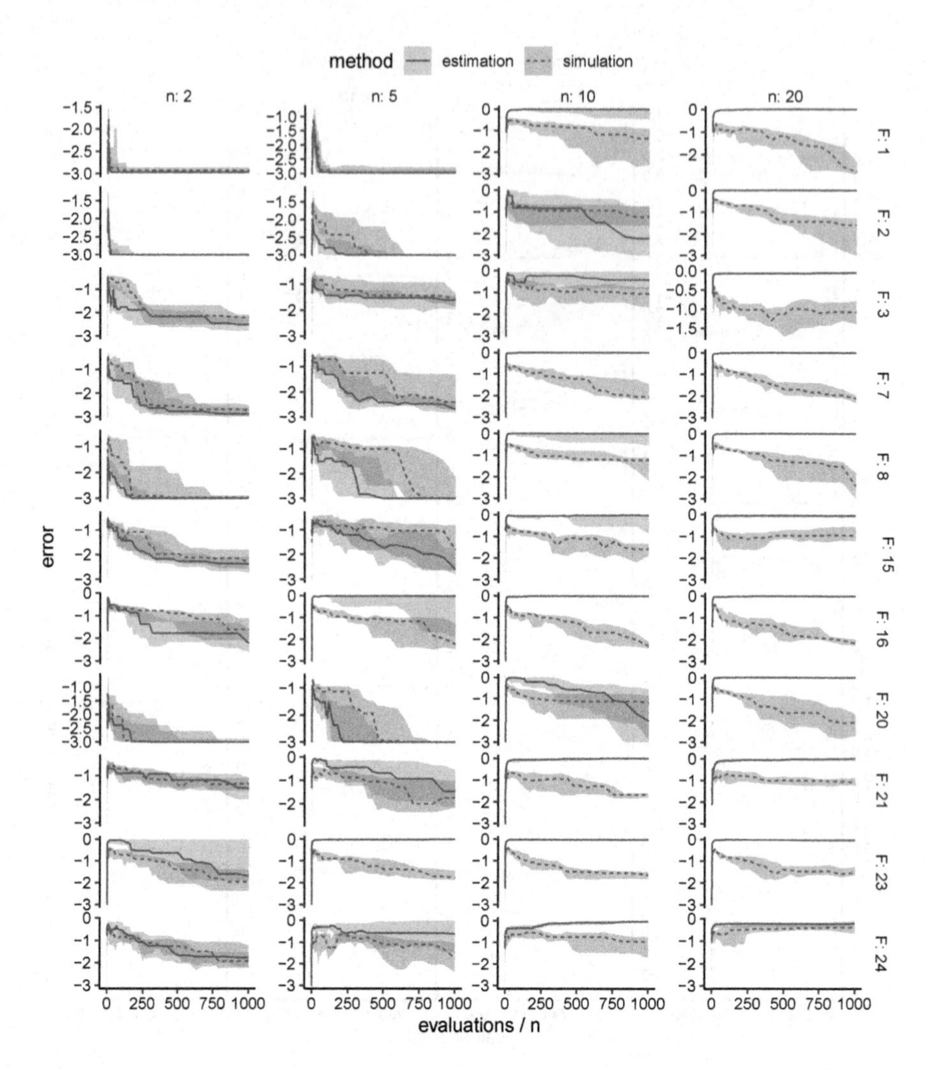

Fig. 3. This is the same plot-type as presented in Fig. 2, but only for the performance of the NM algorithm (instead of DE).

As noted earlier, results between DE, RS, and NM exhibit similar patterns. There is an exception, where the results between DE, NM, and RS differ more strongly: BBOB function 21 and 22. Hence, Fig. 4 shows results for $n = 20$ with function 21 (22 is nearly identical). Here, each plot shows the error for a different algorithm. Coincidentally, this includes the only case where estimation is performing considerably better than simulation for large n (with algorithm RS only). The reason is not perfectly clear. One possibility is a particularly poor model quality. BBOB functions 21 and 22 are both based on a mixture of Gaussian components. Two aspects of this mixture are problematic for GPR: Firstly,

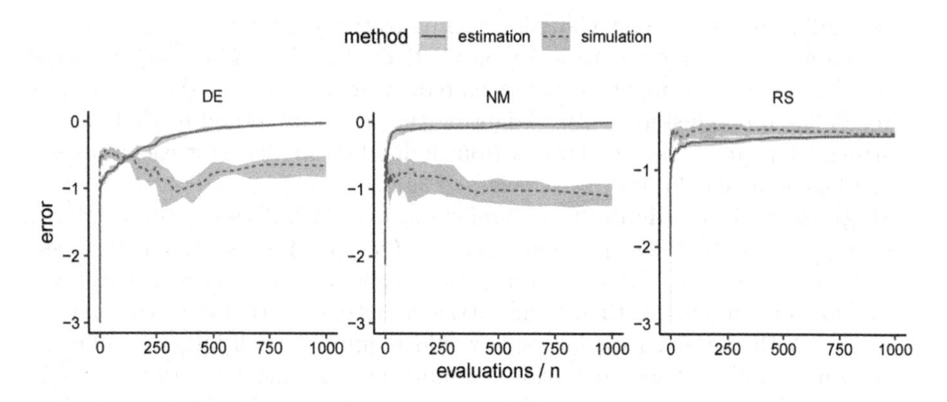

Fig. 4. This is the same plot-type as presented in Fig. 2, but limited to BBOB function 21 and $n = 20$. While Fig. 2 only shows results for DE, this figure compares results for each tested algorithm (DE, NM, RS) for this specific function and dimensionality.

they exhibit a peculiar, localized non-stationarity. The activity of the function may abruptly change direction, depending on the closest Gaussian component. Secondly, overlapping Gaussian components produce discontinuities.

5 Discussion

The results show that the model-based test functions will occasionally deviate considerably from the ground-truth. This has various reasons.

- **Dimensionality:** Clearly, all models are affected by the curse of dimensionality. With 10 or more variables, it becomes increasingly difficult to learn the shape of the real function with a limited number of training samples. Necessarily, this limits how much we can achieve. Despite its more robust performance, simulation also relies on a well-trained model.
- **Continuity:** Our GPR model, or more specifically its kernel, works best if the ground-truth is continuous, i.e., $\lim_{\mathbf{x} \to \mathbf{x}'} \mathrm{f}(\mathbf{x}) = \mathrm{f}(\mathbf{x}')$. Else, model quality decreases. One example is the step ellipsoidal function (BBOB function 7). This weakness could be alleviated, if it is known a-priori: a more appropriate kernel such as the exponential kernel $\mathrm{k}(\mathbf{x}, \mathbf{x}') = \exp(\sum_{i=1}^{n} -\theta_i |x_i - x_i'|)$ could be used. However, this kernel may be less suited for the spectral simulation method [4].
- **Non-stationarity:** Our GPR models assume stationarity of the covariance structure in the data. Yet, some functions are obviously non-stationary. For example, the BBOB variant of the Schwefel function (BBOB function 20) behaves entirely differently close to the search boundaries compared to the optimal region (due to a penalty term). In another way, the two Gallagher's Gaussian functions (BBOB functions 21, 22) show a more localized type of non-stationarity. There, the activity of the function will change direction

depending on the closest Gaussian component. Such functions are particularly difficult to model with common GPR models. Non-stationary variants of GPR exist, and might be better suited. A good choice might be an approach based on clustering [21]. Adapting the spectral method to that case is straight-forward. The simulations from individual models (for each cluster) can be combined (locally) by a weighted sum.
- **Regularity/Periodicity:** Several functions in the BBOB set have some form of regular, symmetric, or periodic behavior. One classical example is the Rastrigin function (e.g., BBOB 3, 4 and 15). While our models seemed to work well for these functions, their regularity, symmetry or periodicity is not reproduced. With the given models, this would require a much larger number of training samples. If such behavior is important (and known a priori), a solution may be to choose a new kernel that, e.g., is itself periodic. This requires that the respective spectral measure of the new kernel is known, to enable the spectral simulation method [4].
- **Extremely fine-grained local structure:** Some functions, such as Schaffer's F7 (BBOB function 17, 18), have an extremely fine-grained local structure. This structure will quickly go beyond even a good model's ability to reproduce accurately. This will be true, even for fairly low-dimensional cases. While there is no easy way out, our results at least suggest one compensation: Many optimization algorithms will not notice such kind of fine-grained ruggedness. For instance, a mutation operator might easily jump across these local bumps, and rather follow the global structure of the function. Hence, an accurate representation of such structures may not be that important in practice, depending on the tested algorithms.
- **Number of samples:** The number of training data samples is one main driver of the complexity for GPR, affecting computational time and memory requirements. The mentioned cluster-GPR approach is one remedy [21].

6 Conclusion

Our first research question was:

Q1. How can simulation with GPR models be used to generate benchmarks for continuous optimization?

As an answer, we use the spectral method for GPR simulation [4]. As this method results in a superposition of cosine functions, it is well suited for continuous search spaces. Conversely, the previously used [23] decomposition-based approach is infeasible due to computational issues. Consecutively, we asked:

Q2. Do simulation-based benchmarks provide better results than estimation-based benchmarks, for continuous optimization?

Our experiments provide evidence that simulation-based benchmarks perform considerably better than estimation-based benchmarks. Only for low-dimensional ($n \leq 5$), unimodal problems did we observe an advantage for estimation.

In practice, if the modality (and dimensionality) of the objective function is known, this may help to select the appropriate approach. In a black-box case, the simulation approach seems to be the more promising choice.

For future research, it would be interesting to investigate how well these results translate to non-stationary GPR models, as discussed in Sect. 5. We also plan to investigate how parameters of the training data generation process affect the generated test functions, perform tests with broader algorithm sets, and demonstrate the approach with real-world applications. Finally, investigating other model types is of importance. Approaches with weaker assumptions than GPR, such as Generative Adversarial Networks [10], may be of special interest.

References

1. Ardia, D., Mullen, K.M., Peterson, B.G., Ulrich, J.: DEoptim: differential evolution in R, Version 2.2-5 (2020). https://CRAN.R-project.org/package=DEoptim. Accessed 25 Feb 2020
2. Bartz-Beielstein, T.: How to create generalizable results. In: Kacprzyk, J., Pedrycz, W. (eds.) Springer Handbook of Computational Intelligence, pp. 1127–1142. Springer, Heidelberg (2015). https://doi.org/10.1007/978-3-662-43505-2_56
3. Bartz-Beielstein, T., et al.: SPOT - Sequential Parameter Optimization Toolbox - v20200429 (2020). https://github.com/bartzbeielstein/SPOT/releases/tag/v20200429. Accessed 29 Apr 2020
4. Cressie, N.A.: Statistics for Spatial Data. Wiley, New York (1993)
5. Dang, N., Pérez Cáceres, L., De Causmaecker, P., Stützle, T.: Configuring irace using surrogate configuration benchmarks. In: Genetic and Evolutionary Computation Conference (GECCO 2017), pp. 243–250. ACM, Berlin, July 2017
6. Daniels, S.J., Rahat, A.A.M., Everson, R.M., Tabor, G.R., Fieldsend, J.E.: A suite of computationally expensive shape optimisation problems using computational fluid dynamics. In: Auger, A., Fonseca, C.M., Lourenço, N., Machado, P., Paquete, L., Whitley, D. (eds.) PPSN 2018. LNCS, vol. 11102, pp. 296–307. Springer, Cham (2018). https://doi.org/10.1007/978-3-319-99259-4_24
7. Fischbach, A., Zaefferer, M., Stork, J., Friese, M., Bartz-Beielstein, T.: From real world data to test functions. In: 26th Workshop Computational Intelligence, pp. 159–177. KIT Scientific Publishing, Dortmund, November 2016
8. Flasch, O.: A modular genetic programming system. Ph.D. thesis, Technische Universität Dortmund, Dortmund, Germany, May 2015
9. Forrester, A., Sobester, A., Keane, A.: Engineering Design via Surrogate Modelling. Wiley, New York (2008)
10. Goodfellow, I.J., et al.: Generative adversarial nets. In: Proceedings of the 27th International Conference on Neural Information Processing Systems, NIPS 2014, vol. 2, pp. 2672–2680. MIT Press, Cambridge (2014)
11. Hansen, N., Auger, A., Mersmann, O., Tusar, T., Brockhoff, D.: COCO: a platform for comparing continuous optimizers in a black-box setting. ArXiv e-prints, August 2016. arXiv:1603.08785v3
12. Hansen, N., et al.: COmparing Continuous Optimizers: numbbo/COCO on Github, March 2019. https://doi.org/10.5281/zenodo.2594848
13. Hansen, N., Finck, S., Ros, R., Auger, A.: Real-parameter black-box optimization benchmarking 2009: noiseless functions definitions. Research report RR-6829, inria-00362633, INRIA, February 2009. https://hal.inria.fr/inria-00362633

14. Journel, A.G., Huijbregts, C.J.: Mining Geostatistics. Academic Press, London (1978)
15. Lantuéjoul, C.: Geostatistical Simulation: Models and Algorithms. Springer, Heidelberg (2002)
16. Nelder, J.A., Mead, R.: A simplex method for function minimization. Comput. J. **7**(4), 308–313 (1965)
17. Preuss, M., Rudolph, G., Wessing, S.: Tuning optimization algorithms for real-world problems by means of surrogate modeling. In: Genetic and Evolutionary Computation Conference (GECCO 2010), pp. 401–408. ACM, Portland, July 2010
18. Rudolph, G., Preuss, M., Quadflieg, J.: Two-layered surrogate modeling for tuning optimization metaheuristics. Technical report TR09-2-005, TU Dortmund, Dortmund, Germany. Algorithm Engineering Report, September 2009
19. Storn, R., Price, K.: Differential evolution – a simple and efficient heuristic for global optimization over continuous spaces. J. Glob. Optim. **11**(4), 341–359 (1997)
20. Volz, V., Naujoks, B., Kerschke, P., Tušar, T.: Single- and multi-objective game-benchmark for evolutionary algorithms. In: Genetic and Evolutionary Computation Conference (GECCO2019). ACM, Prague, July 2019
21. Wang, H., van Stein, B., Emmerich, M., Bäck, T.: Time complexity reduction in efficient global optimization using cluster Kriging. In: Genetic and Evolutionary Computation Conference (GECCO 2017), pp. 889–896. ACM, Berlin, July 2017
22. Ypma, J., Borchers, H.W., Eddelbuettel, D.: NLoptr vers-1.2.1: R interface to NLopt (2019). http://cran.r-project.org/package=nloptr. Accessed 20 Nov 2019
23. Zaefferer, M., Fischbach, A., Naujoks, B., Bartz-Beielstein, T.: Simulation-based test functions for optimization algorithms. In: Genetic and Evolutionary Computation Conference (GECCO 2017), pp. 905–912. ACM, Berlin, July 2017

Comparative Run-Time Performance of Evolutionary Algorithms on Multi-objective Interpolated Continuous Optimisation Problems

Alexandru-Ciprian Zăvoianu(✉), Benjamin Lacroix, and John McCall

School of Computing Science and Digital Media, Robert Gordon University, Aberdeen, Scotland, UK
{c.zavoianu,b.m.e.lacroix,j.mccall}@rgu.ac.uk

Abstract. We propose a new class of multi-objective benchmark problems on which we analyse the performance of four well established multi-objective evolutionary algorithms (MOEAs) – each implementing a different search paradigm – by comparing run-time convergence behaviour over a set of 1200 problem instances. The new benchmarks are created by fusing previously proposed single-objective interpolated continuous optimisation problems (ICOPs) via a common set of Pareto non-dominated seeds. They thus inherit the ICOP property of having tunable fitness landscape features. The benchmarks are of intrinsic interest as they derive from interpolation methods and so can approximate general problem instances. This property is revealed to be of particular importance as our extensive set of numerical experiments indicates that choices pertaining to (i) the weighting of the inverse distance interpolation function and (ii) the problem dimension can be used to construct problems that are challenging to all tested multi-objective search paradigms. This in turn means that the new multi-objective ICOPs problems (MO-ICOPs) can be used to construct well-balanced benchmark sets that discriminate well between the run-time convergence behaviour of different solvers.

Keywords: Multi-objective continuous optimisation · Evolutionary algorithms · Performance analysis · Large-scale benchmarking.

1 Introduction and Motivation

A multi-objective optimisation problem (MOOP) can be defined as:

$$\text{minimize } F(x) = (f_1(x), \ldots, f_m(x))^T, \tag{1}$$

where the search space is multi-dimensional (i.e., $x \in V^d \subset \mathbb{R}^d$) and the $m \in \{2, 3\}$ real-valued objectives of $F(x)$ need to be minimized simultaneously. The conflicting nature of the m objectives means that the general solution of a MOOP is given by a Pareto optimal set (PS) that aggregates all solution candidates

T. Bäck et al. (Eds.): PPSN 2020, LNCS 12269, pp. 287–300, 2020.
https://doi.org/10.1007/978-3-030-58112-1_20

$x^* \in V^d$ with the property that they are not fully dominated – i.e., $\nexists y \in V^d$: $f_i(y) \leq f_i(x^*), \forall i \in \{1, \ldots, m\}$ and $F(y) \neq F(x^*)$. The Pareto front (PF) is the objective space projection of the PS.

Because of their ability to discover high-quality PS approximations called Pareto non-dominated sets (PNs) in single runs, multi-objective evolutionary algorithms (MOEAs) have emerged as some of the most successful MOOP solvers [3]. As an increasing number of MOEA practitioners (e.g., mechatronic engineers [19], industrial designers [16], quality assurance analysts [28]) are tackling ever more challenging real-world problems, difficulties stemming from relying on experimentation/simulation-driven $F(x)$ values are being brought to the forefront. A costly evaluation of solution candidate quality (i.e., fitness) greatly reduces the number of fitness evaluations (*nfe*) that can be computed during an optimisation and runs might be stopped prematurely. Furthermore, running multiple optimisation runs often becomes infeasible and only a single solver (with literature recommended parameter settings) is applied despite the well-known implications of the No Free Lunch Theorems for Optimisation (NFL) [4,22] regarding the benefits of algorithm and/or parameter selection.

To alleviate the aforementioned difficulties of real-world MOEA application, researchers have explored several avenues. Among them, promising results have been delivered by both (i) the integration of surrogate modeling techniques [13,17] that reduce solver dependency on costly fitness evaluations and (ii) the development of multi-method solvers that can deliver a robust performance over large (benchmark) problem sets with a fixed parameterisation [21,23]. While surrogate techniques are often demonstrated on MOOP formulations bound to "closed" simulation or experimentation environments, most MOOP benchmark problems share biases like analytically engineered challenges and strong inter-problem correlations [2]. The present work aims to aid both research streams by introducing a new class of MOOPs with tunable fitness landscapes that:

(i) are intrinsically interesting for benchmark construction as they propose challenges to several state-of-the-art MOEAs, when instantiated randomly;
(ii) can be used to effortlessly generate easy-to-share, lightweight interpolation-based surrogate formulations, when instantiating with real-world data.

2 Multi-objective Interpolated Continuous Optimisation Problems

2.1 Interpolated Continuous Optimisation Problems

First proposed in a single-objective context [12], ICOPs are defined by the following elements:

1. **A search space Ω:** a set, whose elements we refer to as *candidate solutions*, that defines the optimisation problem domain. For continuous problems, this will be a (subset of) real space of chosen dimension. In this paper, we have chosen our search spaces to be the d-dimensional cubes: $\Omega = [-5, 5]^d$.

2. **A distance function**, $e(x, y) : \Omega \times \Omega \rightarrow \mathbb{R}$ defining the distance between two solutions x and y. The pair (Ω, e) with these definitions is a *metric space.* A natural choice of distance function for continuous search spaces is the Euclidean distance.
3. **A set of seeds** $S \subset \Omega$**:** a (generally finite) set of distinct candidate solutions with an assigned fitness. Elements of S and their assigned fitness values will define the entire optimisation problem via interpolation.
4. **An interpolation function** $f_S : \Omega \rightarrow \mathbb{R}$**:** in this paper, we apply the inverse distance weighting method, originally defined by Shepard [18] for use in spatial analysis. Assuming the seed set S contains N seeds, labelled $S = \{s_1, ..., s_N\}$, and with the assigned fitnesses $U = \{u_1, ..., u_N\}$, we define for any solution candidate $x \in \Omega$:

$$f_{S,U}(x) = \begin{cases} \frac{\sum_{j=1}^{N} \frac{u_j}{e(x,s_j)^k}}{\sum_{j=1}^{N} \frac{1}{e(x,s_j)^k}} & \text{if } e(x, s_j) \neq 0 \text{ for all } j \\ u_j, & \text{if } e(x, s_j) = 0 \text{ for some } j \end{cases} \quad (2)$$

where k is a positive real number called the *power parameter.* Higher values of k increase the relative influence of nearby seeds on the interpolated value.
5. **An optimisation objective:** e.g., the minimisation of f_S.

2.2 Multi-objective ICOPs

Multi-Objective ICOPs (MO-ICOPs) can be obtained by associating each single-objective $f_i(x), i \in \{1, \ldots, m\}$ from Eq. 1 with a distinct ICOP. Given that in a real-world optimisation scenario, we would expect each solution candidate x to be evaluated across all m objectives to be optimised, it would make sense that the individual ICOPs share the same seeds (i.e., solution candidate samples), but differ on the fitness values associated to each seed[1]. Hence, a MO-ICOP is defined by the tuple $(S, U_1, \ldots, U_m)$ and the value of k and Eq. 1 can be rewritten:

$$\text{minimize } F_{(S,U_1,\ldots,U_m)}(x) = (f_{S,U_1}(x), \ldots, f_{S,U_m}(x))^T, \quad (3)$$

One major caveat is that the PF of a MO-ICOP cannot be computed analytically and must be estimated in an iterative fashion by aggregating the best (Pareto non-dominated) solutions found during multiple optimisation runs. Sampling-based PF discovery was also required for real-world visualisable (i.e., $x \in V^2$) distance-based MOOPs [9], but for this problem type, restrictions can be imposed to generate artificial instances with prescribed PFs [8].

2.3 The k Parameter

The power parameter k defines the influence of each seed on the interpolation of the rest of the search space as illustrated in Fig. 1. k increase the importance

[1] A similar strategy of composing MOOPs from single-objective problems was recently used for creating the `bbob-biobj` test suite [1].

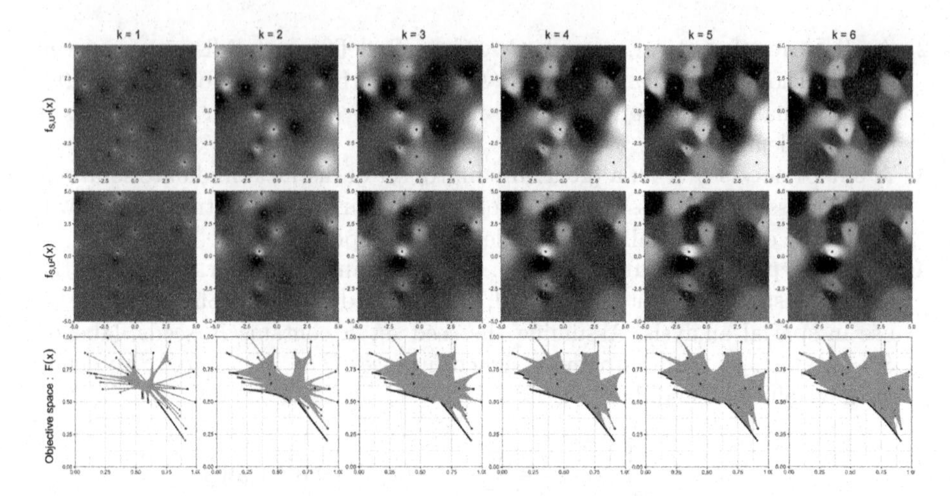

Fig. 1. MO-ICOPs with two objectives for $d = 2$ based on the same tuple (S, U_1, U_2) but different k values. The first two rows show the impact of k on the fitness landscape of each sub-contained ICOP. Non-dominated seeds are marked with red and dominated ones with blue; axes correspond to decision variables x_1 and x_2; lighter shades $\Rightarrow$ lower values. The third row shows how this translates to different shapes of the MO-ICOP objective space (gray) and PF (black) when grid sampling 10^6 candidate solutions. (Color figure online)

of distance weighting in the interpolation function. Lower values of k create flat fitness landscape around the average of the seed's value with sharp *peaks* and *pits* around the seeds. Higher values of k create large basins of attraction around each seed. This in turn influences the PF of resulting MO-ICOPs, as illustrated in the third row plots of Fig. 1.

3 Experimental Setup

3.1 Test Problems

While the properties of MO-ICOPs seeded with real-world data do warrant rigorous examination, in this preliminary study we focus on analysing the run-time convergence behaviour of established MOEA search paradigms with respect to the inherent characteristics of MO-ICOPs by considering a large set of 1200 two-objective random problem instances.

Randomly constructing a MO-ICOP is centred on embedding the seeds in objective space. We define S_{nd} as the set of Pareto non-dominated seeds and S_d as the set of dominated seeds such that $S = S_{nd} \cup S_d$. We start to randomly sample fitness values for each seed in S_{nd} ensuring that no seed in S_{nd} fully dominates another seed in S_{nd}. We then randomly sample fitness values for each seed in S_d, ensuring that every seed in S_d is fully dominated by at least one seed in S_{nd}. The idea behind this construction is to create randomly generated problems that may provide some different PF shapes. Finally, the positions of the seeds in the search space are obtained by random uniform sampling.

For our benchmark set we generated 50 distinct objective-space seed set embeddings and corresponding search space positions for 4 different dimensions $d \in \{5, 10, 20, 30\}$. We obtained the final MO-ICOP instances by associating each of the 200 resulting combinations with 6 values of $k \in \{1, 2, 3, 4, 5, 6\}$. For each problem, the number of non dominated seeds ($|S_{nd}|$) was randomly chosen between 5 and 20 and the number of dominated seeds ($|S_d|$) was randomly chosen between 1 and 80. For each problem, the PF was estimated using 20 million fitness evaluations spread across 400 independent optimisation runs. This data can be accessed at: https://github.com/czavoianu/PPSN_2020.

3.2 Solvers and Parameterisation

The four MOEAs used in our tests were chosen because they exemplify different well-proven strategies for solving multi-objective optimisation problems.

NSGA-II [6] is one of the best-known and most widely applied multi-objective solvers. It has popularised alongside SPEA2 [25] a highly elitist multi-objective evolutionary paradigm in which the population of iteration $t + 1$ is obtained by applying a two-tier selection for survival (i.e., filtering) operator on the union between the population of iteration t and all the offspring generated at iteration t. The filtering relies on a primary Pareto non-dominated sorting criterion and a secondary crowding criterion (for tie-breaking situations). The success of NSGA-II has also popularised two genetic operators for real-valued MOOPs: simulated binary crossover (SBX) and polynomial mutation (PM) [5].

The GDE3 [11] solver maintains the two-tier selection for survival operator introduced by NSGA-II, but aims to also exploit the very good performance of the differential evolution (DE) paradigm [20] on continuous optimisation problems by replacing the SBX and PM operators with a *DE/rand/1/bin* strategy.

MOEA/D-DE with Dynamic Resource Allocation [24] is a state-of-the-art solver that achieves highly competitive solutions for a wide-range of MOOPs. MOEA/D-DE refines the multi-objective search paradigm proposed in MOGLS [10] as it decomposes the original MOOP into several single-objective sub-problems that are the result of weighting-based aggregations of the original MOOP objectives. During the run, individuals are evolved via *DE/rand/1/bin* to become the solution to one or more of the sub-problems. Provided a proper choice of weight vectors, the solver population should provide a very good PS approximation.

The DECMO2++ [23] solver was designed for rapid convergence across a wide range of MOOPs as it integrates and actively pivots between three different search paradigms implemented via coevolved sub-populations. Specifically, while one sub-population implements Pareto-based elitism via the SPEA2 evolutionary model and the associated SBX and PM operators, the other actively co-evolved sub-population uses the DE-centred GDE3 search strategy. Decomposition is implemented via a largely passive archive based on a uniformly weighted

Tschebyscheff distance measure that aims to maintain the best achieved approximation of the PS at each stage of the search process.

We used the standard/literature recommended parameterisation for all four MOEAs and allowed a total computational budget of nfe = 50,000 for each optimisation run. In the case of NSGA-II, GDE3 and DECMO2++ we used a population/archive size of 200. In the case of MOEA/D-DE DRA we used an archive size of 300 – the recommended setting for MOOPs with two objectives.

3.3 Performance Evaluation

The PN quality measure we track during the run-time is Ind_H - a normalised version of the hypervolume [26] that evaluates a PN by relating the size of the objective space it dominates to that of the objective space dominated by the PF.

Hypervolume-ranked performance curves (HRPCs) have been proposed in [27] as a means to quickly estimate the comparative differences in the run-time convergence behaviour of several MOEAs across large benchmark sets. For each MOOP, the strategy is to rank the MOEAs that aim to solve it after every 1000 newly generated individuals (pre-defined comparison stages) using Ind_H values averaged over 100 independent runs. Under a *basic ranking schema*, the solver with the lowest average Ind_H in a set of n_s MOEAs will receive the rank n_s and the best performer will receive the rank 1. A bonus rank of 0 can be given to solvers that are estimated to have fully converged on the problem (i.e., $Ind_H >$ 0.99). By averaging for each MOEA at each comparison stage the ranks achieved on individual MOOPs, one can rapidly obtain an overview of the comparative convergence behaviour across the entire benchmark set.

HRPCs are constructed via two-by-two comparisons between the tested solvers in increasing order of Ind_H-indicated performance. As we also wish to illustrate the magnitude of the differences in run-time convergence behaviours, we complement the basic ranking schema with:

- a *pessimistic ranking schema* under which different ranks are awarded in the stage-wise two-by-two comparison only if the difference between the average Ind_H values of the two MOEAs is higher than a $th = 0.01$, $th = 0.05$ or $th = 0.10$ predefined threshold[2];
- a *statistical ranking schema* under which different ranks are awarded only if the difference between the stage-wise average Ind_H values is statistically significant when using a one-sided Mann-Whitney-Wilcoxon test [14] with a considered significance level of 0.025 after a Bonferroni correction [7].

It is noteworthy that in [27] the ranking was based on the average Ind_H assessment of the run-time MOEA populations at each comparison stage. In

[2] If the difference between the Ind_H-measured qualities of two PNs is larger than $th = 0.05$, the interpretation is that the objective space dominated by the best PN is larger than its counterpart by a size that is equivalent to at least 5% of the size of the objective space dominated by the solution of the MOOP (i.e., the PF).

the present work, motivated by the interactive way in which engineers employ MOEA-based searches in practice [19], the Ind_H averages over independent runs are computed by considering the set of all the Pareto non-dominated individuals that have been discovered by the MOEA till each comparison stage.

4 Results and Analysis

Before going into run-time performance analysis, it worth noting the performance of the four algorithms at *nfe* = 50,000. On the 1200 problems tested, MOEA/D-DE obtained the highest Ind_H 545 times, followed by NSGA-II with 509 end-of-the-run wins. GDE3 and DECMO++ obtained respectively 125 and 21 wins. In the top left plot of Fig. 2 we present the run-time Ind_H-averaged performance of the four tested MOEAs across the entire benchmark of 1200 MO-ICOPs. For *nfe* > 10,000, the plot indicates that MOEA/D-DE DRA achieves the best average performance, ahead of NSGA-II and DECMO2++. GDE3 constantly achieved the lowest run-time Ind_H average values.

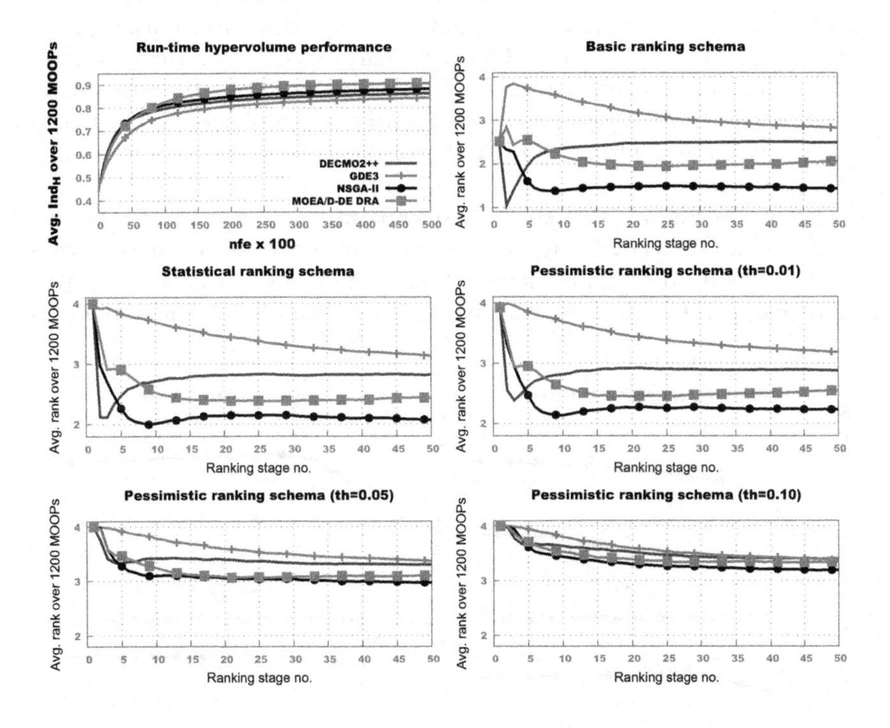

Fig. 2. Results across the entire benchmark of 1200 MO-ICOPs: $d \in \{5, 10, 20, 30\}$ and $k \in \{1, 2, 3, 4, 5, 6\}$. A bonus rank of 0 is awarded for full convergence – i.e., when $Ind_H > 0.99$.

In the early phases of the optimisation runs (i.e., *nfe* < 7,000), both NSGA-II and DECMO2++ achieve slightly better average Ind_H values than MOEA/D.

This is confirmed by the HRPC plots from Fig. 2 as they indicate a slight advantage in early convergence for NSGA-II and DECMO2++ even when considering a pessimistic ranking with a threshold $th = 0.05$. Surprisingly, the general picture of the five rank-based comparisons indicates that NSGA-II edges MOEA/D-DE DRA as the best performer throughout the run-time. Since this is contrasting with the results of the Ind_H-averaged performance plot, it is worthy to focus the analysis on NSGA-II and MOEA/D-DE and thus remove the impact of convergence behaviour cliques on the relative rankings.

Furthermore, by not awarding a bonus rank of 0 for full convergence (i.e., when $Ind_H > 0.99$), the restricted NSGA-II vs. MOEA/D-DE analysis highlights more clearly at each ranking stage the difference between the number of problems on which one solver performs better than the other. Thus, since only ranks of 2 and 1 can be awarded, achieving an average value of 2 at a certain ranking stage is a clear indication that a solver is not better than its counterpart (given the considered ranking criterion) across any MOOPs in the chosen benchmark set. Conversely, HRPC values very close to 1 indicate a clear better performance. More formally, in a one-on-one comparison with no bonuses, a rank value of 1.x associated with a solver, indicates that the solver outperforms its counterpart on $(1 - 0.x) \times 100\%$ of the problems considered in the benchmark.

In light of strong empirical evidence (please see Fig. 1) that the power parameter used in the interpolation function k from Eq. 2 has a significant impact on the geometry of the resulting MO-ICOP problem, it is natural to analyse if and how this translates into divergent MOEA run-time behaviours.

Therefore, in Fig. 3, we present the comparative convergence behaviour of NSGA-II and MOEA/D-DE when only considering a benchmark subset containing the 600 MO-ICOPs with the lowest weights of the inverse distance interpolation function: i.e., $k \in \{1, 2, 3\}$. The average Ind_H values from the top left plot

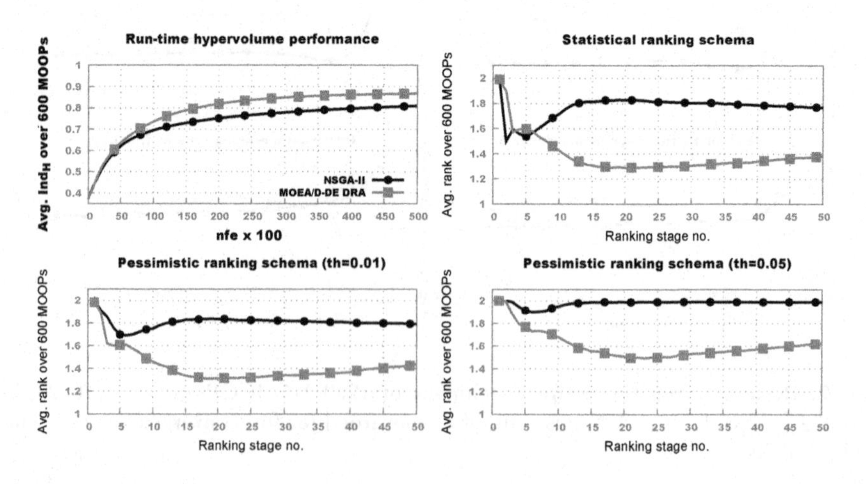

Fig. 3. Results across the benchmark subset with 600 MO-ICOPs with $d \in \{5, 10, 20, 30\}$ and $k \in \{1, 2, 3\}$. No bonus rank is awarded for full convergence.

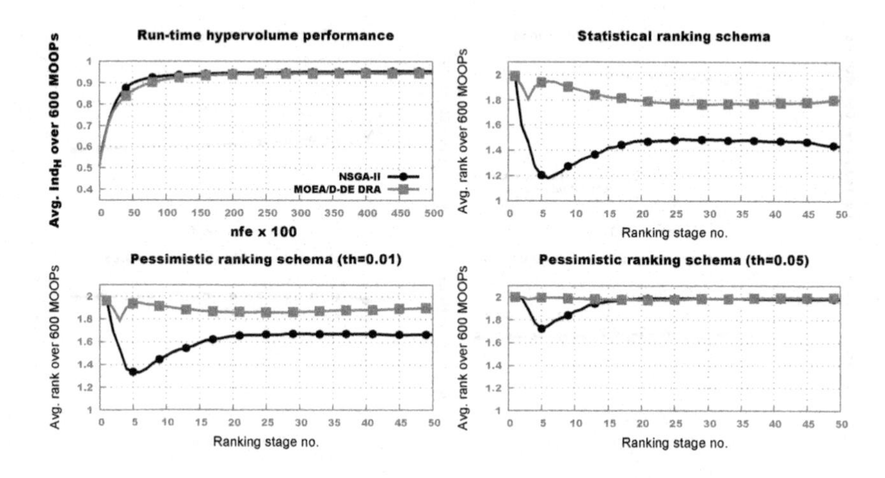

Fig. 4. Results across the benchmark subset with 600 MO-ICOPs with $dim \in \{5, 10, 20, 30\}$ and $k \in \{4, 5, 6\}$. No bonus rank of is awarded for full convergence.

indicate that MOEA/D-DE generally outperforms NSGA-II in all stages of the optimisation runs. Furthermore, this subset of benchmark MO-ICOPs is more challenging as, for both solvers, the achieved average Ind_H values are lower than those reported in Fig. 2 over all 1200 problems.

The different convergence behaviours of the two solvers for low k values is confirmed by all three corresponding HRPC plots. These also highlight that, for $nfe > 15{,}000$, the Ind_H-measured performance of MOEA/D-DE solutions is better across $\approx 40\%$ of the problems in this benchmark subset when considering a Ind_H threshold $th = 0.05$.

The plots in Fig. 4 indicate that for a higher weight of the inverse distance interpolation function (i.e., $k \in \{4, 5, 6\}$), the Ind_H-measured performance of NSGA-II solutions is:

- consistently better across $\approx 50\%$ of the MO-ICOPs in this benchmark subset when considering the one-sided Mann-Whitney-Wilcoxon statistical significance test;
- consistently better across $\approx 30\%$ of the MO-ICOPs, when considering a Ind_H threshold of 0.01.
- only better by large margins (i.e., $th = 0.05$) across $\approx$10–25% of the MO-ICOPs in the early stages of the optimisation runs: $5000 < nfe < 10{,}000$.

It is also noteworthy that on the MO-ICOPs associated with higher values of k, both multi-objective solvers perform well and are able to reach average benchmark-wide Ind_H values higher than 0.9 after 10,000 fitness evaluations.

The plots in Fig. 5 illustrate the run-time convergence behaviour of NSGA-II and MOEA/D-DE on problems with a lower dimension: $d \in \{5, 10\}$. The HRPCs

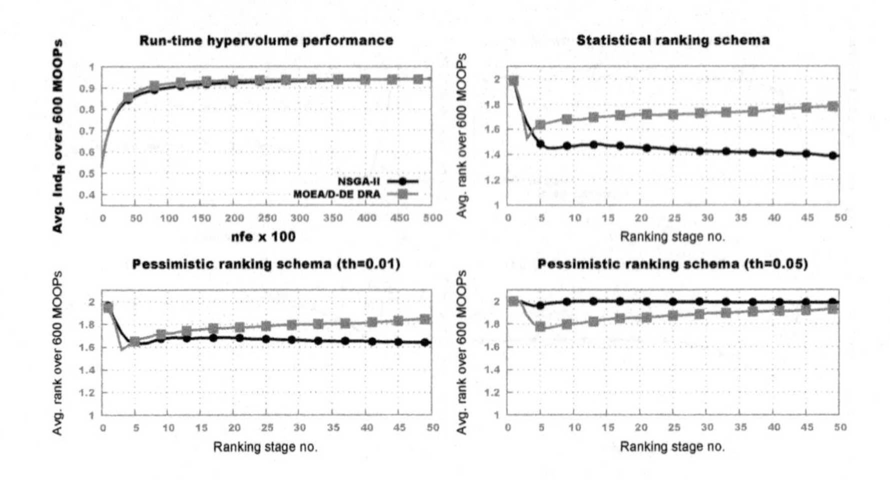

Fig. 5. Results across the benchmark subset with 600 MO-ICOPs with $d \in \{5, 10\}$ and $k \in \{1, 2, 3, 4, 5, 6\}$. No bonus rank is awarded for full convergence.

show that while solver performance is similar early on, for *nfe* > 5,000, when comparing based on statistical significance, NSGA-II performs better across ≈50–60% of the 600 MO-ICOPs and MOEA/D-DE perform better across ≈20–30% of the problems. Smaller differences that progressively favour NSGA-II as the *nfe* increases are also observed when imposing a small Ind_H ranking threshold of $th = 0.01$. However, when looking at the larger thresholds of 0.05, NSGA-II doesn't outperform on any problem in the benchmark while MOEA/D-DE still perform better across ≈10–20% of the problems throughout the entire run-time. The fact that NSGA-II performs better on more problems but with a lower margins (i.e., $th < 0.05$) while MOEA/D performs better on fewer problems but with higher margins (i.e., $th \geq 0.05$) helps to explain the average benchmark-wide Ind_H plot (top-left) associated with this benchmark subset. As a side note, when considering Fig. 4 and 5, the HRPCs provide valuable insight that helps to differentiate convergence behaviours captured by largely similar benchmark-wide Ind_H averaging plots.

The plots in Fig. 6 show that on the benchmark subset of 600 MO-ICOPs with a higher dimension – i.e., $d \in \{20, 30\}$ – MOEA/D-DE consistently outperforms NSGA-II for *nfe* > 15,000 across ≈ 60% of the problems (based on a statistical significance and pessimistic $th = 0.01$ criteria) and ≈30–40% of the problems (when considering the stricter $th = 0.05$ criterion). It is noteworthy that in the early part of the runs (*nfe* < 15,000), NSGA-II performs notably better than MOEA/D-DE across all the considered comparison criteria.

Finally, in order to better understand the interplay between the inverse distance weighting parameter and the MO-ICOP problem dimension on one side and the comparative solver performance on the other, we can compute a (k, d) preference matrix at a fixed point of interest during the run-time. For example, the top-left plot of Fig. 2 indicates that all solvers are past their knee-point

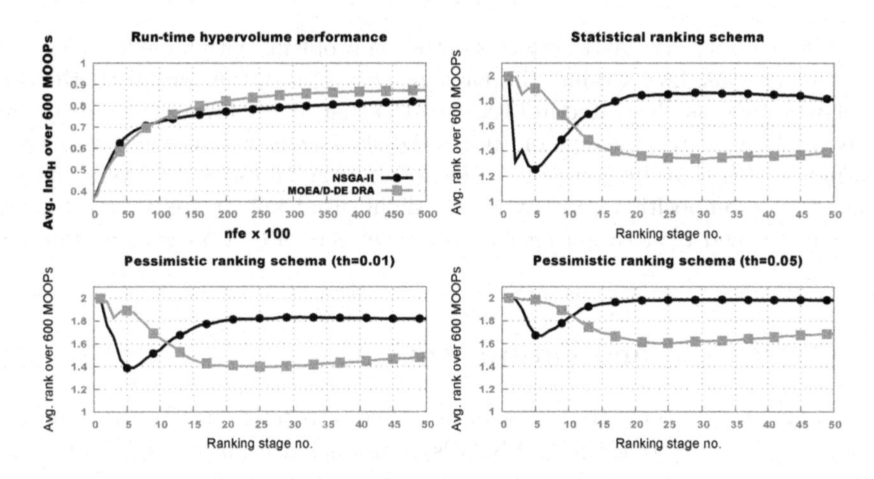

Fig. 6. Results across the benchmark subset with 600 MO-ICOPs with $d \in \{20, 30\}$ and $k \in \{1, 2, 3, 4, 5, 6\}$. No bonus rank is awarded for full convergence.

in convergence at nfe = 20,000 (i.e., ranking stage no. 20). When considering the statistical significance ranking criterion, we can compute the preference for MOEA/D-DE over NSGA-II for the 24 (k, d) combinations by subtracting the percentage of problems on which NSGA-II outperforms (at ranking stage no. 20) from the percentage of problems on which MOEA/D-DE performs better. The resulting (k, d) preference matrix shown in Fig. 7 indicates that while MOEA/D-DE obtains better results than NSGA-II on problems with low k parameters and

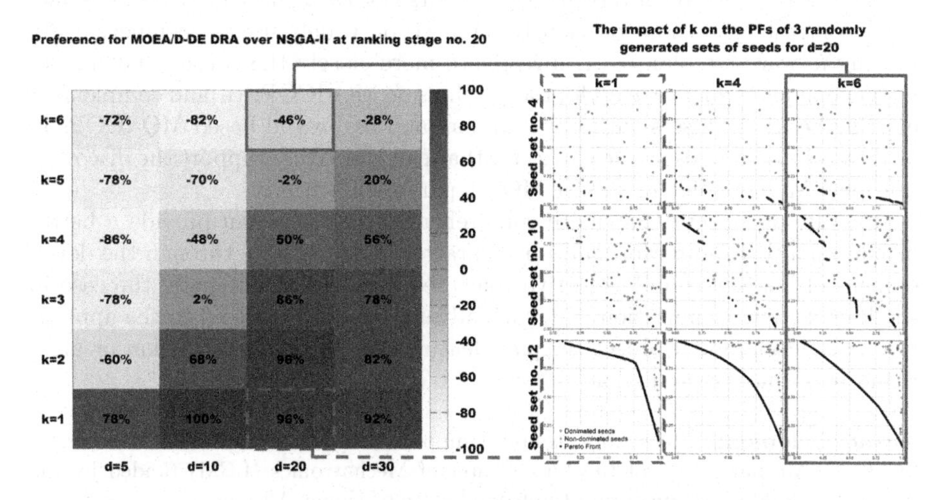

Fig. 7. Preference of MOEA/D-DE over NSGA-II when considering differences in average performance over 100 independent runs confirmed by statistical significance testing (left). Example of the impact of k on the PF shape of 9 MO-ICOPs with $d = 20$ (right).

larger dimensions, NSGA-II performs better on problems with large k values and lower dimensions. Low k values (in particular $k = 1$) result in very discontinuous point-wise PFs (as illustrated in Figs. 1 and 7) that highly favour the directional decomposition search strategy of MOEA/D-DE. Higher k values generate more continuous PFs that are generally easier to converge on for both solvers, but on which the decomposition strategy is at a slight disadvantage (please see Fig. 4) as its more rigid exploration mechanism likely generates PNs with an inferior spread.

5 Conclusions and Future Work

In this paper we (i) describe a new class of multi-objective interpolated continuous optimisation problems (MO-ICOPs) constructed using a weighted inverse distance function and we (ii) proceed to analyse the comparative run-time performance of four established MOEAs on a benchmark of 1200 random MO-ICOP instances using multiple criteria.

The optimisation results indicate that MO-ICOPs propose challenges to all tested multi-objective optimisation paradigms. GDE3 consistently achieves the lowest benchmark-wide average hypervolume attainment levels despite obtaining the best approximation of the PF on 10% of the problems tested. DECMO2++ only demonstrates its characteristic fast converging behaviour during the very start of the run (i.e., $nfe < 5{,}000$) and is outperformed by both MOEA/D-DE and NSGA-II. The comparative performance of NSGA-II and MOEA/D-DE DRA is strongly influenced by the weighting of the inverse distance function and the dimension of the problem.

Since the observed convergence behaviours of GDE3 and DECMO2++ somewhat contrast with those previously reported on widely used benchmarks [23], moving forward we plan to (i) investigate more closely the causes that impact the general performance of all four algorithms on MO-ICOPs (ii)and complement existing benchmark sets with both random and real-world based MO-ICOPs in order to obtain a well-balanced test rig that can effectively support the discovery of robust MOEAs and/or robust MOEA parameterisations.

Finally, we believe that more comprehensive test sets can provide a better insight on algorithm performance by characterising problems through the development of landscape and objective space features. As it is already the case in single-objective optimisation [15], such advancements could lead to the application of landscape and objective space features for algorithm selection or algorithm performance prediction.

Acknowledgments. This work has been supported by the COMET-K2 "Center for Symbiotic Mechatronics" of the Linz Center of Mechatronics (LCM) funded by the Austrian federal government and the federal state of Upper Austria.

References

1. Brockhoff, D., Tusar, T., Auger, A., Hansen, N.: Using well-understood single-objective functions in multiobjective black-box optimization test suites. ArXiv e-prints (2019). arXiv:1604.00359v3
2. Christie, L.A., Brownlee, A.E., Woodward, J.R.: Investigating benchmark correlations when comparing algorithms with parameter tuning. In: Proceedings of the Genetic and Evolutionary Computation Conference Companion, pp. 209–210 (2018)
3. Coello Coello, C.A., Lamont, G.B.: Applications Of Multi-Objective Evolutionary Algorithms. World Scientific, Singapore (2004)
4. Corne, D.W., Knowles, J.D.: No free lunch and free leftovers theorems for multiobjective optimisation problems. In: Fonseca, C.M., Fleming, P.J., Zitzler, E., Thiele, L., Deb, K. (eds.) EMO 2003. LNCS, vol. 2632, pp. 327–341. Springer, Heidelberg (2003). https://doi.org/10.1007/3-540-36970-8_23
5. Deb, K.: Multi-Objective Optimization using Evolutionary Algorithms. Wiley, Hoboken (2001)
6. Deb, K., Pratap, A., Agarwal, S., Meyarivan, T.: A fast and elitist multiobjective genetic algorithm: NSGA-II. IEEE Trans. Evol. Comput. **6**(2), 182–197 (2002)
7. Dunn, O.J.: Multiple comparisons among means. J. Am. Stat. Assoc. **56**(293), 52–64 (1961)
8. Fieldsend, J.E., Chugh, T., Allmendinger, R., Miettinen, K.: A feature rich distance-based many-objective visualisable test problem generator. In: Proceedings of the Genetic and Evolutionary Computation Conference, GECCO 2019, pp. 541–549. Association for Computing Machinery, New York (2019). https://doi.org/10.1145/3321707.3321727
9. Ishibuchi, H., Akedo, N., Nojima, Y.: A many-objective test problem for visually examining diversity maintenance behavior in a decision space. In: Proceedings of the 13th Annual Conference on Genetic and Evolutionary Computation, GECCO 2011, pp. 649–656. Association for Computing Machinery, New York (2011). https://doi.org/10.1145/2001576.2001666
10. Jaszkiewicz, A.: On the performance of multiple-objective genetic local search on the 0/1 knapsack problem - a comparative experiment. IEEE Trans. Evol. Comput. **6**(4), 402–412 (2002)
11. Kukkonen, S., Lampinen, J.: GDE3: the third evolution step of generalized differential evolution. In: IEEE Congress on Evolutionary Computation (CEC 2005), pp. 443–450. IEEE Press (2005)
12. Lacroix, B., Christie, L.A., McCall, J.A.: Interpolated continuous optimisation problems with tunable landscape features. In: Proceedings of the Genetic and Evolutionary Computation Conference Companion, pp. 169–170 (2017)
13. Loshchilov, I., Schoenauer, M., Sebag, M.: A mono surrogate for multiobjective optimization. In: Proceedings of the 12th annual Conference on Genetic and Evolutionary Computation (GECCO), pp. 471–478. ACM (2010)
14. Mann, H.B., Whitney, D.R.: On a test of whether one of two random variables is stochastically larger than the other. Ann. Math. Stat. **18**(1), 50–60 (1947)
15. Muñoz, M.A., Sun, Y., Kirley, M., Halgamuge, S.K.: Algorithm selection for black-box continuous optimization problems: A survey on methods and challenges. Inf. Sci. **317**, 224–245 (2015)
16. Oyama, A., Kohira, T., Kemmotsu, H., Tatsukawa, T., Watanabe, T.: Simultaneous structure design optimization of multiple car models using the k computer. In: IEEE Symposium Series on Computational Intelligence (2017)

17. Pilát, M., Neruda, R.: Hypervolume-based local search in multi-objective evolutionary optimization. In: Proceedings of the 2014 Annual Conference on Genetic and Evolutionary Computation, pp. 637–644. ACM (2014)
18. Shepard, D.: A two-dimensional interpolation function for irregularly-spaced data. In: Proceedings of the 1968 23rd ACM National Conference, pp. 517–524 (1968)
19. Silber, S., et al.: Coupled optimization in MagOpt. Proc. Inst. Mech. Eng. Part I: J. Syst. Control Engineering **230**(4), 291–299 (2016). https://doi.org/10.1177/0959651815593420
20. Storn, R., Price, K.V.: Differential evolution - a simple and efficient heuristic for global optimization over continuous spaces. J. Glob. Optim. **11**(4), 341–359 (1997)
21. Vrugt, J.A., Robinson, B.A.: Improved evolutionary optimization from genetically adaptive multimethod search. Proc. Natl. Acad. Sci. **104**(3), 708–711 (2007)
22. Wolpert, D.H., Macready, W.G.: No free lunch theorems for optimization. IEEE Trans. Evol. Comput. **1**(1), 67–82 (1997)
23. Zăvoianu, A.C., Saminger-Platz, S., Lughofer, E., Amrhein, W.: Two enhancements for improving the convergence speed of a robust multi-objective coevolutionary algorithm. In: Proceedings of the Genetic and Evolutionary Computation Conference, pp. 793–800. ACM (2018)
24. Zhang, Q., Liu, W., Li, H.: The performance of a new version of MOEA/D on CEC09 unconstrained MOP test instances. Technical report, School of CS & EE, University of Essex, February 2009
25. Zitzler, E., Laumanns, M., Thiele, L.: SPEA2: improving the strength Pareto evolutionary algorithm for multiobjective optimization. In: Evolutionary Methods for Design, Optimisation and Control with Application to Industrial Problems (EUROGEN 2001), pp. 95–100. International Center for Numerical Methods in Engineering (CIMNE) (2002)
26. Zitzler, E.: Evolutionary Algorithms for Multiobjective Optimization: Methods and Applications. Ph.D. thesis, Swiss Federal Institute of Technology (1999)
27. Zăvoianu, A.-C., Lughofer, E., Bramerdorfer, G., Amrhein, W., Klement, E.P.: DECMO2: a robust hybrid and adaptive multi-objective evolutionary algorithm. Soft Comput. **19**(12), 3551–3569 (2014). https://doi.org/10.1007/s00500-014-1308-7
28. Zăvoianu, A.C., Lughofer, E., Pollak, R., Meyer-Heye, P., Eitzinger, C., Radauer, T.: Multi-objective knowledge-based strategy for process parameter optimization in micro-fluidic chip production. In: Proceedings of the IEEE SSCI 2017 Conference, pp. 1927–1934. IEEE, Honolulu (2017)

Combinatorial Optimization

On the Design of a Partition Crossover for the Quadratic Assignment Problem

Omar Abdelkafi[1], Bilel Derbel[1(✉)], Arnaud Liefooghe[2], and Darrell Whitley[3]

[1] Univ. Lille, CNRS, Centrale, Inria, UMR 9189 - CRIStAL, 59000 Lille, France
{omar.abdelkafi,bilel.derbel}@univ-lille.fr
[2] JFLI – CNRS IRL 3527, University of Tokyo, Tokyo 113-0033, Japan
arnaud.liefooghe@univ-lille.fr
[3] Colorado State University, Fort Collins, USA
whitley@cs.colostate.edu

Abstract. We conduct a study on the design of a partition crossover for the QAP. On the basis of a bipartite graph representation, we propose to recombine the unshared components from parents, while enabling their fast evaluation using a preprocessing step for objective function decomposition. Besides a formal description and complexity analysis of the proposed crossover, we conduct an empirical analysis on its relative behavior using a number of large-size QAP instances, and a number of baseline crossovers. The proposed operator is shown to have a relatively high intensification ability, while keeping execution time relatively low.

1 Introduction

One of the key ingredients in the success of evolutionary algorithms is the design of effective and efficient crossover operators. In the context of gray-box optimization, problem-specific properties can help in designing dedicated operators. This is the case of *partition crossovers*, developed for a number of combinatorial problems (e.g., TSP [18], SAT [2], NK-landscapes [17]), and allowing to efficiently explore large search spaces. The term *partition* crossover is here used in a general sense, to render the idea of decomposing the variables according to their values in the parents, and then recombining them in such a way that the best improving offspring can be computed efficiently. In other words, a partition crossover is based on the idea of *optimal* recombination of two parents, which requires to find the best possible offspring among all possible ones, while fulfilling the genotype inheritance principle. As such, two issues must be considered.

Firstly, one should specify the recombination mechanism allowing gene transmission while fully exploring the improvement potential of parents. In particular, the shared genes from parents are kept identical and the other genes are to be properly mixed. Secondly, since the set of possible offspring underlying such a process is typically huge, one must rely on some properties to compute the best offspring in the most efficient manner. For instance, the so-called k-bounded binary problems [17] can be decomposed as a linear combination of sub-functions

T. Bäck et al. (Eds.): PPSN 2020, LNCS 12269, pp. 303–316, 2020.
https://doi.org/10.1007/978-3-030-58112-1_21

of at most k variables. This guarantees that the contributions of non interacting variables to the global fitness are additive, and hence, the choice of the optimal gene sequence when performing recombination can be performed greedily in an efficient manner. Another example is TSP [18], where the cost of the global tour is additive with respect to the length of different sub-paths sharing the same cities, but in different order, in the parent solutions.

In this paper, we are interested in the Quadratic Assignment Problem (QAP) [4,7,11,13], for which no partition crossover has been developed so far, despite the broad range of dedicated algorithms. A main difficulty comes from the definition of its objective function, where the contribution of every single variable is sensitive to all other variables. Hence, it becomes a challenging issue to design a decomposition process which enables both parent recombination and fast offspring evaluation. This is precisely the aim of our work, and our contribution is to be viewed as a first step towards the design of an efficient and effective partition crossover operator for the QAP. More specifically, we rely on an intuitive bipartite graph representation for decomposition. We then show how recombination and offspring evaluation can be performed on that basis. Moreover, we conduct an empirical study rendering the performance of the designed crossover comparatively to existing ones, either by itself, or when combined with a fast local search process. Our analysis shows that the proposed crossover has a high intensification power while keeping execution time relatively low.

The paper is organized as follows. In Sect. 2, we introduce the QAP and review some existing crossovers. In Sect. 3, we describe our main contribution towards the design a partition crossover for the QAP. In Sect. 4, we report our experimental findings. In Sect. 5, we conclude the paper.

2 Background

2.1 Problem Definition

The Quadratic Assignment Problem (QAP) [4,13] aims at assigning n facilities I, to n locations J. Let f_{hi} be the flow between facilities h and i, and d_{sj} be the distance between locations s and j. The objective is to minimize the sum of the products between flows and distances, such that each facility is assigned to exactly one location. The solution space can be defined as the set Π of permutations of $\{1, \ldots, n\}$. Given a permutation solution π, the ith element $\pi(i)$ corresponds to assigning facility i to location $\pi(i)$. The QAP is then stated as:

$$\arg\min_{\pi \in \Pi} \sum_{h \in I} \sum_{i \in I} d_{\pi(h)\pi(i)}\, f_{hi} \tag{1}$$

The QAP is NP-hard [13], and is considered as one of the most difficult problems from combinatorial optimization. As such, we have to rely on heuristic approaches such as stochastic local search and evolutionary algorithms [7].

2.2 Representative Crossover Operators

In this paper, we are interested in designing efficient and effective crossover operators. Among the large number of hybrid genetic algorithms for the QAP; see, e.g., [6,9,10,19], crossover appears to be a crucial component. In this respect, we can distinguish two families of crossovers [11]. In *half-structured* crossovers, the offspring preserve only a part of the parent genes, whereas the remaining part is typically generated at random. In *fully-structured* crossovers, the offspring genes are obtained by preserving the ones from parents. In our work, we consider four usual and representative baseline crossovers, two from the first family, and two from the second one. We first describe the half-structured crossovers, namely OPX and UX, and then the fully structured ones, namely CX and SPX. Notice that these crossovers can also find applications in other problems such as TSP [5].

The OPX Crossover. This is a standard one-point crossover, e.g., [6]. Given two parent permutations, a random point is selected to define two parts for each parent. A new offspring is obtained by first preserving the locations from the first parent up to the chosen random point. The remaining part is filled by copying the elements from the second part of the second parent, excluding those that were already copied from the first parent. This may lead to the situation where some facilities are not assigned to any locations. The offspring is hence complemented at random, using the remaining locations that were not yet included.

The UX Crossover. This is a standard uniform crossover operator, e.g., [16]. Some locations are selected at random following a Bernoulli distribution with parameter 1/2. The locations occupied by the selected facilities in the first parent are copied. The locations occupied by the remaining facilities in the second parent are also copied unless they were already included from the first parent. The offspring is complemented by randomly assigning the missing locations.

The CX Crossover. This is the so-called cycle crossover [8,9,12]. For clarity, we consider the example of Fig. 1a. All shared assignments are copied, i.e., facilities 1 and 7 assigned to locations 5 and 9. The crossover starts iterating from the first different facility assignment, which is facility 2 assigned to location 3 in the *first* parent. Looking at the *second* parent, facility 2 is assigned to location 8. This location is occupied by facility 3 in the *first* parent. Similarly, facility 3 is assigned to location 4 in the *second* parent, and so on. A cycle – alternating between the same set of locations in a different order on both parents – is then detected when arriving to facility 8 assigned to location 3 in the second parent. Hence, one parent is selected at random and the so-computed locations are preserved in the offspring. This procedure is repeated until all facilities are assigned.

The SPX Crossover. This is the so-called Swap Path Crossover [1,3]. The crossover is based on iteratively swapping unshared locations. In Fig. 1b, it starts with facility 2 assigned to location 3 (resp. 6) in the first (resp. second) parent. In the first (resp. second) parent, location 6 (resp. 3) is occupied by facility 5 (resp. 8). Hence, a swap is performed between locations 2 and 5 in

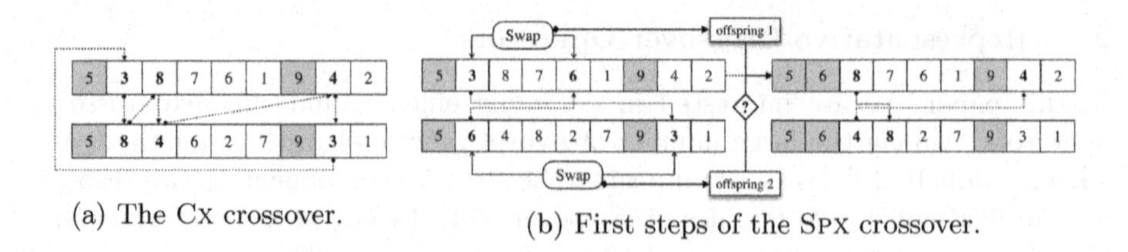

(a) The CX crossover. (b) First steps of the SPX crossover.

Fig. 1. Illustration of the CX and SPX crossovers.

the first parent, *as well as* between locations 2 and 8 in the second parent. Two offspring are obtained, respectively to the first and second parent. Then, the best of the two replaces its corresponding parent, say the first one as in Fig 1b. Notice that after this iteration, the two new parents have one more location in common. The same is then repeated iteratively until both parents become the same. The output offspring is the best ever created during all iterations.

3 A Partition Crossover for the QAP

A partition crossover is based on the idea of decomposing the evaluation function as well as the set of variables during recombination. This decomposition should enable to construct and to evaluate offspring using partial evaluations. The fastest the evaluation step, the fastest the exploration of a large number of offspring, eventually leading to an improving one, hence exploring the so-called dynastic potential of parents at best. It should be clear that none of the previously-described crossovers fulfill this requirement, although CX and SPX attempt to construct an offspring based on the idea of gene transmission.

In the following, we aim at designing a new partition crossover for the QAP. We start by introducing some notations, then we describe a decomposition process for the QAP, and the underlying recombination (Proposition 1) and function evaluation process (Corollary 1), in a formal, but intuitive, manner.

3.1 Recombination Based on a Bipartite Graph Representation

Following standard notations from graph theory, a bipartite graph G is a graph whose nodes can be divided into two disjoint sets $U(G)$ and $V(G)$, such that an edge in $E(G)$ can only connect a node in U to one in V. Given a subset of nodes $Q \subseteq U$ and $R \subseteq V$, we denote by $G[Q, R]$ the subgraph induced by $Q \cup R$ in G.

Let π be a permutation solution for the QAP. Let us define the bipartite graph whose nodes sets are respectively the set of facilities I and locations J, and where every facility i is connected to its unique location j according to permutation π. This corresponds in fact to a very natural representation of a feasible assignment of the facilities to the locations; see Fig. 2.

Definition 1. *Let* $G_\pi = (I, J, E)$ *be the bipartite graph such that* $E(G_\pi) = \{(i, j) \mid \pi(i) = j\}$.

Let us now consider two permutation solutions π_1 and π_2 which will play the role of parents for our target crossover. Let $\widetilde{I}$ (resp., $\overline{I}$) be the set of facilities that are assigned (resp., not assigned) to the same locations, denoted $\widetilde{J}$ (resp., $\overline{J}$).

Definition 2. *Let $\widetilde{I} = \{i \in I \mid \pi_1(i) = \pi_2(i)\}$ and $\overline{I} = I \setminus \widetilde{I}$. Then, let $\widetilde{J} = \{\pi_1(i) \mid i \in \widetilde{I}\}$, and $\overline{J} = J \setminus \widetilde{J}$.*

Notice that for any facility $i \in \overline{I}$, the corresponding locations $\pi_1(i)$ and $\pi_2(i)$ are different, and both belong to $\overline{J}$. Let us now define the bipartite graph $G_{\pi_1\pi_2}$ obtained by merging the edges of G_{π_1} and G_{π_2}; see Fig. 2.

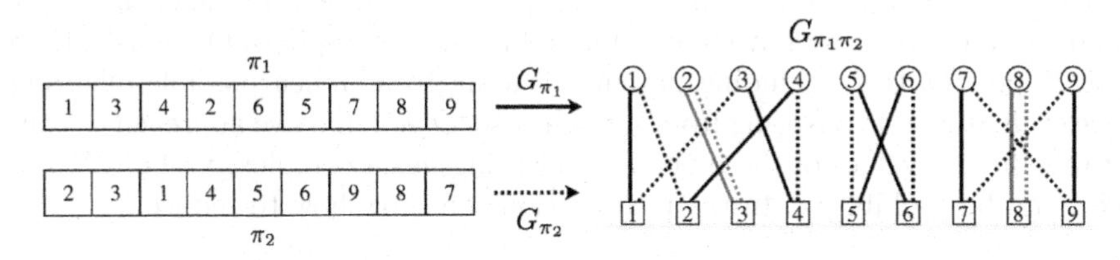

Fig. 2. Illustration of a bipartite graph representation. Solid (resp. dashed) edges are with respect to G_{π_1} (resp. G_{π_2}). We have $\widetilde{I} = \{2, 8\}$, $\widetilde{J} = \{3, 8\}$ and the corresponding edges in gray. There are $k = 3$ connected components C_1, C_2 and C_3 in $G_{\pi_1\pi_2}[\overline{I}, \overline{J}]$, with $U(C_1) = \{1, 3, 4\}$, $U(C_2) = \{5, 6\}$, and $U(C_3) = \{7, 9\}$.

Definition 3. *Let $G_{\pi_1\pi_2} = (I, J, E(G_{\pi_1}) \cup E(G_{\pi_2}))$*

Let us focus on the *connected components* of the so-obtained graph. First, we have that every facility $i \in \widetilde{I}$ is connected, by exactly two edges in $G_{\pi_1\pi_2}$, to a unique location $j = \pi_1(i) = \pi_2(i) \in \widetilde{J}$, and j is not connected to any other facility. Hence, this implies exactly one connected component with these two nodes connected by two parallel edges. Apart from such components (in $\widetilde{I} \cup \widetilde{J}$), the other components of interest connect nodes in $\overline{I}$ to nodes in $\overline{J}$. Then,

Definition 4. *Let k be the number of connected components in $G_{\pi_1\pi_2}[\overline{I}, \overline{J}]$, and let $\mathcal{C} = \{C_1, C_2, \cdots, C_k\}$ be the set of these connected components.*

Notice that $k = 0$ iff $\pi_1 = \pi_2$. In the following, we assume that $\pi_1 \neq \pi_2$, and hence $k \geq 1$. By definition, we also have that for every $\ell \in \{1, \dots, k\}$, C_ℓ is a bipartite graph whose edges form a cycle that are alternatively in $E(G_{\pi_1})$ and in $E(G_{\pi_2})$. An offspring can hence be constructed by: (i) preserving the shared assignments of facilities in $\widetilde{I}$, and (ii) choosing for every set of facilities $U(C_\ell)$, $\ell \in \{1, \dots, k\}$, the locations they are connected to *either* in G_{π_1} (first parent) *or* in G_{π_2} (second parent). More formally, let $m : \{1, \dots, k\} \mapsto \{1, 2\}$ be an *arbitrary* mapping function. Such a mapping can be used to decide which of π_1 or π_2 to consider when choosing the locations of $U(C_\ell)$ for every ℓ, i.e., if $m(\ell) = 1$ then use π_1, otherwise use π_2. This is stated in the following proposition summarizing the proposed crossover recombination mechanism.

Proposition 1. *Given two permutations π_1 and π_2 and an arbitraty mapping function $m : \{1, \ldots, k\} \mapsto \{1, 2\}$, $\overset{\otimes}{\pi}$ as defined in the following is a feasible permutation offspring.*

- $\forall i \in \widetilde{I}$, $\overset{\otimes}{\pi}(i) = \pi_1(i) = \pi_2(i)$
- $\forall i \in \overline{I}$, $\overset{\otimes}{\pi}(i) = \pi_{m(\ell_i)}(i)$, *where* $\ell_i \in \{1, \ldots, k\}$ *is such that* $i \in U(C_{\ell_i})$

The previous proposition is to recall the Cx crossover, where a cycle corresponds to a connected component in our formalism. However, we consider to explore not solely one random offspring, but the whole set of offspring solutions that can be constructed by Proposition 1 to find the best possible one. Since, the recombination process is fully and uniquely determined by the choice of the mapping function m, we have to consider all possible mappings. Obviously, there exist 2^k possibilities (including parents) leading to as much possible offspring. Since evaluating one offspring from scratch costs $\Theta(n^2)$, a naive approach to compute the fitness values of these possible offspring has a complexity of $\Theta(2^k \cdot n^2)$, which can be prohibitive. In the next section, we show how to reduce it.

3.2 Decomposition of the Evaluation Function

Following the previous notations, let $m : \{1, \ldots, k\} \mapsto \{1, 2\}$ be an arbitrary mapping function, and let us define for every ℓ and ℓ' in $\{1, \ldots, k\}$:

$$Q_{\ell\ell'} = \sum_{h \in U(C_{\ell'})} \sum_{i \in U(C_\ell)} d_{\pi_{m(\ell')}(h)\pi_{m(\ell)}(i)} f_{hi} \;;\; Q_\ell = \sum_{h \in \widetilde{I}} \sum_{i \in U(C_\ell)} d_{\pi_1(h)\pi_{m(\ell)}(i)} f_{hi}$$

$$R = \sum_{h \in \widetilde{I}} \sum_{i \in \widetilde{I}} d_{\pi_1(h)\pi_1(i)} f_{hi} \;;\; Q'_\ell = \sum_{i \in \widetilde{I}} \sum_{h \in U(C_\ell)} d_{\pi_{m(\ell)}(h)\pi_1(i)} f_{hi}$$

where $Q_{\ell\ell'}$ represents the contribution of facilities in C_ℓ w.r.t. the facilities in another connected component $C_{\ell'}$. Similarly, Q_ℓ (resp. Q'_ℓ) represents the contribution of facilities in C_ℓ w.r.t. facilities having the same locations in both parents π_1 and π_2. Finally, R represents the pairwise contribution of the facilities that have the same locations in π_1 and π_2. The following proposition shows that the QAP objective function can be decomposed using these contributions.

Proposition 2. *Let $\overset{\otimes}{\pi}$ be a permutation offspring as defined in Proposition 1. Then,*

$$f(\overset{\otimes}{\pi}) = R + \sum_{\ell=1}^{k} \left(Q_\ell + Q'_\ell + \sum_{\ell'=1}^{k} Q_{\ell\ell'} \right) \tag{2}$$

Proof. By decomposing the facilities w.r.t. $\widetilde{I}$, $\overline{I}$, we get:

$$\begin{aligned} f(\overset{\otimes}{\pi}) &= \sum_{h \in I} \sum_{i \in I} d_{\overset{\otimes}{\pi}(h)\overset{\otimes}{\pi}(i)} f_{hi} = \sum_{h \in I} \sum_{i \in \overline{I}} d_{\overset{\otimes}{\pi}(h)\overset{\otimes}{\pi}(i)} f_{hi} + \sum_{h \in I} \sum_{i \in \widetilde{I}} d_{\overset{\otimes}{\pi}(h)\overset{\otimes}{\pi}(i)} f_{hi} \\ &= \sum_{h \in \overline{I}} \sum_{\ell=1}^{k} \sum_{i \in U(C_\ell)} d_{\overset{\otimes}{\pi}(h)\overset{\otimes}{\pi}(i)} f_{hi} + \sum_{h \in \widetilde{I}} \sum_{\ell=1}^{k} \sum_{i \in U(C_\ell)} d_{\overset{\otimes}{\pi}(h)\overset{\otimes}{\pi}(i)} f_{hi} \end{aligned}$$

$$
\begin{aligned}
&+ \sum_{h\in\overline{I}}\sum_{i\in\widetilde{I}} d_{\overset{\otimes}{\pi}(h)\overset{\otimes}{\pi}(i)}\, f_{hi} + \sum_{h\in\widetilde{I}}\sum_{i\in\widetilde{I}} d_{\overset{\otimes}{\pi}(h)\overset{\otimes}{\pi}(i)}\, f_{hi} \\
= &\sum_{\ell'=1}^{k}\sum_{h\in U(C_{\ell'})}\sum_{\ell=1}^{k}\sum_{i\in U(C_\ell)} d_{\overset{\otimes}{\pi}(h)\overset{\otimes}{\pi}(i)}\, f_{hi} + \sum_{h\in\widetilde{I}}\sum_{\ell=1}^{k}\sum_{i\in U(C_\ell)} d_{\overset{\otimes}{\pi}(h)\overset{\otimes}{\pi}(i)}\, f_{hi} \\
&+ \sum_{\ell'=1}^{k}\sum_{h\in U(C_{\ell'})}\sum_{i\in\widetilde{I}} d_{\overset{\otimes}{\pi}(h)\overset{\otimes}{\pi}(i)}\, f_{hi} + \sum_{h\in\widetilde{I}}\sum_{i\in\widetilde{I}} d_{\pi_1(h)\pi_1(i)}\, f_{hi} \\
= &\sum_{\ell'=1}^{k}\sum_{\ell=1}^{k}\sum_{h\in U(C_{\ell'})}\sum_{i\in U(C_\ell)} d_{\pi_{m(\ell')}(h)\pi_{m(l)}(i)}\, f_{hi} \\
&+ \sum_{\ell=1}^{k}\sum_{h\in\widetilde{I}}\sum_{i\in U(C_\ell)} d_{\pi_1(h)\pi_{m(\ell)}(i)}\, f_{hi} + \sum_{\ell'=1}^{k}\sum_{i\in\widetilde{I}}\sum_{h\in U(C_{\ell'})} d_{\pi_{m(\ell')}(h)\pi_1(i)}\, f_{hi} + R \\
= &\sum_{\ell'=1}^{k}\sum_{\ell=1}^{k} Q_{\ell\ell'} + \sum_{\ell=1}^{k} Q_\ell + \sum_{\ell'=1}^{k} Q'_{\ell'} + R
\end{aligned}
$$

□

As a result, we obtain the following corollary which follows from the fact that the contributions appearing in the decomposition of Proposition 2 can *only* have a *constant* number of values, for all possible choices of the mapping m.

Corollary 1. *The whole set of offspring that can be generated by Proposition 1 can be explored and evaluated in $O(n^2 + k^2 \cdot 2^k)$ time.*

Proof. Let us first notice that computing the components of the bipartite graph can be done in $\Theta(n)$ time. Over all the possible mapping functions m, $(m(\ell), m(\ell'))$ can only take 4 different values, namely, $(1,1)$, $(1,2)$, $(2,1)$ and $(2,2)$. Hence, for every *fixed* values of ℓ and ℓ', $Q_{\ell\ell'}$ can only take 4 possible values. These four values depend solely on π_1 and π_2. Therefore, there can only be $\Theta(k^2)$ possible values for $Q_{\ell\ell'}$ over all ℓ and ℓ', and all possible choices of m. All of these $\Theta(k^2)$ values can be precomputed by a simple preprocessing step. Since by definition C_ℓ and $C_{\ell'}$ do not share any nodes for every $\ell \neq \ell'$, this preprocessing step takes obviously $O(n^2)$ time. Similarly, there are only $\Theta(k)$ possible values for Q_ℓ and Q'_ℓ over all possible values of ℓ, and all possible choices of m. Hence, they can also be precomputed in $\Theta(n^2)$ time. Finally, R does not depend on m, and can also be precomputed in $O(n^2)$ time. To summarize, all possible values taken by $Q_{\ell\ell'}$, Q_ℓ, $Q_{\ell'}$, and R can be precomputed in $\Theta(n^2)$ time and stored in $\Theta(k^2)$ memory space before even any specific choice of the mapping function m is made.

Now, let us consider a specific choice for the mapping function m, which fully determines an offspring permutation $\overset{\otimes}{\pi}$ according to Proposition 1. Then, according to Proposition 2, and given the contributions $Q_{\ell\ell'}$, Q_ℓ, $Q_{\ell'}$, and R were already precomputed by the previous discussion, it takes $\Theta(k^2)$ time to compute $f(\overset{\otimes}{\pi})$. The corollary follows since there are 2^k possible mapping functions m. □

Experiment 1. Pseudo-code of the first experimental scenario

1. Let π_1 and π_2 be either (i) two random solutions or (ii) two local optima;
2. Apply crossover on π_1 and π_2 to obtain an offspring $\overset{\otimes}{\pi}$;
3. Apply a local search with $\overset{\otimes}{\pi}$ as initial solution to obtain $\overset{\otimes}{\pi}'$;

4 Experimental Analysis

In the rest of the paper, we provide an empirical analysis of the designed partition crossover, denoted by Px.

4.1 Experimental Setup

We consider the following 8 QAP instances from the literature[1] [14,15]: tai343e0i with $i \in \{0, \dots, 7\}$. They have been selected due to their challenging size of $n = 343$ facilities and locations, which is rarely addressed in the literature. We consider the following scenarios.

Scenario #1. The goal of this scenario is the study the relative ability of the Px crossover to find an improving offspring. As depicted in the high-level template of Experiment 1, we consider two settings where the initial parents are either (i) random solutions, or (ii) local optima. For the latter case, we run a basic hill climbing local search using the standard swap neighborhood to construct the initial local optima. Starting from a random permutation, the best swap move is performed until no improvement is possible. The local search is executed as much times as needed to find as much different local optima as needed. All competing crossovers are applied 100 times using 100 pairs of different initial parents. Once an offspring has been generated by crossover, we also consider to check if it is a local optimum w.r.t. the swap neighborhood by running the local search again.

Scenario #2. The goal of this scenario is to study the relative performance of the Px crossover when plugged into a simple evolutionary algorithm. As depicted in the high level template of Experiment 2, we consider a hybrid evolutionary algorithm embedding the swap-based local search discussed previously. In each generation, a new offspring is generated by performing crossover followed by a local search with probability p. We consider a simple random parent selection and a non-elitist replacement where the newly generated offspring replaces the oldest individual. The value of p is chosen in the set $\{0, 0.05, 0.2\}$. This allows us to study the relative behavior of crossover using a variable amount of local search, ranging from no local search at all ($p = 0$), to a small ($p = 0.05$) and a high ($p = 0.2$) amount. For each configuration, 10 independent runs are performed with a maximum number of generations $G = 1\,000$ and a population size of n.

[1] http://mistic.heig-vd.ch/taillard/problemes.dir/qap.dir/qap.html.

Experiment 2. Pseudo-code of the second experimental scenario

```
  Input: G: maximum number of generations; p ∈ [0,1]: local search ratio;
1 Generate an initial random population P;
2 for g = 1 to G do
3 |   Apply crossover on two randomly selected parents π1 and π2 to obtain π⊗;
4 |   if p < rand(0,1) then
5 |   |   Apply local search with π⊗ as initial solution to obtain π⊗' and let π⊗ = π⊗';
6 |   Replace the oldest individual of the population with π⊗;
```

CPU Running Time. Let us finally notice that we manage to analyze the CPU execution time. It is hence important to recall that the implementations of all algorithms were optimized as much as possible. In particular, computing the best move in a swap-based local search can be performed in a very efficient manner for the QAP. In fact, this can be done in an incremental manner on the basis of the current solution [14,15]. For the sake of fairness when analyzing execution time, this well-established and important consideration is carefully implemented in all experiments. Moreover, when running the Px crossover, we restrict the maximum number of explored offspring to at most 2^{15}, which means that when the number of connected components $k > 15$, not all possible offspring from Proposition 1 are explored. All algorithms are implemented in C++ and run on an Intel Xeon(R) CPU E3-1505M v6 3.00 GHz.

Let us finally notice that more advanced settings including other QAP instances and algorithms, as well as finely tuned operators/components using automated algorithm configuration methods, etc, are left for future investigations.

4.2 Experimental Analysis and Results

We start our analysis by reporting our findings from Experiment 1 in Fig. 3.

Crossover Improvement Ratio. In Fig. 3a, we can see that the Px crossover has a significantly higher improvement rate. Using random parents, it can produce an improving offspring in almost 100% of the cases, whereas all other considered crossovers have a ratio of around 28%. Using local optimal parents, the Px is still able to find improving offspring in around 30% of the cases. This is to contrast with the other crossovers that fail in almost all cases. This first set of observations shows that the Px crossover has a relatively high intensification ability since it is even able to improve over local optima.

Local Optimality. In Fig. 3b, we further show the number of moves performed by the local search initialized with the generated offspring (in line 3), i.e., 0 moves means that the offspring is a local optima w.r.t. the swap neighborhood. For all crossovers except Px, the local search improves the constructed offspring independently of using random or local optima parents. This means that these

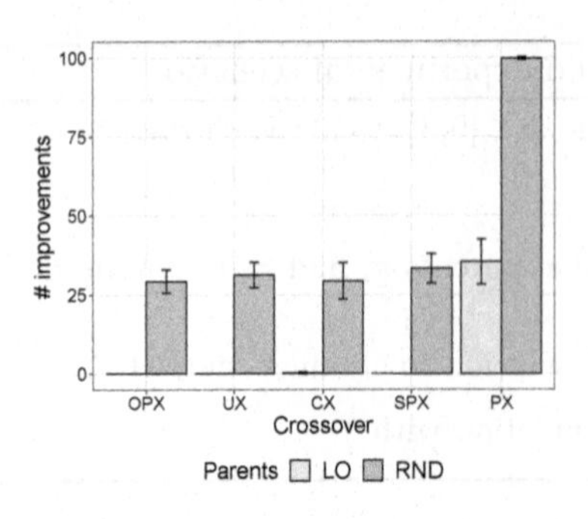

(a) Number of improving offspring

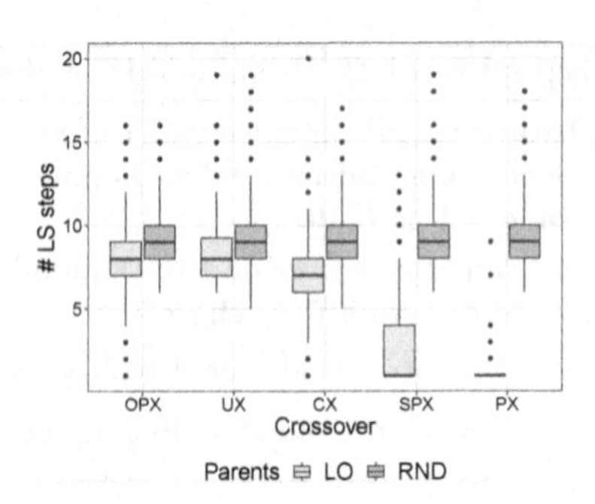

(b) Number of local search moves

Fig. 3. Results from Experiment 1. Results are over all eight instances.

crossovers are more diversification-oriented, since even when using local optimal parents, they are likely to produce an inferior offspring that can be improved by local search. Hence, we can think about these crossovers as acting in a perturbative manner. The situation is completely different for the PX crossover. On one side, with random parents, where the offspring improvement ratio was found to be almost 100%, the local search can still find improvements. On the other side, with local optimal parents, where the offspring improvement ratio is about 30%, the local search cannot find improvements, which indicates that the produced offspring is also likely to be a local optimum. This means that the PX crossover has some ability to act as a tunneling operator allowing to jump from two local optima to a new improving local optimum (with around 30% success rate).

Solution Quality *vs* Execution Time. In the following, we report our findings from Experiment 2. In Table 1, we rank the different algorithms according to: (i) the quality of the best solution that the evolutionary algorithm is able to find during its execution, and (ii) the total CPU execution time. The situation is clearly different depending on whether a local search is used or not.

When performing only crossover with no local search ($p = 0$), the PX crossover is able to find substantially better solutions, followed by SPX, and then by OPX, UX and CX, which do not show any significant difference statistically. Without surprise, these crossovers run however faster than PX and SPX. Notice here the extremely high cost of SPX. Interestingly, the proposed PX crossover implies a relatively reasonable increase in terms of running time (about 3.5 s) compared against SPX (about 287 s).

When performing crossover with local search ($p \in \{0.05, 0.2\}$), we can first see that, compared to not using local search at all, all variants can find much better solutions, while having a higher execution time. Interestingly, the PX and SPX crossovers (which were previously found to perform better than OPX, UX, and CX) now provide slightly worst solutions. On the other side, when analyzing the execution time, we found that using the PX crossover is significantly faster than all other variants, with at least 25% CPU time gain; e.g., for

Table 1. Ranks of crossovers (a lower rank is better). The first (resp. second) part is with respect to solution quality (resp. execution time). For each value of parameter p (local search ratio), a rank c indicates that the corresponding crossover was found to be significantly outperformed by c other ones w.r.t. a Wilcoxon statistical test at a significance level of 0.05. Rows are w.r.t. instances.

	$p=0$					$p=0.05$					$p=0.2$				
#Ins	OPX	UX	CX	SPX	PX	OPX	UX	CX	SPX	PX	OPX	UX	CX	SPX	PX
Mean average deviation to the best (in subscript)															
1	2_{810}	2_{809}	2_{808}	1_{796}	0_{710}	$0_{0.38}$	$0_{0.39}$	$0_{0.37}$	$0_{0.37}$	$0_{0.37}$	$0_{0.26}$	$0_{0.24}$	$0_{0.30}$	$1_{0.32}$	$1_{0.34}$
2	2_{105}	2_{105}	2_{105}	1_{103}	0_{94}	$0_{0.04}$	$0_{0.04}$	$0_{0.03}$	$0_{0.04}$	$1_{0.05}$	$0_{0.02}$	$0_{0.03}$	$0_{0.02}$	$0_{0.03}$	$0_{0.03}$
3	2_{115}	2_{115}	2_{115}	1_{113}	0_{102}	$0_{0.04}$	$0_{0.03}$	$0_{0.03}$	$0_{0.04}$	$1_{0.05}$	$0_{0.02}$	$0_{0.02}$	$0_{0.02}$	$3_{0.03}$	$3_{0.03}$
4	2_{96}	2_{96}	2_{96}	1_{95}	0_{84}	$0_{0.04}$	$0_{0.04}$	$0_{0.04}$	$0_{0.04}$	$1_{0.05}$	$0_{0.03}$	$0_{0.02}$	$0_{0.02}$	$1_{0.03}$	$2_{0.03}$
5	2_{107}	2_{107}	2_{108}	1_{106}	0_{96}	$0_{0.04}$	$0_{0.03}$	$0_{0.03}$	$0_{0.04}$	$1_{0.04}$	$0_{0.02}$	$0_{0.02}$	$0_{0.02}$	$1_{0.03}$	$2_{0.03}$
6	2_{123}	2_{123}	2_{124}	1_{121}	0_{110}	$0_{0.06}$	$0_{0.05}$	$0_{0.05}$	$0_{0.06}$	$1_{0.06}$	$0_{0.04}$	$0_{0.04}$	$0_{0.03}$	$1_{0.05}$	$1_{0.05}$
7	2_{115}	2_{115}	2_{115}	1_{113}	0_{101}	$0_{0.04}$	$0_{0.04}$	$0_{0.04}$	$0_{0.05}$	$0_{0.04}$	$0_{0.03}$	$0_{0.03}$	$0_{0.04}$	$0_{0.04}$	$0_{0.03}$
8	2_{110}	2_{110}	2_{110}	1_{108}	0_{96}	$0_{0.04}$	$0_{0.03}$	$0_{0.03}$	$0_{0.03}$	$1_{0.04}$	$0_{0.02}$	$0_{0.02}$	$0_{0.01}$	$3_{0.03}$	$3_{0.03}$
Mean CPU execution time (in subscript)															
1	$0_{0.61}$	$1_{0.62}$	$2_{0.75}$	4_{286}	$3_{3.39}$	0_{194}	0_{198}	0_{196}	4_{478}	0_{185}	1_{705}	1_{714}	1_{676}	4_{870}	0_{502}
2	$0_{0.61}$	$1_{0.62}$	$2_{0.72}$	4_{287}	$3_{3.44}$	0_{202}	0_{202}	0_{199}	4_{475}	0_{187}	2_{698}	1_{697}	1_{662}	4_{857}	0_{494}
3	$0_{0.60}$	$1_{0.63}$	$2_{0.70}$	4_{286}	$3_{3.38}$	0_{210}	0_{208}	0_{206}	4_{481}	0_{195}	1_{712}	1_{730}	1_{682}	4_{876}	0_{510}
4	$0_{0.60}$	$1_{0.62}$	$2_{0.75}$	4_{287}	$3_{3.41}$	0_{193}	1_{198}	0_{194}	4_{468}	0_{175}	1_{724}	1_{728}	1_{693}	4_{880}	0_{513}
5	$0_{0.60}$	$1_{0.62}$	$2_{0.74}$	4_{287}	$3_{3.39}$	0_{189}	0_{187}	0_{183}	4_{463}	0_{172}	1_{681}	2_{684}	1_{643}	4_{842}	0_{488}
6	$0_{0.61}$	$1_{0.62}$	$2_{0.72}$	4_{286}	$3_{3.39}$	0_{209}	0_{212}	0_{206}	4_{481}	0_{193}	1_{714}	1_{725}	1_{690}	4_{876}	0_{515}
7	$0_{0.61}$	$0_{0.62}$	$2_{0.69}$	4_{286}	$3_{3.37}$	0_{207}	0_{206}	0_{205}	4_{477}	0_{193}	1_{697}	2_{701}	1_{670}	4_{875}	0_{505}
8	$0_{0.61}$	$1_{0.62}$	$2_{0.70}$	4_{287}	$3_{3.40}$	0_{221}	0_{223}	0_{219}	4_{494}	0_{200}	1_{715}	1_{734}	1_{693}	4_{890}	0_{521}

$p = 0.2$, the average execution time over all configurations is of 506 s with PX, against 677, 706, 714, and 871 with PX, CX, OPX, UX, and SPX, respectively. As commented before, knowing that our swap-based hill climbing local search was carefully implemented using a state-of-the-art fast incremental evaluation procedure for finding the best move [14,15], such an observation might be surprising at first sight. However, it can be explained from two perspectives. Firstly, due to Corollary 1, performing the PX crossover is reasonably fast. Secondly, the PX crossover was previously found to have a relatively high intensification power. Thus, it is likely to produce a high-quality offspring, eventually being a local optima. Hence, it is more likely that the local search stops more quickly when attempting to improve the produced offspring. In this case, the cost of the local search is also reduced, hence leading to a decrease in the overall CPU time.

Finally, the previous results about solution quality and execution time are found to hold at any generation, independently of the configuration. This is illustrated in Fig. 4 rendering the convergence profile and the CPU execution time as a function of generations for the first QAP instance. This also confirms that the PX crossover is more intensification-oriented, and should be complemented by other diversification mechanisms when effectively integrated into more advanced evolutionary search processes.

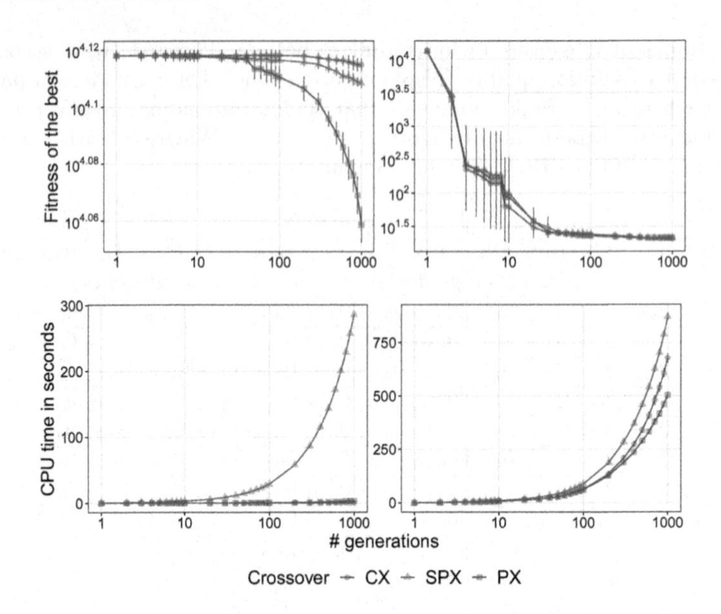

Fig. 4. Experiment 2 on first instance. Top: convergence profile. Bottom: execution time. Left: $p = 0$ (no local search). Right: $p = 0.2$.

QAP Connected Components. To complement our analysis, we provide further observations on the characteristics of the QAP and the relevance of the Px crossover. More precisely, remember that given k connected components implied by the unshared facility assignments $\overline{I}$, the Px crossover is able to provide the best over 2^k possible offspring. In Fig. 5, we report the average value of k (Left), as well the percentage of unshared facility assignments (Right), i.e., $\left|\overline{I}\right| / n$ in %, over all generations and all considered Px runs. We can see that for all considered QAP instances, solutions contain very few shared facility assignments, i.e., less than 1.5%, and the value of k stays relatively low, i.e., 6 in average with the exception of a few outliers exceeding 10. This suggests that the exploration power of the Px crossover can be improved in different ways, since its time complexity guarantees (Corollary 1) should be able to support the fast evaluation of much more offspring solutions. For instance, it would be interesting to investigate the splitting of existing connected components into smaller ones at the aim of processing, and hopefully finding, more improving offspring.

5 Conclusion

In this paper, we presented our first investigations on the design of a partition crossover for the QAP. The proposed recombination and evaluation process is proved to provide a reasonable trade-off between the intensification power and the running time complexity. Our empirical study provides first insights towards the design and integration of more powerful partition crossovers for the QAP. In

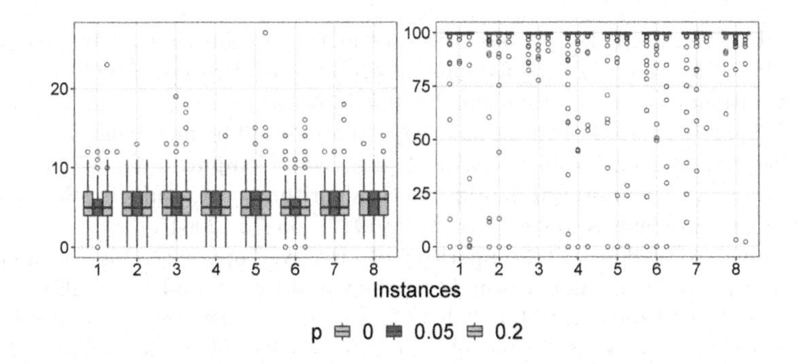

Fig. 5. Left: Number of connected components (k). Right: $|\overline{I}|/n$ in %.

particular, a future challenging issue is to investigate complementary decomposition techniques allowing to break the QAP bipartite graph into further smaller connected components, while maintaining a reasonable recombination and evaluation cost. In this respect, a promising idea would be to investigate the knowledge about the flow and distance values in order to identify the critical facilities and locations when performing decomposition. The challenge is then to guide the recombination process by identifying the most promising components to consider while keeping evaluation cost as low as possible.

Acknowledgments. The three first authors are supported by the French national research agency (ANR-16-CE23-0013-01) and the Research Grants Council of Hong Kong (RGC Project No. A-CityU101/16).

References

1. Ahuja, R.K., Orlin, J.B., Tiwari, A.: A greedy genetic algorithm for the quadratic assignment problem. Comput. Oper. Res. **27**(10), 917–934 (2000)
2. Chicano, F., Ochoa, G., Whitley, D., Tinós, R.: Enhancing partition crossover with articulation points analysis. In: Proceedings of the Genetic and Evolutionary Computation Conference, pp. 269–276 (2018)
3. Glover, F.: Genetic algorithms and scatter search: unsuspected potentials. Stat. Comput. **4**(2), 131–140 (1994)
4. Koopmans, T.C., Beckmann, M.: Assignment problems and the location of economic activities. Econometrica **25**(1), 53–76 (1957)
5. Larrañaga, P., Kuijpers, C.M.H., Murga, R.H., Inza, I., Dizdarevic, S.: Genetic algorithms for the travelling salesman problem: a review of representations and operators. Artif. Intell. Rev. **13**(2), 129–170 (1999)
6. Lim, M.H., Yuan, Y., Omatu, S.: Efficient genetic algorithms using simple genes exchange local search policy for the quadratic assignment problem. Comput. Optim. Appl. **15**(3), 249–268 (2000)
7. Loiola, E.M., Maia, N.M., Boaventura-Netto, P.O., Hahn, P., Querido, T.: A survey for the quadratic assignment problem. Eur. J. Oper. Res. **176**(2), 657–690 (2007)

8. Merz, P., Freisleben, B.: A comparison of memetic algorithms, tabu search, and ant colonies for the quadratic assignment problem. In: Proceedings of the Congress on Evolutionary Computation, pp. 2063–2070 (1999)
9. Merz, P., Freisleben, B.: Fitness landscape analysis and memetic algorithms for the quadratic assignment problem. IEEE Trans. Evol. Comput. **4**(4), 337–352 (2000)
10. Misevicius, A.: An improved hybrid genetic algorithm: new results for the quadratic assignment problem. Knowl.-Based Syst. **17**(2), 65–73 (2004)
11. Misevicius, A., Kilda, B.: Comparison of crossover operators for the quadratic assignment problem. Information Technology and Control **34**(2), (2005)
12. Oliver, I.M., Smith, D.J., Holland, J.R.C.: A study of permutation crossover operators on the traveling salesman problem. In: Proceedings of the Second International Conference on Genetic Algorithms on Genetic Algorithms and their Application, pp. 224–230 (1987)
13. Sahni, S., Gonzalez, T.: P-complete approximation problems. J. ACM **23**(3), 555–565 (1976)
14. Taillard, E.: Robust taboo search for the quadratic assignment problem. Parallel Comput. **17**(4), 443–455 (1991)
15. Taillard, E.: Comparison of iterative searches for the quadratic assignment problem. Location Sci. **3**(2), 87–105 (1995)
16. Tate, D.M., Smith, A.E.: A genetic approach to the quadratic assignment problem. Comput. Oper. Res. **22**(1), 73–83 (1995)
17. Tinós, R., Whitley, D., Chicano, F.: Partition crossover for pseudo-boolean optimization. In: Proceedings of the ACM Conference on Foundations of Genetic Algorithms XIII, pp. 137–149 (2015)
18. Tinós, R., Whitley, D., Ochoa, G.: Generalized asymmetric partition crossover (GAPX) for the asymmetric TSP. In: Proceedings of the 2014 Annual Conference on Genetic and Evolutionary Computation, pp. 501–508 (2014)
19. Vázquez, M., Whitley, L.D.: A hybrid genetic algorithm for the quadratic assignment problem. In: Proceedings of the 2nd Annual Conference on Genetic and Evolutionary Computation, San Francisco, CA, USA, pp. 135–142 (2000)

A Permutational Boltzmann Machine with Parallel Tempering for Solving Combinatorial Optimization Problems

Mohammad Bagherbeik[1(✉)], Parastoo Ashtari[1], Seyed Farzad Mousavi[1], Kouichi Kanda[2], Hirotaka Tamura[2], and Ali Sheikholeslami[1]

[1] University of Toronto, Toronto, ON M5S2E8, Canada
tabrizimo73@gmail.com
[2] Fujitsu Laboratories Limited, Kawasaki, Kanagawa 211-8588, Japan

Abstract. Boltzmann Machines are recurrent neural networks that have been used extensively in combinatorial optimization due to their simplicity and ease of parallelization. This paper introduces the Permutational Boltzmann Machine, a neural network capable of solving permutation optimization problems. We implement this network in combination with a Parallel Tempering algorithm with varying degrees of parallelism ranging from a single-thread variant to a multi-threaded system using a 64-core CPU with SIMD instructions. We benchmark the performance of this new system on Quadratic Assignment Problems, using some of the most difficult known instances, and show that our parallel system performs in excess of 100× faster than any known dedicated solver, including those implemented on CPU clusters, GPUs, and FPGAs.

Keywords: Parallel Boltzmann Machine · Replica exchange Monte-Carlo · Combinatorial optimization · Quadratic Assignment Problem

1 Introduction

Boltzmann Machines (BM), first proposed by Hinton in 1984 [13], are recurrent, fully connected, neural networks that store information within their symmetric edge weights. When combined with Stochastic Local Search methods such as Simulated Annealing (SA) [1] or Parallel Tempering (PT) [10], BMs can be used to perform combinatorial optimization on complex problems such as TSP [3], MaxSAT [8], and MaxCut [16]. In this paper, we present an algorithm for a Permutational Boltzmann Machine (PBM), structured to solve complex, integer based, permutation optimization problems. We combine this PBM with Parallel Tempering and propose both single-threaded and multi-threaded, software implementations of this PBM + PT system using a 64-core CPU along

The authors would like to thank Fujitsu Laboratories Ltd. and Fujitsu Consulting (Canada) Inc. for providing financial support and technical expertise on this research.

T. Bäck et al. (Eds.): PPSN 2020, LNCS 12269, pp. 317–331, 2020.
https://doi.org/10.1007/978-3-030-58112-1_22

with SIMD instructions. As a proof-of-concept, we show how to solve Quadratic Assignment Problems (QAP) [17] using a PBM and present experimental results on some of the hardest QAP instances from QAPLIB [6], Palubeckis [21], and Drezner [9]. We then show that, over the tested instances, our single-threaded and multi-threaded PBM systems can find the best-known-solutions of QAP problems in excess of 10× and 100× faster than the next best solver respectively.

The rest of this paper is organized as follows: Sect. 2 provides background on BMs and the formulation of QAP problems. Section 3 presents the structure of our Permutational Boltzmann Machine and Sect. 4 presents our single and multi-threaded implementations of a PBM + PT system on a multi-core CPU. Section 5 outlines the experiments conducted to benchmark the performance of our PBM + PT system and presents our results. Section 6 concludes this paper.

2 Background

2.1 Boltzmann Machines

BMs, as shown in Fig. 1, are made up of N neurons, $\{x_1, x_2, \ldots, x_N\}$ with binary states represented by vector $\mathbf{S} = [s_1\, s_2\, \ldots\, s_N]^\intercal \in \{0,1\}^N$. Each neuron, x_i, is connected to other neurons, x_j, via symmetric, real-valued weights, $w_{i,j} \in \mathbb{R}$ where $w_{i,j} = w_{j,i}$ and $w_{i,i} = 0$, forming a 2D matrix, $\mathbf{W} \in \mathbb{R}^{N\times N}$. Each neuron also has a bias value, b_i, which forms $\mathbf{B} \in \mathbb{R}^{N\times 1}$. The cumulative inputs to the neurons, also referred to as their *local fields*, h_i, form $\mathbf{H} \in \mathbb{R}^{N\times 1}$ and are calculated using (1).

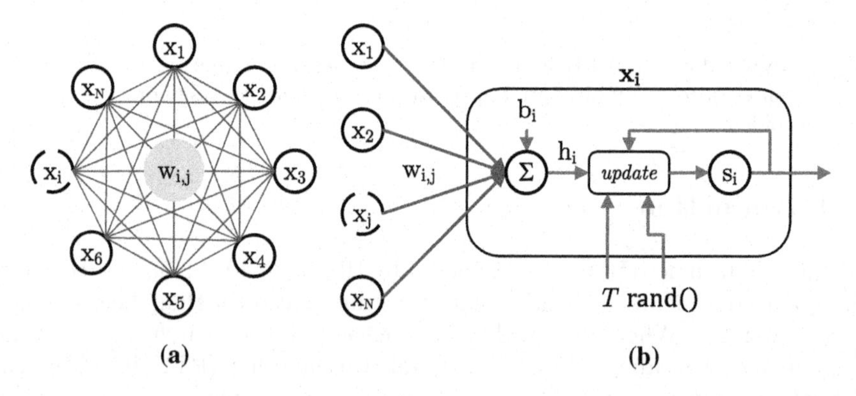

Fig. 1. Structure of a Boltzmann Machine and its neurons. (a) Top level structure of a Boltzmann Machine (b) Detailed structure of a BM neuron, where T is the system temperature and rand() is a uniform random number within [0,1]

$$h_i(\mathbf{S}) = \sum_{j=1}^{N} w_{i,j} s_j + b_i \quad , \quad \mathbf{H}(\mathbf{S}) = \mathbf{W}\mathbf{S} + \mathbf{B} \tag{1}$$

$$E(\mathbf{S}) = -\frac{1}{2}\sum_{i=1}^{N}\sum_{j=1}^{N} w_{i,j}s_is_j - \sum_{i=1}^{N} b_is_i = -\frac{\mathbf{S}^\intercal\mathbf{W}\mathbf{S}}{2} - \mathbf{S}^\intercal\mathbf{B} \tag{2}$$

$$P(\mathbf{S}) = \frac{\exp(-E(\mathbf{S})/T)}{\sum_{\forall \mathbf{S}_j \in \{0,1\}^N} \exp(-E(\mathbf{S}_j)/T)} \tag{3}$$

Each possible state of a BM has an associated *energy* term calculated via (2). The probability of the system being in any state depends on the energy of that state as shown in (3). The lower the energy, the higher the probability that the network will be in that state. BMs create an energy landscape for a problem through the weights that connect their neurons where the state(s) with the lowest energy corresponds to a valid solution. The procedures to convert various optimization problems to the BM format are discussed in [12]. The term T in (3), known as the system *temperature*, flattens or sharpens the energy landscape when increased or decreased respectively, providing a method to maneuver the landscape when searching for the global minimum.

Generally, BMs are combined with Simulated Annealing (SA) to solve optimization problems. Using SA, at a time-step t, where the system is in state $\mathbf{S}^\mathbf{t}$ with temperature T, the local fields $\mathbf{H}(\mathbf{S}^\mathbf{t})$ are calculated using (1). In order to make an update to the system, we must conduct a *trial*. First, a neuron x_i is randomly chosen and the change in energy as a result of flipping its state is calculated via (4). Next, the probability of flipping the neuron's state, P_{move}, is calculated via (5) and is compared against a uniformly distributed random number in [0,1] to determine the change in the neuron's state using (6).

$$\Delta E(\mathbf{S}^\mathbf{t}, i) = \Delta E_{s_i \to !s_i}(\mathbf{S}^\mathbf{t}) = -[1 - 2s_i^\mathbf{t}]h_i(\mathbf{S}^\mathbf{t}) \tag{4}$$

$$P_{move} = min\{1, \exp(-\Delta E/T)\} \tag{5}$$

$$\Delta s_i = \begin{cases} [1 - 2s_i^\mathbf{t}] & \text{if } P_{move} \geq rand() \\ 0 & otherwise \end{cases} \tag{6}$$

After the trial, the system state variables s_i, $\mathbf{H}$, and E need to be updated as shown in (7), (8), and (9) respectively, where $\mathbf{W}_{i,*}$ and $\mathbf{W}_{*,i}$ represent row and column i of $\mathbf{W}$ respectively.

$$s_i^{\mathbf{t}+1} = s_i^\mathbf{t} + \Delta s_i \tag{7}$$

$$\mathbf{H}(\mathbf{S}^{\mathbf{t}+1}) = \mathbf{H}(\mathbf{S}^\mathbf{t}) + \Delta s_i \mathbf{W}_{*,i} \tag{8}$$

$$E(\mathbf{S}^{\mathbf{t}+1}) = E(\mathbf{S}^\mathbf{t}) + \Delta E(\mathbf{S}^\mathbf{t}, i) \tag{9}$$

This procedure is repeated a preset number of times, occasionally *cooling* the system by decreasing T until it goes below a certain threshold, T_{thresh}, at which point the process is terminated and the lowest energy state observed throughout the search is returned. This state may or may not correspond to a valid or optimal solution due to the stochastic nature of the algorithm but, theoretically, if given a long enough cooling schedule, the BM + SA system will eventually converge to an optimal answer [2].

2.2 Quadratic Assignment Problems (QAP)

QAP problems, first formulated in [17], are a class of NP-Hard permutation optimization problems to which many other problems such as the Travelling Salesman Problem can be reduced. While the formulation is relatively simple, QAP remains, to this day, one of the more challenging combinatorial optimization problems. QAP problems entail the task of assigning a set of n facilities to a set of n locations while minimizing the cost of the assignment. QAP problems are comprised of $n \times n$ matrices $\mathbf{F} = (f_{i,j})$ and $\mathbf{D} = (d_{k,l})$ which describe the flows between facilities and distances between locations respectively with the diagonal elements of both matrices being 0. A third $n \times n$ matrix $\mathbf{B_P} = (b_{i,k})$, describes the costs of assigning a facility to a location. All three matrices are comprised of real-valued elements. Given these matrices, each facility must be assigned to a unique location, generating a permutation, $\boldsymbol{\phi} \in \mathbf{S_n}$, where $\mathbf{S_n}$ is the set of all permutations, such that the cost function (10) is minimized.

$$\min_{\boldsymbol{\phi} \in \mathbf{S_n}} cost(\boldsymbol{\phi}) = \min_{\boldsymbol{\phi} \in \mathbf{S_n}} \sum_{i=1}^{n} \sum_{j=1}^{n} f_{i,j} d_{\phi_i,\phi_j} + \sum_{i=1}^{n} b_{i,\phi_i} \tag{10}$$

Generally, there are two variants of the QAP problem: symmetric (*sym*) and asymmetric (*asm*). In the symmetric case, either one or both of $\mathbf{F}$ and $\mathbf{D}$ are symmetric. If one of the matrices is asymmetric, it can be made symmetric by taking the average of an element and its complement. However, if both matrices are asymmetric, we can no longer symmetrize them in this manner. It is important to distinguish between these two cases as they are handled differently by a PBM, as will be shown in Sect. 3.

3 Permutational Boltzmann Machines (PBM)

3.1 Structure and Update Scheme

The PBM's structure is an extension of Clustered Boltzmann Machines (CBM), first proposed by De Gloria [11]. A CBM places neurons that do not have any connections between them into groups called *clusters*. Within a cluster, the states of the neurons have no effect on each other's local fields; simultaneously flipping the states of multiple neurons in the same cluster has the same effect as flipping them in sequence. In a PBM, the neurons are arranged into an $n \times n$ matrix $\mathbf{S_P} = (s_{r,c})$, where each row, r_i, and each column, c_j, forms a cluster, as shown in Fig. 2a. On each cluster, we impose an exactly-1 constraint to ensure that within each row and each column, there is exactly one neuron in the *ON* state.

In the context of a permutation problem, the row-clusters represent a 1-hot encoded integer in [1, n], allowing the neuron states to be represented via the integer permutation vector, ϕ. The column-clusters, in turn, enforce that every integer is unique. The $n^2 \times n^2$ weight matrix is also reshaped into a 4D $(n \times n) \times (n \times n)$ matrix as shown in (11), allowing the generation of the $\mathbf{w}_{r,c}$ sub-matrices via Kronecker Products (denoted by $\otimes$) of rows and columns of $\mathbf{F}$ and $\mathbf{D}$ via (12). The $n \times n$ local field matrix $\mathbf{H_P}$ is calculated via (13).

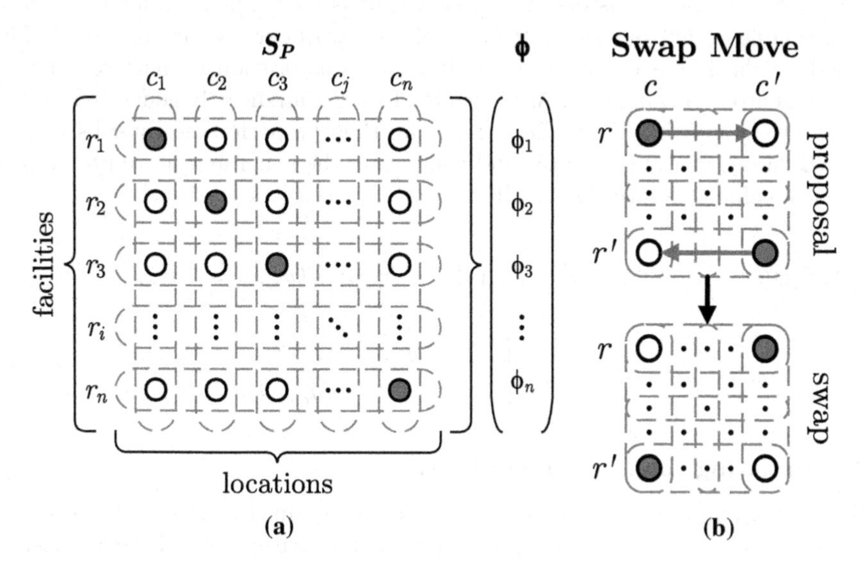

Fig. 2. Structure of a Permutational Boltzmann Machine. (a) The binary neuron state matrix $\mathbf{S_P}$ with row/column cluster structures and the permutation vector ϕ (b) Structure of a permutation Swap Move

$$\mathbf{W_P} = \begin{bmatrix} \mathbf{w}_{1,1} & \mathbf{w}_{1,2} & \cdots & \mathbf{w}_{1,n} \\ \mathbf{w}_{2,1} & \mathbf{w}_{2,2} & \cdots & \mathbf{w}_{2,n} \\ \mathbf{w}_{3,1} & \mathbf{w}_{3,2} & \cdots & \mathbf{w}_{3,n} \\ \vdots & \vdots & \ddots & \vdots \\ \mathbf{w}_{n,1} & \mathbf{w}_{n,2} & \cdots & \mathbf{w}_{n,n} \end{bmatrix}, \quad \mathbf{W_P} \in \mathbb{R}^{(n\times n)\times(n\times n)} \tag{11}$$

$$\mathbf{w}_{r,c} = \begin{cases} -(\mathbf{F}_{r,*})^\intercal \otimes \mathbf{D}_{c,*} & sym \\ -(\mathbf{F}_{r,*})^\intercal \otimes \mathbf{D}_{c,*} - \mathbf{F}_{*,r} \otimes (\mathbf{D}_{*,c})^\intercal & asm \end{cases}, \quad \mathbf{w}_{r,c} \in \mathbb{R}^{n\times n} \tag{12}$$

$$h_{r,c} = \sum_{r'=1}^{n} \sum_{c'=1}^{n} w_{r,c;r',c'} s_{r',c'} + b_{r,c}, \quad \mathbf{H_P} = (h_{r,c}) \in \mathbb{R}^{n\times n} \tag{13}$$

We enforce the $2n$ exactly-1 constraints by not allowing moves that violate the constraints. Assuming that the system is initialized to a valid state that meets all the constraints, we propose trials via moves called *swaps* as shown in Fig. 2b.

A swap proposal involves picking two unique rows, r and r', from the neuron matrix and swapping the states of their ON neurons along columns c and c'. If accepted, this move results in 4 simultaneous bit-flips within the binary neuron matrix. The change in energy as a result of such a move is shown in (14), where the first set of local field terms correspond to the neurons being turned OFF while the second set is due to the neurons being turned ON. The two weights being subtracted are required as we have two pairs of neurons that may be connected across the clusters. The first weight is to compensate for the weight being double added by the local fields of the two neurons turning OFF. The second weight is to account for a coupling that was previously inactive between the two neurons turning ON. As shown in (15), we can directly generate the sum of these weights using **F** and **D**. A trial can then be performed by substituting the ΔE value from (14) into (5) and comparing the generated move probability against a value generated by rand().

$$\Delta E(\boldsymbol{\phi}^{\mathbf{t}}, r, r') = (h^t_{r,c} + h^t_{r',c'}) - (h^t_{r,c'} + h^t_{r',c}) - (w_{r,c;r',c'} + w_{r,c';r',c}) \quad (14)$$

$$w_{r,c;r',c'} + w_{r,c';r',c} = \begin{cases} -2f_{r,r'}d_{c,c'} & sym \\ -(f_{r,r'} + f_{r',r})(d_{c,c'} + d_{c',c}) & asm \end{cases} \quad (15)$$

3.2 Updating the Local Field Matrix

When a swap proposal is accepted, the system state must be updated. Swapping the two values in $\boldsymbol{\phi}$ and adjusting the system energy is simple. However, updating the local field matrix involves a large number of calculations. Attempting to update $\mathbf{H_p}$ via (16) involves fetching four weight sub-matrices from global memory with long access delays. Interestingly, the structure of the weight matrix and the PBM itself allow these calculations to be performed efficiently while storing the majority of required data within L2 or L3 caches. For a symmetric problem, we can generate the required weights with a Kronecker Product operation on the differences between 2 rows of the **F** matrix ($\mathbf{\Delta f}$) and 2 rows of the **D** matrix ($\mathbf{\Delta d}$) using (17). For an asymmetric problem, an additional update using $\mathbf{F}^\intercal$ and $\mathbf{D}^\intercal$ is required. In this manner, the amount of memory required to store the weight data is reduced from n^4 elements for a monolithic weight matrix to $2n^2$ elements to store **F** and **D** when the problem is symmetric. For an asymmetric problem, an additional $2n^2$ elements are needed to store $\mathbf{F}^\intercal$ and $\mathbf{D}^\intercal$. Storing a transposed copy of the matrices, while doubling the required memory, provides significant speedups due to a larger number of cache hits when fetching a small number of rows.

$$\mathbf{H_P^{t+1}} = \mathbf{H_P^t} - (\mathbf{w}_{r,c} + \mathbf{w}_{r',c'}) + (\mathbf{w}_{r,c'} + \mathbf{w}_{r',c}) = \mathbf{H_P^t} + \Delta\mathbf{H_P} \quad (16)$$

$$\Delta\mathbf{H_P} = \begin{cases} \mathbf{\Delta f} \otimes \mathbf{\Delta d} = (\mathbf{F}_{r,*} - \mathbf{F}_{r',*})^\intercal \otimes (\mathbf{D}_{c,*} - \mathbf{D}_{c',*}) & sym \\ \mathbf{\Delta f} \otimes \mathbf{\Delta d} + (\mathbf{F}_{*,r} - \mathbf{F}_{*,r'}) \otimes (\mathbf{D}_{*,c} - \mathbf{D}_{*,c'})^\intercal & asm \end{cases} \quad (17)$$

4 System Overview

4.1 Parallel Tempering

A major weakness of Simulated Annealing in traditional BM optimizers is that it can easily get stuck in a local minimum due to the unidirectional nature of the cooling schedule. Parallel Tempering (PT), first proposed in [23] and developed in [14], provides a means of running M cooperative copies (replicas) of the system, each at a different temperature, in order to search a larger portion of the landscape while allowing a mechanism for escaping from local minima. Replicas are generally arranged in order of increasing T from T_{min} to T_{max} in a temperature ladder. A replica, R_k, operating at temperature T_k, can stochastically exchange temperature with the replica immediately above it on the ladder, R_{k+1}, with an *Exchange Acceptance Probability* (EAP) calculated via (18). Figure 3 outlines the structure of an optimization engine using BM replicas with PT.

$$EAP = min\{1, \exp((1/T_k - 1/T_{k+1})(E_k - E_{k+1}))\} \tag{18}$$

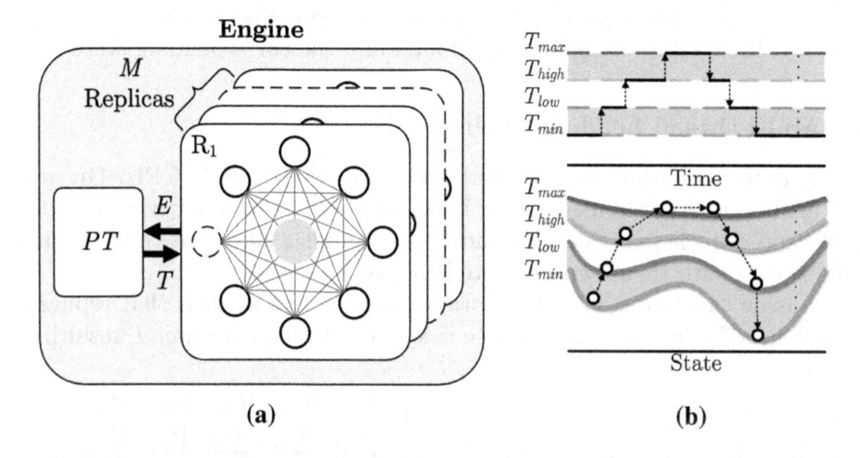

Fig. 3. Overview of a Boltzmann Machine combined with Parallel Tempering (a) Structure of BM + PT Engine (b) Example of a replica escaping from a local minimum and reaching the global optimum via climbing the PT ladder

As implied by (3) and (7), higher T replicas can move around a larger portion of the landscape whereas the moves in lower T replicas are contained to a smaller subspace of the landscape. The ability of replicas to move up or down the ladder, as shown in Fig. 3b, allows for a systematic method of escaping from local minima, making PT a better choice for utilizing parallelism than simply running M disjoint replicas in parallel using SA as proven in [10,14]. In this paper, we implement a PT algorithm based on a modified version of Dabiri's work [7]. One drawback to PT algorithms such as the BM + PT system used in [7] is that their T_{max} and T_{min} must be manually tuned for each problem instance. This requires considerable time and effort while having dramatic effects on the efficacy of the optimization process. We partially address this issue by selecting, from each family of QAP problems, small instances whose solutions can be verified via exact algorithms, to tune a function that automatically selects these parameters for that family within our system.

4.2 Single-Threaded Program

Algorithm 1 presents our proposed PBM + PT system which can be configured for varying levels of multi-threaded operation. A single PT engine ($U = 1$) is used with $M = 32$ replicas. The algorithm starts by initializing the temperature ladder and assigning random permutations to each replica and populating their $\mathbf{H_P}$ matrices and energy values. The system then enters an optimization loop where it runs Y trials for each replica in sequence using the RUN_R() function, updating their states every time a trial is accepted by calling SWAP(). After all replicas have finished their Y trials, temperature exchanges are performed. Similar to QAP solvers such as [15,18–20,24], this process is repeated until the state corresponding to the best-known-solution (BKS) of a problem, E_{BKS}, is reached by one of the replicas, terminating the loop. The system then returns, for each replica, the minimum energy found and the corresponding state.

4.3 Multi-thread Load Balancing

For our implementation, we targeted a 64-core AMD 3990X CPU. Given the structure of a PBM combined with PT, one of the most intuitive ways to extract parallel speedups is to create a thread for each replica such that they all run on a unique core with their own dedicated L1 and L2 caches.

One issue that arises from this form of parallel execution is that replicas at higher T have higher swap acceptance rates than replicas at lower T resulting in

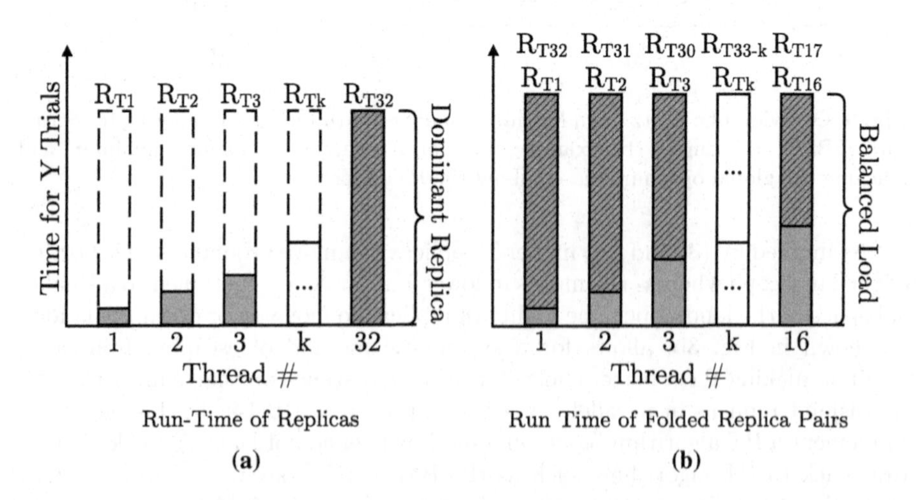

Fig. 4. Load balancing threads via replica folding

Algorithm 1. Permutational Boltzmann Machine with Parallel Tempering

```
Parameters: M, U, C, Y
Input: n, E_BKS, asm, F, D, Fᵀ, Dᵀ, B_P
Output: E_min, φ_min

 1: Engine Variables
 2:     E[U][M]
 3:     E_min[U][M]
 4:     φ[U][M][n]
 5:     φ_min[U][M][n]
 6:     H_P[U][M][n][n]
 7:     T[U][M]
 8:
 9: function RUN_R(u, m)
10:     for y ← 1 to Y do
11:         r, r' ← pick 2 unique rows
12:         ΔE ← (14)
13:         P_swap ← exp(−ΔE/T[u][m])
14:         if P_swap > rand() then
15:             SWAP(u, m, r, r')
16:
17: ▽ Perform Replica Swap Adjustments
18: function SWAP(u,m,r,r')
19:     c, c' ← φ[u][m][r], φ[u][m][r']
20:     fr[n : 1] ← F[r, :] − F[r', :]
21:     dr[n : 1] ← D[c, :] − D[c', :]
22:     if asm then
23:         fc[n : 1] ← Fᵀ[r, :] − Fᵀ[r', :]
24:         dc[n : 1] ← Dᵀ[c, :] − Dᵀ[c', :]
25:     for i ← 1 to n do
26:         H_P[u][m][i, :] +← fr[i] * dr
27:         if asm then
28:             H_P[u][m][i, :] +← fc[i] * dc
29:     φ[u][m][r], φ[u][m][r'] ← c', c
30:     E[u][m] +← ΔE
31:     if E[u][m] < E_min[u][m] then
32:         E_min[u][m] ← E[u][m]
33:         φ_min[u][m] ← φ[u][m]
34: ▽ Exchange Replica Temperatures
35: function PTEXCHANGE(u)
36:     for m ← 1, M − 1 do
37:         δβ ← 1/T[u][m]−1/T[u][m+1]
38:         δE ← E[u][m] − E[u][m + 1]
39:         EAP ← exp(δβ * δE)
40:         if EAP ≥ rand() then
41:             T[u][m], T[u][m + 1] ←
                    T[u][m + 1], T[u][m]
42:
43: ▷ PBM + PT Main Routine
44:
45: ▽ Initialize System
46: for u ← 1, U do
47:     T[u] ← INITFOLDEDLADDER()
48:     for m ← 1, M do
49:         T[m] ← T_min + incr * (m − 1)
50:         φ[u][m] ← RANDOM
51:         φ_min[u][m] ← φ[u][m]
52:         H_P[u][m] ← (13)
53:         E[u][m] ← (2)
54:         E_min[u][m] ← E[u][m]
55:
56: ▽ Run Optimization
57: while E_BKS not in E_min do
58:     #parallel loop threads(U)
59:     for u ← 1, U do
60:         #parallel loop threads(C)
61:         for thread ← 1, C do
62:             for i ← 1, M/C do
63:                 m ← thread*(M/C)+i
64:                 RUN_R(u, m)
65:
66:     #parallel loop threads(U)
67:     for u ← 1, U do
68:         PTEXCHANGE(u)
69:
70: return E_min, φ_min
```

Energies are stored using fp64 while $\mathbf{F}$, $\mathbf{D}$, $\mathbf{B_P}$, and $\mathbf{H_P}$ elements are stored using fp32. Floats allow for the use of fused-multiply-add operations when implementing (17) within SWAP(). To fully utilize our available hardware, SIMD instructions were used wherever possible for significant speed-ups.

more local field updates per trial on average, increasing their run-time. In our experiments, we observed that the number of trials accepted typically increases linearly as T is increased as demonstrated in Fig. 4a. To load-balance the threads, upon initialization of the system, replicas are *folded* and assigned in pairs to threads as shown in Fig. 4b.

The replica-to-thread assignments are static throughout a run to ensure that there is minimal movement of $\mathbf{H_P}$ data between cores. Although the temperature exchanges between replicas can cause load imbalance due to the static folding, their stochastic nature ensures that they are temporary with minimal effects.

4.4 Multi-threaded Configuration Selection

To find the optimal number of engines (U) and *threads-per-engine* (C), we ran all instances within the *sko* and *taiXXb* sets from QAPLIB [6] (excluding *tai*150*b*) 100 times each and recorded the average *time-to-optimum* (TtO) across all 100 runs for each instance. The TtO, reported in seconds, is measured as the average time for a solver to reach the BKS of a problem. We measured TtO values over the selected instances for a system with $U = 1$ across different C values and compared the TtO of each instance against those of a single-threaded system ($U \times C = 1 \times 1$). Figure 5a depicts the average speed-up of a single-engine system as C is varied relative to a 1×1 system, showing that the execution time decreases as the number of threads is increased with diminishing returns. We repeated this experiment, testing different combinations of U and C to find the optimal system configuration. Figure 5b compares the speed-up of different configurations relative to a 1×1 system with the 2×32 configuration having the highest average speed-up despite having no load-balancing. This implies that extra engines, even with load-balancing, cannot make-up for their addtional data movement costs.

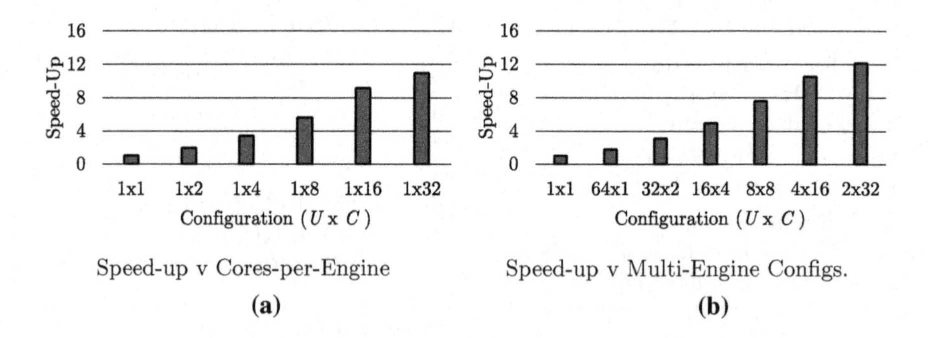

Fig. 5. Speed-up across multi-core configurations

5 Experiments and Results

We benchmark our PBM optimizers using a 64-core AMD Threadripper 3990X system with 128 GB of DDR4 running on CentOS 8.1. Our system was coded in

C++, using the OpenMP API for multi-threading, and compiled with GCC-9. Two separate variants of our solver were benchmarked: PBM ($U = 1$, $C = 1$) and PBM64 ($U = 2$, $C = 32$). We compare the performance of our systems against eight state-of-the-art solvers, described in Table 1. Solvers [4,5,22] use a preset iteration/time limit as their termination criterion while [15,18–20,24] terminate as soon as the *BKS* is reached. All metrics are taken directly from the respective papers. We benchmarked instances from the QAP Library [6] along with ones created by Palubeckis [21] and Drezner [9]. The sets of instances from Palubeckis and Drezner are generated to be difficult with known optima, with the Drezner set being specifically ill-conditioned for meta-heuristics.

Table 1. State-of-the-art solver descriptions

ID	Year	Algorithm	Platform	Hardware
This	2020	Boltzmann Machine + Parallel Tempering	CPU	AMD 3990X
[4]	2019	Hunting Search	CPU	Intel i5-4300
[5]	2017	Break-Out Local Search	CPU	Intel i7-6700
[15]	2017	Break-Out Local Search	FPGA	Xilinx ZCU102
[18]	2018	Genetic Algorithm + Extremal Optimization + Tabu Search	CPU	8 × AMD 6376
[19]	2016	Extremal Optimization + Tabu Search	CPU	8 × AMD 6376
[20]	2016	Extremal Optimization	CPU	8 × AMD 6376
[22]	2018	Multistart Simulated Annealing	GPU	NVidia Titan X
[24]	2012	Ant Colony Optimization	GPU	4 × NVidia GTX480

5.1 Benchmarks: Previously Solved QAP Instances

Table 2 contains TtO values for our two PBM variants and the solvers in Table 1, across some of the most difficult instances from literature that at least one other solver was able to solve with a 100% success rate within a five minute time-out window. The *bur* set, while not difficult, was included as it is the only *asm* set used in literature. The TtO reported for PBM is the average value across 10 consecutive runs with a 5 min time-out window for each run. For the solvers in Table 1, we report only the TtO from the best solver for each instance. The TtOs where PBM or PBM64 outperform the best solver are highlighted in Table 2. In 44 out of 60 instances, the fastest TtO is reported by one or both of our PBM variants with speed-ups in excess of 10× for PBM and 100× for PBM64 on certain instances. Of the remaining 16 instances, PBM64 has either identical or marginally slower performance compared to the best reported solver.

5.2 Benchmarks: Unsolvable QAP Instances

Table 3 contains performance comparisons between PBM64 and the solver from [19], ParEOTS, across QAP instances that no solver to date could consistently solve within a 5 min time-out window. As neither ParEOTS or PBM64 have a 100% success rate on these instances, we also compare their Average Percentage

Table 2. Time-to-optimum (s) comparisons across difficult QAP instances

	ID	Best Solver	PBM	PBM64		ID	Best Solver	PBM	PBM64		ID	Best Solver	PBM	PBM64
bur26a	[20]	0.027	0.017	0.010	**Inst50**	[19]	17	82.55	5.525	**tai20a**	[20]	2.637	0.720	0.035
bur26b	[20]	0.021	0.043	0.013	**Inst60**	[19]	67	58.27	1.606	**tai25a**	[20]	6.330	0.441	0.029
bur26c	[20]	0.009	0.022	0.011	**Inst70**	[19]	127	101.49	10.67	**tai30a**	[22]	0.76	0.560	0.093
bur26d	[4]	7.951	0.109	0.022	**Inst80**	[19]	116	X	27.66	**tai35a**	[5]	19.20	2.620	0.149
bur26e	[20]	0.010	0.042	0.012	**sko42**	[15]	0.330	0.130	0.023	**tai40a**	[19]	64.00	X	93.68
bur26f	[20]	0.009	0.176	0.012	**sko49**	[15]	2.460	0.972	0.081	**tai12b**	[20]	0.001	0.004	0.009
bur26g	[20]	0.006	0.052	0.012	**sko56**	[19]	0.600	0.891	0.082	**tai15b**	[20]	0.001	0.012	0.010
bur26h	[20]	0.010	0.037	0.011	**sko64**	[19]	1.300	0.568	0.046	**tai20b**	[18]	~0.0	0.010	0.010
dre15	[19]	~0.0	0.016	0.010	**sko72**	[19]	8.700	2.733	0.205	**tai25b**	[18]	~0.0	0.030	0.012
dre18	[19]	~0.0	0.017	0.026	**sko81**	[18]	22.40	4.125	0.210	**tai30b**	[18]	0.100	0.076	0.017
dre21	[19]	~0.0	0.054	0.023	**sko90**	[19]	92.00	8.016	0.487	**tai35b**	[19]	0.200	0.168	0.021
dre24	[19]	~0.0	0.139	0.028	**sko100a**	[19]	69.00	9.597	0.561	**tai40b**	[20]	0.061	0.149	0.016
dre28	[19]	0.1	0.299	0.053	**sko100b**	[19]	45.00	6.772	0.574	**tai50b**	[24]	0.200	1.938	0.111
dre30	[19]	0.1	0.483	0.087	**sko100c**	[18]	56.00	16.02	0.747	**tai60b**	[24]	0.400	4.064	0.222
dre42	[19]	0.7	3.069	0.276	**sko100d**	[19]	37.00	9.579	0.841	**tai80b**	[24]	5.500	9.332	0.594
dre56	[19]	5.6	29.25	1.590	**sko100e**	[19]	47.00	6.138	0.609	**tai100b**	[24]	10.10	6.766	0.481
dre72	[19]	26	152.7	11.01	**sko100f**	[19]	57.00	10.78	0.647	**tho30**	[20]	0.235	0.093	0.019
Inst20	[19]	~0.0	0.144	0.024	**tai12a**	[20]	0.011	0.004	0.009	**tho40**	[19]	0.400	1.113	0.245
Inst30	[19]	0.1	3.504	0.226	**tai15a**	[20]	0.089	0.040	0.012	**wil50**	[20]	27.54	0.390	0.036
Inst40	[19]	4.0	21.92	0.812	**tai17a**	[20]	0.292	0.053	0.012	**wil100**	[19]	97.00	15.49	0.770

Table 3. Performance across unsolvable instances

	BKS	ParEOTS [19]			PBM64				BKS	ParEOTS [19]			PBM64		
		#bks	APD	Time	#bks	APD	Time			#bks	APD	Time	#bks	APD	Time
dre90	1838	9	0.968	167	9	0.870	69.33	**tai60a**	7205962	3	0.146	255	0	0.997	300.0
dre110	2264	6	6.334	223	6	4.487	229.9	**tai80a**	13499184	0	0.364	300	0	1.831	300.0
dre132	2744	1	22.78	294	3	7.048	244.6	**tai100a**	21052466	0	0.298	300	0	1.709	300.0
Inst100	15008994	1	0.120	300	0	0.185	300.0	**tai150b**	498896643	0	0.061	300	9	~0.0	111.4
Inst150	58352664	0	0.126	300	0	0.171	300.0	**tai256c**	44759294	0	0.178	300	0	0.139	300.0
Inst200	75405684	0	0.125	300	0	0.153	300.0	**tho150**	8133398	1	0.007	291	0	0.031	300.0
tai50a	4938796	3	0.077	264	3	0.151	262.9								

Deviation, calculated as $APD = 100 \times (Avg - BKS)/BKS$. Avg is calculated as the average of the best cost found in each run. We benchmarked PBM64 using the same procedure reported in [19], running each instance 10 times with a time-out window of 5 min and reporting the average time of the 10 runs along with the number of runs that reached the BKS, *#bks*

PBM64 displays better performance on the *dre* instances and has a near 100% success rate on *tai150b*. Across other instances, ParEOTS reports equal or better performance despite PBM64 performing better on smaller instances from the same family of problems. Further testing is required to compare the TtO of ParEOTS and PBM64 if ran without a time-out limit.

6 Conclusion

We demonstrated a Permutational Boltzmann Machine with Parallel Tempering, that is capable of solving NP-Hard problems such as QAP in excess of 100× faster than other state-of-the-art solvers. The speed of the PBM is attributed to its simple structure where we can utilize parallelism through the parallel execution of its replicas on dedicated computational units along with using SIMD instructions when performing local field updates. Though our PBM + PT system, which uses a 64-core CPU, was the fastest in solving the majority of the QAP test cases by a wide margin, its flexibility allows it to be scaled to match the user's available hardware while maintaining competitive performance with other state-of-the-art solvers.

References

1. Aarts, E., Korst, J.: Simulated Annealing and Boltzmann Machines (1988)
2. Aarts, E.H.L., Korst, J.H.M.: Boltzmann machines as a model for parallel annealing. Algorithmica **6**(1–6), 437–465 (1991). https://doi.org/10.1007/bf01759053
3. Aarts, E.H., Korst, J.H.: Boltzmann machines for travelling salesman problems. Eur. J. Oper. Res. **39**(1), 79–95 (1989). https://doi.org/10.1016/0377-2217(89)90355-x
4. Agharghor, A., Riffi, M., Chebihi, F.: Improved hunting search algorithm for the quadratic assignment problem. Indonesian J. Electr. Eng. Comput. Sci. **14**, 143 (2019). https://doi.org/10.11591/ijeecs.v14.i1.pp143-154
5. Aksan, Y., Dokeroglu, T., Cosar, A.: A stagnation-aware cooperative parallel breakout local search algorithm for the quadratic assignment problem. Comput. Ind. Eng. **103**, 105–115 (2017). https://doi.org/10.1016/j.cie.2016.11.023
6. Burkard, R.E., Karisch, S.E., Rendl, F.: QAPLIP-a quadratic assignment problem library. J. Global Optim. **10**(4), 391–403 (1997). https://doi.org/10.1023/A:1008293323270
7. Dabiri, K., Malekmohammadi, M., Sheikholeslami, A., Tamura, H.: Replica exchange MCMC hardware with automatic temperature selection and parallel trial. IEEE Trans. Parallel Distrib. Syst. **31**(7), 1681–1692 (2020). https://doi.org/10.1109/TPDS.2020.2972359
8. d'Anjou, A., Grana, M., Torrealdea, F., Hernandez, M.: Solving satisfiability via Boltzmann machines. IEEE Trans. Pattern Anal. Mach. Intell. **15**(5), 514–521 (1993). https://doi.org/10.1109/34.211473
9. Drezner, Z., Hahn, P.M., Taillard, É.D.: Recent advances for the quadratic assignment problem with special emphasis on instances that are difficult for metaheuristic methods. Ann. Oper. Res. **139**(1), 65–94 (2005). https://doi.org/10.1007/s10479-005-3444-z
10. Earl, D.J., Deem, M.W.: Parallel tempering: theory, applications, and new perspectives. Phys. Chem. Chem. Phys. **7**(23), 3910 (2005). https://doi.org/10.1039/b509983h
11. Gloria, A.D., Faraboschi, P., Olivieri, M.: Clustered Boltzmann machines: massively parallel architectures for constrained optimization problems. Parallel Comput. **19**(2), 163–175 (1993). https://doi.org/10.1016/0167-8191(93)90046-n

12. Glover, F., Kochenberger, G., Du, Yu.: Quantum bridge analytics I: a tutorial on formulating and using QUBO models. 4OR **17**(4), 335–371 (2019). https://doi.org/10.1007/s10288-019-00424-y
13. Hinton, G.E., Sejnowski, T.J., Ackley, D.H.: Boltzmann machines: constraint satisfaction networks that learn. Carnegie-Mellon University, Department of Computer Science Pittsburgh (1984)
14. Hukushima, K., Nemoto, K.: Exchange Monte Carlo method and application to spin glass simulations. J. Phys. Soc. Jpn. **65**(6), 1604–1608 (1996)
15. Kanazawa, K.: Acceleration of solving quadratic assignment problems on programmable SoC using high level synthesis. In: FSP 2017; Fourth International Workshop on FPGAs for Software Programmers, pp. 1–8 (2017)
16. Korst, J.H., Aarts, E.H.: Combinatorial optimization on a Boltzmann machine. J. Parallel Distrib. Comput. **6**(2), 331–357 (1989). https://doi.org/10.1016/0743-7315(89)90064-6
17. Lawler, E.L.: The quadratic assignment problem. Manage. Sci. **9**(4), 586–599 (1963). https://doi.org/10.1287/mnsc.9.4.586
18. López, J., Múnera, D., Diaz, D., Abreu, S.: Weaving of metaheuristics with cooperative parallelism. In: Auger, A., Fonseca, C.M., Lourenço, N., Machado, P., Paquete, L., Whitley, D. (eds.) PPSN 2018. LNCS, vol. 11101, pp. 436–448. Springer, Cham (2018). https://doi.org/10.1007/978-3-319-99253-2_35
19. Munera, D., Diaz, D., Abreu, S.: Hybridization as cooperative parallelism for the quadratic assignment problem. In: Blesa, M.J., et al. (eds.) HM 2016. LNCS, vol. 9668, pp. 47–61. Springer, Cham (2016). https://doi.org/10.1007/978-3-319-39636-1_4
20. Munera, D., Diaz, D., Abreu, S.: Solving the quadratic assignment problem with cooperative parallel extremal optimization. In: Chicano, F., Hu, B., García-Sánchez, P. (eds.) EvoCOP 2016. LNCS, vol. 9595, pp. 251–266. Springer, Cham (2016). https://doi.org/10.1007/978-3-319-30698-8_17
21. Palubeckis, G.: An algorithm for construction of test cases for the quadratic assignment problem. Informatica Lith. Acad. Sci. **11**, 281–296 (2000)
22. Sonuc, E., Sen, B., Bayir, S.: A cooperative GPU-based parallel multistart simulated annealing algorithm for quadratic assignment problem. Eng. Sci. Technol. Int. J. **21**(5), 843–849 (2018). https://doi.org/10.1016/j.jestch.2018.08.002
23. Swendsen, R.H., Wang, J.S.: Replica Monte Carlo simulation of spin-glasses. Phys. Rev. Lett. **57**(21), 2607–2609 (1986). https://doi.org/10.1103/physrevlett.57.2607
24. Tsutsui, S.: ACO on multiple GPUs with CUDA for faster solution of QAPs. In: Coello, C.A.C., Cutello, V., Deb, K., Forrest, S., Nicosia, G., Pavone, M. (eds.) PPSN 2012. LNCS, vol. 7492, pp. 174–184. Springer, Heidelberg (2012). https://doi.org/10.1007/978-3-642-32964-7_18

Solution Repair by Inequality Network Propagation in LocalSolver

Léa Blaise[1,2(✉)], Christian Artigues[1], and Thierry Benoist[2]

[1] LAAS-CNRS, Université de Toulouse, CNRS, INP, Toulouse, France
[2] LocalSolver, 36 Avenue Hoche, 75008 Paris, France
lblaise@localsolver.com

Abstract. This paper focuses on optimization problems whose constraints comprise a network of binary and ternary linear inequalities. These constraints are often encountered in the fields of scheduling, packing, layout, and mining. Alone, small-neighborhood local search algorithms encounter difficulties on these problems. Indeed, moving from a good solution to another requires small changes on many variables, due to the tight satisfaction of the constraints.

The solution we implemented in LocalSolver is a kind of constraint propagation: when the solution obtained after a local transformation is infeasible, we gradually repair it, one constraint at a time. In order to extend the local transformation rather than cancel it, we impose never to go back on the decision to increase or decrease the value of a variable. We show that the success of this repair procedure is guaranteed for a large class of constraints.

We apply this method to several scheduling problems, characterized by precedences and disjunctive resource constraints. We give numerical results on the Job Shop, Open Shop and Unit Commitment Problems, and show that our repair algorithm dramatically improves the performance of our local search algorithms.

Keywords: Constraint propagation · Local search · Repair algorithm · Solver · Disjunctive scheduling

1 Introduction

LocalSolver is a global mathematical programming solver, whose goal is to offer a model-and-run approach to optimization problems, including combinatorial, continuous, and mixed problems, and to offer high quality solutions in short running times, even on large instances. It allows practitioners to focus on the modeling of the problem using a simple formalism, and then to defer its resolution to a solver based on efficient and reliable optimization techniques, including local search (but also linear, non-linear, and constraint programming). As described in [5] and [11], the local search components in LocalSolver are mostly based on small neighborhoods (flips, shifts, insertions, ...).

T. Bäck et al. (Eds.): PPSN 2020, LNCS 12269, pp. 332–345, 2020.
https://doi.org/10.1007/978-3-030-58112-1_23

In this paper, we mainly target constraints defined by linear inequalities between two or three variables, and disjunctions and chains of such inequalities. Many scheduling problems comprise such a network of binary or ternary inequalities. For instance, the Job Shop Problem [10] is characterized by precedences and disjunctive resource constraints. However, this kind of structure is also typical of packing, layout, and mining problems.

These problems are highly constrained: in a good solution of a Job Shop instance, the precedences and disjunctive resource constraints are often very tight. Because of that, moving from a solution of makespan x to a solution of makespan $x - 1$ requires a lot of small changes on many integer variables – the start times of the tasks. Being able to move from a good feasible solution to another using random small neighborhoods is then very unlikely: one would have to randomly target the right set of integer variables and to randomly shift them all by the right amount. For these reasons, the algorithms described in [11] encounter serious difficulties on these problems. In the vast literature on job-shop scheduling by local search (for example [4,16] and [19]), these difficulties are overcome by exploiting higher level dedicated solution representations, such as the disjunctive graph. In this work, we aim at keeping the modeling elements simple and we wish to target other problems as well. Hence, we focus on the direct integer variable representation.

To tackle this problem, we designed a solution repair algorithm based on constraint propagation: a promising but infeasible solution is gradually repaired, one constraint at a time. This gradual procedure can be compared to ejection chains algorithms, originally proposed by Glover (1996) [12] to generate neighborhoods of compound moves for the Traveling Salesman Problem, and more recently studied by Ding *et al.* (2019) [7] in the context of scheduling problems. With powerful moves enhancing the local search, these methods yield great results. Our repair mechanism can also be compared to the min-conflicts heuristic introduced by Minton *et al.* (1992) [15] to solve CSPs, and more recently applied to the field of scheduling by Ahmeti and Musliu (2018) [1]. This method consists in creating a complete but inconsistent assignment for the variables of a CSP, and to repair the constraint violations until the assignment is consistent. The min-conflicts heuristic corresponds to selecting a variable involved in a violated constraint, and setting its value to the one minimizing the number of outstanding constraint violations.

This paper is organized as follows. Section 2 formally introduces our repair mechanism. Our method is simpler than the ones mentioned above, since we repair the violated constraints in the order we encounter them, by changing the values of the involved variables just enough to repair the current constraint. However, it yields satisfactory results in practice, has strong theoretical properties, and has the advantage of being very fast, which is crucial to integrate it in a high-performance solver. In Sects. 3 and 4, we apply our method to binary constraints on Boolean and numeric variables respectively. We show that some constraints have very strong properties ensuring that the repair mechanism will succeed, and we characterize such constraints. In Sect. 5, we introduce some more complex constraints: disjunctions and chains, for which we adapt our repair mechanism. In Sect. 6, we introduce ternary constraints. Finally, we give numerical results in Sect. 7. We apply our method to the Job Shop, Open Shop, and Unit Commitment Problems, on which the repair procedure dramatically improves the performance of our local search algorithms.

2 Repair Mechanism: Definitions and General Algorithm

In this section, we formally describe the repair mechanism implemented in LocalSolver to supplement its local search algorithms, which consists in gradually repairing an infeasible solution, one constraint at a time.

Definition 1 (Constraint). *A constraint is a relation on variables of an optimization problem that a solution must satisfy. It is characterized by its feasible set, which is a subset of the Cartesian product of the domains of its variables. The set of variables involved in a constraint* $\mathcal{C}$ *is denoted* $var(\mathcal{C})$.

In the remainder of the paper, when the context is clear, a constraint can refer either to its defining equation or to the related feasible set.

We consider any iteration of the local search, and we assume that this iteration starts from an initial feasible solution, denoted $\mathcal{S}_0$, in which each variable X has an initial value x_0, and an initial domain $\mathcal{D}_X = [\underline{x}, \overline{x}]$. A local transformation is applied to $\mathcal{S}_0$: we now have a solution $\mathcal{S}$, that we assume to be infeasible. The value of each variable X in the current solution is denoted x.

Property 1 (No backtracking). To ensure that the repair mechanism extends the local transformation rather than cancel it, we impose never to go back on a previous decision. In other words, if a variable has already been modified (in the local transformation, or when repairing a constraint), it can be modified again in the same direction, but it cannot be modified in the opposite direction. Then, a modification of a variable's value is equivalent to a domain reduction: when it increases (resp. decreases), its lower (resp. upper) bound is adjusted accordingly.

Because of Property 1, the repair phase is equivalent to a kind of constraint propagation, whose filtering algorithm will be referred to as "half bound consistency" (HBC) in the remainder of the paper. Indeed, while the initial domain of each variable X is $\mathcal{D}_X = [\underline{x}, \overline{x}]$, it will be reduced by successive modifications of one of its bounds only throughout the propagation. The first modification of X's value x (in the local transformation, or during the propagation) determines which of X's bounds will have its modifications propagated. Therefore, throughout the iteration, a variable X's non empty domain is always of the form $[\underline{x}, b]$ or $[b, \overline{x}]$, with $b \in [\underline{x}, \overline{x}]$.

We now describe the repair procedure. After the local transformation is performed, all the constraints involving a variable that has just been modified (variables X such that $x \neq x_0$) are put in the propagation queue.

While the queue is non empty, we apply the HBC filtering algorithm on its front constraint $\mathcal{C}$, as follows. If $\mathcal{C}$ is already verified, it is skipped. Otherwise, we compute a new domain for each variable $x \in var(\mathcal{C})$, whose tighter bounds are chosen to ensure its consistency towards $\mathcal{C}$. If at least one of these reduced domains is empty, then $\mathcal{C}$ cannot be repaired: the propagation fails, and the constraints remaining in the propagation queue are ignored. Else, each variable $X \in var(\mathcal{C})$ is assigned a new value x'. If x is still in X's reduced domain, then $x' = x$. If not, x' is the value of the reduced domain that is closest to x: it is one of X's reduced bounds. Thus, x' is the projection of x on X's reduced domain.

We now consider the solution $\mathcal{S}'$, in which each variable $X \in var(\mathcal{C})$ takes the value x'. If this solution satisfies $\mathcal{C}$, the domain reductions are propagated: for each variable $X \in var(\mathcal{C})$ such that $x \neq x'$, its new bound x' is propagated (its other bound being left unchanged), and every other constraint involving X is added to the propagation queue. If $\mathcal{S}'$ violates the constraint, then there exists several ways to repair it, and several ways to choose a new valid value for the variables, none of which is supposedly better than the others. We detail how we deal with this situation in the following sections: we show that this situation never occurs for certain classes of constraints, and we explain how we randomly repair the more complex constraints.

When the queue is empty, either the propagation fails because there exists no feasible solution respecting the decisions of the local transformation, and the algorithm reverts back to its initial solution $\mathcal{S}_0$, or a feasible solution is found. In the latter case, for each variable, if its domain has been reduced, then each of these reductions corresponded to a modification of the same bound, and its current value is equal to this modified bound.

3 Repair of Binary Constraints on Boolean Variables

We first focus on the simple case of binary constraints on Boolean variables (*leq*, *nand*, *or*, *xor*):

$$X \leq Y; \quad X + Y \leq 1; \quad X + Y \geq 1; \quad X + Y = 1$$

When considering Boolean variables, Property 1 implies that each variable can only be modified once during each iteration of the local search. This implies that every time a violated constraint is propagated, there exists at most one way to repair it. Indeed, since the constraint is in the propagation queue, at least one of its variables must have been modified. Then, this variable cannot be modified again, and the constraint may only be repaired by changing the other one.

This procedure is similar to one iteration of the limited backtracking algorithm [9] for the 2-SAT Problem. We start from a solution where the variables that have already been modified by the local transformation have a definitive value, and the others only have a temporary value. When applying the HBC filtering algorithm on a violated constraint, two configurations arise. Either only one of the two Boolean variables already has a definitive value, and the other one is assigned the definitive value that repairs the constraint, or both variables already have a definitive value, and the propagation fails because the constraint cannot be repaired. The difference between the limited backtracking algorithm and ours is that we never take arbitrary decisions on which we can backtrack if we encounter a constraint that we cannot repair. Indeed, these arbitrary decisions were taken during the local transformation, which we choose not to reconsider.

Proposition 1. *If there exists a feasible solution that respects the decisions of the local transformation, the repair algorithm is guaranteed to find it.*

Sketch of Proof. This ensues from the similarity between our repair algorithm on binary Boolean constraints and the limited backtracking algorithm. See [9]. □

Example 1. We consider a small problem with three Boolean variables X, Y and Z, and three binary Boolean constraints $\mathcal{C}_1 : X + Y \leq 1$, $\mathcal{C}_2 : Y + Z \geq 1$ and $\mathcal{C}_3 : X \geq Z$. We assume that the initial feasible solution $\mathcal{S}_0$ is such that $x_0 = 0$, $y_0 = 1$ and $z_0 = 0$. We assume that after the local transformation, the current solution $\mathcal{S}$ is infeasible and verifies $x = 1$, $y = y_0 = 1$ and $z = z_0 = 0$. The propagation takes place as follows:

- Modification of X (local transformation) $\Rightarrow q = \{\mathcal{C}_1, \mathcal{C}_3\}$.
- Propagate $\mathcal{C}_1 : X + Y \leq 1$. Repair: $y = 0$. Modification of $Y \Rightarrow q = \{\mathcal{C}_3, \mathcal{C}_2\}$.
- Propagate $\mathcal{C}_3 : X \geq Z$. Already verified $\Rightarrow q = \{\mathcal{C}_2\}$.
- Propagate $\mathcal{C}_2 : Y + Z \geq 1$. Repair: $z = 1$. Modification of $Z \Rightarrow q = \{\mathcal{C}_3\}$.
- Propagate $\mathcal{C}_3 : X \geq Z$. Already verified $\Rightarrow q = \emptyset$.

A feasible solution was found: $x = 1$, $y = 0$, $z = 1$.

4 Repair of Binary Constraints on Numeric Variables

We now consider binary constraints on numeric variables. First, we describe the specific binary constraints actually propagated in LocalSolver. We then characterize the more general form of binary constraints verifying useful properties.

4.1 Repair of Binary Linear Inequalities on Numeric Variables

In addition to binary Boolean constraints, we consider inequalities of the form

$$aX + bY \leq c$$

where X and Y are integer or real variables, and a, b and c are any constants.

Remark 1. The special case where $a = 1$ and $b = -1$ corresponds to the generalized precedence constraints encountered in scheduling problems.

Repair Mechanism. The propagation of these binary linear constraints follows the general repair algorithm described in Sect. 2. In more concrete terms, the application of the HBC filtering algorithm to a constraint $\mathcal{C}$: $aX + bY \leq c$ takes place as follows. If the inequality is already verified, its propagation is skipped. If the variables cannot be shifted enough in the right direction to repair $\mathcal{C}$, the propagation fails. Else, if only one of the variables can be shifted in the right direction (X by symmetry), the algorithm applies the only necessary and sufficient change on X to repair $\mathcal{C}$: $X \leftarrow \frac{c-by}{a}$. We show in Sect. 4.2 that the initial solution $\mathcal{S}_0$ being feasible ensures that the last possible case – both variables being able to move in the right direction – never actually happens.

4.2 Properties on Binary Linear Inequalities

Property 2. If the initial solution $\mathcal{S}_0$ is feasible, then there exists at most one way to repair the constraint $\mathcal{C}$: $aX + bY \leq c$.

Proof. Let $aX + bY \leq c$ be a violated constraint. Then $ax + by > c$. Let us assume that there exists several ways to repair the constraint, which implies that both X and Y can be shifted in the repair direction. Therefore $ax \leq ax_0$, $by \leq by_0$, and $ax_0 + by_0 > c$: the constraint was also violated in the initial solution. □

Property 3. If the initial solution $\mathcal{S}_0$ is feasible, and if there exists a feasible solution compatible with the local transformation's decisions, then the algorithm is guaranteed to find it.

Sketch of Proof. The proof comes from Property 2: whenever a violated constraint is encountered during the propagation, there exists exactly one necessary and sufficient way to repair it. Then, the algorithm can never take a "wrong" decision that would prevent it from finding a feasible solution at the end. □

4.3 Characterization of Repairable Binary Constraints

In this section, we describe a condition on any binary constraint that ensures that, if there exists a feasible solution that respects the decisions of the local transformation, the propagation will succeed. Since this property ensures a high efficiency in the repair procedure, it is greatly desirable.

As seen in Sect. 4.2, this property is verified if and only if there always exists a necessary way to repair each encountered constraint. Indeed, if not, there is no indication as to which possible repair is the "right" one.

As described in Sect. 2, when applying the HBC filtering algorithm on a constraint, we consider the solution in which the value of each variable X is the projection of its current value x on its reduced domain. If this solution is feasible, then it is a necessary repair. We then introduce a condition on any binary constraint that ensures that if there exists a solution, then the projection of their current value on their reduced domains is a feasible solution.

Definition 2 (Minimal repair). *Let us consider a repair of an infeasible solution (x, y) into a feasible solution (x', y'). This repair is minimal if any intermediate solution $(\lambda x + (1-\lambda)x', \mu y + (1-\mu)y')$ with $\lambda, \mu \in]0, 1[$ is infeasible.*

Remark 2. A minimal repair is not always necessary. For example, two solutions (x, y_1) and (x, y_2) with $y_1 < y < y_2$ can both correspond to a minimal repair.

Definition 3 (Biconvex constraint). *A binary constraint $\mathcal{C}$ on real variables X and Y is called biconvex if*

$$\begin{cases} \forall x \in \mathcal{D}_X,\ \mathcal{C}^x = \{y \in \mathcal{D}_Y : (x, y) \in \mathcal{C}\} \text{ is convex} \\ \forall y \in \mathcal{D}_Y,\ \mathcal{C}^y = \{x \in \mathcal{D}_X : (x, y) \in \mathcal{C}\} \text{ is convex} \end{cases}$$

A binary constraint $\mathcal{C}$ on integer variables X and Y is called biconvex if

$$\begin{cases} (x_1, y_1) \text{ and } (x_2, y_1) \text{ feasible} \Rightarrow \forall \tilde{x} \in [x_1, x_2],\ (\tilde{x}, y_1) \text{ feasible} \\ (x_1, y_1) \text{ and } (x_1, y_2) \text{ feasible} \Rightarrow \forall \tilde{y} \in [y_1, y_2],\ (x_1, \tilde{y}) \text{ feasible} \end{cases}$$

Definition 4 (Path-connected constraint). *A binary constraint $\mathcal{C}$ on real variables X and Y is path-connected if for any two feasible solutions (x, y) and (x', y'), there exists a continuous feasible path between them. That is:*

$$\forall (x, y), (x', y') \in \mathcal{C},\ \exists f : [0, 1] \to \mathcal{C} \text{ continuous}, f(0) = (x, y), f(1) = (x', y')$$

A binary constraint $\mathcal{C}$ on integer variables X and Y is path-connected if for any two feasible solutions (x, y) and (x', y'), there exists a path of feasible "neighbor"[1] *solutions between them. That is:*

$$\forall (x, y), (x', y') \in \mathcal{C},\ \exists \{f_0, ..., f_n\} \in \mathcal{C}, \begin{cases} f_0 = (x, y), f_n = (x', y') \\ \forall i < n,\ \|f_i - f_{i+1}\|_\infty \leq 1 \end{cases}$$

Proposition 2. *If a binary constraint $\mathcal{C}$ on integer or real variables X and Y is biconvex and path-connected (and closed in the case of real variables), and if it is possible to repair it, then it admits a necessary repair. This repair consists in projecting the variables on their reduced domains.*

Sketch of proof.[2] We assume that the current infeasible solution (x, y) verifies $x \geq x_0$ and $y \geq y_0$ by symmetry, and that there exists $x' \geq x$ and $y' \geq y$ such that (x', y') is feasible.

As illustrated in Fig. 1, the intersection of $\mathcal{C}$ with any rectangle is still biconvex and path-connected. Then, there exists a feasible path f between (x_0, y_0) and (x', y'), inside the rectangle that these two points define. As depicted in Fig. 2, this path must admit a point (x, y_1) of abscissa x and a point (x_1, y) of ordinate y, only one of which respects Proposition 1 ((x, y_1) by symmetry).

From this, we show that the projection of (x, y) on the reduced domains yields a feasible solution $(x, y^\star)$: it is a necessary repair. □

[1] Two solutions (x_1, y_1) and (x_2, y_2) are neighbors when $\|(x_1, y_1) - (x_2, y_2)\|_\infty \leq 1 \Leftrightarrow |x_1 - x_2| \leq 1$ and $|y_1 - y_2| \leq 1$

[2] For reasons of space, we only give a sketch of proof here. For the full proof, see [6].

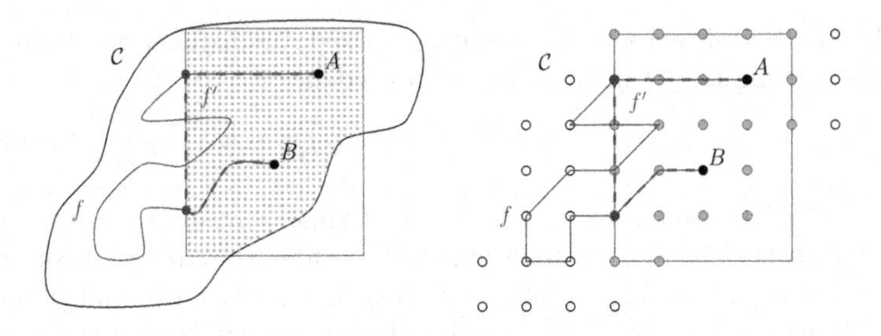

Fig. 1. Path-connectedness of the intersection – real (left) and integer (right) variables

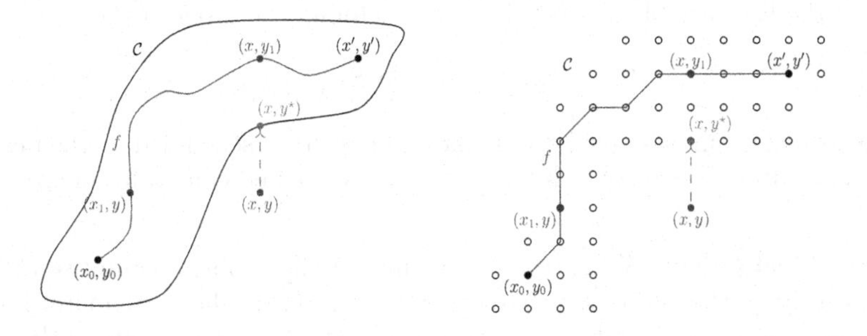

Fig. 2. Necessary repair – real (left) and integer (right) variables

5 Repair of Disjunctions and Chains of Binary Linear Inequalities on Numeric Variables

In addition to the previously mentioned constraints, we now consider disjunctions of inequalities of the form

$$\bigvee_i (a_i X_i + b_i Y_i \leq c_i)$$

where the X_i and Y_i are integer or real variables, and the a_i, b_i and c_i are any constants, as well as chains of inequalities of the form

$$\bigwedge_i \left(aX_{L[f(i)]} + bX_{L[g(i)]} \leq c_{L[h(i)]}\right)$$

where X is an array of integer or real variables, L is a list variable (representing an ordering), and a, b and the elements of the array c are any constants.

Remark 3. When the a_i (resp. a) equal 1 and the b_i (resp. b) equal −1, these constraints describe packing or disjunctive resource constraints (tasks taken two by two in a disjunction, or ordered in regard of the list variable in a chain).

Example 2. Disjunctive resource constraints – non-overlapping of tasks scheduled on the same machine – are often encountered in the field of scheduling.

When the number of tasks n on a resource is fixed, its disjunctive nature can be described by using $O(n^2)$ disjunctions of precedences:

$$\forall 1 \leq i < j \leq n,\ (S_j \geq S_i + d_i) \vee (S_i \geq S_j + d_j)$$

where S_i and d_i, respectively, denote the start time (variable) and the duration (fixed) of a task i. In this formulation, we consider that if two tasks i and j are scheduled on the same machine, either j starts after the end of i, or the opposite. However, by using a chained constraint, we can reduce the number of constraints to $O(n)$. Indeed, in any feasible solution, the tasks are scheduled in a certain order. We can then formulate the constraint with a list variable L: the value of the list variable is a permutation, defining the order of the tasks.

$$\forall 1 \leq i < n,\ S_{L[i+1]} \geq S_{L[i]} + d_{L[i]}$$

In this formulation, we consider that the $(i+1)$-th task scheduled on the machine must start after the end of i-th task, creating a chained constraint of size $O(n)$.

Repair Mechanism. When the constraints of the problem comprise disjunctions or chains, the properties listed in Sect. 4.2 do not hold anymore. Indeed, these constraints rarely admit a necessary repair. In order to efficiently repair them, we use a filtering algorithm slightly different from HBC: the one we apply is based on random choices rather than on projections alone.

Repair of a Disjunction. We assume that a disjunction is violated. Since *a priori* none of the inequalities of the disjunction should prevail over the others, the algorithm chooses one at random and tries to repair it. If it cannot be repaired, it tries to repair the subsequent one, and so forth. If none of them can be repaired, the propagation fails.

Let $aX + bY \leq c$ be the inequality that was randomly chosen for repair in the disjunction. If only one of its variables can be shifted in the repair direction, then the constraint is repaired as described in Sect. 4.1. It is also possible that both variables can be shifted in the repair direction, since the chosen inequality may not have been the one that was respected in the initial solution $\mathcal{S}_0$. If so, the algorithm randomly chooses how to shift them. Let Δ be the distance to feasibility: $\Delta = aX + bY - c$, and let δ_X and δ_Y be the shares of the repair respectively attributed to X and Y, verifying $\delta_X + \delta_Y = \Delta$. The algorithm has four equiprobable ways to repair the constraint: either X repairs it alone ($\delta_X = \Delta$), or Y repairs it alone ($\delta_Y = \Delta$), or X and Y equitably share the repair ($\delta_X = \delta_Y = \frac{\Delta}{2}$), or X and Y randomly share the repair ($\delta_X = random(1, \Delta - 1)$ and $\delta_Y = \Delta - \delta_X$).

Repair of a Chain. Violated chains are repaired one index at a time. When considering one inequality of the chain, the repair procedure is very similar to that of an inequality of a disjunction.

Property 4. If there exists a solution that respects the decisions of the move, there is always a non-zero probability for the propagation to succeed, depending on whether the algorithm always takes the "right" random decisions.

Example 3 (Scheduling problem). We consider a scheduling problem with three tasks. Task t, of duration 3 and release date 1, and task t', of duration 2, must not overlap, and must both be scheduled before task t'', of duration 4. There are three integer variables S, S' and S'' (start times of the tasks), two precedences $\mathcal{P}_1$: $S - S'' \leq -3$ and $\mathcal{P}_2$: $S' - S'' \leq -2$, and one disjunctive resource constraint $\mathcal{R}$: $(S - S' \leq -3) \vee (S' - S \leq -2)$. We assume that the initial feasible solution $\mathcal{S}_0$ is such that $s_0 = 1$, $s'_0 = 4$ and $s''_0 = 6$. We assume that after the local transformation, the current solution $\mathcal{S}$ is infeasible and verifies $s = s_0 = 1$, $s' = s'_0 = 4$ and $s'' = 5$. The propagation can take place as follows:

- Modification of S'' (local transformation) $\Rightarrow q = \{\mathcal{P}_1, \mathcal{P}_2\}$.
- Propagation of $\mathcal{P}_1$: $S - S'' \leq -3$. Already verified $\Rightarrow q = \{\mathcal{P}_2\}$.
- Propagation of $\mathcal{P}_2$: $S' - S'' \leq -2$. Repair: $s' = 3$. Modif. of $S' \Rightarrow q = \{\mathcal{R}\}$.
- Propagation of $\mathcal{R}$: $(S - S' \leq -3) \vee (S' - S \leq -2)$. Repair (randomly chosen): $s = 2$, $s' = 0$. Modification of S and $S' \Rightarrow q = \{\mathcal{P}_1, \mathcal{P}_2\}$.
- Propagation of $\mathcal{P}_1$: $S - S'' \leq -3$. Already verified $\Rightarrow q = \{\mathcal{P}_2\}$.
- Propagation of $\mathcal{P}_2$: $S' - S'' \leq -3$. Already verified $\Rightarrow q = \emptyset$.

A feasible solution was found: $s = 2$, $s' = 0$, $s'' = 5$, as illustrated in Fig. 3.

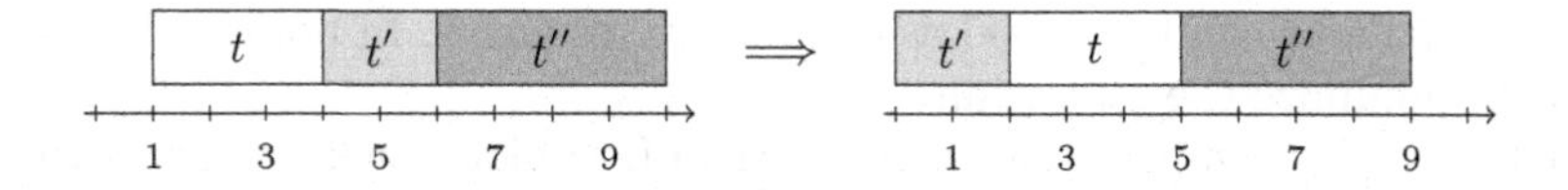

Fig. 3. Initial and repaired solutions

6 Repair of Ternary Linear Inequalities on Numeric Variables

We extend the repair mechanism to ternary linear inequalities, and to disjunctions and chains of such inequalities:

$$aX + bY + cZ \leq d$$

$$\bigvee_i (a_i X_i + b_i Y_i + c_i Z_i \leq d_i)$$

$$\bigwedge_i \left(aX_{L[f_1(i)]} + bX_{L[f_2(i)]} + cY_{L[f_3(i)]} \leq d_{L[f_4(i)]}\right)$$

Remark 4. If all as and cs equal 1 and all bs equal -1, these constraints describe precedences and disjunctive resource constraints on tasks of variable duration.

Repair Mechanism. As in the previous section, when the current constraint is a ternary linear inequality, or a disjunction or chain of such inequalities, there likely exists several minimal ways to repair it. The filtering algorithm applied to these constraints is then non deterministic and based on random choices as well.

Let $aX+bY+cZ \leq d$ be a violated constraint. If only one or two variables can be shifted in the repair direction, the constraint is repaired like a binary linear inequality in a disjunction (see Sect. 5). If all three variables can be shifted in the repair direction, the algorithm chooses between eleven different repair methods. It can choose to repair the constraint by shifting one of the variables only, or by shifting two variables: either equitably or randomly. Finally, it can choose to repair it by shifting all three variables: either equitably or randomly.

The repair of disjunctions and chains of ternary inequalities is similar to that of disjunctions and chains of binary inequalities, described in Sect. 5.

7 Numerical Results

7.1 Repair of Binary Constraints in Shop Scheduling Problems

In this section, we consider the Job Shop and Open Shop Problems. In both problems, n jobs are divided into m activities each – one activity per machine. The machines are disjunctive, which can be modeled using either $O(mn^2)$ disjunctions or $O(m)$ chains of size $O(n)$. In the Job Shop Problem, the activities of each job are ordered: there are $O(n^2)$ precedences, whereas in the Open Shop Problem, the jobs can be viewed as disjunctive resources. In both problems, the goal is to minimize the makespan.

In Table 1, we compare the performance of LocalSolver with and without our repair mechanism[1] on three classic Job Shop instance classes: the FT class by Fisher and Thompson [10], the LA class by Lawrence [14], and the ORB class by Applegate and Cook [3], as well as on a classic Open Shop instance class by Taillard [18]. We give the average optimality gap with and without repairs, after 10 and 60 s of search.

While positive, it can be noted that these results remain behind those of dedicated scheduling algorithms or specialized methods (constraint-based scheduling, disjunctive graph-based local search) presented for example in [17] and [20] for the Job Shop Problem, and in [13] for the Open Shop Problem. However, as already mentioned, we are interested in more general forms of disjunctions, and we wish to keep our modeling elements simple and non-specialized.

On average, 77% of the improving moves on the Job Shop instances, and 62% on the Open Shop instances, gave a solution that was initially infeasible, but was successfully repaired by our algorithm.

[1] So far, we always assumed that the iteration's initial solution $\mathcal{S}_0$ was feasible, so as to guarantee certain desirable properties. Yet, the repair mechanism can also be applied when $\mathcal{S}_0$ is infeasible, with a lower probability of success. Thus, it is also called in the early stages of LocalSolver's local search, before a feasible solution is found.

Table 1. Optimality gap – Job Shop and Open Shop Problems

Instance class	No of instances	Gap 10s		Gap 60s	
		No repairs	Repairs	No repairs	Repairs
Job Shop					
FT	3	73%	5%	15%	2%
LA	40	246%	8%	91%	3%
ORB	10	120%	6%	22%	3%
Open Shop					
4×4	10	41%	0.4%	41%	0%
5×5	10	65%	2.7%	65%	1.3%
7×7	10	102%	5.0%	93%	2.8%
10×10	10	569%	7.2%	261%	3.8%

7.2 Repair of Ternary Constraints in the Unit Commitment Problem

In this section, we focus on the Unit Commitment Problem, recently studied in [8]. We study a simplified version of the problem, where the level of production of each turned on unit is fixed to its average production rate. We model the problem as follows: each production range on any unit represents a task of variable duration. The constraints are chained disjunctive resource constraints: two consecutive tasks scheduled on a same unit must be separated by at least the unit's setup time. Since both the start time and the duration of each task are decision variables, the constraints are ternary.

In Table 2, we measure the performance improvement given by our repair algorithm in LocalSolver (compared to results obtained when turning repairs off), in 10 and 60 s. We also give the percentage of improving local transformations which needed repairing. We used the instances from [2]: the number of units varies from 10 to 100, and the number of time steps is 24.

Table 2. Performance improvement – simplified Unit Commitment Problem

No units	No instances	Improv. 10s	Improv. 60s	% repaired moves
10	30	4%	3%	3%
20	27	7%	6%	3%
50	25	13%	10%	3%
75	19	12%	12%	4%
100	25	7%	14%	4%

8 Conclusion

In this paper, we considered a family of optimization problems, characterized by a network of binary and ternary linear inequalities. We introduced a solution repair algorithm based on constraint propagation, overcoming the difficulties met by small-neighborhood search algorithms. The two main specificities of our propagation algorithm are that a domain reduction is only propagated if it excludes the current value of the variable, and that each variable must always be shifted in the same direction. We also described some desirable properties on the constraints, which ensure the success of the repair procedure.

The main limitation of our approach is the possibility of failure on complex constraints repairs, which incurs the withdrawal of the tentative move. However, this is largely tempered in practice by the rapidity of the overall iteration process, which allows a huge number of moves to be tested in a short time frame, and generally leads to successful repairs. As a result, its integration into LocalSolver dramatically improves its performance on the targeted problems, not only on classic scheduling problems such as the Job Shop, Open Shop and Unit Commitment Problems, but also on some 3D packing and mining industrial instances.

References

1. Ahmeti, A., Musliu, N.: Min-conflicts heuristic for multi-mode resource-constrained projects scheduling. In: Proceedings of the Genetic and Evolutionary Computation Conference, GECCO 2018, pp. 237–244. Association for Computing Machinery (2018)
2. Angulo, A., Espinoza, D., Palma, R.: Thermal unit commitment instances for paper: a polyhedral-based approach applied to quadratic cost curves in the unit commitment problem. http://www.dii.uchile.cl/~daespino/UC_instances_archivos/portada.htm
3. Applegate, D., Cook, W.: A computational study of the job-shop scheduling problem. ORSA J. Comput. **3**, 149–156 (1991)
4. Beck, J., Feng, T., Watson, J.P.: Combining constraint programming and local search for job-shop scheduling. INFORMS J. Comput. **23**, 1–14 (2011)
5. Benoist, T., Estellon, B., Gardi, F., Megel, R., Nouioua, K.: LocalSolver 1.x: a black-box local-search solver for 0–1 programming. 4OR **9**, 299–316 (2011)
6. Blaise, L., Artigues, C., Benoist, T.: Solution repair by inequality network propagation in LocalSolver. Technical report, LAAS-CNRS (2020). https://hal.laas.fr/hal-02866559
7. Ding, J., Shen, L., Lü, Z., Peng, B.: Parallel machine scheduling with completion-time-based criteria and sequence-dependent deterioration. Comput. Oper. Res. **103**, 35–45 (2019)
8. Dupin, N., Talbi, E.G.: Parallel matheuristics for the discrete unit commitment problem with min-stop ramping constraints. Int. Trans. Oper. Res. **27**(1), 219–244 (2020)
9. Even, S., Itai, A., Shamir, A.: On the complexity of timetable and multicommodity flow problems. SIAM J. Comput. **5**, 691–703 (1976)

10. Fisher, H., Thompson, G.L.: Probabilistic Learning Combinations of Local Job-Shop Scheduling Rules, pp. 225–251. Prentice-Hall, Englewood Cliffs (1963)
11. Gardi, F., Benoist, T., Darlay, J., Estellon, B., Megel, R.: Mathematical Programming Solver Based on Local Search. Wiley, Hoboken (2014)
12. Glover, F.: Ejection chains, reference structures and alternating path methods for traveling salesman problems. Discrete Appl. Math. **65**(1), 223–253 (1996)
13. Grimes, D., Hebrard, E., Malapert, A.: Closing the open shop: contradicting conventional wisdom. In: Gent, I.P. (ed.) CP 2009. LNCS, vol. 5732, pp. 400–408. Springer, Heidelberg (2009). https://doi.org/10.1007/978-3-642-04244-7_33
14. Lawrence, S.: Resource-constrained project scheduling: an experimental investigation of heuristic scheduling techniques (supplement). Graduate School of Industrial Administration, Carnegie-Mellon University, Pittsburgh, Pennsylvania, Technical report (1984)
15. Minton, S., Johnston, M.D., Philips, A.B., Laird, P.: Minimizing conflicts: a heuristic repair method for constraint satisfaction and scheduling problems. Artif. Intell. **58**(1), 161–205 (1992)
16. Peng, B., Lü, Z., Cheng, T.: A tabu search/path relinking algorithm to solve the job shop scheduling problem. Comput. Oper. Res. **53**, 154–164 (2015)
17. Siala, M., Artigues, C., Hebrard, E.: Two clause learning approaches for disjunctive scheduling. In: Pesant, G. (ed.) CP 2015. LNCS, vol. 9255, pp. 393–402. Springer, Cham (2015). https://doi.org/10.1007/978-3-319-23219-5_28
18. Taillard, E.: Benchmarks for basic scheduling problems. Eur. J. Oper. Res. **64**(2), 278–285 (1993)
19. Vaessens, R., Aarts, E., Lenstra, J.: Job shop scheduling by local search. Eindhoven University of Technology, Computing Science Notes (1994)
20. Zhang, J., Ding, G., Zou, Y., Qin, S., Fu, J.: Review of job shop scheduling research and its new perspectives under industry 4.0. J. Intell. Manuf. **30**, 1809–1830 (2017)

Optimising Tours for the Weighted Traveling Salesperson Problem and the Traveling Thief Problem: A Structural Comparison of Solutions

Jakob Bossek(✉), Aneta Neumann, and Frank Neumann

Optimisation and Logistics, The University of Adelaide, Adelaide, Australia
{jakob.bossek,aneta.neumann,frank.neumann}@adelaide.edu.au

Abstract. The Traveling Salesperson Problem (TSP) is one of the best-known combinatorial optimisation problems. However, many real-world problems are composed of several interacting components. The Traveling Thief Problem (TTP) addresses such interactions by combining two combinatorial optimisation problems, namely the TSP and the Knapsack Problem (KP). Recently, a new problem called the node weight dependent Traveling Salesperson Problem (W-TSP) has been introduced where nodes have weights that influence the cost of the tour. In this paper, we compare W-TSP and TTP. We investigate the structure of the optimised tours for W-TSP and TTP and the impact of using each others fitness function. Our experimental results suggest (1) that the W-TSP often can be solved better using the TTP fitness function and (2) final W-TSP and TTP solutions show different distributions when compared with optimal TSP or weighted greedy solutions.

Keywords: Evolutionary algorithms · Traveling Thief Problem · Node weight dependent TSP

1 Introduction

The Traveling Salesperson Problem (TSP) is one of the most prominent combinatorial optimisation problems and has been widely studied in the literature. It also serves as a basis for many more complex vehicle routing problems. Often real-world optimisation problems involve multiple interacting components that have to be optimised simultaneously. Moreover, due to the interactions the different silo problems can not be optimised separately in order to come up with an overall good solution [4].

The Traveling Thief Problem introduced in [3] is a multi-component problem that has recently gained significant attention in the evolutionary computation literature [6,7,9,15–19]. It combines the TSP and the classical Knapsack Problem by assigning items with profits and weights to the cities. The goal is to maximise the difference of profits of the collected items and the costs of a tour

T. Bäck et al. (Eds.): PPSN 2020, LNCS 12269, pp. 346–359, 2020.
https://doi.org/10.1007/978-3-030-58112-1_24

where the weights of items collected while visiting the cities increase the cost of moving from one city to the next one. More precisely, the weights of the items collected so far reduce the speed of the vehicle in a linear fashion and the cost of moving from city i to city j is determined by the current speed and the distance $d(i, j)$ of i and j. A wider range of benchmark instances have been introduced [11] and various competitions have been carried out at evolutionary computation conferences.

Understanding the interactions within the TTP is difficult. If the given tour is fixed and only the remaining (still $\mathcal{NP}$-hard) packing problem has to be solved, then this can be done by dynamic programming and also approximation algorithms are available [10]. However, optimising the tour for the TTP when the packing part is fixed seems to be significantly more difficult. In order to gain a better understanding on how node weights that influence the cost of a tour impact the optimisation, the node weight dependent Traveling Salesperson Problem (W-TSP) has been introduced recently [5]. Here each node has a weight and the cost of going from city i to city j is their distance $d(i, j)$ times the weight of the nodes visited so far. For special cases approximation algorithms have been designed in [5] that establish a relation to the minimum latency problem [2]. Furthermore, experimental investigations have been carried out to examine the impact of the node weights on the optimised salesperson tour.

With this paper, we continue this line of research and further bridge the gap in understanding the impact of node weights on salesperson tours. We examine and compare TTP and W-TSP in a systematic study. We consider a variant to TTP where the packing plan – and in consequence the total profit – is fixed and the goal is to minimise the cost of the weighted TTP tour length. We call this problem W-TTP. In our experimental investigations, we investigate instances where each item of a given TTP benchmark is present with probability p. Our study suggests, that with increasing p, i.e. increasing average number of nodes with strictly positive node weight, for the simple randomised search heuristic considered in this paper, it is advantageous to use the W-TTP objective as a driver for the search process instead of the W-TSP objective in order to find good solutions for the W-TSP. Furthermore, we consider the difference in terms of the structure of solutions obtained using the different problem formulation. In terms of structural similarity of W-TTP and W-TSP solutions produced by our simple heuristic, good W-TSP on average show higher similarity with the solutions obtained by a naive weighted greedy approach (WGR) then this is the case for W-TTP solutions with respect to a similarity measure based on the inversion number. In contrast, good W-TTP solutions on average share more edges with optimal TSP solutions. We hope that in future such findings can be leveraged to develop more sophisticated heuristic search algorithms for both the W-TSP and the Traveling Thief Problem.

The paper is structured as follows. We introduce the problems examined in this paper in Sect. 2, and we carry out our experimental investigations in Sect. 3. Afterwards, in Sect. 4, we investigate the relation of solutions among the two problems in terms of objective value ratios. In Sect. 5 we perform a structural similarity analysis of solutions with optimal TSP tours and weighted greedy solutions. We finish with some concluding remarks and avenues for future research.

2 Problem Formulation

The classical Traveling Salesperson problem is one of the most studied $\mathcal{NP}$-hard combinatorial optimisation problems. Given a set of n cities $V = \{1, \ldots, n\}$ and distances $d(i, j)$ between them, the goal is to find a permutation π which minimizes the tour length given by

$$\text{TSP}(\pi) = d(\pi_n, \pi_1) + \sum_{i=1}^{n-1} d(\pi_i, \pi_{i+1}).$$

Motivated by the TTP, we study variants of this problem where node weights influence the cost of a tour.

2.1 The Traveling Thief Problem

The Travelling Thief Problem (TTP) was first introduced in [3]. Given is a set of n cities $V = \{1, \ldots, n\}$ with pairwise distances $d(i, j)$ between them and a set $E_i = \{e_{i1}, \ldots, e_{im_i}\}$ of $m_i = |E_i|$ items at city i, $1 \leq i \leq n$. We denote by $E = \cup E_i$ the overall set of items. There is a profit $p \colon E \to \mathbb{R}^+$ and weight function $w \colon E \to \mathbb{R}^+$ on the items and knapsack capacity C which limits the total weight of a selection of items.

The goal in the TTP is to find a tour $\pi = (\pi_1, \ldots, \pi_n)$ and a packing plan $x = (x_{11}, \ldots, x_{nm_n})$ such that their combination π and x maximises the sum of the profits minus the travel cost associated with π and x. Note that in the classical TTP, there is usually no item available at city 1.

We indicate by a bitstring $x = (x_{11}, \ldots, x_{nm_n})) \in \{0, 1\}^m$, where $m = \sum_{i=1}^{n} m_i$, the items present in a problem instance. Item e_{ij} is present iff $x_{ij} = 1$ holds.

We denote by

$$w(\pi_i, x) = \sum_{k=1}^{m_{\pi_i}} w(e_{\pi_i k}) x_{\pi_i k}$$

the weight of the items taken in city π_i with packing plan x. The number of present items at city π_i is

$$\eta(\pi_i) = \sum_{k=1}^{m_{\pi_i}} x_{\pi_i k}.$$

In our experiments, we consider the case where all cities have the same number of items and use the notion IPN for *items per node*.

Let $\omega(i) = \sum_{j=1}^{i} w(\pi_j, x)$ be the sum of the weights of the cities in permutation π up to the ith city. The cost of a tour is given by the time the vehicle takes to complete the tour. Here the weight of the items present when going from city i to city j depends on the distance $d(i,j)$ and the speed $\upsilon \in [\upsilon_{\min}, \upsilon_{\max}]$, where $\upsilon_{\min}$ is the minimum speed and $\upsilon_{\max}$ is the maximum speed of the vehicle. The tour has to start and city 1 and therefore $\pi_1 = 1$ is required.

The goal in the standard formulation of TTP is to maximize

$$\mathrm{TTP}(\pi, x) = \sum_{e \in E} p(e) x_e - R \left(\frac{d(\pi_n, \pi_1)}{\upsilon_{max} - \nu\omega(n)} + \sum_{i=1}^{n-1} \frac{d(\pi_i, \pi_{i+1})}{\upsilon_{max} - \nu\omega(i)} \right)$$

where $\sum_{e \in E} p(e)$ is the sum over all packed items' profits, $\nu = (\upsilon_{max} - \upsilon_{min})/C$ is a constant value defined by the input and R is a constant called the renting rate.

We assume that the packing plan is fixed x for a given instance. If x is fixed then the profits and the weights at the cities are completely determined. We ignore the profit part and the renting rate as both are constant and do not have any impact on the order of solutions with respect to the fitness function TTP. In our study, we investigate the following cost function which depends on the weights of the items determined by x and the chosen permutation π:

$$\text{W-TTP}(\pi, x) = \left(\frac{d(\pi_n, \pi_1)}{\upsilon_{max} - \nu\omega(n)} + \sum_{i=1}^{n-1} \frac{d(\pi_i, \pi_{i+1})}{\upsilon_{max} - \nu\omega(i)} \right)$$

We call the problem of finding a tour which minimizes this goal function the weighted TTP-problem (W-TTP).

2.2 The Node Weight Dependent TSP

We also consider the node weight dependent TSP problem (W-TSP) recently introduced in [5]. In addition to the input of the TSP, we have a set of possible items E_i available at each city i. Following the notation for W-TTP, we indicate by a bitstring $x \in \{0,1\}^m$ whether an item e_{ij} is present.

Given a set of n cities $V = \{1, \ldots, n\}$ with distances $d(i,j)$ between the cities and a weight function $w\colon E \to \mathbb{R}^+$ on the set of items, the goal is to find a permutation π that minimizes the weighted TSP cost. The tour has to start and city 1 and therefore $\pi_1 = 1$ is required. We denote by

$$w(\pi_i, x) = \sum_{k=1}^{m_{\pi_i}} w(e_{\pi_i k}) x_{\pi_i k}$$

the weight of the items presents at city π_i. The fitness of a given tour π and a given set of present items indicated by x is given as

$$\text{W-TSP}(\pi, x) = d(\pi_n, \pi_1) \left(\sum_{j=1}^{n} w(\pi_j, x) \right) + \sum_{i=1}^{n-1} d(\pi_i, \pi_{i+1}) \left(\sum_{j=1}^{i} w(\pi_j, x) \right).$$

Note, that the standard TSP is the special case where $w(\pi_1) = 1$ and $w(\pi_i) = 0$, $2 \leq i \leq n$.

Our fitness function definitions for W-TTP and W-TSP work with a set of present items which can also be defined in terms of the input items without using the bitstring x. We use the notation of present items indicated by x as we will use TTP benchmarks where different subsets of items of a given TTP instance have to be collected in the computed tour.

2.3 Problem Comparison

The TSP, W-TTP, and W-TSP place different emphasize on the weight of nodes. The TSP can be considered as the special case of W-TSP where only the first node receives a weight of 1. Furthermore, TSP is a special case of the tour optimisation variant of TTP where no item is collected, and the vehicle always travels at maximum speed $v_{\max}$. W-TSP allows for a very drastic and high weightening of distance costs as the weights are collected during the route and each distance is multiplied with the weight of the cities visited. TTP in more limited in terms of the impact of the weightening as the weight of the items reduces the speed from $v_{\max}$ to $v_{\min}$ in a linear fashion. Using the interval $[v_{\min}, v_{\max}]$ for the speed also ensures that the weighted distance for going from city i to j is always in the interval $[d(i,j)/v_{\max}, d(i,j)/v_{\min}]$ where as in the case of W-TSP this can be in the range $[0, W \cdot d(i,j)]$ where W is the total weight amount all cities.

3 Experimental Setup

The focus of this paper is on understanding interactions between solutions for the W-TTP and the recently introduced W-TSP. To study these effects, we consider a subset of instances from the TTP 2017 CEC Competition[1] for our experiments [11]. We choose all instances which are based on the following classical TSPlib [12] instances: a280, berlin52, ch130, ch150, eil101, eil51, eil76, kroA100, kroC100, kroD100, lin105, pcb442, pr1002, pr2392, pr76, rd100, st70. Therein, all three weight/profit classes are covered: bounded strongly correlated (bsc), uniform similar weights (usw) and uncorrelated (u). Furthermore, the number of items per node (IPN) is either one or five. In total our benchmark set contains 102 instances. The subset is a cross-section of the TTP benchmark set with instances of few nodes up to instances with several thousand nodes. In addition, optimal tours for the classical TSP are known for these instances. This will be of essential for structural similarity analysis in Sect. 5. Recall that in our setup the packing plan is initially fixed and so are the weights at the nodes; no changes to the packing are made in the course of optimisation. To account for the stochasticity in the packing and the influence of the fraction of active items, for each instance and each $p \in \{0.01, 0.05, 0.1, 0.2, 0.3, 0.4, 0.6, 0.8, 1.0\}$ we generated 31 random packings from a $\text{Bin}(m, p)$-distribution where m is the number of items

[1] https://cs.adelaide.edu.au/~optlog/TTP2017Comp/.

of the TTP instance at hand, i.e. each items is packed with probability p and not packed with inverse probability $(1-p)$. In order to make all generated packings feasible, we set the knapsack capacity C to the sum of all item weights (not just the packed ones).[2] Note that this choice for the knapsack capacity allows us to explore different degrees of filling of the vehicle. In consequence a transition from the classical TSP (p close to zero) and the TTP with a fully loaded vehicle (p close to one) is possible.

We consider the classical (1 + 1)-EA with inversion mutation on permutations. Preliminary benchmarking with swap and insertion mutation showed its superiority; this confirms the experimental results in [5] on the W-TSP. We urge the reader to carefully read the following sentences as they convey a crucial aspect of our study: we run (1+1)-EA with either the W-TTP or the W-TSP for driving the evolutionary search process (EA driver). In addition, the best so far solution in every iteration and in particular the final best solution is evaluated with both W-TTP and W-TSP resulting in four different relevant combinations.

(1 + 1)-EA is applied each one time on each instance and each of the 31 associated packings plans. Note, that we do not perform additional independent runs for each fixed packing plan. Instead, the 31 runs already account for the stochasticity. Our implementation and data is available in a public GitHub repository.[3]

4 Comparison in Terms of Solution Quality

We first approach the following research question: is it beneficial to use each others fitness function for optimisation purposes? More precisely, if we aim to optimise the W-TTP (W-TSP), should we use the actual objective function as EA driver or is it of benefit to use the W-TSP (W-TTP) objective function instead? One might argue that it certainly makes no sense to use another fitness function as a surrogate. However, our results prove this assumption wrong in many cases. Figure 1 show the distribution of objective value ratios across all runs on all considered instances separated by the instance property IPN and the packing probability p. The ratios are to be interpreted as follows: when the objective is W-TSP we divide the W-TSP objective value of the final solution determined with the W-TSP-driver by the W-TSP objective value of the final solution obtained by optimising with the W-TTP-driver and vice versa. Since both objectives are to be minimised a ratio below 1.0 indicates that it is advantageous to use the actual objective function to guide the EA; the result one would expect. Returning to Fig. 1 we actually see that this assumption does not always hold true; at least in one direction. The data shows that it is consistently advisable to use the W-TTP objective function to optimise the W-TTP. However, a closer look shows that the W-TTP-related box-plots show a characteristic U-shape with peaks in the area of $p \approx 0.5$. In contrast, with W-TSP

[2] Note that this step is relevant for the W-TTP only; the W-TSP objective function does not cope with a knapsack limit.

[3] GitHub repository: http://github.com/jakobbossek/ttp.

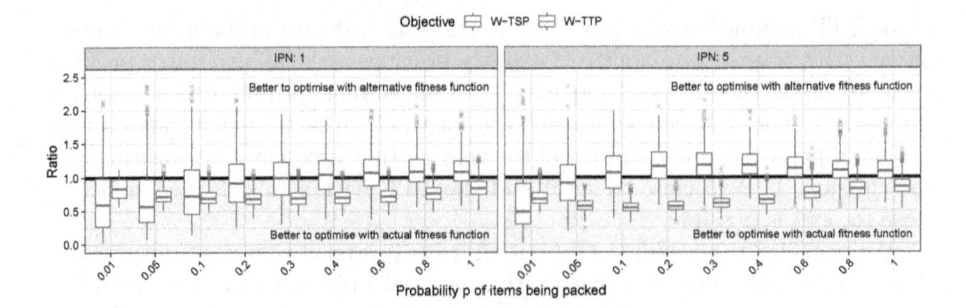

Fig. 1. Distribution of objective value ratios of final tours. Ratios are calculated by the following rule: if W-TSP is to be minimised we divided the W-TSP tour-length obtained by optimising with the actual W-TSP driver with the W-TSP tour-length of the solution calculated when the algorithm is run with the W-TTP driver instead. Ratios for W-TTP optimisation are calculated analogously. Ratios below zero indicate a benefit for the actual objective function.

being in the focus of optimisation we observe a very different pattern. Here, with $p \to 1$, the median ratio increases. The median surpasses 1.0 for the first time at a level of $p = 0.4$ with one item per node and $p = 0.1$ for IPN $= 5$. Our assumption is that for IPN $= 1$ and given $p \in [0, 1]$ in expectation np nodes have a strictly positive weight. In contrast, if there are multiple items per node, due to independence of the item activation in the packing plan generation, in each node $m_i p$ are expected to be active. Hence, in expectation, there will be more nodes with strictly positive weight assigned in this setting. Either case it seems as with increasing p oftentimes the W-TTP-driver leads to better W-TSP tours. The results suggest that using the W-TSP objective produces large basins of attraction for qualitatively bad local optima. Figure 2 shows a less aggregated view. Here, the ratios are shown for three representative instances from the benchmark set (still aggregated across weight/profit types bsc, usw and u since the type does not reveal any different patterns). Here, in particular the largest pr2392-based instances with $n = 2\,392$ nodes stands out from the crowd: here the aforementioned U-shape observed for the W-TTP is inverse for the W-TSP at least for IPN $= 5$. For this particular instance the difference between median ratios is highest and using the W-TTP EA-driver for moderate p leads to median quality gains of ≥ 1.5 which is massive.

Figure 3 visualises the trajectories/development of incumbent solutions for two representative instances. In particular for $p = 0.3$ (second column) we see that for these particular runs in fact the final W-TSP objective is better when the EA driver is W-TTP. Moreover, occasional decrease in fitness values can be observed even though the general optimisation goal is still purchased.

In order to make sense out the data we trained a simple decision tree to decide which EA-driver to use in order to solve the W-TSP. Since the W-TTP is best solved by adopting the W-TTP driver (beside few outliers) we did not perform this step for the other direction. Our goal was a simple binary classification task.

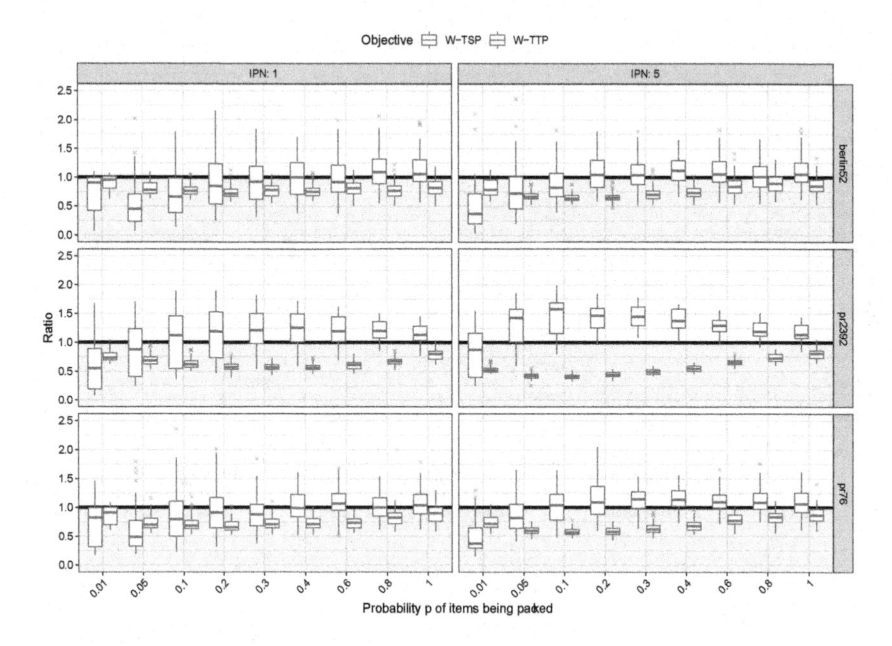

Fig. 2. More fine-grained objective ratios for three representative instances (rows) and different item counts (columns).

I.e. the target is to decide which EA-driver is preferable while predictor variables are the instance size n, the IPN value and the probability p. We used 10-fold cross-validation and the R-package `rpart` [13] interfaced by package `mlr` [1] to train the model and access its performance. The cross-validation results report a mean miss-classification test error of 18.5% and thus an accuracy of 81.5% in predicting the best EA-driver. This is not overwhelming, though admittedly higher than tossing a coin. The final decision tree is depicted in Fig. 4. The splits used by the model, i.e. decisions made when we follow the nodes from the root down to leaf level, very much reflect our previous observations where the W-TTP driver is advantageous for larger p and IPN > 1.

5 Structural Similarity Analysis of Solutions

In the following we conduct a similarity analysis of solutions. To be more concise we investigate the similarity of final W-TTP and W-TSP solutions calculated in our study with two types of permutations: (1) optimal TSP solutions for the underlying TSP instance and (2) tours calculated by a greedy algorithm which favors visiting "heavy" nodes, i.e. nodes of high weight, later in the tour. In a nutshell the algorithm termed *weighted greedy* (WGR) works as follows. In a first step nodes are sorted in ascending order of their node weight. The second step is about tour construction. Here, nodes are visited in ascending order of node weight. In case of ties, i.e. several nodes with the same node weight, these nodes

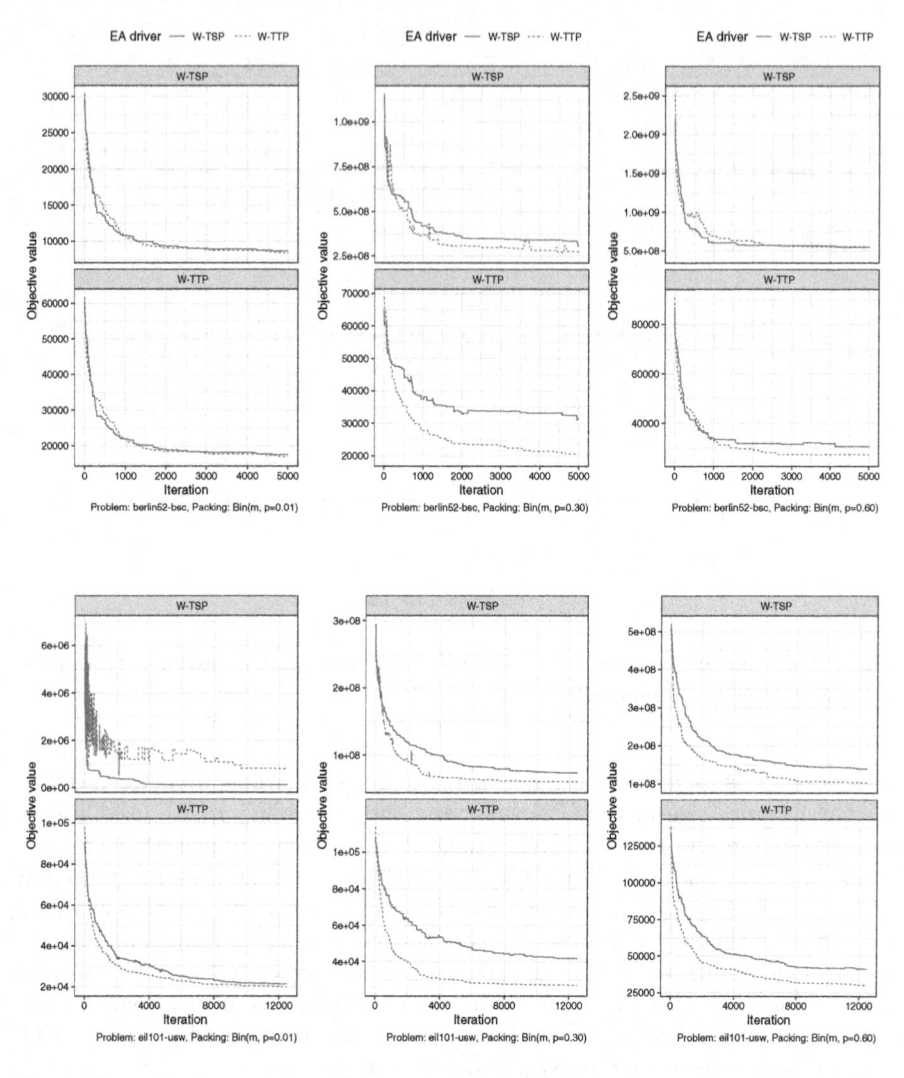

Fig. 3. Exemplary trajectories for instance berlin52 (two top rows) and eil101 (two bottom rows) with bounded strongly correlated weights and 5 items per node. The EA was run with both W-TSP and W-TTP as driver (indicated by color and line type). Likewise, incumbent solutions were evaluated with both objective functions (W-TSP in top and W-TTP in bottom row). (Color figure online)

are visited following the nearest neighbor heuristic [8]. This construction method can be seen as a naive approach to solve the W-TTP or W-TSP respectively where one might assume that nodes with a high weight loading should be visited later on even if this requires to take some long distance edges beforehand. Note that the optimal TSP tours and WGR tours pose two extremes: the TSP tour is focused on the distances only neglecting node weights completely. In contrast,

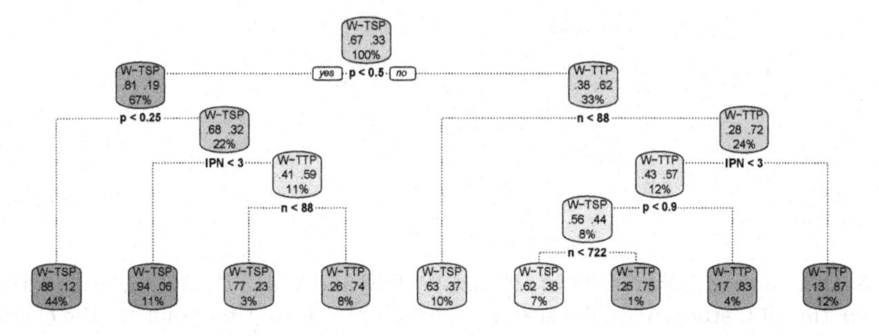

Fig. 4. Decision tree for the machine learning task of determining which objective function should be used in order to optimise the W-TSP. Within the splits p is the probability of items being active, n is the number of nodes and IPN is the number of items per node. Values within the nodes indicate the majority decision (top), the fraction of data processed by the left/right branch respectively (center) and the percentage of overall data points processed at that node.

WGRs' focus, though not able to guarantee optimality, is mainly on late heavy node placement in the tour.

For the purpose of measuring similarity we use two metrics for the comparison of two tours (permutations) π^1 and π^2. The first is termed *common edges* (CE) and is defined as the proportion of edges shared by both tours. The second metric is based on the mathematical term of *inversion* which – in the classical sense – is a measure of the sortedness of a sequence: for a permutation π, if $1 \leq i < j \leq n$ and $\pi_i > \pi_j$ the pair (i, j) is called an inversion [14]. The total count of inversions $\text{IN}(\pi^1, \pi^2)$ is termed the *inversion number* which is at most $n(n-1)/2$ with higher values indicating stronger dissimilarity with respect to sortedness. In our setting though we are given two permutations π^1, π^2 and we call (i, j) an inversion, if node i is visited before (after) node j in π^1 and after (before) j in π^2. In order to obtain a normalised similarity version we define our second measure as follows:

$$\text{INV}(\pi^1, \pi^2) := 1 - \left(\frac{2 \cdot \text{IN}(\pi^1, \pi^2)}{n(n-1)}\right) \in [0, 1].$$

We want to stress that with a simple heuristic like the (1 + 1)-EA it is unlikely to get optimal solutions to our problems. In consequence, the following observations are based on sub-optimal approximations to the W-TTP and W-TSP respectively. Nevertheless, we believe that our insights are valuable first steps towards a better understanding of tour composition.

Figure 5 shows the distribution of the similarity of W-TSP and W-TTP solutions with optimal TSP tours and WGR tours by means of the two measures CE and INV throughout the whole benchmark set. For ease of reference, we denote the similarity with CE[TSP], CE[WGR], INV[TSP] and INV[WGR]. Regarding CE[TSP]-similarity we observe a U-shape with increasing probability p for W-TSP. The box-plots for W-TTP however show a clear downward trend, i.e. the more items have to be collected by the thief, the less similar the tour gets to the

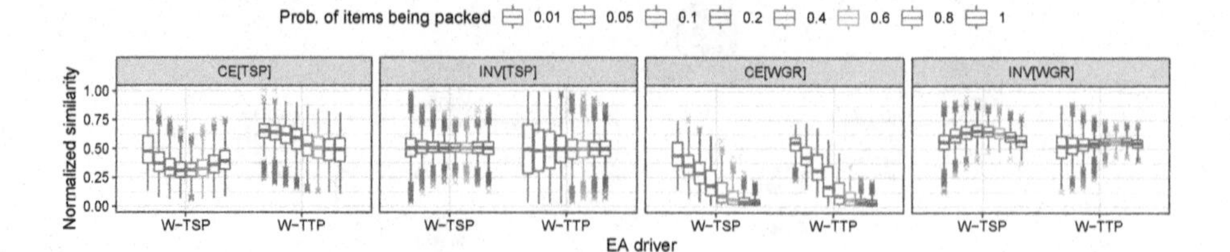

Fig. 5. Distribution of similarity of all final W-TSP and W-TTP solutions calculated in our experimental study. We calculate the similarity to the optimal tour for the classical TSP and the weighted greedy tours (WGR) respectively.

TSP. Nevertheless, for both W-TTP and W-TSP the median similarity is larger than 25% for all values of p and even above 50% for the W-TTP. Compared with this for both considered optimisation problems the CE[WGR]-similarity strongly decreases with increasing p. Here, median values close to 0% with low variance are reached if on average at least 60% of the items are active. The CE-measure is plain simple and kind of binary in the sense that an edge is either shared or not. However, even if the number of shared edges approaches zero the INV-similarity can show different patterns as it measures the number of swaps needed to transform one tour into another. In fact, median INV[WGR]-values are $> 50\%$ for all considered settings and both W-TSP and W-TTP. Moreover, with increasing p there is trend towards a narrowed outlier distribution, i.e. outliers are less frequent indicating a lower total range of similarity values. In addition, for the W-TSP we observe an inverted U-shape with its median peak at about $p = 0.2$. This suggests that for the W-TSP and a relatively low number of active items it is in fact advisable to place these heavy nodes in the end of the permutation. All observations made so far are valid for all considered instances and IPN values (see Fig. 6 for a less aggregated view for three representative instances). We clearly observe the same patterns even though the actual similarity values can differ substantially (cf. the CE[TSP]-similarity in Fig. 6). In particular pr2392-based instances stand out. This is partly explained by its size (2 392 nodes) which is much bigger than the majority of our benchmark instances and the fact that we use a very simple heuristic. Therefore, our W-TSP and W-TTP solutions for those instances are likely far away from optimal.

Coming back to the actual measures: the only measure which shows strong variance throughout the instance set is INV[TSP]. This observation can be visually derived from Fig. 5 where we see many partly extreme outliers and is backed up by the representative more fine-grained plots in Fig. 6. The strong variance is even more pronounced for the W-TTP solutions. To be honest, at this point we have no clear explanation to this phenomenon.

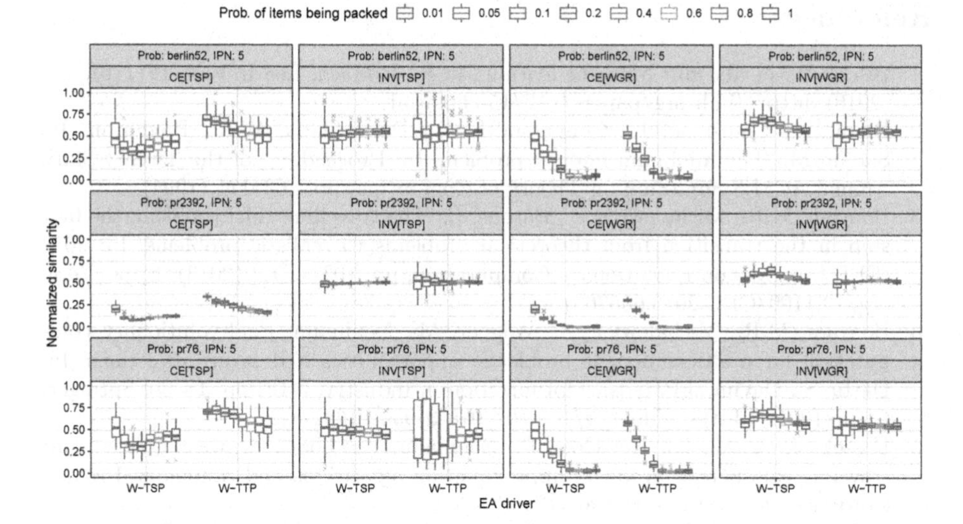

Fig. 6. Distribution of similarity of all W-TSP and TTP solutions calculated for instances of type berlin52 (top row), pr2392 (middle row) and pr75 (bottom row) to the respective optimal TSP tours and weighted greedy tours (WGR).

6 Conclusion

Multi-component problems appear frequently in real-world applications and the TTP (combining the TSP and KP) has been introduced as a benchmark problem to study such problem in greater depth. Understanding the interaction of the two components is still a challenging task and we focused in this paper on the weighted TSP part of the problem. We have carried out a structural comparison of TSP variants called W-TTP and W-TSP where the weight on nodes determined by a collection of items plays a crucial role in determining an optimal tour when the to be collected set of items is fixed. Our results show that W-TTP is closer to the TSP than the W-TSP and that using the fitness function of W-TTP can surprisingly lead to better results when the goal is to optimise W-TSP.

Future work will investigate the similarity of high quality solutions of W-TTP and W-TSP. Furthermore, evolving instances that show a significant performance difference for optimised tours of W-TTP and W-TSP and their characterization in terms of problem features would help to push forward the understanding of the these problems.

Acknowledgment. This work has been supported by the Australian Research Council (ARC) through grants DP160102401 and DP190103894, and by the South Australian Government through the Research Consortium "Unlocking Complex Resources through Lean Processing".

References

1. Bischl, B., et al.: mlr: Machine learning in R. J. Mach. Learn. Res. **17**(170), 1–5 (2016). http://jmlr.org/papers/v17/15-066.html
2. Blum, A., Chalasani, P., Coppersmith, D., Pulleyblank, W.R., Raghavan, P., Sudan, M.: The minimum latency problem. In: Proceedings of the Twenty-Sixth Annual ACM Symposium on Theory of Computing, pp. 163–171 (1994)
3. Bonyadi, M.R., Michalewicz, Z., Barone, L.: The travelling thief problem: the first step in the transition from theoretical problems to realistic problems. In: 2013 IEEE Congress on Evolutionary Computation, pp. 1037–1044 (2013). https://doi.org/10.1109/CEC.2013.6557681
4. Bonyadi, M.R., Michalewicz, Z., Wagner, M., Neumann, F.: Evolutionary computation for multicomponent problems: opportunities and future directions. In: Datta, S., Davim, J.P. (eds.) Optimization in Industry. MIE, pp. 13–30. Springer, Cham (2019). https://doi.org/10.1007/978-3-030-01641-8_2
5. Bossek, J., Casel, K., Kerschke, P., Neumann, F.: The node weight dependent traveling salesperson problem: approximation algorithms and randomized search heuristics (2020). to appear at GECCO 2020
6. El Yafrani, M., Ahiod, B.: Population-based vs. single-solution heuristics for the travelling thief problem. In: Genetic and Evolutionary Computation Conference (GECCO), pp. 317–324. ACM (2016)
7. Faulkner, H., Polyakovskiy, S., Schultz, T., Wagner, M.: Approximate approaches to the traveling thief problem. In: Conference on Genetic and Evolutionary Computation (GECCO), pp. 385–392. ACM (2015)
8. Lawler, E.: The Travelling Salesman Problem: A Guided Tour of Combinatorial Optimization. Wiley-Interscience Series in Discrete Mathematics and Optimization. Wiley, Hoboken (1985). https://books.google.com.au/books?id=qbFlMwEACAAJ
9. Mei, Y., Li, X., Yao, X.: On investigation of interdependence between sub-problems of the travelling thief problem. Soft Comput. **20**(1), 157–172 (2016)
10. Neumann, F., Polyakovskiy, S., Skutella, M., Stougie, L., Wu, J.: A fully polynomial time approximation scheme for packing while traveling. In: Disser, Y., Verykios, V.S. (eds.) ALGOCLOUD 2018. LNCS, vol. 11409, pp. 59–72. Springer, Cham (2019). https://doi.org/10.1007/978-3-030-19759-9_5
11. Polyakovskiy, S., Bonyadi, M.R., Wagner, M., Michalewicz, Z., Neumann, F.: A comprehensive benchmark set and heuristics for the traveling thief problem. In: Proceedings of the Genetic and Evolutionary Computation Conference (GECCO), pp. 477–484. ACM (2014). https://doi.org/10.1145/2576768.2598249
12. Reinelt, G.: TSPLIB-a traveling salesman problem library. ORSA J. Comput. **3**(4), 376–384 (1991)
13. Therneau, T., Atkinson, B.: rpart: recursive partitioning and regression trees (2018). https://CRAN.R-project.org/package=rpart. R package version 4.1-13
14. Vitter, J.S., Flajolet, P.: Average-Case Analysis of Algorithms and Data Structures, pp. 431–524. MIT Press, Cambridge (1991)
15. Wagner, M., Lindauer, M., Mısır, M., Nallaperuma, S., Hutter, F.: A case study of algorithm selection for the traveling thief problem. J. Heuristics **24**(3), 295–320 (2017). https://doi.org/10.1007/s10732-017-9328-y
16. Wu, J., Polyakovskiy, S., Neumann, F.: On the impact of the renting rate for the unconstrained nonlinear knapsack problem. In: Proceedings of the Genetic and Evolutionary Computation Conference (GECCO), pp. 413–419. ACM (2016). https://doi.org/10.1145/2908812.2908862

17. Wu, J., Polyakovskiy, S., Wagner, M., Neumann, F.: Evolutionary computation plus dynamic programming for the bi-objective travelling thief problem. In: Proceedings of the Genetic and Evolutionary Computation Conference (GECCO), pp. 777–784. ACM (2018). https://doi.org/10.1145/3205455.3205488
18. Wu, J., Wagner, M., Polyakovskiy, S., Neumann, F.: Exact approaches for the travelling thief problem. In: Proceedings of the 11th International Conference on Simulated Evolution and Learning (SEAL), pp. 110–121 (2017)
19. Yafrani, M.E., Ahiod, B.: Efficiently solving the traveling thief problem using hill climbing and simulated annealing. Inf. Sci. **432**, 231–244 (2018)

Decentralized Combinatorial Optimization

Lee A. Christie(✉)

School of Computing, Robert Gordon University, Aberdeen, Scotland
l.a.christie@rgu.ac.uk

Abstract. Combinatorial optimization is a widely-studied class of computational problems with many theoretical and real-world applications. Optimization problems are typically tackled using hardware and software controlled by the user. Optimization can be competitive where problems are solved by competing agents in isolation, or by groups sharing hardware and software in a distributed manner.

Blockchain technology enables *decentralized applications* (DApps). Optimization as a DApp would be run in a trustless manner where participation in the system is voluntary and problem-solving is incentivized with bitcoin, ether, or other fungible tokens. Using a purpose-built blockchain introduces the problem of bootstrapping robust immutability and token value. This is solved by building a DApp as a smart-contract on top of an existing Turing-complete blockchain platform such as Ethereum.

We propose a means of using Ethereum Virtual Machine smart contracts to automate the payout of cryptocurrency rewards for market-based voluntary participation in the solution of combinatorial optimization problems without trusted intermediaries.

We suggest use of this method for optimization-as-a-service, automation of contests, and long-term recording of best-known solutions.

1 Introduction

1.1 Motivation

Combinatorial optimization is an important class of computational problem, and may be tackled using some form of incentivized cooperative optimization. However, this requires solving the cheating problem. It also requires management by a centralizing party, and some degree of trust between parties.

Smart contracts on an open, decentralized blockchain allow us to automate the issuance of reward without the need for a trusted third party. This can enable us to co-ordinate problem solving between multiple parties. The existing infrastructure provided by open, public blockchains such as Ethereum provides an opportunity to build a decentralised application for combinatorial optimization.

T. Bäck et al. (Eds.): PPSN 2020, LNCS 12269, pp. 360–372, 2020.
https://doi.org/10.1007/978-3-030-58112-1_25

1.2 Cooperative Optimization

In a traditional approach to optimization, multiple participants may tackle the same problem, either at the same time or at different times with no co-ordination. Each attempt to optimize the function is done without regard to other attempts. Optimization in isolation can be seen when solving benchmark problems, which is a competitive activity, with results published at the end of optimization process.

We define *cooperative optimization* as multiple *hosts* working together in collaboration to solve an optimization problem. In contrast to competitive optimization, the hosts are allied in a team and do not gain from the loss of another host. Hosts could all be under the control of a single entity, the *client*, or could be part of a volunteer computing project where the problem is specified by the client. Cooperative optimization may be practiced by 'grid search' over a search space, 'parameter sweep' varying the parameters of an algorithm, or by parallel algorithms such as a genetic algorithm with island models. The degree to which hosts share information varies by technique, however it is typically not the case that a host loses by helping another host succeed.

1.3 Incentivized Cooperative Optimization

When multiple independent participants take part in cooperative optimisation, some form of incentivization may be introduced. Incentivization may take the form of financial reward, or non-monetary reputation/scoring.

Typically under cooperative optimization systems, hosts are expected to carry out a set program/algorithm and report the result. This program could be, for example, exhaustive search over a small subspace, or running a particular algorithm with a given set of parameters. It may be the case that carrying out these instructions yields no useful result, for example if the subspace contained no good or viable solutions, or the parameter set was sub-optimal for the problem. Participants are rewarded for the amount of computation done.

It may be possible for hosts to 'cheat' by falsely reporting that the work was carried out. Verification is possible if the host is required to perform some 'residual' side-calculation, but often the only way to ensure the residual is computing correctly is for the client to redundantly re-issue the same work to another participant, assuming the verifier is incentivized to perform the verification diligently and not cheat at verifying.

This issue is more broadly described as *the cheating problem* in volunteer computing projects, and solutions have been proposed [4,6]. In this work, we propose a novel system of distributing optimization without central control, and without the need to control for cheating.

1.4 Outline

In Sect. 2, we give background on decentralized applications (DApps), the concept on which the proposed system is built. In Sect. 3, we outline the concept of decentralized optimization (DOpt) proposed. In Sect. 4, we discuss the work which has been done to implement a proof-of-concept for the DOpt system. In Sect. 5, we propose potential practical applications which further motivate the development of decentralised optimisation. In Sect. 6, we outline the further work which needs to be done.

2 Decentralized Applications (DApps)

2.1 Bitcoin

There has long been a desire to implement a peer-to-peer digital currency. Traditionally, a system without the need for trusted intermediaries faces the issue of *double-spending* - reversing a transaction after receiving a good or service. Many systems had been proposed and attempted, however the first widely-adopted decentralized digital currency has been Bitcoin.

The Bitcoin white paper [8] put forth a solution to the double-spending problem of digital currencies in the form of proof-of-work by partial hash inversion, a solution to the Byzantine generals problem [7]. This allows multiple parties in a distributed peer-to-peer network to reach a consensus about the current state of the system without any trusted parties or identification. Bitcoin is regarded as the first successful decentralised application (DApp).

2.2 Proof of Work (PoW)

Distributed ledgers are a means of coordinating on the state of a system. They need a means of agreeing on the state. A naive voting system would be vulnerable to Sybil attack [3], whereby a malicious agent gains multiple votes by adopting multiple identities. PoW serves as a way of randomly selecting a participant in the network who decides on which new transactions to append to a ledger.

Transactions are grouped into *blocks*. Proof of work requires miners to solve a satisfaction problem to mine a new block and attach it to the *blockchain*, a cryptography-hashed backwards-singly linked list of blocks. Rewriting old state in the blockchain is not possible without redoing the work at a rate faster than the honest miners, which provides security to the blockchain. The method used by Bitcoin is *partial hash inversion*, where miners update a nonce value in the block header until, by chance, the hash of the block header is less than a target value. If the target begins with n binary zeros, the probability of random data hash being satisfactory is 2^{-n}. The target is automatically adjusted such that the average period of block mining is 10 min and a transaction is considered practically irreversible after 6 blocks (1 h).

It is important to note that each solve attempt is a statistically independent event. This is a key property known as *zero progress*. The proof of work algorithm is designed to have the following properties[1]:

[1] Agreement on the exact desirable properties and their relative importance is debatable, and may vary between blockchain designs.

- Zero Progress - There should be no learnable structure in the problem, so that each attempt is a statistically independent event.
- Asymmetric - It must be computationally expensive to solve the puzzle, but trivially easy to verify.
- Scalable Difficulty - The difficulty of the puzzle should be able to be automatically scaled to set the target amount of time to solve.
- Not Predictable - No user should be able to start on the next puzzle until the current is solved.
- Not Beneficial - The computation should serve no purpose other than proving that the miners consumed electrical energy to secure the blockchain.

Given that the current energy consumption of the bitcoin network use to proof of work is estimated at around 71 TWh/year [2], it is often suggested that proof-of-work be replaced by useful computation such as protein folding, or in our case, optimization. Such useful work fails to meet most of the desirable properties of a proof-of-work algorithm given above.

2.3 Smart Contracts

A *smart contract* is an agreement between two or more parties which is enforced, not by law, but by software. [10] Parties involved may be persons, companies, or autonomous software agents.

Platforms such as Bitcoin allow for smart contracts. In the context of Bitcoin, a smart contract is code embedded in the blockchain as a script and replicated across all nodes on the network. The code can be executed based on transactions and can be used to pay out native tokens in the form of *bitcoin* (BTC) when pre-defined criteria are met.

The script comes in two parts: the *locking script* and an *unlocking script*. The locking script is attached to a bitcoin output. The unlocking script redeems the value to use as an input to another transaction. If the two scripts concatenated together forms a valid execution, the spend is valid, otherwise is it rejected.

The *Bitcoin scripting language* (Script) is intentionally Turing-incomplete. [12] This is a limitation with implementing arbitrary functions. The rationale is to ensure that contracts are executed in a deterministic number of operations, preventing denial of service attacks. Additionally, scripts are stateless: the blockchain only records whether a given output has been *spent*, and spending is all-or-nothing. The following subsection will cover the basic concept of using scripts to control cryptocurrency ownership, whereas Subsect. 2.5 we will discuss a Turing-complete system, which would be necessary for distributed optimization.

2.4 Standard Payments Vs Transaction Puzzles

Most Bitcoin transactions are simple payment of value (bitcoins) from one user to another using the standard Pay-to-Public-Key-Hash [11] (P2PKH) script.[2]

[2] Other scripts such as P2SH-wapped and native bech32 are also common.

The details of the script are omitted for brevity. The effect of this script is checking that the following two conditions hold of a future unlocking transaction which attempts to spend the given transaction output:

1. reveals the public key whose double-hash (known as an *address*) matches the previously specified recipient address; **and**
2. is signed by the private key corresponding to the revealed public key.

Note that both of these conditions are encoded in the locking script specified by the sender (constructed as standard by the sender's wallet software). The script is written to ensure the value is transferred to the intended recipient.

One uncommon use-case of Bitcoin scripts is known as a *transaction puzzle.* A transaction puzzle does *not* specify a recipient, and therefore is of the class of *anyone-can-spend* transaction outputs.

Instead of the locking script being designed to target a specific recipient, it pays anyone who can provide the solution to a satisfaction problem, such as providing a blob of data whose SHA256 hash is equal to the predetermined value.

One potential application of a Bitcoin transaction puzzle could be set so as to require a solution to a satisfaction problem, however f must be implementable in the Turing-incomplete Bitcoin scripting language.

Other limitations exist, such as the requirement for the entire puzzle reward to be paid out to one single solver. The reward cannot be shared for partial solutions or best-so-far solutions.

Additional details need to be considered. For example, if a time-limit is to be set on the problem (after which the problem-setter recovers their funds), a time-lock will need to be set, which is an extra complexity on the script. This was not used in the hash puzzle example above.

One major limitation is that any anyone-can-spend transactions are highly vulnerable to *mempool attacks.* A mempool attack is one in which in which an attacker intercepts the solution (which is broadcast publicly) and re-transmits their own solution with a higher network priority. The result is that the attacker is rewarded with the entire prize amount and the honest participant receives nothing. Such attacks can be automated anonymously on the network.

2.5 Ethereum

Ethereum [1,14] is a platform for Turing-complete DApps. It uses a native token called *ether* (ETH) which functions as a currency and also is used to pay for execution time of smart contracts. Contracts execute on the global, decentralized *Ethereum Virtual Machine* (EVM).

A smart contract account can 'hold' ether just as a user can, and interactions with the contract can result in the contract sending and receiving ether. A simple example is a *faucet* contract containing two methods, one default method marked 'payable' into which users may donate ether, and another from which another user may specify a desired amount, and the contract will automatically payout the requested amount. Arbitrarily complex programming logic and statefulness can be used in the design of EVM DApps in determining payout conditions and amounts.

2.6 Game Theory and DApps

Participation in a DApp is typically voluntary, incentivized, and pseudonymous. Identities with negative reputation can be abandoned and replaced. This means that iterated interactions with the same bad actor may not be detectable, and participants with high reputation my be engaged in negative behaviour under an alternate identity. It may be assumed that participants will use any possible exploits in the system for self-profit. DApps requires careful, and explicit consideration of game theory in their design and security auditing of their implementation.

It is important that all rules of the DApp be enforced either by the terms of smart contract directly, or indirectly. Rules enforced directly by smart contract cannot be broken as the software does not allow it. Rules enforced indirectly may used a 'watchtower' system of enforcing a financial penalty in the event that one participant cheats and another participant detects the cheating. This may be used where it is infeasible to enforce a given rule directly. Such penalties are only possible when potentially-cheating actions require collateralizing tokens, typically under some time-lock.

3 Decentralized Optimization (DOpt)

3.1 Decentralized Optimization

In this work we introduce the concept of *decentralized optimization* (DOpt), a system in which participants (*hosts*) race to solve an optimization problem in real time without central co-ordination or requirement for trust. Distributed optimization differs from cooperative optimization with incentivisation in that limited sharing of information (which could help competitors) is practiced in exchange for rewards. Hosts are not required to share information as the optimization proceeds, but do so out of incentivized self-interest.

DOpt does not face the issue of cheating in the same way as traditional volunteer computing projects. Part of the novelty of this approach is that following a prescribed process is not a required behaviour. In fact, participants are able to use any algorithm without the need to disclose the details of their algorithm. Hosts are given the freedom to tackle the problem how they wish, and are rewarded for progress, not process. This incentivizes the hosts to use methods and computational resources that will be competitive.

3.2 Collective Optimization Trajectory

In implementing a distributed optimization system, the proposed solution is to use *collective optimization trajectory*, a ledger which tracks the best candidate solution found so far over time.

Each host will receive reward proportional to the time spend in the lead. This reward structure has the following properties, we state here with justifications for why each is desirable:

- Verifiability - It is easy to check that a given solution has a given newly-leading fitness. This enables the system to run without the need for a centralized verifier.
- Domain Invariance - Improvement can always be noted regardless of the scale of the domain. This is desirable for maximum generalizability to arbitrary optimization problems.
- Codomain Invariance - Improvement can always be noted regardless of the scale of the codomain. Again, this is for generalizability.
- Sybil Resistance - Hosts do not gain from manufacturing alternate identities, removing incentive to Sybil attack. This is necessary, since vulnerability to Sybil attacks is a severe security issue.
- Divisible - Not all of the reward will go to one host, as lead changes between hosts. This is desirable to having smaller participants receive zero reward, desensitizing participation.
- Predictable Payout - The total cost to the client can be allocated ahead of time. This is required since under an EVM system, the client will need to have the funds available and deposited in the contract.

The payout is distributed to host as shown in Fig. 1. (1) Client submits problem. (2) First activity recorded when red (solid line) host submits candidate, red's reign as leader begins. (3) First improvement when blue (dotted line) host submits improving candidate, blue is now leader and red's reign ends. (4) At end of optimization (N blocks after start), blue (dotted line) host is the final leader. (5) Reward distributed proportionately to each host's total reign duration.

The process of updating new best fitness and candidate is outlined in Algorithm 1.

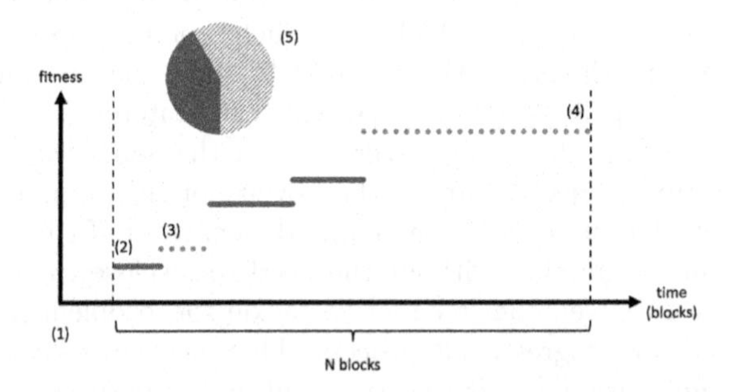

Fig. 1. Payout proportion based on time in the lead. (Color figure online)

This proposed system does not perfectly reward effort, however it is easy to construct obvious alternatives which would fail to satisfy these properties. For example, rewarding improving solutions by a set amount (e.g. 10% higher fitness), or towards a target (closeness to 0) fails codomain invariance and predictable payout. Rewarding finding new good solutions within a distance to old solutions fails domain invariance and predictable payout. Rewarding detecting learnable structure in problems, or fully evaluating subspaces fails verifiability. Rewarding each host for discovering a good solution over certain fitnesses, or rewarding hosts checking each other's work fails Sybil resistance. Rewarding only the best candidate fails divisibility.

3.3 Trajectory Broadcaster Vs Recipient

When a host discovers a new best-fitness candidate and broadcasts it to the network, we call this host the *broadcaster*. For distributed optimization to work as a system distinct, collective optimization trajectory must posses certain properties. It must be more beneficial from a game theory perspective for hosts to broadcast a newly-found best candidate than it is to hoard the information for themselves. The problem of mempool attacks must be considered, as when a new candidate solution with the next best-so-far candidate, there is a period of time in which the transaction has not been included in the blockchain.

When hosts hear of a new candidate broadcast via the blockchain, we call these host the *recipients*. This process is illustrated in Fig. 2. (1) The current best candidate on the blockchain. (2) Candidate is imported by a new host. (3) Host algorithm(s) explores the space. (4) New solution is broadcast and host picks it up, redirecting the search. (5) Eventually this individual host finds a new best solution, broadcasting this and becoming the leading host.

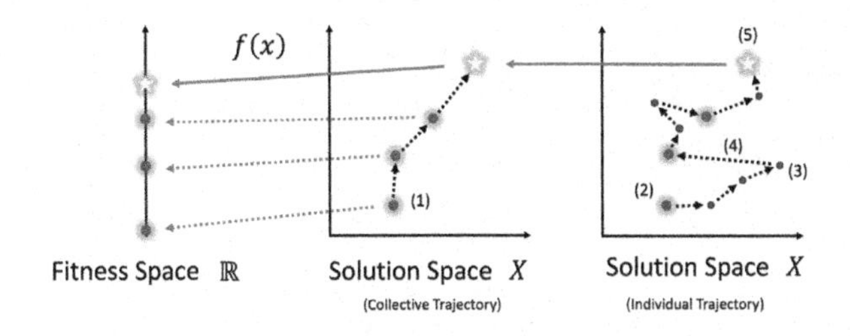

Fig. 2. Collective Optimization Trajectory.

It must be possible for a recipient's algorithm to learn from collective trajectory despite the information flow being severely limited. If algorithms cannot benefit from trajectory information, then the first mover with the fastest algorithm and most computational power has an unmitigated advantage and will not be overtaken, the reward will all go to one host and not be subdivided.

Proposed examples of algorithms benefiting from trajectory are[3]:

- GA - Treat received candidates migration events; add as elites.
- EDA - Contrast received candidates to candidates generated by current probabilistic model; adjust model.
- Hill-climber - Hill-climb to refine received candidate to local optimum; perturb from this local optimum to find adjacent local optima.
- PSO - Use received candidate to update global best information to swarm.

[3] Algorithms may be adapted, or specifically-designed for decentralised optimisation.

4 Proof-of-Concept

4.1 Implementation

A smart contract for decentralised solving of the OneMax problem has been implemented, with a Web 3.0 GUI using NodeJS and the React framework.[4] The function may be swapped out by replacing an arbitrary EVM function.

The current smart contract which has been built is using the Solidity [13] smart contract language for the EVM. Development and testing is running on the Truffle development framework [5] consisting of the *Truffle* development tools, *Ganche* development blockchain, and *Drizzle* Redux components for front-end.

The current implementation allows a client to pseudonymously connect to the DApp using a Web 3.0 enabled browser. Hosts can submit a candidate solution, and if it is an improvement, the new candidate and fitness will be recorded, and the broadcaster is recorded as the current leader.

A screenshot of the GUI is shown in Fig. 3. The two left boxes allow candidates to be submitted though a Web 3.0 enabled browser. The right box displays the current status of the contract, showing the current best candidate and fitness registered on the blockchain, with the address of the reigning leader. The UI runs entirely on the browser client without need for a back-end server. ♦ denotes a state-mutating action associated with a transaction fee.

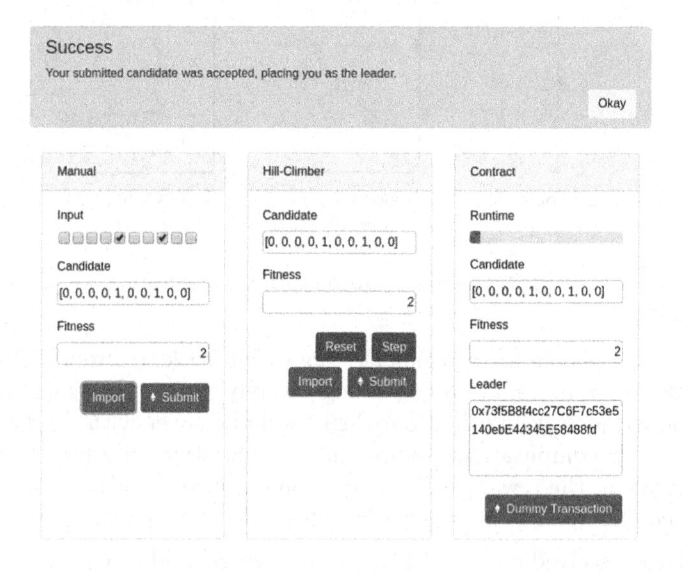

Fig. 3. A screenshot of the front-end for the prototype of DOpt.

[4] Source available at https://github.com/leechristie/dopt-concept.

4.2 Security Considerations - Reentrancy

Payout for satisfying criteria is best done using a withdrawal system in which funds are allocated for a host and then the host initiates a withdrawal. This is a 'best-practice' design pattern in smart contracts which avoids triggering sending of funds with arbitrary conditions and reduces the potential for certain classes of security exploit known as 'reentrancy bugs' [9].

The 'current reign' duration is defined as the number of blocks since the last time an improving candidate was submitted to the contract. The 'total reign' for a given host is the sum of all reigns held by that host, i.e. the amount of time (in number of blocks) that the host was the leading host.

Currently, runtime is displayed from the block in which the first candidate was submitted until a fixed runtime of 15 blocks has elapsed. When a host is the leader for a contiguous n blocks, they are said to have a *reign* of duration n blocks. The client sets the *total reward* amount as the amount they are willing to pay for the optimization. The *total reign* is calculated as the sum of their reigns, and the reward is calculated is as

$$\text{assigned reward} = \text{sum of reigns} \times \text{total reward} \;/\; \text{total runtime}$$

for example if the total runtime is 15 blocks with a total reward of 0.0015 ether and the host has held the lead twice, for a duration of 4 then 3 blocks, the assigned reward is 0.0007 ether. A host may withdraw ether from the smart contract provided the requested amount is less than or equal to the assigned reward less the amount already withdrawn. At any time during the optimization of the function, the total reward may be increased by anyone (usually the client) who sends funds to the payable method. The method of computing balance given above will retroactively update all balances without additional effort, since the balance the tracked variables are sum of reign and amount withdrawn. The contract requires donated amounts be divisible by the runtime.

5 Applications

5.1 Optimization-as-a-Service

The primary application of DOpt is to set up a network whereby an optimization problem can be submitted to the DOpt network by anyone and solved by anyone. Both problem submitters and solvers may be anonymous on the network and do not need trust or communication between them.

Those submitting problems are in the role of *client*. The client designated the total reward amount along with a predefined total runtime (counted by N blocks mined on the blockchain). Optimization runtime begins when the first host submits a candidate and ends N blocks after the start time.

Those with spare CPU cycles which may be used to run optimization problems are in the role of *host*. The host decides which of the available active problems to attempt to optimize, and what algorithm(s) to run, which improving candidates to submit to the network, and when to stop and switch problems.

```
State: best, leader, max_runtime, sub_count, start, sub_block, completed
Data: candidate, fitness, sender, block
Result: updates host reign
if fitness ≠ eval(candidate) then
  | revert;
if fitness ≤ best.fitness then
  | revert;
if runtime ≥ max_runtime then
  | revert;
best = fitness, candidate;
sub_count++;
if start = null then
  | start ← block;
else if block ≥ submission block then
  | reign ← block - sub_block;
  | completed[leader] += reign;
  | allocated += reign;
leader ← sender;
sub_block ← block;
```

Algorithm 1: Algorithm to update state on receiving new candidate.

Some potential decision criteria are:

- Age of Problem - The amount of time the problem has already been running before the host joins;
- Reward per Runtime - The rate at which reward is issued for being the leader;
- Competition - The number of other candidates which have been submitted and the rate they are being submitted; and
- Success - Whether a problem currently being attempted by the host is yielding successful improvement

A host may run as many or few processes and as much or as little compute power they wish on a given problem or problems.

As with many decentralized systems, price could be expected to be dictated by the market without the need to add pricing infrastructure. If the network contains a high number of active optimization problems and a low number of hosts, the price could be expected to trend higher. Clients submitting problems whose payout per unit time is low with respect to the market average will find little-to-no submitted candidates unless they increase the total reward amount to a competitive level. If the total number of hosts is high and the total number of active problems is low, most problems will be tackled by a high number of competitive hosts. However, deployment in a real-money scenario would be required to see how participants and pricing behaves in practice.

A client may set their reward level above market rate to increase the level of prioritization, resulting in more hosts devoting more CPU time to their problem. A client may set their reward level lower and increase the runtime if the importance of the problem is lower and they are prepared to wait longer.

DOpt may pay participants in ether (ETH). If the price volatility of ether as a currency disincentives use as a unit of account, it is possible to allow exchange of value using another token, such as tokens under the ERC-20 token standard, allowing use of application-specific utility tokens, or stable-coin.

5.2 Automation of Contests

In the optimization community, optimization contests are run wherein a set of benchmarks are given out and the most efficient algorithms provided for this benchmark set are awarded. The system could be used for regular (such as annual) or ad-hoc run contests.

DOpt is *not* a suitable direct replacement for this design of contest. As with PoW systems, as it does not attempt to maintain a level playing field for participants since advantage is given to both quality of algorithm, and level of compute power. Participants may still reveal the algorithms used however the system does not require this. Participants could compete for financial reward or prestige given a fixed time window in which to apply all algorithms and hardware resources they can to a given problem. The contest would be automatically run by the DOpt smart contract system.

5.3 Long-Term Recording of Best-Known Solutions

Some large optimization benchmark problems have best-known solutions which are improved over long periods of time. The same DOpt system run with a very long duration runtime, or modified to run endlessly, may be used to record the best-known solutions on the blockchain.

These current record-holder could receive a small payout over the span of holding the record. Note that if the system is modified to allow endless runtime, the client would need to top-up the total reward with additional deposits over time to allow the runtime to continue rewarding solvers.

Alternatively, the system could be run without financial incentive, the reward being only to have their record publicly available in the blockchain. Anyone citing the best-known solution to a given problem could cite the solution in the blockchain, which is verifiable by all.

6 Further Work

Work must be done to identify existing algorithms which can run competitively using the collective trajectory concept. In addition, which classes of problems are best suited to this approach needs to be identified. An API will be required to allow hosts to connect their algorithms to the DOpt network, automatically discover problems, submit candidates, and receive broadcast trajectory. A process will be required to prevent mempool attacks for secure deployment of DOpt.

References

1. Buterin, V.: A next-generation smart contract and decentralized application platform (2015). https://github.com/ethereum/wiki/wiki/White-Paper
2. Digiconomist: Bitcoin energy consumption index. https://digiconomist.net/bitcoin-energy-consumption
3. Douceur, J.R.: The sybil attack. In: Druschel, P., Kaashoek, F., Rowstron, A. (eds.) IPTPS 2002. LNCS, vol. 2429, pp. 251–260. Springer, Heidelberg (2002). https://doi.org/10.1007/3-540-45748-8_24
4. Du, W., Jia, J., Mangal, M., Murugesan, M.: Uncheatable grid computing. Electrical Engineering and Computer Science, 26 2004. https://surface.syr.edu/cgi/viewcontent.cgi?article=1025&context=eecs
5. Group, T.B.: Sweet tools for smart contracts (2019). https://www.trufflesuite.com/
6. Hine, J.H., Dagger, P.: Securing distributed computing against the hostile host. In: Proceedings of the 27th Australasian Conference on Computer Science ACSC 2004, Vol. 26, pp. 279–286. Australian Computer Society Inc., Darlinghurst (2004). http://dl.acm.org/citation.cfm?id=979922.979956
7. Lamport, L., Shostak, R., Pease, M.: The byzantine generals problem. ACM Trans. Program. Lang. Syst. (TOPLAS) **4**(3), 382–401 (1982)
8. Nakamoto, S.: Bitcoin: A peer-to-peer electronic cash system (2008). https://bitcoin.org/bitcoin.pdf
9. Solidity: Security considerations - re-entrancy (2019). https://solidity.readthedocs.io/en/v0.5.11/security-considerations.html#re-entrancy
10. Szabo, N.: Formalizing and securing relationships on public networks. First Monday, **2**(9) (1997)
11. Wiki, B.: Pay-to-pubkey hash (P2PKH) - Bitcoin Wiki (2016). https://en.bitcoin.it/wiki/Hashlock
12. Wiki, B.: Script (2019). https://en.bitcoin.it/wiki/Script
13. Wood, G.: Solidity the Contract-Oriented Programming Language. https://github.com/ethereum/solidity
14. Wood, G.: Ethereum: A secure decentralised generalised transaction ledger eip-150 revision (2017). http://gavwood.com/paper.pdf

PbO-CCSAT: Boosting Local Search for Satisfiability Using Programming by Optimisation

Chuan Luo[1,2(✉)], Holger Hoos[2], and Shaowei Cai[3]

[1] Microsoft Research, Beijing, China
chuan.luo@microsoft.com
[2] Leiden Institute of Advanced Computer Science, Leiden University, Leiden, The Netherlands
hh@liacs.nl
[3] State Key Laboratory of Computer Science, Institute of Software, Chinese Academy of Sciences, Beijing, China
caisw@ios.ac.cn

Abstract. Propositional satisfiability (SAT) is a prominent problem in artificial intelligence with many important applications. Stochastic local search (SLS) is a well-known approach for solving SAT and known to achieve excellent performance on randomly generated, satisfiable instances. However, SLS solvers for SAT are usually ineffective in solving application instances. Here, we propose a highly configurable SLS solver dubbed *PbO-CCSAT*, which leverages a powerful technique known as configuration checking (CC) in combination with the automatic algorithm design paradigm of programming by optimisation (PbO). Our *PbO-CCSAT* solver exposes a large number of design choices, which are automatically configured to optimise the performance for specific classes of SAT instances. We present extensive empirical results showing that our *PbO-CCSAT* solver significantly outperforms state-of-the-art SLS solvers on SAT instances from many applications, and further show that *PbO-CCSAT* is complementary to state-of-the-art complete solvers.

Keywords: Stochastic local search · Satisfiability · Programming by optimisation.

1 Introduction

Given a formula in conjunctive normal form (CNF), the propositional satisfiability (SAT) problem is to decide whether there exists an assignment to all variables under which the formula evaluates to true. SAT is one of the most prominent NP-complete problems and plays a critical role in many important application problems [10,24].

There are two major families of SAT solvers: *complete* and *incomplete* solvers. Complete solvers can solve any instance if they run sufficiently long, but in practice cannot solve certain types of instances within reasonable time. Incomplete

T. Bäck et al. (Eds.): PPSN 2020, LNCS 12269, pp. 373–389, 2020.
https://doi.org/10.1007/978-3-030-58112-1_26

solvers, most of which are based on stochastic local search (SLS), cannot prove unsatisfiability but have been shown surprisingly effective for solving hard random SAT instances (see, *e.g.*, [19]).

However, application SAT instances usually exhibit complex structure; for solving such instances, SLS solvers are considered to be less effective than state-of-the-art complete solvers based on conflict-driven clause learning (CDCL). While the grand challenge of devising SLS-based solvers that reach the performance of CDCL solvers in solving application satisfiable instances remains open, recent advances in SLS solvers for SAT have produced a series of promising results:

1. An idea called configuration checking (CC) has led to a series of highly effective SLS solvers for SAT (see, *e.g.*, [6,19]). Among these, *DCCASat* [19] has shown efficacy in solving structured SAT instances.
2. A novel paradigm for algorithm design called *programming by optimisation (PbO)* [12] has provided a basis for automatically constructing SLS solvers that perform well on particular classes of SAT instances. By automatically configuring the parametric *SATenstein* framework, good performance was achieved on several classes of structured SAT instances [15]; however, these instances were not particularly hard and the improvements were relatively modest.

As we will demonstrate in this work, leveraging PbO in combination with CC, much better performance can be achieved on substantially harder structured SAT instances (see Sect. 5.5). Our main contributions are as follows:

Firstly, we introduce a new PbO-based design for SLS solvers for SAT that integrates variants of CC strategies [6,18], and related techniques. This forms the core of *PbO-CCSAT*, our new framework for SLS-based SAT solving.

Secondly, we augment the resulting flexible hierarchical CC framework with other effective diversification strategies [1,22]. As a result, a better balance can be achieved between intensification and diversification within *PbO-CCSAT*, which is important for achieving high performance in SLS solvers.

Finally, we use an algorithm configurator called *SMAC* [13] to determine effective configurations of *PbO-CCSAT* for a variety of structured SAT instances. We compare *PbO-CCSAT* with 13 state-of-the-art SAT solvers, including SLS and CDCL ones. Our experiments show that *PbO-CCSAT* not only outperforms existing SLS solvers on all classes of instances, but also performs better than state-of-the-art CDCL solvers on several types of instances; for example, *PbO-CCSAT* can solve 34 SAT-encoded spectrum repacking instances whose satisfiability status was previously unknown and which could not be solved by any other competing solver.

2 Preliminaries

Given a set of n Boolean *variables* $\{x_1, x_2, \cdots, x_n\}$ and a set of $2{\cdot}n$ corresponding *literals* $\{x_1, \neg x_1, x_2, \neg x_2, \cdots, x_n, \neg x_n\}$, where each literal is either a Boolean variable or its negation, a *clause* is a disjunction of literals, *i.e.*, $c_i = l_{i,1} \vee l_{i,2} \vee \cdots \vee l_{i,k_i}$, where k_i is the length of clause c_i. In what follows, we consider formulae in conjunctive normal form (CNF): $F = c_1 \wedge c_2 \wedge \cdots \wedge c_m$, where each c_i $(1 \leq i \leq m)$ is a clause, and m is the number of clauses.

A complete *assignment* is a mapping that assigns a truth value (*true* or *false*) to each variable in a given formula F. Given a formula F in CNF and a complete assignment α, each clause $c \in C(F)$ under assignment α has two possible *states*: *satisfied* or *unsatisfied*; c is satisfied under α if it contains either a literal x that is assigned *true* under α or a literal $\neg x$ such that x is assigned *false*; otherwise, c is unsatisfied; F is satisfied under α if all clauses $c \in C(F)$ are satisfied under α, and unsatisfied otherwise. The *SAT problem* (for CNF) is to determine whether there exists a complete assignment under which the given formula is satisfied.

Since our work focuses on improving the performance of SLS solvers, it is useful to recall the basic operation of SLS-based SAT solvers: After generating a complete assignment, the key idea is to repeatedly modify (or *flip*) the truth value of a single variable, which is chosen according to a variable selection heuristic *pickVarHeur* (the most critical part of the algorithm), until a satisfying assignment is reached or a given bound on running time has been exceeded.

High-performance SLS algorithms usually operate in two modes: intensification and diversification. In intensification mode, SLS solvers prefer to select a variable in a greedy manner, *e.g.*, focus on flipping variables such that the number of unsatisfied clauses is maximally decreased. In diversification mode, SLS solvers select variables in a way intended to move the search process to a different part of the search space.

SLS algorithms for SAT pick the variables to be flipped in each search iteration based on various variable properties. For a variable x, the *score* of x, denoted $score(x)$, is the increment in the number (or total weight if using clause weighting scheme) of satisfied clauses by flipping x [29]; the *age* of x, denoted $age(x)$, is the number of steps that has occurred since x's last flip [8]. Due to the space limit, the definitions of other variable properties (such as *subscore* [6], *hscore* [4], $hscore_2$ [4]) can be found online.[1]

3 Related Work

The performance of stochastic local search can be seriously degraded by stagnation – situations in which a limited set of candidate solutions is frequently revisited. Configuration checking (CC) is an effective idea for addressing this issue, and the main idea of CC is to prevent SLS solvers from visiting a candidate solution whose context has not changed since it was last visited [6]. CC has been shown effective in solving a variety of combinatorial problems. In the context of SAT, a hierarchical combination of CC variants has given rise to *DCCASat* [19], a novel SLS solver that achieves good performance on several classes of structured SAT instances.

[1] https://github.com/chuanluocs/PbO-CCSAT.

The key idea behind programming by optimisation (PbO) is to significantly expand the design space; the resulting flexibility is later exploited by automatically making design choices in a way that optimises performance for specific classes of inputs [12]. The PbO paradigm usually involves exposing all design choices as configurable parameters, and configuring the resulting flexible solver framework using an algorithm configurator. PbO has led to major improvements in SLS-based solvers for structured SAT instances [15] and in the state of the art for solving a broad range of NP-hard problems (see, *e.g.*, [20]).

4 The *PbO-CCSAT* Solver

We now introduce *PbO-CCSAT* – our highly configurable SLS framework for SAT solving. *PbO-CCSAT* works in two major phases: initialisation and local search. For the former, an initial assignment will be generated as the starting solution for local search. For the latter, since each local search step of SLS for SAT typically picks one variable and flips its value, the design space mainly includes different heuristics for selecting the variable whose value is to be flipped. After providing a design space consisting of different heuristics for generating initial assignments and variable selection in local search, we utilise an automated configurator to determine an effective configuration for a given instance set.

4.1 Initialisation

PbO-CCSAT includes a random initialisation mechanism and a greedy strategy, yielding the following 2 instantiations of the initialisation component:

1. ***randomAsgnGenHeur:*** Given a formula in CNF, assign *True* or *False* to each variable independently and uniformly at random.
2. ***greedyAsgnGenHeur:*** Given a formula in CNF, count the number of appearances of all positive literals (denoted as *#allPosLit*) and the number of appearances of all negative literals (denoted as *#allNegLit*). If *#allPosLit* is greater than *#allNegLit*, then all variables are assigned *True*; otherwise, all variables are assigned *False*.

To select which heuristic is used within a given configuration of *PbO-CCSAT*, we use a categorical parameter with value 1 for *randomAsgnGenHeur* and 2 for *greedyAsgnGenHeur*. All choices between different instantiations for a specific component of *PbO-CCSAT* discussed in the following are handled analogously.

4.2 Variable Selection in Local Search

The variable selection heuristic is critical in any SLS-based SAT solver. In our *PbO-CCSAT* solver, it makes use of several components, as seen in Algorithm 1.

Algorithm 1: Variable selection heuristic of *PbO-CCSAT*

```
   Output: variable v
 1 if performRW then
 2     with probability rw_prob do
 3         return v ← selVarRWHeur();
 4 if performProbDiv then
 5     with probability div_prob do
 6         c ← selUnsatClause();
 7         return v ← selVarFromUnsatClause(c);
 8 if performCSCC then
 9     if CSDvars ≠ ∅ then
10         return v ← selVarFromSet(CSDvars);
11 if NVDvars ≠ ∅ then
12     return v ← selVarFromSet(NVDvars);
13 if performAspiration then
14     if SDvars ≠ ∅ then
15         return v ← selVarFromSet(SDvars);
16 if performCWS then
17     activate clause weighting scheme clauseWeightScheme();
18 c ← selUnsatClause();
19 return v ← selVarFromUnsatClause(c);
```

1. **Random Walk Component:** *PbO-CCSAT* can optionally activate this component, in order to ensure probabilistic approximate completeness of the stochastic search process [11]. (Lines 1–3 in Algorithm 1)
2. **Probabilistic Diversification Component:** *PbO-CCSAT* has an option to activate this component before the CC-based greedy search, in order to achieve stronger diversification. (Lines 4–7 in Algorithm 1)
3. **CC-based Intensification Component:** *PbO-CCSAT* employs Boolean-valued parameters to realise a flexible hierarchical combination of different configuration checking mechanisms [6, 18]. (Lines 8–15 in Algorithm 1)
4. **Diversification Component:** *PbO-CCSAT* employs a diversification component after the greedy search, if no suitable variable is selected by the CC-based greedy search; this is useful for balancing intensification and diversification. (Lines 16–19 in Algorithm 1)

Random Walk Component. Different from existing CC-based SAT solvers, *PbO-CCSAT* employs a specific random walk (RW) component to achieve probabilistic approximate completeness [11], which is important for incomplete solvers; this component is activated using Boolean-valued parameter *performRW*. If activated (*i.e., performRW* = *True*), with probability *rw_prob*, the RW component selects the variable to be flipped according to a heuristic *selVarRWHeur* (Line 3 in Algorithm 1) that randomly selects a variable in a randomly chosen clause.

Probabilistic Diversification Component. To provide stronger diversification, *PbO-CCSAT* integrates a probabilistic diversification component before its intensification component; this optional diversification mechanism is activated by means of the Boolean-valued parameter *performProbDiv*. If activated (*performProbDiv* = *True*), with probability *div_prob*, the component first picks a currently unsatisfied clause c via *selUnsatClause*, and then selects a variable in clause c via *selVarFromUnsatClause*. *PbO-CCSAT* supports 2 instantiations of *selUnsatClause* (Line 6 in Algorithm 1):

1. Select an unsatisfied clause uniformly at random.
2. Select an unsatisfied clause with a probability proportional to its weight if a clause weighting scheme has been activated.

PbO-CCSAT supports 7 instantiations of heuristic *selVarFromUnsatClause* (Line 7 in Algorithm 1):

1. Select a variable appearing in c uniformly at random.
2. Select the variable in c with the highest *age*.
3. Select the variable in c with the highest *score*, breaking ties in favour of variables with higher *age*.
4. Select the variable in c with the highest *hscore*, breaking ties in favour of variables with higher *age*.
5. Select the variable in c with the highest $hscore_2$, breaking ties in favour of variables with higher *age*.
6. Select the variable in c according to the *Novelty* heuristic [22].
7. Select the variable in c with a probability proportional to its *score* and *age* (which resembles the diversification heuristic from *Sparrow* [1]).

CC-Based Intensification Component. Various successful SLS solvers have been developed based on configuration checking (CC). Among these CC-based solvers, *DCCASat* [19], which uses a hierarchical combination of three mechanisms, *i.e.*, clause-states-based configuration checking (CSCC) [18], neighbouring-variables-based configuration checking (NVCC) [6] and aspiration [5], achieves state-of-the-art performance on solving hard SAT instances [19]. The hierarchical combination approach utilised by *DCCASat* is deterministic and executes the underlying mechanisms (*i.e.*, CSCC, NVCC and aspiration) in a fixed sequence.

Our design for the CC-based intensification component of *PbO-CCSAT* is based on the idea that additional flexibility in a hierarchical CC approach, such as the one found in *DCCASat*, can yield further performance improvements. During the search process, *PbO-CCSAT* will maintain three sets of candidate variables (*i.e.*, *CSDvars*, *NVDvars* and *SDvars*) conforming to the criteria of the CSCC, NVCC and aspiration mechanisms, respectively. The definitions of these sets can be found online (See footnote 1).

Based on the canonically deterministic CC heuristic, we propose the flexibly hierarchical CC heuristic, which can optionally decide which candidate sets need to be checked, by introducing Boolean-valued parameters. Our flexible hierarchical CC heuristic works as follows (Lines 8–15 in Algorithm 1): If *performCSCC* = *True* and *CSDvars* $\neq \varnothing$, heuristic *selVarFromSet* is used to select a variable from set *CSDvars*; otherwise, if *NVDvars* $\neq \varnothing$, *selVarFromSet* is used to select a variable from set *NVDvars*; if still no variable has been selected at this point and *performAspiration* = *True* as well as *SDvars* $\neq \varnothing$, *selVarFromSet* is used to select a variable from set *SDvars*.

We support 4 instantiations of the heuristic *selVarFromSet* (Lines 10, 12 and 15 of Algorithm 1):

1. Given a variable set S, select $x \in S$ with the highest *score*, breaking ties uniformly at random.
2. Given a variable set S, select $x \in S$ with the highest *score*, breaking ties in favour of ones with higher *age*.
3. Given a variable set S, select $x \in S$ with the highest *score*, breaking ties in favour of ones with higher *hscore*.
4. Given a variable set S, select $x \in S$ with the highest *score*, breaking ties in favour of ones with higher $hscore_2$.

Diversification Component. Our *PbO-CCSAT* solver uses clause weighting for diversification. Whether or not clause weighting is used is controlled by a Boolean-valued parameter, *performCWS*. *PbO-CCSAT* integrates two clause weighting schemes, *SWT* [6] and *PAWS* [30]. As outlined in Lines 16–19 of Algorithm 1, the diversification component of *PbO-CCSAT* works as follows: If *performCWS* = *True*, clause weights are updated according to a clause weighting scheme, *clauseWeightScheme*; then, an unsatisfied clause c is selected according to heuristic *selUnsatClause*, and a variable from c is chosen according to heuristic *selVarFromUnsatClause*.

PbO-CCSAT supports 2 instantiations of *clauseWeightScheme* (Line 17 in Algorithm 1):

1. Activating the *SWT* scheme [6].
2. Activating the *PAWS* scheme [30].

The details of *SWT* and *PAWS* can be found online (See footnote 1). The instantiations of heuristics *selUnsatClause* (Line 18 in Algorithm 1) and *selVarFromUnsatClause* (Line 19 in Algorithm 1) in the diversification component are the same ones used in the probabilistic diversification component described earlier.

4.3 Configuration Space and Default Configuration

The configuration space of *PbO-CCSAT* indicated in Sects. 4.1 and 4.2, and represented by a set of parameters, is used by an automated configurator for performance optimisation. For the default configuration of *PbO-CCSAT*, we used a version of *DCCASat* that is known to perform well on structured SAT instances [19], to provide a strong starting point for our automatic configuration process.

PbO-CCSAT has 23 parameters in total, including 5 Boolean parameters, 5 categorical parameters with 3.4 possible values on average, 7 real-number parameters and 6 integer parameters. For the overview of the full configuration space of *PbO-CCSAT* (*i.e.*, all heuristics and parameters, as well as the conditions when heuristics and parameters are activated) as well as the default settings for all parameters and design choices, please refer to the supplementary materials available online (See footnote 1).

Table 1. Comparative results on `FCC-SAT [Test]`.

Solver	`FCC-SAT [Test]`	
	#SAT	PAR10
PbO-CCSAT [SLS]	9 019	1 978.6
GNovelty+PCL [SLS]	8 802	3 162.0
PbO-CCSAT (Default)* [SLS]	8 192	6 554.0
Sparrow [SLS]	8 138	6 722.1
COMiniSatPS_Pulsar [CDCL]	8 031	7 323.3
MapleLCMDistChrBt-DL-v3 [CDCL]	7 925	7 928.0
Maple_LCM_Dist [CDCL]	7 780	8 628.3
MapleLCMDistChronoBT [CDCL]	7 638	9452.5
SATenstein [SLS]	7 587	9 619.8
MapleCOMSPS [CDCL]	7 464	10 238.2
DDFW [SLS]	7 448	10 490.5
Lingeling [CDCL]	7 359	10 856.2
YalSAT [SLS]	7 319	11 040.1
*Sattime** [SLS]	6 997	12 761.7
Dimetheus [Hybrid]	5 858	19 008.7
*Dimetheus (Random)** [Hybrid]	5 202	22 329.2
*Dimetheus (APP)** [Hybrid]	1 174	43 780.8

5 Experiments

In this section, we describe a series of experiments we conducted to evaluate *PbO-CCSAT* on a broad set of SAT benchmarks from various applications.

5.1 Benchmarks

We selected a broad set of 7 well-studied applications of SAT. The entire benchmark collection process and details of our benchmarks are publicly available online (See footnote 1). Overall, we used 26 932 instances – far more than the number of instances typically used in empirical studies of SAT algorithms.

Table 2. Comparative results on FCC-UNKNOWN.

Solver	FCC-UNKNOWN		
	#SAT	#UNSAT	PAR10
MapleLCMDistChrBt-DL-v3 [CDCL]	14	95	36 646.5
COMiniSatPS_Pulsar [CDCL]	13	93	36 969.7
PbO-CCSAT [SLS]	100	0	37 580.9
Maple_LCM_Dist [CDCL]	14	87	37 720.9
MapleLCMDistChronoBT [CDCL]	6	92	38 057.8
GNovelty+PCL [SLS]	64	0	42 048.5
MapleCOMSPS [CDCL]	5	42	44 138.3
Sparrow [SLS]	20	0	47 546.9
PbO-CCSAT (Default)* [SLS]	20	0	47 572.7
Dimetheus [Hybrid]	0	10	48 809.7
DDFW [SLS]	7	0	49 145.1
Lingeling [CDCL]	0	4	49 528.4
SATenstein [SLS]	2	0	49 761.8
*Dimetheus (APP)** [Hybrid]	0	0	50 000.0
*Dimetheus (Random)** [Hybrid]	0	0	50 000.0
*Sattime** [SLS]	0	0	50 000.0
YalSAT [SLS]	0	0	50 000.0

FCC-SAT + FCC-UNKNOWN. SAT-encoded instances from a spectrum repacking project by the US Federal Communications Commission (FCC) [24]. FCC-SAT is the set of 9 482 instances which are known to be satisfiable, and FCC-UNKNOWN denotes the set of 397 instances whose satisfiability status is unknown.

PTN + PTN-More. SAT-encoded instances for tackling a long-standing open problem in mathematics known as Boolean Pythagorean Triples (PTN) [10]. PTN denotes the set of 23 instances publicly available, while PTN-More is a large set of 556 instances generated by the PTN encoder.

SMT-QF-BV. A set of 16 434 SAT-encoded instances of prominent satisfiability-modulo-theories (SMT) problems [25].

Community. A set of 20 instances with community structure [9].

SC17-Main-mp1-9. A set of 20 instances from the 2017 SAT Competition [2].

We note that, in our experiments, each of these benchmarks (*i.e.*, FCC-SAT, PTN, SMT-QF-BV, Community and SC17-Main-mp1-9) is split into a training set and a testing set. The training sets are used to determine optimised configurations, and the testing sets are used for performance evaluation.

Table 3. Comparative results on PTN [Test].

Solver	PTN [Test]	
	#SAT	PAR10
PbO-CCSAT [SLS]	12	4.5
DDFW [SLS]	12	40.9
SATenstein [SLS]	12	41.7
GNovelty+PCL [SLS]	12	64.0
Sparrow [SLS]	12	159.7
PbO-CCSAT (Default)* [SLS]	11	4672.7
Dimetheus [Hybrid]	9	12538.5
YalSAT [SLS]	9	12554.6
*Sattime** [SLS]	4	34105.0
*Dimetheus (Random)** [Hybrid]	2	41676.9
MapleCOMSPS [CDCL]	1	45973.8
COMiniSatPS_Pulsar [CDCL]	0	50000.0
*Dimetheus (APP)** [Hybrid]	0	50000.0
Lingeling [CDCL]	0	50000.0
Maple_LCM_Dist [CDCL]	0	50000.0
MapleLCMDistChrBt-DL-v3 [CDCL]	0	50000.0
MapleLCMDistChronoBT [CDCL]	0	50000.0

5.2 Competitors

We assessed our new approach against a baseline comprised of 13 state-of-the-art SAT solvers, including 6 SLS, 6 CDCL and 1 hybrid solver. We briefly describe these competitors below.

`SLS competitors.` ***SATenstein*** [15] is a unified SLS framework that integrates components from a broad range of prominent SLS-based SAT algorithms. ***YalSAT*** [3] is the winner of the Random Track of the 2017 SAT Competition. ***DDFW*** [14], ***Sparrow*** [1], ***Sattime*** [16] and ***GNovelty+PCL*** [27] are known to be quite effective in solving structured SAT instances.

`CDCL competitors.` ***Lingeling*** [3] is an efficient CDCL solver that won a number of awards in SAT competitions. ***COMiniSatPS_Pulsar*** [26] is the winner of the NoLimit Track of the 2017 SAT Competition. ***MapleCOMSPS*** [17], ***Maple_LCM_Dist*** [21], ***MapleLCMDistChronoBT*** [23] and ***MapleLCMDistChrBt-DL-v3*** are the winners of the Main Tracks of the 2016–2019 SAT Competitions, respectively.

`Hybrid competitor.` ***Dimetheus*** [7] is a complex solver that effectively hybridises preprocessing, CDCL, SLS and message passing techniques.

Table 4. Comparative results on `PTN-More`.

Solver	PTN-More	
	#SAT	PAR10
PbO-CCSAT [SLS]	556	6.1
DDFW [SLS]	556	28.2
SATenstein [SLS]	554	295.9
*Sattime** [SLS]	554	420.5
GNovelty+PCL [SLS]	542	1 404.7
PbO-CCSAT (Default)* [SLS]	527	3 121.3
Sparrow [SLS]	497	5 798.9
Dimetheus [Hybrid]	288	24 467.2
*Dimetheus (Random)** [Hybrid]	45	46 066.3
YalSAT [SLS]	16	48 583.4
COMiniSatPS_Pulsar [CDCL]	0	50 000.0
*Dimetheus (APP)** [Hybrid]	0	50 000.0
Lingeling [CDCL]	0	50 000.0
MapleCOMSPS [CDCL]	0	50 000.0
Maple_LCM_Dist [CDCL]	0	50 000.0
MapleLCMDistChrBt-DL-v3 [CDCL]	0	50 000.0
MapleLCMDistChronoBT [CDCL]	0	50 000.0

5.3 Configuration Protocol

PbO-CCSAT has been designed as a highly parametric SLS framework for SAT, to be automatically configured to perform well on different classes of SAT instances. In this context, we used *SMAC* (version: 2.10.03), a widely used, state-of-the-art automated configurator [13]. We now describe the protocol we used for configuring *PbO-CCSAT* and all other solvers described in Sect. 5.2.

Following standard practice, we used *SMAC* to minimise PAR10 (*i.e.*, average running time, where unsuccessful runs are counted as 10 times the cutoff time), which aims to achieve the best average performance on a given benchmark. We used an overall time budget of 36 000 s (= 10 h) for each run of *SMAC*, and a cutoff of 60 CPU seconds per solver run during configuration. For each training set, we performed 25 independent runs of *SMAC*, resulting in 25 optimised configurations. Each of these was then evaluated on the entire training set, with one solver run per instance and a cutoff time of 60 CPU seconds per run, and the configuration with the lowest PAR10 was selected as the result of the configuration process. As per *SMAC*'s default settings, we performed one run per problem instance during this validation phase when configuring algorithms. The configurations of *PbO-CCSAT* obtained in this way for all our training sets are reported online (See footnote 1).

Table 5. Comparative results on `SMT-QF-BV [Test]`.

Solver	`SMT-QF-BV [Test]`		
	#SAT	#UNSAT	PAR10
Maple_LCM_Dist [CDCL]	15 441	228	2 100.5
COMiniSatPS_Pulsar [CDCL]	15 437	221	2 117.9
MapleLCMDistChrBt-DL-v3 [CDCL]	15 420	231	2 164.1
MapleLCMDistChronoBT [CDCL]	15 414	226	2 196.9
MapleCOMSPS [CDCL]	15 414	224	2 198.4
Lingeling [CDCL]	15 211	236	2 803.4
Dimetheus [Hybrid]	13 930	213	6 847.8
*Dimetheus (APP)** [Hybrid]	9 429	197	20 668.7
PbO-CCSAT [SLS]	7 012	0	28 591.3
YalSAT [SLS]	6 015	0	31 617.5
SATenstein [SLS]	6 009	0	31 633.9
*Sattime** [SLS]	5 831	175	31 665.7
PbO-CCSAT (Default)* [SLS]	5 869	0	32 051.8
*Dimetheus (Random)** [Hybrid]	5 644	173	32 252.4
GNovelty+PCL [SLS]	5 555	0	33 000.8
DDFW [SLS]	5 506	0	33 179.9
Sparrow [SLS]	5 420	0	33 415.1

5.4 Experimental Setup

All our experiments have been performed on a cluster of computers with dual 16-core, 2.10 GHz Intel Xeon E5-2683 CPUs, 40 MB L3 cache and 94 GB RAM.

For each benchmark, *PbO-CCSAT* was compared against the 13 state-of-the-art competitors described in Sect. 5.2. All competitors (except *Sattime*, which does not expose any configurable parameters) were automatically configured using *SMAC* on the respective training sets. Furthermore, we also report results for *PbO-CCSAT (Default)* (*i.e., DCCASat*) and two specific instantiations of *Dimetheus*, namely *Dimetheus (Random)* and *Dimetheus (APP)* (for solving random and application SAT instances, respectively).

Each solver was run once on each instance, with a cutoff time of 5000 CPU seconds, following the rules of SAT competitions. All runs were performed using *runsolver* [28] to record running time. For each benchmark and solver, we report the number of satisfiable instances solved ('#SAT') and the PAR10 score in CPU seconds. Additionally, for `FCC-UNKNOWN` and `SMT-QF-BV [Test]`, we also report the number of unsatisfiable instances solved by the complete solvers used in our experiments ('#UNSAT').

Table 6. Comparative results on `Community [Test]`.

Solver	`Community [Test]`	
	#SAT	PAR10
PbO-CCSAT [SLS]	14	3 435.8
DDFW [SLS]	14	3 470.6
MapleCOMSPS [CDCL]	14	3 477.4
SATenstein [SLS]	14	3 630.5
MapleLCMDistChronoBT [CDCL]	14	3 631.8
Maple_LCM_Dist [CDCL]	13	6 886.1
MapleLCMDistChrBt-DL-v3 [CDCL]	13	6 972.4
YalSAT [SLS]	13	6 973.6
COMiniSatPS_Pulsar [CDCL]	12	10 259.4
*Sattime** [SLS]	12	10 863.1
Lingeling [CDCL]	10	16 799.8
Dimetheus [Hybrid]	10	17 006.1
Sparrow [SLS]	9	20 378.1
GNovelty+PCL [SLS]	8	23 351.8
*Dimetheus (Random)** [Hybrid]	8	23 572.4
PbO-CCSAT (Default)* [SLS]	4	36 779.4
*Dimetheus (APP)** [Hybrid]	0	50 000.0

5.5 Experimental Results

Tables 1, 2, 3, 4, 5, 6 and 7 show the results for *PbO-CCSAT* and its competitors on all testing benchmarks. All solvers except those marked with '*' were automatically configured using *SMAC* using the protocal described previously. Those solvers marked with '*' use a specific configuration or do not expose configurable parameters. In each of the tables, we list the solvers ranked according to their PAR10 scores in ascending order, so the best-performing solvers appear at the top. We also use **boldface** to highlight the results for *PbO-CCSAT* and *PbO-CCSAT (Default)*.

On all benchmarks, *PbO-CCSAT* performs significantly better than other SLS solvers, which indicates that *PbO-CCSAT* improves the state of the art in SLS for solving SAT. Also, it is clear that *PbO-CCSAT* is complementary in performance to state-of-the-art CDCL solvers.

Notably, on the `FCC-UNKNOWN` benchmark, where the satisfiability status of each instance was previously unknown, *PbO-CCSAT* solved 34 satisfiable instances that could not be solved by any of the competitors, while there are only 2 instances that were solved by at least one of the competitors, but not by our *PbO-CCSAT* solver. We note that for this real-world application on spectrum repacking, better solver performance brings immediate, tangible benefits.

Table 7. Comparative results on SC17-mp1-9 [Test].

Solver	SC17-mp1-9 [Test]	
	#SAT	PAR10
PbO-CCSAT [SLS]	15	476.0
*Sattime** [SLS]	14	3 423.8
YalSAT [SLS]	14	3 457.4
SATenstein [SLS]	14	3 810.8
Lingeling [CDCL]	13	6 856.7
MapleCOMSPS [CDCL]	13	7 110.7
Sparrow [SLS]	12	10 344.1
COMiniSatPS_Pulsar [CDCL]	12	10 537.8
MapleLCMDistChrBt-DL-v3 [CDCL]	12	10 934.5
GNovelty+PCL [SLS]	11	13 633.2
Maple_LCM_Dist [CDCL]	10	16 917.1
MapleLCMDistChronoBT [CDCL]	9	20 776.6
Dimetheus [Hybrid]	6	30 537.8
PbO-CCSAT (Default)* [SLS]	5	33 977.1
DDFW [SLS]	2	43 718.2
*Dimetheus (APP)** [Hybrid]	0	50 000.0
*Dimetheus (Random)** [Hybrid]	0	50 000.0

Furthermore, on the SC17-Main-mp1-9 benchmark, our *PbO-CCSAT* solver was able to solve one difficult competition instance (mp1-9_21.cnf) that could not be solved by any competitor; none of the solvers that participated in the 2017 SAT Competition solved this instance. This clearly demonstrates the improvement to the state of the art in solving SAT instances from real-world applications achieved by *PbO-CCSAT*.

5.6 The Effect of Automatic Configuration

We recall that the default configuration of *PbO-CCSAT* is equivalent to the *DCCASat* solver (the version used for solving structured SAT instances), which inspired much of the design of our CC-based SLS framework. As seen from the results presented in Tables 1, 2, 3, 4, 5, 6 and 7, automatically configuring the flexible *PbO-CCSAT* framework leads to substantial improvements of the average performance on the testing sets for all benchmarks, up to a factor of over 1000 in terms of PAR10 score for PTN [Test]. To further investigate these performance gains, it is instructive to look at them on a per-instance basis; further per-instance based analysis can be found online (See footnote 1). In a nutshell, automatic configuration leads to performance improvements on the large majority of instances, including the prominent and challenging FCC, PTN and SC17-Main-mp1-9 instances.

6 Conclusions and Future Work

In this work, we presented a substantial improvement in the state of the art in SLS solvers for structured SAT instances from a wide range of applications. We designed a new, highly configurable SLS solver dubbed *PbO-CCSAT*, by applying the PbO paradigm to configuration checking. Through automatically configuring *PbO-CCSAT* for 7 prominent benchmarks, we obtained significant improvements over previous SLS solvers in all cases. Furthermore, for solving satisfiable instances, *PbO-CCSAT* performs better than state-of-the-art complete solvers on 6 of our 7 satisfiable benchmarks, and narrows the performance gap compared to cutting-edge CDCL solvers on the remaining benchmark. Our results indicate that SLS solvers can be very effective in solving highly structured SAT instances from real-world applications.

Different from *SATenstein* which is an earlier parametric SAT solver, our new *PbO-CCSAT* solver is strongly based on configuration checking. It is also substantially less complex, since it is focussed on a smaller set of relatively simple mechanisms that complement each other well and can be easily integrated into a common framework. Nevertheless, in our experiments, *PbO-CCSAT* performs much better on all benchmarks we considered. In future work, we intend to integrate *PbO-CCSAT* with neural-network-based approaches [31], in order to achieve even better performance on challenging real-world SAT instances.

The source code, configuration space, default configuration and optimised configurations of *PbO-CCSAT*, as well as details about definitions, related work, benchmarks, competitors and further experimental analysis can be found at https://github.com/chuanluocs/PbO-CCSAT.

References

1. Balint, A., Fröhlich, A.: Improving stochastic local search for SAT with a new probability distribution. In: Strichman, O., Szeider, S. (eds.) SAT 2010. LNCS, vol. 6175, pp. 10–15. Springer, Heidelberg (2010). https://doi.org/10.1007/978-3-642-14186-7_3
2. Balyo, T., Heule, M.J.H., Järvisalo, M. (eds.): Proceedings of SAT Competition 2017: Solver and Benchmark Descriptions. University of Helsinki (2017)
3. Biere, A.: CaDiCaL, Lingeling, Plingeling, Treengeling and YalSAT entering the SAT competition 2017. In: Proceedings of SAT Competition 2017: Solver and Benchmark Descriptions, pp. 14–15 (2017)
4. Cai, S., Luo, C., Su, K.: Scoring functions based on second level score for k-SAT with long clauses. J. Artif. Intell. Res. **51**, 413–441 (2014)
5. Cai, S., Su, K.: Configuration checking with aspiration in local search for SAT. In: 2012 Proceedings of AAAI, pp. 434–440 (2012)
6. Cai, S., Su, K.: Local search for Boolean satisfiability with configuration checking and subscore. Artif. Intell. **204**, 75–98 (2013)
7. Gableske, O.: On the interpolation between product-based message passing heuristics for SAT. In: Järvisalo, M., Van Gelder, A. (eds.) SAT 2013. LNCS, vol. 7962, pp. 293–308. Springer, Heidelberg (2013). https://doi.org/10.1007/978-3-642-39071-5_22

8. Gent, I.P., Walsh, T.: Towards an understanding of hill-climbing procedures for SAT. In: 1993 Proceedings of AAAI, pp. 28–33 (1993)
9. Giráldez-Cru, J., Levy, J.: Generating SAT instances with community structure. Artif. Intell. **238**, 119–134 (2016)
10. Heule, M.J.H., Kullmann, O., Marek, V.W.: Solving and verifying the boolean pythagorean triples problem via Cube-and-Conquer. In: Creignou, N., Le Berre, D. (eds.) SAT 2016. LNCS, vol. 9710, pp. 228–245. Springer, Cham (2016). https://doi.org/10.1007/978-3-319-40970-2_15
11. Hoos, H.H.: On the run-time behaviour of stochastic local search algorithms for SAT. In: 1999 Proceedings of AAAI, pp. 661–666 (1999)
12. Hoos, H.H.: Programming by optimization. Commun. ACM **55**(2), 70–80 (2012)
13. Hutter, F., Hoos, H.H., Leyton-Brown, K.: Sequential model-based optimization for general algorithm configuration. In: Coello, C.A.C. (ed.) LION 2011. LNCS, vol. 6683, pp. 507–523. Springer, Heidelberg (2011). https://doi.org/10.1007/978-3-642-25566-3_40
14. Ishtaiwi, A., Thornton, J., Sattar, A., Pham, D.N.: Neighbourhood clause weight redistribution in local search for SAT. In: 2005 Proceedings of CP, pp. 772–776 (2005)
15. KhudaBukhsh, A.R., Xu, L., Hoos, H.H., Leyton-Brown, K.: SATenstein: automatically building local search SAT solvers from components. Artif. Intell. **232**, 20–42 (2016)
16. Li, C.M., Li, Yu.: Satisfying versus falsifying in local search for satisfiability. In: Cimatti, A., Sebastiani, R. (eds.) SAT 2012. LNCS, vol. 7317, pp. 477–478. Springer, Heidelberg (2012). https://doi.org/10.1007/978-3-642-31612-8_43
17. Liang, J.H., Ganesh, V., Poupart, P., Czarnecki, K.: Learning rate based branching heuristic for SAT solvers. In: Creignou, N., Le Berre, D. (eds.) SAT 2016. LNCS, vol. 9710, pp. 123–140. Springer, Cham (2016). https://doi.org/10.1007/978-3-319-40970-2_9
18. Luo, C., Cai, S., Su, K., Wu, W.: Clause states based configuration checking in local search for satisfiability. IEEE Trans. Cybern. **45**(5), 1014–1027 (2015)
19. Luo, C., Cai, S., Wu, W., Su, K.: Double configuration checking in stochastic local search for satisfiability. In: 2014 Proceedings of AAAI, pp. 2703–2709 (2014)
20. Luo, C., Hoos, H.H., Cai, S., Lin, Q., Zhang, H., Zhang, D.: Local search with efficient automatic configuration for minimum vertex cover. In: 2019 Proceedings of IJCAI, pp. 1297–1304 (2019)
21. Luo, M., Li, C., Xiao, F., Manyà, F., Lü, Z.: An effective learnt clause minimization approach for CDCL SAT solvers. In: 2017 Proceedings of IJCAI, pp. 703–711 (2017)
22. McAllester, D.A., Selman, B., Kautz, H.A.: Evidence for invariants in local search. In: 1997 Proceedings of AAAI, pp. 321–326 (1997)
23. Nadel, A., Ryvchin, V.: Chronological backtracking. In: Beyersdorff, O., Wintersteiger, C.M. (eds.) SAT 2018. LNCS, vol. 10929, pp. 111–121. Springer, Cham (2018). https://doi.org/10.1007/978-3-319-94144-8_7
24. Newman, N., Fréchette, A., Leyton-Brown, K.: Deep optimization for spectrum repacking. Commun. ACM **61**(1), 97–104 (2018)
25. Niemetz, A., Preiner, M., Biere, A.: Propagation based local search for bit-precise reasoning. Formal Methods Syst. Des. **51**(3), 608–636 (2017). https://doi.org/10.1007/s10703-017-0295-6
26. Oh, C.: COMiniSatPS Pulsar and GHackCOMSPS. In: Proceedings of SAT Competition 2017: Solver and Benchmark Descriptions, pp. 12–13 (2017)
27. Pham, D.N., Duong, T., Sattar, A.: Trap avoidance in local search using pseudo-conflict learning. In: 2012 Proceedings of AAAI, pp. 542–548 (2012)

28. Roussel, O.: Controlling a solver execution with the runsolver tool. J. Satisfiability Boolean Mode. Comput. **7**(4), 139–144 (2011)
29. Selman, B., Levesque, H.J., Mitchell, D.G.: A new method for solving hard satisfiability problems. In: 1992 Proceedings of AAAI, pp. 440–446 (1992)
30. Thornton, J., Pham, D.N., Bain, S., Ferreira Jr., V.: Additive versus multiplicative clause weighting for SAT. In: 2004 Proceedings of AAAI, pp. 191–196 (2004)
31. Yolcu, E., Póczos, B.: Learning local search heuristics for Boolean satisfiability. In: 2019 Proceedings of NeurIPS, pp. 7990–8001 (2019)

Evaluation of a Permutation-Based Evolutionary Framework for Lyndon Factorizations

Lily Major[1(✉)], Amanda Clare[1], Jacqueline W. Daykin[1,2], Benjamin Mora[3], Leonel Jose Peña Gamboa[1], and Christine Zarges[1]

[1] Department of Computer Science, Aberystwyth University, Aberystwyth, UK
{jam86,afc,jwd6,lep31,chz8}@aber.ac.uk
[2] Department of Information Science, Stellenbosch University, Stellenbosch, South Africa
[3] Computer Science Department, Swansea University, Swansea, UK
b.mora@swansea.ac.uk

Abstract. String factorization is an important tool for partitioning data for parallel processing and other algorithmic techniques often found in the context of big data applications such as bioinformatics or compression. Duval's well-known algorithm uniquely factors a string over an ordered alphabet into Lyndon words, i.e., patterned strings which are strictly smaller than all of their cyclic rotations. While Duval's algorithm produces a pre-determined factorization, modern applications motivate the demand for factorizations with specific properties, e.g., those that minimize the number of factors or consist of factors with similar lengths. In this paper, we consider the problem of finding an alphabet ordering that yields a Lyndon factorization with such properties. We introduce a flexible evolutionary framework and evaluate it on biological sequence data. For the minimization case, we also propose a new problem-specific heuristic, Flexi-Duval, and a problem-specific mutation operator for Lyndon factorization. Our results show that our framework is competitive with Flexi-Duval for minimization and yields high quality and robust solutions for balancing where no problem-specific algorithm is available.

Keywords: Alphabet ordering · Biosequences · Duval's algorithm · Lyndon words · Permutations · String factorization

1 Introduction

Many application areas, including text compression and bioinformatics, benefit from first breaking a string into factors with mathematically interesting properties. Such a factorization can suggest a mechanism to skip over irrelevant regions of a string, to partition computation or to construct local indices into the string.

The first author is the main contributor. The ordering of the remaining authors is alphabetical with respect to the last name and does not imply any kind of weighting.

T. Bäck et al. (Eds.): PPSN 2020, LNCS 12269, pp. 390–403, 2020.
https://doi.org/10.1007/978-3-030-58112-1_27

For example, Mantaci et al. [17] propose that a factorization into Lyndon words will permit the separate processing of local suffixes when building a suffix array, so that internal or external memory can be used, or to allow online processing. The Burrows-Wheeler Transform (BWT) [4] is at the heart of compression algorithms such as bzip2 [20], and bioinformatics algorithms such as Bowtie2 and BWA [15,16]. The bijective BWT is based on Lyndon factorization [14] and recent work has shown [2] that when using this transformation, the number of search steps for a pattern is based on the number of distinct Lyndon factors in a particular suffix of the pattern. Lyndon words have also been used to discover tandem approximate repeats in biosequences [10] and in musicology [6]. A Lyndon word is a string which is minimal in the lexicographic order of its cyclic rotations. Duval's Algorithm (DA) [11] takes a string over an ordered alphabet, such as the Roman alphabet, and factors it into Lyndon factors in linear time and constant space. While the Lyndon factorization of a string for a given alphabet is unique [7], it may not yield the most suitable factors for a given application. Hence interest arises in seeking different factorizations via alphabet ordering techniques. For example, the string *parallelproblem* factors into the three Lyndon words $(p)(ar)(allelproblem)$ when using the Roman alphabet ordering. Using the alphabet ordering $m < b < o < e < l < r < a < p$ increases the number of Lyndon factors to $(p)(a)(ra)(l)(l)(elpr)(o)(ble)(m)$, and similarly, the ordering $p < r < o < l < a < b < e < m$ approximately balances the lengths as $(parallel)(problem)$. However, choosing an alphabet ordering can be computationally non-trivial. Regarding the BWT, the problem of deciding whether there exists an alphabet ordering satisfying a parameterized number of runs (maximal unary substrings) is NP-complete, while the corresponding minimization problem is APX-hard [3].

We wish to attempt two tasks: the production of the minimal number of Lyndon factors achievable over all possible alphabet orderings, and alternatively, the production of factors that are evenly balanced in size. We have developed an Evolutionary Algorithm (EA) to search for alphabet orderings for both tasks. EAs are *anytime* algorithms, i.e., they can be terminated at any time and will produce a solution, though the solution might be improved later if time permits. Finding improved permutations for a variety of tasks is an ideal use case for EAs [22].

To the best of our knowledge, alphabet ordering has not been considered other than for natural language text [5]. Furthermore, no consideration has been made in the domain for a heuristic setting, or in the context of Evolutionary Computation other than preliminary work in [8], where the impact of alphabet permutations produced significant variations in the numbers of factors obtained. Four fitness functions (maximize/minimize/parameterize/balance) for the number of factors demonstrated different factorizations suitable for various downstream tasks. This previous work mostly concentrated on random sequences; now we consider protein sequences and look at factorizations of proteins across many genomes. Here, we concentrate on comparing different mutation operators to develop the algorithm in a principled way. We also present a new problem-specific algorithm for the task of minimizing the number of factors.

2 Preliminaries

Let an alphabet Σ be a non-empty set of unique symbols also known as letters or characters. We denote Σ^+ as the set of all non-empty finite strings over the alphabet Σ and $\Sigma^* = \Sigma^+ \cup \{\varepsilon\}$ where ε is the empty string with length of zero.[1]

For a string $\boldsymbol{w} = w_0 w_1 \ldots w_{n-1}$ with each $w_i \in \Sigma$ and length n (or $|\boldsymbol{w}|$), we will represent $\boldsymbol{w}$ as an array $\boldsymbol{w}[0..n-1]$ with entries $\boldsymbol{w}[0]$, $\boldsymbol{w}[1]$, ..., $\boldsymbol{w}[n-1]$. A string is also known as a word.

Repetitions in strings will be denoted with a superscript as follows: for $\Sigma = \{a\}$ and $\boldsymbol{w} = aaa$ then $\boldsymbol{w} = a^3$. Lyndon words involve cyclic rotations (cyclic shifts or conjugates) of a string. For example, the cyclic rotations of the string *abc* are: *abc*, *bca*, *cab*. For a string $\boldsymbol{w} = \boldsymbol{w_p w_f w_s}$ of non-zero length, and strings $\boldsymbol{w_p}, \boldsymbol{w_f}, \boldsymbol{w_s} \in \Sigma^*$, then $\boldsymbol{w_p}$ is said to be a prefix of $\boldsymbol{w}$ while $\boldsymbol{w_s}$ is said to be a suffix of $\boldsymbol{w}$ and $\boldsymbol{w_f}$ is said to be a substring or factor.

For any non-empty string $\boldsymbol{w}$, $\boldsymbol{w} \in \Sigma^+$, there exists a unique factorization of $\boldsymbol{w}$ into Lyndon words, such that $\boldsymbol{w} = (\boldsymbol{u_1}) \geq (\boldsymbol{u_2}) \geq \cdots \geq (\boldsymbol{u_k})$ [7]. We denote these Lyndon factors as strings between brackets. For $\Sigma = \{a, b, c, d\}$, and a string $\boldsymbol{w} = dcba$, then the Lyndon factorization is written as $(d)(c)(b)(a)$.

We refer to the number of Lyndon factors produced from Duval's Algorithm for a given string $\boldsymbol{w} \in \Sigma^*$ and alphabet ordering π as $L(\Sigma, \pi, \boldsymbol{w})$.

3 Algorithms

In this section, we introduce our mutation-based evolutionary framework including the problem representation and different fitness functions. Afterwards, we derive a problem-specific mutation operator based on a modified version of Duval's Algorithm [11].

3.1 Representation and Fitness Function

We consider the problem of finding an alphabet ordering that induces a Lyndon factorization with a given property when used as input for Duval's Algorithm. [11]. Since an ordering of an alphabet Σ is equivalent to a permutation π of the letters in Σ, a permutation-based representation is the most natural choice for the problem at hand. Formally, given an alphabet Σ with size σ, the objective is to find a permutation π^* of Σ that optimizes a fitness function $f\colon S_\sigma \to \mathbb{R}^+$, where S_σ denotes the permutation space of the alphabet Σ.

As before, let $L(\Sigma, \pi, \boldsymbol{w})$ denote the number of Lyndon factors for a word $\boldsymbol{w} \in \Sigma^*$ and alphabet ordering π for Σ. Furthermore, let ℓ_j be the length of the j-th factor in a given factorization. In this paper, we consider three of the fitness functions introduced in [8], which deal with two aims for the Lyndon factorizations:

[1] We denote strings using boldface and characters in plainface.

Algorithm 1: (1+1) EA

```
1   Input: Word w ∈ Σ*; Fitness function f
2   Output: Best ordering of Σ found for f
3 select x ∈ S_σ uniformly at random;
4 while termination criterion not met do
5 |   y ← mutate(x);
6 |   if f(y) ≥ f(x) then
7 |   |   x ← y;
```

- Minimization of the number of factors, i.e., we minimize $f(\pi) = L(\Sigma, \pi, \boldsymbol{w})$.
- Balancing the length of the factors, i.e., all factors should be similar in length:
  1. difference between the maximum and the minimum length:

$$f(\pi) = \max_{1 \leq j \leq L(\Sigma,\pi,\boldsymbol{w})} \ell_j - \min_{1 \leq j \leq L(\Sigma,\pi,\boldsymbol{w})} \ell_j \tag{1}$$

  2. the standard deviation of the factor length:

$$f(\pi) = \sqrt{\frac{\sum_{j=1}^{L(\Sigma,\pi,\boldsymbol{w})} (\mu - \ell_j)^2}{L(\Sigma, \pi, \boldsymbol{w})}}, \tag{2}$$

  where μ is the mean length of the factors.

Note that for the balancing version we additionally require at least two factors. We do this by penalizing individuals with only one factor so that they are never accepted. All three fitness functions use Duval's [11] linear time and constant space algorithm to compute a unique Lyndon factorisation for a given string and alphabet ordering.

3.2 Mutation-Based Evolutionary Framework

We consider a simple mutation-based evolutionary framework in the spirit of a (1+1) EA (see Algorithm 1). The algorithm starts with a random permutation. It uses mutation to create an offspring and keeps the offspring if it is at least as good as its parent.

We compare three common mutation operators for permutation problems [12]:

- *Swap*: Selects two random letters and swaps them.
- *Insert*: Selects a random letter and moves it to a new random position.
- *Scramble*: Selects a random interval in the permutation and randomizes it.

For minimization, we additionally use a fourth, problem-specific mutation operator, the *LF-inspired* operator, that is proposed in Sect. 3.4. For each operator, we consider three variants: one application of the operator, three applications in succession and a random number of applications in succession where the random number is determined by a Poisson distribution with parameter $\lambda = 1$. This gives a total of twelve mutation operators for minimization and nine for balancing. Note that the EA (Algorithm 1) is formulated in a very general way and without defining specific stopping criteria. We provide details for the different cases we consider in Sect. 4.

We remark that we conducted preliminary experiments with Boltzmann selection (as used in Simulated Annealing [12,13]) instead of elitist selection in lines 4–5 of Algorithm 1. However, the results are not competitive and are therefore not included in this paper.

3.3 Flexi-Duval Algorithm

Here, we present a modification to Duval's Algorithm (DA) [11] for computing the Lyndon factorization (LF) of a string. For DA, the input string is over a totally ordered alphabet, such as the Roman or integer alphabet. When comparing a pair of letters during its linear scan, a non-fortuitous order such as $b > a$, or a repetition, will cause a factor(s) to be created. The new heuristic algorithm Flexi-Duval (FDA, Algorithm 2) is formulated from DA, with the goal of minimizing the number of factors. In contrast to DA, FDA does not assume that the underlying alphabet is ordered but rather induces an ordering while scanning the string. In the above case of comparing b and a, if currently unassigned, the reverse order $b < a$ would be given thus causing the current factor to be extended. The result is a Lyndon factorization over a partially ordered alphabet. This ordering may produce fewer Lyndon factors (and therefore an increase in length for some factors) than the original DA. Note however, for a repetition such as a^k, $a \in \Sigma$, both DA and FDA will produce k factors.

Table 1 shows examples of factoring strings using both DA and FDA. We see that input strings exist for both algorithms such that neither is necessarily optimal for minimizing the number of Lyndon factors. This depends of course on the alphabet ordering, however, for some applications there is no inherent ordering of the letters, such as nucleotide or amino acids letters in molecular biology. This allows flexibility in specifying suitable factors for a task.

3.4 Problem-Specific Mutation Operator

We introduce a problem-specific mutation operator (Algorithm 3) which is based on Duval's Algorithm and the concept of inducing the alphabet ordering in Flexi-Duval (Algorithm 2). By partially applying Duval's Algorithm to the input string using our alphabet ordering, the values of i and j can be found when a new Lyndon factor would be created. From this, we can find the characters in the alphabet ordering which cause the Lyndon factor to be created and then move the character at position j in the string in the alphabet ordering so that $\boldsymbol{w}[j] > \boldsymbol{w}[i]$. The effect of this is that, typically, a Lyndon factor is lengthened.

Algorithm 2: Algorithm FDA

```
1      Input: A string w of length n on an unordered Σ
2      Output: End positions of the Lyndon words
3  h ← 0;                                 // start of the current Lyndon factor
4  while h < n − 1 do
5      i ← h;               // search for the start of the next Lyndon factor
6      j ← h + 1;             // search for the end of the next Lyndon factor
7      while j < n and (isNotAssigned(w[j], w[i]) or w[j] ≥ w[i]) do
8          if w[i] ≠ w[j] and isNotAssigned(w[j], w[i]) then
9              assign(w[j] > w[i]);
10             i ← h;
11         else
12             if w[j] > w[i] then
13                 i ← h;
14             else
15                 i ← i + 1;
16         j ← j + 1;
17     repeat
18         h ← h+(j − i);
19         output(h − 1);
20     until h ≥ i;
```

Table 1. Example factorizations for DA and FDA.

String	Algorithm	Alphabet ordering	Factorization
$bab^{O(n)}$	Duval's	$a < b$	$(b)(abb...b)$
$bab^{O(n)}$	Flexi-Duval	$b < a$	$(ba)(b)...(b)$
$dcba$	Duval's	$a < b < c < d$	$(d)(c)(b)(a)$
$dcba$	Flexi-Duval	$d > c > b > a$	$(dcba)$

4 Methodology

To provide a large dataset of strings, protein sequences from a collection of genomes of prokaryotic organisms were obtained from NCBI RefSeq [19]. We are interested in producing Lyndon factoring algorithms that either reduce the number of factors (EA & FDA - Sect. 5.2) or balance their length (EA only - Sect. 5.3). Our *LF-inspired* operator is specifically designed for minimization, and thus is only analyzed in this case. We select one genome (GCF_000064305.2_ASM6430v2) to find the most suitable mutation operator for both the minimization and balanced factor fitness functions (Sect. 5.1). Each fitness function (Sect. 3.1) is evaluated using each of the 2446 proteins in this genome for each of the mutation operators (Sects. 3.2 and 3.4). We use two methods of finding the best mutation operator for the EA. For minimization the best operator has a minimal average of the best fitness found over all iterations for all

Algorithm 3: LF-inspired Mutation Operator

```
1      Input: A string w of length n, an alphabet Σ and alphabet ordering π
2      Output: A new alphabet ordering π
3  if L(Σ, π, w) > 1 then
4  |   i ← 0;
5  |   j ← 1;
6  |   while true do
7  |   |   if j = n or w[j] < w[i] then
8  |   |   |   break;
9  |   |   else
10 |   |   |   if w[j] > w[i] then
11 |   |   |   |   i ← 0;
12 |   |   |   else
13 |   |   |   |   i ← i + 1;
14 |   |   |   j ← j + 1;
15 |   c = w[j − i];
16 |   remove c from π;
17 |   p ← index of w[i] in π or |π| if missing;
18 |   if p = |π| then
19 |   |   append c to π;
20 |   else
21 |   |   m ← |π| − 1 − p;
22 |   |   r ← 0;
23 |   |   if m > 0 then
24 |   |   |   r ← uniformly at random from {0, 1, ..., m};
25 |   |   insert c into π at position p + 1 + r, shifting elements to the right;
26 output(π);
```

proteins. For balancing the best operators have minimal average standard deviation. We remark that the best operator does not change if the least difference between the average longest and average shortest Lyndon factor is used instead.

We select a set of 90 genomes to evaluate the performance of the selected mutation operator for the minimization fitness function with DA and FDA (Sect. 5.2) and balancing fitness functions (Sect. 5.3). As a baseline for minimization of the number of factors we consider two additional, very simple heuristics for each protein to produce an alphabet ordering: sorting unique characters by decreasing frequency for (a) each single protein and (b) for entire genomes.

We stop at a predefined number of iterations or if the optimum is found (a value of 1 for the minimization of the number of Lyndon factors or 0 for the balancing factor length fitness functions). Our preliminary experiments[2] show the majority of the improvement in fitness for each fitness function to occur after a relatively low number of iterations. Improvement after this generally takes many more iterations for a slight improvement. We therefore used 3000

[2] Code is available at: https://github.com/jam86/Evaluation-of-a-Permutation-Based-Evolutionary-Framework-for-Lyndon-Factorizations.

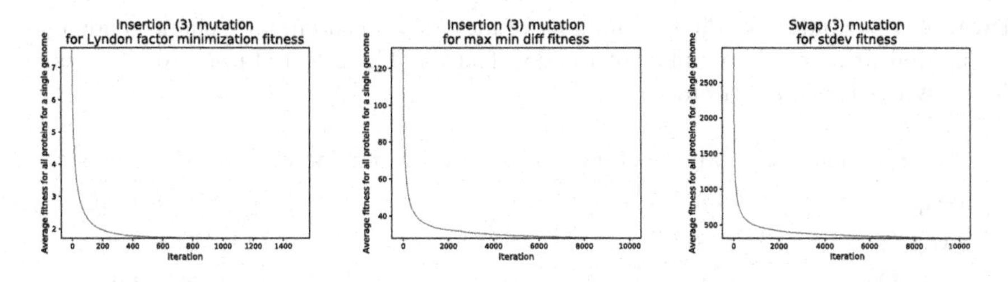

Fig. 1. Average fitness over time for all proteins in GCF_000064305.2_ASM6430v2 for each fitness function of the EA for the best mutation operators for each fitness function.

iterations for minimization of the number of Lyndon factors and 10000 iterations for balancing the length of factors (Fig. 1).

5 Results and Discussion

5.1 Single Genome

We present results for the EA (Algorithm 1) using the proteins in GCF_000064305.2_ASM6430v2 as the input. For minimization of the number of Lyndon factors, we find that *Insertion* mutation applied three times in succession gives the lowest average of the best found number of Lyndon factors (Table 2). We observe that all settings involving standard mutation operators perform somewhat similarly with averages ranging from 1.709 (*Insertion (3)*) to 1.933 (*Scramble*). In general, *Insertion* performs slightly better than *Swap* which in turn performs slightly better than *Scramble*, no matter how often these operators are applied. However, applying an operator three times performs slightly better than applying it only once or a random number of times. All standard operators achieve min and mode number of Lyndon factors of 1 and max number of Lyndon factors of 8 or 9.

All standard mutation operators yield averages smaller than 2. The best average achieved by *LF-inspired (3)* is 5.460, but this clearly improves over DA. Similar observations can be made for min, mode, and max number of Lyndon factors, no matter how often the operator is applied in succession. Due to the bias introduced by *LF-inspired* mutation, the EA quickly converges to a local optimum and is unable to escape. Combining the mutation with crossover as done previously for other (permutation-based) combinatorial optimization problems [21] is a promising direction for future research.

For the balance fitness functions we find again the mutation operators which give the lowest average standard deviation to be *Insertion* mutation applied three times in succession for Eq. 1 (Table 3). Further, we find the best mutation operator for Eq. 2 to be *Swap* mutation applied three times (Table 4). For both balancing fitness functions we make very similar observations in terms of the standard mutation operators and the problem-specific operator: All standard

Table 2. Distribution of the best number of factors per protein for a single genome for the minimization of the number of Lyndon factors. The selected operator is in bold. We show DA for comparison.

Mutation operator	Max factors	Min factors	Mode factors	Mean ± stdev
Swap	8	1	1	1.877 ± 1.002
Swap (3)	8	1	1	1.727 ± 1.043
Scramble	8	1	1	1.933 ± 1.136
Scramble (3)	9	1	1	1.911 ± 1.155
Insertion	8	1	1	1.753 ± 1.042
Insertion (3)	**8**	**1**	**1**	**1.709 ± 1.047**
LF-inspired	14	1	5	5.601 ± 2.057
LF-inspired (3)	13	1	5	5.460 ± 2.020
Swap Poisson	8	1	1	1.796 ± 1.027
Scramble Poisson	8	1	1	1.913 ± 1.156
Insertion Poisson	8	1	1	1.739 ± 1.044
LF-inspired Poisson	13	1	5	5.596 ± 2.107
DA	16	2	7	7.284 ± 2.296

Table 3. Balance factor length for Eq. 1 average maximum, minimum, and average standard deviation of Lyndon factor length per protein for a single genome. The selected operator is in bold.

Mutation operator	Avg. max	Avg. min	Avg. mean ± Avg. stdev
Swap	112.956	79.516	96.033 ± 13.227
Swap (3)	123.076	95.123	109.110 ± 11.365
Scramble	119.588	85.857	102.228 ± 13.391
Scramble (3)	121.521	90.147	105.445 ± 12.592
Insertion	112.007	78.131	94.885 ± 13.271
Insertion (3)	**122.953**	**95.378**	**109.198 ± 11.182**
Swap Poisson	120.794	90.632	105.524 ± 12.199
Scramble Poisson	120.842	89.093	104.725 ± 12.702
Insertion Poisson	115.520	84.570	99.866 ± 12.229

operators perform very similarly while our *LF-inspired* operator performs worse. However, we do not show *LF-inspired* for balancing as the operator is designed to minimize rather than balance. We remark that it performs poorly with both balancing fitness functions. Moreover, applying standard operators three times in succession outperforms the other considered variants. When looking into the different standard operators *Swap* and *Insertion* are slightly better than *Scramble* for minimizing the difference while the picture is mixed when minimizing the standard deviation.

Table 4. Balance factor length for Eq. 2 average maximum, minimum, and average standard deviation of Lyndon factor length per protein for a single genome. The selected operator is in bold.

Mutation operator	Avg. max	Avg. min	Avg. mean ± Avg. stdev
Swap	96.988	47.251	63.456 ± 14.708
Swap (3)	**108.813**	**75.157**	**89.153 ± 11.112**
Scramble	106.498	63.335	79.080 ± 13.606
Scramble (3)	108.308	72.090	86.853 ± 11.889
Insertion	94.676	40.298	56.681 ± 15.567
Insertion (3)	107.425	71.074	84.978 ± 11.571
Swap Poisson	104.742	62.679	78.067 ± 13.160
Scramble Poisson	106.829	67.627	82.797 ± 12.611
Insertion Poisson	98.418	49.903	65.407 ± 14.254

Table 5. Number of Lyndon factors for methods that minimize the number of factors on the test set of 90 genomes.

Alphabet ordering	Max factors	Min factors	Mode factors	Mean ± stdev
Duval's Algorithm	41	2	6	6.802 ± 2.169
Protein based frequency	31	1	7	7.227 ± 2.307
Genome based frequency	25	2	7	7.164 ± 2.238
EA with Insertion (3)	12	1	1	1.725 ± 1.102
Flexi-Duval Algorithm	6	1	1	1.197 ± 0.443

5.2 Minimizing the Number of Lyndon Factors

We find that FDA produces fewer Lyndon factors across our entire set of 90 test genomes than DA (Table 5). On average, the EA with the fitness function of the number of Lyndon factors almost performs as well as FDA. Given that FDA is a problem-specific heuristic developed for the purpose of minimizing the number of Lyndon factors, it is encouraging to see that our simple EA yields comparable performance in terms of solution quality. With respect to the runtime, FDA is far faster than the EA. For the same set of 90 genomes, FDA computed the factorization of all proteins in minutes compared to hours for the EA.

As the output of the FDA is a partial order, we convert each alphabet ordering result to a total order using depth first traversal, for inspection of the relevance of the order to biology. The positions of the characters within the orderings, averaged over each genome, are shown in Fig. 2. Proteins are not random, and usually begin with a methionine (M) [1]. This is reflected in its positions across the bottom of the graph. Some of the more common amino acids (L, K) are found next, and some of the less common amino acids (C, W) are towards the top of the graph, later in the ordering. We therefore compare the FDA results to the baseline of using a frequency-based ordering (Table 5).

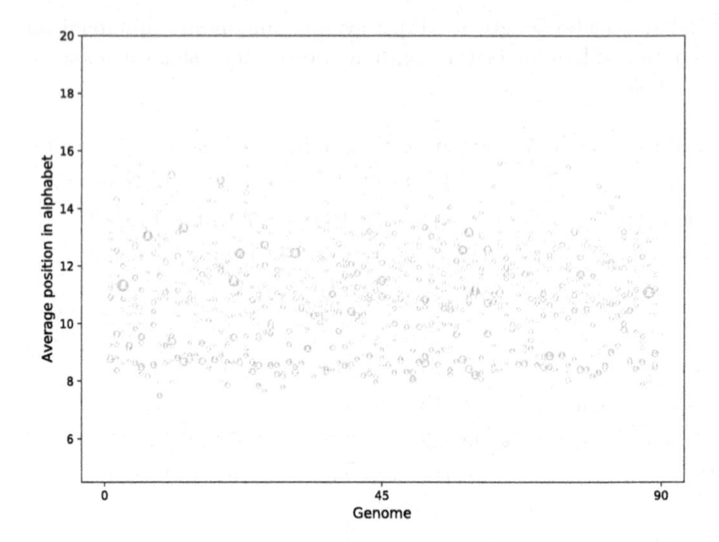

Fig. 2. Average position of each character in the FDA alphabet ordering across all genomes in the test set of 90 genomes. The size of each point is proportional to the relative frequency of each amino acid in the entire genome. The y axis is truncated to begin at 4.5.

FDA is significantly better than using an ordering based on the frequency of amino acids in each protein or genome.[3] Surprisingly, some proteins are factored into fewer Lyndon factors when using the whole genome frequency for alphabet ordering than when using the protein-specific frequency. Of the total 303,007 proteins in the test set of 90 genomes, 114,975 (37.94%) are factored into more Lyndon factors when using the protein frequency-based ordering than when using the genome frequency-based ordering.

5.3 Balancing the Length of Lyndon Factors

While there exists a problem-specific heuristic for the minimization problem, to the best of our knowledge no such method exists for balancing the lengths of the factors. Applying our EA for this use case is therefore an important step to demonstrate its usefulness. Based on our results in Sect. 5.1, we consider *Insertion (3)* for minimizing the difference of the factor lengths and *Swap (3)* for minimizing the standard deviation. We find that the fitness functions for balancing the length of Lyndon factors are a great improvement over the regular output of DA (Table 6). The fitness function that uses the standard deviation of factor length produces smaller factors than the fitness function that minimizes the difference in factor length. However both produce equivalently stable variation in balancing the factor lengths (as seen by the average standard deviations).

[3] Using a Kolmogorov-Smirnov [18] (KS) two-sample test we obtained extremely low p-values calculated as zero.

Table 6. Average maximum, minimum, mean, and standard deviation of Lyndon factor lengths for the EA and DA per protein in 90 test genomes.

Alphabet ordering	Avg. max	Avg. min	Avg. mean ± Avg. stdev
Duval's Algorithm	197.195	1.384	48.197 ± 70.204
Max Min Diff – Insertion (3)	116.816	91.474	104.110 ± 10.309
Stdev – Swap (3)	104.220	72.955	86.114 ± 10.486

Comparing results in Table 6 for the 90-genome set and Tables 3 and 4 for a single genome, we observe that the average standard deviation values on the 90-genome test set are very similar to the single genome (e.g., 10.309 (Table 6) vs 11.182 (Table 3)). This indicates that the genome we used to determine the best mutation operator is representative for the whole dataset. It is also important to remark, that the optimal solutions for all genomes are currently unknown, i.e., we do not know what the smallest possible difference/standard deviation is. However, achieving similar results across different sets of genomes indicates that our proposed method is successful in finding good solutions in a robust way.

6 Conclusion and Future Work

We have proposed a flexible mutation-based evolutionary framework for the problem of finding string factorizations with specific properties. We have considered two optimization goals (minimizing the number of factors and balancing the length of factors) and evaluated our framework on biological sequence data. We find that for the minimization problem our framework is competitive with a simple problem-specific heuristic in terms of solution quality while it yields high quality and robust solutions for the balancing variant where no such problem-specific algorithm is available. Moreover, we observe that a novel problem-specific mutation operator for minimization is not yet competitive with standard permutation-based operators from the literature.

Finding alphabet orderings with desired effects has been considered previously in [8] for random sequences as well as proteins from a single genome. Biological sequences have structure [9,23] that is not necessarily present in totally random, fixed length sequences. This work more fully explores the factoring of proteins from a large number of genomes.

Future work will deal with the improvement and further evaluation of our framework, particularly the analysis of other fitness functions, problem-specific mutation operators and a comparison with population-based algorithms. Moreover, we plan to apply our framework to other types of strings such as a natural language corpus. While the EA produces a total alphabet ordering, FDA only determines a partial ordering, which is more flexible and may prove useful for finding consensus orderings in the context of sets of genomes. Future work will also investigate how such consensus orderings can be determined and how these could inform us about the biological structures of the collection of proteins.

Acknowledgment. This work is supported by the UKRI AIMLAC CDT, http://cdt-aimlac.org, grant no. EP/S023992/1, and was part-funded by the European Regional Development Fund through the Welsh Government, grant 80761-AU-137 (West).

References

1. Adams, J.M., Capecchi, M.R.: N-formylmethionyl-sRNA as the initiator of protein synthesis. Proc. Natl. Acad. Sci. USA **55**(1), 147–155 (1966). https://doi.org/10.1073/pnas.55.1.147
2. Bannai, H., Kärkkäinen, J., Köppl, D., Piatkowski, M.: Indexing the bijective BWT. In: 30th Annual Symposium on Combinatorial Pattern Matching (CPM 2019), pp. 17:1–17:14. Leibniz International Proceedings in Informatics (LIPIcs) (2019)
3. Bentley, J., Gibney, D., Thankachan, S.V.: On the complexity of BWT-runs minimization via alphabet reordering. arXiv:1911.03035v2 (2019)
4. Burrows, M., Wheeler, D.J.: A block-sorting lossless data compression algorithm. Technical report, Digital Systems Research Center, Palo Alto (1994). https://www.hpl.hp.com/techreports/Compaq-DEC/SRC-RR-124.pdf
5. Chapin, B., Tate, S.R.: Higher compression from the Burrows-Wheeler Transform by modified sorting. In: Data Compression Conference, DCC 1998, Snowbird, Utah, USA, 30 March–1 April 1998, p. 532. IEEE Computer Society (1998). https://doi.org/10.1109/DCC.1998.672253
6. Chemillier, M.: Periodic musical sequences and Lyndon words. Soft. Comput. **8**(9), 611–616 (2004). https://doi.org/10.1007/s00500-004-0387-2
7. Chen, K.T., Fox, R.H., Lyndon, R.C.: Free differential calculus, IV - the quotient groups of the lower central series. Ann. Math. **68**(1), 81–95 (1958). https://doi.org/10.2307/1970044
8. Clare, A., Daykin, J.W., Mills, T., Zarges, C.: Evolutionary search techniques for the Lyndon factorization of biosequences. In: Proceedings of the Genetic and Evolutionary Computation Conference Companion, GECCO 2019, Prague, Czech Republic, 13–17 July 2019, pp. 1543–1550. ACM (2019). https://doi.org/10.1145/3319619.3326872
9. De Lucrezia, D., Slanzi, D., Poli, I., Polticelli, F., Minervini, G.: Do natural proteins differ from random sequences polypeptides? Natural vs. random proteins classification using an evolutionary neural network. PloS One **7**(5), e36634–e36634 (2012). https://doi.org/10.1371/journal.pone.0036634
10. Delgrange, O., Rivals, E.: STAR: an algorithm to search for tandem approximate repeats. Bioinform. (Oxford, Engl.) **20**(16), 2812–2820 (2004). https://doi.org/10.1093/bioinformatics/bth335
11. Duval, J.P.: Factorizing words over an ordered alphabet. J. Algorithms **4**(4), 363–381 (1983). https://doi.org/10.1016/0196-6774(83)90017-2
12. Eiben, A.E., Smith, J.E.: Introduction to Evolutionary Computing. NCS, 2nd edn. Springer, Heidelberg (2015). https://doi.org/10.1007/978-3-662-44874-8
13. Kirkpatrick, S., Gelatt Jr., C.D., Vecchi, M.P.: Optimization by simulated annealing. Science **220**(4598), 671–680 (1983)

14. Kufleitner, M.: On bijective variants of the Burrows-Wheeler Transform. In: Holub, J., Zdárek, J. (eds.) Proceedings of the Prague Stringology Conference 2009, Prague, Czech Republic, pp. 65–79 (2009)
15. Langmead, B., Salzberg, S.L.: Fast gapped-read alignment with Bowtie 2. Nat. Methods **9**, 357–359 (2012). https://doi.org/10.1038/nmeth.1923
16. Li, H., Durbin, R.: Fast and accurate short read alignment with Burrows-Wheeler transform. Bioinformatics **25**(14), 1754–1760 (2009). https://doi.org/10.1093/bioinformatics/btp324
17. Mantaci, S., Restivo, A., Rosone, G., Sciortino, M.: Suffix array and Lyndon factorization of a text. J. Discret. Algorithms **28**, 2–8 (2014)
18. Massey, F.J.: The Kolmogorov-Smirnov test for goodness of fit. J. Am. Stat. Assoc. **46**(253), 68–78 (1951). https://doi.org/10.2307/2280095
19. O'Leary, N.A., et al.: Reference sequence (RefSeq) database at NCBI: current status, taxonomic expansion, and functional annotation. Nucleic Acids Res. **44**(D1), D733–D745 (2016). https://doi.org/10.1093/nar/gkv1189
20. Seward, J.: bzip2 and libbzip2 (1996). http://sourceware.org/bzip2/
21. Tinós, R., Helsgaun, K., Whitley, D.: Efficient recombination in the Lin-Kernighan-Helsgaun traveling salesman heuristic. In: Auger, A., Fonseca, C.M., Lourenço, N., Machado, P., Paquete, L., Whitley, D. (eds.) PPSN 2018. LNCS, vol. 11101, pp. 95–107. Springer, Cham (2018). https://doi.org/10.1007/978-3-319-99253-2_8
22. Whitley, D.: Permutations. In: Bäck, T., Fogel, D.B., Michalewicz, Z. (eds.) Handbook of Evolutionary Computation, p. C1.4. IOP Publishing and Oxford University Press (1997)
23. Yu, J.-F., et al.: Natural protein sequences are more intrinsically disordered than random sequences. Cell. Mol. Life Sci. **73**(15), 2949–2957 (2016). https://doi.org/10.1007/s00018-016-2138-9

Optimising Monotone Chance-Constrained Submodular Functions Using Evolutionary Multi-objective Algorithms

Aneta Neumann(✉) and Frank Neumann(✉)

Optimisation and Logistics, School of Computer Science, The University of Adelaide, Adelaide, SA, Australia
{aneta.neumann,frank.neumann}@adelaide.edu.au

Abstract. Many real-world optimisation problems can be stated in terms of submodular functions. A lot of evolutionary multi-objective algorithms have recently been analyzed and applied to submodular problems with different types of constraints. We present a first runtime analysis of evolutionary multi-objective algorithms for chance-constrained submodular functions. Here, the constraint involves stochastic components and the constraint can only be violated with a small probability of α. We show that the GSEMO algorithm obtains the same worst case performance guarantees as recently analyzed greedy algorithms. Furthermore, we investigate the behavior of evolutionary multi-objective algorithms such as GSEMO and NSGA-II on different submodular chance constrained network problems. Our experimental results show that this leads to significant performance improvements compared to the greedy algorithm.

1 Introduction

Evolutionary algorithms have been widely applied to solve complex optimisation problems. They are well suited for broad classes of problems and often achieve good results within a reasonable amount of time. The theory of evolutionary computation aims to explain such good behaviours and also point out the limitations of evolutionary computing techniques. A wide range of tools and techniques have been developed in the last 25 years and we point the reader to [2,7,14,27] for comprehensive presentations.

Stochastic components play a crucial role in many real-world applications and chance constraints allow to model constraints that can only be violated with a small probability. A chance constraint involves random components and it is required that the constraint is violated with a small probability of at most α. We consider chance constraints where the weight $W(S)$ of a possible solution S can violate a given constraint bound C with probability at most α, i.e. $\Pr[W(S) > C] \leq \alpha$ holds.

T. Bäck et al. (Eds.): PPSN 2020, LNCS 12269, pp. 404–417, 2020.
https://doi.org/10.1007/978-3-030-58112-1_28

Evolutionary algorithms have only recently been considered for chance constrained problems and we are pushing forward this area of research by providing a first runtime analysis for submodular functions. In terms of the theoretical understanding and the applicability of evolutionary algorithms, it is desirable to be able to analyse them on a broad class of problems and design appropriate evolutionary techniques for such classes. Submodular functions model a wide range of problems where the benefit of adding solution components diminishes with the addition of elements. They have extensively been studied in the literature [3,9,13,18,21,24,25,33] and allow to model a variety of real-word applications [12,19,20,22]. In recent years, the design and analysis of evolutionary algorithms for submodular optimisation problems has gained increasing attention. We refer to the recent book of Zhou et al. [37] for an overview. Such studies usually study evolutionary algorithms in terms of their runtime and approximation behaviour and evaluate the performance of the designed algorithms on classical submodular combinatorial optimisation problems.

To our knowledge, there is so far no runtime analysis of evolutionary algorithms for submodular optimisation with chance constraints and the runtime analysis of evolutionary algorithms for chance constrained problems has only started recently for very special cases of chance-constrained knapsack problem [26]. Chance constraints are in general hard to evaluate exactly, but well-known tail inequalities such as Chernoff bounds and Chebyshev's inequality may be used to estimate the probability of a constraint violation. We provide a first runtime analysis by analysing GSEMO together with multi-objective formulations that use a second objective taking the chance constraint into account. These formulations based on tail inequalities are motivated by some recent experimental studies of evolutionary algorithms for the knapsack problem with chance constraints [1,34,35]. The GSEMO algorithm has already been widely studied in the area of runtime analysis in the field of evolutionary computation [10] and more broadly in the area of artificial intelligence where the focus has been on submodular functions and Pareto optimisation [29–32]. We analyse this algorithm in the chance constrained submodular optimisation setting investigated in [6] in the context of greedy algorithms. Our analyses show that GSEMO is able to achieve the same approximation guarantee in expected polynomial time for uniform IID weights and the same approximation quality in expected pseudo-polynomial time for independent uniform weights having the same dispersion.

Furthermore, we study GSEMO experimentally on the influence maximization problem in social networks and the maximum coverage problem. Our results show that GSEMO significantly outperforms the greedy approach [6] for the considered chance constrained submodular optimisation problems. Furthermore, we use the multi-objective problem formulation in a standard setting of NSGA-II. We observe that NSGA-II is outperformed by GSEMO in most of our experimental settings, but usually achieves better results than the greedy algorithm.

The paper is structured as follows. In Sect. 2, we introduce the problem of optimising submodular functions with chance constraints, the GSEMO algorithm and tail inequalities for evaluating chance constraints. In Sect. 3, we provide a runtime analysis for submodular functions where the weights of the constraints are either identically uniformly distributed or are uniformly distributed and have the same dispersion. We carry out experimental investigations that compare the performance of greedy algorithms, GSEMO, and NSGA-II in Sect. 4 and finish with some concluding remarks.

2 Preliminaries

Given a set $V = \{v_1, \ldots, v_n\}$, we consider the optimization of a monotone submodular function $f\colon 2^V \rightarrow \mathbb{R}_{\geq 0}$. We call a function monotone iff for every $S, T \subseteq V$ with $S \subseteq T$, $f(S) \leq f(T)$ holds. We call a function f submodular iff for every $S, T \subseteq V$ with $S \subseteq T$ and $x \notin T$ we have

$$f(S \cup \{x\}) - f(S) \geq f(T \cup \{x\}) - f(T).$$

Here, we consider the optimization of a monotone submodular function f subject to a chance constraint where each element $s \in V$ takes on a random weight $W(s)$. Precisely, we examine constraints of the type

$$\Pr[W(S) > C] \leq \alpha.$$

where $W(S) = \sum_{s \in S} w(s)$ is the sum of the random weights of the elements and C is the given constraint bound. The parameter α specifies the probability of exceeding the bound C that can be tolerated for a feasible solution S.

The two settings, we investigate in this paper assume that the weight of an element $s \in V$ is $w(s) \in [a(s) - \delta, a(s) + \delta]$, $\delta \leq \min_{s \in V} a(s)$, is chosen uniformly at random. Here $a(s)$ denotes the expected weight of items s. For our investigations, we assume that each item has the same dispersion δ. We call a feasible solution S a γ-approximation, $0 \leq \gamma \leq 1$, iff $f(S) \geq \gamma \cdot f(OPT)$ where OPT is an optimal solution for the given problem.

2.1 Chance Constraint Evaluation Based on Tail Inequalities

As the probability $(Pr(W(X) > C)$ used in the objective functions is usually computational expensive to evaluate exactly, we use the approach taken in [34] and compute an upper bound on this probability using tail inequalities [23]. We assume that $w(s) \in [a(s) - \delta, a(s) + \delta]$ holds for each $s \in V$ which allows to use Chebyshev's inequality and Chernoff bounds.

The approach based on (one-sided) Chebyshev's inequality used in [34] upper bounds the probability of a constraint violation by

$$\hat{\Pr}(W(X) > C) \leq \frac{\delta^2 |X|}{\delta^2 |X| + 3(C - E_W(X))^2} \tag{1}$$

The approach based on Chernoff bounds used in [34] upper bounds the probability of a constraint violation by

$$\hat{\Pr}[W(X) > C] \leq \left(\frac{e^{\frac{C - E_W(X)}{\delta |X|}}}{\left(\frac{\delta|X| + C - E_W(X)}{\delta|X|}\right)^{\frac{\delta|X| + C - E_W(X)}{\delta|X|}}} \right)^{\frac{1}{2}|X|} \quad (2)$$

We use $\hat{\Pr}(W(X) > C)$ instead of $\Pr(W(X) > C)$ for our investigations usng multi-objective models of the problem.

2.2 Multi-objective Formulation

Following the approach of Yue et al. [34] for the chance constrained knapsack problem, we evaluate a set X by the multi-objective fitness function $g(X) = (g_1(X), g_2(X))$ where g_1 measures the tightness in terms of the constraint and g_2 measures the quality of X in terms of the given submodular function f.

We define

$$g_1(X) = \begin{cases} E_W(X) - C & \text{if} \quad (C - E_W(X))/(\delta \cdot |X|) \geq 1 \\ \hat{Pr}(W(X) > C) & \text{if } (E_W(X) < C) \wedge (C - E_W(X))/(\delta|X| < 1) \\ 1 + (E_W(X) - C) & \text{if} \quad E_W(X) \geq C \end{cases} \quad (3)$$

and

$$g_2(X) = \begin{cases} f(X) & \text{if} \quad g_1(X) \leq \alpha \\ -1 & \text{if } \hat{Pr}(W(X) > C) > \alpha \end{cases} \quad (4)$$

where $E_W(X) = \sum_{s \in X} a(s)$ denotes the expected weight of the solution. The term $(C - E_W(X))/(\delta \cdot |X|) \geq 1$ in g_1 implies that a set X of cardinality $|X|$ has probability 0 of violating the chance constraint due to the upper bound on the intervals.

We say a solution Y dominates a solution X (denoted by $Y \succcurlyeq X$) iff $g_1(Y) \leq g_1(X) \wedge g_2(Y) \geq g_2(X)$. We say that Y strongly dominates X (denoted by $Y \succ X$) iff $Y \succcurlyeq X$ and $g(Y) \neq g(X)$ The dominance relation also translates to the corresponding search points used in GSEMO. Comparing two solutions, the objective function guarantees that a feasible solution strongly dominates every infeasible solution. The objective function g_1 ensures that the search process is guided towards feasible solutions and that trade-offs in terms of the probability of a constraint violation and the function value of the submodular function f are computed for feasible solutions.

2.3 Global SEMO

Our multi-objective approach is based on a simple multi-objective evolutionary algorithm called Global Simple Evolutionary Multi-Objective Optimizer (GSEMO, see Algorithm 1) [11]. The algorithm encodes sets as bitstrings of

Algorithm 1: Global SEMO

```
1 Choose x ∈ {0,1}^n uniformly at random;
2 P ← {x};
3 repeat
4 |  Choose x ∈ P uniformly at random;
5 |  Create y by flipping each bit x_i of x with probability 1/n;
6 |  if ∄w ∈ P : w ≻ y then
7 |  |  S ← (P ∪ {y})\{z ∈ P | y ≽ z};
8 until stop;
```

length n and the set X corresponding to a search point x is given as $X = \{v_i \mid x_i = 1\}$. We use x when referring to the search point in the algorithm and X when referring to the set of selected elements and use applicable fitness measure for both notations in an interchangeable way. GSEMO starts with a random search point $x \in \{0,1\}^n$. In each iteration, an individual $x \in P$ is chosen uniformly at random from the current population P In the mutation step, it flips each bit with a probability $1/n$ to produce an offspring y. y is added to the population if it is not strongly dominated by any other search point in P. If y is added to the population, all search points dominated by y are removed from the population P.

We analyze GSEMO in terms of its runtime behaviour to obtain a good approximation. The expected time of the algorithm required to achieve a given goal is measured in terms of the number of iterations of the repeat loop until a feasible solution with the desired approximation quality has been produced for the first time.

3 Runtime Analysis

In this section, we provide a runtime analysis of GSEMO which shows that the algorithm is able to obtain a good approximation for important settings where the weights of the constraint are chosen according to a uniform distribution with the same dispersion.

3.1 Uniform IID Weights

We first investigate the case of uniform identically distributed (IID) weights. Here each weight is chosen uniformly at random in the interval $[a - \delta, a + \delta]$, $\delta \leq a$. The parameter δ is called the dispersion and models the uncertainty of the weight of the items.

Theorem 1. *Let $k = \min\{n + 1, \lfloor C/a \rfloor\}$ and assume $\lfloor C/a \rfloor = \omega(1)$. Then the expected time until GSEMO has computed a $(1-o(1))(1-1/e)$-approximation for a given monotone submodular function under a chance constraint with uniform iid weights is $O(nk(k + \log n))$.*

Proof. Every item has expected weight a and uncertainty δ. This implies $g_1(X) = g_1(Y)$ iff $|X| = |Y|$ and $E_W(X) = E_W(Y) < C$. As GSEMO only stores for each fixed g_1-value one single solution, the number of solutions with expected weight less than C is at most $k = \min\{n+1, C/a\}$. Furthermore, there is at most one individual X in the population $g_2(X) = -1$. Hence, the maximum population that GSEMO encounters during the run of the algorithm is at most $k+1$.

We first consider the time until GSEMO has produced the bitstring 0^n. This is the best individual with respect to g_1 and once included will always stay in the population. The function g_1 is strictly monotone decreasing with the size of the solution. Hence, selecting the individual in the population with the smallest number of elements and removing one of them least to a solution with less elements and therefore with a smaller g_1-value. Let $\ell = |x|_1$ be the number of elements of the solution x with the smallest number of elements in P. Then flipping one of the 1-bits corresponding these elements reduces k by one and happens with probability at least $\ell/(en)$ once x is selected for mutation. The probability of selecting x is at least $1/(k+1)$ as there are at most $k+1$ individuals in the population. Using the methods of fitness-based partitions, the expected time to obtain the solution 0^n is at most

$$\sum_{\ell=1}^{n} \left(\frac{\ell}{e(k+1)n}\right)^{-1} = O(nk \log n).$$

Let $k_{opt} = \lfloor C/a \rfloor$, the maximal number of elements that can be included in the deterministic version of the problem.

The function g_1 is strictly monotonically increasing with the number of elements and each solution with same number of elements has the same g_1-value.

We consider the solution X with the largest k for which

$$f(X) \geq (1 - (1 - 1/k_{opt})^k) \cdot f(OPT)$$

holds in the population and the mutation which adds an element with the largest marginal increase $g_2(X \cup \{x\}) - g_2(X)$ to X. The probability for such a step picking X and carrying the mutation with the largest marginal gain is $\Omega(1/kn)$ and its waiting time is $O(kn)$.

This leads to a solution Y for which

$$f(X) \geq (1 - (1 - 1/k_{opt})^{k+1}) \cdot f(OPT)$$

holds. The maximal number of times such a step is required after having included the search point 0^n into the population is k which gives the runtime bound of $O(k^2 n)$.

For the statement on the approximation quality, we make use of the lower bound on the maximal number of elements that can be included using the Chernoff bound and Chebyshev's inequality given in [6].

Using Chebyshev's inequality (Eq. 1) at least

$$k_1^* = \max\left\{k \mid k + \frac{\sqrt{(1-\alpha)k\delta^2}}{\sqrt{3\alpha}a} \le k_{opt}\right\}$$

elements can be included and when using Chernoff bound (Eq. 2), at least

$$k_2^* = \max\left\{k \;\middle|\; k + \frac{\sqrt{3\delta k \ln(1/\alpha)}}{a} \le k_{opt}\right\}$$

elements can be included.

Including k^* elements in this way least to a solution X^* with

$$f(X^*) \ge (1 - (1 - 1/k_{opt})^{k^*}) \cdot f(OPT).$$

As shown in [6], both values of k_1^* and k_2^* yield $(1 - o(1))(1 - 1/e) \cdot f(OPT)$ if $\lfloor C/a \rfloor = \omega(1)$ which completes the proof. □

3.2 Uniform Weights with the Same Dispersion

We now assume that the expected weights do not have to be the same, but still require the same dispersion for all elements, i.e. $w(s) \in [a(s) - \delta, a(s) + \delta]$ holds for all $s \in V$.

We consider the (to be minimized) objective function $\hat{g}_1(X) = E_W(X)$ (instead of g_1) together with the previously defined objective function g_2 and evaluate a set X by $\hat{g}(X) = (\hat{g}_1(X), g_2(X))$. We have $Y \succeq X$ iff $\hat{g}_1(Y) \le \hat{g}_1(X)$ and $g_2(Y) \ge g_2(X)$

Let $a_{\max} = \max_{s \in V} a(s)$ and $a_{\min} = \min_{s \in V} a(s)$, and $\delta \le a_{\min}$. The following theorem shows that GSEMO is able to obtain a $(1/2 - o(1))(1 - 1/e)$-approximation if $\omega(1)$ elements can be included in a solution.

Theorem 2. *If $C/a_{\max} = \omega(1)$ then GSEMO obtains a $(1/2 - o(1))(1 - 1/e)$-approximation for a given monotone submodular function under a chance constraint with uniform weights having the same dispersion in expected time $O(P_{\max} \cdot n(C/a_{\min} + \log n + \log(a_{\max}/a_{\min})))$.*

Proof. We first consider the time until the search point 0^n is included in the population. We always consider the individual x with the smallest $\hat{g}_1$-value. Flipping every 1-bit of $x \neq 0^n$ leads to an individual with a smaller $\hat{g}_1$-value and is therefore accepted. Furthermore, the total weight decrease of these 1-bit flips is $\hat{g}_1(x)$ which also equals the total weight decrease of all single bit flip mutation when taking into account that 0-bit flips give decrease of the $\hat{g}_1$-value of zero. A mutation carrying out a single bit flip happens each iteration with probability at least $1/e$. The expected decrease in $\hat{g}_1$ is therefore at least by a factor of $(1 - 1/(P_{\max}en))$ and the expected minimal $\hat{g}_1$-value in the next generation is at most

$$(1 - 1/(P_{\max} \cdot en)) \cdot \hat{g}_1(x).$$

We use drift analysis, to upper bound the expected time until the search point 0^n is included in the population. As $a_{\min} \leq \hat{g_1}(x) \leq na_{\max}$ holds for any search point $x \neq 0^n$, the search point 0^n is included in the population after an expected number of $O(P_{\max}n(\log n + \log(a_{\max}/a_{\min})))$ steps.

After having include the search point 0^n in the population, we follow the analysis of POMC for subset selection with general deterministic cost constraints [29] and always consider the individual x with the largest $\hat{g_1}$-value for which

$$g_2(x) \geq \left[1 - \prod_{k=1}^{n}\left(1 - \frac{a(k)x_k}{C}\right)\right] \cdot f(OPT).$$

Note that the search point 0^n meets this formula. Furthermore, we denote by $\hat{g_1}^*$, the maximal $\hat{g_1}$-value for which $\hat{g_1}(x) \leq \hat{g_1}^*$ and

$$g_2(x) \geq \left[1 - \left(1 - \frac{\hat{g_1}^*}{Cr}\right)^r\right] \cdot f(OPT).$$

for some r, $0 \leq r \leq n-1$, holds. We use $\hat{g_1}^*$ to track the progress of the algorithm and it has been shown in [29] that $\hat{g_1}^*$ does not decrease during the optimisation process of GSEMO.

Choosing x for mutation and flipping the 0-bit of x corresponding to the largest marginal gain in terms of $g_2/\hat{g_1}$ gives a solution y for which

$$\begin{aligned} g_2(y) &\geq \left[1 - \left(1 - \frac{a_{\min}}{C}\right) \cdot \left(1 - \frac{\hat{g_1}^*}{Cr}\right)^r\right] \cdot f(OPT) \\ &\geq \left[1 - \left(1 - \frac{\hat{g_1}^* + a_{\min}}{C(r+1)}\right)^{r+1}\right] \cdot f(OPT) \end{aligned}$$

holds and $\hat{g_1}^*$ increases by at least $a_{\min}$. The $\hat{g_1}^*$-value for the considered solution, can increase at most $C/a_{\min}$ times and therefore, once having included the search point 0^n, the expected time until such improvements have occurred is $O(P_{\max}nC/a_{\min})$.

Let x^* be the feasible solution of maximal cost included in the population after having increased the $\hat{g_1}^*$ at most $C/a_{\min}$ times as described above. Furthermore, let v^* be the element with the largest g_2-value not included in x^* and $\hat{x}$ be the solution containing the single element with the largest g_2-value. $\hat{x}$ is produced from the search point 0^n in expected time $O(P_{\max}n)$.

Let r be the number of elements in a given solution. According to [6], the maximal $\hat{g_1}^*$ deemed as feasible is at least

$$C_1^* = C - \sqrt{\frac{(1-\alpha)r\delta^2}{3\alpha}}$$

when using Chebyshev's inequality (Eq. 1) and at least

$$C_2^* = C - \sqrt{3\delta r \ln(1/\alpha)}$$

when using the Chernoff bound (Eq. 2). For a fixed C^*-value, we have

$$\left[1-\left(1-\frac{C^*}{C(r+1)}\right)^{r+1}\right] \cdot f(OPT)$$

We have $\hat{g_1}(x^*) + a(v^*) > C_1^*$ when working with Chebyshev's inequality and $\hat{g_1}(x^*) + a(v^*) > C_2^*$ when using Chernoff bound. In addition, $f(\hat{x}) \geq f(v^*)$ holds. According to [6], x^* or $\hat{x}$ is therefore a $(1/2-o(1))(1-1/e)$-approximation which completes the proof. □

For the special case of uniform IID weights, we have $a = a_{\max} = a_{\min}$ and $P_{max} \leq C/a+1$. Furthermore, the solution x^* already gives a $(1-o(1))(1-1/e)$-approximation as the element with the largest f-value is included in construction of x^*. This gives a bound on the expected runtime of $O(nk(k+\log n))$ to obtain a $(1-o(1))(1-1/e)$-approximation for the uniform IID case when working with the function $\hat{g_1}$ instead of g_1. Note that this matches the result given in Theorem 1.

4 Experimental Investigations

In this section, we investigate the GSEMO and the NSGA-II algorithm on important submodular optimisation problems with chance constraints and compare them to the greedy approach given in [6].

4.1 Experimental Setup

We examine GSEMO and NSGA-II for constraints with expected weights 1 and compare them to the greedy algorithm (GA) given in [6]. Our goal is to study different chance constraint settings in terms of the constraint bound C, the dispersion δ, and the probability bound α. We consider different benchmarks for chance constrained versions of the maximum influence problems and the maximum coverage problem.

For each benchmark set, we study the performance of the GSEMO and the NSGA-II algorithms for different budgets. We consider $C = 20, 50, 100$ for influence maximization and $C = 10, 15, 20$ for maximum coverage. We consider all combinations of $\alpha = 0.1, 0.001$, and $\delta = 0.5, 1.0$ for the experimental investigations of the algorithms and problems. Chebyshev's inequality leads to better results when α is relatively large and the Chernoff bounds gives better results for small α (see [6,34]). Therefore, we use Eq. 1 for $\alpha = 0.1$ and Eq. 2 for $\alpha = 0.001$ when computing the upper bound on the probability of a constraint violation. We allow 5 000 000 fitness evaluations for each evolutionary algorithm run. We run NSGA-II with parent population size 20, offspring size 10, crossover probability 0.90 and standard bit mutation for 500 000 generations. For each tested instance, we carry out 30 independent runs and report the minimum, maximum, and average results. In order to test the statistical significance of the results, we use the Kruskal-Wallis test with 95% confidence in order to measure the statistical validity of our results. We apply the Bonferroni post-hoc statistical

Table 1. Results for Influence Maximization with uniform chance constraints.

C	α	δ	GA (1)	GSEMO (2)					NSGA-II (3)				
				Mean	Min	Max	Std	Stat	Mean	Min	Max	Std	Stat
20	0.1	0.5	51.51	**55.75**	54.44	56.85	0.5571	$1^{(+)}$	55.66	54.06	56.47	0.5661	$1^{(+)}$
	0.1	1.0	46.80	**50.65**	49.53	51.68	0.5704	$1^{(+)}$	50.54	49.61	52.01	0.6494	$1^{(+)}$
50	0.1	0.5	90.55	**94.54**	93.41	95.61	0.5390	$1^{(+)}, 3^{(+)}$	92.90	90.75	94.82	1.0445	$1^{(+)}, 2^{(-)}$
	0.1	1.0	85.71	**88.63**	86.66	90.68	0.9010	$1^{(+)}, 3^{(+)}$	86.89	85.79	88.83	0.8479	$1^{(+)}, 2^{(-)}$
100	0.1	0.5	144.16	**147.28**	145.94	149.33	0.8830	$1^{(+)}, 3^{(+)}$	144.17	142.37	146.18	0.9902	$2^{(-)}$
	0.1	1.0	135.61	**140.02**	138.65	142.52	0.7362	$1^{(+)}, 3^{(+)}$	136.58	134.80	138.21	0.9813	$2^{(-)}$
20	0.001	0.5	48.19	**50.64**	49.10	51.74	0.6765	$1^{(+)}$	50.33	49.16	51.25	0.5762	$1^{(+)}$
	0.001	1.0	39.50	**44.53**	43.63	45.55	0.4687	$1^{(+)}$	44.06	42.18	45.39	0.7846	$1^{(+)}$
50	0.001	0.5	75.71	**80.65**	78.92	82.19	0.7731	$1^{(+)}$	80.58	79.29	81.63	0.6167	$1^{(+)}$
	0.001	1.0	64.49	69.79	68.89	71.74	0.6063	$1^{(+)}$	**69.96**	68.90	71.05	0.6192	$1^{(+)}$
100	0.001	0.5	116.05	**130.19**	128.59	131.51	0.7389	$1^{(+)}, 3^{(+)}$	127.50	125.38	129.74	0.9257	$1^{(+)}, 2^{(-)}$
	0.001	1.0	96.18	**108.95**	107.26	109.93	0.6466	$1^{(+)}, 3^{(+)}$	107.91	106.67	110.17	0.7928	$1^{(+)}, 2^{(-)}$

procedure, that is used for multiple comparison of a control algorithm, to two or more algorithms [5]. $X^{(+)}$ is equivalent to the statement that the algorithm in the column outperformed algorithm X. $X^{(-)}$ is equivalent to the statement that X outperformed the algorithm given in the column. If algorithm X does not appear, then no significant difference was determined between the algorithms.

4.2 The Influence Maximization Problem

The influence maximization problem (IM) (see [16,21,29,36] detailed descriptions) is a key problem in the area of social influence analysis.

IM aims to find the set of the most influential users in a large-scale social network. The primary goal of IM is to maximize the spread of influence through a given social network i.e. a graph of interactions and relationships within a group of users [4,15]. However, the problem of influence maximization has been studied subject to a deterministic constraint which limits the cost of selection [29].

The social network is modeled as a directed graph $G = (V, E)$ where each node represents a user, and each edge $(u, v) \in E$ has been assigned an edge probability $p_{u,v}$ that user u influences user v. The aim of the IM problem is to find a subset $X \subseteq V$ such that the expected number of activated nodes $E[I(X)]$ of X is maximized. Given a cost function $c\colon V \to \mathbb{R}^+$ and a budget $C \geq 0$, the corresponding submodular optimization problem under chance constraints is given as

$$\arg\max_{X \subseteq V} E[I(X)] \text{ s.t. } \Pr[c(X) > C] \leq \alpha.$$

For influence maximization, we consider uniform cost constraints where each node has expected cost 1. The expected cost of a solution is therefore $E_W(X) = |X|$.

In order to evaluate the algorithms on the chance constrained influence maximization problem, we use a synthetic data set with 400 nodes and 1 594 edges [29].

Table 2. Results for Maximum Coverage with uniform chance constraints for graphs frb30-15-01 (rows 1–12) and frb35-17-01 dataset (rows 13–24).

C	α	δ	GA (1)	GSEMO (2)					NSGA-II (3)				
				Mean	Min	Max	Std	Stat	Mean	Min	Max	Std	Stat
10	0.1	0.5	371.00	**377.23**	371.00	379.00	1.8323	$1^{(+)}$	376.00	371.00	379.00	2.5596	$1^{(+)}$
	0.1	1.0	321.00	**321.80**	321.00	325.00	1.5625	$1^{(+)}$	321.47	321.00	325.00	1.2521	
15	0.1	0.5	431.00	**439.60**	435.00	442.00	1.7340	$1^{(+)}, 3^{(+)}$	437.57	434.00	441.00	1.7555	$1^{(+)}, 2^{(-)}$
	0.1	1.0	403.00	**411.57**	408.00	414.00	1.7750	$1^{(+)}$	410.67	404.00	414.00	2.5098	$1^{(+)}$
20	0.1	0.5	446.00	**450.07**	448.00	451.00	0.8277	$1^{(+)}, 3^{(+)}$	448.27	445.00	451.00	1.3113	$1^{(+)}, 2^{(-)}$
	0.1	1.0	437.00	**443.87**	441.00	446.00	1.2794	$1^{(+)}, 3^{(+)}$	441.37	438.00	444.00	1.6914	$1^{(+)}, 2^{(-)}$
10	0.001	0.5	348.00	**352.17**	348.00	355.00	2.4081	$1^{(+)}$	350.80	348.00	355.00	2.8935	$1^{(+)}$
	0.001	1.0	321.00	**321.67**	321.00	325.00	1.5162	$1^{(+)}$	321.33	321.00	325.00	1.0613	
15	0.001	0.5	414.00	**423.90**	416.00	426.00	2.4824	$1^{(+)}$	422.67	419.00	426.00	2.2489	$1^{(+)}$
	0.001	1.0	371.00	**376.77**	371.00	379.00	1.8134	$1^{(+)}$	376.33	371.00	379.00	2.6824	$1^{(+)}$
20	0.001	0.5	437.00	**443.53**	440.00	445.00	1.1958	$1^{(+)}, 3^{(+)}$	440.23	437.00	443.00	1.6955	$1^{(+)}, 2^{(-)}$
	0.001	1.0	414.00	**424.00**	420.00	426.00	1.7221	$1^{(+)}$	422.50	417.00	426.00	2.5291	$1^{(+)}$
10	0.1	0.5	448.00	**458.80**	451.00	461.00	3.3156	$1^{(+)}$	457.97	449.00	461.00	4.1480	$1^{(+)}$
	0.1	1.0	376.00	**383.33**	379.00	384.00	1.7555	$1^{(+)}$	382.90	379.00	384.00	2.0060	$1^{(+)}$
15	0.1	0.5	559.00	**559.33**	555.00	562.00	2.0057	$3^{(+)}$	557.23	551.00	561.00	2.4309	$1^{(-)}, 2^{(-)}$
	0.1	1.0	503.00	**507.80**	503.00	509.00	1.1567	$1^{(+)}$	507.23	502.00	509.00	1.8323	$1^{(+)}$
20	0.1	0.5	587.00	**587.20**	585.00	589.00	1.2149	$3^{(+)}$	583.90	580.00	588.00	1.9360	$1^{(-)}, 2^{(-)}$
	0.1	1.0	569.00	**569.13**	566.00	572.00	1.4559	$3^{(+)}$	565.30	560.00	569.00	2.1520	$1^{(-)}, 2^{(-)}$
10	0.001	0.5	413.00	**423.67**	418.00	425.00	1.8815	$1^{(+)}$	422.27	416.00	425.00	2.6121	$1^{(+)}$
	0.001	1.0	376.00	**383.70**	379.00	384.00	1.1492	$1^{(+)}$	381.73	377.00	384.00	2.6514	$1^{(+)}$
15	0.001	0.5	526.00	**527.97**	525.00	532.00	2.1573	$1^{(+)}$	527.30	520.00	532.00	2.7436	
	0.001	1.0	448.00	**458.87**	453.00	461.00	2.9564	$1^{(+)}$	457.10	449.00	461.00	4.1469	$1^{(+)}$
20	0.001	0.5	568.00	**568.87**	565.00	572.00	1.5025	$3^{(+)}$	564.60	560.00	570.00	2.7618	$1^{(-)}, 2^{(-)}$
	0.001	1.0	526.00	**528.03**	525.00	530.00	1.8843	$1^{(+)}$	527.07	522.00	530.00	2.2427	

Table 1 shows the results obtained by GA, GSEMO, and NSGA-II for the combinations of α and δ. The results show that GSEMO obtains the highest mean values compared to the results obtained by GA and NSGA-II. Furthermore, the statistical tests show that for most of the combinations of α and δ GSEMO and NSGA-II significantly outperform GA. The solutions obtained by GSEMO have significantly better performance than NSGA-II in the case of a high budget i.e. for $C = 100$. A possible explanation for this is that the relatively small population size of NSGA-II does not allow one to construct solutions in a greedy fashion, as is possible for GA and GSEMO.

4.3 The Maximum Coverage Problem

The maximum coverage problem [8,17] is an important NP-hard submodular optimisation problem. We consider the chance constrained version of the problem. Given a set U of elements, a collection $V = \{S_1, S_2, \ldots, S_n\}$ of subsets of U, a cost function c: $2^V \rightarrow \mathbb{R}^+$, and a budget C, the goal is to find

$$\arg\max_{X \subseteq V} \{f(X) = |\cup_{S_i \in X} S_i| \text{ s.t. } \Pr(c(X) > C) \leq \alpha\}.$$

We consider linear cost functions. For the uniform case each set S_i has an expected cost of 1 and we have $E_W(X) = |\{i \mid S_i \in X\}|$.

For our experiments, we investigate maximum coverage instances based on graphs. The U elements consist of the vertices of the graph and for each vertex, we generate a set which contains the vertex itself and its adjacent vertices. For the chance constrained maximum coverage problem, we use the graphs frb30-15-01 (450 nodes, 17 827 edges) and frb35-17-01 (595 nodes and 27 856 edges) from [28].

The experimental results are shown in Table 2. It can be observed that GSEMO obtains the highest mean value for each setting. Furthermore, GSEMO statistically outperforms GA for most of the settings. For the other settings, there is no statistically significant difference in terms of the results for GSEMO and GA. NSGA-II is outperforming GA for most of the examined settings and the majority of the results are statistically significant. However, NSGA-II performs worse than GA for frb35-17-01 when $C = 20$ and $\alpha = 0.1$.

5 Conclusions

Chance constraints involve stochastic components and require a constraint only to be violated with a small probability. We carried out a first runtime analysis of evolutionary algorithms for the optimisation of submodular functions with chance constraints. Our results show that GSEMO using a multi-objective formulation of the problem based on tail inequalities is able to achieve the same approximation guarantee as recently studied greedy approaches. Furthermore, our experimental results show that GSEMO computes significantly better solutions than the greedy approach and often outperforms NSGA-II.

For future work, it would be interesting to analyse other probability distributions for chance constrained submodular functions. A next step would be to examine uniform weights with a different dispersion and obtain results for uniform weights with the same dispersion when using the fitness function g instead of $\hat{g}$.

Acknowledgment. This work has been supported by the Australian Research Council (ARC) through grant DP160102401 and by the South Australian Government through the Research Consortium "Unlocking Complex Resources through Lean Processing".

References

1. Assimi, H., Harper, O., Xie, Y., Neumann, A., Neumann, F.: Evolutionary bi-objective optimization for the dynamic chance-constrained knapsack problem based on tail bound objectives. CoRR abs/2002.06766 (2020). to appear at ECAI 2020
2. Auger, A., Doerr, B.: Theory of Randomized Search Heuristics: Foundations and Recent Developments. World Scientific Publishing Co., Inc. (2011)
3. Bian, A.A., Buhmann, J.M., Krause, A., Tschiatschek, S.: Guarantees for Greedy maximization of non-submodular functions with applications. In: Proceedings of the 34th International Conference on Machine Learning, ICML 2017, vol. 70, pp. 498–507. PMLR (2017)

4. Chen, W., Wang, Y., Yang, S.: Efficient influence maximization in social networks. In: Proceedings of the 15th ACM SIGKDD International Conference on Knowledge Discovery and Data Mining, pp. 199–208 (2009)
5. Corder, G.W., Foreman, D.I.: Nonparametric Statistics for Non-statisticians: A Step-by-Step Approach. Wiley, Hoboken (2009)
6. Doerr, B., Doerr, C., Neumann, A., Neumann, F., Sutton, A.M.: Optimization of chance-constrained submodular functions. In: The Thirty-Fourth AAAI Conference on Artificial Intelligence, AAAI 2020, pp. 1460–1467. AAAI Press (2020). https://www.aaai.org/Papers/AAAI/2020GB/AAAI-DoerrB.6164.pdf
7. Doerr, B., Neumann, F.: Theory of Evolutionary Computation - Recent developments in discrete optimization. Natural Computing Series. Springer, Heidelberg (2020). https://doi.org/10.1007/978-3-030-29414-4
8. Feige, U.: A threshold of ln n for approximating set cover. J. ACM **45**(4), 634–652 (1998)
9. Feldman, M., Harshaw, C., Karbasi, A.: Greed is good: Near-optimal submodular maximization via greedy optimization. In: COLT. Proceedings of Machine Learning Research, vol. 65, pp. 758–784. PMLR (2017)
10. Friedrich, T., Neumann, F.: Maximizing submodular functions under matroid constraints by evolutionary algorithms. Evol. Comput. **23**(4), 543–558 (2015)
11. Giel, O., Wegener, I.: Evolutionary algorithms and the maximum matching problem. In: Alt, H., Habib, M. (eds.) STACS 2003. LNCS, vol. 2607, pp. 415–426. Springer, Heidelberg (2003). https://doi.org/10.1007/3-540-36494-3_37
12. Golovin, D., Krause, A.: Adaptive submodularity: theory and applications in active learning and stochastic optimization. J. Artif. Intell. Res. **42**, 427–486 (2011)
13. Harshaw, C., Feldman, M., Ward, J., Karbasi, A.: Submodular maximization beyond non-negativity: guarantees, fast algorithms, and applications. In: Proceedings of the 36th International Conference on Machine Learning, ICML 2019, vol. 97, pp. 2634–2643. PMLR (2019)
14. Jansen, T.: Analyzing Evolutionary Algorithms - The Computer Science Perspective. Natural Computing Series. Springer, Heidelberg (2013). https://doi.org/10.1007/978-3-642-17339-4
15. Kempe, D., Kleinberg, J.M., Tardos, É.: Maximizing the spread of influence through a social network. In: Proceedings of the Ninth ACM SIGKDD International Conference on Knowledge Discovery and Data Mining, pp. 137–146. ACM (2003)
16. Kempe, D., Kleinberg, J.M., Tardos, É.: Maximizing the spread of influence through a social network. Theory Comput. **11**, 105–147 (2015)
17. Khuller, S., Moss, A., Naor, J.: The budgeted maximum coverage problem. Inf. Process. Lett. **70**(1), 39–45 (1999)
18. Krause, A., Golovin, D.: Submodular function maximization. In: Tractability: Practical Approaches to Hard Problems, pp. 71–104. Cambridge University Press (2014)
19. Krause, A., Guestrin, C.: Near-optimal observation selection using submodular functions. In: Proceedings of the Twenty-Second Conference on Artificial Intelligence, AAAI 2007, pp. 1650–1654. AAAI Press (2007)
20. Lee, J., Mirrokni, V.S., Nagarajan, V., Sviridenko, M.: Non-monotone submodular maximization under matroid and knapsack constraints. In: Proceedings of the 41st Annual ACM Symposium on Theory of Computing, STOC 2009, pp. 323–332. ACM (2009)

21. Leskovec, J., Krause, A., Guestrin, C., Faloutsos, C., VanBriesen, J.M., Glance, N.S.: Cost-effective outbreak detection in networks. In: Proceedings of the 13th ACM SIGKDD International Conference on Knowledge Discovery and Data Mining 2007, pp. 420–429. ACM (2007)
22. Mirzasoleiman, B., Jegelka, S., Krause, A.: Streaming non-monotone submodular maximization: personalized video summarization on the fly. In: Proceedings of the Thirty-Second AAAI Conference on Artificial Intelligence, AAAI 2018, pp. 1379–1386. AAAI Press (2018)
23. Motwani, R., Raghavan, P.: Randomized Algorithms. Cambridge University Press, Cambridge (1995)
24. Nemhauser, G.L., Wolsey, L.A.: Best algorithms for approximating the maximum of a submodular set function. Math. Oper. Res. **3**(3), 177–188 (1978)
25. Nemhauser, G.L., Wolsey, L.A., Fisher, M.L.: An analysis of approximations for maximizing submodular set functions - I. Math. Program. **14**(1), 265–294 (1978). https://doi.org/10.1007/BF01588971
26. Neumann, F., Sutton, A.M.: Runtime analysis of the (1 + 1) evolutionary algorithm for the chance-constrained knapsack problem. In: Proceedings of the 15th ACM/SIGEVO Conference on Foundations of Genetic Algorithms, FOGA 2019, pp. 147–153. ACM (2019)
27. Neumann, F., Witt, C.: Bioinspired Computation in Combinatorial Optimization. Natural Computing Series. Springer, Heidelberg (2010). https://doi.org/10.1007/978-3-642-16544-3
28. Nguyen, T.H., Bui, T.: Benchmark instances. https://turing.cs.hbg.psu.edu/txn131/
29. Qian, C., Shi, J., Yu, Y., Tang, K.: On subset selection with general cost constraints. In: International Joint Conference on Artificial Intelligence, IJCAI 2017, pp. 2613–2619 (2017)
30. Qian, C., Shi, J., Yu, Y., Tang, K., Zhou, Z.: Subset selection under noise. In: Advances in Neural Information Processing Systems 30: Annual Conference on Neural Information Processing Systems, NIPS 2017, pp. 3563–3573 (2017)
31. Qian, C., Yu, Y., Zhou, Z.: Subset selection by Pareto optimization. In: Proceedings of the 28th International Conference on Neural Information Processing Systems, NIPS 2015, vol. 1, pp. 1774–1782 (2015)
32. Roostapour, V., Neumann, A., Neumann, F., Friedrich, T.: Pareto optimization for subset selection with dynamic cost constraints. In: The Thirty-Third AAAI Conference on Artificial Intelligence, AAAI 2019, pp. 2354–2361. AAAI Press (2019)
33. Vondrák, J.: Submodularity and curvature: the optimal algorithm. RIMS Kôkyûroku Bessatsu **B23**, 253–266 (2010)
34. Xie, Y., Harper, O., Assimi, H., Neumann, A., Neumann, F.: Evolutionary algorithms for the chance-constrained knapsack problem. In: Proceedings of the Genetic and Evolutionary Computation Conference, GECCO 2019, pp. 338–346. ACM (2019). https://doi.org/10.1145/3321707.3321869
35. Xie, Y., Neumann, A., Neumann, F.: Specific single- and multi-objective evolutionary algorithms for the chance-constrained knapsack problem. CoRR abs/2004.03205 (2020). to appear at GECCO 2020
36. Zhang, H., Vorobeychik, Y.: Submodular optimization with routing constraints. In: Proceedings of the 30th AAAI Conference on Artificial Intelligence, AAAI 2016, pp. 819–826. AAAI Press (2016)
37. Zhou, Z., Yu, Y., Qian, C.: Evolutionary Learning: Advances in Theories and Algorithms. Springer, Heidelberg (2019). https://doi.org/10.1007/978-981-13-5956-9

Parameter-Less Population Pyramid for Permutation-Based Problems

Szymon Wozniak, Michal W. Przewozniczek(✉), and Marcin M. Komarnicki

Department of Computational Intelligence,
Wroclaw University of Science and Technology, Wroclaw, Poland
{michal.przewozniczek,marcin.komarnicki}@pwr.edu.pl

Abstract. Linkage learning is frequently employed in state-of-the-art methods dedicated to discrete optimization domains. Information about *linkage* identifies a subgroup of genes that are found dependent on each other. If such information is precise and properly used, it may significantly improve a method's effectiveness. The recent research shows that to solve problems with so-called overlapping blocks, it is not enough to use linkage of high quality – it is also necessary to use many different linkages that are diverse. Taking into account that the overlapping nature of problem structure is typical for practical problems, it is important to propose methods that are capable of gathering many different linkages (preferably of high quality) to keep them diverse. One of such methods is a Parameter-less Population Pyramid (P3) that was shown highly effective for overlapping problems in binary domains. Since P3 does not apply to permutation optimization problems, we propose a new P3-based method to fill this gap. Our proposition, namely the Parameter-less Population Pyramid for Permutations (P4), is compared with the state-of-the-art methods dedicated to solving permutation optimization problems: Generalized Mallows Estimation of Distribution Algorithm (GM-EDA) and Linkage Tree Gene-pool Optimal Mixing Evolutionary Algorithm (LT-GOMEA) for Permutation Spaces. As a test problem, we use the Permutation Flowshop Scheduling problem (Taillard benchmark). Statistical tests show that P4 significantly outperforms GM-EDA for almost all considered problem instances and is superior compared to LT-GOMEA for large instances of this problem.

Keywords: Genetic algorithms · Linkage learning · Linkage diversity · Parameter-less · Estimation of Distribution Algorithms · Permutation problems · Random keys · Scheduling

1 Introduction

The decomposition of problem structure is an important group of techniques that improve the effectiveness of the evolutionary methods. Among all, these techniques are employed in the Estimation of Distribution Algorithms (EDAs) that construct a probabilistic model of the solved problem during their run.

T. Bäck et al. (Eds.): PPSN 2020, LNCS 12269, pp. 418–430, 2020.
https://doi.org/10.1007/978-3-030-58112-1_29

Another way of incorporating the knowledge about the problem structure is the so-called Gray Box Optimization that employs the knowledge of the user [6,17]. Finally, the linkage learning techniques that concentrate on detecting the inter-gene dependencies are frequently used in various optimization methods [5,8,10,13,16]. Introducing the problem decomposition techniques into the methods dedicated to solving the permutation-based problems is not an easy task. The nature of the dependencies between genes encoding permutations is different from the dependencies in a typical genotype-like encoding. Nevertheless, an EDA, namely the GM-EDA [4] and Linkage Tree GOMEA (LT-GOMEA) that employs linkage learning [3], were recently proposed and shown effective in solving the permutation-based problems.

The recent research concerning the evolutionary methods dedicated to solving problems in discrete domains shows the importance of linkage quality [12]. Intuitively, the more precise linkage (problem decomposition information) we possess, the more effective the method that uses it will be. The less intuitive issue of linkage diversity was pointed in [13]. The presented results and their analysis show that to effectively solve an optimization problem with so-called overlaps (inter-gene blocks dependencies), it is favorable to use many different linkage models. An up-to-date method that maintains many different linkages is P3. To obtain linkage diversity, P3 employs a population structure that is significantly different from other evolutionary methods. P3 was shown highly competitive in solving overlapping problems [9,13,18]. Since the nature of overlapping problems is typical for problems encountered in practice [14], the objective of this paper is to propose a method based on the P3 idea that would be dedicated to solving permutation problems. The novelty of the proposed method, namely the Parameter-less Population Pyramid for Permutations (P4), consists of re-composing the already known mechanisms and ideas rather than on the P3 modifications (we introduce some adjustments, but they are slight). Nevertheless, the proposed P4 significantly outperforms GM-EDA on almost all test case groups and is more effective than LT-GOMEA. The supremacy of P4 is most visible for the largest and most complicated test cases. As a benchmark problem, we use the Permutation Flowshop Scheduling Problem (PFSP) [15] that was also employed in the papers proposing GM-EDA and LT-GOMEA [3,4].

The rest of this paper is organized as follows. In the next section, we present the related work. In Sect. 3, we describe the details of PFSP. The fourth section presents the details of P4. The results of the experiments and the discussion are given in Sect. 5. Finally, the last section concludes the paper and points the most promising future research directions.

2 Related Work

In this section, we present the related work. First, we describe the idea behind linkage learning using the Dependency Structure Matrix (DSM). Then, we present the ideas behind methods employing problem decomposition.

2.1 Linkage Learning Using DSM

DSM is a square matrix that indicates dependencies between components. Its concept is derived from the organization theory [8]. Each DSM entry $d_{i,j} \in R$ refers to the relationship between the i-th and j-th component. Higher values indicate higher dependency. Many ways are available to compute the value of each DSM entry. The mutual information is one of the most frequently used in the field of evolutionary computation [5,8,16], and is defined as follows.

$$I(X;Y) = \sum_{x \in X} \sum_{y \in Y} p(x,y) \log_2 \frac{p(x,y)}{p(x)p(y)} \quad (1)$$

where both X and Y are random variables. Note that $\forall_{X,Y}\ I(X;Y) \geq 0$ and $I(X;Y)$ equals 0 when X and Y are independent, because then $p(x,y) = p(x)p(y)$.

In evolutionary methods, each gene is related to a single component in the DSM matrix. The DSM is computed on the base of pairwise gene values frequencies that exist in the population. Such linkage information is the key to the effectiveness of the evolutionary methods applied to solve hard computational problems [5,8,16]. Some of the recent research investigates the dependency between linkage quality and method effectiveness. Moreover, some problems may be easy to decompose by the DSM-based linkage learning, while for some problems, the DSM-based linkage quality remains low. More details may be found in [12].

The information represented by a DSM matrix may be utilized in many ways. Here, we concentrate on the creation of so-called Linkage Trees (LT). LT is a result of using a hierarchical clustering algorithm and the DSM matrix. LT is built from nodes that are clusters. Each cluster shall contain genes that were found dependent on each other. The procedure of LT creation is as follows. First, each gene index is assigned to a single LT leaf. Then, the nodes that contain the most dependent genes are joined together until only one cluster (the root of LT) remains. More details concerning the LT creation process and LT examples may be found in [13,16].

One of the ways of using LT is the Optimal Mixing (OM) operator [5,16]. OM involves two individuals (called *donor* and *source*) and a single LT node (the cluster of dependent genes). LT node marks a group of genes in the donor. These marked genes from the donor replace the appropriate genes in the source individual. This operation is reverted if it causes a decrease in the fitness of the source individual. Otherwise, the source remains modified. Methods employing OM, for each source usually consider all LT nodes (except the root) in the random order. For each source and node pair, the donor is selected randomly. More information about OM may be found in [3,5,16].

Note that recently the DSM-like matrix was employed by the linkage learning technique that avoids using the statistical measures to detected inter-gene dependencies. Instead of the gene dependency prediction, it uses an empirical check. The fundamental difference is that the prediction-based linkage learning (e.g., employing statistical measures) may detect dependencies that are false. The so-called empirical check does not suffer for this flaw (however, it may still miss detecting the true dependencies). More information may be found in [13].

2.2 Methods Employing Problem Decomposition Techniques

LT-GOMEA is an improved version of the Linkage Tree Genetic Algorithm (LTGA) [16]. LTGA maintains a population of individuals for which it constructs DSM at every method iteration. At each method iteration, each individual is updated by OM. The mutation operator is not used. LTGA requires specifying the population size that is its only parameter. To transform LTGA into a parameter-less method, LT-GOMEA employs the population-sizing scheme proposed in [7]. LT-GOMEA maintains multiple instances of LTGA. It starts with an instance containing only one individual. Then, the instance with a doubled population size is executed at each 4th iteration. During its run, LT-GOMEA drops useless LTGAs. A single LTGA population is found useless if all of its individuals are the same, or its average population fitness is lower than the average population fitness of at least one LTGA with larger population size. If any LTGA is found useless, then all LTGAs with smaller populations are found useless too. All LTGA instances are independent, they during the so-called Forced Improvements phase they may interact with the globally best-found individual. LT-GOMEA proposed in [3] also introduces mechanisms dedicated to solving the permutation-based problems. These mechanisms are discussed in Sect. 4.

GM-EDA is an EDA recently proposed for permutation-based problems optimization [4]. It uses a model that is dedicated to permutation spaces and may be found equivalent to Gaussian distribution. GM-EDA samples the model to create new individuals. The probabilistic model employed by GM-EDA allowed it to outperform other EDAs. Note that EDAs were originally designed to solve problems in discrete and real-valued spaces.

3 Permutation Flowshop Scheduling Problem

The Permutation Flowshop Scheduling Problem (PFSP) [15] is a well-known permutation optimization problem, frequently used as a benchmark [3,4]. In PFSP, we consider J jobs and M machines. Each job i consists of M operations. All operations must be in the same order $j = 1, 2, ..., M$. Each operation j can only be processed on the j-th machine. Each machine can process only one job at a time. If the j-th operation j of the i-th job is to be processed on the particular machine, but the machine is busy (it processes another operation from a different job), then the operation must wait until the machine finishes its work. Each test case is defined by the number of jobs J, the number of machines M, and the operation execution times on each machine for each job. Let $\pi = \{\pi_1, \pi_2, ..., \pi_J\}$ be a job-processing sequence, where π_1 and π_J indicate the first and the last job respectively. By $c(\pi_i, j)$ we define the completion time of the i-th job on the j-th machine. $c(\pi_i, j)$ may be computed as follows.

$$c(\pi_1, 1) = p(\pi_1, 1),$$
$$c(\pi_1, j) = c(\pi_1, j-1) + p(\pi_1, j), \text{ for } j = 2, ..., M,$$
$$c(\pi_i, 1) = c(\pi_{i-1}, 1) + p(\pi_i, 1), \text{ for } i = 2, ..., J,$$
$$c(\pi_i, j) = \max\{c(\pi_{i-1}, j), c(\pi_i, j-1)\} + p(\pi_i, j), \text{ for } i = 2, ..., J;\ j = 2, ..., M. \tag{2}$$

where $p(\pi_i, j)$ is the processing time of job i on machine j.

The Total Flow Time (TFT) of a sequence π can be computed as folows.

$$TFT(\pi) = \sum_{i=1}^{J} c(\pi_i, M) \tag{3}$$

In PFSP considered in this paper, the objective is to find the job-processing schedule $\pi*$ that minimizes the TFT value.

4 Parameter-Less Population Pyramid for Permutation Problems

In this section, we describe the proposed P4 method. As stated in the Introduction, we do not propose new mechanisms. Instead, we propose a different composition of the already proposed ideas. The results reported in Sect. 5 show that the proposed P4 is competitive to state-of-the-art methods.

P3 is a DSM-using evolutionary method that was proposed to solve binary-encoded problems [5]. It employs the same linkage learning technique as LTGA and the OM operator. The main difference between P3 and other evolutionary methods is the structure of its population. Individuals form subpopulations called *levels*. Each level has its separate linkage information. Thus, during its run, P3 maintains and uses many different linkages. The recent research shows that even a high-quality linkage may not be enough to solve overlapping problems. In such problems, the groups of tightly dependent genes also have inter-gene dependencies with other groups. The other requirement is to use a diverse linkage (use many different linkages at the same time) [13]. The nature of overlapping problems is typical for problems encountered in practice [14]. Thus, the method that effectively solves problems with overlaps is expected to effectively solve practical problems as well. P3 was shown to outperform other state-of-the-art methods in solving so-called overlapping problems [9,13,18]. Therefore, the main motivation behind this paper is to use the P3 concept that allows for using a diverse linkage and apply it to permutation-based search spaces.

Same as P3, P4 uses the pyramid-like population structure, each pyramid level has its separate DSM and a separate linkage tree. The idea of P4 is presented in Fig. 1. At each method iteration, a new individual is added to the population (pyramid). In P3, each new individual is initialized by a local optimization procedure [5,13]. In P4, this operation is abandoned due to its high computational cost. A new individual is added to the first level of a pyramid.

After this operation, a new individual is crossed with each individual in the pyramid using the OM operator (individuals that are already a part of the pyramid act as the donors). If the OM operation improves the individual, then its improved version is added to the higher level of the pyramid. The pyramid level appropriate for the new individual is found using the following formula $newLevel = max(Level(source)Level(donor)) + 1$, where $Level(source)$ and $Level(donor)$ are the levels of source and donor individual, respectively. If an individual is to be added to the level that does not exist, then the new level is created. Thus, the better fitting individuals (with a higher number of improvements) shall be found in the higher levels of the pyramid. Moreover, the number of levels usually increases during the method run.

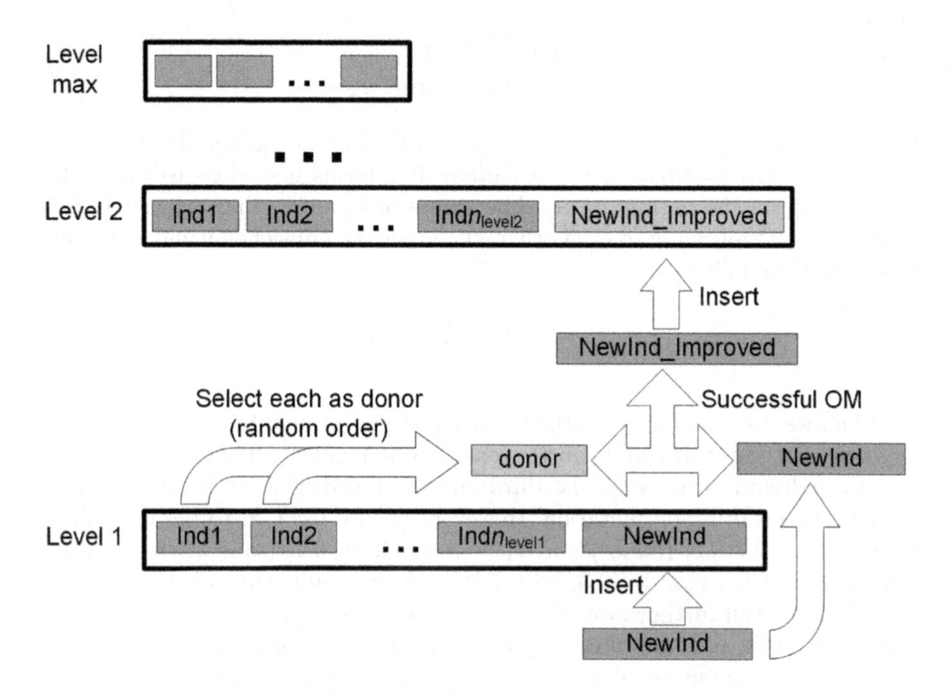

Fig. 1. Pyramid-like population structure in P4

In P4, we use *random keys*, to encode solutions [1,3]. When random keys are used, typical evolutionary operators yield feasible solutions and no repair operator is necessary. Each solution (permutation) is represented by a n-dimensional vector of real values: $\overrightarrow{r} = [r_0, ..., r_{n-1}]$. To transform $\overrightarrow{r}$ into a permutation $\mathbf{x}$, the vector values are sorted in an ascending order, so, $r_{x_0} < r_{x_1} < ... < r_{x_{n-1}}$ holds. For instance, $\overrightarrow{r} = [0.1, 0.4, 0.2]$ encodes the permutation $x = (1, 3, 2)$. As in [3], we limit the available r_i values to the range $[0; 1]$.

For each of its pyramid levels, P4 must be capable of computing a DSM matrix that represents gene dependencies. To obtain it, we use the dependency measure proposed in [3]:

$$D(i,j) = 1 - \delta(i,j). \tag{4}$$

where

$$\delta(i,j) = \delta(j,i) = \delta_1(i,j)\delta_2(i,j). \tag{5}$$

δ_1 is denoted as *relative ordering information*. It is the entropy of the probability that $r_i < r_j$. The maximum of $\delta_1(i,j)$ is 1 and takes place when the relation ordering is found random. The minimum value is 0 and is equivalent to the certainty that one value will always take place before the other. Thus, $\delta_1(i,j)$ is defined as follows.

$$\delta_1(i,j) = 1 - \{-[p_{ij}\log_2(p_{ij}) + (1.0 - p_{ij})\log_2(1.0 - p_{ij})]\} \tag{6}$$

where

$$p_{ij} = \frac{1}{n}\sum_{k=0}^{n-1}\begin{cases}1 & \text{if } r_i^k < r_j^k\\ 0 & \text{otherwise}\end{cases} \tag{7}$$

where r_i^k is the value of ith gene of the kth individual in the population.

δ_2 is denoted as *adjacency information*. It informs how close to each other r_i and r_j are in the permutations. The values of δ_2 are in the range $[0;1]$. The closer, the two values are to each other, the lower the adjacency value is. Finally, δ_2 is defined as follows.

$$\delta_2 = 1 - \frac{1}{n}\sum_{k=0}^{n-1}(r_i^k - r_j^k)^2 \tag{8}$$

When we use random keys, the following situation may take place. We have detected that the third and fourth genes are dependent. Therefore, when we mix two individuals, we wish the third and the fourth gene to be close to each other in the resulting permutation. However, when if use OM to the individuals $\overrightarrow{r_d} = [0.1, 0.8, 0.7, 0.2]$ (donor) and $\overrightarrow{r_s} = [0.3, 0.6, 0.5, 0.1]$ (source) the resulting individual will be: $\overrightarrow{r_{res}} = [0.3, 0.6, 0.7, 0.2]$. As a result, the third and fourth genes will be even further away from each other in r_{res} than in the r_s, which is counter-intuitive and unwanted. To overcome this issue, we employ the *random rescaling* [2]. Random rescaling takes place with a given probability p_δ. When it is triggered, the values of the random keys that are transferred to the resulting individual are scaled to the randomly selected range. For instance, if the values of the third and fourth gene from the previous example are rescaled to the range $[0.5; 0.55]$, then the third and the fourth gene would have values 0.55 and 0.5, respectively. The resulting individual would be: $\overrightarrow{r_{res}} = [0.3, 0.6, 0.55, 0.5]$ and the sequence would not be interrupted by other genes. As in [3], we use $p_\delta = 0.1$ and n equal value ranges. Note that the method proposed in [2] was unable to solve hard overlapping problems without using the random rescaling.

As pointed in [3], the nature of OM may quickly decrease the population diversity, even if random rescaling is used. Therefore, we adopt the re-encoding operator. At the beginning of each iteration, the random keys of all individuals are randomly chosen in the way that the order of random keys is preserved.

As stated above, P4 is the re-composition of already known mechanisms. The objective of the re-composition is to use the pyramid-like population structure idea proposed in P3. The motivation is to use the linkage diversity preserving framework proposed in P3 to effectively solve hard practical permutation-based problems. Nevertheless, P4 introduces some changes in the general P3 framework. These changes might be found minor, but are important for the method effectiveness. The first difference, already mentioned above, is that P4 does not use local optimization to initialize new individuals. The second difference, is that during the OM operation, P4 does not prioritize the shorter linkage tree nodes (which is done by P3). The reason behind this change is to omit a potential bias that may influence the method performance on a particular problem [3]. Finally, the third difference between P4 and P3 is that P3 accepts a new individual in the pyramid only if it is unique (the same individual does not exist on any of the pyramid levels). In P4, we abandon this mechanism due to the following reasons. Since we use random keys, the uniqueness check would be reasonable only on the permutation level. However, any permutation may be encoded in many different ways, and the relations between random key values are useful as well.

5 The Results

5.1 Experiment Setup

The objective of the experiments was to compare the effectiveness of P4 and the state-of-the-art methods dedicated to solving permutation-based problems. As in [3,4], we use PFSP as the comparison base. As competing methods, we use LT-GOMEA with random rescaling and re-coding and GM-EDA. LT-GOMEA was chosen because P4 adopts many of its mechanisms. GM-EDA [4] was chosen since it is an up-to-date proposition for permutation-based optimization. Similarly to P4 and LT-GOMEA, GM-EDA also models the optimized problem. However, instead of linkage learning, it uses a probabilistic model. Thus, it is interesting to compare the performance of P4 with an up-to-date method that employs a different concept for problem decomposition.

We use the Fitness Function Evaluation number (FFE) as the stop condition. The number of available FFE was the same as suggested in [15] and the same as used in [3,4] and is presented in Table 1. For each job and machine number combination, we consider ten different test cases. For each test case, we consider the results from 20 independent runs. The detailed results and the source codes are available at https://github.com/przewooz/P4.

As a quality measures we use the *median of relative percentage deviation* (MRPD) and the *average relative percentage deviation* (ARPD) defined in (9) and (10), respectively. The same quality measures were used in [3,4].

$$MRPD = \text{median}\left(\bigcup_{i=0}^{n-1}\left\{\frac{100(TFT_i - UB)}{UB}\right\}\right), \quad (9)$$

Table 1. Available FFE per test case [4]

J jobs × M machines	Max. FFE	J jobs × M machines	Max. FFE
20 × 5	182224100	100 × 5	235879800
20 × 10	224784800	100 × 10	266211000
20 × 20	256896400	100 × 20	283040000
50 × 5	220712150	200 × 10	272515500
50 × 10	256208100	200 × 20	287728850
50 × 20	275954150	500 × 20	260316750

where TFT_i is the total flow time of the best solution found in the i-th run, and UB is the best-known result for the considered TFT.

$$ARPD = \frac{1}{N}\left(\sum_{i=1}^{n}\left\{\frac{100(TFT_i - UB)}{UB}\right\}\right). \tag{10}$$

Additionally, to check the statistical significance of the median differences, we use the unpaired Wilcoxon test and a significance level of 5%.

5.2 Experiments Results and Discussion

In this section, we present the detailed results of the experiments. The summarized results for each experiment group are presented in Table 2. Since every group consists of 10 test cases, we check the statistical significance of median result differences. Thus, for each test case, we can check which method performs better. In Table 3, we present the detailed results for most of the considered test cases. Complete results are available in the data pack.

P4 and LT-GOMEA find the best-known result in every run for experiments with 20 jobs. GM-EDA fails to do so. Thus, it is outperformed by P4 (and LT-GOMEA) for every test case group employing 20 jobs. For test cases concerning 50 jobs and 5 machines, LT-GOMEA performs significantly better than P4. However, the LT-GOMEA dominance vanishes with the increase of machine number. The explanation of this observation may be as follows. The number of subpopulations maintained at the same time by LT-GOMEA is usually much lower than the number of levels (subpopulations) in P4. Thus, LT-GOMEA maintains a smaller number of linkages (one per each subpopulation). Therefore, P4 preserves linkage diversity better than LT-GOMEA. On the other hand, since LT-GOMEA uses a smaller number of subpopulations, the subpopulations in LT-GOMEA contain more individuals than levels in P4. Since LT-GOMEA computes the DSM for larger groups of individuals, then its linkage should be of higher quality. Thus, it is allowed to assume that LT-GOMEA uses the linkage of lower diversity but higher quality. This assumption is also justified by the results presented in [12] – LT-GOMEA is shown to use a linkage of higher quality than P3 that is the

Table 2. Summarized results

Test case group	P4 vs LT-GOMEA			P4 vs GM-EDA		
	Better	Equal	Worse	Better	Equal	Worse
20×5	0	10	0	**10**	0	0
20×10	0	10	0	**10**	0	0
20×20	0	10	0	**10**	0	0
50×5	1	1	**8**	**10**	0	0
50×10	2	4	**4**	**10**	0	0
50×20	4	2	4	**10**	0	0
100×5	4	2	4	5	0	5
100×10	**5**	5	0	**10**	0	0
100×20	**9**	1	0	**10**	0	0
200×10	**4**	4	2	**5**	3	2
200×20	**10**	0	0	**8**	1	1
500×20	**10**	0	0	**10**	0	0

base of P4. Additionally, selection pressure in LT-GOMEA improves the convergence. P4 does not use the selection pressure (all the individuals are stored in the pyramid). Therefore, its convergence is slower. To summarize, for test cases in 50×5 group LT-GOMEA maintains linkage that is diverse enough and the selection pressure that allows it to find better solutions than those proposed by P4.

The situation changes when the number of machines and/or jobs increases. For 50×10 and 50×20 test case groups, the results are close to tie or tied. A reasonable explanation is as follows. When the test cases become more complicated, the diverse linkage starts to be more important. Therefore, the larger and the more complicated the test case is, the more effective P4 is when compared with LT-GOMEA. Note that for the largest test case groups, P4 outperforms LT-GOMEA significantly. The results are even more decisive if MRPD and ARPD measures reported in Table 3 for 200×20 and 500×20 test cases are taken into account. The above observations correspond with the results for overlapping problems reported in [13] – the larger is the test case, the more significant is the supremacy of P3 over the competing methods.

GM-EDA is outperformed by P4 significantly for most of the considered test case groups. The results are tied only for the 100×5 group the results are tied. Most likely, for these test cases, GM-EDA is capable of converging faster than P4. Note that GM-EDA maintains only one probabilistic model of a problem at a time. Thus, the diversity of problem decomposition is significantly lower than in the P4 and LT-GOMEA cases. Concerning GM-EDA, we can not discuss the linkage diversity because GM-EDA does not use linkage. Nevertheless, it decomposes the problem in the other way. The results presented here may

Table 3. The comparison between P4, LT-GOMEA, and GM-EDA on the base of MRPD and ARPD

	P4	LT-GOMEA	GM-EDA		P4	LT-GOMEA	GM-EDA
20 × 20	**0.00 (0.00)**	**0.00 (0.00)**	0.57 (0.65)	**50 × 5**	0.18 (0.18)	**0.06 (0.07)**	0.74 (0.79)
	0.00 (0.00)	**0.00 (0.00)**	0.22 (0.29)		0.41 (0.40)	**0.18 (0.16)**	0.88 (0.94)
	0.00 (0.00)	**0.00 (0.00)**	0.00 (0.04)		0.72 (0.70)	**0.71 (0.75)**	1.32 (1.34)
	0.00 (0.00)	**0.00 (0.00)**	0.24 (0.28)		**0.50 (0.48)**	0.67 (0.66)	1.18 (1.27)
	0.00 (0.00)	**0.00 (0.00)**	0.33 (0.26)		0.42 (0.37)	**0.25 (0.26)**	0.84 (0.89)
	0.00 (0.00)	**0.00 (0.00)**	0.10 (0.30)		0.40 (0.38)	**0.31 (0.31)**	0.80 (0.82)
	0.00 (0.00)	**0.00 (0.00)**	0.58 (0.61)		0.27 (0.30)	**0.18 (0.20)**	0.99 (0.96)
	0.00 (0.00)	**0.00 (0.00)**	0.58 (0.52)		0.38 (0.38)	**0.23 (0.22)**	0.95 (0.97)
	0.00 (0.00)	**0.00 (0.00)**	0.33 (0.56)		0.35 (0.33)	**0.26 (0.25)**	0.84 (0.81)
	0.00 (0.00)	**0.00 (0.00)**	0.18 (0.41)		**0.40 (0.41)**	0.44 (0.40)	0.93 (1.00)
50 × 10	0.68 (0.69)	**0.62 (0.67)**	2.16 (2.10)	**50 × 20**	**0.49 (0.53)**	0.54 (0.51)	1.71 (1.76)
	0.74 (0.66)	**0.53 (0.50)**	2.41 (2.45)		0.37 (0.35)	**0.15 (0.18)**	1.55 (1.58)
	0.42 (0.43)	0.43 (0.41)	1.78 (1.84)		**0.56 (0.60)**	0.58 (0.54)	2.17 (2.24)
	0.54 (0.54)	0.58 (0.64)	1.78 (1.83)		0.57 (0.53)	**0.53 (0.49)**	1.79 (1.92)
	0.51 (0.51)	**0.45 (0.47)**	2.09 (2.01)		**0.67 (0.64)**	0.72 (0.73)	2.38 (2.30)
	0.36 (0.36)	**0.28 (0.29)**	1.48 (1.55)		0.41 (0.46)	**0.39 (0.40)**	1.81 (1.78)
	0.64 (0.62)	**0.57 (0.62)**	1.94 (1.97)		**0.54 (0.54)**	1.05 (1.03)	2.06 (2.10)
	0.62 (0.57)	0.75 (0.69)	1.95 (2.03)		0.69 (0.68)	0.69 (0.65)	2.26 (2.24)
	0.63 (0.63)	0.63 (0.67)	2.09 (2.10)		0.56 (0.59)	**0.38 (0.41)**	1.78 (1.79)
	0.63 (0.59)	0.65 (0.62)	2.00 (2.00)		**0.55 (0.55)**	1.07 (1.03)	1.98 (1.95)
100 × 5	0.86 (0.88)	**0.56 (0.58)**	0.83 (0.82)	**200 × 10**	1.09 (1.11)	**0.99 (1.07)**	1.16 (1.19)
	0.89 (0.88)	**0.62 (0.63)**	1.00 (1.00)		**1.44 (1.45)**	2.12 (1.95)	1.47 (1.46)
	0.82 (0.84)	**0.73 (0.76)**	0.78 (0.80)		1.14 (1.13)	1.17 (1.18)	**1.13 (1.12)**
	0.68 (0.70)	0.69 (0.74)	0.78 (0.78)		1.16 (1.15)	1.24 (1.19)	**1.13 (1.13)**
	0.68 (0.68)	0.73 (0.71)	0.82 (0.80)		**1.08 (1.10)**	1.47 (1.44)	1.31 (1.32)
	0.97 (0.94)	1.04 (1.03)	**0.79 (0.80)**		1.32 (1.31)	**1.26 (1.28)**	1.36 (1.39)
	0.85 (0.85)	**0.75 (0.77)**	1.01 (1.00)		1.36 (1.40)	1.29 (1.37)	**1.18 (1.17)**
	0.97 (0.94)	0.94 (1.02)	**0.88 (0.90)**		**1.02 (1.04)**	1.50 (1.43)	1.25 (1.26)
	0.93 (0.93)	1.12 (1.06)	**0.92 (0.87)**		1.25 (1.29)	1.14 (1.16)	**1.13 (1.11)**
	0.76 (0.76)	0.95 (0.91)	0.92 (0.96)		**1.20 (1.23)**	1.45 (1.44)	1.32 (1.29)
100 × 10	**1.01 (1.01)**	1.61 (1.61)	1.71 (1.69)	**200 × 20**	**1.28 (1.28)**	1.59 (1.62)	1.60 (1.59)
	1.15 (1.13)	1.24 (1.17)	2.06 (2.07)		**1.27 (1.29)**	2.18 (2.12)	1.46 (1.45)
	1.08 (1.09)	**1.06 (1.09)**	1.68 (1.71)		**1.18 (1.21)**	1.67 (1.73)	1.35 (1.33)
	0.94 (1.06)	1.28 (1.27)	1.96 (1.89)		1.48 (1.48)	2.04 (2.10)	**1.47 (1.43)**
	0.89 (0.91)	1.57 (1.62)	1.72 (1.73)		**1.34 (1.31)**	2.17 (2.14)	1.61 (1.64)
	1.08 (1.11)	1.35 (1.31)	1.59 (1.70)		**1.31 (1.29)**	2.19 (2.06)	1.51 (1.52)
	0.78 (0.77)	**0.70 (0.75)**	1.48 (1.51)		**1.26 (1.25)**	1.67 (1.62)	1.27 (1.27)
	1.21 (1.11)	**1.17 (1.10)**	1.83 (1.88)		**1.37 (1.40)**	2.00 (1.92)	1.55 (1.57)
	0.97 (0.92)	**0.93 (0.96)**	1.80 (1.76)		**1.33 (1.39)**	1.90 (1.93)	1.47 (1.48)
	0.93 (0.91)	1.31 (1.35)	1.46 (1.50)		**1.44 (1.43)**	1.85 (1.91)	1.47 (1.44)
100 × 20	**0.91 (0.95)**	1.56 (1.53)	2.04 (2.03)	**500 × 20**	**1.12 (1.12)**	1.49 (1.49)	9.11 (8.90)
	0.85 (0.88)	**0.82 (0.84)**	1.79 (1.80)		**1.03 (1.02)**	1.50 (1.47)	8.45 (8.58)
	0.83 (0.78)	1.82 (1.77)	1.97 (1.93)		**1.07 (1.14)**	1.49 (1.50)	8.57 (8.46)
	1.05 (1.06)	1.92 (1.78)	1.89 (1.86)		**0.93 (0.96)**	1.43 (1.46)	8.78 (8.75)
	0.86 (0.89)	1.10 (1.09)	1.76 (1.77)		**0.90 (0.88)**	1.29 (1.32)	8.78 (8.72)
	0.97 (1.02)	2.14 (2.08)	2.20 (2.17)		**1.10 (1.11)**	1.41 (1.42)	8.32 (8.58)
	1.14 (1.10)	1.58 (1.69)	1.90 (1.90)		**1.18 (1.19)**	1.63 (1.64)	9.25 (9.15)
	1.17 (1.12)	2.04 (1.82)	1.96 (1.96)		**1.17 (1.16)**	1.58 (1.55)	8.54 (8.62)
	1.02 (0.98)	1.77 (1.72)	1.85 (1.82)		**0.99 (1.00)**	1.43 (1.45)	8.81 (8.69)
	1.05 (1.03)	1.68 (1.55)	2.09 (2.05)		**1.19 (1.12)**	1.45 (1.46)	8.60 (8.51)

indicate that no matter what kind of problem decomposition is employed, this information should be diverse.

6 Conclusions

In this paper, we propose a new method for permutation-based problem optimization, namely P4. Our proposition is based on the P3 idea, and, except for the slight changes in the P3 framework, it is a composition of many, but already known, mechanisms. Nevertheless, we show that the intuitions behind P4 are accurate. It significantly outperforms GM-EDA and is more effective than LT-GOMEA, especially for the largest and most complicated test cases. The differences in the absolute values of chosen quality measures indicate that the advantage of P4 for the largest test cases is more significant than for the medium-sized test cases for which it is outperformed by LT-GOMEA. Thus, it is allowed to state that we propose an up-to-date method that properly fills the gap in the field of permutation-based problem optimization.

Our proposition is at its early stage. The main directions of further research are as follows. The behavior of P4 for the test cases on which it was outperformed by LT-GOMEA shall be analyzed, and the appropriate updates to P4 shall be introduced. The issue of linkage and problem decomposition diversity should be investigated in detail to improve the understanding of evolutionary methods behavior. Recently the linkage quality measures are proposed for problems in different domains [11,12]. Similar propositions should be formulated for permutation-based problems. Finally, another interesting issue is to try to introduce the problem decomposition diversity into EDAs.

Acknowledgments. We thank Peter Bosman for supporting us with outcomes of all runs of GM-EDA and LT-GOMEA on the Taillard problem and the LT-GOMEA source codes.

This work was supported by the Polish National Science Centre (NCN) under Grant 2015/19/D/ST6/03115 and the statutory funds of the Department of Computational Intelligence, Wroclaw University of Science and Technology.

References

1. Bean, J.C.: Genetic algorithms and random keys for sequencing and optimization. INFORMS J. Comput. **6**(2), 154–160 (1994)
2. Bosman, P.A.N., Thierens, D.: Crossing the road to efficient ideas for permutation problems. In: Proceedings of the 3rd Annual Conference on Genetic and Evolutionary Computation, GECCO 2001, pp. 219–226. Morgan Kaufmann Publishers Inc., San Francisco (2001)
3. Bosman, P.A., Luong, N.H., Thierens, D.: Expanding from discrete Cartesian to permutation gene-pool optimal mixing evolutionary algorithms. In: Proceedings of the Genetic and Evolutionary Computation Conference 2016, GECCO 2016, pp. 637–644. Association for Computing Machinery, New York (2016)

4. Ceberio, J., Irurozki, E., Mendiburu, A., Lozano, J.A.: A distance-based ranking model estimation of distribution algorithm for the flowshop scheduling problem. IEEE Trans. Evol. Comput. **18**(2), 286–300 (2014)
5. Goldman, B.W., Punch, W.F.: Parameter-less population pyramid. In: Proceedings of the 2014 Annual Conference on Genetic and Evolutionary Computation, GECCO 2014, pp. 785–792. ACM, New York (2014). https://doi.org/10.1145/2576768.2598350
6. Gomes, T.M., de Freitas, A.R.R., Lopes, R.A.: Multi-heap constraint handling in gray box evolutionary algorithms. In: Proceedings of the Genetic and Evolutionary Computation Conference, GECCO 2019, pp. 829–836. Association for Computing Machinery, New York (2019). https://doi.org/10.1145/3321707.3321872
7. Harik, G.R., Lobo, F.G.: A parameter-less genetic algorithm. In: Proceedings of the 1st Annual Conference on Genetic and Evolutionary Computation - Volume 1, GECCO 1999, pp. 258–265 (1999)
8. Hsu, S.H., Yu, T.L.: Optimization by pairwise linkage detection, incremental linkage set, and restricted/back mixing: DSMGA-II. In: Proceedings of the 2015 Annual Conference on Genetic and Evolutionary Computation, GECCO 2015, pp. 519–526. ACM, New York (2015). https://doi.org/10.1145/2739480.2754737
9. Komarnicki, M.M., Przewozniczek, M.W.: Comparative mixing for DSMGA-II. In: Proceedings of the 2020 Annual Conference on Genetic and Evolutionary Computation, GECCO 2020 (2020, in press)
10. Kwasnicka, H., Przewozniczek, M.: Multi population pattern searching algorithm: a new evolutionary method based on the idea of messy genetic algorithm. IEEE Trans. Evol. Comput. **15**, 715–734 (2011)
11. Omidvar, M.N., Yang, M., Mei, Y., Li, X., Yao, X.: DG2: a faster and more accurate differential grouping for large-scale black-box optimization. IEEE Trans. Evol. Comput. **21**(6), 929–942 (2017). https://doi.org/10.1109/TEVC.2017.2694221
12. Przewozniczek, M.W., Frej, B., Komarnicki, M.M.: On measuring and improving the quality of linkage learning in modern evolutionary algorithms applied to solve partially additively separable problems. In: Proceedings of the 2020 Annual Conference on Genetic and Evolutionary Computation, GECCO 2020 (2020, in press)
13. Przewozniczek, M.W., Komarnicki, M.M.: Empirical linkage learning. IEEE Trans. Evol. Comput. (2020, in press)
14. Watson, R.A., Hornby, G.S., Pollack, J.B.: Hierarchical building-block problems for GA evaluation. In: Proceedings of the 1998 International Conference on Parallel Problem Solving from Nature, vol. 2, pp. 97–106 (1998)
15. Taillard, E.: Benchmarks for basic scheduling problems. Eur. J. Oper. Res. **64**(2), 278–285 (1993). Project Management anf Scheduling
16. Thierens, D., Bosman, P.A.: Hierarchical problem solving with the linkage tree genetic algorithm. In: Proceedings of the 15th Annual Conference on Genetic and Evolutionary Computation, GECCO 2013, pp. 877–884. ACM, New York (2013). https://doi.org/10.1145/2463372.2463477
17. Whitley, L.D., Chicano, F., Goldman, B.W.: Gray box optimization for Mk landscapes (NK landscapes and MAX-kSAT). Evol. Comput. **24**(3), 491–519 (2016)
18. Zielinski, A.M., Komarnicki, M.M., Przewozniczek, M.W.: Parameter-less population pyramid with automatic feedback. In: Proceedings of the Genetic and Evolutionary Computation Conference Companion, GECCO 2019, pp. 312–313. Association for Computing Machinery, New York (2019). https://doi.org/10.1145/3319619.3322052

Connection Between Nature-Inspired Optimization and Artificial Intelligence

Biologically Plausible Learning of Text Representation with Spiking Neural Networks

Marcin Białas[1(✉)], Marcin Michał Mirończuk[1], and Jacek Mańdziuk[2]

[1] National Information Processing Institute, al. Niepodległości 188 b, 00-608 Warsaw, Poland
{marcin.bialas,marcin.mironczuk}@opi.org.pl

[2] Faculty of Mathematics and Information Sciences, Warsaw University of Technology, Koszykowa 75, 00-662 Warsaw, Poland
mandziuk@mini.pw.edu.pl

Abstract. This study proposes a novel biologically plausible mechanism for generating low-dimensional spike-based text representation. First, we demonstrate how to transform documents into series of spikes (*spike trains*) which are subsequently used as input in the training process of a spiking neural network (SNN). The network is composed of biologically plausible elements, and trained according to the unsupervised Hebbian learning rule, Spike-Timing-Dependent Plasticity (STDP). After training, the SNN can be used to generate low-dimensional spike-based text representation suitable for text/document classification. Empirical results demonstrate that the generated text representation may be effectively used in text classification leading to an accuracy of 80.19% on the *bydate* version of the *20 newsgroups* data set, which is a leading result amongst approaches that rely on low-dimensional text representations.

Keywords: Spiking neural network · STDP · Hebbian learning · Text processing · Text representation · Spike-based representation · Representation learning · Feature learning · Text classification · 20 newsgroups bydate

1 Introduction

Spiking neural networks (SNNs) are an example of biologically plausible artificial neural networks (ANNs). SNNs, like their biological counterparts, process sequences of discrete events occurring in time, known as spikes. Traditionally, spiking neurons, due to their biological validity, have been studied mostly by theoretical neuroscientists, and have become a standard tool for modeling brain processes on a micro scale. However, recent years have shown that spiking computation can also successfully address common machine learning challenges [35]. Another interesting aspect of SNNs is the adaptation of such algorithms to neuromorphic hardware which is a brain-inspired alternative to the traditional von

T. Bäck et al. (Eds.): PPSN 2020, LNCS 12269, pp. 433–447, 2020.
https://doi.org/10.1007/978-3-030-58112-1_30

Neumann machine. Thanks to mimicking processes observed in brain synaptic connections, neuromorphic hardware is a highly fault-tolerant and energy-efficient substitute for classical computation [27].

Recently we have witnessed significant growth in the volume of research into SNNs. Researchers have successfully adapted SNNs for the processing of images [35], audio signals [9,39,40], and time series [18,31]. However, to the best of the authors knowledge, there is only one work related to text processing with SNNs [37]. This state of affairs is caused by the fact that text, due to its structure and high dimensionality, presents a significant challenge to tackle by the SNN approach. The motivation of this study is to broaden the current knowledge of the application of SNNs to text processing. More specifically, we have developed and evaluated a novel biologically inspired method for generation of spike-based text representation that may be used in text/document classification task [25].

1.1 Objectives and Summary of Approach

This paper proposes an *Spike Encoder for Text* (SET) which generates spike-based text representation suitable for classification task. Text data is highly dimensional (the most common text representation is in the form of a vector with many features) which, due to the *curse of dimensionality* [2,17,20,26], usually leads to overfitted classification models with poor generalisation [4,13,30,32,38].

Processing highly dimensional data is also computationally expensive. Therefore, researchers have sought text representations which may overcome this drawback [3]. One of possible approaches is based on transformation of high dimensional feature space to low-dimensional representation [5,6,36].

In the above context we propose the following two-phase approach to SNN based text classification. Firstly, the text is transformed into *spike trains*. Secondly, spike trains representation is used as the input in the SNN training process performed according to biologically plausible unsupervised learning rule, and generating the spike-based text representation. This representation has significantly lower dimensionality than the spike trains representation and can be used effectively in subsequent SNN text classification. The proposed solution has been empirically evaluated on the publicly available version, *bydate* [21] of the real data set known as *20 newsgroups*, which contains 18 846 text documents from twenty different newsgroups of *Usenet*, a worldwide distributed discussion system.

Both the input and output of the SNN rely on spike representations, though of very different forms. For the sake of clarity, throughout the paper the former representation (SNN input) will be referred to as *spike trains*, and the latter one (SNN output) as *spike-based*, or *spiking encoding*, or *low-dimensional.*

1.2 Contribution

The main contribution of this work can be summarized as follows:

- To propose an original approach to document processing using SNNs and its subsequent classification based on generated spike-based text representation;

- To experimentally evaluate the influence of various parameters on the quality of generated representation, which leads to better understanding of the strengths and limitations of SNN-based text classification approaches;
- To propose an SNN architecture which may potentially contribute to development of other SNN based approaches. We believe that the solution presented may serve as a building block for larger SNN architectures, in particular deep spiking neural networks (DSNNs) [35];

1.3 Related Work

As mentioned above, we are aware of only one paper related to text processing in the context of SNNs context [37] which, nevertheless, differs significantly from our approach. The authors of [37] focus on transforming word embeddings [23,28] into spike trains, whilst our focus is not only on representation of text in the form of spike trains, but also on training the SNN encoder which generates low-dimensional text representation. In other words, our goal is to generate a low-dimensional text representation with the use of SNN base, whereas in [37] the transformation of an existing text embedding into spike trains is proposed.

This remainder of the paper is structured as follows. Section 2 presents an overview of the proposed method; Sect. 3 describes the evaluation process of the method and experimental results; and Sect. 4 presents the conclusions.

2 Proposed Spiking Neural Method

The proposed method transforms input text to spike code and uses it as training input for the SNN to achieve a meaningful spike-based text representation. The method is schematically presented in Fig. 1. In phase *I*, text is transformed into a vector representation and afterwards each vector is encoded as spike trains. Once the text is encoded in the form of neural activity, it can be used as input to the core element of our method - *a spiking encoder*. The encoder is a two-layered SNN with adaptable synapses. During the learning phase (*II*), the spike trains are propagated through the encoder in a feed-forward manner and synaptic weights are modified simultaneously according to unsupervised learning rule. After the learning process, the output layer of the spiking encoder provides spike-based representation of the text presented to the system.

In the remainder of this section all elements of the system described above are discussed in more detail.

2.1 Input Transformation

2.1.1 Text Vectorization

During a *text to spike transformation* phase like the one illustrated in Fig. 1 text is preprocessed for further spiking computation. Text input data (*data corpus*) is organized as a set D of documents $d_i, i = 1, \ldots, K$. In the first step a dictionary T containing all unique words $t_j, j = 1, \ldots, |T|$ from the corpus data is built.

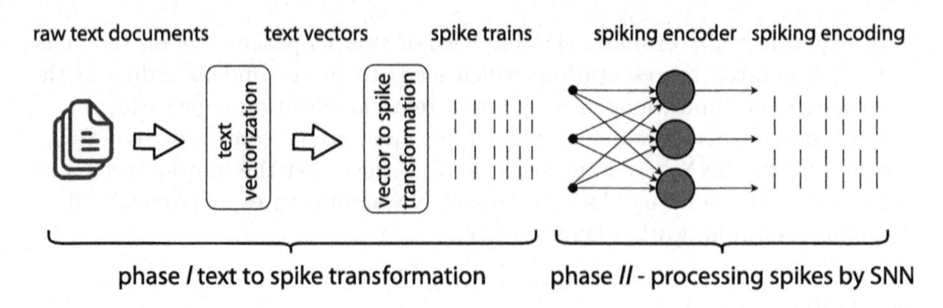

Fig. 1. A schema of the proposed method for generating spike-based low-dimensional text representation.

Next, each document d_i is transformed into an M-dimensional ($M = |T|$) vector W_i, the elements of which, $W_i[j] := w_{ij}, j = 1, \ldots, M$ represent the relevance of words t_j to document d_i. In effect, the corpus data is represented by a real-valued matrix $W_{K \times M}$ also called *document-term matrix.*

The typical weighting functions are *term-frequency* (TF), *inverse document frequency* (IDF), or their combination TF-IDF [12,22]. In TF the weight w_{ij} is equal to the number of times the j-th word appears in d_i with respect to the length of $|d_i|$ (the number of all non-unique words in d_i). IDF takes into account the whole corpus D and sets w_{ij} as the logarithm of a ratio between $|D|$ and the number of documents containing word t_j. Consequently, IDF mitigates the impact of words that occur very frequently in a given corpus and are presumably less informative from the point of view of document classification than the words occurring in a small fraction of the documents. TF-IDF sets w_{ij} as a product of TF and IDF weights. In this paper we use TF-IDF weighting which is the most popular approach in text processing domain.

2.1.2 Vector to Spike Transformation

In order to transform a vector representation to spike trains one, presentation time t_p which establishes for how long each document is presented to the network, and the time gap between two consecutive presentations Δt_p, must be defined. A time gap period, without any input stimuli is necessary to eliminate interference between documents and allow dynamic parameters of the system to decay and "be ready" for the next input.

Technically, for a given document d_i, represented as M dimensional vector of weights w_{ij}, for each weight w_{ij} in every millisecond of document presentation a spike is generated with probability proportional to w_{ij}. Thanks to this procedure, we ultimately derive a spiking representation of the text.

In our experiments each document is presented for $t_p = 600[ms]$ and $\Delta t_p = 300[ms]$, and proportionality coefficient α is set to 1.5.

For a better clarification, let's consider a simple example and assume that for a word *baseball* the corresponding weight w_{ij} in some document d_i is equal to 0.1. Then for each millisecond of a presentation time a probability of emitting a spike $P(spike|baseball)$ equals $\alpha \cdot 0.1 = 0.15$. Hence, 90 spikes during $600[ms]$ presentation time are expected, on average, to be generated.

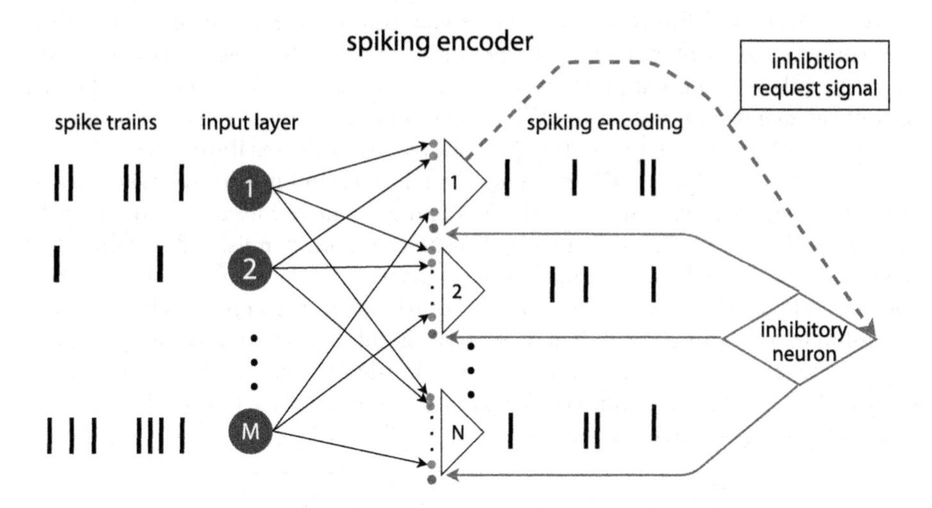

Fig. 2. Spiking encoder architecture.

2.2 Spiking Encoder Architecture and Dynamics

A spiking encoder is the key element of the proposed method. The encoder, presented in Fig. 2, is a two layered SNN equipped with an additional inhibitory neuron. The first layer contains M neurons (denoted by blue circles) and each of them represents one word t_j from the dictionary T. Neuron dynamics is defined by the *spike trains* generated based on weights w_{ij} corresponding to documents $d_i, i = 1, \ldots, K$. Higher numbers of spikes are emitted by neurons representing words which are statistically more relevant for a particular document, according to the chosen TF-IDF measure. The *spike trains* for each neuron are presented in Fig. 2 as a row of short vertical lines.

In the brain spikes are transmitted between neurons via synaptic connections. A neuron which generates a spike is called a *presynaptic neuron*, whilst a target neuron (spike receiver) is a *postsynaptic neuron*. In the proposed SNN architecture (cf. Fig. 2) two different types of synaptic connections are utilised: *excitatory* ones and *inhibitory* ones. Spikes transmitted through excitatory connections (denoted by green circles in Fig. 2) leads to firing of postsynaptic neuron, while impulses traveling through inhibitory ones (red circles in Fig. 2) hinder *postsynaptic neuron* activity. Each time an encoder neuron fires its weights are modified according to the proposed learning rule. The neuron simultaneously sends an *inhibition request signal* to the *inhibitory neuron* and activates it. Then the inhibitory neuron suppresses the activity of all encoder output layer neurons using *recursive inhibitory connection* (red circles). The proposed architecture satisfies the competitive learning paradigm [19] with a winner-takes-all (WTA) strategy.

In this work we consider a biologically plausible neuron model known as leaky integrate and fire (LIF) [11]. The dynamics of such a neuron is described in terms of changes of its membrane potential (MP). If the neuron is not receiving any spikes its potential is close to the value of $u_{rest} = -65[mV]$ known as resting membrane potential. When the neuron receives spikes transmitted through excitatory synapses, the MP moves towards excitatory equilibrium potential, $u_{exc} = 0[mV]$. When many signals are simultaneously transmitted through excitatory synapses the MP rises and at some point can reach a threshold value of $u_{th} = -52[mV]$, in which case the neuron fires. After firing, the neuron resets its MP to u_{rest} and becomes inactive for $t_{ref} = 3[ms]$ (the refractory period). In the opposite scenario, when the neuron receives spikes through the inhibitory synapse, its MP moves towards inhibitory equilibrium potential $u_{inh} = -90[mV]$, i.e. further away from its threshold value, which decreases the chance of firing. The dynamics of the membrane potential u in the LIF model is described by the following equation:

$$\tau \frac{du}{dt} = (u_{rest} - u) + g_e(u_{exc} - u) + g_i(u_{inh} - u) \tag{1}$$

where g_e and g_i denote excitatory and inhibitory conductance, resp. and $\tau = 100[ms]$ is membrane time constant. The values of g_e and g_i depend on presynaptic activity. Each time a signal is transmitted through the synapse the conductance is incremented by the value of weight corresponding to that synapse, and decays with time afterwards according to Eq. (4)

$$\tau_e \frac{dg_e}{dt} = -g_e, \quad \tau_i \frac{dg_i}{dt} = -g_i, \tag{2}$$

where $\tau_e = 2[ms]$, $\tau_i = 2[ms]$ are decay time constants. In summary, if there is no presynaptic activity the MP converges to u_{rest}. Otherwise, its value changes according to the signals transmitted through the neuron synapses.

2.3 Hebbian Synaptic Plasticity

We utilise a modified version of the Spike-Timing-Dependent Plasticity (STDP) learning process [33]. STDP is a biologically plausible unsupervised learning protocol belonging to the family of Hebbian learning (HL) methods [14]. In short, the STDP process results in an increase of the synaptic weight if the postsynaptic spike is observed soon after presynaptic one ('pre-before-post'), and in a decrease of the synaptic weight in the opposite scenario ('post-before-pre'). The above learning scheme increases the relevance of those synaptic connections which contribute to the activation of the postsynaptic neuron, and decreases the importance of the ones which do not. We modify STDP in a manner similar to [8,29], i.e. by skipping the weight modification in the post-before-pre scenario and introducing an additional scaling mechanism. The plasticity of the excitatory synapse s_{ij} connecting a presynaptic neuron i from the input layer with postsynaptic neuron j from the encoder layer can be expressed as follows:

$$\Delta s_{ij} = \eta(A(t) - (R(t) + 0.1)s_{ij}) \tag{3}$$

where

$$A(t) = -\tau_A \frac{dA(t)}{dt}, \quad R(t) = -\tau_R \frac{dR(t)}{dt} \tag{4}$$

and $\eta = 0.01$ is a small learning constant. In Eqs. (3)–(4) $A(t)$ represents a presynaptic trace and $R(t)$ is a scaling factor which depends on the history of postsynaptic neuron activity. Every time the presynaptic neuron i fires $A(t)$ is set to 1 and exponentially decays in time ($\tau_A = 5[ms]$). If the postsynaptic neuron fires just after the presynaptic one ('pre-before-post') $A(t)$ is close to 1 and the weight increase is high. The other component of Eq. (3) $(R(t) + 0.1)s_{ij}$ is a form of synaptic scaling [24]. Every time the postsynaptic neuron fires $R(t)$ is incremented by 1 and afterwards decays with time ($\tau_R = 70[ms]$). The role of the small constant factor 0.1, is to maintain scaling even when activity is relatively small. The overall purpose of synaptic scaling is to decrease the weights of the synapses which are not involved in firing the postsynaptic neuron. Another benefit of synaptic scaling is to restrain weights from uncontrolled growth which can be observed in HL [1].

2.4 Learning Procedure

For a given *data corpus* (set of documents) D the training procedure is performed as follows. Firstly, we divide D into s subsets $u_i, i = 1, \ldots, s$ in the manner described in Sect. 3.1. Secondly, each subset u_i is transformed to spike trains and used as input for a separate SNN encoder $H_i, i = 1, \ldots, s$ composed of N neurons. Please note that each encoder is trained with the use of one subset only. Such a training setup allows the processing of the data in parallel manner. Another advantage is that this limits the number of excitatory connections per neuron, which reduces computational complexity (the number of differential equations that need to be evaluated for each spike) as during training encoder H_i is exposed only to the respective subset, T_i of the training set dictionary T and the number of its excitatory connections is limited to $|T_i| < |T|$. Spike trains are presented to the network four times (in four training epochs).

Once the learning process is completed, the *connection pruning* procedure is applied. Please observe that HL combined with competitive learning should lead to highly specialised neurons which are activated only for some subset of the inputs. The specialisation of a given neuron depends on the set of its connection weights. If the probability of firing should be high for some particular subset of the inputs, the weights representing words from those inputs must be high. The other weights should be relatively low due to the synaptic scaling mechanism. Based on this assumption, after training, for each output layer neuron we prune θ per cent of its incoming connections with the lowest weights. θ is a hyper parameter of the method empirically evaluated in the experimental section.

3 Empirical Evaluation and Results Comparison

This section presents experimental evaluation of the method proposed. In Subsect. 3.1 the technical issues related to the setup of experiment and implementation of the training and evaluation procedures are discussed. The final two subsections focus respectively on the experimental results and compare them with the literature.

3.1 Experiment Setup

3.1.1 Data Set and Implementation Details

The *bydate* version[1] of *20 newsgroups* is a well known benchmark set in the text classification domain. The set contains newsgroups post related to different categories (topics) gathered from *Usenet*, in which each category corresponds to one newsgroup. Categories are organised into a hierarchical structure with the main categories being *computers*, *recreation and entertainment*, *science*, *religion*, *politics*, and *forsale*. The corpus consists of 18 846 documents nearly equally distributed among twenty categories and explicitly divided into two subsets: the training one (60%) and the test one (40%).

The dynamics of the spiking neurons (including the plasticity mechanism) was implemented using the *BRIAN 2* simulator [34]. *Scikit-learn Python library*[2] was used for processing the text and creating the TF-IDF matrix.

3.1.2 Training

As mentioned in Sect. 2.4 the training set was divided into $s = 11$ subsets u_i each of which, except for u_{11}, contained 1 500 documents. Division was performed randomly. Firstly the entire training set was shuffled, and then consecutively assigned to the subsets according to the resulting order with a 500 document redundancy (overlap) between the neighbouring subsets, as described in Table 1.

The overlap between subsequent subsets resulted from preliminary experiments which suggested that such an approach improves classification accuracy. While we found the concept of partial data overlap to be reasonably efficient, it by no means should be regarded as an optimal choice. The optimal division of data into training subsets remains an open question and a subject of our future research.

3.1.3 Evaluation Procedure

The outputs of all SNNs $H_i, i = 1, \ldots, s$, i.e. spike-based encodings represented as sums of spikes per document were joined to form a single matrix (a final low-dimensional text representation) which was evaluated in the context of a classification task. The joined matrix of spike rates was used as an input to the Logistic Regression (LR) [15,17] classifier with *accuracy* as the performance measure.

[1] http://qwone.com/~jason/20Newsgroups/20news-bydate.tar.gz.
[2] https://scikit-learn.org/.

Table 1. Division of the training set into subsets u_1–u_{11}.

Subset	u_1	u_2	u_3	u_4	u_5	u_6	u_7	u_8	u_9	u_{10}	u_{11}
First index	0	1000	2000	3000	4000	5000	6000	7000	8000	9000	10000
Last index	1500	2500	3500	4500	5500	6500	7500	8500	9500	10500	11314
Size	1500	1500	1500	1500	1500	1500	1500	1500	1500	1500	1314

3.2 Experimental Results

In the first experiment we looked more closely at the weights of neurons after training and the relationship between the inhibition mechanism and the quality/efficacy of resulting text representation. We trained eleven SNN encoders with 50 neurons each according to the procedure presented above. After training, 5 neurons from the first encoder (H_1) was randomly sampled and their weights were used for further analysis. Figure 3 illustrates the highest 200 weights sorted in descending order.

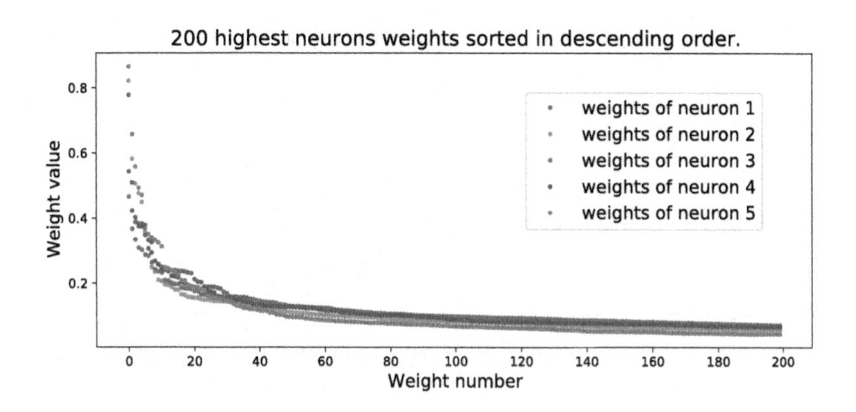

Fig. 3. Weights extracted from the encoder's neurons.

The weights of each neuron are presented with a different colour. The plots show that every neuron has a group of dominant connections represented by the weights with the highest values - the first several dozen connections. It means that each neuron will be activated more easily by the inputs that contain words corresponding to these weights. For example neuron 4 will potentially produce more spikes for documents related to religion because its 10 highest weights corresponds to words *'jesus'*, *'god'*, *'paul'*, *'faith'*, *'law'*, *'christians'*, *'christ'*, *'sabbath'*, *'sin'*, *'jewish'*. A different behaviour is expected from neuron 2 whose 10 highest weights corresponds to words *'drive'*, *'scsi'*, *'disk'*, *'hard'*, *'controller'*, *'ide'*, *'drives'*, *'help'*, *'mac'*, *'edu'*. This one will be more likely activated for computer related documents. On the other hand, not all neurons can be classified so easily. For instance 10 highest weights of neuron 5 are linked to words *'cs'*, *'serial', 'ac', 'edu', 'key', 'bit', 'university', 'windows', 'caronni', 'uk'*, hence a designation of this neuron is less obvious. We have repeated the above sampling and weigh inspection procedure several times and the observations are qualitatively the same. For the sake of space savings we do not report them in detail.

Hence, a question arises as to how well documents can be encoded with the use of neurons trained in the manner described above? Intuitively, in practice the quality of encoding may be related to the level of competition amongst neurons in the evaluation phase. If the inhibition value is kept high enough to satisfy WTA strategy then only a few neurons will be activated and the others will be immediately suppressed. This scenario will lead to highly sparse representations of the input documents, with just a few or (in extreme cases) only one neuron dominating the rest. Since differences between documents belonging to different classes may be subtle, such a sparse representation may not be the optimal setup. In order to check the influence of the inhibition level on the resulting spike-based representation we tested the performance of the trained SNNs H_1–H_{11} for various inhibition levels by adjusting the value of the neurons' inhibitory synapses. The results are illustrated in Fig. 4 (top).

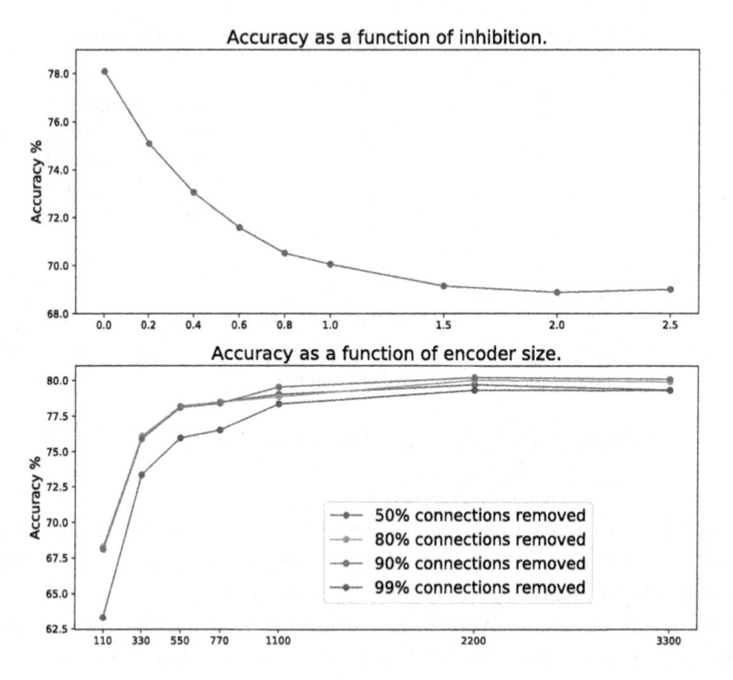

Fig. 4. Accuracy for various inhibition levels (top) and encoder sizes (bottom).

Clearly the accuracy strongly depends on the inhibition level. The best outcomes ($\approx$ 78%) are accomplished with inhibition set to 0 and rapidly decrease along with the inhibition raise. For the inhibition values higher than 1.5 the accuracy plot enters a plateau at the level of approximately 68%. The results show that the most effective representation of documents is generated with the absence of inhibition during the evaluation phase, i.e. when all neurons have the same chance of being activated and contribute to the document representation.

The second series of experiments aimed at exploring the relationship between the efficacy of document representation and the size of the encoders. Furthermore, the sensitivity of the trained encoders to *connection pruning*, with respect to their efficiency, was verified. The results of both experiments are shown in the bottom plot of Fig. 4. Seven encoders of various sizes (between 110 and 3 300 neurons) were trained, and once the training was completed the before the *connection pruning* procedure took place.

In the plot, four colored curves illustrate particular pruning scenarios and their impact on classification accuracy for various encoder sizes. 99%, 90%, 80%, and 50% of the weakest weights were respectively removed in the four discussed cases. Overall, for smaller SNN encoders (between 110 and 1 100 neurons) the accuracy rises rapidly along with the encoder size increase. For larger SSNs, changes in the accuracy are slower and for all four curves stay within the range $[77.5\%, 80.19\%]$.

In terms of the *connection pruning* degree the biggest changes in accuracy (between 63% and 79%) are observed when 99% of connections have been deleted (the red curve). In particular, the results of the encoders with fewer than 1 100 neurons demonstrate that this range of pruning heavily affects classification accuracy. In larger networks additional neurons compensate the features removed by the *connection pruning* mechanism and the results are getting closer to other pruning setups.

Interestingly, for the networks smaller than 770 neurons, the differences in accuracy between 50%, 80%, and 90% pruning setups are negligible, which suggests that relatively high redundancy of connections still exist in the networks pruned in the range of 50% to 80%. Apparently, retaining as few as 10% of the weights does not impact the quality of representation and does not cause deterioration of results. This result well correlates with the outcomes of the weight analysis reported above and confirms that a meaningful subset of connections is sufficient for proper encoding the input. The best overall classification result (80.19%) was achieved by the SNN encoder with 2 200 neurons and level of pruning set to 90% (the green curve). It proves that SEM can effectively reduce dimensionality of the text input from initial $\approx$ 130 000 (the size of *20 newsgroups* training vocabulary) to the size of $550 - 2\ 200$, and maintain classification accuracy above 77.5%.

3.3 Results Analysis. Comparison with the Literature

Since to our knowledge this paper presents the first attempt of using SNN architecture to text classification, in order to make some comparisons we selected results reported for other neural networks trained with similar input (*document-term matrix*) and yielding low-dimensional text representation as the output. The results are presented in Table 2. SET achieved 80.19% accuracy and outperformed the remaining shallow approaches. While this result looks promising we believe that there is still room for improvement with further tuning of the method (in particular a division of samples into training subsets), as well as extension of the SNN encoder by adding more layers. Another interesting direction would be to learn semantic relevance between different words and documents [10,41].

Table 2. Accuracy [%] comparison for several low-dimensional text representation methods on *bydate* version of *20 newsgroups* data set.

Method	Accuracy
SET (this paper)	**80.19**
K-competitive Autoencoder for TExt (KATE) [7]	76.14
Class Preserving Restricted Boltzmann Machine (CPr-RBM) [16]	75.39
Variational Autoencoder [7]	74.30

4 Conclusions

This work offers a novel approach to text representation relying on Spiking Neural Networks. Using the proposed low-dimensional text representation the LR classifier accomplished 80.19% accuracy on a standard benchmark set (*20 newsgroups bydate*) which is a leading result among shallow approaches relying on low-dimensional representations.

We have also examined the influence of the inhibition mechanism and synaptic connections sparsity on the quality of the representation showing that (i) it is recommended that inhibition be disabled during the SNN evaluation phase, and (ii) pruning out as many as 90% of connections with lowest weights did not affect the representation quality while heavily reducing the SNN computational complexity, i.e. the number of differential equations describing the network.

There are a few lines of potential improvement that we plan to explore in the further work. Most notably, we aim to expand the SNN encoder towards Deep SNN architecture by adding more layers of spiking neurons which should possibly allow to learn more detailed features of the input data.

References

1. Abbott, L.F., Nelson, S.B.: Synaptic plasticity: taming the beast. Nat. Neurosci. **3**(Suppl 1), 1178–1183 (2000)
2. Aggarwal, C.C.: Data Mining. Springer, Cham (2015). https://doi.org/10.1007/978-3-319-14142-8
3. Aggarwal, C.C.: Machine Learning for Text, 1st edn. Springer, Cham (2018). https://doi.org/10.1007/978-3-319-73531-3_9
4. Asif, M., Ishtiaq, A., Ahmad, H., Aljuaid, H., Shah, J.: Sentiment analysis of extremism in social media from textual information. Telematics Inform. **48**, 101345 (2020). https://doi.org/10.1016/j.tele.2020.101345

5. Ayesha, S., Hanif, M.K., Talib, R.: Overview and comparative study of dimensionality reduction techniques for high dimensional data. Inf. Fusion **59**, 44–58 (2020)
6. Bengio, Y., Courville, A., Vincent, P.: Representation learning: a review and new perspectives. IEEE Trans. Pattern Anal. Mach. Intell. **35**(8), 1798–1828 (2013). https://doi.org/10.1109/tpami.2013.50
7. Chen, Y., Zaki, M.J.: KATE: k-competitive autoencoder for text. In: Proceedings of the 23rd ACM SIGKDD International Conference on Knowledge Discovery and Data Mining, Halifax, NS, Canada, 13–17 August 2017, pp. 85–94. ACM (2017). https://doi.org/10.1145/3097983.3098017
8. Diehl, P., Cook, M.: Unsupervised learning of digit recognition using spike-timing-dependent plasticity. Front. Comput. Neurosci. **9**, 99 (2015). https://doi.org/10.3389/fncom.2015.00099
9. Dominguez-Morales, J.P., et al.: Deep spiking neural network model for time-variant signals classification: a real-time speech recognition approach. In: 2018 International Joint Conference on Neural Networks (IJCNN). IEEE, July 2018. https://doi.org/10.1109/ijcnn.2018.8489381
10. Gao, Y., Wang, W., Qian, L., Huang, H., Li, Y.: Extending embedding representation by incorporating latent relations. IEEE Access **6**, 52682–52690 (2018). https://doi.org/10.1109/ACCESS.2018.2866531
11. Gerstner, W., Kistler, W.M.: Spiking Neuron Models. Cambridge University Press, Cambridge (2002). https://doi.org/10.1017/cbo9780511815706
12. Haddoud, M., Mokhtari, A., Lecroq, T., Abdeddaïm, S.: Combining supervised term-weighting metrics for SVM text classification with extended term representation. Knowl. Inf. Syst. **49**(3), 909–931 (2016). https://doi.org/10.1007/s10115-016-0924-1
13. Hartmann, J., Huppertz, J., Schamp, C., Heitmann, M.: Comparing automated text classification methods. Int. J. Res. Market. **36**(1), 20–38 (2019). https://doi.org/10.1016/j.ijresmar.2018.09.009
14. Hebb, D.O.: The organization of behavior: A neuropsychological theory. New York (1949)
15. Hosmer, D.W., Lemeshow, S.: Applied Logistic Regression. Second edn. Wiley, Hoboken (2000). https://doi.org/10.1002/0471722146
16. Hu, J., Zhang, J., Ji, N., Zhang, C.: A new regularized restricted boltzmann machine based on class preserving. Knowl.-Based Syst. **123**, 1–12 (2017). https://doi.org/10.1016/j.knosys.2017.02.012
17. James, G., Witten, D., Hastie, T., Tibshirani, R.: An Introduction to Statistical Learning: with Applications in R. Springer, Heidelberg (2013). https://doi.org/10.1007/978-1-4614-7138-7
18. Kasabov, N., Capecci, E.: Spiking neural network methodology for modelling, classification and understanding of EEG spatio-temporal data measuring cognitive processes. Inf. Sci. **294**, 565–575 (2015). https://doi.org/10.1016/j.ins.2014.06.028
19. Kaski, S., Kohonen, T.: Winner-take-all networks for physiological models of competitive learning. Neural Netw. **7**, 973–984 (1994). https://doi.org/10.1016/S0893-6080(05)80154-6
20. Keogh, E., Mueen, A.: Curse of Dimensionality, pp. 314–315. Springer, Boston (2017). https://doi.org/10.1007/978-1-4899-7687-1_192
21. Lang, K.: NewsWeeder: learning to filter netnews. In: Proceedings of the Twelfth International Conference on Machine Learning, pp. 331–339 (1995)
22. Manning, C.D., Raghavan, P., Schütze, H.: Introduction to Information Retrieval. Cambridge University Press, New York (2008)

23. Mikolov, T., Sutskever, I., Chen, K., Corrado, G.S., Dean, J.: Distributed representations of words and phrases and their compositionality. In: Burges, C.J.C., Bottou, L., Ghahramani, Z., Weinberger, K.Q. (eds.) Advances in Neural Information Processing Systems 26: 27th Annual Conference on Neural Information Processing Systems 2013. Proceedings of a meeting held December 5–8, 2013, Lake Tahoe, Nevada, United States, pp. 3111–3119 (2013)
24. Miller, K.D., MacKay, D.J.C.: The role of constraints in Hebbian learning. Neural Comput. **6**(1), 100–126 (1994). https://doi.org/10.1162/neco.1994.6.1.100
25. Mladenić, D., Brank, J., Grobelnik, M.: Document Classification, pp. 372–377. Springer, Boston (2017). https://doi.org/10.1007/978-1-4899-7687-1_75
26. Murphy, K.P.: Machine Learning - A Probabilistic Perspective. Adaptive Computation and mAchine Learning Series. MIT Press, Cambridge (2012)
27. Nawrocki, R.A., Voyles, R.M., Shaheen, S.E.: A mini review of neuromorphic architectures and implementations. IEEE Trans. Electron Devices **63**(10), 3819–3829 (2016). https://doi.org/10.1109/TED.2016.2598413
28. Pennington, J., Socher, R., Manning, C.D.: Glove: global vectors for word representation. In: Moschitti, A., Pang, B., Daelemans, W. (eds.) Proceedings of the 2014 Conference on Empirical Methods in Natural Language Processing, EMNLP 2014, October 25–29, 2014, Doha, Qatar, A Meeting of SIGDAT, a Special Interest Group of the ACL, pp. 1532–1543. ACL (2014). https://doi.org/10.3115/v1/d14-1162
29. Querlioz, D., Bichler, O., Dollfus, P., Gamrat, C.: Immunity to device variations in a spiking neural network with memristive nanodevices. IEEE Trans. Nanotechnol. **12**(3), 288–295 (2013). https://doi.org/10.1109/TNANO.2013.2250995
30. Raza, M., Hussain, F.K., Hussain, O.K., Zhao, M., ur Rehman, Z.: A comparative analysis of machine learning models for quality pillar assessment of SaaS services by multi-class text classification of users' reviews. Future Gen. Comput. Syst. **101**, 341–371 (2019)
31. Reid, D., Hussain, A.J., Tawfik, H.: Financial time series prediction using spiking neural networks. PLoS ONE **9**(8), e103656 (2014). https://doi.org/10.1371/journal.pone.0103656
32. Silva, R.M., Almeida, T.A., Yamakami, A.: MDLText: an efficient and lightweight text classifier. Knowl.-Based Syst. **118**, 152–164 (2017). https://doi.org/10.1016/j.knosys.2016.11.018
33. Song, S., Miller, K., Abbott, L.: Competitive Hebbian learning through spike timing-dependent plasticity. Nat. Neurosci. **3**, 919–926 (2000). https://doi.org/10.1038/78829
34. Stimberg, M., Goodman, D., Benichoux, V., Brette, R.: Equation-oriented specification of neural models for simulations. Front. Neuroinform. **8**, 6 (2014). https://doi.org/10.3389/fninf.2014.00006
35. Tavanaei, A., Ghodrati, M., Kheradpisheh, S.R., Masquelier, T., Maida, A.: Deep learning in spiking neural networks. Neural Netw. **111**, 47–63 (2019). https://doi.org/10.1016/j.neunet.2018.12.002
36. Vlachos, M.: Dimensionality Reduction, pp. 354–361. Springer, Boston (2017). https://doi.org/10.1007/978-1-4899-7687-1_71
37. Wang, Y., Zeng, Y., Tang, J., Xu, B.: Biological neuron coding inspired binary word embeddings. Cogn. Comput. **11**(5), 676–684 (2019). https://doi.org/10.1007/s12559-019-09643-1
38. Webb, G.I.: Overfitting, pp. 947–948. Springer, Boston (2017). https://doi.org/10.1007/978-1-4899-7687-1_960

39. Wu, J., Chua, Y., Zhang, M., Li, H., Tan, K.C.: A spiking neural network framework for robust sound classification. Front. Neurosci. **12** (2018). https://doi.org/10.3389/fnins.2018.00836
40. Wysoski, S.G., Benuskova, L., Kasabov, N.: Evolving spiking neural networks for audiovisual information processing. Neural Netw. **23**(7), 819–835 (2010). https://doi.org/10.1016/j.neunet.2010.04.009
41. Zheng, S., Bao, H., Xu, J., Hao, Y., Qi, Z., Hao, H.: A bidirectional hierarchical skip-gram model for text topic embedding. In: 2016 International Joint Conference on Neural Networks, IJCNN 2016, Vancouver, BC, Canada, 24–29 July 2016, pp. 855–862. IEEE (2016). https://doi.org/10.1109/IJCNN.2016.7727289

Multi-Objective Counterfactual Explanations

Susanne Dandl(✉), Christoph Molnar, Martin Binder, and Bernd Bischl

Department of Statistics, LMU Munich, Ludwigstr. 33, 80539 Munich, Germany
susanne.dandl@stat.uni-muenchen.de

Abstract. Counterfactual explanations are one of the most popular methods to make predictions of black box machine learning models interpretable by providing explanations in the form of 'what-if scenarios'. Most current approaches optimize a collapsed, weighted sum of multiple objectives, which are naturally difficult to balance a-priori. We propose the Multi-Objective Counterfactuals (MOC) method, which translates the counterfactual search into a multi-objective optimization problem. Our approach not only returns a diverse set of counterfactuals with different trade-offs between the proposed objectives, but also maintains diversity in feature space. This enables a more detailed post-hoc analysis to facilitate better understanding and also more options for actionable user responses to change the predicted outcome. Our approach is also model-agnostic and works for numerical and categorical input features. We show the usefulness of MOC in concrete cases and compare our approach with state-of-the-art methods for counterfactual explanations.

Keywords: Interpretability · Interpretable machine learning · Counterfactual explanations · Multi-objective optimization · NSGA-II

1 Introduction

Interpretable machine learning methods have become very important in recent years to explain the behavior of black box machine learning (ML) models. A useful method for explaining *single* predictions of a model are counterfactual explanations. ML credit risk prediction is a common motivation for counterfactuals. For people whose credit applications have been rejected, it is valuable to know why they have not been accepted, either to understand the decision making process or to assess their actionable options to change the outcome. Counterfactuals provide these explanations in the form of "if these features had different values, your credit application would have been accepted". For such explanations to be plausible, they should only suggest small changes in a few features.

This work has been partially supported by the German Federal Ministry of Education and Research (BMBF) under Grant No. 01IS18036A and by the Bavarian State Ministry of Science and the Arts in the framework of the Centre Digitisation.Bavaria (ZD.B). The authors of this work take full responsibility for its content.

T. Bäck et al. (Eds.): PPSN 2020, LNCS 12269, pp. 448–469, 2020.
https://doi.org/10.1007/978-3-030-58112-1_31

Therefore, counterfactuals can be defined as close neighbors of an actual data point, but their predictions have to be sufficiently close to a (usually quite different) desired outcome. Counterfactuals explain why a certain outcome was not reached, can offer potential reasons to object against an unfair outcome and give guidance on how the desired prediction could be reached in the future [35]. Note that counterfactuals are also valuable for predictive modelers on a more technical level to investigate the pointwise robustness and the pointwise bias of their model.

2 Related Work

Counterfactuals are closely related to adversarial perturbations. These have the aim to deceive ML models instead of making the models interpretable [30]. Attribution methods such as Local Interpretable Model-agnostic Explanations (LIME) [27] and Shapley Values [22] explain a prediction by determining how much each feature contributed to it. Counterfactual explanations differ from feature attributions since they generate data points with a different, desired prediction instead of attributing a prediction to the features.

Counterfactual methods can be model-agnostic or model-specific. The latter usually exploit the internal structure of the underlying ML model, such as the trained weights of a neural network, while the former are based on general principles which work for arbitrary ML models - often by only assuming access to the prediction function of an already fitted model. Several model-agnostic counterfactual methods have been proposed [8,11,16,18,25,29,37]. Apart from Grath et al. [11], these approaches are limited to classification. Unlike the other methods, the method of Poyiadzi et al. [25] can obtain plausible counterfactuals by constructing feasible paths between data points with opposite predictions.

A model-specific approach was proposed by Wachter et al. [35], who also introduced and formalized the concept of counterfactuals in predictive modeling. Like many model-specific methods [15,20,24,28,33] their approach is limited to differentiable models. The approach of Tolomei et al. [32] generates explanations for tree-based ensemble binary classifiers. As with [35] and [20], it only returns a single counterfactual per run.

3 Contributions

In this paper, we introduce Multi-Objective Counterfactuals (MOC), which to the best of our knowledge is the first method to formalize the counterfactual search as a multi-objective optimization problem. We argue that the mathematical problem behind the search for counterfactuals should be naturally addressed as multi-objective. Most of the above methods optimize a collapsed, weighted sum of multiple objectives to find counterfactuals, which are naturally difficult to balance a-priori. They carry the risk of arbitrarily reducing the solution set to a single candidate without the option to discuss inherent trade-offs – which should be especially relevant for model interpretation that is by design very hard to precisely capture in a (single) mathematical formulation.

Compared to Wachter et al. [35], we use a distance metric for mixed feature spaces and two additional objectives: one that measures the number of feature changes to obtain sparse and therefore more interpretable counterfactuals, and one that measures the closeness to the nearest observed data points for more plausible counterfactuals. MOC returns a Pareto set of counterfactuals that represents different trade-offs between our proposed objectives, and which are constructed to be diverse in feature space. This seems preferable because changes to different features can lead to a desired counterfactual prediction[1] and it is more likely that some counterfactuals meet the (hidden) preferences of a user. A single counterfactual might even suggest a strategy that is interpretable but not actionable (e.g., 'reduce your number of pregnancies') or counterproductive in more general contexts (e.g., 'increase your age to reduce the risk of diabetes'). In addition, if multiple otherwise quite different counterfactuals suggest changes to the same feature, the user may have more confidence that the feature is an important lever to achieve the desired outcome. We refer the reader to Appendix A for two concrete examples illustrating the above.

Compared to other counterfactual methods, MOC is model-agnostic and handles classification, regression and mixed feature spaces, which furthermore increases its practical usefulness in general applications. Together with [16], our paper also includes one of the first benchmark studies that compares multiple counterfactual methods on multiple, heterogeneous datasets.

4 Methodology

[35] loosely define counterfactuals as:

> "You were denied a loan because your annual income was £30,000. If your income had been £45,000, you would have been offered a loan. Here the statement of decision is followed by a counterfactual, or statement of how the world would have to be different for a desirable outcome to occur. Multiple counterfactuals are possible, as multiple desirable outcomes can exist, and there may be several ways to achieve any of these outcomes."

We now formalize this statement by stating four objectives, which a counterfactual should adhere to. In the subsequent section we provide detailed definitions of these objectives and tie them together as a multi-objective optimization problem in order to generate a diverse set of different trade-off solutions.

4.1 Multi-Objective Counterfactuals

Definition 1 (Counterfactual Explanation). *Let $\hat{f} : \mathcal{X} \rightarrow \mathbb{R}$ be a prediction function, $\mathcal{X}$ the feature space and $Y' \subset \mathbb{R}$ a set of desired outcomes. The latter*

[1] Rashomon effect [5].

can either be a single value or an interval of values. We define a counterfactual explanation $\mathbf{x}'$ *for an observation* $\mathbf{x}^*$ *as a data point fulfilling the following: (1) its prediction* $f(\mathbf{x}')$ *is close to the desired outcome set* Y'*, (2) it is close to* $\mathbf{x}^*$ *in the* $\mathcal{X}$ *space, (3) it differs from* $\mathbf{x}^*$ *only in a few features, and (4) it is a* plausible *data point according to the probability distribution* $\mathbb{P}_{\mathcal{X}}$*. For classification models, we assume that* $\hat{f}$ *returns the probability for a user-selected class and* Y' *has to be the desired probability (range).*

This can be translated into a multi-objective minimization task:

$$\min_{\mathbf{x}} \mathbf{o}(\mathbf{x}) := \min_{\mathbf{x}} \left(o_1(\hat{f}(\mathbf{x}), Y'),\, o_2(\mathbf{x}, \mathbf{x}^*), o_3(\mathbf{x}, \mathbf{x}^*), o_4(\mathbf{x}, \mathbf{X}^{obs})\right), \tag{1}$$

with $\mathbf{o} : \mathcal{X} \to \mathbb{R}^4$ and $\mathbf{X}^{obs}$ as the observed (i.e. training) data. The first component o_1 quantifies the distance between $\hat{f}(\mathbf{x})$ and Y'. We define it as:[2]

$$o_1(\hat{f}(\mathbf{x}), Y') = \begin{cases} 0 & \text{if } \hat{f}(\mathbf{x}) \in Y' \\ \inf\limits_{y' \in Y'} |\hat{f}(\mathbf{x}) - y'| & \text{else} \end{cases}.$$

The second component o_2 quantifies the distance between $\mathbf{x}^*$ and $\mathbf{x}$ using the Gower distance to account for mixed features [10]:

$$o_2(\mathbf{x}, \mathbf{x}^*) = \frac{1}{p} \sum_{j=1}^{p} \delta_G(x_j, x_j^*) \in [0, 1]$$

with p being the number of features. The value of δ_G depends on the feature type:

$$\delta_G(x_j, x_j^*) = \begin{cases} \frac{1}{\widehat{R}_j} |x_j - x_j^*| & \text{if } x_j \text{ is numerical} \\ \mathbb{I}_{x_j \neq x_j^*} & \text{if } x_j \text{ is categorical} \end{cases}$$

with $\widehat{R}_j$ as the value range of feature j, extracted from the observed dataset.

Since the Gower distance does not take into account how many features have been changed, we introduce objective o_3, which counts the number of changed features using the L_0 norm:

$$o_3(\mathbf{x}, \mathbf{x}^*) = ||\mathbf{x} - \mathbf{x}^*||_0 = \sum_{j=1}^{p} \mathbb{I}_{x_j \neq x_j^*}.$$

The fourth objective o_4 measures the weighted average Gower distance between $\mathbf{x}$ and the k nearest observed data points $\mathbf{x}^{[1]}, ..., \mathbf{x}^{[k]} \in \mathbf{X}^{obs}$ as an empirical approximation of how likely $\mathbf{x}$ originates from the distribution of $\mathcal{X}$:

$$o_4(\mathbf{x}, \mathbf{X}^{obs}) = \sum_{i=1}^{k} w^{[i]} \frac{1}{p} \sum_{j=1}^{p} \delta_G(x_j, x_j^{[i]}) \in [0, 1] \text{ where } \sum_{i=1}^{k} w^{[i]} = 1.$$

[2] We chose the L_1 norm over the L_2 norm for a natural interpretation. Its non-differentiability is negligible for evolutionary optimization.

Throughout this paper, we set k to 1. Further procedures to increase the plausibility of the counterfactuals are integrated into the optimization algorithm and are described in Sect. 4.3.

Balancing the four objectives is difficult since the objectives contradict each other. For example, minimizing the distance between counterfactual outcome and desired outcome Y' (o_1) becomes more difficult when we require counterfactual feature values close to $\mathbf{x}^*$ (o_2 and o_3) and to the observed data (o_4).

4.2 Counterfactual Search

Our proposed method MOC uses the *Nondominated Sorting Genetic Algorithm II* (NSGA-II) [7] with modifications specific to the problem considered. First, unlike the original NSGA-II, it uses *mixed integer evolutionary strategies* (MIES) [19] to work with the mixed discrete and continuous search space. Furthermore, a different crowding distance sorting algorithm is used, and we propose some optional adjustments tailored to the counterfactual search in the upcoming section.

For MOC, each candidate is described by its feature vector (the 'genes') and the objective values of the candidates are evaluated by Eq. (1). Features of candidates are recombined and mutated with predefined probabilities – some of the control parameters of MOC. Numerical features are recombined by the simulated binary crossover recombinator [6], all other feature types by the uniform crossover recombinator [31]. Based on [19], numerical features are mutated by the scaled Gaussian mutator. Categorical features are altered by uniformly sampling from their admissible levels, while binary and logical features are simply flipped. After recombination and mutation, some feature values are randomly set to the values of $\mathbf{x}^*$ with a given (low) probability – another control parameter – to prevent all features from deviating from $\mathbf{x}^*$.

Contrary to NSGA-II, the crowding distance is computed not only in the objective space $\mathbb{R}^4$ (L_1 norm) but also in the feature space $\mathcal{X}$ (Gower distance), and the distances are summed up with equal weighting. As a result, candidates are more likely kept if they differ greatly from another candidate in their feature values although they are similar in the objective values. Diversity in $\mathcal{X}$ is desired because the chances of obtaining counterfactuals that meet the (hidden) preferences of users are higher. This approach is based on Avila et al. [2].

MOC stops if either a predefined number of generations is reached (default) or the performance no longer improves for a given number of successive generations.

4.3 Further Modifications

Initialization. Naively, we could initialize a population by uniformly sampling some feature values from their full range of possible values, while randomly setting other features to the values of $\mathbf{x}^*$ to induce sparsity. However, if a feature has a large influence on the prediction, it should be more likely that the counterfactual values differ from $\mathbf{x}^*$. The importance of a feature for an entire dataset can be measured as the standard deviation of the partial dependence plot [12]. Analogously, we propose to measure the feature importance for a single prediction with the standard deviation of the Individual Conditional Expectation (ICE) curve of $\mathbf{x}^*$. ICE curves show for one observation and for one feature how the prediction changes when the feature is changed, while other features are fixed to the values of the considered observation [9]. The greater the standard deviation of the ICE curve, the higher we set the probability that the feature value is initialized with a different value than the one of $\mathbf{x}^*$. Therefore, the standard deviation σ_j^{ICE} of each feature x_j is transformed into probabilities within $[p_{min}, p_{max}] \cdot 100\%$:

$$P(value\ differs) = \frac{(\sigma_j^{ICE} - min(\sigma^{ICE})) \cdot (p_{max} - p_{min})}{max(\sigma^{ICE}) - min(\sigma^{ICE})} + p_{min}$$

with $\boldsymbol{\sigma}^{ICE} := (\sigma_1^{ICE}, ..., \sigma_p^{ICE})$. p_{min} and p_{max} are control parameters with default values 0.01 and 0.99.

Actionability. To get more actionable counterfactuals, extreme values of numerical features outside a predefined range are capped to the upper or lower bound after recombination and mutation. The ranges can either be derived from the minimum and maximum values of the features in the observed dataset or users can define these ranges. In addition, users can identify non-actionable features such as the country of birth or gender. The values of these features are permanently set to the values of $\mathbf{x}^*$ for all candidates within MOC.

Penalization. Furthermore, candidates whose predictions are further away from the target than a predefined distance $\epsilon \in \mathbb{R}$ can be penalized. After the candidates have been sorted into fronts F_1 to F_K using nondominated sorting, the candidate that violates the constraint least will be reassigned to front F_{K+1}, the candidate with the second smallest violation to F_{K+2}, and so on. The concept is based on Deb et al. [7]. Since the constraint violators are in the last fronts, they are less likely to be selected for the next generation.

Mutation. Since the aforementioned mutators do not take the data distribution into account and can potentially generate unlikely new candidates, we suggest a conditional mutator. It generates plausible feature values conditional on the values of the other features. For each input feature, we trained a transformation tree [14] on X^{obs}, which is then used to sample values from the conditional distribution. We mutate the feature in randomized order since a feature mutation now depends on the previous changes.

How our proposed strategies for initialization and mutation affect MOC is later examined in a benchmark study (Sects. 6 and 7).

4.4 Evaluation Metric

We use the popular hypervolume indicator (HV) [38] to evaluate the quality of our estimated Pareto front, with reference point $\mathbf{s} = (\inf_{y' \in Y'} |\hat{f}(\mathbf{x}^*) - y'|, 1, p, 1)$, representing the maximal values of the objectives. We compute the HV always over the complete archive of evaluated solutions.

4.5 Tuning of Parameters

We also use HV, when we tune MOC's control parameters – population size, the probabilities for recombining and mutating a feature of a candidate – with iterated F-racing [21]. Furthermore, we let iterated F-racing decide whether our proposed strategies for initialization and mutation of Sect. 4.3 are preferable. Tuning is performed on six binary classification datasets from OpenML [34] – which were not used in the benchmark. A summary of the tuning setup and results can be found in Table 5 in Appendix B. Iterated F-racing found both our initialization and mutation strategy to be advantageous. The tuned parameters were used for the credit data application and the benchmark study.

5 Credit Data Application

This section demonstrates the usefulness of MOC to explain the prediction of credit risk using the German credit dataset [13]. The dataset has 522 complete observations and nine features containing credit and customer information. Categories with few case numbers were combined. The binary target indicates whether a customer has a 'good' or 'bad' credit risk. We chose the first observation of the dataset as $\mathbf{x}^*$ with the following feature values:

Age	Sex	Job	Housing	Saving accounts	Checking account	Credit amount	Duration	Purpose
22	Female	2	Own	Little	Moderate	5951	48	Radio/TV

We tuned a support vector machine (with radial-basis (RBF) kernel) on the remaining data with the same tuning setup as for the benchmark (Appendix C). To obtain a single numerical outcome, only the predicted probability for the class 'good' credit risk was returned. We obtained an accuracy of 0.64 for the model using two nested cross-validations (CV) (5-fold CV in outer and inner loop) and a predicted probability for 'good' credit risk of 0.41 for $\mathbf{x}^*$.

We set the desired outcome interval to $Y' = [0.5, 1]$, which indicates a change to a 'good' credit risk. We generated counterfactuals using MOC with the parameter setting selected by iterated F-racing. Candidates with a prediction below 0.5 were penalized.

A total of 136 counterfactuals were found by MOC. In the following, we focus upon the 82 of them with predictions within $[0.5, 1]$. Credit *duration* was changed

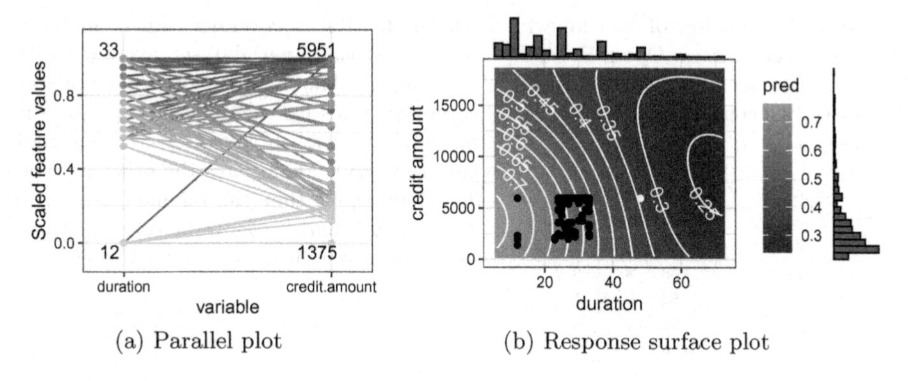

(a) Parallel plot

(b) Response surface plot

Fig. 1. Visualization of counterfactuals for the first data point $\mathbf{x}^*$ of the credit dataset. **(a)** Feature values of the counterfactuals. Only changed features are shown. The given numbers indicate the minimum and maximum feature values of the counterfactuals. **(b)** Response surface plot for the model prediction along features duration and credit amount, holding other feature values constant at the value of $\mathbf{x}^*$. Colors and contour lines indicate the predicted value. The white point is $\mathbf{x}^*$ and the black points are the counterfactuals that only proposed changes in duration and/or credit amount. The histograms show the marginal distributions of the features in the observed dataset.

for all counterfactuals, followed by *credit amount* (86%). Since a user might not want to investigate all returned counterfactuals individually (in feature space), we provide a visual summary of the Pareto set in Fig. 1, either as a parallel coordinate plot or a response surface plot[3] along two features. All counterfactuals had values equal to or smaller than the values of $\mathbf{x}^*$ for *duration* and *credit amount*. The response surface plot illustrates why these feature changes were recommended. The color gradient and contour lines indicate that either *duration* or both *credit amount* and *duration* must be decreased to reach the desired outcome. Due to the fourth objective and the conditional mutator, we obtained counterfactuals in high density areas (indicated by histograms). Counterfactuals in the lower left corner seem to be in a less favorable region far from $\mathbf{x}^*$, but they are close to the training data.

6 Experimental Setup

In this section, the performance of MOC is evaluated in a benchmark study for binary classification. The datasets are from the OpenML platform [34] and are briefly described in Table 1. We selected datasets with no missing values, with up to 3500 observations and a maximum of 40 features. We randomly selected ten observed data points per dataset as $\mathbf{x}^*$ and excluded them from the training data. For each dataset, we tuned and trained the following models: logistic regression, random forest, xgboost, RBF support vector machine and a

[3] This is equivalent to a 2-D ICE-curve through $\mathbf{x}^*$ [9]. We refer to Sect. 4.3 for a general definition of ICE curves.

Table 1. Description of benchmark datasets. Legend: *task:* OpenML task id; *Obs:* Number of rows; *Cont/Cat:* Number of continuous/categorical features.

Task	Name	Obs	Cont	Cat
3718	boston	506	12	1
3846	cmc	1473	2	7
145976	diabetes	768	8	0
9971	ilpd	583	9	1
3913	kc2	522	21	0
3	kr-vs-kp	3196	0	36
3749	no2	500	7	0
3918	pc1	1109	21	0
3778	plasma_retinol	315	10	3
145804	tic-tac-toe	958	0	9

Table 2. MOC's coverage rate of methods to be compared per dataset averaged over all models. The number of nondominated counterfactuals for each method are given in parentheses. Higher values of coverage indicate that MOC dominates the other method. The * indicates that the binomial test with $H_0 : p < 0.5$ that a counterfactual is covered by MOC is significant at the 0.05 level.

	DiCE	Recourse	Tweaking
boston	1* (36)	0.92* (24)	0.9* (10)
cmc	1* (17)		0.75 (8)
diabetes	1* (64)	0.45 (40)	1 (3)
ilpd	1* (26)	1* (37)	0.83 (6)
kc2	1* (53)	0.31 (55)	1 (2)
kr-vs-kp	1* (8)		0.2 (10)
no2	1* (58)	0.5 (12)	0.9* (10)
pc1	1* (60)	0.66* (38)	
plasma_retinol	1* (7)		0.89* (9)
tic-tac-toe	1* (20)		0.75 (8)

one-hidden-layer neural network. The tuning parameter set and the performance using nested resampling are in Table 8 in Appendix C. Each model returned only the probability for one class. The desired target for each $\mathbf{x}^*$ was set to the opposite of the predicted class:

$$Y' = \begin{cases}]0.5, 1] & \text{if } \hat{f}(\mathbf{x}^*) \leq 0.5 \\ [0, 0.5] & \text{else} \end{cases}.$$

The benchmark study aimed to answer two research questions:

Q1) How does MOC perform compared to other state-of-the-art methods for counterfactuals?
Q2) How do our proposed strategies for initialization and mutation of Sect. 4.3 influence the performance of MOC?

For the first one, we compared MOC – once with and once without our proposed strategies for initialization and mutation – with 'DiCE' by Mothilal et al. [24], 'Recourse' by Ustun et al. [33] and 'Tweaking' by Tolomei et al. [32]. We chose DiCE, Recourse and Tweaking because they are implemented in general open source code libraries.[4] The methods are only applicable to certain models: DiCE can handle neural networks and logistic regressions, Recourse can handle logistic regressions and Tweaking can handle random forests. Since Recourse can only process binary and numerical features, we did not train logistic regression on cmc, tic-tac-toe, kr-vs-kp and plasma_retinol. As a baseline, we selected the

[4] Most other counterfactual methods are implemented for specific examples, but cannot be easily used for other datasets.

closest observed data point to $\mathbf{x}^*$ (according to the Gower distance) that has a prediction equal to our desired outcome. Since this approach is part of the *What-If Tool* [36], we call this approach 'Whatif'.

The parameters of DiCE, Recourse and Tweaking were set to the default values recommended by the authors (Appendix D). To allow for a fair comparison, we initialized MOC with the parameters of iterated F-racing which were tuned on other binary classification datasets (Appendix B). While MOC can potentially return several hundreds of counterfactuals, the other methods are designed to either return one or a few. We have therefore limited the maximum number of counterfactuals to ten for all approaches.[5] Tweaking and Whatif generated only one counterfactual by design. For MOC we reduced the number of counterfactuals by preferring the ones that achieved the target prediction Y' and/or the highest HV contribution.

For all methods, only nondominated counterfactuals were considered for the evaluation. Since we are interested in a diverse set of counterfactuals, we evaluate the methods based on the size of their counterfactual set, its objective values, and the coverage rate derived from the coverage indicator by Zitzler and Thiele [38]. The coverage rate is the relative frequency with which counterfactuals of a method are dominated by MOC's counterfactuals for a certain model and $\mathbf{x}^*$. A counterfactual covers another counterfactual if it dominates it, and it does not cover the other if both have the same objective values or the other has lower values in at least one objective. A coverage rate of 1 implies that for each generated counterfactual of a method MOC generated at least one dominating counterfactual. We only computed the coverage rate over counterfactuals that met the desired target Y'.

To answer the second research question, we compared the dominated HV over the generations of MOC with and without our proposed strategies for initialization and mutation. As a baseline, we used a random search approach that has the same population size (20) and number of generations (175) as MOC. In each generation, some feature values were uniformly sampled from their set of possible values derived from the observed data and $\mathbf{x}^*$, while other features were set to the values of $\mathbf{x}^*$. The HV for one generation was computed over the newly generated candidates combined with the candidates of the previous generations.

7 Results

Q1) MOC vs. State-of-the-Art Counterfactual Methods

Table 2 shows the coverage rate of each method (to be compared) by the tuned MOC per dataset. Some fields are empty because Recourse could not process features with more than two classes and Tweaking never achieved the desired outcome for pc1. MOC's counterfactuals dominated all counterfactuals of DiCE for all datasets. The same holds for Tweaking except for kr-vs-kp and tic-tac-toe because the counterfactuals of Tweaking had the same objective values as

[5] Note that this artificially penalizes our approach in the benchmark comparison.

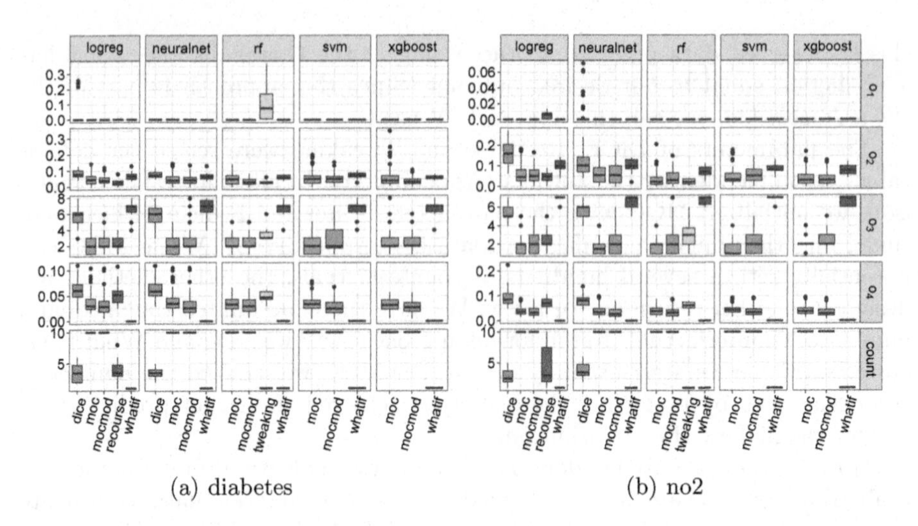

(a) diabetes

(b) no2

Fig. 2. Boxplots of the objective values and number of nondominated counterfactuals (*count*) per model for MOC with our proposed strategies for initialization and mutation (*mocmod*), MOC without these modifications, Whatif, DiCE, Recourse and Tweaking for the datasets diabetes and no2. Lower values are better except for *count*.

the ones of MOC. MOC's coverage rate of Recourse only exceeded 90% for boston and ilpd since Recourse's counterfactuals often deviated less from $\mathbf{x}^*$ (but performed worse in other objectives).

Figure 2 compares MOC (with (*mocmod*) and without (*moc*) our proposed strategies for initialization and mutation) with the other methods for the datasets diabetes and no2 and for each model separately. The resulting boxplots for all other datasets are shown in Figs. 4 and 5 in the Appendix. They agree with the results shown here. Compared to the other methods, both versions of MOC found the most nondominated solutions, which met the target and changed the least features. DiCE performed worse than MOC in all objectives. Tweaking's counterfactuals were often closer to $\mathbf{x}^*$, but they were further away from the nearest training data point and more features were changed. Tweaking's counterfactuals often did not reach the desired outcome because they stayed too close to $\mathbf{x}^*$. The MOC with our proposed modifications found counterfactuals closer to $\mathbf{x}^*$ and the observed data, but required more feature changes compared to MOC without the modifications.

Q2) MOC Strategies for Initialization and Mutation

Figure 3 shows the ranks of the dominated HVs for MOC without modifications, for each modification of MOC and random search. Ranks were calculated per dataset, model, $\mathbf{x}^*$ and generation, and were averaged over all datasets, models and $\mathbf{x}^*$. We transformed HVs to ranks because the HVs are not comparable across $\mathbf{x}^*$. It can be seen that the MOC with our proposed modifications clearly

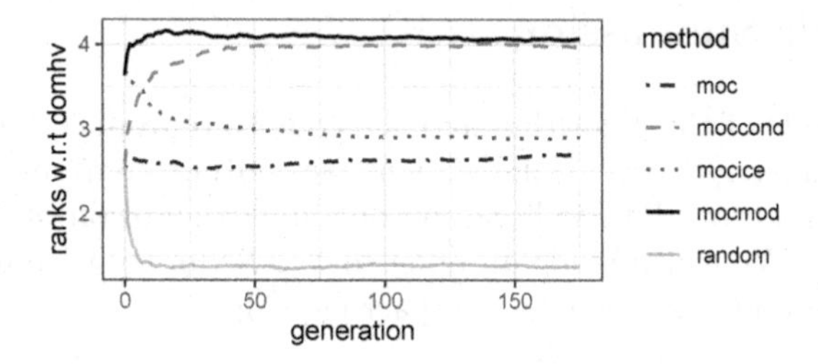

Fig. 3. Comparison of the ranks w.r.t. the dominated HV (*domhv*) per generation averaged over all models and datasets. For each approach, the population size of each generation was 20. A higher HV and therefore a higher rank is better. Legend: *moc*: MOC without our proposed modifications; *moccond*: MOC with the conditional mutator; *mocice*: MOC with the ICE curve variance initialization; *mocmod*: MOC with both modifications; *random*: random search.

outperforms the MOC without these modifications. The ranks of the initial population were higher when the ICE curve variance was used to initialize the candidates. The use of the conditional mutator led to higher dominated HVs over the generations. We received the best performance over the generations when both modifications were used. At each generation, all versions of MOC outperformed random search. Figure 6 in the Appendix shows the ranks over the generations for each dataset separately. They largely agree with the results shown here. The performance gains of MOC compared to random search were particularly evident for higher-dimensional datasets.

8 Conclusion and Outlook

In this paper, we introduced Multi-Objective Counterfactuals (MOC), which to the best of our knowledge is the first method to formalize the counterfactual search as a multi-objective optimization problem. Compared to state-of-the-art approaches, MOC returns a diverse set of counterfactuals with different trade-offs between our proposed objectives. Furthermore, MOC is model-agnostic and suited for classification, regression and mixed feature spaces. We demonstrated the usefulness of MOC to explain a prediction on the German credit dataset and showed in a benchmark study that MOC finds more counterfactuals than other counterfactual methods that are closer to the training data and required fewer feature changes. Our proposed initialization strategy (based on ICE curve variances) and our conditional mutator resulted in higher performance in fewer evaluations and in counterfactuals that were closer to the data point we were interested in and to the observed data.

MOC has only been evaluated on binary classification, and only with respect to the dominated HV and the individual objectives. It is an open question how to let users select the counterfactuals that meet their – a-priori unknown – trade-off between the objectives. We leave these investigations to future research.

9 Electronic Submission

The complete code of the algorithm and the code to reproduce the experiments and results of this paper are available at https://github.com/susanne-207/moc. The implementation of MOC is based on our implementation of [19], which we also used for [3]. We will provide an open source R library with our implementation of the method based on the `iml` package [23].

A Illustration of MOC's Benefits

This section illustrates the benefits of having a *diverse set* of counterfactuals using the diabetes dataset of the benchmark study (Sect. 6). We will compare the counterfactuals returned by MOC with the ones of Recourse [33] and Tweaking [32]. Due to space constraints, we only show the six counterfactuals of MOC with the highest HV contribution for both examples.

Table 3. Counterfactuals and corresponding objective values of MOC and Recourse for the prediction of a logistic regression for observation 741 of the diabetes dataset. Shaded fields indicate values that differ from the value of observation 741 in brackets.

Feature ($\mathbf{x}^*$)	MOC_1	MOC_2	MOC_3	MOC_4	MOC_5	MOC_6	$Recourse_1$	$Recourse_2$	$Recourse_3$
preg (11)	11.00	6.35	11.00	11.00	11.00	6.35	11.00	11.00	10.92
plas (120)	27.78	3.29	79.75	94.85	79.75	3.18	57.00	57.00	57.00
pres (80)	80.00	80.00	80.00	80.00	80.00	80.00	80.00	80.00	80.00
skin (37)	37.00	37.00	37.00	37.00	37.00	37.00	37.00	36.81	37.00
insu (150)	150.00	150.00	17.13	150.00	40.61	150.00	150.00	150.00	150.00
mass (42.3)	42.30	42.30	29.17	15.36	29.17	42.30	42.30	42.30	42.30
pedi (0.78)	0.78	0.78	0.31	0.78	0.17	0.78	0.78	0.78	0.78
age (48)	48.00	41.61	44.42	48.00	48.00	48.00	28.36	28.36	28.36
o_1	0.00	0.00	0.00	0.00	0.00	0.00	0.00	0.00	0.00
o_2	0.06	0.12	0.10	0.07	0.10	0.11	0.08	0.08	0.08
o_3	1.00	3.00	5.00	2.00	4.00	2.00	2.00	3.00	3.00
o_4	0.10	0.05	0.03	0.07	0.04	0.07	0.09	0.09	0.09

Table 3 contrasts MOC's counterfactuals with the three counterfactuals of Recourse for the prediction of observation 741. A logistic regression predicted a probability of having diabetes of 0.89 for this observation. The desired target is a prediction of less than 0.5, which indicates having no diabetes. All counterfactuals of Recourse suggest the same reduction in *age* and plasma concentration (*plas*), with two counterfactuals additionally suggesting a minimal reduction in the number of pregnancies (*preg*) or the skin fold thickness (*skin*).[6] Apart from that a reduction in *age* or *preg* is impossible, they do not offer many options

[6] By reclassifying *age* and *preg* as integers (instead of decimals), integer changes would be recommended by MOC, Recourse and Tweaking.

Table 4. Counterfactuals and corresponding objective values given by MOC and Tweaking for the prediction of a random forest for observation 268 of the cmc dataset. Shaded fields indicate values that differ from the value of observation 268 in brackets.

Feature ($\mathbf{x}^*$)	MOC_1	MOC_2	MOC_3	MOC_4	MOC_5	MOC_6	$Tweaking_1$
preg (2)	2.00	2.00	2.00	2.00	2.00	2.00	1.53
plas (128)	121.50	90.21	126.83	128.00	88.44	120.64	119.71
pres (64)	64.00	64.00	64.00	64.00	64.00	64.00	64.00
skin (42)	42.00	42.00	42.00	42.00	42.00	42.00	42.00
insu (0)	0.00	0.00	0.00	0.00	0.00	90.93	0.00
mass (40)	40.00	40.00	40.00	40.00	40.00	40.00	40.00
pedi (1.1)	1.10	0.48	1.10	0.17	0.46	1.10	1.10
age (24)	24.00	24.00	24.00	24.00	25.85	24.00	28.29
o_1	0.00	0.00	0.00	0.00	0.00	0.00	0.00
o_2	0.00	0.06	0.00	0.05	0.06	0.02	0.02
o_3	1.00	2.00	1.00	1.00	3.00	2.00	3.00
o_4	0.05	0.02	0.05	0.04	0.01	0.03	0.06

for users. Instead, MOC returned a larger set of counterfactuals that provide more options for actionable user responses and are closer to the observed data than Recourse's counterfactuals (o_4). Counterfactual MOC_1 has overall lower objective values than all counterfactuals of Recourse. MOC_3 suggested changes to five features so that it is especially close to the nearest training data point (o_4).

Table 4 compares the set of counterfactuals found by MOC with the single counterfactual found by Tweaking for the prediction of observation 268. A random forest classifier predicted a probability of having diabetes of 0.62 for this observation. Again, the desired target is a prediction of less than 0.5. Tweaking suggested reducing the number of children and plasma glucose concentration (*plas*) while increasing the *age* so that the probability of diabetes decreases. This is contradictory and not plausible. In contrast, MOC's counterfactuals suggest various strategies, e.g., only a decrease of *plas*, which is easier to realize. In addition, MOC_1, MOC_3 and MOC_6 dominate the counterfactual of Tweaking. Since five of six counterfactuals suggest changes to *plas*, the user may have more confidence that *plas* is an important lever to achieve the desired outcome.

B Iterated F-racing

We used iterated F-racing (irace) [21] to tune the parameters of MOC for binary classification. The parameters and considered ranges are given in Table 5. The number of generations was not part of the parameter set because it would be always tuned to the upper bound. Instead, the number of generations was determined after the other parameters were tuned with irace. Irace was initialized with a maximum budget of 3000 evaluations equal to 3000 runs of MOC. In every step, irace randomly selected one of 300 instances. Each instance consisted of a trained model, a randomly selected data point from the observed data as $\mathbf{x}^*$

Table 5. Parameter space investigated with iterated F-racing, as well as the resulting optimized configuration (*Result*).

Name	Description	Range	Result
M	Population size	[20, 100]	20
initialization	Initialization strategy	[Random, ICE curve]	ICE curve
conditional	Whether to use the conditional mutator	[TRUE, FALSE]	TRUE
p.rec	Probability a pair of parents is chosen to recombine	[0.3, 1]	0.57
p.rec.gen	Probability a feature is recombined	[0.3, 1]	0.85
p.rec.use.orig	Probability the indicator for feature changes is recombined	[0.3, 1]	0.88
p.mut	Probability a child is chosen to be mutated	[0.05, 0.8]	0.79
p.mut.gen	Probability one feature is mutated	[0.05, 0.8]	0.56
p.mut.use.orig	Probability indicator for a feature change is flipped	[0.05, 0.5]	0.32

and a desired outcome. The desired target for each $\mathbf{x}^*$ was the opposite of the predicted class:

$$Y' = \begin{cases}]0.5, 1] & \text{if } \hat{f}(\mathbf{x}^*) \leq 0.5 \\ [0, 0.5] & \text{else} \end{cases}.$$

The trained model was either logistic regression, random forest, xgboost, RBF support vector machine or a two-hidden-layer neural network. Each model estimated only the probability for one class. The models were trained on datasets obtained from the OpenML platform [34] (without the sampled $\mathbf{x}^*$) and are briefly described in Table 7. While these datasets were not used in the benchmark study (Sect. 6), the same preprocessing steps were conducted and the models were tuned with the same setup (see Sect. C for details).

In each step of irace, parameter configurations were evaluated by running MOC on the same selected instance. MOC stopped after evaluating 8000 candidates with Eq. (1), which should be enough to ensure convergence of the HV in most cases. The integral of the first order spline approximation of the dominated HV over the evaluations was the performance criterion as recommended by [26]. The integral takes into account not only the extent but also the rate of convergence of the dominated HV. A Friedman test was used to discard less promising configurations. The first Friedman test was conducted after initial configurations were evaluated on 15 instances; afterward, the test was conducted after evaluating the remaining configurations on a single instance to accelerate the exclusion process. The best configuration returned is given in Table 5.

To obtain a default parameter for the number of generations for the benchmark study, we determined for the 300 instances after how many generations of the tuned MOC the dominated HV has not increased for 10 generations. We chose the maximum of 175 generations as a default for the study.

Table 6. Tuning search space per model. The hyperparameters *ntrees* and *nrounds* were log-transformed.

Model	Hyperparameter	Range
randomforest	ntrees	[0, 1000]
xgboost	nrounds	[0, 1000]
svm	cost	[0.01, 1]
logreg	lr	[0.0005, 0.1]
neuralnet	lr	[0.0005, 0.1]
	layer_size	[1, 6]

Table 7. Description of datasets for tuning with iterated F-racing. Legend: *Task:* OpenML task id; *Obs:* Number of rows; *Cont/Cat:* Number of continuous/categorical features.

Task	Name	Obs	Cont	Cat
3818	tae	151	3	2
3917	kc1	2109	21	0
52945	breastTumor	277	0	6
3483	mammography	11183	6	0
3822	nursery	12960	0	8
3586	abalone	4177	7	1

C Model Hyperparameters for the Benchmark Study

We used random search (with 200 iterations for neural networks and 100 iterations for all other models) and 5-fold CV (with misclassification error as performance measure) to tune the hyperparameters of the models on the training data. The tuning search space was the same as for iterated F-racing and is shown in Table 6. Numerical features were scaled (standardization (Z-score) for random forest, min-max-scaling (0–1-range) for all other models) and categorical features were one-hot encoded. For neural network and logistic regression, ADAM [17] was the optimizer, the batch size was 32 with a 1/3 validation split and early stopping was conducted after 5 patience steps. Logistic regression needed these configurations because we constructed the model as a zero-hidden-layer neural network. For all other hyperparameters of the models, we chose the default values of the `mlr` [4] and `keras` [1] R packages. Table 8 shows the accuracies of the trained models using nested resampling (5-fold CV in outer and inner loop).

Table 8. Accuracy using nested resampling per benchmark dataset and model. Legend: *Name:* OpenML task name; *rf:* random forest. Logistic regression (*logreg*) was only trained on datasets with numerical or binary features.

Name	rf	xgboost	svm	logreg	neuralnet
boston	0.90	0.89	0.87	0.86	0.87
cmc	0.70	0.72	0.67		0.68
diabetes	0.76	0.74	0.75	0.63	0.68
ilpd	0.69	0.67	0.65	0.53	0.58
kc2	0.81	0.80	0.79	0.75	0.72
kr-vs-kp	0.99	0.99	0.97		0.99
no2	0.63	0.59	0.58	0.55	0.54
pc1	0.93	0.93	0.91	0.91	0.88
plasma_retinol	0.53	0.52	0.58		0.55
tic-tac-toe	0.99	0.99	0.98		0.97

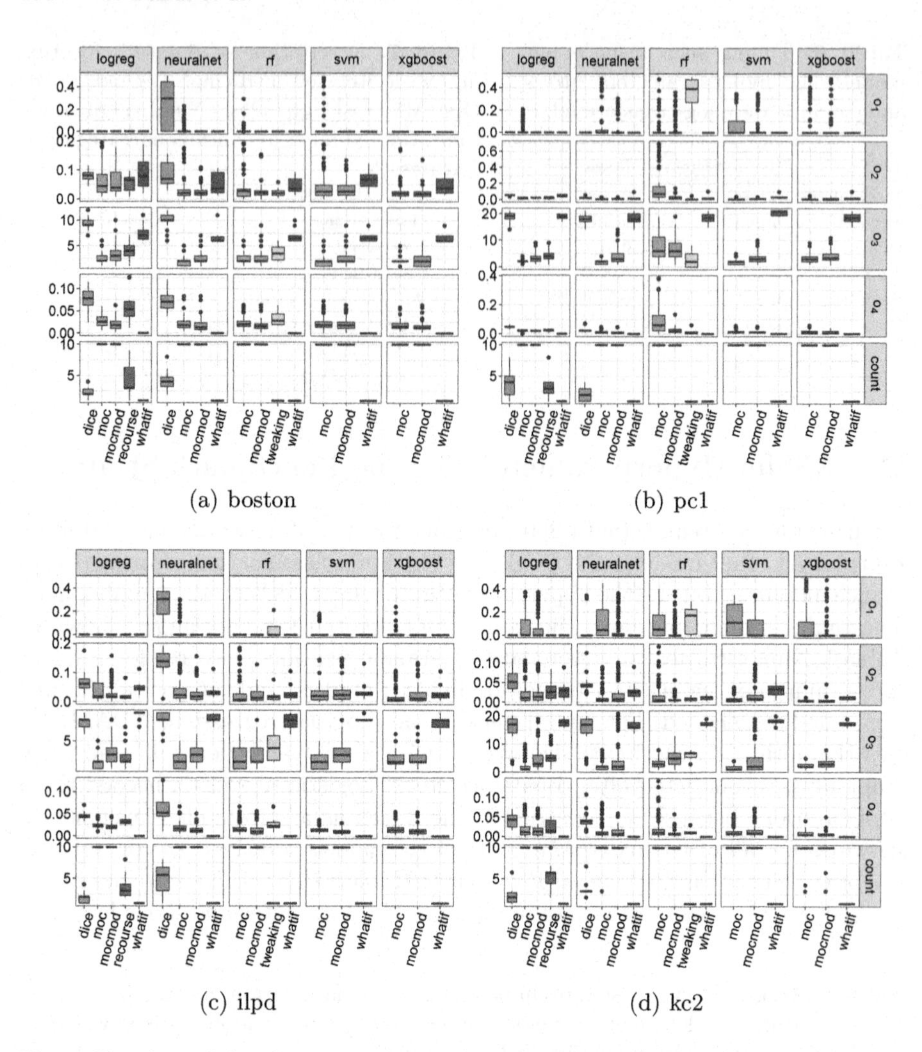

(a) boston (b) pc1

(c) ilpd (d) kc2

Fig. 4. Boxplots of the objective values and number of nondominated counterfactuals (*count*) per dataset and model for MOC with our proposed strategies for initialization and mutation (*mocmod*), MOC without these modifications, Whatif, DiCE, Recourse and Tweaking. Lower values are better except for *count*.

D Control Parameters of Counterfactual Methods

For Tweaking [32], we only changed ϵ, a positive threshold that limits the tweaking of each feature. It was set to 0.5 because it obtained better results for the authors on their data example on Ad Quality in comparison to the default value 0.1. We used the R implementation of Tweaking on Github: https://github.com/katokohaku/featureTweakR (commit `6f3e614`). For Recourse [33], we left all parameters at their default settings. We used the Python implementation of Recourse on Github: https://github.com/ustunb/actionable-recourse (com-

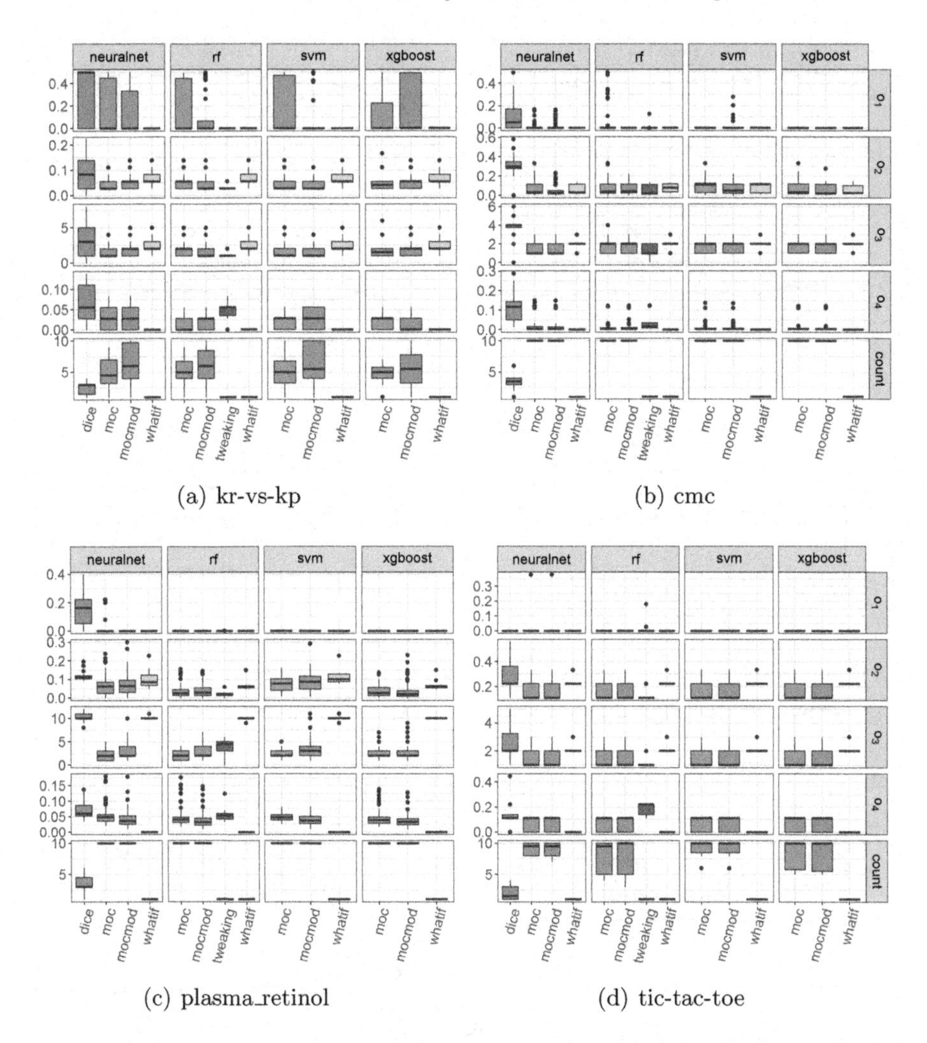

(a) kr-vs-kp

(b) cmc

(c) plasma_retinol

(d) tic-tac-toe

Fig. 5. Boxplots of the objective values and number of nondominated counterfactuals (*count*) per dataset and model for MOC with our proposed strategies for initialization and mutation (*mocmod*), MOC without these modifications, Whatif, DiCE, Recourse and Tweaking. Lower values are better except for *count*.

mit `aaae8fa`). For DiCE [24], we used the 'DiverseCF' version proposed by the authors [24] and left the control parameters at their defaults. We used the inverse mean absolute deviation for the feature weights. For datasets where the mean absolute deviation of a feature was zero, we set the feature weight to 10. We used the Python implementation of DiCE available on Github: https://github.com/microsoft/DiCE (commit `fed9d27`).

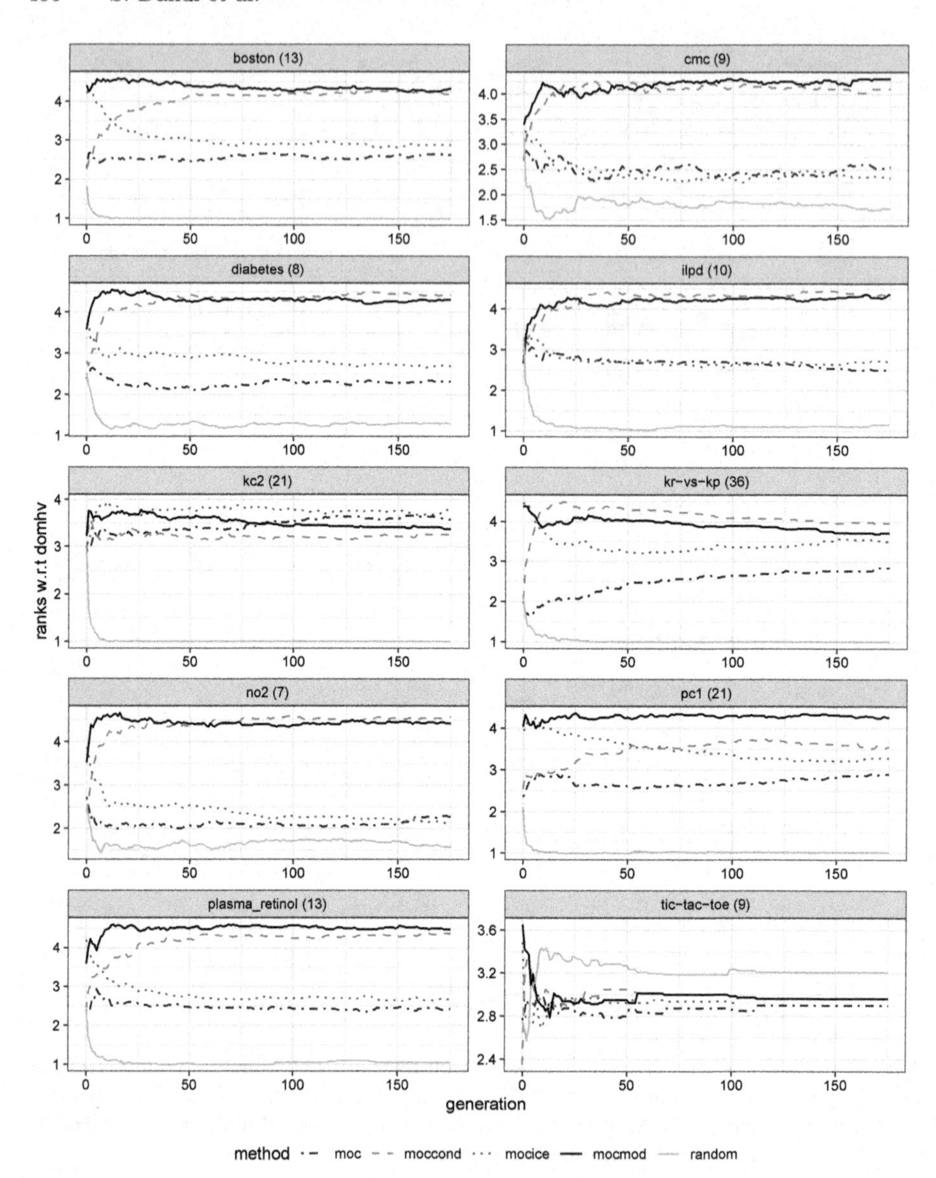

Fig. 6. Comparison of the ranks w.r.t. the dominated HV (*domhv*) per generation and per benchmark dataset averaged over all models. The numbers in parentheses indicate the number of features. For each approach, the population size of each generation was 20. Higher ranks are better. Legend: *moc*: MOC without modifications; *moccond*: MOC with the conditional mutator; *mocice*: MOC with the ICE curve variance initialization; *mocmod*: MOC with both modifications; *random*: random search.

References

1. Allaire, J., Chollet, F.: keras: R Interface to 'Keras' (2019). https://keras.rstudio.com, R package version 2.3.0
2. Avila, S.L., Krähenbühl, L., Sareni, B.: A multi-niching multi-objective genetic algorithm for solving complex multimodal problems. In: OIPE. Sorrento, Italy (2006). https://hal.archives-ouvertes.fr/hal-00398660
3. Binder, M., Moosbauer, J., Thomas, J., Bischl, B.: Multi-Objective Hyperparameter Tuning and Feature Selection using Filter Ensembles (2019). Accepted at GECCO 2020
4. Bischl, B., et al.: mlr: Machine Learning in R. J. Mach. Learn. Res. **17**(170), 1–5 (2016). http://jmlr.org/papers/v17/15-066.html, R package version 2.17
5. Breiman, L.: Statistical modeling: the two cultures. Stat. Sci. **16**(3), 199–231 (2001). https://doi.org/10.1214/ss/1009213726
6. Deb, K., Agarwal, R.B.: Simulated binary crossover for continuous search space. Complex Syst. **9**, 115–148 (1995)
7. Deb, K., Pratap, A., Agarwal, S., Meyarivan, T.: A fast and elitist multiobjective genetic algorithm: NSGA-II. IEEE Trans. Evol. Comput. **6**(2), 182–197 (2002). https://doi.org/10.1109/4235.996017
8. Dhurandhar, A., Pedapati, T., Balakrishnan, A., Chen, P., Shanmugam, K., Puri, R.: Model Agnostic Contrastive Explanations for Structured Data. CoRR abs/1906.00117 (2019). http://arxiv.org/abs/1906.00117
9. Goldstein, A., Kapelner, A., Bleich, J., Pitkin, E.: Peeking inside the black box: visualizing statistical learning with plots of individual conditional expectation. J. Comput. Graph. Stat. **24**(1), 44–65 (2015). https://doi.org/10.1080/10618600.2014.907095
10. Gower, J.C.: A general coefficient of similarity and some of its properties. Biometrics **27**(4), 857–871 (1971)
11. Grath, R.M., et al.: Interpretable Credit Application Predictions With Counterfactual Explanations. CoRR (abs/1811.05245) (2018). http://arxiv.org/abs/1811.05245
12. Greenwell, B.M., Boehmke, B.C., McCarthy, A.J.: A simple and effective model-based variable importance measure. arXiv preprint arXiv:1805.04755 (2018)
13. Hofmann, H.: German Credit Risk (2016). https://www.kaggle.com/uciml/german-credit. Accessed 25 Jan 2020
14. Hothorn, T., Zeileis, A.: Transformation Forests (2017)
15. Joshi, S., Koyejo, O., Vijitbenjaronk, W., Kim, B., Ghosh, J.: Towards Realistic Individual Recourse and Actionable Explanations in black-box decision making systems. CoRR abs/1907.09615 (2019). http://arxiv.org/abs/1907.09615
16. Karimi, A., Barthe, G., Balle, B., Valera, I.: Model-Agnostic Counterfactual Explanations for Consequential Decisions. CoRR (abs/1905.11190) (2019). http://arxiv.org/abs/1905.11190
17. Kingma, D., Ba, J.: Adam: a method for stochastic optimization. In: International Conference on Learning Representations, December 2014
18. Laugel, T., Lesot, M.J., Marsala, C., Renard, X., Detyniecki, M.: Comparison-Based Inverse Classification for Interpretability in Machine Learning. CoRR (abs/1712.08443) (2017). http://arxiv.org/abs/1712.08443
19. Li, R., et al.: Mixed integer evolution strategies for parameter optimization. Evol. Comput. **21**(1), 29–64 (2013)

20. Looveren, A.V., Klaise, J.: Interpretable Counterfactual Explanations Guided by Prototypes. CoRR abs/1907.02584 (2019). http://arxiv.org/abs/1907.02584
21. López-Ibáñez, M., Dubois-Lacoste, J., Cáceres, L.P., Birattari, M., Stützle, T.: The irace package: iterated racing for automatic algorithm configuration. Oper. Res. Perspect. **3**, 43–58 (2016). https://doi.org/10.1016/j.orp.2016.09.002, http://www.sciencedirect.com/science/article/pii/S2214716015300270, R package version 3.4.1
22. Lundberg, S.M., Lee, S.I.: A unified approach to interpreting model predictions. In: Advances in Neural Information Processing Systems, pp. 4765–4774 (2017)
23. Molnar, C., Bischl, B., Casalicchio, G.: iml: an R package for interpretable machine learning. JOSS **3**(26), 786 (2018). https://doi.org/10.21105/joss.00786
24. Mothilal, R.K., Sharma, A., Tan, C.: Explaining Machine Learning Classifiers through Diverse Counterfactual explanations. CoRR (abs/1905.07697) (2019). http://arxiv.org/abs/1905.07697
25. Poyiadzi, R., Sokol, K., Santos-Rodriguez, R., Bie, T.D., Flach, P.: FACE: Feasible and Actionable Counterfactual Explanations (2019)
26. Radulescu, A., López-Ibáñez, M., Stützle, T.: Automatically improving the anytime behaviour of multiobjective evolutionary algorithms. In: Purshouse, R.C., Fleming, P.J., Fonseca, C.M., Greco, S., Shaw, J. (eds.) EMO 2013. LNCS, vol. 7811, pp. 825–840. Springer, Heidelberg (2013). https://doi.org/10.1007/978-3-642-37140-0_61
27. Ribeiro, M.T., Singh, S., Guestrin, C.: "Why should i trust you?" Explaining the predictions of any classifier. In: Proceedings of the 22nd ACM SIGKDD International Conference on Knowledge Discovery and Data Mining, pp. 1135–1144 (2016)
28. Russell, C.: Efficient Search for Diverse Coherent Explanations. CoRR (abs/1901.04909) (2019). http://arxiv.org/abs/1901.04909
29. Sharma, S., Henderson, J., Ghosh, J.: CERTIFAI: Counterfactual Explanations for Robustness, Transparency, Interpretability, and Fairness of Artificial Intelligence models. CoRR abs/1905.07857 (2019). http://arxiv.org/abs/1905.07857
30. Su, J., Vargas, D.V., Sakurai, K.: One pixel attack for fooling deep neural networks. IEEE Trans. Evol. Comput. **23**, 828–841 (2017)
31. Syswerda, G.: Uniform crossover in genetic algorithms. In: Proceedings of the 3rd International Conference on Genetic Algorithms, pp. 2–9. Morgan Kaufmann Publishers Inc., San Francisco (1989)
32. Tolomei, G., Silvestri, F., Haines, A., Lalmas, M.: Interpretable predictions of tree-based ensembles via actionable feature tweaking. In: Proceedings of the 23rd ACM SIGKDD International Conference on Knowledge Discovery and Data Mining, KDD 2017, pp. 465–474. ACM, New York (2017). https://doi.org/10.1145/3097983.3098039
33. Ustun, B., Spangher, A., Liu, Y.: Actionable recourse in linear classification. In: Proceedings of the Conference on Fairness, Accountability, and Transparency, FAT* 2019, pp. 10–19. ACM, New York (2019). https://doi.org/10.1145/3287560.3287566
34. Vanschoren, J., van Rijn, J.N., Bischl, B., Torgo, L.: OpenML: networked science in machine learning. SIGKDD Explor. **15**(2), 49–60 (2013). https://doi.org/10.1145/2641190.2641198
35. Wachter, S., Mittelstadt, B.D., Russell, C.: Counterfactual Explanations without Opening the Black Box: Automated Decisions and the GDPR. CoRR (abs/1711.00399) (2017). http://arxiv.org/abs/1711.00399
36. Wexler, J., Pushkarna, M., Bolukbasi, T., Wattenberg, M., Viégas, F.B., Wilson, J.: The What- If Tool: Interactive Probing of Machine Learning Models. CoRR abs/1907.04135 (2019). http://arxiv.org/abs/1907.04135

37. White, A., d'Avila Garcez, A.: Measurable Counterfactual Local Explanations for Any Classifier (2019)
38. Zitzler, E., Thiele, L.: Multiobjective optimization using evolutionary algorithms—a comparative case study. In: Eiben, A.E., Bäck, T., Schoenauer, M., Schwefel, H.-P. (eds.) PPSN 1998. LNCS, vol. 1498, pp. 292–301. Springer, Heidelberg (1998). https://doi.org/10.1007/BFb0056872

Multi-objective Magnitude-Based Pruning for Latency-Aware Deep Neural Network Compression

Wenjing Hong[1,2,3], Peng Yang[1], Yiwen Wang[4], and Ke Tang[1(✉)]

[1] Guangdong Provincial Key Laboratory of Brain-Inspired Intelligent Computation, Department of Computer Science and Engineering, Southern University of Science and Technology, Shenzhen 518055, China
{hongwj,yangp,tangk3}@sustech.edu.cn

[2] Department of Management Science, University of Science and Technology of China, Hefei 230027, China

[3] Guangdong-Hong Kong-Macao Greater Bay Area Center for Brain Science and Brain-Inspired Intelligence, Guangzhou 510515, China

[4] Department of Electronic and Computer Engineering, Department of Chemical and Biological Engineering, Hong Kong University of Science and Technology, Hong Kong, China
eewangyw@ust.hk

Abstract. Layer-wise magnitude-based pruning is a popular method for Deep Neural Network (DNN) compression. It has the potential to reduce the latency for an inference made by a DNN by pruning connects in the network, which prompts the application of DNNs to tasks with real-time operation requirements, such as self-driving vehicles, video detection and tracking. However, previous methods mainly use the compression rate as a proxy for the latency, without explicitly accounting for latency in the training of the compressed network. This paper presents a new layer-wise magnitude-based pruning method, namely Multi-objective Magnitude-based Latency-Aware Pruning (MMLAP). MMLAP captures latency directly and incorporates a novel multi-objective evolutionary algorithm to optimize both accuracy of a DNN and its latency efficiency when designing compressed networks, i.e., when tuning hyper-parameters of LMP. Empirical studies show the competitiveness of MMLAP compared to well-established LMP methods and show the value of multi-objective optimization in yielding Pareto-optimal compressed networks in terms of accuracy and latency.

This work was supported in part by the National Key Research and Development Program of China under Grant 2017YFB1003102, the Natural Science Foundation of China under Grant 61672478 and Grant 61806090, the Guangdong Provincial Key Laboratory under Grant 2020B121201001, the Shenzhen Peacock Plan under Grant KQTD2016112514355531, the Guangdong-Hong Kong-Macao Greater Bay Area Center for Brain Science and Brain-Inspired Intelligence Fund (NO. 2019028), and the National Leading Youth Talent Support Program of China.

T. Bäck et al. (Eds.): PPSN 2020, LNCS 12269, pp. 470–483, 2020.
https://doi.org/10.1007/978-3-030-58112-1_32

Keywords: Compression · Magnitude-based pruning · Latency-aware · Multi-objective optimization · Multi-objective evolutionary algorithm

1 Introduction

Deep Neural Networks (DNNs) have demonstrated impressive performance on various machine learning tasks such as visual recognition [7], knowledge search [27] and natural language processing [14]. However, modern DNNs often contain millions of parameters or hundreds of layers, which can lead to prohibitive computational costs [5,13]. Thus, the development of algorithms for reducing computational costs of DNN inference is therefore essential for applications with high real-time requirements, such as video detection and tracking [3], and self-driving vehicles [16]. For example, it is immensely critical to minimize the latency for an inference made by accident detection networks on self-driving vehicles, while maintaining an acceptable level of task performance (e.g., accuracy). Hence, a variety of approaches have been developed in recent years to compress pre-trained networks into sparse networks [21,30,31].

Layer-wise Magnitude-based Pruning (LMP) is a powerful DNN compression approach that has achieved significant success in many applications [18,28]. It has the potential to bring a computationally expensive DNN to latency-critical applications by removing redundant connections in the network. To be more specific, LMP prunes connections in each layer by removing connections with absolute values lower than a layer-specific threshold. Therefore, it is possible for LMP approaches to achieve rapid DNN compression simply by assigning a threshold value to each layer, even for DNNs with enormous parameters.

However, previous work on LMP mainly uses the compression rate as a proxy for the latency metric, without explicitly accounting for the real latency performance of a DNN deployed in a specific hardware [18,22]. Such practice is shown to be unable to yield the latency-optimal DNN compression [20,25]. A similar phenomenon can also be observed in our experiments in Sect. 4. Hence, to further improve the latency performance, it is critical to explicitly consider latency when designing network pruning methods.

In this paper, we propose a new LMP approach by directly incorporating latency into the training for DNN compression. It is worthy noting that the design of latency-aware LMP is non-trivial for three reasons. First, the latency performance is hardware-related and relies heavily on various aspects including memory management mechanisms and compute mode, thus may not be expressible in closed form and could be sensitive to hardware environment. Hence, unlike proxies such as the compression ratio, to obtain the latency performance of a pruned DNN, the model needs to be actually deployed on the given hardware platform, imposing higher requirements on engineering capabilities. Second, due to the complexity of operating environment on real-world hardware platforms, the latency may not be evaluated precisely, which brings challenges to optimization algorithms. Third, not only the latency performance is important, but also the accuracy of the obtained pruned network determines whether a network can be used for task solving. Nevertheless, latency and accuracy usually conflict, and thus there exists no single network that can be both latency-optimal and accuracy-optimal. To solve these issues, a multi-objective optimization-based framework for latency-aware LMP, namely Multi-objective Magnitude-based Latency-Aware Pruning (MMLAP), is proposed. Evolutionary algorithms that are shown to be effective for optimization problems that are typically not differentiable and with noise [18,23] are adopted. Furthermore, considering the expensive computation costs of network retraining, an efficient instantiation of MMLAP based on memetic algorithms and an iterative pruning and adjusting strategy is developed. The central idea is to achieve a dedicated MMLAP algorithm for DNN compression by combining fast heuristic methods with powerful multi-objective evolutionary algorithms.

Our contributions are summarized as follows.

1. A novel MMLAP framework for optimizing both latency and accuracy is proposed for DNN compression. To the best of our knowledge, this is the first pruning method to directly minimize the latency and also the first multi-objective optimization-based framework for latency-aware LMP in the specialized literature.
2. An efficient MMLAP approach based on memetic algorithms is developed. It employs multi-objective evolutionary algorithms for Pareto-optimal networks, and further improves search efficiency by integrating LMP heuristics delicately.
3. Experimental studies show that the instantiated MMLAP approach can not only achieve the best performance in terms of both latency and accuracy compared with state-of-the-art LMP methods, but also exhibit more flexibility by offering a set of networks with different trade-offs.

The remainder of this paper is organized as follows. Section 2 reviews related work. Section 3 details the proposed MMLAP framework as well as its instantiation based on memetic algorithms. The experimental studies are presented in Sect. 4. Section 5 finally concludes this article.

2 Related Work

With the immense application potential in real-world scenarios with high real-time requirements, network pruning for DNN compression has become a research hotspot [9,17,31]. Current pruning methods can be classified into two broad categories: connection pruning [8,18] and neuron pruning [19,32]. The former prunes networks at element-wise granularity, often referred to as fine-grained pruning, while the latter is often referred to as coarse-grained pruning, which produce denser but larger networks than the former. In this work, we focus on connection pruning.

LMP is a popular connection pruning method that removes the connections in each layer with absolute values lower than a layer-specific threshold, with the intuition that these connections are likely to have less impact on accuracy but increase the computational complexity [8,10,26,28]. By taking the advantage of the multi-layer nature of DNNs, LMP can achieve a rapid pruning, especially for DNNs with millions or billions connections [18]. Thus, it has been widely studied recent years and have achieved significant results in many applications [31]. Early works of LMP are mainly handcrafted and tune the layer-specific thresholds by experience, e.g., iterative pruning and retraining [10], dynamic network surgery [8], and layer-wise optimal brain surgeon [6]. In recent years, some optimization-based methods are proposed to implement automated LMP, e.g., optimization-based LMP [18]. Nevertheless, these methods mainly focus on achieving sparse networks in terms of the compression rate, ignoring the real latency performance of a network. As what will be shown in our experiments, such methods could hardly achieve latency-optimal networks. As a consequence, it is difficult for these methods to guarantee that the latency for an inference made by the pruned network meets application requirements.

Latency-aware LMP has rarely been considered before [20]. The only work in this direction that we found in the specialized literature is the constraint-aware LMP method presented in [2] which incorporates latency directly in the pruning process. However, it assumes that the latency constraint value should be feasible given the pre-trained network and compression algorithm, which can be hardly satisfied in real-world applications. In contrast, it is more realistic to require a minimal task performance (e.g., accuracy in a classification task), because an acceptable task performance is a prerequisite for a DNN to be adopted. Thus, in this work, we propose to directly minimize the latency during the DNN pruning, while preserving accuracy of the network as closely as possible.

3 The Proposed Approach

Although directly capturing latency brings in advantages when designing LMP for tasks with high real-time requirements, it also leads to a challenging problem. That is, the conflict between latency and accuracy results in that no single compressed network can achieve optimal accuracy and optimal latency simultaneously. Therefore, optimizing for both accuracy and latency is crucial when designing latency-aware LMP, and the complexity in measuring the latency performance poses further challenges to the design of the optimization algorithm. In this section, we present the proposed multi-objective optimization-based latency-aware LMP, namely MMLAP. The MMLAP framework and an efficient instantiation of MMLAP based on memetic algorithms are presented in the following, respectively.

3.1 MMLAP Framework

Given a pre-trained DNN, the problem of MMLAP is to minimize the latency, while preserving the original accuracy of the network as closely as possible. More specifically, suppose a pre-trained DNN W_0 with $(L + 1)$ layers, i.e.,

$W_0 = \{W_{i,j}^l | W_{i,j}^l \neq 0, l \in [1, L], i \in [1, n_l], j \in [1, n_{l+1}]\}$, where $W_{i,j}^l$ denotes the connection weight between the i-th neuron in layer l and the j-th neuron in layer $(l+1)$, n_l denotes the number of neurons in layer l, and $[a, b]$ denotes the set of integers between a and b. Let $\Phi(W_0, \phi)$ denote the network compression algorithm and $W = \Phi(W_0, \phi)$ denote the compressed network, where ϕ denotes a vector specifying the magnitude thresholds for pruning each layer in the network. Here, $W_{i,j}^l = 0$ indicates that the corresponding connection is pruned. Let $f_1(W)$ denote the evaluation function measuring the accuracy of network W and let $f_2(W)$ denote the evaluation function measuring the latency of network W. The MMLAP problem is defined as

$$\begin{aligned} \min F(W) = \{f_1(W), f_2(W)\} \\ s.t. f_1(W_0) - f_1(W) \leq \delta, \end{aligned} \tag{1}$$

where δ denotes the accuracy constraint.

This problem poses several challenges to optimization algorithms: 1) the latency performance is hardware-related and relies heavily on various aspects including memory management mechanisms and compute mode, thus may not be expressible in closed form and may not be easily differentiable, 2) the actual latency could be sensitive to the current state of the deployed hardware, thus may leading to noisy evaluation results, 3) the conflict between latency and accuracy makes it difficult to achieve an effective network for real-world applications.

To approach the above challenges, multi-objective evolutionary optimization is employed in this work. Multi-objective evolutionary optimization provides a general framework for optimizing black-box multi-objective optimization problems that are typically not differentiable and may not be evaluated precisely [11,24], thus may be promising for network LMP. In problem (1), multi-objective evolutionary optimization is employed to search for a set of pruning vectors that produce a set of Pareto-optimal compressed networks in terms of latency and accuracy. The search process maintains a population of pruning vectors ϕ and iteratively updates the population considering both latency and accuracy of the compressed networks. At each generation, one or more new pruning vectors are generated to produce new compressed networks. The new networks are then compared with the networks produced in the previous generation, and the worse networks will be discarded according to a multi-objective selection strategy. The search terminates when a predefined halting condition is satisfied.

3.2 An Instantiation of MMLAP

In this section, an instantiation of MMLAP for DNN compression is presented to illustrate the detailed steps of a MMLAP algorithm.

A key issue in designing a MMLAP for DNN compression is its efficiency. On one hand, multi-objective evolutionary optimization needs to search the whole objective space spanned by accuracy and latency, and thus its convergence speed may be sacrificed to some extent for diversity across different objectives [33]. On the other hand, retraining compressed networks over the whole data set in the iterative search process may be prohibitively expensive.

To overcome the above issues, an efficient memetic MMLAP algorithm is proposed for DNN compression. The main idea is to incorporate fast heuristic algorithms into multi-objective evolutionary optimization and estimate the objectives using a small set of the data set. The algorithm contains three key components: (1) a local search based on weighted Negatively Correlated Search (NCS) [29], (2) a global search based on non-dominated sorting and indicator-based search, and (3) an iterative pruning and adjusting strategy.

Firstly, the algorithm is a hybrid of local search and global search procedures. The global search considers the overall performance over different objectives and the local search focuses on a local improvement towards a specified objective direction. For the global search, the non-dominated sorting [4] and indicator-based search [12] commonly used in multi-objective evolutionary optimization are adopted. In each generation, the population is first ranked into front layers based on the non-dominated sorting mechanism and then sorted based on the hypervolume contribution [1] in the same front layer. The worst individual in the population is then compared with a new individual generated by SBX crossover and polynomial mutation, which are commonly used in continuous optimization [4]. The better one between these two individuals in terms of dominance relation will be reserved into the next generation. For the local search, NCS is a heuristic algorithm that has shown to be effective and efficient in pruning for DNN compression [18]. Here, each individual in the population runs a separate NCS and each NCS is modified to search for a latency-optimal compressed network while satisfying a weighted accuracy constraint. Note that, the initial thresholds are set conservatively so that there would be some individuals feasible for the accuracy constraint. When all individuals in the current population finish the NCS search process, the local search process terminates. Algorithm 1 presents the details of this memetic multi-objective evolutionary algorithm. Generally, the algorithm maintains a population of pruning vectors. The population is randomly initialized, and then the above global and local search strategies are adopted alternately until a predefined stopping condition is satisfied.

Secondly, a recently proposed iterative pruning and adjusting strategy [18] is integrated to improve the training efficiency. Each iteration consists of pruning followed by adjusting, allowing the remaining connections to learn to compensate for the pruning loss. This approach has been shown to effectively reduce the number of epochs required for retraining [8,18]. More specifically, the whole data set is split into multiple small sets, and then, the pruning phase is carried out on only one batch of the split data set and the adjusting phase utilizes the rest batches to recover the accuracy of compressed networks. This pruning and adjusting strategy as well as the above memetic multi-objective evolutionary algorithm are integrated into MMLAP.

The flowchart of the resultant MMLAP for DNN compression is shown in Fig. 1. Concretely, the whole data set D is split into $|D|/K$ small sets, where $|D|$ denotes the size of D and each set contains K batches. In each iteration, one

Algorithm 1: An instantiation of MMLAP

Input : The network files, the layers to be pruned, the accuracy constraint δ, the number of memetic rounds $maxR$, the generations for multi-objective optimization $maxGen$, the generations for NCS $ncsGen$, the population size N, epoch number $numE$

Output: An approximation set P

1 Initialize the population P randomly based on the layers to be pruned;
2 Initialize accuracy constraints $\{cons_1, \cdots, cons_N\}$ uniformly based on δ;
3 **for** $round = 1 : maxR$ **do**
4   **for** $t = 1 : ncsGen$ **do**
5   **for** $i = 1 : N$ **do**
6   Optimize the i-th individual using NCS, where the fitness is set to the latency value when accuracy constraint $cons_i$ is satisfied, and the worst value when the $cons_i$ is unsatisfied;
7   **end**
8   Retrain each individual for $numE$ epochs;
9   **end**
10   **for** $g = 1 : maxGen$ **do**
11   Rank the population based on non-dominated sorting;
12   Sort the last front of the rank based on hypervolume contribution and select the worst individual;
13   Generate an offspring by applying crossover and mutation to two parents randomly selected from the population;
14   Evaluate the multi-objective functions of the offspring;
15   Update the above-obtained worst individual using the offspring, i.e., remove the dominated one or randomly remove one of the non-dominated;
16   **end**
17   Retrain each individual for $numE$ epochs;
18 **end**
19 **return** non-dominated individuals in P;

split data set is adopted for pruning and adjusting. One batch extracted from the selected split data set is used for searching pruning vectors in the memetic multi-objective evolutionary algorithm, and the rest batches are used to adjust the compressed networks to recover their accuracy. Let $T = \{T^l_{i,j} | l \in [1, L], i \in [1, n_l], j \in [1, n_{l+1}]\}$ denote a binary matrix, where $T^l_{i,j}$ indicates the states of whether the corresponding connection is currently pruned or not. The memetic multi-objective evolutionary algorithm is to adjust T by optimizing parameter importance $c = \{c_l | l \in [1, L]\}$ based on

$$T^l_{i,j} = \begin{cases} 0 & \text{if } a_l > |W^l_{i,j}| \\ T^l_{i,j} & \text{if } a_l \leq |W^l_{i,j}| < b_l \\ 1 & \text{if } b_l \leq |W^l_{i,j}|, \end{cases} \tag{2}$$

where $a_l = 0.9 \times \max\{\theta_l + c_l\sigma_l, 0\}$, $b_l = 1.1 \times \max\{\theta_l + c_l\sigma_l, 0\}$, θ_l and σ_l denote the mean of absolute values and standard deviation of weights in layer l, which is borrowed from that in [8].

After the current pruning and adjusting, the resultant compressed networks will be reserved into the next iteration. The search terminates when a predefined halting condition is satisfied. The non-dominated compressed networks in the final population or the compressed network with the minimum latency that satisfies the task performance constraint will be kept as candidates to be deployed.

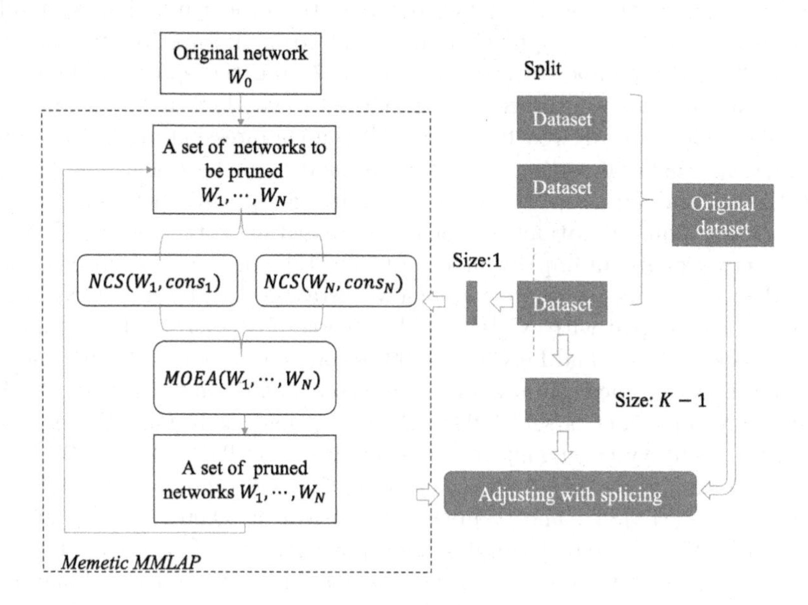

Fig. 1. MMLAP for DNN compression.

4 Experimental Studies

In this section, experimental studies are conducted to examine the performance of MMLAP. The experiments aim at illustrating the performance from three aspects: (1) the best Pareto-optimal performance in terms of latency and accuracy achieved by the instantiated MMLAP compared with well-established LMP methods for DNN compression, (2) the value of latency awareness in designing LMP when preferring real-time performance, and (3) the effectiveness of multi-objective optimization of MMLAP compared with multiple separate optimization processes.

4.1 Experiment Settings

To assess the potential of MMLAP, computational studies have been carried out to compare the instantiated MMLAP against a number of well-established LMP methods, including Dynamic Network Surgery (DNS) [8] and Optimization-based LMP (OLMP) [18]. Besides, a variant of OLMP that runs multiple OLMPs separately, denoted as multi-OLMP, is created in order to verify the impact of multi-objective optimization where collaboration exists in the population. The parameters of the above-mentioned compared algorithms are set following suggestions in their original literature. For the instantiated MMLAP, the setting of NCS is same as that in OLMP except that the fitness function is modified as shown in Sect. 3, the population size is set to 5, the maximum generation is set to 2000, the number of rounds is set to 7, and the epoch number is set to 1. Parameters to be optimized are initialized to random numbers between 0 and 0.001, which is borrowed from [18]. All the compression algorithms follow the same experimental settings for SGD method (e.g., base learning rate and batch size) in training networks when retraining compressed networks, and the additional retraining iterations for each compressed network is set to 15,000.

The networks are all implemented on Caffe [15] and the released projects of DNS. All experiments are conducted on a workstation with Intel XEON E5-2620 v4 processor equipped with NVIDIA GTX-1080Ti GPU cards. The commonly used networks LeNet-5 and LeNet-300-100 [10] are employed as the original networks. For LeNet-5, the accuracy constraint is set to 0.05, and for LeNet-300-100, the accuracy constraint is set to 0.08, following that set in OLMP. For multi-OLMP, the accuracy constraint vector is set to $\{0.032, 0.048, 0.064, 0.08, 0.096\}$ on LeNet-5 and is set to $\{0.02, 0.03, 0.04, 0.05, 0.06\}$ on LeNet-300-100. The dataset used is MNIST. The training set consists of 60,000 samples and the testing set consists of 10,000 samples. The split batch size K is set to 1000. The stabilized (after a network is loaded into memory) forwarding time of one batch over 100 trials is used to measure the latency of a network.

4.2 Results and Discussions

In this section, the performance of MMLAP is examined in terms of latency and accuracy. Firstly, the compressed networks obtained by the instantiation MMLAP is compared with the original networks and that obtained by two well-established LMP methods, under pre-defined accuracy constraints. The results show the advantages of MMLAP and latency awareness. Secondly, the Pareto-optimal networks in terms of latency and accuracy obtained by the compared algorithms are illustrated together, and show the advantages of multi-objective optimization of MMLAP. Thirdly, further analysis of the evolutionary behavior of MMLAP shows that despite the noisy estimation of objectives, MMLAP as a whole could still tend to achieve higher accuracy and lower latency. The detailed results and discussions are shown below.

The results of the compared algorithms under the pre-defined accuracy constraints are shown in Table 1. For MMLAP, the best network satisfying the specified constraint in the final population is shown in the table. The Origin method refers to the original network. The best results are highlighted in bold. In general, it can be observed that the instantiated MMLAP achieves the best performance in terms of both latency and accuracy. More specifically, it not only reduces the latency significantly, but also is able to achieve slightly better accuracy, which implies the effectiveness of multi-objective optimization to some extent. In particular, when examined on LeNet-5, MMLAP can reduce the latency by 104.8 ms with a slightly better accuracy. Besides, we would like to note the comparison between the compressed networks obtained by DNS and OLMP. The results in Table 1 show the superiority of DNS over OLMP in terms of latency. Nevertheless, according to the results in terms of the compression rate shown in [18], OLMP yielded better compression rates than DNS. Thus, this implies to some extent that the compression rate might not be a good proxy for latency evaluation.

Table 1. Comparison results in terms of accuracy and latency.

Network	Method	Top-1 Error (%)	Latency (ms)	Accuracy Improvement (%)	Latency Improvement (ms)
LeNet-300-100	Origin	2.28	102.9	0.00	0.0
	DNS	1.99	100.7	0.29	2.2
	OLMP	2.18	102.5	0.10	0.4
	MMLAP	1.78	91.7	**0.50**	**11.2**
LeNet-5	Origin	0.91	760.9	0.00	0.0
	DNS	0.91	660.8	0.00	100.1
	OLMP	0.91	669.2	0.00	91.7
	MMLAP	0.87	656.1	**0.04**	**104.8**

The Pareto-optimal networks found by the compared algorithms are illustrated in Fig. 2. Generally, MMLAP achieves the best performance in terms of multi-objective optimization. To be specific, the networks generated by MMLAP yield better or competitive performance compared with the compressed networks obtained by other algorithms, i.e., belonging to the non-dominated front. Furthermore, MMLAP maintains a population of networks with different trade-offs, thereby being capable of offering more options in a single run. This could be helpful in practical applications. On the other hand, as the multi-OLMP algorithm is created by running multiple OLMP separately, we would like to take a look at the comparison between MMLAP and multi-OLMP to study the effect of multi-objective optimization. Note that, multi-OLMP actually consumed much more epochs than MMLAP, since each OLMP follows the default setting except the accuracy constraint. In spite of that, the results show that MMLAP generally performed better than multi-OLMP, implying the value of multi-objective optimization for latency-aware LMP. In addition, multi-OLMP needs to manually set the value of accuracy constraints, and thus could be inefficient in practical use. As a consequence, MMLAP not only achieves the best performance in terms of latency and accuracy, but also offers more flexibility.

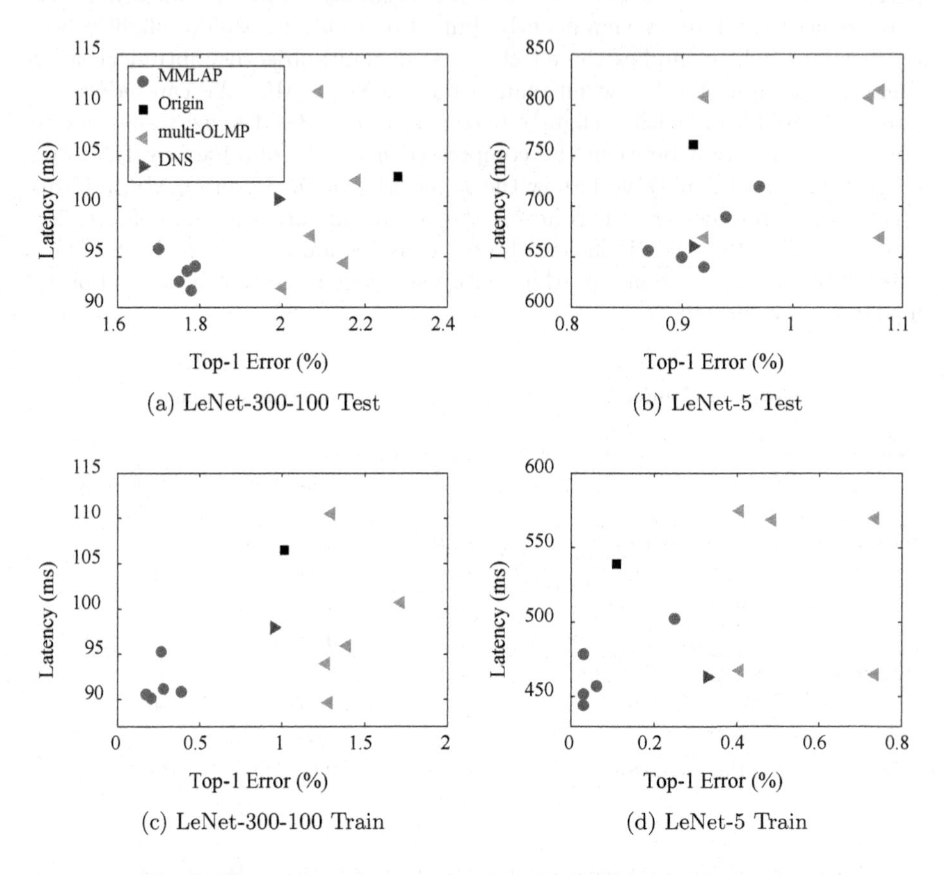

Fig. 2. Illustration of comparison results in the objective space.

The evolutionary behavior of MMLAP is further examined by illustrating the networks obtained at different rounds during the optimization, as shown in Fig. 3. Here, the performance of MMLAP is examined on one batch for one trial on the training set, which could be rather noisy but fast. According to the figures, the population of MMLAP progressively approaches towards the Pareto front.

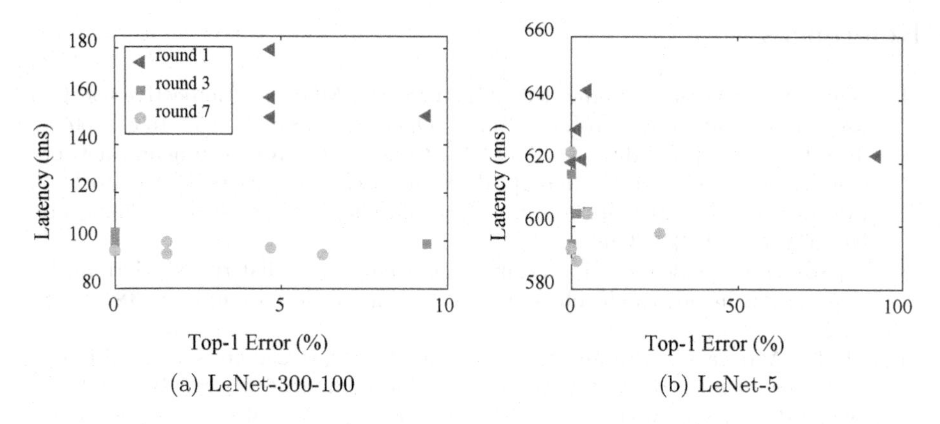

(a) LeNet-300-100 (b) LeNet-5

Fig. 3. Evolutionary behaviors of MMLAP during the optimization.

5 Conclusion

In this paper, a new LMP approach based on multi-objective latency-aware framework, namely MMLAP, is proposed for DNN compression. In general, the advantage of MMLAP over existing LMP methods contains three aspects. First, it directly captures the latency performance during the designing of compressed networks, and thus may be of important interest to latency-critical applications. Second, it optimizes for task performance (e.g., accuracy) and latency simultaneously and is capable of offering more network options for practical use. Third, it provides an efficient instantiation which integrates fast heuristic algorithms and multi-objective evolutionary algorithm. To the best of our knowledge, this is the first multi-objective optimization-based framework for latency-aware LMP. The experimental studies further show the superiority of the resultant MMLAP approach over state-of-the-art LMP approaches.

There are several future directions. First, to better understand the value of MMLAP, it is worthy of making an comprehensive investigation about the strengths and weaknesses of existing representative multi-objective evolutionary algorithms when applied to DNN compression. Second, the performance of MMLAP needs further analysis by taking into account an in-depth analysis of the noise in latency measurements, as well as more complex tasks and networks. Third, problem specific search operators are expected to further improve the effectiveness and efficiency, thus it is worth in-depth study. Fourth, there exists various types of compression methods, it is relevant to study the integration of these methods and MMLAP, leading to a more effective compression method.

References

1. Beume, N., Naujoks, B., Emmerich, M.T.M.: SMS-EMOA: multiobjective selection based on dominated hypervolume. Eur. J. Oper. Res. **181**(3), 1653–1669 (2007)
2. Chen, C., Tung, F., Vedula, N., Mori, G.: Constraint-aware deep neural network compression. In: Ferrari, V., Hebert, M., Sminchisescu, C., Weiss, Y. (eds.) ECCV 2018. LNCS, vol. 11212, pp. 409–424. Springer, Cham (2018). https://doi.org/10.1007/978-3-030-01237-3_25
3. Ciaparrone, G., Sánchez, F.L., Tabik, S., Troiano, L., Tagliaferri, R., Herrera, F.: Deep learning in video multi-object tracking: a survey. Neurocomputing **381**, 61–88 (2020)
4. Deb, K., Agrawal, S., Pratap, A., Meyarivan, T.: A fast and elitist multiobjective genetic algorithm: NSGA-II. IEEE Trans. Evol. Comput. **6**(2), 182–197 (2002)
5. Dong, J.-D., Cheng, A.-C., Juan, D.-C., Wei, W., Sun, M.: DPP-Net: device-aware progressive search for pareto-optimal neural architectures. In: Ferrari, V., Hebert, M., Sminchisescu, C., Weiss, Y. (eds.) ECCV 2018. LNCS, vol. 11215, pp. 540–555. Springer, Cham (2018). https://doi.org/10.1007/978-3-030-01252-6_32
6. Dong, X., Chen, S., Pan, S.J.: Learning to prune deep neural networks via layer-wise optimal brain surgeon. In: Advances in Neural Information Processing Systems 30, Long Beach, CA, pp. 4857–4867 (2017)
7. Esteva, A., et al.: A guide to deep learning in healthcare. Nat. Med. **25**(1), 24–29 (2019)
8. Guo, Y., Yao, A., Chen, Y.: Dynamic network surgery for efficient DNNs. In: Advances in Neural Information Processing Systems 29, Barcelona, Spain, pp. 1379–1387 (2016)
9. Han, S., Mao, H., Dally, W.J.: Deep compression: Compressing deep neural networks with pruning, trained quantization and Huffman coding (2015). arXiv preprint arXiv:1510.00149
10. Han, S., Pool, J., Tran, J., Dally, W.J.: Learning both weights and connections for efficient neural network. In: Advances in Neural Information Processing Systems 28, Montreal, Quebec, Canada, pp. 1135–1143 (2015)
11. Hong, W., Tang, K.: Convex hull-based multi-objective evolutionary computation for maximizing receiver operating characteristics performance. Memetic Comput. **8**(1), 35–44 (2015). https://doi.org/10.1007/s12293-015-0176-8
12. Hong, W., Tang, K., Zhou, A., Ishibuchi, H., Yao, X.: A scalable indicator-based evolutionary algorithm for large-scale multiobjective optimization. IEEE Trans. Evol. Comput. **23**(3), 525–537 (2019)
13. Huang, G., Liu, Z., van der Maaten, L., Weinberger, K.Q.: Densely connected convolutional networks. In: IEEE Conference on Computer Vision and Pattern Recognition, Honolulu, HI, pp. 2261–2269 (2017)
14. Huang, P., He, X., Gao, J., Deng, L., Acero, A., Heck, L.P.: Learning deep structured semantic models for web search using clickthrough data. In: 22nd ACM International Conference on Information and Knowledge Management, San Francisco, CA, pp. 2333–2338 (2013)
15. Jia, Y., et al.: Caffe: convolutional architecture for fast feature embedding. In: Proceedings of the ACM International Conference on Multimedia, Orlando, FL, pp. 675–678 (2014)
16. Kim, J., Misu, T., Chen, Y., Tawari, A., Canny, J.F.: Grounding human-to-vehicle advice for self-driving vehicles. In: IEEE Conference on Computer Vision and Pattern Recognition, Long Beach, CA, pp. 10591–10599 (2019)

17. LeCun, Y., Denker, J.S., Solla, S.A.: Optimal brain damage. In: Advances in Neural Information Processing Systems 2, Colorado, USA, pp. 598–605 (1989)
18. Li, G., Qian, C., Jiang, C., Lu, X., Tang, K.: Optimization based layer-wise magnitude-based pruning for DNN compression. In: Proceedings of the 27th International Joint Conference on Artificial Intelligence, Stockholm, Sweden, pp. 2383–2389 (2018)
19. Li, H., Kadav, A., Durdanovic, I., Samet, H., Graf, H.P.: Pruning filters for efficient ConvNets. In: 5th International Conference on Learning Representations, Toulon, France (2017)
20. Marculescu, D., Stamoulis, D., Cai, E.: Hardware-aware machine learning: modeling and optimization. In: Proceedings of the International Conference on Computer-Aided Design, San Diego, CA, p. 137 (2018)
21. Molchanov, D., Ashukha, A., Vetrov, D.P.: Variational dropout sparsifies deep neural networks. In: Proceedings of the 34th International Conference on Machine Learning, Sydney, Australia, pp. 2498–2507 (2017)
22. Qi, H., Sparks, E.R., Talwalkar, A.: Paleo: a performance model for deep neural networks. In: 5th International Conference on Learning Representations, Toulon, France (2017)
23. Rakshit, P., Konar, A., Das, S.: Noisy evolutionary optimization algorithms - a comprehensive survey. Swarm Evol. Comput. **33**, 18–45 (2017)
24. Real, E., et al.: Large-scale evolution of image classifiers. In: Proceedings of the 34th International Conference on Machine Learning, Sydney, Australia, pp. 2902–2911 (2017)
25. Sandler, M., Howard, A.G., Zhu, M., Zhmoginov, A., Chen, L.: MobileNetV2: inverted residuals and linear bottlenecks. In: IEEE Conference on Computer Vision and Pattern Recognition, Salt Lake City, UT, pp. 4510–4520 (2018)
26. See, A., Luong, M., Manning, C.D.: Compression of neural machine translation models via pruning. In: Proceedings of the 20th Conference on Computational Natural Language Learning, Berlin, Germany, pp. 291–301 (2016)
27. Silver, D., et al.: Mastering the game of Go with deep neural networks and tree search. Nature **529**(7587), 484 (2016)
28. Sun, Y., Wang, X., Tang, X.: Sparsifying neural network connections for face recognition. In: 2016 IEEE Conference on Computer Vision and Pattern Recognition, Las Vegas, NV, pp. 4856–4864 (2016)
29. Tang, K., Yang, P., Yao, X.: Negatively correlated search. IEEE J. Sel. Areas Commun. **34**(3), 542–550 (2016)
30. Ullrich, K., Meeds, E., Welling, M.: Soft weight-sharing for neural network compression. In: 5th International Conference on Learning Representations, Toulon, France (2017)
31. Wang, E., et al.: Deep neural network approximation for custom hardware: where we've been, where we're going. ACM Comput. Surv. **52**(2), 40:1–40:39 (2019)
32. Yu, R., et al.: NISP: pruning networks using neuron importance score propagation. In: IEEE Conference on Computer Vision and Pattern Recognition, Salt Lake City, UT, pp. 9194–9203 (2018)
33. Zhang, H., Sun, J., Liu, T., Zhang, K., Zhang, Q.: Balancing exploration and exploitation in multiobjective evolutionary optimization. Inf. Sci. **497**, 129–148 (2019)

Network Representation Learning Based on Topological Structure and Vertex Attributes

Shengxiang Hu, Bofeng Zhang(✉), Ying Lv, Furong Chang, and Zhuocheng Zhou

School of Computer Engineering and Science, Shanghai University, Shanghai, China
{mathripper,bfzhang}@shu.edu.cn

Abstract. Network Representation Learning (NRL) is an essential task in the field of network data analysis, which tries to learn the distributed representation of each vertex in the network for downstream vector-based data mining tasks. NRL is helpful in solving the computationally expensive or intractable problems of large-scale network analysis. Most related NRL methods only focus on encoding the network topology information into vertex representation. However, vertices may contain rich attributes that directly impact the network formation and measure the attribute-level similarity between vertices. Additionally, encoding the vertex attributes information into the representation vector may improve the performance of the representation. This paper proposes a general NRL framework TAFNE that can effectively retain both network topology and vertex attributes information. For complex types of vertex attributes, we design two different information fusion methods that take both training efficiency and generality into account. The proposed TAFNE framework is extensively evaluated through various data analysis tasks, including clustering, visualization and node classification, and achieves superior performance compared with baseline methods.

Keywords: Network Representation Learning · Vertex attributes · Network topology · Information fusion

1 Introduction

Deep Learning has achieved great success in analyzing Euclidean data such as natural languages, images, and audio. However, the non-Euclidean data is also valuable in daily life, effectively analyzing such data to extract favorable information seems to be a challenge. Many network-based data mining tasks, e.g. node classification [21], link prediction [10], recommendation [27], and key user discovery [7] are applied in various fields. For example, in the biological protein network, link prediction task is applied to study the correlation between proteins.

Supported by National Key R&D Program of China grant (NO. 2017YFC0907505).

T. Bäck et al. (Eds.): PPSN 2020, LNCS 12269, pp. 484–497, 2020.
https://doi.org/10.1007/978-3-030-58112-1_33

Many studies used a discrete matrix to represent network data and perform network analysis tasks based on matrix spectral decomposition [12]. However, such methods have at least a quadratic time complexity respect to the number of vertices, which makes them difficult to generalize to large networks. Besides, it is also a tedious task to design specific algorithms for different networks.

To solve such problems, NRL aims to learn low-dimensional potential representations of network vertices that encode the network topology, vertex attribute, and other related information. The learned representation vectors can be used in subsequent vector-based machine learning tasks [29]. Most of the NRL methods focus on how to preserve network topology, such as DeepWalk [17], LINE [23], Node2Vec [5] and DNE [22] etc. These methods expect to keep vertices with similar topological contexts adjacent to each other in the new low-dimensional representation space to retain network topology information. However, the vertices representations learned by these methods may loss some information about the original network. For example, in a citation network, each vertex represents a paper with the abstract as its attribute, edges refer to the citation relationship among vertices. Two papers may study similar problems but are not directly connected or have no common neighbors. Considering only topology information cannot clearly explain the similarity between them, but their attributes can. So, vertex attribute information should also be taken into account to augment the network representation performance.

This paper proposes a Topology and Vertex Attributes Fusion Network Embedding (TAFNE) which can effectively encode both network topology and vertex attribute information into representations. Firstly, we train a Transformer-decoder [25] to capture topology information from the random walk vertex sequences. In this way, we can improve feature extraction capabilities through the self-attention mechanism. Then we suggest two different information fusion methods to preserve vertex attributes. Vertex attributes do not have to be numerical or nominal, but can also be, for example, full-text descriptions. We evaluate our approach on three data mining tasks: node classification, clustering, and visualization on attributed and non-attributed networks. The experimental results show that TAFNE outperforms state-of-the-art models.

The major contributions of this paper are summarized as follows.

- To preserve the topology information, we design an embedding model that combines a network structure embedding layer with the Transformer-decoder to take advantage of its strong feature extraction capabilities.
- To preserve vertex attribute information, we propose two information infusion methods that make vertices with similar attributes adjacent to each other in the low dimensional representation space.
- To evaluate the proposed TAFNE framework, we have widely conducted node classification, clustering, and visualization tasks on different types of datasets and achieved excellent performance.

The rest of this paper is organized as follows: In Sect. 2, we summarize the related works. In Sect. 3, we explain the research question and some essential concepts. In Sect. 4, we introduce the TAFNE framework in detail. In Sect. 5, we evaluate the TAFNE framework through various data mining tasks and present the experimental results. Section 6, we summarize the work in this paper.

2 Related Work

According to the different implementations of the algorithms, we divide the existing NRL methods into two categories: matrix factorization based methods and neural network based methods.

The matrix factorization based methods use different types of matrices to preserve the network information, such as adjacency matrix, k-step transition probability matrix, and context matrix [28], then leverage matrix factorization to obtain the network representations. Spectrum Embedding [2] is a method for calculating non-linear embeddings, which uses a Laplacian spectral decomposition to find low-dimensional representations for the input network data. Maximize Modularity [24] performs a decomposition of modularity matrix to learn community-oriented vertex representation [15]; TADW [28] executes inductive matrix decomposition [14] on the vertex context matrix to retain both network structure and vertex text feature in the representation vector. However, due to the fact that matrix factorization demands a lot of memory and computing resources, these matrix factorization based methods are difficult to extend to large networks. It is no longer suitable for the current Internet environment. What is more, the design of relational matrices will directly affect the matrix factorization performance, which brings additional contingency.

The neural network based methods have been proposed in order to extract sophisticated structural features and learn highly non-linear vertex representation recently. Such methods can usually be well extended to large networks. DeepWalk [17] performs truncated random walk on the network to generate vertex sequence sets. The frequency of the occurrence of vertex context pairs reflects their relevance. It learns from the experience of word representation learning and introduces the word embedding algorithm (Skip-Gram) [13] to learn the vertex representation over the vertex sequences. In the Struc2vec [19] method, vertex structural role proximity is encoded into a multilayer graph, and then DeepWalk is performed on the multilayer graph to learn vertex representations. SDNE [26] uses first and second-order similarity to preserve the network structure and designs a deeply embedding model for capturing highly non-linear network structures while retaining global and local structure information. DNE [22] utilizes LSTM [4] to keep the transfer possibilities among the nodes and designs a Laplacian-supervised embedding space optimization to preserve network local structure information. Compared with the matrix factorization based methods, the neural network based methods are easy to generalize and more robust because they are not affected by artificially designed relation matrices. However, most of the above works consider only network topology information. In this paper, we aim to propose a general NRL framework that encodes both network topology and the rich types of vertex attributes into the representation.

3 Problem Definition

In this section, we define the research problem and some important concepts.

Definition 1 (Network). *The network is defined as* $G = \{V, E, A\}$, *where* $V = \{v_1, \ldots, v_n\}$ *is the vertex set of network* G *and* n *is the number of vertices.* $E = \{e_{ij}\}_{i,j=1}^{n}$ *is the edge set of network* G, e_{ij} *denotes the edge from vertex* v_i *to* v_j, *which is attached with a weight* $w_{ij} \in \mathbb{R}$, $w_{ij} = 1$ *in unweighted network or* $w_{ij} = 0$ *if vertex* v_i, v_j *are not linked directly.* $A = \{\boldsymbol{a}_1, \ldots, \boldsymbol{a}_n\}$ *denotes the attributes set of the vertices,* $\boldsymbol{a}_i$ *means the attributes of vertex* v_i.

Definition 2 (Topology and Vertex Attributes Fusion Network Embedding). *Given a network* $G = \{V, E, A\}$, *the Topology and Vertex Attributes Fusion Network Embedding task aims to learn a representation matrix* $R \in \mathbb{R}^{|V| \times d}$, *where* $d \ll |V|$ *and the i_th row of* $R(R_i)$ *is the low-dimensional representation vector of vertex* v_i. *Meanwhile the representation vector* R_i *preserves both the context structural information and vertex attributes, which means the vertices with similar attributes or similar context structure in the source network are also adjacent to each other in the embedding space.*

4 Topology and Vertex Attributes Fusion Network Embedding

In this section, we present details of the proposed NRL framework TAFNE as shown in Fig. 1. The TAFNE framework consists of three parts: (a) A random walk process guided by second-order biased proximity, (b) A multi-layer

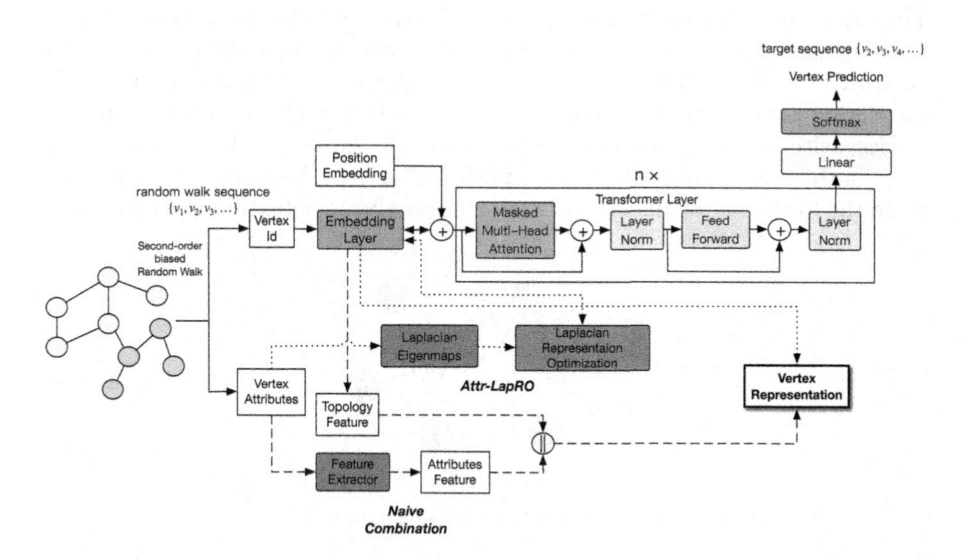

Fig. 1. The proposed Topology and Vertex Attributes Fusion Network Embedding (TAFNE) framework.

Transformer-decoder [25] for capturing network topology information preserved in the vertex sequences, (c) A information fusion model for preserving the vertex attributes.

4.1 Second-Order Biased Random Walk

In real networks, edge usually represents the similarity or some kinds of interactions between two vertices. Some related works, e.g., LINE [23], leverage the first-order similarity to guide the random walk process. The shortcoming is that, in unweighted networks, the degree of pairwise proximity and the social relationship information are left out. As shown in Fig. 2, since vertex i and j share more neighbors, i is closer with j than k in terms of social relationship. Here we utilize the second-order biased proximity to characterize the vertex similarity and guide the random walk process. For each pair of directly connected vertices, the similarity s_{ij} and the walk probability P are calculated as follows.

$$\begin{aligned} s_{ij} &= w_{ij}\frac{\left|N_{v_i}\bigcap N_{v_j}\right|+1}{max(d_i,d_j)} \\ P(v_i\rightarrow v_j) &= \frac{e^{s_{ij}}}{\sum_{k=0}^{d_i} e^{s_{ik}}} \end{aligned} \tag{1}$$

where w_{ij} is the weight of edge e_{ij}, d_i is the degree of vertex v_i, N_{v_i} is the set of one-hop neighbors of vertex v_i.

4.2 Topology Information Preservation

The sequences generated by the random walk can be treated as sentences in natural language. The context information of the tokens in the sentences reflects the network's topology. The pairwise co-occur frequency of the vertices reflects the correlation between them. Following the idea of DeepWalk, we assume that the vertices with similar contexts (neighbors) are also similar, and are close to each other in the target embedding space. With such a hypothesis, we expect to maximize the likelihood of the central vertex when given contextual vertices. Given

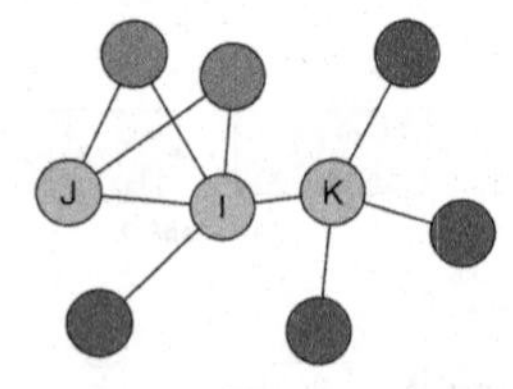

Fig. 2. A toy example of networks. First-order similarity and second-order biased similarity is quite different among vertices i, j and k.

a random walk sequence $S = \{v_1, \ldots, v_n\}$, the likelihood function is defined as follows.

$$D(S) = \sum_i P(v_i \mid v_1, ..., v_{i-1}; W) \tag{2}$$

where W means the model parameters.

In related researches, DNE [22] chose LSTM as its prediction model, but LSTM cannot train in parallel and lack the ability to resolve long-distance dependencies. Since Transformer [25] proposed in 2017, Transformer and its variants [9,18] have achieved encouraging results on various natural language processing tasks. The self-attention mechanism can not only capture longer distance language structures but also train in parallel to speed up the training process. Therefore, we employ a multi-layer Transformer-decoder [25] as our prediction model to take advantage of its powerful feature extraction capabilities. The Transformer-decoder applies a masked multi-head self-attention operation, followed by a position-wise feedforward layer, over the input context vertices to produce an output distribution over target vertex. This process and the loss function L_{lm} can be formulated as follows:

$$\begin{aligned} h_0 &= T_S R^e + R^p \\ h_l &= transformer\_decoder(h_{l-1}) \; \forall l \in [1, n] \\ \hat{y} &= softmax(h_n) \\ L_{lm}(\hat{y}, y) &= -\sum y * log(\frac{1}{\hat{y}}) \end{aligned} \tag{3}$$

where $T_S \in \mathbb{R}^{|V| \times |V|}$ is the tokenized matrix of vertex sequence S, T_i is the one-hot vector for vertex v_i, $R^e \in \mathbb{R}^{|V| \times d_e}$ is the vertex embedding matrix, $R^p \in \mathbb{R}^{|V| \times d_e}$ is the position embedding matrix, n is the number of layers, y is the target predicted vertex. We only want to get the vertex embedding feature through the prediction process, so the output of the embedding layer $R^e \in \mathbb{R}^{|V| \times d_e}$ is what we expect.

4.3 Information Fusion

In this subsection, we present the details of the information fusion model, as shown in the lower part of Fig. 1, which is designed to encode vertex attributes information into vertex representation. We expect vertices with similar attributes to be adjacent to each other in the low-dimensional target embedding space as well. Because of the rich types of vertex attributes, we propose two ways to incorporate the features extracted from the attributes of network vertices into the embedding process, which guarantees that the proposed NRL framework can not only train more efficiently but also be applied to various kinds of networks.

Naive Combination. For textual attributes or numerical attributes with sufficient dimensions, we utilize a pre-designed feature extractor to obtain the attributes feature $R^a \in \mathbb{R}^{|V| \times d_a}$ and directly combine it with the embedding feature R^e as a final vertex representation. So the dimension of vertex attributes must be greater than d_a to avoid introducing noise in the process of feature extraction. In this paper, we utilize the BOW (bag of words) model to obtain text vectors. According to the types of vertex attributes, users can flexibly design feature extractor. In our experiments, we take the hidden layer output of an autoencoder [16] as the attribute feature. The feature extractor and Transformer-decoder can be trained in parallel without affecting each other, which can greatly speed up the training process. The above process can be formulated as follows:

$$\begin{aligned} R^a &= feature\_extractor(A) \\ R &= [R^e|R^a] \end{aligned} \tag{4}$$

where $R \in \mathbb{R}^{|V| \times d}$ is the vertex representation matrix, $R^e \in \mathbb{R}^{|V| \times d_e}$ is the vertex embedding feature matrix, $R^a \in \mathbb{R}^{|V| \times d_a}$ is the attribute feature matrix, $[R^e|R^a]$ represents the operation of horizontally concatenating R^e and R^a. We assume that vertex attributes and network structure contribute equally to the vertex representation and set $d_a = d_e = \frac{d}{2}$, d is the dimension of vertex representation vector.

Attribute-Driven Laplacian Representation Optimization. In real-world networks, vertex attributes may have complex types, which do not apply to the Navie Combination described above. Some related works [1,22] use Laplacian Eigenmaps to enhance the ability of NRL models to retain network structure. Inspired by above works, we propose an Attributes-driven Laplacian Representation Optimization (Attr-LapRO) to make the proposed framework more general. We define the loss function of the Attr-LapRO as follows.

$$L_{lap} = \sum_{ij} (\mathbf{r_i} - \mathbf{r_j})^2 I_{ij} = 2 * Tr(R^T L R) \tag{5}$$

Where $R \in \mathbb{R}^{|V| \times d}$ is the vertex representation matrix, $I \in \mathbb{R}^{|V| \times |V|}$ is the vertex attribute similarity matrix, $I_{ij} \in [0, 1]$ represents the attribute cosine similarity score of vertex v_i and v_j, $L = D - I$ is Laplacian eigenmap, $D \in \mathbb{R}^{n \times n}$ is a diagonal matrix, $D_{ii} = \sum_j I_{ij}$. In this way, Attr-LapRO and Transformer-decoder share one output, so $R = R^e$, $d_e = d$.

We alternately and iteratively optimize the loss functions L_{lm} and L_{lap}, which is mainly for two reasons. One is that the parameters of the Transformer-decoder become challenging to update with two loss functions. The other reason is that the training of each stage can speed up the other's convergence process since vertices representation vectors are shared between the two stages.

Based on the two information fusion methods introduced above, we propose two types of TAFNE models: TAFNE$_{vanilla}$ and TAFNE$_{lap}$. TAFNE$_{vanilla}$ uses Naive Combination as the information fusion method, TAFNE$_{lap}$ utilizes Attr-LapRO to incorporate vertex attributes into the embedding process.

5 Evaluation

In this section, we first introduce the datasets used in this work and then validate the performance of our model compared to various state-of-the-art NRL algorithms through three downstream data mining tasks (i.e., clustering visualization, and node classification) on five datasets. Finally, we analyze the sensitivity of parameters.

5.1 Datasets

In order to fully evaluate the proposed method, we conduct experiments on three citation networks and two social networks with different sizes. Table 1 presents detailed information of the five datasets.

- Facebook [20] is a page-page graph of verified Facebook sites. Vertices represent official Facebook pages while the links are mutual likes between sites. Vertex attributes are extracted from the site descriptions that the page owners created to summarize the purpose of the site. All the vertices are divided into four categories, which are defined by Facebook.
- BlogCatalog [24] is a social network. Vertices represent bloggers, and categories of blogs written by bloggers are used as vertex labels. Each vertex has one or more labels. Edges represent friendship relationships between bloggers.
- Cora, CiteSeer, and PubMed [21] are citation networks, where each vertex represents a scientific publication, and the edge represents the citation relationship between vertices. Vertices are divided into different categories according to their research field, and each vertex has an abstract as its attribute.

5.2 Baseline Methods

We compare our method with several baseline methods, including DeepWalk [17], SDNE [26], Struc2Vec [19], DNE [22], TADW [28] and Attributes. In the Attributes method, the vertex attribute feature is treated as the vertex representation. Although there are other NRL methods, we can not list all of them. The methods mentioned above all have great innovations and conduct various verification experiments compared with other methods in the corresponding papers.

5.3 Parameter Settings

For all datasets, we set the dimension of the learned representation vector to $d = 128$. For the baseline methods, we follow the best parameter settings recommended in the original paper. In DeepWalk method, window size is $w = 10$, walk length is $l = 40$, and walks per vertex $\gamma = 40$. In Struc2Vec method, window size is $w = 10$, walk length is $l = 80$, walks per vertex is $\gamma = 10$. In SDNE method, the number of model layers is 3, and the hyperparameter $\alpha = 0.1$, $\beta = 10$. In DNE method, walk length is $l = 100$, walks per vertex is $\gamma = 100$, and the LSTM learning rate is 0.001. In TADW method, we set the parameters to the same as given in the corresponding paper. In Attributes method, the dimension of the vertex attribute feature is reduced to 128 via SVD [6]. In TAFNE method, we set the walk length to $l = 100$, walks per vertex $\gamma = 100$. Transformer-decoder has 8 layers and 4 masked self-attention heads per layer. At the stage of optimizing the loss function L_{lm}, we take advantage of Adam optimization scheme [8] with a max learning rate $lr_{lm_{max}} = 1e-4$. The learning rate was raised linearly from zero over the first 5000 updates and annealed to $1e-5$ using an exponential scheduler.

Table 1. Statistics of the datasets

Dataset	Nodes	Edges	Categories
Facebook	22470	171002	4
BlogCatalog	10312	333983	39
Cora	2707	5429	7
Citeseer	3311	4732	6
PubMed	19717	44338	3

Table 2. Clustering performance (NMI)

Methods	3-Cora	Cora
DeepWalk	0.045	0.012
SDNE	0.027	0.027
Struc2Vec	0.001	0.005
DNE	0.387	0.309
Attributes	0.451	0.286
TADW	0.613	0.411
$\text{TAFNE}_{vanilla}$	**0.655**	**0.468**
TAFNE_{lap}	0.648	0.447

5.4 Experiments Results

Clustering. In the real world, most data is unlabeled, and annotating data manually costs a lot, so learning the representation of a network is very important for unsupervised learning tasks. We perform clustering tasks on the 3-Cora and Cora datasets. 3-Cora is separated from Cora with 3 different categories.

In the clustering task, we use each baseline method to generate the vertices representation vectors that are used as features for clustering. The vertices are divided into several categories using the K-Means algorithm, and we evaluate the performance with NMI (Normalized Mutual Information)[3] score. Table 2 records the result of clustering. This result shows that TAFNE is significantly better than other methods. TADW achieves the best performance among baseline methods because of the consideration of both network topology and vertex attributes, but TAFNE still outperforms it. $\text{TAFNE}_{vanilla}$ is absolutely 0.042 and 0.057 better than TADW on the 3-Cora dataset and Cora dataset, respectively. Benefit from the combination with Transformer-decoder and information fusion model, TAFNE can retain network information more comprehensively. Therefore TAFNE is more robust in the clustering task.

Visualization. In the visualization task, we expect to reveal the network by visualizing the learned representations intuitively. We apply our method and baseline methods to the 3-Cora dataset, which has nearly 1300 vertices, and each vertex belongs to one of the three categories: Neural Network, Rule Learning, and Reinforcement Learning. We exploit t-SNE [11] to map the vertex representation learned by different methods to 2-dimension space. Figure 3 shows the visualization on the 3-Cora dataset, each point represents a scientific publication, and the different color represents a different category. The visualization of DeepWalk, SDNE, and Struc2Vec are not meaningful because the points of the same class are not clustered together. Although DNE and TADW can cluster most points of the same label together, the boundaries are not visible enough. The performance of TAFNE is much better than the baseline methods. Our method can not only cluster the points of the same category together, but also the clusters can be clearly separated from each other. This experiment indicates that TAFNE can learn more robust and informative representations.

Classification. In the node classification task, we perform multi-label and multi-class classification tasks on two types of datasets: one without vertex attributes (i.e., BlogCatalog) and one with vertex attributes (i.e., Facebook, Citeseer and PubMed). We treat the representations learned by various methods as the vertices feature vectors. For each dataset, a portion (L_V) of the labeled vertices are randomly sampled as the training data to train an MLP (Multilayer Perceptron) as the classifier, the rest of the vertices are the test data. We repeat this process 10 times and record the average performance in terms of classification accuracy and F1-score.

On the BlogCatalog dataset, we only focus on learning network topology feature to verify the effectiveness of the Transformer-decoder in capturing the

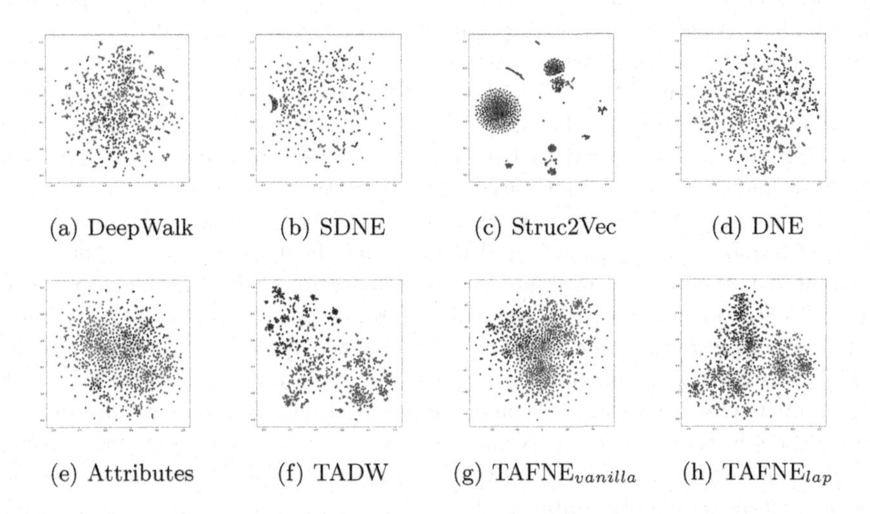

(a) DeepWalk (b) SDNE (c) Struc2Vec (d) DNE

(e) Attributes (f) TADW (g) TAFNE$_{vanilla}$ (h) TAFNE$_{lap}$

Fig. 3. Visualization of the 3-Cora dataset. Each point represents one scientific publication. Different colors correspond to different categories, i.e., Red: Nerual Network, Blue: Rule Learning, Green: Reinforcement Learning (Color figure online)

Table 3. Multi-label classification performance (Micro-F1/Macro-F1)

L_V	10%	20%	30%
DeepWalk	33.17/17.40	35.84/20.39	37.28/21.96
SDNE	31.25/11.54	32.27/14.33	32.24/16.31
Struc2Vec	11.16/5.24	11.20/5.57	12.86/4.83
DNE	34.13/17.38	37.25/21.41	37.56/23.06
TAFNE	**38.43/19.04**	**42.05/23.28**	**42.27/25.66**

Table 4. Multi-class classification performance (accuracy) on three datasets

Dataset	Facebook			CiteSeer			PubMed		
L_V	10%	20%	30%	10%	20%	30%	10%	20%	30%
DeepWalk	72.09	74.83	76.14	50.94	51.54	53.28	69.94	71.37	72.51
SDNE	53.44	54.81	55.81	30.41	32.12	32.75	39.04	39.37	39.83
Struc2Vec	35.58	35.13	35.82	25.01	26.76	27.75	47.08	48.15	49.32
DNE	68.06	70.62	73.06	50.56	52.97	54.04	73.11	73.04	74.72
TADW	79.48	80.50	83.08	60.89	62.32	64.78	80.63	81.76	84.32
Attributes	74.27	76.62	78.41	57.31	59.20	61.09	75.45	77.52	78.79
TAFNE$_{vanilla}$	**82.23**	**82.96**	**85.28**	**63.20**	**65.75**	**67.01**	**83.19**	**84.67**	**86.94**
TAFNE$_{lap}$	80.14	81.45	83.23	62.18	64.13	65.61	81.83	84.27	85.23

network structure information without vertex attributes. Since TADW performs DeepWalk to preserve the network structure, so we exclude TADW. Then we perform a multi-label classification task on the vertices representations and report the performance with Micro-F1 and Macro-F1 scores of the classification results. From Table 3, one can see that TAFNE is 4.71% (Micro-F1) and 2.6% (Macro-F1) with 30% labeled vertices better than the best baseline DNE. The self-attention mechanism takes into account more contextual information, by which the model can better retain network global and local structure information.

Furthermore, we conduct multi-class classification tasks on the three attributed networks and evaluate the representations with classification accuracy. Table 4 documents the classification results. As we can see, TAFNE consistently outperforms other baseline methods. Compared to the best baseline TADW, TAFNE$_{vanilla}$ absolutely improves the accuracy by 2.20% on Facebook, 2.27% on CiteSeer, and 2.62% on PubMed. The classification results demonstrate the effectiveness of TAFNE to consider the network topology and vertex attribute information comprehensively, and the quality of the learned representation vectors are greatly improved.

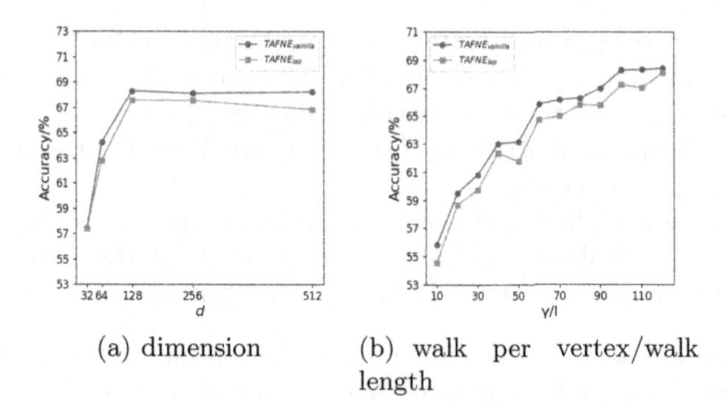

(a) dimension (b) walk per vertex/walk length

Fig. 4. Parameters sensitivity analysis of TAFNE on CiteSeer with train ratio as 50%

5.5 Paramter Sensitivity

In this section, we investigate the sensitivity of the method to the choice of parameter values. TAFNE has three main parameters: the dimension of the learned representation vector d, walk per vertex γ and walk length l. We test the classification accuracy with different parameters above CiteSeer dataset, and randomly sample 50% vertices as the trainset and the rest as the testset. We fix other parameters when inspecting each parameter.

Figure 4(a) shows the effect of different d on classification results. As the representation vector dimension d increases, the accuracy increases first and then decreases. The classification accuracy is highest when $d = 128$. For convenience we set $\gamma = l$. Figure 4(b) shows that with the increase of γ and l, the classification accuracy also improves, because γ and l correspond to the richness of the corpus. The bigger γ and l are, the richer the corpus is. Since the self-attention mechanism can solve the problem of long-distance dependency well, the increase in walk length not only does not have a negative impact but also better captures the global structure of the network.

6 Conclusion

To solve the problems of analyzing large-scale networks with computational overhead or intractability, and generate higher-quality vertices representations, this paper proposes an NRL framework TAFNE that can simultaneously incorporate network topology and vertex attribute information into the embedding process. The advantages of TAFNE can be summarized as follows:

(1) By utilizing the Transformer-decoder with masked self-attention mechanism, TAFNE can take more contextual information into account than previous models when extracting structural features from random walk vertex sequences. Thus TAFNE preserves global and local structural information better;

(2) For the networks where vertex contains textual attributes or numerical attributes of sufficient dimensions, TAFNE encodes the vertex attributes by an autoencoder and treats the attributes feature as part of the vertex representation. In this way, attributes feature directly measures the attribute-level similarity among vertices.
(3) For the networks where vertex contains attributes of complex types, TAFNE constructs an attribute similarity matrix to constrain the vertex representation. Such an approach makes TAFNE more generic.

Experimental results on the clustering, visualization, and node classification tasks over five datasets demonstrate the effectiveness of TAFNE.

References

1. Belkin, M., Niyogi, P.: Laplacian eigenmaps and spectral techniques for embedding and clustering. In: Advances in Neural Information Processing Systems, pp. 585–591 (2002)
2. Belkin, M., Niyogi, P.: Laplacian eigenmaps for dimensionality reduction and data representation. Neural Comput. **15**(6), 1373–1396 (2003)
3. Estévez, P.A., Tesmer, M., Perez, C.A., Zurada, J.M.: Normalized mutual information feature selection. IEEE Trans. Neural Netw. **20**(2), 189–201 (2009)
4. Graves, A.: Generating sequences with recurrent neural networks. arXiv preprint arXiv:1308.0850 (2013)
5. Grover, A., Leskovec, J.: node2vec: scalable feature learning for networks. In: Proceedings of the 22nd ACM SIGKDD International Conference on Knowledge Discovery and Data Mining, pp. 855–864 (2016)
6. Halko, N., Martinsson, P.G., Tropp, J.A.: Finding structure with randomness: probabilistic algorithms for constructing approximate matrix decompositions. SIAM Rev. **53**(2), 217–288 (2011)
7. Henderson, K., et al.: RolX: structural role extraction & mining in large graphs. In: Proceedings of the 18th ACM SIGKDD International Conference on Knowledge Discovery and Data Mining, pp. 1231–1239 (2012)
8. Kingma, D.P., Ba, J.: Adam: a method for stochastic optimization. arXiv preprint arXiv:1412.6980 (2014)
9. Krause, B., Kahembwe, E., Murray, I., Renals, S.: Dynamic evaluation of transformer language models. arXiv preprint arXiv:1904.08378 (2019)
10. Liben-Nowell, D., Kleinberg, J.: The link-prediction problem for social networks. J. Am. Soc. Inf. Sci. Technol. **58**(7), 1019–1031 (2007)
11. van der Maaten, L., Hinton, G.: Visualizing data using t-SNE. J. Mach. Learn. Res. **9**(Nov), 2579–2605 (2008)
12. Malliaros, F.D., Vazirgiannis, M.: Clustering and community detection in directed networks: a survey. Phys. Rep. **533**(4), 95–142 (2013)
13. Mikolov, T., Sutskever, I., Chen, K., Corrado, G.S., Dean, J.: Distributed representations of words and phrases and their compositionality. In: Advances in Neural Information Processing Systems, pp. 3111–3119 (2013)
14. Natarajan, N., Dhillon, I.S.: Inductive matrix completion for predicting gene-disease associations. Bioinformatics **30**(12), i60–i68 (2014)
15. Newman, M.E.: Finding community structure in networks using the eigenvectors of matrices. Phys. Rev. E **74**(3), 036104 (2006)

16. Ng, A., et al.: Sparse autoencoder. CS294A Lecture notes **72**(2011), 1–19 (2011)
17. Perozzi, B., Al-Rfou, R., Skiena, S.: DeepWalk: online learning of social representations. In: Proceedings of the 20th ACM SIGKDD International Conference on Knowledge Discovery and Data Mining, pp. 701–710 (2014)
18. Rae, J.W., Potapenko, A., Jayakumar, S.M., Lillicrap, T.P.: Compressive transformers for long-range sequence modelling. arXiv preprint arXiv:1911.05507 (2019)
19. Ribeiro, L.F., Saverese, P.H., Figueiredo, D.R.: struc2vec: learning node representations from structural identity. In: Proceedings of the 23rd ACM SIGKDD International Conference on Knowledge Discovery and Data Mining, pp. 385–394 (2017)
20. Rozemberczki, B., Allen, C., Sarkar, R.: Multi-scale attributed node embedding (2019)
21. Sen, P., Namata, G., Bilgic, M., Getoor, L., Galligher, B., Eliassi-Rad, T.: Collective classification in network data. AI Mag. **29**(3), 93–93 (2008)
22. Sun, X., Song, Z., Dong, J., Yu, Y., Plant, C., Böhm, C.: Network structure and transfer behaviors embedding via deep prediction model. In: Proceedings of the AAAI Conference on Artificial Intelligence, vol. 33, pp. 5041–5048 (2019)
23. Tang, J., Qu, M., Wang, M., Zhang, M., Yan, J., Mei, Q.: LINE: large-scale information network embedding. In: Proceedings of the 24th International Conference on World Wide Web, pp. 1067–1077 (2015)
24. Tang, L., Liu, H.: Relational learning via latent social dimensions. In: Proceedings of the 15th ACM SIGKDD International Conference on Knowledge Discovery and Data Mining, pp. 817–826 (2009)
25. Vaswani, A., et al.: Attention is all you need. In: Advances in Neural Information Processing Systems, pp. 5998–6008 (2017)
26. Wang, D., Cui, P., Zhu, W.: Structural deep network embedding. In: Proceedings of the 22nd ACM SIGKDD International Conference on Knowledge Discovery and Data Mining, pp. 1225–1234 (2016)
27. Wang, X., He, X., Wang, M., Feng, F., Chua, T.S.: Neural graph collaborative filtering. In: Proceedings of the 42nd International ACM SIGIR Conference on Research and Development in Information Retrieval, pp. 165–174 (2019)
28. Yang, C., Liu, Z., Zhao, D., Sun, M., Chang, E.: Network representation learning with rich text information. In: Twenty-Fourth International Joint Conference on Artificial Intelligence (2015)
29. Zhang, D., Yin, J., Zhu, X., Zhang, C.: Network representation learning: a survey. IEEE Trans. Big Data **6**(1), 3–28 (2018)

A Committee of Convolutional Neural Networks for Image Classification in the Concurrent Presence of Feature and Label Noise

Stanisław Kaźmierczak(✉) and Jacek Mańdziuk

Faculty of Mathematics and Information Science, Warsaw University of Technology, Warsaw, Poland
{s.kazmierczak,mandziuk}@mini.pw.edu.pl

Abstract. Image classification has become a ubiquitous task. Models trained on good quality data achieve accuracy which in some application domains is already above human-level performance. Unfortunately, real-world data are quite often degenerated by noise existing in features and/or labels. There are numerous papers that handle the problem of either feature or label noise separately. However, to the best of our knowledge, this piece of research is the first attempt to address the problem of concurrent occurrence of both types of noise. Basing on the MNIST, CIFAR-10 and CIFAR-100 datasets, we experimentally prove that the difference by which committees beat single models increases along with noise level, no matter whether it is an attribute or label disruption. Thus, it makes ensembles legitimate to be applied to noisy images with noisy labels. The aforementioned committees' advantage over single models is positively correlated with dataset difficulty level as well. We propose three committee selection algorithms that outperform a strong baseline algorithm which relies on an ensemble of individual (nonassociated) best models.

Keywords: Committee of classifiers · Ensemble learning · Label noise · Feature noise · Convolutional neural networks

1 Introduction

A standard image classification task consists in assigning a correct label to an input sample picture. In the most widely-used supervised learning approach, one trains a model to recognize the correct class by providing input-output image-label pairs (training set). In many cases, the achieved accuracy is very high [6,38], close to or above human-level performance [7,15].

The quality of real-world images is not perfect. Data may contain some noise defined as anything that blurs the relationship between the attributes of an instance and its class [16]. There are mainly two types of noise considered in the literature: feature (attribute) noise and class (label) noise [11,27,36].

T. Bäck et al. (Eds.): PPSN 2020, LNCS 12269, pp. 498–511, 2020.
https://doi.org/10.1007/978-3-030-58112-1_34

Despite the fact that machine algorithms (especially those based on deep architectures) perform on par with humans or even better on high-quality pictures, their performance on distorted images is noticeably worse [10]. Similarly, label noise may potentially result in many negative consequences, e.g. deterioration of prediction accuracy along with an increase of model's complexity, size of a training set, or length of a training process [11]. Hence, it is necessary to devise methods that reduce noise or are able to perform well in its presence. The problem is furthermore important considering the fact that the acquisition of accurately labeled data is usually time-consuming, expensive and often requires a substantial engagement of human experts [2].

There are many papers in the literature which tackle the problem of label noise. Likewise, a lot of works have been dedicated to studying attribute noise. However, to the best of our knowledge, there are no papers that consider the problem of feature and label noise occurring simultaneously in the computer vision domain. In this paper, we present the method that successfully deals with the concurrent presence of attribute noise and class noise in image classification problem.

1.1 The Main Contribution

Encouraged by the promising results of ensemble models applied to label noise and Convolutional Neural Network (CNN) based architectures utilized to handle noisy images we examine how a committee of CNN classifiers (each trained on the whole dataset) deal with noisy images marked with noisy labels. With regard to the common taxonomy, there are four groups of ensemble methods [22]. The first one relates to *data selection mechanisms* aiming to provide different subset for every single classifier to be trained on. The second one refers to *the feature level.* Methods among this group select features that each model uses. The third one, *the classifier level group*, comprises algorithms that have to determine the base model, the number of classifiers, other types of classifiers, etc. The final one refers to *the combination of classifiers level* where an algorithm has to decide how to combine models' individual decisions to make a final prediction. In this study, we assume having a set of well-trained CNNs which make the ultimate decision by means of soft voting (averaging) scheme [26]. We concentrate on the task of finding an optimal or near-optimal model committee that deals with concurrent presence of attribute and label noise in the image classification problem. In summary, the main contribution of this work is threefold:

- addressing the problem of simultaneously occurring feature and label noise which, to the best of our knowledge, is a novel unexplored setting;
- designing three methods of building committees of classifiers which outperform a strong baseline algorithm that employs a set of individually best models;
- proving empirically that a margin of ensembles gain over the best single model rises along with an increase of both noise types, as well as dataset difficulty, which makes the proposed approaches specifically well-suited to the case of noisy images with noisy labels.

The remainder of this paper is arranged as follows. Section 2 provides a literature review, with considerable emphasis on methods addressing label noise and distorted images. Section 3 introduces the proposed novel algorithms for classifiers' selection. Sections 4 and 5 describe the experimental setup and analysis of results, respectively. Finally, brief conclusions and directions for further research are presented in the last section.

2 Related Literature

Many possible sources of label noise have been identified in the literature [11], e.g. insufficient information provided to the expert [3,11], expert (human or machine) mistakes [25,32], the subjectivity of the task [17] or communication problems [3,36]. Generally speaking, there are three main approaches to dealing with label noise [11]. The first one is based on algorithms that are naturally robust to class noise. This includes ensemble methods like bagging and boosting. It has been shown in [9] that bagging performs generally better than boosting in this task. The second group of methods relies on data cleansing. In this approach, corrupted instances are identified before the training process starts off and some kind of filter (e.g. voting or partition filter [4,37] which is deemed easy, cheap and relatively solid) is applied to them. Ultimately, the third group consists of methods that directly model label noise during the learning phase or were specifically designed to take label noise into consideration [11].

In terms of images, the key reasons behind feature noise are faults in sensor devices, analog-to-digital converter errors [30] or electromechanical interferences during the image capturing process [14]. State-of-the-art approaches to deal with feature noise are founded on deep architectures. In [24] CNNs (LeNet-5 for MNIST and an architecture similar to the base model C of [33] for CIFAR-10 and SVHN datasets) were used to handle noisy images. Application of a denoising procedure (Non-Local Means [5]) before the training phase improves classification accuracy for some types and levels of noise. In [28] several combinations of denoising autoencoder (DAE) and CNNs were proposed, e.g. DAE-CNN, DAE-DAE-CNN, etc. The paper states that properly combined DAE and CNN achieve better results than individual models and other popular methods like Support Vector Machines [35], sparse rectifier neural network [13] or deep belief network [1].

3 Proposed Algorithms

As mentioned earlier, classification results are obtained via a soft voting scheme. More specifically, probabilities of particular classes from single CNNs are summed up and the class with the highest cumulative value is ultimately selected (see Fig. 1).

Many state-of-the-art results in image classification tasks are achieved by an ensemble of well-trained networks that were not selected in any way [7,18,21]. In [31] the authors went further and noticed that limiting ensemble size just to

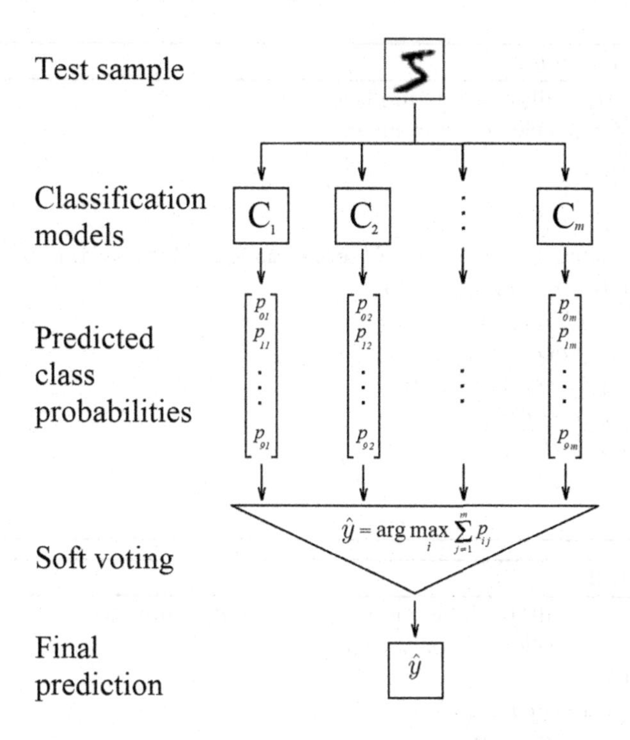

Fig. 1. The concept of the soft voting approach for the 10-class problem.

two best-performing models increased accuracy in their case. We adopted that idea to the algorithm called (for the purpose of this study) *top-n*, which serves as a benchmark in our experiments. First, all models are sorted in descending order according to their accuracies. Then, ensembles constituted by k best networks where k ranges from 1 to the number of available models are created. Finally, the committee with the best score on the validation dataset is chosen. Algorithm 1 summarizes the procedure.

The first algorithm proposed in this paper, called *2-opt-c*, was inspired by the local search technique commonly applied to solving the Traveling Salesman Problem (TSP) [8]. The original *2-opt* formulation looks for any two nodes of the current salesman's route which, if swapped, would shorten the route length. It works until no improvement is made within a certain number of sampling trials. The *2-opt-c* algorithm receives an initial committee as an input and in each step modifies it by either *adding* or *subtracting* or *exchanging* one or two elements. A single modification that maximizes accuracy on the validation set is performed. Thus, there are eight possible atomic operations listed in Algorithm 2. The procedure operates until neither of the operations improves performance. The *1-opt-c* works in a very similar way but is limited to three operations that modify only one element in a committee (adding, removal and swap). The *top-n-2-opt-c* and *top-n-1-opt-c* operate likewise besides being initialized with the output of the *top-n* procedure, not an empty committee.

Algorithm 1. *top-n*

```
   input : M_all–all available models
   output: C_best–selected committee
 1 C_best ← ∅;
 2 C_curr ← ∅;
 3 acc_best = 0;
 4 M_sorted = sort(M_all);        // sort models descending by accuracy
 5 for i ← 1 to size(M_sorted) do
 6 |   C_curr ← C_curr ∪ M_sorted[i];
 7 |   acc_curr ← accuracy(C_curr);
 8 |   if acc_curr > acc_best then
 9 |   |   acc_best ← acc_curr;
10 |   |   C_best ← C_curr;
11 |   end
12 end
```

Algorithm 2. *2-opt-c*

```
   input : M_all–all available models, C_0–initial committee
   output: C_best–selected committee
 1 C_best ← C_0;
 2 acc_best ← accuracy(C_best);
 3 while acc_best rises do
 4 |   acc_curr = 0;
 5 |   acc_curr, C_curr ← add(C_best, M_all, acc_curr);
 6 |   acc_curr, C_curr ← remove(C_best, acc_curr);
 7 |   acc_curr, C_curr ← swap(C_best, M_all, acc_curr);
 8 |   acc_curr, C_curr ← addTwo(C_best, M_all, acc_curr);
 9 |   acc_curr, C_curr ← removeTwo(C_best, acc_curr);
10 |   acc_curr, C_curr ← addAndSwap(C_best, M_all, acc_curr);
11 |   acc_curr, C_curr ← removeAndSwap(C_best, M_all, acc_curr);
12 |   acc_curr, C_curr ← swapTwice(C_best, M_all, acc_curr);
13 |   if acc_curr > acc_best then
14 |   |   acc_best ← acc_curr;
15 |   |   C_best ← C_curr;
16 |   end
17 end
```

Another algorithm, called *stochastic*, relies on the fact that the performance of the entire ensemble depends on diversity among individual component classifiers, on the one hand, and the predictive performance of single models, on the other hand [29]. The pseudocode of the algorithm is presented in Algorithm 3. In the ith epoch, a committee size ranges from i to $i + range$ with the expected value equal to $i + \frac{range}{2}$. This property is assured by the formula in line 7 which increases the probability of adding a new model along with decreasing ensemble size and vice versa. Figure 2 illustrates this relationship. In each step, one model

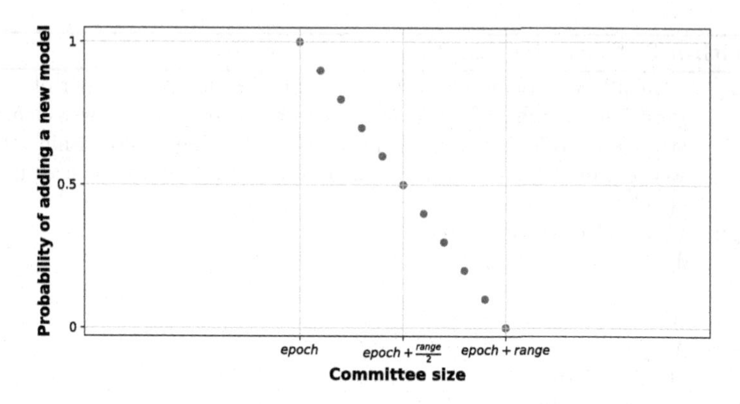

Fig. 2. Probability of adding a new model to the committee in the *stochastic* algorithm.

is either added or removed. In the first scenario a model with the best individual performance is appended to the committee with probability t_a (lines 11–14) or a model which minimizes the maximum correlation between any model from the committee and itself is added with probability $1 - t_a$ (lines 15–18). Analogously, if the algorithm decides to decrease a committee size it removes the weakest model with probability t_r (lines 22–25) or a model which minimizes the highest correlation between any two models in the committee with probability $1 - t_r$ (lines 26–29). In each epoch, the algorithm performs N_i iterations to explore the solution space. A correlation between two models is measured by the Pearson correlation coefficient calculated on probability vectors obtained from predictions on the validation set.

4 Experimental Setup

4.1 MNIST, CIFAR-10 and CIFAR-100 Datasets

As a benchmark, we selected three datasets with a diversified difficulty level. MNIST database contains a large set of 28×28 grayscale images of handwritten digits (10 classes) and is commonly used in machine learning experiments [23]. The training set and the test set are composed of 60 000 and 10 000 images, respectively.

CIFAR-10 [20] is another popular image dataset broadly used to assess machine learning/computer vision algorithms. It contains 60 000 32×32 color images in 10 different classes. The training set includes 50 000 pictures, while the test set 10 000 ones.

The CIFAR-100 dataset is similar to CIFAR-10. It comprises 60 000 images with the same resolution and three color channels as well. The only difference is the number of classes–CIFAR-100 has 100 of them, thus yielding 600 pictures per class.

Algorithm 3. *Stochastic* algorithm

input : M_{all}–all available models, N–number of epochs, N_i–number of iterations within an epoch, t_a–probability threshold below which the strongest model is added, t_r–probability threshold below which the weakest model is removed, r–range of possible committee sizes in an epoch

output: C_{best}–selected committee

```
1  C_curr ← ∅;
2  C_best ← ∅;
3  acc_best ← 0;
4  M_left ← M_all;
5  for i ← 0 to N − 1 do
6  |  for j ← 1 to N_i do
7  |  |  p_a ← 1 − (size(C_curr) − i)/r;
8  |  |  u ← generate from the uniform distribution U(0, 1);
9  |  |  if u < p_a then
10 |  |  |  u_a ← generate from the uniform distribution U(0, 1);
11 |  |  |  if size(C_curr) == 0 or u_a < t_a then
12 |  |  |  |  m_a ← getStrongestModel(M_left);
13 |  |  |  |  C_curr ← C_curr ∪ {m_a};
14 |  |  |  |  M_left ← M_left \ {m_a};
15 |  |  |  else
   |  |  |  |  // select model from M_left which minimizes maximum
   |  |  |  |  // correlation between any model from C_curr and itself
16 |  |  |  |  m_a ← getMarginallyCorrelatedModel(C_curr, M_left);
17 |  |  |  |  C_curr ← C_curr ∪ {m_a};
18 |  |  |  |  M_left ← M_left \ {m_a};
19 |  |  |  end
20 |  |  else
21 |  |  |  u_r ← generate from the uniform distribution U(0, 1);
22 |  |  |  if size(C_curr) == 1 or u_r < t_r then
23 |  |  |  |  m_r ← getWeakestModel(C_curr);
24 |  |  |  |  C_curr ← C_curr \ {m_r};
25 |  |  |  |  M_left ← M_left ∪ {m_r};
26 |  |  |  else
   |  |  |  |  // select model from M_curr which minimizes maximum
   |  |  |  |  // correlation between any two models in C_curr
27 |  |  |  |  m_r ← getMaximallyCorrelatedModel(C_curr);
28 |  |  |  |  C_curr ← C_curr \ {m_r};
29 |  |  |  |  M_left ← M_left ∪ {m_r};
30 |  |  |  end
31 |  |  end
32 |  |  acc_curr ← accuracy(C_curr);
33 |  |  if acc_curr > acc_best then
34 |  |  |  acc_best ← acc_curr;
35 |  |  |  C_best ← C_curr;
36 |  |  end
37 |  end
38 end
```

Table 1. CNN architectures used for MNIST, CIFAR-10 and CIFAR-100.

Layer	Type	#maps & neurons	Kernel/pool size
1	Convolutional	32 maps of 32 × 32 neurons (CIFAR) 32 maps of 28 × 28 neurons (MNIST)	3 × 3
2	Batch normalization		
3	Convolutional	32 maps of 32 × 32 neurons (CIFAR) 32 maps of 28 × 28 neurons (MNIST)	3 × 3
4	Batch normalization		
5	Max pooling		2 × 2
6	Dropout (20%)		
7	Convolutional	64 maps of 16 × 16 neurons (CIFAR) 64 maps of 14 × 14 neurons (MNIST)	3 × 3
8	Batch normalization		
9	Convolutional	64 maps of 16 × 16 neurons (CIFAR) 64 maps of 14 × 14 neurons (MNIST)	3 × 3
10	Batch normalization		
11	Max pooling		2 × 2
12	Dropout (20%)		
13	Convolutional	128 maps of 8 × 8 neurons (CIFAR) 128 maps of 7 × 7 neurons (MNIST)	3 × 3
14	Batch normalization		
15	Convolutional	128 maps of 8 × 8 neurons (CIFAR) 128 maps of 7 × 7 neurons (MNIST)	3 × 3
16	Batch normalization		
17	Max pooling		2 × 2
18	Dropout (20%)		
19	Dense	128 neurons	
20	Batch normalization		
21	Dropout (20%)		
22	Dense	10 neurons (MNIST, CIFAR-10) 100 neurons (CIFAR-100)	

4.2 CNN Architectures

Individual classifiers are in the form of a convolutional neural network composed of VGG blocks (i.e. a sequence of convolutional layers, followed by a max pooling layer) [31], additionally enhanced by adding dropout [34] and batch normalization [18]. All convolutional layers and hidden dense layer have ReLU as an activation function and their weights were initialized with *He normal initializer* [15]. Softmax was applied in the output layer while the initial weights were drawn from the *Glorot uniform distribution* [12]. Table 1 summarizes the CNNs architectures. Please note that slight differences in architectures for MNIST, CIFAR-10 and

CIFAR-100 are caused by distinct image sizes and numbers of classes in the datasets. Without any attribute or label noise, a single CNN achieved approximately 99%, 83% and 53% accuracy on MNIST, CIFAR-10 and CIFAR-100, respectively.

4.3 Training Protocol

The following procedure was applied to all three datasets. Partition of a dataset into training and testing subsets was predefined as described in Sect. 4.1. At the very beginning, all features (RGB values) were divided by 255 to fit the $[0, 1]$ range. Images were neither preprocessed nor formatted in any other way. From the training part, we set aside 5 000 samples as a validation set for a single models training and another 5 000 samples for a committee performance comparison. From now on when referring to the training set we would mean all training samples excluding the above-mentioned 10 000 samples used for validation purposes.

To create noisy versions of the datasets we degraded features of the three copies of each dataset by adding Gaussian noise with standard deviation $\sigma = 0.1, 0.2, 0.3$, respectively. The above distortion was applied to the training set, two validation sets and the test set. All affected values were then clipped to $[0, 1]$ range. Next, for the original datasets and each of the three copies influenced by the Gaussian noise, another three copies were created and their training set labels were altered with probability $p = 0.1, 0.2, 0.3$, respectively. If a label was selected to be modified, a new value was chosen from the discrete uniform distribution $\mathcal{U}\{0,\ 9\}$. If the new label value equaled the initial value, then a new label was drawn again, until the sampled label was different from the original one. Hence, we ended up with 16 different versions of each dataset in total (no feature noise plus three degrees of feature noise multiplied by analogous four options regarding the label noise).

The second step was to train CNNs on each of the above-mentioned dataset versions. We set the maximum number of epochs to 30 and batch size to 32. In the case of four consecutive epochs with no improvement on the validation set, training was stopped and weights from the best epoch were restored. The Adam optimizer [19] was used to optimize the cross-entropy loss function:

$-\sum_{c=1}^{K} y_{o,c} \log(p_{o,c})$ where K is the number of classes, $y_{o,c}$ – a binary indicator whether c is a correct class for observation o, and $p_{o,c}$ – a predicted probability that o is from class c. The learning rate was fixed to 0.001.

4.4 Algorithms Parametrization

The *stochastic* algorithm was run with the following parameters: the number of available models – 25, the number of epochs – 16, the number of iterations within an epoch – 1000, probability threshold below which the strongest model is added – 0.5, probability threshold below which the weakest model is removed – 0.5, range of possible committee sizes in each epoch – 10. The above parametrization allows the algorithm to consider any possible committee size from 1 to 25. Other analyzed algorithms do not require setting of any steering parameters.

5 Experimental Results

This section presents experimental results of testing various ensemble selection algorithms. For each pair $(\sigma, p) \in \{0, 0.1, 0.2, 0.3\} \times \{0, 0.1, 0.2, 0.3\}$ 50 CNNs were independently trained from which 25 were drawn to create one instance of experiment. Each experiment was repeated 20 times to obtain reliable results. In the whole study, we assume not having any knowledge regarding either the type or the level of noise the datasets are affected by.

Figure 3 depicts the relative accuracy margin that committees gained over the *top-1* algorithm which selects the best individual model from the whole library of models. Scores are averaged over *top-n, 2-opt-c, 1-opt-c, top-n-2-opt-c, top-n-1-opt-c* and *stochastic* algorithms. For example, if the best individual model achieves 80% accuracy while the mean accuracy of ensembles found by analyzed algorithms equals 88% then the relative margin of ensembles over *top-1* is equal to 10%. *Attribute* curves refer to computations where all scores within particular attribute noise level are averaged over label noise (four values for every attribute noise level). *Label* curves are created analogously–for particular label noise all scores within specific label noise are averaged over attribute noise. *Both* curves concern increasing noise level concurrently on both attributes and labels by the same amount (i.e. with $\sigma = p$). For example, 0.2 value on the x-axis refers to $\sigma = 0.2$ attribute noise and $p = 0.2$ label noise.

Two main conclusions can be drawn from the plots. First, a committee margin rises along with an increase of both noise types (separately and jointly as well). Secondly, a difference increases further for more demanding datasets. In case of MNIST, the difference is less than 1% while when concerning CIFAR-100 the margin amounts to more than 20% for 0.1 and 0.2 noise level and around 30% for 0.3 noise level. Figure 4 illustrates in a concise, aggregated way how algorithms perform in comparison with *top-n* for various noise levels. Values on the y-axis indicate how much (percentage-wise) the margin achieved by *top-n* over *top-1* is better/worse than the margin attained by the rest of the algorithms. For example, if accuracies of *top-1*, *top-n* and *stochastic* methods are 80%, 85% and 85.5%, respectively then the value for *stochastic* algorithm amounts to 10% in that case since 0.5% constitutes 10% of 5%. Line $y = 0$ refers to *top-n*. For each dataset the leftmost plot, for the given level of *attribute* noise, presents scores averaged over the *label* noise (four values for each level). Likewise, in the middle plot, for the given level of *label* noise, the scores averaged over the four values of *attribute* noise are depicted. In the third plot, the scores are not averaged since x-values refer to both *attribute* and *label* noise. From the first row of plots, which refers to the MNIST dataset, it stems that there are huge relative differences in results achieved by the algorithms which, furthermore, vary a lot between noise levels. This phenomenon is caused by the fact that all errors for all noise levels are below 1% in MNIST. Thus, even very little absolute difference between scores may be reflected in high relative value (one instance constitutes 0.01% of test set size). Therefore, it is hard to draw any vital conclusions for this dataset other than a general observation that for a relatively easy dataset the results of all algorithms are close to each other.

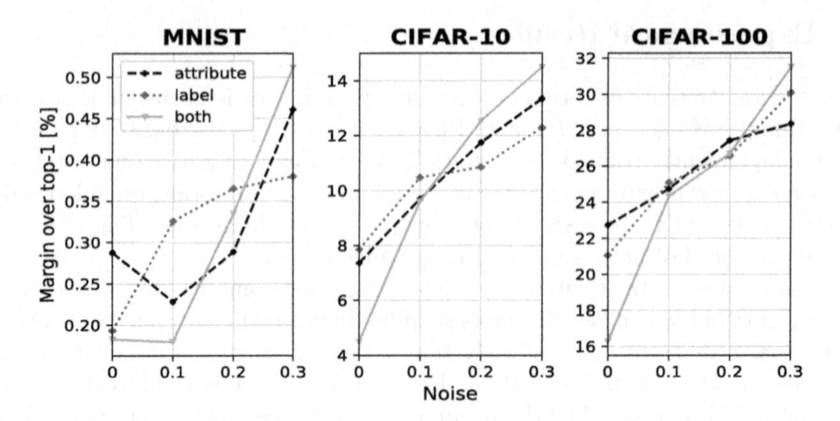

Fig. 3. Relative accuracy margin that committees gained over the *top-1* algorithm.

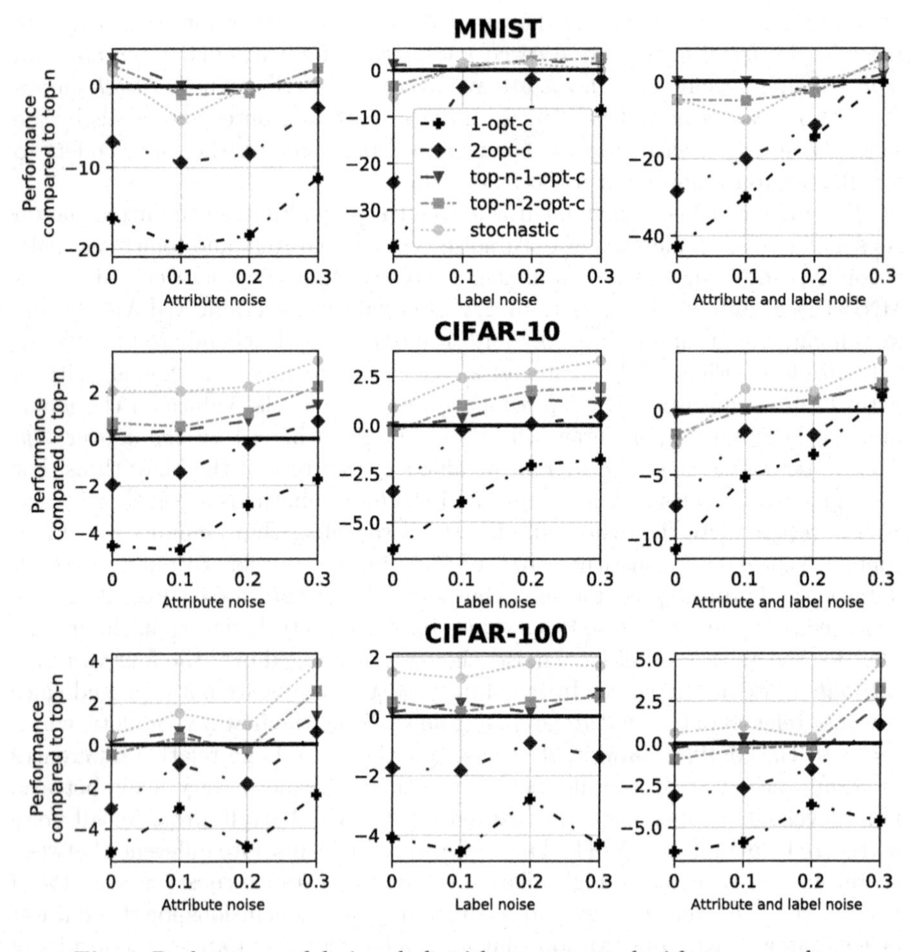

Fig. 4. Performance of designed algorithms contrasted with *top-n* results.

From the plots related to CIFAR-10 and CIFAR-100, one can see that three of our algorithms noticeably surpassed the *top-n* one. The *stochastic* method achieved better results on all noise levels. The only yellow dot below zero refers to no noise case on either attributes and labels. Both *top-n-2-opt-c* and *top-n-1-opt-c* also beat *top-n* in most of the cases. Another observation is that our algorithms are positively correlated with a noise level in the sense that the attained margin rises along with increasing noise.

We have also analyzed 35-sized libraries of the models. The relationships between results achieved by the algorithms remain similar to those with 25 models, only the absolute accuracy values are slightly higher. It is not surprising since algorithms have a wider choice of models and may keep more of them in a committee. As the last remark, we noticed that *2-opt-c* and *1-opt-c* obtained very high accuracy on validation sets (greater than *top-n-2-opt-c* and *top-n-1-opt-c*, respectively) however it was not reflected on test sets. This observation suggests that one has to be careful when dealing with methods whose performance is measured solely on the validation set with neglecting models' diversity, as such committees tend to overfit. We also noticed that sizes of committees found by respective algorithms form the following order: *top-n* > *top-n-1-opt-c* > *top-n-2-opt-c* > *stochastic* > *2-opt-c* > *1-opt-c*. Moreover, the harder dataset is the more numerous ensembles are.

6 Conclusions and Future Work

The main goal of this paper is to address the problem of concurrently occurring feature and label noise in the image classification task which, to the best of our knowledge, has not been considered in the computer vision literature. To this end, we propose five novel ensemble selection algorithms among which four are inspired by the local optimization algorithm derived from the TSP and one employs a stochastic search. Three out of five methods outperform the strong baseline reference algorithm that applies a set of individually selected best models (*top-n*). We have also empirically proven that a margin gained by the committees over the best single model rises along with an increase of both types of noise as well as with raising dataset difficulty, thus making proposed ensembles specifically well-suited to noisy images with noisy labels.

There are a couple of lines of inquiry worth pursuing in the future. Firstly, one may experiment with the parametrization of the *stochastic* algorithm (a range of possible committee sizes in an epoch, a tradeoff between individual performance and ensemble diversity, etc.). Analysis of other correlation measures could be insightful as well. Secondly, all our algorithms operate on probability vectors, which allows us to assume that they would achieve similar results in other domains and are not limited to noisy images only. Finally, this paper addresses only one aspect of ensembling - models selection. Other areas which could be considered when forming a committee model, briefly mentioned in Sect. 1.1, are also worth investigation.

References

1. Bengio, Y., Lamblin, P., Popovici, D., Larochelle, H.: Greedy layer-wise training of deep networks. In: Advances in NIPS, pp. 153–160 (2007)
2. Breve, F.A., Zhao, L., Quiles, M.G.: Semi-supervised learning from imperfect data through particle cooperation and competition. In: The 2010 International Joint Conference on Neural Networks (IJCNN), pp. 1–8. IEEE (2010)
3. Brodley, C.E., Friedl, M.A.: Identifying mislabeled training data. J. Artif. Intell. Res. **11**, 131–167 (1999)
4. Brodley, C.E., Friedl, M.A., et al.: Identifying and eliminating mislabeled training instances. In: Proceedings of the National Conference on Artificial Intelligence, pp. 799–805 (1996)
5. Buades, A., Coll, B., Morel, J.M.: A non-local algorithm for image denoising. In: 2005 IEEE Computer Society Conference on Computer Vision and Pattern Recognition (CVPR 2005), vol. 2, pp. 60–65. IEEE (2005)
6. Chan, T.H., Jia, K., Gao, S., Lu, J., Zeng, Z., Ma, Y.: PCANet: a simple deep learning baseline for image classification? IEEE Trans. Image Process. **24**(12), 5017–5032 (2015)
7. Ciregan, D., Meier, U., Schmidhuber, J.: Multi-column deep neural networks for image classification. In: 2012 IEEE Conference on Computer Vision and Pattern Recognition, pp. 3642–3649. IEEE (2012)
8. Croes, G.A.: A method for solving traveling-salesman problems. Oper. Res. **6**(6), 791–812 (1958)
9. Dietterich, T.G.: An experimental comparison of three methods for constructing ensembles of decision trees: bagging, boosting, and randomization. Mach. Learn. **40**(2), 139–157 (2000). https://doi.org/10.1023/A:1007607513941
10. Dodge, S., Karam, L.: A study and comparison of human and deep learning recognition performance under visual distortions. In: 2017 26th International Conference on Computer Communication and Networks (ICCCN), pp. 1–7. IEEE (2017)
11. Frénay, B., Verleysen, M.: Classification in the presence of label noise: a survey. IEEE Trans. Neural Netw. Learn. Syst. **25**(5), 845–869 (2013)
12. Glorot, X., Bengio, Y.: Understanding the difficulty of training deep feedforward neural networks. In: Proceedings of the Thirteenth International Conference on Artificial Intelligence and Statistics, pp. 249–256 (2010)
13. Glorot, X., Bordes, A., Bengio, Y.: Deep sparse rectifier neural networks. In: Proceedings of the Fourteenth International Conference on Artificial Intelligence and Statistics, pp. 315–323 (2011)
14. González, R., Woods, R.: Digital Image Processing, vol. 60. Prentice Hall, Upper Saddle River (2008). ISBN 9780131687288
15. He, K., Zhang, X., Ren, S., Sun, J.: Delving deep into rectifiers: surpassing human-level performance on imagenet classification. In: Proceedings of the IEEE International Conference on Computer Vision, pp. 1026–1034 (2015)
16. Hickey, R.J.: Noise modelling and evaluating learning from examples. Artif. Intell. **82**(1–2), 157–179 (1996)
17. Hughes, N.P., Roberts, S.J., Tarassenko, L.: Semi-supervised learning of probabilistic models for ECG segmentation. In: The 26th Annual International Conference of the IEEE Engineering in Medicine and Biology Society, vol. 1, pp. 434–437. IEEE (2004)
18. Ioffe, S., Szegedy, C.: Batch normalization: accelerating deep network training by reducing internal covariate shift. arXiv preprint arXiv:1502.03167 (2015)

19. Kingma, D.P., Ba, J.: Adam: a method for stochastic optimization. arXiv preprint arXiv:1412.6980 (2014)
20. Krizhevsky, A., Hinton, G., et al.: Learning multiple layers of features from tiny images (2009)
21. Krizhevsky, A., Sutskever, I., Hinton, G.E.: ImageNet classification with deep convolutional neural networks. In: Advances in NIPS, pp. 1097–1105 (2012)
22. Kuncheva, L.I.: Combining Pattern Classifiers: Methods and Algorithms. Wiley, Hoboken (2014)
23. LeCun, Y., Cortes, C., Burges, C.: MNIST handwritten digit database, **2**. ATT Labs (2010). http://yann.lecun.com/exdb/mnist
24. Nazaré, T.S., da Costa, G.B.P., Contato, W.A., Ponti, M.: Deep convolutional neural networks and noisy images. In: Mendoza, M., Velastín, S. (eds.) CIARP 2017. LNCS, vol. 10657, pp. 416–424. Springer, Cham (2018). https://doi.org/10.1007/978-3-319-75193-1_50
25. Pechenizkiy, M., Tsymbal, A., Puuronen, S., Pechenizkiy, O.: Class noise and supervised learning in medical domains: the effect of feature extraction. In: 19th IEEE Symposium on Computer Based Medical Systems (CBMS 2006), pp. 708–713. IEEE (2006)
26. Polyak, B.T., Juditsky, A.B.: Acceleration of stochastic approximation by averaging. SIAM J. Control Optim. **30**(4), 838–855 (1992)
27. Quinlan, J.R.: Induction of decision trees. Mach. Learn. **1**(1), 81–106 (1986). https://doi.org/10.1007/BF00116251
28. Roy, S.S., Hossain, S.I., Akhand, M., Murase, K.: A robust system for noisy image classification combining denoising autoencoder and convolutional neural network. Int. J. Adv. Comput. Sci. Appl. **9**(1), 224–235 (2018)
29. Sagi, O., Rokach, L.: Ensemble learning: a survey. Wiley Interdisc. Rev. Data Min. Knowl. Discov. **8**(4), e1249 (2018)
30. Shapiro, L.G., Stockman, G.C.: Computer Vision. Prentice Hall, Upper Saddle River (2001)
31. Simonyan, K., Zisserman, A.: Very deep convolutional networks for large-scale image recognition. arXiv preprint arXiv:1409.1556 (2014)
32. Snow, R., O'connor, B., Jurafsky, D., Ng, A.Y.: Cheap and fast-but is it good? Evaluating non-expert annotations for natural language tasks. In: Proceedings of the 2008 Conference on Empirical Methods in Natural Language Processing, pp. 254–263 (2008)
33. Springenberg, J.T., Dosovitskiy, A., Brox, T., Riedmiller, M.: Striving for simplicity: the all convolutional net. arXiv preprint arXiv:1412.6806 (2014)
34. Srivastava, N., Hinton, G., Krizhevsky, A., Sutskever, I., Salakhutdinov, R.: Dropout: a simple way to prevent neural networks from overfitting. J. Mach. Learn. Res. **15**(1), 1929–1958 (2014)
35. Vincent, P., Larochelle, H., Lajoie, I., Bengio, Y., Manzagol, P.A.: Stacked denoising autoencoders: learning useful representations in a deep network with a local denoising criterion. J. Mach. Learn. Res. **11**, 3371–3408 (2010)
36. Zhu, X., Wu, X.: Class noise vs. attribute noise: a quantitative study. Artif. Intell. Rev. **22**(3), 177–210 (2004). https://doi.org/10.1007/s10462-004-0751-8
37. Zhu, X., Wu, X., Chen, Q.: Bridging local and global data cleansing: identifying class noise in large, distributed data datasets. Data Min. Knowl. Disc. **12**(2–3), 275–308 (2006). https://doi.org/10.1007/s10618-005-0012-8
38. Zoph, B., Le, Q.V.: Neural architecture search with reinforcement learning. arXiv preprint arXiv:1611.01578 (2016)

Improving Imbalanced Classification by Anomaly Detection

Jiawen Kong[1(✉)], Wojtek Kowalczyk[1], Stefan Menzel[2], and Thomas Bäck[1]

[1] Leiden University, Leiden, The Netherlands
{j.kong,w.j.kowalczyk,t.h.w.baeck}@liacs.leidenuniv.nl
[2] Honda Research Institute Europe GmbH, Offenbach, Germany
stefan.menzel@honda-ri.de

Abstract. Although the anomaly detection problem can be considered as an extreme case of class imbalance problem, very few studies consider improving class imbalance classification with anomaly detection ideas. Most data-level approaches in the imbalanced learning domain aim to introduce more information to the original dataset by generating synthetic samples. However, in this paper, we gain additional information in another way, by introducing additional attributes. We propose to introduce the outlier score and four types of samples (safe, borderline, rare, outlier) as additional attributes in order to gain more information on the data characteristics and improve the classification performance. According to our experimental results, introducing additional attributes can improve the imbalanced classification performance in most cases (6 out of 7 datasets). Further study shows that this performance improvement is mainly contributed by a more accurate classification in the overlapping region of the two classes (majority and minority classes). The proposed idea of introducing additional attributes is simple to implement and can be combined with resampling techniques and other algorithmic-level approaches in the imbalanced learning domain.

Keywords: Class imbalance · Anomaly detection · Borderline samples

1 Introduction

The imbalanced classification problem has caught growing attention from many fields. In the field of computational design optimization, product parameters are modified to generate digital prototypes and the performances are usually evaluated by numerical simulations which often require minutes to hours of computation time. Here, some parameter variations (minority number of designs) would result in valid and producible geometries but violate given constraints in the final steps of the optimization. Under this circumstance, performing proper imbalanced classification algorithms on the design parameters could save computation time. In the imbalanced learning domain, many techniques have proven to be efficient in handling imbalanced datasets, including resampling techniques

This project has received funding from the European Union's Horizon 2020 research and innovation programme under grant agreement number 766186 (ECOLE).

T. Bäck et al. (Eds.): PPSN 2020, LNCS 12269, pp. 512–523, 2020.
https://doi.org/10.1007/978-3-030-58112-1_35

and algorithmic-level approaches [5,9,15], where the former aims to produce balanced datasets and the latter aims to make classical classification algorithms appropriate for handling imbalanced datasets. The resampling techniques are standard techniques in imbalance learning since they are simple and easily configurable and can be used in synergy with other learning algorithms [4]. The main idea of most oversampling approaches is to introduce more information to the original dataset by creating synthetic samples. However, very few studies consider the idea of introducing additional attributes to the imbalanced dataset.

The anomaly detection problem can be considered as an extreme case of the class imbalance problem. In this paper, we propose to improve the imbalanced classification with some anomaly detection techniques. We propose to introduce the outlier score, which is an important indicator to evaluate whether a sample is an outlier [2], as an additional attribute of the original imbalanced datasets. Apart from this, we also introduce the four types of samples (safe, borderline, rare and outlier), which have been emphasized in many studies [14,16], as another additional attribute. In our experiments, we consider four scenarios, i.e. four different combinations using the additional attributes and performing resampling techniques. The results of our experiments demonstrate that introducing the two proposed additional attributes can improve the imbalanced classification performance in most cases. Further study shows that this performance improvement is mainly contributed by a more accurate classification in the overlapping region of the two classes (majority and minority classes).

The remainder of this paper is organized as follows. In Sect. 2, the research related to our work is presented, also including the relevant background knowledge on four resampling approaches, outlier score and the four types of samples. In Sect. 3, the experimental setup is introduced in order to understand how the results are generated. Section 4 gives the results and further discussion of our experiments. Section 5 concludes the paper and outlines further research.

2 Related Work

As mentioned in the Introduction, we propose to introduce two additional attributes into the imbalanced datasets in order to gain more information on the data characteristics and improve the classification performance. Introducing additional attributes can be regarded as a data preprocessing method, which is independent of resampling techniques and algorithmic-level approaches, and can also be combined with these two approaches. In this section, the background knowledge related to our experiment is given, including resampling techniques (Sect. 2.1), the definition of four types of samples in the imbalance learning domain (Sect. 2.3) and the outlier score (Sect. 2.2).

2.1 Resampling Techniques

In the following, we introduce two oversampling techniques (SMOTE and ADASYN) and two undersampling techniques (NCL and OSS).

Oversampling Techniques. The synthetic minority oversampling technique (SMOTE) is the most famous resampling technique [3]. SMOTE produces synthetic minority samples based on the randomly chosen minority samples and their K-nearest neighbours. The new synthetic sample can be generated by interpolation between the selected minority sample and one of its K-nearest neighbours. The main improvement in the adaptive synthetic (ADASYN) sampling technique is that the samples which are harder to learn are given higher importance and will be oversampled more often in ADASYN [7].

Undersampling Techniques. One-Sided Selection (OSS) is an undersampling technique which combines Tomek Links and the Condensed Nearest Neighbour (CNN) Rule [4,11]. In OSS, noisy and borderline majority samples are removed with so-called Tomek links [17]. The safe majority samples which have limited contribution for building the decision boundary are then removed with CNN. Neighbourhood Cleaning Rule (NCL) emphasizes the quality of the retained majority class samples after data cleaning [12]. The cleaning process is first performed by removing ambiguous majority samples through Wilson's Edited Nearest Neighbour Rule (ENN) [19]. Then, the majority samples which have different labels from their three nearest neighbours are removed. Apart from this, if a minority sample has different labels from its three nearest neighbours, then the three neighbours are removed.

2.2 Four Types of Samples in the Imbalance Learning Domain

Napierala and Stefanowski proposed to analyse the local characteristics of minority examples by dividing them into four different types: safe, borderline, rare examples and outliers [14]. The identification of the type of an example can be done through modeling its k-neighbourhood. Considering that many applications involve both nominal and continuous attributes, the HVDM metric (Appendix A) is applied to calculate the distance between different examples. Given the number of neighbours k (odd), the label to a minority example can be assigned through the ratio of the number of its minority neighbours to the total number of neighbours ($R_{\frac{min}{all}}$) according to Table 1. The label for a majority example can be assigned in a similar way.

Table 1. Rules to assign the four types of minority examples.

Type	Safe (S)	Borderline (B)	Rare (R)	Outlier (O)
Rule	$\frac{k+1}{2k} < R_{\frac{min}{all}} \leqslant 1$	$\frac{k-1}{2k} \leqslant R_{\frac{min}{all}} \leqslant \frac{k+1}{2k}$	$0 < R_{\frac{min}{all}} < \frac{k-1}{2k}$	$R_{\frac{min}{all}} = 0$
E.G. given the neighbourhood of a fixed size $k = 5$				
Rule	$\frac{3}{5} < R_{\frac{min}{all}} \leqslant 1$	$\frac{2}{5} \leqslant R_{\frac{min}{all}} \leqslant \frac{3}{5}$	$0 < R_{\frac{min}{all}} < \frac{2}{5}$	$R_{\frac{min}{all}} = 0$

2.3 Outlier Score

Many algorithms have been developed to deal with anomaly detection problems and the experiments in this paper are mainly performed with the nearest-neighbour based local outlier score (LOF). Local outlier factor (LOF), which indicates the degree of a sample being an outlier, was first introduced by Breunig et al. in 2000 [2]. The LOF of an object depends on its relative degree of isolation from its surrounding neighbours. Several definitions are needed to calculate the LOF and are summarized in the following Algorithm 1.

Algorithm 1: Local Outlier Factor (LOF) algorithm [2]

Input : $\mathbf{X}$ - input data $\mathbf{X} = (X_1, ..., X_n)$
n - the number of input examples
k - the number of neighbours
Output: LOF score of every X_i

1 initialization;
2 calculate the distance $d(\cdot)$ between every two data points;
3 **for** $i = 1$ **to** n **do**
4 calculate *k-distance*(X_i): the distance between X_i and its kth neighbour;
5 find out k-distance neighbourhood $N_k(X_i)$: the set of data points whose distance from X_i is not greater than *k-distance*(X_i);
6 **for** $j = 1$ **to** n **do**
7 calculate reachability distance:

$$reach\text{-}dist_k(X_i, X_j) = \max\{k\text{-}distance(X_j), d(X_i, X_j)\};$$

8 calculate local reachability density:

$$lrd_k(X_i) = 1/avg\text{-}reach\text{-}dist_k(X_i) = 1/\left(\frac{\sum_{o \in N_k(X_i)} reach\text{-}dist_k(X_i, X_j)}{|N_k(X_i)|}\right);$$

intuitively, the local reachability density of X_i is the inverse of the average reachability distance based on the k-nearest neighbours of X_i;
9 calculate LOF:

$$LOF_k(X_i) = \frac{\sum_{o \in N_k(X_i)} lrd_k(X_j)}{|N_k(X_i)| \cdot lrd_k(X_i)} = \frac{\sum_{o \in N_k(X_i)} \frac{lrd_k(X_j)}{lrd_k(X_i)}}{|N_k(X_i)|}$$

the LOF of X_i is the average local reachability density of X_i's k-nearest neighbours divided by the local reachability density of X_i.
10 **end**
11 **end**

According to the definition of LOF, a value of approximately 1 indicates that the local density of data point X_i is similar to its neighbours. A value below 1 indicates that data point X_i locates in a relatively denser area and does not seem to be an anomaly, while a value significantly larger than 1 indicates that data point X_i is alienated from other points, which is most likely an outlier.

3 Experimental Setup

In this paper, we propose to introduce the *four types of samples* and the *outlier score* as additional attributes of the original imbalanced dataset, where the former can be expressed as $R_{\frac{min}{all}}$ (Table 1) and the latter can be calculated through Python library `PyOD` [20].

The experiments reported in this paper are based on 7 two-class imbalanced datasets, including 6 imbalanced benchmark datasets (given in Table 2) and a 2D imbalanced chess dataset, which is commonly used for visualising the effectiveness of the selected techniques in the imbalanced learning domain [4]. Imbalance ratio (IR) is the ratio of the number of majority class samples to the number of minority class samples. For each dataset, we consider four scenarios, whether to perform resampling techniques on the original datasets and whether to perform resampling techniques on the datasets with additional attributes. For each scenario of each dataset, we repeat the experiments 30 times with different random seeds. After that, the paired t-tests were performed on each of the 30 performance metric values to test if there is significant difference between the results of each scenario on a 5% significance level. Each collected dataset is divided into 5 stratified folds (for cross-validation) and only the training set is oversampled, where the stratified fold is to ensure that the imbalance ratio in the training set is consistent with the original dataset and only oversampling the training set is to avoid over-optimism problem [15]. Our code is available on Github (https://github.com/FayKong/PPSN2020) for the convenience of reproducing the main results and figure.

Table 2. Information on benchmark datasets [1].

Datasets	#Attributes	#Samples	Imbalance ratio (IR)
glass1	9	214	1.82
ecoli4	7	336	15.8
vehicle1	18	846	2.9
yeast4	8	1484	28.1
wine quality	11	1599	29.17
page block	10	5472	8.79

In this paper, we evaluate the performance through several different measures, including Area Under the ROC Curve (AUC), precision, recall, F-Measure

(F1) and Geometric mean (Gmean) [13]. These performance measures can be calculated as follows.

$$AUC = \frac{1 + TP_{rate} - FP_{rate}}{2}, \quad TP_{rate} = \frac{TP}{TP + FN}, \quad FP_{rate} = \frac{FP}{FP + TN};$$

$$GM = \sqrt{\frac{TP}{TP + FN} \times \frac{TN}{FP + TN}}; \quad FM = \frac{(1 + \beta)^2 \times Recall \times Precision}{\beta^2 \times Recall + Precision}$$

$$Recall = TP_{rate} = \frac{TP}{TP + FN}, \quad Precision = \frac{TP}{TP + FP}, \quad \beta = 1;$$

where TP, FN, FP, TN indicate True Positives, False Negatives, False Negatives and True Negatives in the standard confusion matrix for binary classification.

4 Experimental Results and Discussion

Like other studies [7,13], we also use `SVM` and `Decision Tree` as the base classifiers in our experiments to compare the performance of the proposed method and the existing methods. Please note that we did not tune the hyperparameters for the classification algorithms and the resampling techniques [9]. The experimental results with the two additional attributes (four types of samples and LOF score) are presented in Table 3. We can observe that introducing outlier score and four types of samples as additional attributes can significantly improve the imbalanced classification performance in most cases. For 5 out of 7 datasets (*2D chess dataset, glass1, yeast4, wine quality and page block*), only introducing additional attributes (with no resampling) gives better results than performing resampling techniques.

According to our experimental setup, we notice that introducing the outlier score focuses on dealing with the minority samples since the outlier score indicates the degree of a sample being an outlier. Meanwhile, introducing four types of samples (safe, borderline, rare and outlier) puts emphasis on separating the overlapping region and safe region. The visualisation of different scenarios for the *2D chess* dataset is given in Fig. 1 in order to further study the reason for the performance improvement.

From both the experimental results in Table 3 and the visualisation in Fig. 1, we can conclude that, for the *2D chess* dataset, the experiment with the two additional attributes outperforms the experiment with the classical resampling technique SMOTE. The figure also illustrates that the proposed method has a better ability to handle samples in the overlapping region.

Table 3. Experimental results with `SVM` and `Decision Tree`. "**Add** = YES" means we introduce the two additional attributes to the original datasets; gray cells indicate that the proposed method (**Add** = YES) significantly outperforms the existing methods (**Add** = NO); "—" means that TP + FN = 0 or TP + FP = 0 and the performance metric cannot be computed.

2D chess dataset											
Methods	Add	Decision Tree					SVM				
		AUC	Precision	Recall	F1	Gmean	AUC	Precision	Recall	F1	Gmean
NONE	NO	0.8482	0.5743	0.6992	0.6208	0.8047	0.8285	—	—	—	—
	YES	0.9771	0.9557	0.9070	0.9226	0.9469	0.9859	0.9846	0.9485	0.9643	0.9723
SMOTE	NO	0.8584	0.6422	0.7102	0.6646	0.8183	0.5921	0.1636	0.5004	0.2437	0.5855
	YES	0.9704	0.9191	0.9061	0.9064	0.9453	0.9933	0.9633	0.9667	0.9622	0.9801
ADASYN	NO	0.8482	0.5743	0.6992	0.6208	0.8047	0.6172	0.1434	0.5904	0.2299	0.5892
	YES	0.9771	0.9557	0.9070	0.9226	0.9469	0.9925	0.8546	0.9667	0.8999	0.9721
NCL	NO	0.5786	0.1245	0.6652	0.2092	0.5541	0.5290	0.1076	0.4212	0.1693	0.4802
	YES	0.9715	0.8542	0.9667	0.8988	0.9716	0.9946	0.9119	0.9667	0.9337	0.9766
OSS	NO	0.7569	0.4197	0.5227	0.4554	0.6813	0.6262	0.3050	0.0295	0.0535	0.0958
	YES	0.9743	0.9321	0.9391	0.9316	0.9640	0.9937	0.9532	0.9564	0.9524	0.9745
glass1 dataset											
Methods	Add	Decision Tree					SVM				
		AUC	Precision	Recall	F1	Gmean	AUC	Precision	Recall	F1	Gmean
NONE	NO	0.7029	0.6099	0.6235	0.6044	0.6806	0.6779	0.6394	0.5533	0.5828	0.6633
	YES	0.7328	0.6283	0.6344	0.6227	0.6956	0.7779	0.6506	0.65917	0.6430	0.7089
SMOTE	NO	0.7008	0.5750	0.6561	0.6060	0.6782	0.7140	0.5125	0.7236	0.5785	0.6111
	YES	0.7595	0.6393	0.6988	0.6589	0.7273	0.8288	0.6537	0.8802	0.7369	0.7760
ADASYN	NO	0.7095	0.5922	0.6728	0.6187	0.6842	0.7338	0.5159	0.7982	0.6103	0.6271
	YES	0.7799	0.6614	0.7106	0.6780	0.7419	0.8388	0.6545	0.8996	0.7456	0.7845
NCL	NO	0.5926	0.4401	0.9302	0.5843	0.3761	0.6750	0.4124	1.0000	0.5765	0.2177
	YES	0.5897	0.3976	0.9239	0.5527	0.3806	0.7790	0.4299	1.0000	0.5948	0.3403
OSS	NO	0.7010	0.5688	0.6841	0.6132	0.6804	0.6810	0.5850	0.5837	0.5683	0.6444
	YES	0.7611	0.6342	0.7136	0.6637	0.7295	0.7784	0.6085	0.7382	0.6543	0.7128
ecoli4 dataset											
Methods	Add	Decision Tree					SVM				
		AUC	Precision	Recall	F1	Gmean	AUC	Precision	Recall	F1	Gmean
NONE	NO	0.8446	0.7241	0.6433	0.6432	0.7694	0.9919	0.8889	0.8000	0.7993	0.8797
	YES	0.8525	0.6435	0.6017	0.5734	0.6920	0.9889	0.9143	0.7500	0.7835	0.8512
SMOTE	NO	0.8824	0.7938	0.7233	0.7102	0.8328	0.9894	0.8290	0.8000	0.7268	0.8457
	YES	0.8629	0.8315	0.7300	0.7262	0.8303	0.9931	0.8824	0.9500	0.8881	0.9639
ADASYN	NO	0.8719	0.8407	0.7083	0.7221	0.8236	0.9903	0.7813	0.8000	0.7034	0.8389
	YES	0.8747	0.7833	0.6717	0.6623	0.7822	0.9934	0.8800	0.9500	0.8857	0.9634
NCL	NO	0.8007	0.6080	0.6333	0.5651	0.7380	0.9869	0.8258	0.9000	0.7886	0.8976
	YES	0.8523	0.7297	0.7550	0.6499	0.7982	0.9914	0.8533	0.9500	0.8556	0.9549
OSS	NO	0.8398	0.6284	0.7250	0.5958	0.7872	0.9877	0.8458	0.8133	0.7580	0.8668
	YES	0.9115	0.6858	0.8350	0.6787	0.8586	0.9890	0.8830	0.9117	0.8626	0.9408
vehicle1 dataset											
Methods	Add	Decision Tree					SVM				
		AUC	Precision	Recall	F1	Gmean	AUC	Precision	Recall	F1	Gmean
NONE	NO	0.6699	0.5018	0.4301	0.4575	0.6004	0.8673	0.7074	0.3593	0.4747	0.5824
	YES	0.7385	0.5855	0.5329	0.5573	0.6794	0.9081	0.6873	0.6266	0.6536	0.7500
SMOTE	NO	0.7241	0.5398	0.5557	0.5458	0.6796	0.8945	0.5538	0.9237	0.6913	0.8264
	YES	0.7403	0.5825	0.5629	0.5704	0.6938	0.9204	0.5808	0.9745	0.7272	0.8582
ADASYN	NO	0.7211	0.5359	0.5570	0.5446	0.6791	0.8995	0.5485	0.9465	0.6937	0.8303
	YES	0.7481	0.5842	0.5789	0.5797	0.7025	0.9206	0.5800	0.9809	0.7284	0.8597
NCL	NO	0.7411	0.4153	0.9506	0.5769	0.7093	0.8411	0.4108	0.9768	0.5776	0.7059
	YES	0.7781	0.4560	0.9392	0.6118	0.7529	0.8752	0.5076	1.0000	0.6728	0.8139
OSS	NO	0.7125	0.4857	0.6066	0.5370	0.6837	0.8702	0.5745	0.7014	0.6293	0.7560
	YES	0.7531	0.5524	0.6286	0.5859	0.7174	0.9062	0.6088	0.9117	0.7290	0.8515
yeast4 dataset											
Methods	Add	Decision Tree					SVM				
		AUC	Precision	Recall	F1	Gmean	AUC	Precision	Recall	F1	Gmean
NONE	NO	0.6736	0.3619	0.2217	0.2653	0.4482	0.8469	—	—	—	—
	YES	0.8647	0.8320	0.6708	0.7260	0.8132	0.9910	0.8628	0.8036	0.8270	0.8920
SMOTE	NO	0.7320	0.2632	0.4029	0.3082	0.6082	0.9052	0.2112	0.6769	0.3160	0.7773
	YES	0.9115	0.7665	0.6892	0.7171	0.8235	0.9922	0.7096	0.9442	0.8079	0.9639
ADASYN	NO	0.7226	0.2494	0.3958	0.2963	0.6041	0.9011	0.2061	0.6902	0.3104	0.7815
	YES	0.9114	0.7531	0.6553	0.6906	0.8036	0.9923	0.6951	0.9618	0.8051	0.9727
NCL	NO	0.8176	0.1929	0.6819	0.2992	0.7772	0.9063	0.2552	0.5745	0.3516	0.7256
	YES	0.9785	0.6733	0.9772	0.7928	0.9791	0.9917	0.7512	0.9436	0.8337	0.9649
OSS	NO	0.7066	0.2899	0.3561	0.3020	0.5713	0.8488	0.2094	0.0258	0.0447	0.0781
	YES	0.9130	0.7637	0.7699	0.7532	0.8708	0.9892	0.8312	0.8390	0.8310	0.9121
wine quality dataset											
Methods	Add	Decision Tree					SVM				
		AUC	Precision	Recall	F1	Gmean	AUC	Precision	Recall	F1	Gmean
NONE	NO	0.5844	0.1180	0.1275	0.1132	0.2817	0.9790	0.9653	0.9113	0.9333	0.9525
	YES	0.9790	0.9653	0.9113	0.9333	0.9525	0.9944	0.9636	0.8274	0.8761	0.9031
SMOTE	NO	0.5597	0.0648	0.1801	0.0930	0.3704	0.6935	0.1065	0.4223	0.1680	0.5941
	YES	0.9685	0.9715	0.8630	0.9031	0.9239	0.9942	0.8809	0.9055	0.8890	0.9488
ADASYN	NO	0.5601	0.0654	0.1909	0.0953	0.3800	0.6920	0.1039	0.4231	0.1650	0.5933
	YES	0.9859	0.9709	0.8467	0.8917	0.9141	0.9944	0.8805	0.9055	0.8888	0.9488
NCL	NO	0.5922	0.1037	0.2593	0.1423	0.4817	0.7207	0.2582	0.1891	0.1818	0.3755
	YES	0.9845	0.8567	0.9492	0.8949	0.9703	0.9939	0.9359	0.8818	0.8890	0.9308
OSS	NO	0.5733	0.0729	0.2158	0.1054	0.4135	0.5078	—	—	—	—
	YES	0.9859	0.9636	0.9818	0.9723	0.9901	0.9941	0.9282	0.9424	0.9307	0.9690
page block dataset											
Methods	Add	Decision Tree					SVM				
		AUC	Precision	Recall	F1	Gmean	AUC	Precision	Recall	F1	Gmean
NONE	NO	0.9083	0.8108	0.7442	0.7687	0.8519	0.9723	0.8743	0.7046	0.7663	0.8304
	YES	0.9369	0.8535	0.8289	0.8350	0.9014	0.9880	0.8481	0.8460	0.8379	0.9091
SMOTE	NO	0.9122	0.7485	0.7910	0.7620	0.8735	0.9646	0.6815	0.8792	0.7536	0.9099
	YES	0.9300	0.8216	0.8404	0.8245	0.9051	0.9847	0.7404	0.9496	0.8251	0.9533
ADASYN	NO	0.9130	0.7302	0.7990	0.7558	0.8763	0.9613	0.5716	0.9277	0.6983	0.9194
	YES	0.9328	0.8452	0.8321	0.8356	0.9032	0.9843	0.7529	0.9726	0.8435	0.9661
NCL	NO	0.9338	0.6528	0.9091	0.7502	0.9223	0.9669	0.6628	0.8960	0.7412	0.9127
	YES	0.9563	0.7318	0.9400	0.8156	0.9474	0.9844	0.7355	0.9606	0.8255	0.9577
OSS	NO	0.9071	0.7297	0.7936	0.7473	0.8711	0.9555	0.8375	0.6755	0.7310	0.8107
	YES	0.9248	0.7820	0.8349	0.7957	0.8972	0.9808	0.7845	0.8655	0.8111	0.9137

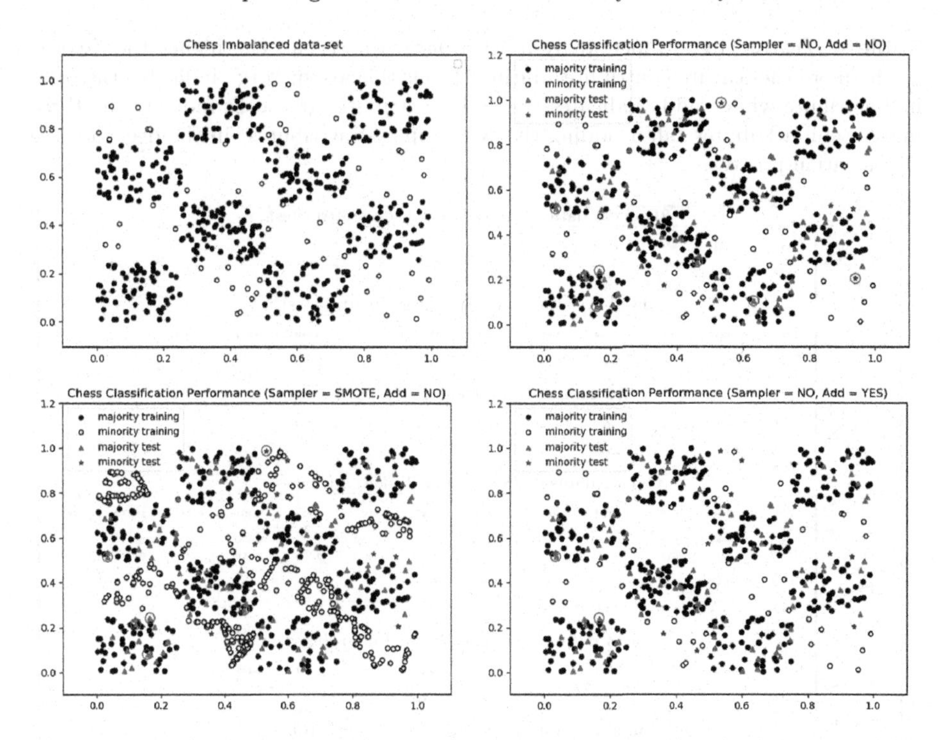

Fig. 1. [top left]. Original imbalanced 2D chess dataset. [top right]. Classification performance for original chess dataset. The red-circled points indicate the misclassified points. [bottom left]. Classification performance for SMOTE-sampled chess dataset. [bottom right]. Classification performance for chess dataset with additional attributes. (Color figure online)

Apart from the visualisation, the feature importance (with `Decision Tree`) is also analysed in order to get an additional insight into the usefulness of the new attributes. Detailed importance score of each attribute is shown in Table 4. According to the feature importance analysis, we can conclude that the introduced "four types of samples" attribute plays an important role in the decision tree classification process for all datasets in our experiment. For 3 out of 7 datasets, the introduced "outlier score" attribute provides useful information during the classification process. The conclusions above show that the two introduced attributes are actually used in the decision process and the "four types of samples" attribute is more important than the "outlier score" attribute.

Table 4. Feature importance analysis with `Decision Tree`. The higher the "score" is, the more the feature contributes during the classification; "org" indicates the original attribute while "add" indicates the added attribute; grey cells indicate the three most useful attributes (after adding the two proposed attributes) in the decision tree classification process.

2D chess dataset (2 original & 2 added attributes)													
Score \ Attr / Add	org1	org2	add1	add2	—	—	—	—	—	—	—	—	—
NO	0.4636	0.5364	—	—	—	—	—	—	—	—	—	—	—
YES	0.0101	0.0097	0.8152	0.1650	—	—	—	—	—	—	—	—	—
glass1 dataset (9 original & 2 added attributes)													
Score \ Attr / Add	org1	org2	org3	org4	org5	org6	org7	org8	org9	add1	add2	—	—
NO	0.2063	0.0213	0.2354	0.1291	0.0302	0.0418	0.2634	0.0000	0.0726	—	—	—	—
YES	0.1770	0.0056	0.1527	0.1099	0.0000	0.0110	0.1892	0.0000	0.0056	0.2413	0.1077	—	—
ecoli4 dataset (7 original & 2 added attributes)													
Score \ Attr / Add	org1	org2	org3	org4	org5	org6	org7	add1	add2	—	—	—	—
NO	0.1093	0.0587	0.0000	0.0000	0.6591	0.1729	0.0000	—	—	—	—	—	—
YES	0.0000	0.0337	0.0000	0.0000	0.6119	0.0000	0.0808	0.1742	0.0994	—	—	—	—
vehicle1 dataset (18 original & 2 added attributes)													
Score \ Attr / Add	org1	org2	org3	org4	org5	org6	org7	org8	org9	org10	org11	org12	org13
NO	0.1304	0.0654	0.0892	0.0403	0.0563	0.0233	0.0028	0.0707	0.0000	0.0635	0.0172	0.0416	0.0438
YES	0.0248	0.0216	0.1317	0.0426	0.0227	0.0205	0.0179	0.0024	0.0000	0.0338	0.0260	0.1291	0.0828
Score \ Attr / Add	org14	org15	org16	org17	org18	add1	add2	—	—	—	—	—	—
NO	0.0862	0.0414	0.0498	0.0516	0.1265	—	—	—	—	—	—	—	—
YES	0.0146	0.0310	0.0325	0.0291	0.0471	0.2413	0.0485	—	—	—	—	—	—
yeast4 dataset (8 original & 2 added attributes)													
Score \ Attr / Add	org1	org2	org3	org4	org5	org6	org7	org8	add1	add2	—	—	—
NO	0.3301	0.2446	0.1839	0.0720	0.0106	0.0000	0.1233	0.0355	—	—	—	—	—
YES	0.0385	0.0297	0.0483	0.0116	0.0000	0.0000	0.0248	0.0053	0.7771	0.0646	—	—	—
wine quality dataset (11 original & 2 added attributes)													
Score \ Attr / Add	org1	org2	org3	org4	org5	org6	org7	org8	org9	org10	org11	add1	add2
NO	0.0466	0.1402	0.1215	0.1194	0.0806	0.0635	0.0428	0.1287	0.0483	0.0841	0.1244	—	—
YES	0.0000	0.0098	0.0000	0.0000	0.0000	0.0000	0.0263	0.0000	0.0000	0.0000	0.0000	0.9639	0.0000
page block dataset (10 original & 2 added attributes)													
Score \ Attr / Add	org1	org2	org3	org4	org5	org6	org7	org8	org9	org10	add1	add2	—
NO	0.5452	0.0096	0.0117	0.1899	0.0530	0.0285	0.0983	0.0382	0.0122	0.0134	—	—	—
YES	0.5282	0.0006	0.0036	0.1745	0.0205	0.0223	0.0833	0.0129	0.0007	0.0082	0.1288	0.0164	—

5 Conclusions and Future Research

In this paper, we propose to introduce additional attributes to the original imbalanced datasets in order to improve the classification performance. Two additional attributes, namely four types of samples and outlier score, and the resampling techniques (SMOTE, ADASYN, NCL and OSS) are considered and experimentally tested on seven imbalanced datasets. According to our experimental results, two main conclusions can be derived:

1) In most cases, introducing these two additional attributes can improve the class imbalance classification performance. For some datasets, only introducing additional attributes gives better classification results than only performing resampling techniques.
2) An analysis of the experimental results also illustrates that the proposed method has a better ability to handle samples in the overlapping region.

In this paper, we only validate our idea with four resampling techniques and seven benchmark datasets. As future work, other anomaly detection techniques, such as the clustering-based local outlier score (CBLOF) [8] and histogram-based outlier score (HBOS) [6] could be included in the analysis. Future work could also consider an extension of this research for engineering datasets [10], especially for the design optimization problems mentioned in our Introduction. Detailed analysis of the feature importance and how the proposed method affects the classification performance in the overlapping region would also be worth studying.

Appendix A

Heterogeneous Value Difference Metric (HVDM)

HVDM is a heterogeneous distance function that returns the distance between two vectors **x** and **y** [18], where the vectors can involve both nominal and numerical attributes. The HVDM distance can be calculated by [18]:

$$HVDM(\mathbf{x}, \mathbf{y}) = \sqrt{\sum_{a=1}^{n} {d_a}^2(x_a, y_a)}, \tag{1}$$

where n is the number of attributes. The function $d_a(\cdot)$ returns the distance between x_a and y_a, where x_a, y_a indicate the ath attribute of vector x and y respectively. It is defined as follows:

$$d_a(x, y) = \begin{cases} 1, & \text{if } x \text{ or } y \text{ is unknown} \\ \text{norm\_vdm}_a(x, y), & \text{if } a\text{th attribute is nominal} \\ \text{norm\_diff}_a(x, y), & \text{if } a\text{th attribute is continuous} \end{cases} \tag{2}$$

where

$$\text{norm\_vdm}_a(x, y) = \sqrt{\sum_{c=1}^{C} \left| \frac{N_{a,x,c}}{N_{a,x}} - \frac{N_{a,y,c}}{N_{a,y}} \right|^2}, \quad \text{norm\_diff}_a(x, y) = \frac{|x - y|}{4\sigma_a}, \tag{3}$$

where

- C is the number of total output classes,
- $N_{a,x,c}$ is the number of instances which have value x for the ath attribute and output class c and $N_{a,x} = \sum_{c=1}^{C} N_{a,x,c}$,
- σ_a is the standard deviation of values of the ath attribute.

References

1. Alcalá-Fdez, J., et al.: KEEL data-mining software tool: data set repository, integration of algorithms and experimental analysis framework. J. Multiple-Valued Log. Soft Comput. **17** (2011)

2. Breunig, M.M., Kriegel, H.P., Ng, R.T., Sander, J.: LOF: identifying density-based local outliers. In: Proceedings of the 2000 ACM SIGMOD International Conference on Management of Data, pp. 93–104 (2000)
3. Chawla, N.V., Bowyer, K.W., Hall, L.O., Kegelmeyer, W.P.: SMOTE: synthetic minority over-sampling technique. J. Artif. Intell. Res. **16**, 321–357 (2002)
4. Fernández, A., García, S., Galar, M., Prati, R.C., Krawczyk, B., Herrera, F.: Learning from Imbalanced Data Sets. Springer, Cham (2018). https://doi.org/10.1007/978-3-319-98074-4
5. Ganganwar, V.: An overview of classification algorithms for imbalanced datasets. Int. J. Emerg. Technol. Adv. Eng. **2**(4), 42–47 (2012)
6. Goldstein, M., Dengel, A.: Histogram-based outlier score (HBOS): a fast unsupervised anomaly detection algorithm. KI-2012: Poster and Demo Track, pp. 59–63 (2012)
7. He, H., Bai, Y., Garcia, E.A., Li, S.: ADASYN: adaptive synthetic sampling approach for imbalanced learning. In: 2008 IEEE International Joint Conference on Neural Networks (IEEE World Congress on Computational Intelligence), pp. 1322–1328. IEEE (2008)
8. He, Z., Xu, X., Deng, S.: Discovering cluster-based local outliers. Pattern Recogn. Lett. **24**(9–10), 1641–1650 (2003)
9. Kong, J., Kowalczyk, W., Nguyen, D.A., Bäck, T., Menzel, S.: Hyperparameter optimisation for improving classification under class imbalance. In: 2019 IEEE Symposium Series on Computational Intelligence (SSCI), pp. 3072–3078. IEEE (2019)
10. Kong, J., Rios, T., Kowalczyk, W., Menzel, S., Bäck, T.: On the performance of oversampling techniques for class imbalance problems. In: Lauw, H.W., Wong, R.C.-W., Ntoulas, A., Lim, E.-P., Ng, S.-K., Pan, S.J. (eds.) PAKDD 2020. LNCS (LNAI), vol. 12085, pp. 84–96. Springer, Cham (2020). https://doi.org/10.1007/978-3-030-47436-2_7
11. Kubat, M., Matwin, S., et al.: Addressing the curse of imbalanced training sets: one-sided selection. In: ICML, Nashville, USA, vol. 97, pp. 179–186 (1997)
12. Laurikkala, J.: Improving identification of difficult small classes by balancing class distribution. In: Quaglini, S., Barahona, P., Andreassen, S. (eds.) AIME 2001. LNCS (LNAI), vol. 2101, pp. 63–66. Springer, Heidelberg (2001). https://doi.org/10.1007/3-540-48229-6_9
13. López, V., Fernández, A., García, S., Palade, V., Herrera, F.: An insight into classification with imbalanced data: empirical results and current trends on using data intrinsic characteristics. Inf. Sci. **250**, 113–141 (2013)
14. Napierala, K., Stefanowski, J.: Types of minority class examples and their influence on learning classifiers from imbalanced data. J. Intell. Inf. Syst. **46**(3), 563–597 (2015). https://doi.org/10.1007/s10844-015-0368-1
15. Santos, M.S., Soares, J.P., Abreu, P.H., Araujo, H., Santos, J.: Cross-validation for imbalanced datasets: avoiding overoptimistic and overfitting approaches [research frontier]. IEEE Comput. Intell. Mag. **13**(4), 59–76 (2018)
16. Skryjomski, P., Krawczyk, B.: Influence of minority class instance types on SMOTE imbalanced data oversampling. In: First International Workshop on Learning with Imbalanced Domains: Theory and Applications, pp. 7–21 (2017)
17. Tomek, I.: Two modifications of CNN. IEEE Trans. Syst. Man Cybern. **6**, 769–772 (1976)
18. Wilson, D.R., Martinez, T.R.: Improved heterogeneous distance functions. J. Artif. Intell. Res. **6**, 1–34 (1997)

19. Wilson, D.L.: Asymptotic properties of nearest neighbor rules using edited data. IEEE Trans. Syst. Man Cybern. **SMC-2**(3), 408–421 (1972)
20. Zhao, Y., Nasrullah, Z., Li, Z.: PyOD: a python toolbox for scalable outlier detection. arXiv preprint arXiv:1901.01588 (2019)

BACS: A Thorough Study of Using Behavioral Sequences in ACS2

Romain Orhand[1,2(✉)], Anne Jeannin-Girardon[1,2], Pierre Parrend[1,3], and Pierre Collet[1,2]

[1] Icube Laboratory - UMR 7357, 300 bd Sébastien Brant, 67412 Illkirch, France
[2] University of Strasbourg, 4 rue Blaise Pascal, 67081 Strasbourg, France
[3] ECAM Strasbourg-Europe, 2 rue de Madrid, 67300 Schiltigheim, France
{rorhand,anne.jeannin,pierre.parrend,pierre.collet}@unistra.fr

Abstract. This papers introduces BACS, a learning classifier system that integrates Behavioral Sequences to ACS2 (Anticipatory Classifier System 2), in order to address the Perceptual Aliasing Issue: this issue occurs when systems can not differentiate situations that are truly distinct in partially observable environments. In order to permit this integration, BACS implements (1) an aliased-state detection algorithm allowing the system to build behavioral classifiers and (2) an evolved Anticipatory Learning Process. A study of the capabilities of BACS is presented through a thorough benchmarking on 23 mazes. The obtained results show that Behavioral Sequences are a suitable approach to address the perceptual aliasing issue.

Keywords: Anticipatory Learning Classifier System · Machine learning · Behavioral Sequences · Perceptual aliasing issue

1 Introduction

Learning Classifier Systems (LCS) are rule-based machine learning algorithms that combine several components to build a population of rules (also called classifiers) according either to the environment in which they evolve, or to association patterns within some data [24]. Those components rely on different learning paradigms to (1) generate classifiers making local approximations of the problem they face and (2) evolve their population of classifiers so as to form effectively distributed and accurate approximations [6]. Hence, they can be applied to a wide variety of areas such as classification problems, function approximation problems or reinforcement learning problems [5].

The work presented in this paper focuses on reinforcement learning problems, especially those where only partial observations are provided, as this is the case in many real-world environments in applications such as for self-driving cars [23] or robotics [22]. Such environment are challenging for LCS when they cannot differentiate each state of their environment with respect to their current perception, leading them to consider different states as identical due to their

T. Bäck et al. (Eds.): PPSN 2020, LNCS 12269, pp. 524–538, 2020.
https://doi.org/10.1007/978-3-030-58112-1_36

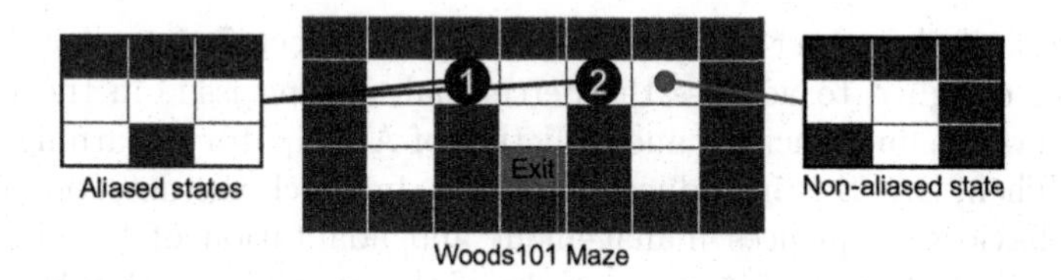

Fig. 1. Illustration of the perceptual aliasing issue within the Woods101 maze, if the provided observations are the eight squares adjacent to each position. Positions 1 and 2 are aliased states. (Color figure online)

perceptive capabilities: these states are said to be aliased. LCS are then said to suffer from the *perceptual aliasing issue* [14] that can prevent the system from achieving its task or from building a relevant population of classifiers.

Maze environments have been widely used as reinforcement learning benchmarks within the learning classifier system literature [2]. Figure 1 depicts an example of a maze environment, called Woods101. In this maze, two states (depicted in red) are represented by the same perception, given that the perceptive capabilities are limited to the eight squares adjacent to each position. The two positions circled in red are aliased because they are indistinguishable from the agent's perception, in comparison with the state circled in blue.

The work hereby presented targets **Anticipatory Learning Classifier Systems** (ALCS), because they are based on the cognitive mechanism of Anticipatory Behavioral Control (ABC) [25]. Anticipatory Behavioral Control consists in learning to perform new tasks by *anticipating* the consequences of an action while being in a specific state. This learning is possible according to the association of "conditions", "actions" and "effects" described in [13]. This mechanisms allows ALCS to learn directly from the cause and effect relationship they anticipates, instead of relying on an accuracy measure of the reward predictions during a mostly stochastic process, as done in XCS [31]. Furthermore, ABC provides more flexibility for implementing other cognitive mechanisms such as Latent Learning (learning without reinforcement or motivation), One-Step Mental Acting (learning from the execution of hypothetical actions) or the Lookahead-Winner Algorithm (a look-ahead planning combined with a reward learning process) [26].

ALCS can tackle the perceptual aliasing issue by learning to predict the next state given actions carried out under specific conditions: the predicted state can become a condition to be considered for further actions, making it possible for the system to build specific sequences of actions called **Behavioral Sequences** [25]. These behavioral sequences consist in a succession of actions that have been triggered by a unique classifier and executed subsequently, enabling the system to deal with aliased states.

This paper presents BACS (Behavioral Anticipatory Classifier System), a Learning Classifier System based on an enhanced version of Stolzmann's Anticipatory Classifier System (ACS2) [10], in which Behavioral Sequences have been integrated in order to allow the system to tackle the perceptual aliasing issue.

The rest of the paper is organized as follows: Sect. 2, introduces different approaches designed to address the perceptual aliasing issue in the LCS field, along with the main principles and structure of Anticipatory Learning Classifier Systems. Then, BACS is introduced in Sect. 3, in which the detection of aliased states, behavioral sequences management and adaptation of learning policies are described. Section 4 presents a study of the capabilities of BACS through a thorough benchmarking on 23 mazes used as testbeds in the literature. The advantages and the limits of BACS are presented in Sect. 5, before concluding in Sect. 6.

2 Related Works

ALCS observe the principles of the cognitive mechanism of Anticipatory Behavioral Control that constitutes the core of its behavior. This cognitive mechanism is implemented thanks to two machine learning paradigms: the Anticipatory Learning Process (ALP) and Reinforcement Learning (RL). These two paradigms are briefly described as guidelines for designing ACS in Subsect. 2.2, after giving more specific details regarding the resolution of perceptual aliasing issues by LCS in Subsect. 2.1.

2.1 LCS and the Perceptual Aliasing Issue

An intuitive approach to deal with the perceptual aliasing issue is to implement *a memory mechanism based on the encountered states* in order to determine the next optimal action. For instance, the non-aliased states encountered before an aliased one in the Woods101 maze (Fig. 1) can provide additional information helping disambiguating the state. This idea has been implemented in LCS such as [11,16,18,32,33], none of them being ALCS. [33] is quite different from [11, 16,18,32] as this system associates its perceptions of the environment in a single image to uniquely identifying encountered states rather than using some internal states represented by memory bits or immediate memory from the past states.

Another approach involves *chaining classifiers*, which can be done in two different ways. On the first hand, each classifier *Cl* has a probability to reference another classifier that has been used before or after *Cl*: hence, chains of classifier called *corporations* are built by the learning classifier system in order to bridge the aliased states. Such corporations were only used in [27–29], which are not ALCS. On the other hand, *sequences of actions* called **Behavioral Sequences** were only built in ALCS [20,25]. If an aliased state is detected, a new classifier, related to a state preceding this aliased state and anticipating a state following this aliased state, is built by chaining all the required actions in a sequence.

As stated earlier, the work hereby presented targets ALCS because they are based on the cognitive mechanism of Anticipatory Behavioral Control: the learning process of such systems, based on anticipation, mainly operates on the perceptions they get from their environment, rather than on stochastic components as [11,16,18,27–29,32]: when these systems encounter aliased states, they use stochastic components to detect the aliasing or to set up their additional parameters (memory or references management) of their classifiers. Stochastic components in ALCS are only used to find more general classifiers (classifiers which accurately correspond to as many states as possible). Thus, a distinction is made between finding accurate classifiers and generalizing these classifiers, giving more flexibility to the system as the roles of the different components are separated.

Furthermore, [33] has shown very promising results in aliased environments, although it does not include generalization mechanism during its learning process, thus preventing the system from discovering underlying rules in an environment. Moreover, modifications were implemented in the reinforcement mechanisms, restraining the system to specific environments (for instance, it cannot deal with multiple goal tasks).

Hence, Behavioral Sequences appear to be an appealing approach to deal with the perceptual aliasing issue, but for some reasons, they have never been integrated to ACS2, one of the most well-known ALCS. The main principles and structures of ALCS are introduced in the next section as BACS relies on them.

2.2 Principles of Anticipatory Learning Classifier Systems

The two machine learning paradigms used within ALCS are complementary and dedicated to two distinct purposes.

On the first hand, the ALP is used to learn the interactions between an environment and the ALCS. These interactions are described as $\{C, A, E\}$ tuples made up of a condition component C, an action component A and an effect component E [25]. Such a tuple describes the consequences E of an action A under states that meet the prerequisites defined by C. These tuples are built by comparing the expected effects of the ALCS with the actual effects occurring in the environment, thus enabling the ALCS to fit its classifiers to the environment.

On the other hand, RL provides ALCS more adaptative abilities to carry out specific tasks, by strengthening or weakening classifiers suitable to the situational context with environmental rewards. Hence, more pieces of information from the situational context to solve a task are integrated thanks to RL.

Each classifier of an ALCS consists of the following components: a $\{C, A, E\}$ tuple, a mark that specifies states for which the effects were not anticipated, a measurement of the quality q of anticipation (computed by the ALP), a prediction of expected rewards r (computed by RL) and finally, a fitness value that is the product of q by r. The ALCS manage their population of classifiers according to the following cycle (for further details, refer to [10,20]):

1. The system perceives new sensory inputs from the environment.
2. The ALP is applied on the previous *action set* (the set of all classifiers whose action component corresponds to the last action performed). Newly generated classifiers are added to this *action set* and to the population of all classifiers.
3. RL is applied on the previous *action set.*

4. A *matching set* is formed out of the current population of all classifiers, by selecting the classifiers whose condition component matches the current sensory inputs.
5. An action is selected from this matching set. An action is randomly selected with an exploration probability (it could also be selected by roulette-wheel [12] or by choosing the action related to the classifier having the best fitness with ϵ-greedy policy [17]).
6. An *action set* is formed out the current *matching set* by selecting the classifiers whose action component matches the selected action.
7. The system interacts with its environment by performing the selected action.
8. If the system achieves its goal, the ALP and RL are applied to the last *action set*. Otherwise, this cycle loops once again.

Therefore, LCS that rely on and exploit Anticipatory Behavioral Control within their inner components belong to the ALCS. ALCS vary among each other in the sense that the steps of the presented cycle may slightly differ or be enhanced.

One of the most well-known ALCS is ACS2 [10], an enhanced version of Stolzmann's Anticipatory Classifier System [25]. It exploits a generalization mechanism by Genetic Algorithm (GA), aiming at generalizing over-specialized classifiers created by the ALP [8]. It also uses an extension of the ALP known as Specification of Unchanging Components (SUC) that enables the system to cope with environmental states that are not affected by actions [25]. ACS2 also includes two exploratory biases within its action selection policies to efficiently adapt the classifier population to the environment: the *action delay bias* is used to select the earliest executed action while the *knowledge array bias* is used to select the action for which the system has the least anticipation quality [4]. Recently, an action planning mechanism [9] has been introduced to ACS2 [30] in order to speed up the learning process. BACS, hereby presented, is an evolution of ACS2 that integrates the Behavioral Sequences, in order to deal with the perceptual aliasing issue.

3 BACS: Behavioral Anticipatory Classifier System

Behavioral sequences expand the $\{C, A, E\}$ tuple of each classifier, permitting the execution of a sequence of actions. Integrating such sequences in ACS2 requires (1) the introduction of a mechanism for aliased state detection (Sect. 3.1) that triggers the generation of *behavioral classifiers* (classifiers whose action component is now built with a behavioral sequence –section 3.2), as well as (2) the adjustment of both the ALP (Sect. 3.3) and the action selection policies (Sect. 3.4).

3.1 Detection of Aliased States

Detecting aliased state is possible, during the ALP, by using the agent's perception for both the state $s(t-1)$ and the current state $s(t)$, jointly with classifier marks that corresponds to states for which the effects were not anticipated. If a classifier correctly anticipates the state $s(t)$ and if its mark matches a unique state identical to the state $s(t-1)$, then the state $s(t-1)$ is identified as aliased [20], because this classifier both anticipates *and* fails to anticipate this particular state.

3.2 Building and Using Behavioral Classifiers

Once an aliased state is detected, a *behavioral classifier* is generated, using both the classifier Cl_{t-1} that correctly anticipated the state $s(t)$ in the aliased state $s(t-1)$ and the penultimate classifier Cl_{t-2} selected in the previous state $s(t-2)$. In order to cap the size of behavioral sequences, a parameter is introduced, B_{seq}, that determines the maximal length of all sequences: if $B_{seq} = 1$, then the classifier is a "regular" classifier (such as ACS2 classifiers: one classifier for one action), otherwise the length of the behavioral sequence in the classifier is at most B_{seq}.

The generation of a behavioral classifier fails if the length of its behavioral sequences exceeds B_{seq} or, if neither Cl_{t-2} nor Cl_{t-1} anticipates a change. Otherwise, the new classifier is generated as follows: (1) its condition is that of Cl_{t-2}, (2) its behavioral sequence is the chaining of the actions of first Cl_{t-2} and then Cl_{t-1}, and (3) its effect is built by applying the *passthrough* operator [20] on the effects of both Cl_{t-1} and Cl_{t-2} and by deleting all parts of the built effect that match same parts of the condition. Once built, the new behavioral classifier is only added to the population of all classifiers, according to the insertion process described in [10], as its action component is different from that of the classifiers of the current action set.

When BACS uses a behavioral classifier, it has to execute each action of the behavioral sequence before applying the learning processes. These actions can lead to a loop: as an example, in maze environments, a loop can be characterized by taking a step forward, then a step backward, then a step forward, and so forth. These loops are described as *looping behavioral sequences* in [20]. In order to prevent looping behavioral sequences, [20] proposed a mechanism that was integrated to BACS that records the states encountered thanks to the behavioral sequence so as to detect previously encountered states. Upon such a detection, the quality of the behavioral classifiers that led to this loop is decreased.

3.3 Adapting the Anticipatory Learning Process

The ALP used in BACS is directly inspired by the ALP described in [10] and in [20]. The process depends on whether the action component of the selected classifier is an *action* or a *behavioral sequence*.

If it is an action, then the ALP is the same as the ALP used in ACS2 [10], except regarding the mechanisms implemented for the detection of aliased states and the building of behavioral classifiers.

Otherwise, a *useless case* is added to the ALP as in [20] to prevent the use of behavioral sequences when the classifier does not anticipate any changes, by decreasing the quality of such behavioral classifiers, as the behavioral sequences should enable the system to bridge the aliased states. Moreover, BACS does not build behavioral classifiers from this part of the ALP (when the action component of the selected classifier is a behavioral sequence) because the probability to build such classifiers with a number of actions that exceeds B_{seq} is higher: as a matter of fact, the number of actions in a sequence increases more quickly if a sequence is built from two behavioral classifiers.

3.4 Adapting the Classifier Selection Policies

Action selection policies are used in order to pick an action among the set of classifiers matching the current agent perception. Thereupon, action selection policies of ACS2 (commonly an ϵ-greedy policy) are no longer suitable since they are not dedicated to the selection of a *sequence* of actions: to preserve the benefits introduced by biasing these policies, BACS now returns classifiers instead of actions through its ϵ-greedy policy. BACS is not permitted to select sequences of actions *per se*, because behavioral sequences are not required for dealing with non-aliased states.

In the case of aliased states, the *knowledge array bias* has another benefit in BACS: given the nature of such states (the system cannot predict the next state), the quality of classifiers cannot efficiently increase. **As such, this exploratory bias promotes actions that enable the system to bridge this aliased state, and BACS can thus exploit them while building behavioral sequences.**

The integration of Behavioral Sequences into ACS2, resulting in BACS, requires the detection of aliased states to trigger the creation of behavioral classifiers. The action selection policies, that include the exploratory biases introduced within ACS2, have been updated to enable the selection of a classifier instead of an action. And finally, the Anticipatory Learning Process has been split depending on the action component of the selected classifier (1) to promote behavioral classifiers that can bridge the aliased states and (2) to avoid the triggering of the building of behavioral classifiers having too much actions in their sequences.

The validation of BACS was done using a maze benchmark containing both aliased and non-aliased environments: the obtained results are presented in the following section.

4 BACS Performance in Maze Environments

In the following, Sect. 4.1 characterizes the maze environments that constitute the proposed benchmark. The experimental protocol set for the experiments and the metrics used for evaluating BACS are described in 4.2, before presenting the achieved results in Sect. 4.3.

4.1 Maze Characterization

The maze used for experiments can be characterized by the *average distance to the exit*, their *type of aliasing* and their *complexity* [2].

The *average distance to the exit* corresponds to the average number of actions to take, from any position in the maze, to reach the exit.

The *aliasing type* specifies an aliased state by its distance to the exit from this state and the optimal action that allows the agent to get closer to the exit. The authors of [2] have proposed four aliased types (ranked by increasing complexity) by comparing the distances and actions of two distinct aliased states:

- *Pseudo-aliasing* if the distances and the actions are identical for both states.
- *Type I aliasing* if the distances are different but the actions are identical.
- *Type II aliasing* if the distances are identical and the actions are different.
- *Type III aliasing* if the distances and the actions are different.

Aliased states introduce an extra challenge for anticipatory systems, since such systems must also anticipate the state following an aliased one. In the end, the aliasing type of a maze is defined as the more complex aliasing type found in the environment (this does not exclude the presence of lower types of aliasing).

Lastly, the *complexity* of a maze quantifies how difficult it is for an agent to learn in this environment, and mainly depends on the number of aliased states as well as their type, on whether or not some aliased states are adjacent, on the distance to the exit or on the size of the maze. The complexity is computed as the ratio of how long an agent trained by Q-Learning takes to reach the exit to the average distance to the exit (for further details, refer to [2]). The higher the complexity, the higher the complexity of the maze.

The benchmark constituted for these experiments is made up of 23 mazes environments [1,2,4,7,8,15,19–21,25,32]. These mazes are sorted first by their aliasing type, then by their complexity, following a decreasing order, in Table 1. This set of maze environments is non-exhaustive: the mazes that were selected are the ones for which the topology was clearly described in the literature. This benchmark can be further extended with other maze environments thanks to the integration of these environments to *OpenAI Gym* [3,15].

This benchmark, including maze environments of varying complexity, was used to validate the mechanisms implemented in BACS.

4.2 Experimental Protocol

The goal of BACS in a maze environment is to move from its starting position, one grid-cell at a time and in either eight adjacent positions, until it reaches the exit. Its perceptive capabilities are limited to the eight squares adjacent to each position. Its starting position in the mazes is a random position, distinct to the exit.

For each maze, 30 experiments were run. An experiment is a succession of trials (attempts to reach the exit within a minimal number of actions). A maximal number of actions (100) beyond which BACS is stopped was set to cap the

Table 1. Main characteristics of the maze environments used in the benchmark from [2]. $*$ indicates that no values were provided because corresponding mazes were not aliased ones.

Maze	Distance to exit	Complexity	Aliasing type
MazeE2 [2]	2.73	251.2	III
Woods101.5 [32]	3.1	251	III
Maze10 [32]	5.11	171	III
MazeE1 [20]	3.07	167	III
Woods102 [32]	3.31	167	III
Woods100 [20]	2.33	166	III
Woods101 [2]	2.9	149	III
Maze7 [32]	4.33	82	II
MazeF4 [25]	4.5	47	II
MiyazakiB [21]	3.33	1.03	II
Littman57 [2]	3.71	154	I
MiyazakiA [21]	3.05	69	I
Littman89 [19]	3.77	61	I
MazeB [1]	3.5	1.26	I
MazeD [1]	2.75	1.03	I
Cassandra4×4 [2]	2.27	1	I
Maze4 [7]	3.5	$*$	Not aliased
Maze5 [4]	4.61	$*$	Not aliased
MazeA [1]	4.23	$*$	Not aliased
MazeF1 [25]	1.8	$*$	Not aliased
MazeF2 [25]	2.5	$*$	Not aliased
MazeF3 [25]	3.38	$*$	Not aliased
Woods14 [8]	9.5	$*$	Not aliased

trials. For each experiment, the first 1000 trials are used to explore the maze ($\epsilon = 0.8$ for the ϵ-greedy policy), using both the ALP and RL ($\beta = 0.05$). Then, BACS is switched to pure exploitation ($\epsilon = 0$, no ALP), having 100 more trials to bootstrap the expected rewards of classifiers that anticipate changes with RL ($\beta = 0.05$), before recording the number of actions required by BACS to reach the exit for 500 more trials. Other parameters (not described here) are initialized to the values provided in [10].

In order to evaluate BACS, these experiments were run using an ACS2 [4] (control experiment) in order to address specific questions:

- BACS behaviors using a single action ($B_{seq} = 1$) have been compared to that of the control experiment to check its ability to have similar behaviors in non-aliased mazes, with and without the genetic algorithms (GA). The

ratios of correct transitions learned by at least one reliable classifier of the population to all possible transitions (knowledge ratio) were collected, along with the number of actions required to reach the exit.
- To study the effect of the length of the Behavioral Sequences for solving the task, two different lengths for each experiment ($B_{seq} = 2$ or 3) were used. BACS-2 stands for BACS with $B_{seq} = 2$, etc. The number of actions required to reach the exit were compared to that of the control experiment in aliased mazes. The GA was not used because its coupling with behavioral sequences is not desired: generalizing the behavioral classifiers should be prevented as such classifiers should solely be used to bridge aliased states.

In the following, μ and σ are, respectively, the average and standard deviations of each metric. All averages are statistically compared using Welsh t-test or One Sample T-Test (significance threshold 0.05).

4.3 BACS Performance

In non-aliased mazes, BACS-1 and the control experiment achieve the average optimal number of steps to reach the exit both with and without GA (all $p \ll 0.05$). In each maze, both systems achieved full knowledge ratios ($\mu = 100\%, \sigma = 0$) with and without GA: it means that both systems were able to build a full representation of their environment within their populations of classifiers. **Hence, BACS-1 behaves similarly as ACS2 in non-aliased environments**.

Figure 2 depicts the average number of steps required to reach the exit in aliased mazes achieved by BACS-2 and BACS-3 as compared to ACS2. BACS-2 and BACS-3 equivalently perform in 9 out 16 mazes (all $p \gg 0.05$), while both of them significantly achieve better results than ACS2 (all $p \ll 0.05$). BACS-3 significantly achieves the best results in two mazes (MazeD and Maze10, both $p \ll 0.05$), while BACS-2 significantly achieves the best results in three mazes (MazeB, MazeE1, Woods102, all $p \ll 0.05$). Finally, ACS2 achieves the best results in the two remaining mazes: Littman57 and MiyazakiB (both $p \ll 0.05$). **Hence, the behavioral sequences enable BACS to tackle the perceptual aliasing issue in most mazes** (14 out 16). Optimal performance over the 30 experiments is achieved by BACS in three aliased mazes (Woods100, MazeF4 and Maze7, $p \ll 0.05$), while ACS2 does not reach optimal performance in any aliased mazes.

5 Discussion

Behavioral Sequences length

The length of the Behavioral Sequences has an impact on BACS ability to deal with aliased states. This is especially true if there are successions of such aliased states in the environment: when the case arises, BACS must build a sequence of actions as long as this succession is. This can be illustrated with Maze10 in Fig. 2.

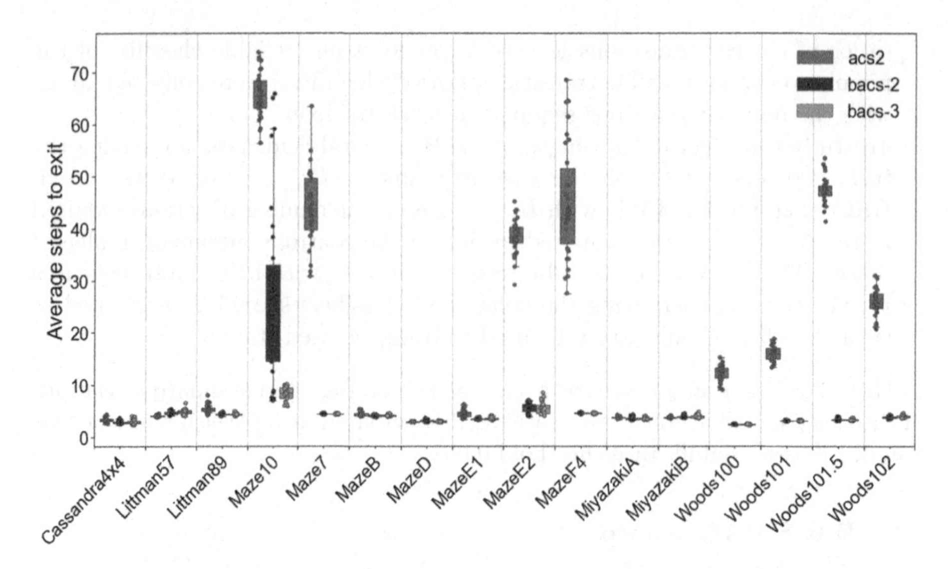

Fig. 2. BACS reaches the exit more efficiently in aliased mazes thanks to the Behavioral Sequences. The lower the average number of steps, the better the performance.

Conversely, using long Behavioral Sequences can impair BACS performance, as shown for example by the Woods102 maze, resulting in an increase in the number of actions to take to fulfil the objective.

Using Behavioral Sequences can increase the number of actions to take to fulfil an objective
The results show that, in the case of the aliased environments Littman57 and MiyazakiB, BACS performs slightly less well than ACS2, regardless of the B_{seq} parameter. This slight performance drop is caused by the behavioral classifiers since they do not guarantee that sequence of actions will be optimal for each state of the environment, especially when an environment has adjacent type-I aliased states (the optimal action is the same among different aliased states) as this is the case in Littman57 and MiyazakiB. This point can be illustrated by considering the state in-between both aliased state of Woods101 (illustrated on Fig. 1): from this state, to reach the exit, BACS can either go South using a regular classifier (single action) or go East first then South-West using a Behavioral Sequence, as both will have really close fitness value. In the latter case, it means that an additional action would be counted. However, BACS does not yet include a mechanism that would promote the minimal number of actions to take between two states, but such a mechanism would allow BACS to favor classifiers that are the most adapted to the context it evolves in (*e.g.* for reaching the exit within a minimal number of actions).

Building Behavioral Sequences tends to increase the number of classifiers in the population
BACS classifier population represents the knowledge acquired by the system through its interactions with its environment. As such, building behavioral classifiers is equivalent to have the system learn different ways of dealing with an aliased state. Then, the more aliased is an environment and the longer the maximal length of the Behavioral Sequences, the more the population of classifiers will increase: for instance, the average number of classifiers in MazeE1 increases to 1230 with BACS-3, as compared to the average 450 classifiers with ACS2. Setting up constraints during the generation of behavioral classifiers would alleviate the growth of the population of classifiers.

On the ratio of aliased states to non-aliased states
Environments having a high ratio of aliased states to non-aliased state are more challenging for BACS, as it requires non-aliased states to be able to bridge the aliased states through a Behavioral Sequence. This difficulty can be observed by considering the performance of BACS in Woods100 (2 aliased states and 4 non-aliased states) and MazeE2 (36 aliased states and 12 non aliased states): BACS could optimally reach the exit of Woods100 within the minimal number of actions while it could not in MazeE2.

Comparing ALCS
A comparison between BACS and the results presented in [20,25] shows that BACS performs as well as [25] in MazeF4 and [20] in Woods100 (both systems achieved optimal performance) and BACS performs better than [20] in MazeE1 and MazeE2 (both $p \ll 0.05$). These results could be mainly explained by the roulette-wheel selection used in [20]: it enables their system to potentially select an inadequate classifier during exploitation. Performing a more thorough comparison, especially with other systems dealing with the perceptual aliasing issue, is very challenging as experimental protocols may differ, the evaluations of the performance are very diverse, source code for a system is simply not available or more complex mazes are needed. Carrying out a detailed comparison of methods enabling LCS to deal with the perceptual aliasing issue is also a future research direction.

6 Conclusion

When only partial observations of the environment are available, Anticipatory Learning Classifier Systems face the perceptual aliasing issue and thus, they cannot differentiate states that actually are truly distinct. In order to cope with this issue, BACS (Behavioral Anticipatory Classifier System) has been built upon the state-of-the-art ACS2 in order to integrate Behavioral Sequences, used to build sequences of actions. To use such sequences, BACS required the implementation of (1) a new classifier selection policies, (2) a mechanism to detect aliased states in environments as well as (3) adjustments to the Anticipatory Learning Process. A benchmark of 23 mazes, along with a thorough experimental protocol, has been set up to validate the use of Behavioral Sequence for dealing with the perceptual aliasing issue. The achieved results show that (1) BACS using a single action behaves similarly to ACS2 and (2) using the Behavioral Sequences with BACS is a suitable approach to deal with aliased states.

A finer management of the behavioral classifiers would further enhance BACS capabilities by restraining the growth of the classifier population and by fitting the use of the sequences to the environment in which the system evolves. Furthermore, BACS actually relies on non-aliased states to build behavioral classifiers: reducing this dependence would allow BACS to evolve more easily in environment having a high ratio of aliased states to non-aliased states.

References

1. Arai, S., Sycara, K.: Credit assignment method for learning effective stochastic policies in uncertain domains. In: Proceedings of the 3rd Annual Conference on Genetic and Evolutionary Computation, pp. 815–822 (2001)
2. Bagnall, A.J., Zatuchna, Z.V.: On the classification of maze problems. In: Bull, L., Kovacs, T. (eds.) Foundations of Learning Classifier Systems, Studies in Fuzziness and Soft Computing, vol. 183, pp. 305–316. Springer, Heidelberg (2005). https://doi.org/10.1007/11319122_12
3. Brockman, G., et al.: Openai gym (2016)
4. Butz, M.V.: Biasing exploration in an anticipatory learning classifier system. In: Lanzi, P.L., Stolzmann, W., Wilson, S.W. (eds.) IWLCS 2001. LNCS (LNAI), vol. 2321, pp. 3–22. Springer, Heidelberg (2002). https://doi.org/10.1007/3-540-48104-4_1
5. Butz, M.V.: Combining gradient-based with evolutionary online learning: an introduction to learning classifier systems. In: 7th International Conference on Hybrid Intelligent Systems (HIS 2007), pp. 12–17. IEEE (2007)
6. Butz, M.V.: Learning classifier systems. In: Kacprzyk, J., Pedrycz, W. (eds.) Springer Handbook of Computational Intelligence, pp. 961–981. Springer, Heidelberg (2015). https://doi.org/10.1007/978-3-662-43505-2_47
7. Butz, M.V., Goldberg, D.E., Stolzmann, W.: Probability-enhanced predictions in the anticipatory classifier system. In: Luca Lanzi, P., Stolzmann, W., Wilson, S.W. (eds.) IWLCS 2000. LNCS (LNAI), vol. 1996, pp. 37–51. Springer, Heidelberg (2001). https://doi.org/10.1007/3-540-44640-0_4
8. Butz, M.V., Goldberg, D.E., Stolzmann, W.: The anticipatory classifier system and genetic generalization. Nat. Comput. **1**(4), 427–467 (2002)
9. Butz, M.V., Stolzmann, W.: Action-planning in anticipatory classifier systems. In: Proceedings of the 1999 Genetic and Evolutionary Computation Conference Workshop Program, pp. 242–249 (1999)
10. Butz, M.V., Stolzmann, W.: An algorithmic description of ACS2. In: Lanzi, P.L., Stolzmann, W., Wilson, S.W. (eds.) IWLCS 2001. LNCS (LNAI), vol. 2321, pp. 211–229. Springer, Heidelberg (2002). https://doi.org/10.1007/3-540-48104-4_13
11. Cliff, D., Ross, S.: Adding temporary memory to ZCS. Adapt. Behav. **3**(2), 101–150 (1994)
12. Goldberg, D.: Genetic Algorithms in Search, Optimization, and Machine Learning. Addison-wesley, Reading, Boston (1989). Schraudolph, N.N. J **3**(1) (1989)

13. Hoffmann, J.: Anticipatory behavioral control. In: Butz, M.V., Sigaud, O., Gérard, P. (eds.) Anticipatory Behavior in Adaptive Learning Systems. LNCS (LNAI), vol. 2684, pp. 44–65. Springer, Heidelberg (2003). https://doi.org/10.1007/978-3-540-45002-3_4
14. Kaelbling, L.P., Littman, M.L., Moore, A.W.: Reinforcement learning: a survey. J. Artif. Intell. Res. **4**, 237–285 (1996)
15. Kozlowski, N., Unold, O.: Integrating anticipatory classifier systems with openai gym. In: Proceedings of the Genetic and Evolutionary Computation Conference Companion, pp. 1410–1417 (2018)
16. Lanzi, P.L.: Adding memory to XCS. In: 1998 IEEE International Conference on Evolutionary Computation Proceedings. IEEE World Congress on Computational Intelligence (Cat. No. 98TH8360), pp. 609–614. IEEE (1998)
17. Lanzi, P.L.: Adaptive Agents with Reinforcement Learning and Internal Memory, pp. 333–342. MIT Press (2000)
18. Lanzi, P.L., Wilson, S.W.: Toward optimal classifier system performance in non-markov environments. Evol. Comput. **8**(4), 393–418 (2000)
19. Loch, J., Singh, S.P.: Using eligibility traces to find the best memoryless policy in partially observable markov decision processes. In: ICML, pp. 323–331 (1998)
20. Métivier, M., Lattaud, C.: Anticipatory classifier system using behavioral sequences in non-markov environments. In: Lanzi, P.L., Stolzmann, W., Wilson, S.W. (eds.) IWLCS 2002. LNCS (LNAI), vol. 2661, pp. 143–162. Springer, Heidelberg (2003). https://doi.org/10.1007/978-3-540-40029-5_9
21. Miyazaki, K., Kobayashi, S.: Proposal for an algorithm to improve a rational policy in pomdps. In: IEEE SMC1999 Conference Proceedings. 1999 IEEE International Conference on Systems, Man, and Cybernetics (Cat. No. 99CH37028), vol. 5, pp. 492–497. IEEE (1999)
22. Oliehoek, F.A., Amato, C., et al.: A Concise Introduction to Decentralized POMDPs, vol. 1. Springer, Heidelberg (2016). https://doi.org/10.1007/978-3-319-28929-8
23. Qiao, Z., Muelling, K., Dolan, J., Palanisamy, P., Mudalige, P.: POMDP and hierarchical options MDP with continuous actions for autonomous driving at intersections. In: 2018 21st International Conference on Intelligent Transportation Systems (ITSC), pp. 2377–2382. IEEE (2018)
24. Sigaud, O., Wilson, S.W.: Learning classifier systems: a survey. Soft Comput. **11**(11), 1065–1078 (2007)
25. Stolzmann, W.: An introduction to anticipatory classifier systems. In: Lanzi, P.L., Stolzmann, W., Wilson, S.W. (eds.) IWLCS 1999. LNCS (LNAI), vol. 1813, pp. 175–194. Springer, Heidelberg (2000). https://doi.org/10.1007/3-540-45027-0_9
26. Stolzmann, W., Butz, M., Hoffmann, J., Goldberg, D.: First cognitive capabilities in the anticipatory classifier system, February 2000
27. Tomlinson, A., Bull, L.: A corporate classifier system. In: Eiben, A.E., Bäck, T., Schoenauer, M., Schwefel, H.-P. (eds.) PPSN 1998. LNCS, vol. 1498, pp. 550–559. Springer, Heidelberg (1998). https://doi.org/10.1007/BFb0056897
28. Tomlinson, A., Bull, L.: A corporate XCS. In: Lanzi, P.L., Stolzmann, W., Wilson, S.W. (eds.) IWLCS 1999. LNCS (LNAI), vol. 1813, pp. 195–208. Springer, Heidelberg (2000). https://doi.org/10.1007/3-540-45027-0_10
29. Tomlinson, A., Bull, L.: Cxcs: Improvements and corporate generalization. In: Proceedings of the 3rd Annual Conference on Genetic and Evolutionary Computation, pp. 966–973 (2001)

30. Unold, O., Rogula, E., Kozłowski, N.: Introducing action planning to the anticipatory classifier system ACS2. In: Burduk, R., Kurzynski, M., Wozniak, M. (eds.) CORES 2019. AISC, vol. 977, pp. 264–275. Springer, Cham (2020). https://doi.org/10.1007/978-3-030-19738-4_27
31. Wilson, S.W.: Classifier fitness based on accuracy. Evol. Comput. **3**(2), 149–175 (1995)
32. Zang, Z., Li, D., Wang, J.: Learning classifier systems with memory condition to solve non-markov problems. Soft Comput. **19**(6), 1679–1699 (2015)
33. Zatuchna, Z.V., Bagnall, A.: Learning mazes with aliasing states: an LCS algorithm with associative perception. Adapt. Behav. **17**(1), 28–57 (2009)

Nash Equilibrium as a Solution in Supervised Classification

Mihai-Alexandru Suciu and Rodica Ioana Lung(✉)

Centre for the Study of Complexity, Babeş-Bolyai University, Cluj-Napoca, Romania
mihai-suciu@cs.ubbcluj.ro, rodica.lung@econ.ubbcluj.ro
http://csc.centre.ubbcluj.ro/

Abstract. The supervised classification problem offers numerous challenges to algorithm designers, most of them stemming from the size and type of the data. While dealing with large data-sets has been the focus of many studies, the task of uncovering subtle relationships within data remains an important challenge, for which new solutions concepts have to be explored. In this paper, we propose a general framework for the supervised classification problem based on game theory and the Nash equilibrium concept. The framework is used to estimate parameters of probabilistic classification models to approximate the equilibrium of a game as the optimum of a function. CMA-ES is adapted to compute such model parameters; a noise mechanism is used to enhance the diversity of the search. To illustrate the approach we use Probit regression; numerical experiments indicate that the game-theoretic approach may provide a better insight into data than other models.

Keywords: Binary classification · Game theory · Nash equilibrium · CMA-ES

1 Introduction

Game theory is a field of mathematics that models different types of interactions among agents and provides a variety of solution concepts to conflicting objectives. In theory these solutions can be adapted for complex problems that exhibit multiple, conflicting, interacting objectives and require trade-off solutions. The purpose of this paper is to explore the use of the popular Nash equilibrium concept as a novel way to approach supervised classification, based on the assumption that converting the classification problem into a normal form game and considering its corresponding equilibrium as solution may provide a better model for the interaction among instances in the same class than other approaches, as decisions made in a game take into account the actions of all other players.

There are approaches that use game theory to enhance machine learning techniques. In [2] a linear regression classifier is modeled as a public good game,

T. Bäck et al. (Eds.): PPSN 2020, LNCS 12269, pp. 539–551, 2020.
https://doi.org/10.1007/978-3-030-58112-1_37

in which players choose the precision of the output data they reveal while features are public. Schuurmans and Zinkevich investigate a reduction of supervised learning to game playing [11], and they establish a bijection between the Nash equilibria and critical (or KKT) points of the problem. Another approach that uses game theory proposes a method for explaining individual predictions of classification models with contributions of individual feature values [12]. In [7] for each feature an importance value for a prediction is assigned in order to determine feature importance and understand model predictions.

In our endeavour the classification problem is converted into a large game in which instances of the data set are players that have to choose their class in order to maximize their payoff computed as the F_1 score corresponding to their class. The correct classification corresponds to the Nash equilibrium of this game. In mixed form, this Nash equilibrium can be computed as the minimum of a real valued function. This function can be used to estimate parameters of probabilistic classification models. The general framework that constructs the game and the function yielding the equilibrium is called FROG: Framework based on Optimization and Game theory. FROG has three main components: the game, presented in Sect. 2, a classification model, which in this paper is the Probit model, presented in Sect. 3 and the optimization model (Sect. 4). Preliminary numerical experiments presented in Sect. 5 indicate the potential of this approach to solve the binary classification problem in comparison with other methods.

2 FROG - Game

The binary classification problem [5, p. 11] can be expressed as follows: given a set of input data $X = (x_1, \ldots x_N)^\top$, with $x_i \in \mathbb{R}^p$, $p \geq 1$, and labels $Y = (y_1, \ldots, y_N)^\top$, with $y_i \in \{0, 1\}$, such that label y_i corresponds to x_i, $i = 1, \ldots, N$, find a model that makes a good prediction of Y from X.

In what follows we will construct a normal form N-player game such that its equilibrium represents the correct classification for X and use the function that computes the Nash equilibrium of the game to estimate model parameters in order to provide a classification model with properties close to the equilibrium of the game.

The game $\Gamma = ([N], S, U|X, Y)$ is defined as:

- $[N]$ is the set of **players**, $[N] = \{1, \ldots, N\}$ representing the N instances in X; a player i is an observation x_i from X;
- $S = \{0, 1\}^N$ is the set of pure **strategy** profiles of the game; each player chooses a class; an element $s \in S$ is $s = (s_1, \ldots, s_N)$, with $s_i \in \{0, 1\}$ representing the class chosen by player i;
- $U = (u_1, \ldots, u_N)$ is the **payoff** function, with $u_i : S \to \mathbb{R}$ the payoff of player i, $i = 1, \ldots N$, computed as:

$$u_i(s) = \begin{cases} F_1(s, y_i) \text{ if } s_i \neq y_i \\ \quad 1 \quad\;\; \text{ if } s_i = y_i \end{cases}. \tag{1}$$

$F_1(s, y_i)$ is computed as the F_1 measure [13] corresponding to situation s and label y_i, which in this case can be expressed as:

$$F_1(s, y_i) = \frac{2T(s, y_i)}{2T(s, y_i) + F(s)}, \tag{2}$$

with

$$T(s, y_i) = |\{k \in [N] | s_k = y_i\}|, \text{ and } F(s) = |\{k \in [N] | s_k \neq y_k\}|, \tag{3}$$

where $|\{\cdot\}|$ denotes the cardinality of a set, $T(s, y_i)$ counts the number of instances with label y_i in s that have chosen strategy corresponding to label y_i, i.e. are correctly classified, and $F(s)$ counts the number of players that do not choose their label, i.e. are incorrectly classified in s.

The payoff is computed such that if a player has chosen the correct class it will get the maximum value of 1, and otherwise the F_1 score corresponding to its class.

Example 1. Consider a data set X having 10 instances with two attributes $X = (x_1, \dots, x_{10})^\top$, $x_i \in \mathbb{R}^2$ and the labels $Y = (0, 0, 0, 0, 1, 1, 1, 1, 1, 1)^\top$. There are $N = 10$ players, each one chooses a label: 0 or 1. An example of strategy profile of the game can be $s \in S$, $s = (1, 0, 1, 1, 1, 0, 1, 1, 1, 1)$, in which the first player chooses the label 1, the second one chooses 0, and so on. We have $T(s, 0) = 1$, $T(s, 1) = 5$, and $F(s) = 4$. Then, for example, the payoff function for the first player in situation s is:

$$u_1(s) = F_1(s, 0) = \frac{2T(s, 0)}{2T(s, 0) + F(s)} = \frac{2 \cdot 1}{2 \cdot 1 + 4} = \frac{1}{3},$$

and the payoff for player 6:

$$u_1(s) = F_1(s, 1) = \frac{2T(s, 1)}{2T(s, 1) + F(s)} = \frac{2 \cdot 5}{2 \cdot 5 + 4} = \frac{5}{7}.$$

The Nash equilibrium of game Γ represents the correct classification. The assumption is that a model that approximates the Nash equilibrium of this game should yield also a good classification for the problem that would present some properties of the Nash equilibrium, such as stability against unilateral deviations. A model that captures a Nash equilibrium should be also more robust to outliers as interactions among instances are captured by the game equilibrium and not by their actual attributes.

To approximate the Nash equilibrium, the game can be converted into an optimization problem for which mixed Nash equilibria are optimal values. Mixed strategies of the game represent probability distributions over the strategies sets of the players. We will denote by σ_{ij} the probability that player i chooses strategy j, $j \in \{0, 1\}$. Since we have $\sigma_{i0} + \sigma_{i1} = 1$ we will use only σ_i to denote the probability σ_{i1} that player i chooses class 1; then $\sigma_{i0} = 1 - \sigma_i$. We can refer to

$\sigma = (\sigma_1, \ldots, \sigma_N)$ as to a mixed strategy profile for Γ. We will use s_{ij} to denote player i's pure strategy in which $\sigma_{ij} = 1$. If $\sigma = (\sigma_1, \ldots, \sigma_N)$ and $\sigma'_i \in [0, 1]$, the notation (σ'_i, σ_{-i}) refers to the strategy profile in which all players except i use σ_i and player i uses σ'_i.

The payoff function u is extended to mixed strategies by considering the expected payoff:

$$u_i(\sigma) = \sum_{s \in S} \left(u_i(s) \cdot \prod_{j \in [N], s_j = 0} (1 - \sigma_j) \cdot \prod_{l \in [N], s_l = 1} \sigma_l \right) \tag{4}$$

Formally, a strategy profile σ is a Nash equilibrium in mixed form for game Γ if $\forall i \in [N]$ and $\forall q_i \in [0, 1]$ we have $u_i((q_i, \sigma_{-i})) \leq u_i(\sigma)$. In the classification context, the Nash equilibrium σ assigns to each x_i in X a probability to belong to class 1 such that there is no unilateral change in this probability that a player can make to improve its expected payoff, i.e. $u_i(\sigma)$.

According to [8], and adapting the intermediate functions to game Γ, the Nash equilibria in mixed form can be computed as the global minima of function $v : [0, 1]^N \to \mathbb{R}$:

$$v(\sigma) = \sum_{i \in [N]} (1 - u_i(\sigma))^2. \tag{5}$$

Evaluation of Payoff Functions $u_i(\sigma)$. Payoff functions $u_i(\sigma)$ in (4) sum products of N probabilities over all 2^N elements of S . However, if we consider the nature of game Γ we can rewrite formula (4):

$$u_i(\sigma) = \begin{cases} 1 - \sigma_i + \mathcal{S}_{i0}(\sigma) \cdot \sigma_i, & \text{if } y_i = 0 \\ \sigma_i + \mathcal{S}_{i1}(\sigma) \cdot (1 - \sigma_i), & \text{if } y_i = 1 \end{cases}, \tag{6}$$

where $\mathcal{S}_{i0}(\sigma)$ and $\mathcal{S}_{i1}(\sigma)$ are comput ed as sums over distinct possible values for the corresponding F_1 score when considering all possible combinations between the rest of $[N] \setminus \{i\}$ players and probabilities in σ (see (7) and (8)). Let m_0 and m_1 denote the number of instances having class 0 and class 1 respectively. The F_1 score for a label is computed considering the number of instances correctly classified with that label, called true positive/negative (TP/TN) depending on the convention regarding which label is identified as positive/negative, and the number of instances incorrectly classified with each label, called false positive (FP) and false negative (FN). In our approach we consider the label "0" as positive and "1" as negative. Then $TP \in \{0, \ldots, m_0\}$ and $TN \in \{0, \ldots, m_1\}$, $FP = m_1 - TN$ and $FN = m_0 - TP$. Then $\mathcal{S}_{i0}$ and $\mathcal{S}_{i1}$ are:

$$\mathcal{S}_{i0}(\sigma) = \sum_{TP=0}^{m_0 - 1} \sum_{TN=0}^{m_1} \frac{2TP}{TP + N - TN} \mathcal{P}_0(\sigma_{-i}; TP, m_0 - 1 - TP) \cdot \\ \cdot \mathcal{P}_1(\sigma; m_1 - TN, TN), \tag{7}$$

and

$$\mathcal{S}_{i1}(\sigma) = \sum_{TP=0}^{m_0}\sum_{TN=0}^{m_1-1} \frac{2TN}{TN+N-TP}\mathcal{P}_0(\sigma;TP,m_0-TP)\cdot$$
$$\cdot\mathcal{P}_1(\sigma_{-i};m_1-1-TN,TN), \quad (8)$$

where for a given $Y = (y_1, y_2, \ldots, y_d)^\top$ with $y_j \in \{0,1\}$ such that $k_1+k_2 \leq h, h := |\{j \in [d]|y_j = l\}|$ and $k_1, k_2 \in \mathbb{N} \cup \{0\}$ and $l \in \{0,1\}$, $\mathcal{P}_l(\sigma; k_1, k_2)$ is defined as the sum of all products of (k_1+k_2) factors which are σ_j or $1-\sigma_j$ such that j satisfies $(y_j = l \wedge (1 \leq j \leq d))$ and exactly k_1 are of the form $1-\sigma_j$ and exactly k_2 are of the form σ_j. When σ_{-i} is used $k_1+k_2 < h$ is required. The notation σ_{-i} represents σ with σ_i removed from it, i.e. $\sigma_{-i} = (\sigma_1, \ldots, \sigma_{i-1}, \sigma_{i+1}, \ldots, \sigma_N)$. For example $P_0(\sigma_{-i}; TP, m_0-1-TP)$ in (7) takes the sum over all products of probabilities σ_j and $1-\sigma_j$ where $j \neq i$ and $y_j = 0$ such that TP factors are $1-\sigma_j$ and m_0-1-TP are σ_j.

Example 2. Let us consider the payoff u_1 for the first player in Example 1 and $\sigma = (0.4, 0.3, 0.2, 0.2, 0.6, 0.1, 0.4, 0.7, 0.3, 0.6)$. We have:

$$u_1(\sigma) = 1 - \sigma_1 + \mathcal{S}_{10}(\sigma)\sigma_1$$

where

$$\mathcal{S}_{10}(\sigma) = \sum_{TP=0}^{3}\sum_{TN=0}^{6} \frac{2TP}{TP+N-TN}\mathcal{P}_0(\sigma_{-1}, TP, 3-TP)\cdot\mathcal{P}_1(\sigma, 6-TN, TN).$$

Let us consider, for example, $TP = 1$ and $TN = 3$. There are $\binom{3}{1} = 3$ possible ways in which the rest of players having class 0 can choose their class in such a way that $TP = 1$ and $\binom{6}{3} = 20$ ways in which players having class one can choose their class such that $TN = 3$. For all these $\binom{3}{1}\cdot\binom{6}{3} = 60$ situations, the corresponding value of F_1 is the same: $\frac{2TP}{TP+N-TN} = \frac{1}{4}$. The sum $P_0(\sigma_{-1}, 1, 2)$ has 3 terms:

$$\mathcal{P}_0(\sigma_{-1}, 1, 2) = (1-\sigma_2)\sigma_3\sigma_4 + \sigma_2(1-\sigma_3)\sigma 4 + \sigma_2\sigma_3(1-\sigma_4) = 0.124,$$

and $P_1(\sigma, 3, 3)$ has 20 terms:

$$\begin{aligned}\mathcal{P}_1(\sigma,3,3) = \sigma_5\sigma_6\sigma_7(1-\sigma_8)(1-\sigma_9)(1-\sigma_{10}) + \\ + \sigma_5\sigma_6(1-\sigma_7)\sigma_8(1-\sigma_9)(1-\sigma_{10}) + \ldots \\ \ldots + (1-\sigma_5)(1-\sigma_6)(1-\sigma_7)\sigma_8\sigma_9\sigma_{10} = 0.33772.\end{aligned}$$ □

3 FROG - Classification Model

FROG constitutes a framework in which we can estimate parameters for probabilistic classification models by considering an equilibrium type of solution. A general classification model seeks to find a function $\phi(x;\theta)$, $\phi : \mathbb{R}^{p+1} \to \mathbb{R}$ that associates to each $\theta \in \mathbb{R}^{p+1}$ either a probability, or a value that can be further converted into a probability, that instance $x \in X$ belongs to a class - usually to the one labeled 1.

Classification Model: Probit. We use the Probit classification model to experiment with FROG. Within this model the probability that an instance x belongs to class 1 is estimated by using the cumulative distribution function Φ of the standard normal distribution:

$$\phi(x;\theta) = \Phi(x \cdot \theta) \tag{9}$$

where $x \cdot \theta$ denotes the dot product of $(1, x_1, \ldots, x_p)$ and $\theta \in \mathbb{R}^{p+1}$.

Since $\phi(x;\theta)$ are probabilities, we can pass them over directly to function $v()$: let $\sigma_i(\theta) = \phi(x_i;\theta)$ and $\sigma(\theta) = (\sigma_i(\theta))_{i=\overline{1,N}}$. Within FROG, the goal is to find θ that minimizes $v(\sigma(\theta))$. Since we use the Probit classification model, we will call this instance of our framework **P-FROG**.

4 FROG - Optimization

Optimization Method: CMA-ES. Covariance Matrix Adaptation Evolution Strategy (CMA-ES) [4] is a search heuristic that uses a covariance matrix adaptation mechanism over a normally distributed population of individuals in order to find the optima of non-linear or non-convex optimization problems. It advertises as a robust, fast method, recommended for model parameter estimation. Its implementation is available in various programming languages. FROG uses CMA-ES to compute model parameters that minimize an approximation of function $v(\cdot)$ in (5), which is presented in what follows.

CMA-ES: Fitness Function Evaluation. Even considering Eq. (6) the computational complexity of evaluating $v(\sigma(\phi))$ remains prohibitive. In order to reduce it further, within CMA-ES the evaluation of an individual θ takes an approximation for u_i for each $i \in [N]$ by considering the the following average probabilities from σ: $\tilde{\sigma}_j$ is the average of all σ_i having $y_i = j$. Then we replace $\mathcal{S}_{i0}$ and $\mathcal{S}_{i1}$ in (7) and (8) with:

$$\begin{aligned}\widetilde{\mathcal{S}}_{i0}(\sigma) &= \sum_{TP=0}^{m_0-1}\sum_{TN=0}^{m_1} \frac{2TP}{TP+N-TN} C_{m_0-1}^{TP}(1-\tilde{\sigma}_0)^{TP}\tilde{\sigma}_0^{m_0-1-TP} \cdot \\ &\qquad \cdot C_{m_1}^{m_1-TN}(1-\tilde{\sigma}_1)^{m_1-TN}\tilde{\sigma}_1^{TN},\end{aligned} \tag{10}$$

$$\begin{aligned}\widetilde{\mathcal{S}}_{i1}(\sigma) &= \sum_{TP=0}^{m_0}\sum_{TN=0}^{m_1-1} \frac{2TN}{TN+N-TP} C_{m_0}^{TP}(1-\tilde{\sigma}_0)^{TP}\tilde{\sigma}_0^{m_0-TP} \cdot \\ &\qquad \cdot C_{m_1-1}^{m_1-1-TN}(1-\tilde{\sigma}_1)^{m_1-1-TN}\tilde{\sigma}_1^{TN}.\end{aligned} \tag{11}$$

In Eq. (6) we will have actually only two distinct payoff functions:

$$\begin{aligned}\tilde{u}_0(\sigma) &= 1 - \tilde{\sigma}_0 + \widetilde{\mathcal{S}}_{i0}(\sigma)\cdot\tilde{\sigma}_0, \quad \text{if } y_i = 0\\ \tilde{u}_1(\sigma) &= \tilde{\sigma}_1 + \widetilde{\mathcal{S}}_{i1}(\sigma)\cdot(1-\tilde{\sigma}_1), \text{ if } y_i = 1,\end{aligned} \tag{12}$$

and $v()$ in (5) is replaced with:

$$\tilde{v}(\sigma) = m_0\tilde{u}_0(\sigma) + m_1\tilde{u}_1(\sigma) \tag{13}$$

CMA-ES: Noise. The search of CMA for this problem is hindered by the plateaux of the fitness function $\tilde{v}()$. In order to escape these plateaux and to preserve search diversity, a simple noise inducing mechanism is added to CMA-ES. Thus, if at the end of an iteration the algorithm decides that it encountered a *flat function* a small normal noise $N(0,\omega)$ is added to X in order to modify the fitness function and move the search towards another region of the space. $N(0,\omega)$ represents the normal distribution with mean 0 and standard deviation ω. While ω is constant during the entire search, the noise is generated randomly within the fitness function such that each individual is evaluated using a different noise. The noise is also triggered at random with probability 0.1 in order to further diversify the search. As the noise is part of the optimization framework, we call this Probit Framework for Optimization with Game theory Plus: **P-FROG+**.

5 Numerical Experiments

Numerical experiments are used to illustrate the potential of P-FROG+ to approach classification problems compared with other standard classification methods.

Data Sets. We use four real world data sets[1] and eight synthetic ones (Table 1). All problems require the classification of the instances into two classes (binary classification problem), the attributes are categorical, integer or real, and their number varies depending on the data set. The real world data sets are: iris data set (from which we removed the setosa instances in order to obtain a linear non separable binary classification problem), the acute inflammations data set [1], the Statlog (heart disease) data set and the Somerville Happiness survey data set. The synthetic data sets are generated such that the problems are non linearly separable. We use the *make_classification* function from *scikit-learn*[2] Python library [9] to generate the data, for reproducibility we report the number of instances, attributes and seed used to generate the data set. All synthetic data sets are generated for a binary classification problem. Table 1 presents the number of instances and attributes for each data set. All data sets included can be considered challenging considering the results provided by standard classifiers. The ability of FROG to find the correct classification for data-sets that are correctly classified by other methods has been successfully tested but due to space limitations we cannot include those results here, and choose focus only on the more difficult ones.

Comparisons with Other Models. We compare the performance of our approach with six well known classifiers: Decision Tree, k-nearest-neighbour classifier, support vector classifier with a linear and a radial kernel in order to classify linearly and non linear separable classes, Logistic Regression and a Stochastic Gradient

[1] UCI Machine Learning Repository https://archive.ics.uci.edu/ml/index.php, accessed January 2020.

[2] Version 0.21.1.

Table 1. Data sets.

Id	Data set	Instances	Attributes
1	Iris	100	4
2	Acute Inflamations	120	6
3	Statlog	270	13
4	Somerville	143	7
5	$Synthetic_1$	100	2 (seed = 4)
6	$Synthetic_2$	100	2 (seed = 5)
7	$Synthetic_3$	200	2 (seed = 4)
8	$Synthetic_4$	200	2 (seed = 5)
9	$Synthetic_5$	500	2 (seed = 4)
10	$Synthetic_6$	500	2 (seed = 5)
11	$Synthetic_7$	1000	2 (seed = 4)
12	$Synthetic_8$	1000	2 (seed = 5)

Descent classifier [6]. They can handle data with different characteristics and thus are suited for comparisons. We used their implementation in *scikit-learn* to allow for reproducibility of the results.

Classification Metrics. As performance metric we use the area under the receiver operating characteristic curve AUC [6]. AUC is a performance measure of the accuracy of the classifier in distinguishing between classes. The measure is equivalent to the probability that a randomly selected positive sample is ranked higher than a randomly selected negative sample [10]. A maximum value of 1 is an ideal situation where the model can correctly classify all instances.

To estimate prediction error we use the standard approach of K-Fold Cross-Validation [6] where part of the data is used to fit the model and another part is used to assess the model performance. For K-Folding we divide the instances in K cross-validation folds at random. K runs are performed such that in each run one fold is used as test data to validate the model fitted to the other $K-1$ folds. We use a stratified K-Folding approach - i.e. each fold contains approximately the same proportion of samples from the two classes.

For our experiments we use $K = 10$, we fit all compared models on the training instances, and report the area under the ROC curve (AUC) for each fold on the test instances. We repeat the stratified 10-fold cross-validation fourteen times with different random number generator seeds for splitting the data into folds, thus performing a 14×10 folds cross-validation.

Results. Results are presented in Fig. 1 and Tables 2 and 3. Table 2 presents statistical comparisons between P-FROG and the other classification models, and Table 3 presents the effect of noise magnitude on P-FROG+.

Figure 1 presents boxplots for obtained AUC over the 14×10 folds for P-FROG+, P-FROG and compared models. For the real data sets on P-FROG reports the best results when compared with other models for data sets D1, D3, and D4 and classifies correctly all instances of data set D2. The noise variant, P-FROG+, also outperforms all the other methods for D1, D3, and D4. For the synthetic data sets our approach, P-FROG, is able to better classify the data than the compared models. P-FROG+ obtains the best classification results for 5 out of 8 synthetic tables.

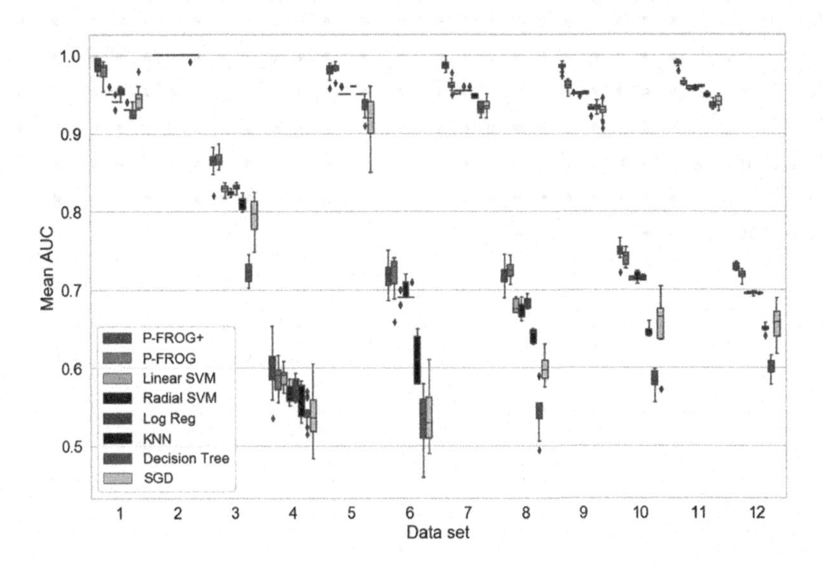

Fig. 1. Boxplots of AUC for the 14 runs of 10 fold cross-validation. Results for Probit FROG with noise $N(0, 0.1)$, Probit FROG without noise, and compared classifiers.

Table 2 presents corresponding Wilcoxon singed rank test results comparing P-FROG+ and P-FROG results with the other methods. The comparison is performed based on fold results, with a significance level of 0.05. P-FROG+ uses a noise $N(0, 0.1)$. A $\bullet$ indicates that the method in the heading reports significantly better results that the method indicated by the line for these data sets. We find P-FROG+ to be better than P-FROG in 5 instances and except one result - Table 2, compared with Linear SVM - better than all the others. Without using the noise, P-FROG can be considered better in 68 out of the 72 cases. These results, while not generalizable, do indicate that P-FROG+ offers a competitive framework that can be used in practical applications.

Computational Time. The duration of one evaluation of function $\tilde{v}()$ for a random generated vector σ for different number of instances (players) is illustrated in Figure 2, indicating that P-FROG can be scaled to larger data sets using this approach.

CMA-ES. CMA-ES is a robust optimization method designed to be used as it is, recommended for problems where classical approaches fail. We have used

Table 2. Wilcoxon signed rank test comparing medians of AUC values for each fold for results represented in Fig. 1. A • indicates that the method indicated in the heading reports significantly better results ($\alpha = 0.05$). A - indicates no statistical difference. P-FROG+ uses a noise $N(0, 0.1)$.

Method	P-FROG+												P-FROG											
Data set	1	2	3	4	5	6	7	8	9	10	11	12	1	2	3	4	5	6	7	8	9	10	11	12
Linear SVM	•	•	•	–	•	•	•	•	•	•	•	•	•	•	•	–	•	•	•	•	•	•	•	•
Radial SVM	•	•	•	•	•	•	•	•	•	•	•	•	•	•	•	–	•	•	•	•	•	•	•	•
Log Reg	•	•	•	•	•	•	•	•	•	•	•	•	•	•	•	–	•	•	•	•	•	•	–	•
KNN	•	•	•	•	•	•	•	•	•	•	•	•	•	•	•	•	•	•	•	•	•	•	•	•
Decision tree	•	•	•	•	•	•	•	•	•	•	•	•	•	•	•	•	•	•	•	•	•	•	•	•
SGD	•	•	•	•	•	•	•	•	•	•	•	•	•	•	•	•	•	•	•	•	•	•	•	•
P–FROG	–	•	–	–	•	–	•	–	•	–	•	–												

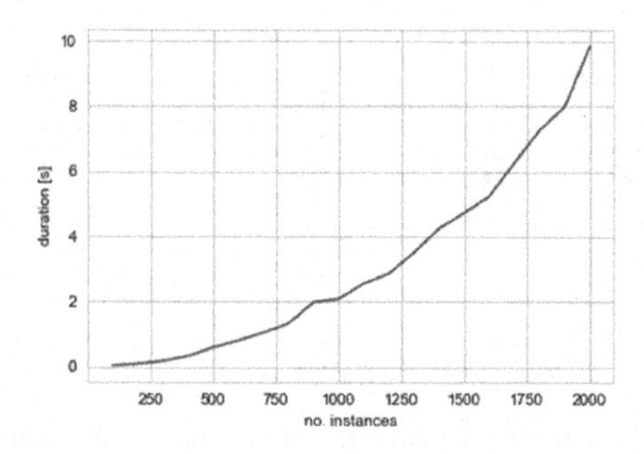

Fig. 2. Duration (s) for of one evaluation of function v for different number of instances (players).

the *cma* library in Python [3]. The initial solution was set constant to 0.1 for all variables, $\sigma = 0.3$. Function $fmin$ was modified to trigger the noise if a plateau is detected or at random with probability 0.1. The standard deviation ω of the noise $N(0, \omega)$ is a parameter of the method. Table 3 presents results obtained with different values of ω. For each data set, the average and standard deviation over the total number of folds (140) of the AUC values and minimum v values reported by CMA-ES are presented. A wilcoxon sing rank test was used to assess difference between results. A * marks for each data set the variant that has significantly outperformed most of the others. We find that, while differences among results vary from data to data, a 0.1 value of ω reports the best solutions in most of the cases. Considering that the all data are standardized to have mean 0 and variance 1, we can consider this value to be reasonably independent from data.

Table 3. Results obtained by P-FROG+ with different noise magnitudes ω. Average and standard deviations for AUC values and minimum values of v reported by CMA-ES for each fold. A * indicates that the result significantly outperformed most other results for the data set according to a Wilcoxon sign rank test with $\alpha = 0.05$.

Noise / Data	ω	0.001		0.01		0.1		0.2		0.3		0	
		Mean	StDev	Mean	StDev	Mean	StDev	Mean	StDev	Mean	StDev	Mean	StDev
1	AUC	0.967	0.059	0.97243	0.0578	0.9872*	0.03519	0.97929	0.04288	0.97171	0.06324	0.97957	0.04902
	MIN	0.00004	0.00006	0.00003	0.00005	0.00076	0.00103	0.00767	0.01171	0.0175	0.01991	0.00003	0.00007
6	AUC	1*	0	1*	0	1*	0	1*	0	0.99643	0.02977	1*	0
	MIN	0	0	0	0	0.00001	0.00004	0.00134	0.0024	0.0102	0.01735	0	0
7	AUC	0.82141	0.07191	0.82353	0.07097	0.8623*	0.07491	0.87534	0.07601	0.85281	0.10837	0.86607	0.07284
	MIN	0.03664	0.01139	0.03355	0.01179	0.07818	0.02972	0.13373	0.05581	0.1946	0.08986	0.03851	0.01211
8	AUC	0.5967	0.13083	0.57646	0.13065	0.5979*	0.12	0.55672	0.11877	0.53756	0.1157	0.5867	0.12159
	MIN	1.15115	0.25178	1.08642	0.25494	0.78711	0.31044	0.51895	0.29657	0.3579	0.21575	1.18954	0.21679
10	AUC	0.96757	0.0534	0.96086	0.06251	0.9801*	0.04523	0.988	0.03322	0.98871	0.02989	0.98329	0.03381
	MIN	0.00005	0.00003	0.00006	0.00007	0.00075	0.00113	0.00758	0.01096	0.023	0.03124	0.00046	0.00026
11	AUC	0.711	0.1284	0.7178*	0.13096	0.71871	0.13956	0.70771	0.15075	0.67186	0.14223	0.71986	0.14586
	MIN	0.4048	0.08922	0.40282	0.09257	0.53237	0.16369	0.64174	0.25417	0.5658	0.25792	0.5059	0.54748
17	AUC	0.9595	0.04093	0.95693	0.04504	0.9872*	0.02464	0.9894*	0.02271	0.9905*	0.02158	0.96154	0.04486
	MIN	0.00027	0.00012	0.00017	0.00011	0.00212	0.00251	0.01587	0.02198	0.0624	0.09515	0.00049	0.00026
18	AUC	0.70579	0.10716	0.70914	0.10362	0.71611	0.11571	0.7152*	0.11814	0.70014	0.11707	0.72507	0.10626
	MIN	1.0191	0.19661	1.02661	0.22888	1.24023	0.29218	1.44148	0.47829	1.431	0.37988	1.10787	0.47824
24	AUC	0.95023	0.03139	0.95471	0.03182	0.9849*	0.01916	0.9883*	0.01512	0.9887*	0.01283	0.96133	0.02921
	MIN	0.00133	0.00043	0.00116	0.00046	0.00923	0.0091	0.06244	0.07479	0.1763	0.2013	0.00142	0.00054
25	AUC	0.71118	0.05781	0.71363	0.05914	0.74809	0.06705	0.75398	0.06256	0.7500*	0.07665	0.74093	0.0743
	MIN	2.6003	0.24798	2.52968	0.25468	3.0298	0.431	3.71034	0.7724	4.0502	1.03084	2.94932	1.74796
31	AUC	0.96393	0.01815	0.96572	0.01814	0.9895*	0.0129	0.9909*	0.00819	0.99072	0.00845	0.96492	0.01861
	MIN	0.00147	0.00035	0.00181	0.00057	0.01765	0.01563	0.11515	0.12406	0.3505	0.43161	0.00164	0.00051
32	AUC	0.69944	0.04278	0.70185	0.04545	0.72916	0.05551	0.7396*	0.05501	0.7426*	0.04818	0.7188*	0.05167
	MIN	6.7225	0.45564	6.63487	0.5145	7.77233	0.99537	9.50447	1.59769	10.5024	2.01597	6.84834	0.53202

6 Conclusions

FROG is a novel framework for classification based on game theory and optimization. The main idea behind FROG is to construct a game for which the equilibrium represents a correct classification and use it to approximate model parameters. The equilibrium of the game can be computed by minimizing a real valued function. FROG uses CMA-ES as minimizer. Because the function yielding the game equilibrium may present plateaux, a noise inducing mechanism is added to CMA-ES to enhance the search. When noise is used, we refer to the framework as FROG+. The existence of plateaux depends on the classification model.

P-FROG uses FROG to estimate equilibrium parameters for the Probit classification model. P-FROG+ denotes the version that uses the noise mechanism during the search of CMA-ES. We find that P-FROG outperforms other standard classification methods for the tested data-sets, while using the noise mechanism in P-FROG+ can further improve results.

While results indicate the potential of the approach, the general framework can be further improved and explored. The computational complexity of the minimizing function, which is the main drawback of FROG - may be reduced by using other approximation methods in order to be able to tackle larger data sets. However, in spite drawbacks, exploring the use of solutions concepts from game theory in the field of classification may provide a more detailed insight into data than other models.

Acknowledgments. The first author would like to acknowledge the financial support provided for this research by Babeş-Bolyai University grant GTC 31379/2020.

References

1. Czerniak, J., Zarzycki, H.: Application of rough sets in the presumptive diagnosis of urinary system diseases. In: Sołdek, J., Drobiazgiewicz, L. (eds.) Artificial Intelligence and Security in Computing Systems, vol. 752, pp. 41–51. Springer, Boston (2003). https://doi.org/10.1007/978-1-4419-9226-0_5
2. Gast, N., Ioannidis, S., Loiseau, P., Roussillon, B.: Linear regression from strategic data sources. arXiv:1309.7824 [cs, math, stat], July 2019
3. Hansen, N., Akimoto, Y., Baudis, P.: CMA-ES/pycma on Github. Zenodo, February 2019. https://doi.org/10.5281/zenodo.2559634
4. Hansen, N., Ostermeier, A.: Completely derandomized self-adaptation in evolution strategies. Evol. Comput. **9**(2), 159–195 (2001). https://doi.org/10.1162/106365601750190398
5. Hastie, T., Tibshirani, R., Friedman, J.: The Elements of Statistical Learning. Springer Series in Statistics. Springer, New York (2001). https://doi.org/10.1007/978-0-387-21606-5
6. Hastie, T., Tibshirani, R., Friedman, J.: The Elements of Statistical Learning: Data Mining, Inference and Prediction. Springer, Heidelberg (2009). https://doi.org/10.1007/978-0-387-84858-7. http://www-stat.stanford.edu/ tibs/ElemStatLearn/
7. Lundberg, S.M., Lee, S.I.: A unified approach to interpreting model predictions. In: Guyon, I., Luxburg, U.V., Bengio, S., Wallach, H., Fergus, R., Vishwanathan, S., Garnett, R. (eds.) Advances in Neural Information Processing Systems, vol. 30, pp. 4765–4774. Curran Associates, Inc. (2017). http://papers.nips.cc/paper/7062-a-unified-approach-to-interpreting-model-predictions.pdf
8. McKelvey, R.D., McLennan, A.: Chapter 2 Computation of equilibria in finite games. In: Handbook of Computational Economics, vol. 1, pp. 87–142. Elsevier (1996). https://doi.org/10.1016/S1574-0021(96)01004-0. http://www.sciencedirect.com/science/article/pii/S1574002196010040
9. Pedregosa, F., et al.: Scikit-learn: machine learning in Python. J. Mach. Learn. Res. **12**, 2825–2830 (2011)
10. Rosset, S.: Model selection via the AUC. In: Proceedings of the Twenty-First International Conference on Machine Learning, ICML 2004. p. 89. Association for Computing Machinery, New York (2004). https://doi.org/10.1145/1015330.1015400
11. Schuurmans, D., Zinkevich, M.A.: Deep learning games. In: Lee, D.D., Sugiyama, M., Luxburg, U.V., Guyon, I., Garnett, R. (eds.) Advances in Neural Information Processing Systems, vol. 29, pp. 1678–1686. Curran Associates, Inc. (2016). http://papers.nips.cc/paper/6315-deep-learning-games.pdf

12. Strumbelj, E., Kononenko, I.: An efficient explanation of individual classifications using game theory. J. Mach. Learn. Res. **11**, 1–18 (2010)
13. Tan, P.N., Steinbach, M., Karpatne, A., Kumar, V.: Introduction to Data Mining, 2nd edn. Pearson, London (2018)

Analyzing the Components of Distributed Coevolutionary GAN Training

Jamal Toutouh(✉), Erik Hemberg, and Una-May O'Reilly

Massachusetts Institute of Technology, CSAIL, Cambridge, MA, USA
toutouh@mit.edu,
{hembergerik,unamay}@csail.mit.edu

Abstract. Distributed coevolutionary Generative Adversarial Network (GAN) training has empirically shown success in overcoming GAN training pathologies. This is mainly due to diversity maintenance in the populations of generators and discriminators during the training process. The method studied here coevolves sub-populations on each cell of a spatial grid organized into overlapping Moore neighborhoods. We investigate the impact on performance of two algorithm components that influence the diversity during coevolution: the performance-based selection/replacement inside each sub-population and the communication through migration of solutions (networks) among overlapping neighborhoods. In experiments on MNIST dataset, we find that the combination of these two components provides the best generative models. In addition, migrating solutions without applying selection in the sub-populations achieves competitive results, while selection without communication between cells reduces performance.

Keywords: Generative Adversarial Networks · Coevolution · Diversity · Selection pressure · Communication

1 Introduction

Machine learning with Generative Adversarial Networks (GANs) is a powerful method for generative modeling [9]. A GAN consists of two neural networks, a generator and a discriminator, and applies adversarial learning to optimize their parameters. The generator is trained to transform its inputs from a random latent space into "artificial/fake" samples that approximate the true distribution. The discriminator is trained to correctly distinguish the "natural/real" samples from the ones produced by the generator. Formulated as a minmax optimization problem through the definitions of generator and discriminator loss, training can converge on an optimal generator that is able to fool the discriminator.

GANs are difficult to train. The adversarial dynamics introduce convergence pathologies [5,14]. This is mainly because the generator and the discriminator are differentiable networks, their weights are updated by using (variants of) simultaneous gradient-based methods to optimize the minmax objective, that rarely

T. Bäck et al. (Eds.): PPSN 2020, LNCS 12269, pp. 552–566, 2020.
https://doi.org/10.1007/978-3-030-58112-1_38

converges to an equilibrium. Thus, different approaches have been proposed to improve the convergence and the robustness in GAN training [7,15,18,20,26].

A promising research line is the application of distributed competitive coevolutionary algorithms (Comp-COEA). Fostering an arm-race of a population of generators against a population of discriminators, these methods optimize the minmax objective of GAN training. Spatially distributed populations (cellular algorithms) are effective at mitigating and resolving the COEAs pathologies attributed to a lack of diversity [19], which are similar to the ones observed in GAN training. `Lipizzaner` [1,22] is a spatial distributed Comp-COEA that locates the individuals of both populations on a spatial grid (each cell contains a GAN). In a cell, each generator is evaluated against all the discriminators of its neighborhood, the same with the discriminator. It uses neighborhood communication to propagate models and foster diversity in the sub-populations. Moreover, the selection pressure helps the convergence in the sub-populations [2].

Here, we evaluate the impact of neighborhood communication and selection pressure on this type of GAN training. We conduct an ablation analysis to evaluate different combinations of these two components. We ask the following research questions: **RQ1:** *What is the effect on the quality of the generators when training with communication or isolation and the presence or absence of selection pressure?*. The quality of the generators is evaluated in terms of the accuracy of the samples generated and their diversity. **RQ2:** *What is the effect on the diversity of the network parameters when training with communication or isolation and the presence or absence of selection pressure?* **RQ3:** *What is the impact on the computational cost of applying migration and selection/replacement?*

The main contributions of this paper are: *i)* proposing distributed Comp-COEA GAN training methods by applying different types of ablation to `Lipizzaner`, *ii)* evaluating the impact of the communication and the selection pressure on the quality of the returned generative model, and *iii)* analyzing their computational cost.

The paper is organized as follows. Section 2 presents related work. Section 3 describes `Lipizzaner` and the ablations the methods analyzed. The experimental setup is in Sect. 4 and results in Sect. 5. Finally, conclusions are drawn and future work is outlined in Sect. 6.

2 Related Work

In 2014, Goodfellow introduced GAN training [9]. Robust GAN training methods are still investigated [5,14]. Competitive results have been provided by several practices that stabilize the training [7]. These methods include different strategies, such as using different cost functions to the generator or discriminator [4,15,18,29,32] and decreasing the learning rate through the iterations [20].

GAN training that involves multiple generators and/or discriminators empirically show robustness. Some examples are: iteratively training and adding new generators with boosting techniques [25]; combining a cascade of GANs [30]; training an array of discriminators on different low-dimensional projections of the data [17]; training the generator against a dynamic ensemble of discriminators [16]; independently optimizing several "local" generator-discriminator pairs so that a "global" supervising pair of networks can be trained against them [6].

Evolutionary algorithms (EAs) may show limited effectiveness in high dimensional problems because of runtime [24]. Parallel/distributed implementations allow EAs to keep computation times at reasonable levels [3,8]. There has been an emergence of large-scale distributed evolutionary machine learning systems. For example: EC-Star [11], which runs on hundreds of desktop machines; a simplified version of Natural Evolution Strategies [31] with a novel communication strategy to address a collection of reinforcement learning benchmark problems [21] or deep convolutional networks trained with genetic algorithms [23].

Theoretical studies and empirical results demonstrate that the spatially distributed COEA GAN training mitigates convergence pathologies [1,22,26]. In this study, we focus on spatially distributed Comp-COEA GAN training, such as `Lipizzaner` [1,22]. `Lipizzaner` places the individuals of the generator and discriminator populations on each cell (i.e., each cell contains a generator-discriminator pair). Overlapping Moore neighborhoods determine the communication among the cells to propagate the models through the grid. Each generator is evaluated against all the discriminators of its neighborhood and the same happens with each discriminator. This intentionally fosters diversity to address GAN training pathologies. Mustangs [26], a `Lipizzaner` variant, uses randomly selected loss functions to train each cell for each epoch to increase diversity. Moreover, training the GANs in each cell with different subsets of the training dataset has been demonstrated effective in increasing diversity across the grid [27]. These three approaches return an ensemble/mixture of generators defined by the best neighborhood (sub-population of generators) built using evolutionary ensemble learning [28]. They have shown competitive results on standard benchmarks.

Here, we investigate the impact of two key components for diversity in Comp-COEA GAN training using ablations.

3 Comp-COEA GAN Training Ablations

Here, we evaluate the impact of communication and selection pressure on the performance of distributed Comp-COEA GAN training. One goal of the training is to maintain diversity as a means of resilience to GAN training pathologies. The belief is that a lack of sufficient diversity results in convergence to pathological training states [22] and too much diversity results in divergence. The use of selection and communication is one way of regulating the diversity.

This section presents `Lipizzaner` [22] and describes the ablations applied to gauge the impact of communication/isolation and selection pressure.

3.1 Lipizzaner Training

`Lipizzaner` adversarially trains a population of generators $\mathbf{g} = \{g_1, ..., g_N\}$ and a population of discriminators $\mathbf{d} = \{d_1, ..., d_N\}$, where N is the size of the

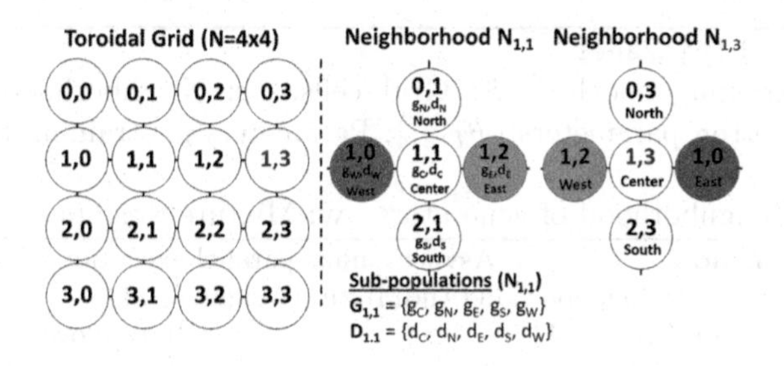

Fig. 1. A 4×4 grid (left) and the neighborhoods $\mathrm{N}_{1,1}$ and $\mathrm{N}_{1,3}$ (right).

populations. It defines a toroidal grid in whose cells a pair generator-discriminator, i.e., a GAN, is placed (called *center*). This allows the definition of neighborhoods with sub-populations $\mathbf{g}^k$ and $\mathbf{d}^k$, of $\mathbf{g}$ and of $\mathbf{d}$, respectively. The size of these sub-populations is denoted by s ($s \leq N$). `Lipizzaner` uses the five-cell Moore neighborhood ($s = 5$), i.e., the neighborhoods include the cell itself (*center*) and the cells in the *West*, *North*, *East*, and *South* (see Fig. 1).

Each cell asynchronously executes in parallel its own learning algorithm. Cells interact with the neighbors at the beginning of each training epoch. This communication is carried out by gathering the latest updated *center* generator and discriminator of its overlapping neighborhoods. Figure 1 illustrates some examples of the overlapping neighborhoods on a *4x4* toroidal grid. The updates in cell $\mathrm{N}_{1,0}$ and $\mathrm{N}_{1,2}$ will be communicated to the $\mathrm{N}_{1,1}$ and $\mathrm{N}_{1,3}$ neighborhood.

`Lipizzaner` returns an ensemble of generators that consist of a sub-population of generators $\mathbf{g}^k$ and a mixture weight vector $\mathbf{w} \subset \mathbb{R}^s$, where $w_i \in \mathbf{w}$ represents the probability that a data point is drawn from $\mathbf{g}^k_i$, i.e., $\sum_{w_i \in \mathbf{w}^k} w_i = 1$. Thus, an Evolutionary Strategy, `ES`-(1+1), evolves $\mathbf{w}$ to optimize the weights in order to get the most accurate generative model according to a given metric [22], such as Fréchet Inception Distance (FID) [10].

`Algorithm` 1 summarizes the main steps of `Lipizzaner`. It starts the parallel execution of the training process in each cell. First, the cell randomly initializes single generator and discriminator models. Then, the training process consist of a loop with three main steps: *i)* gathering the GANs (neighbors) to update the neighborhood; *ii)* updating the *center* by applying the training method presented in `Algorithm` 2; and *iii)* evolving mixture of weights by applying `ES`-(1+1) (line 7 in `Algorithm` 1). These three steps are repeated T (training epochs) times. The returned generative model is the best performing neighborhood with its optimal mixture weights, i.e., the best ensemble of generators.

The Comp-COEA training applied to the networks of each sub-population is shown in `Algorithm` 2. The output is the sub-population n with the *center* updated. This method is built on the basis of *fitness evaluation*, *selection and replacement*, and *reproduction* based on gradient descent updates of the weights.

The *fitness* of each network is evaluated in terms of the average *loss function* $\mathcal{L}$ when it is evaluated against all the adversaries. After evaluating all the

Algorithm 1. `Lipizzaner`

Input: T: Training epochs, E: Grid cells, s: Neighborhood size, θ_{EA}: `mixtureEvolution` parameters, θ_{Train}: Parameters for training the models in each cell

Return: g: Neighborhood of generators, $\mathbf{w}$: Mixture weights

1: **parfor** $c \in E$ **do** ▷ Asynchronous parallel execution of all cells in grid
2: $n, \mathbf{w} \leftarrow$ initializeNeighborhoodAndMixtureWeights(c, s)
3: **for** epoch **do** $\in [1, \ldots, \mathrm{T}]$ ▷ Iterate over training epochs
4: $n \leftarrow$ copyNeighbours(c, s) ▷ Gather neighbour networks
5: $n \leftarrow$ trainModels(n, θ_{Train}) ▷ Update GANs weights
6: $g \leftarrow$ getNeighborhoodGenerators(n) ▷ Get the generators
7: $\mathbf{w} \leftarrow$ `mixtureEvolution` $(\mathbf{w}, g, \theta_{EA})$ ▷ Evolve mixture weights by `ES`-(1+1)
8: **end parfor**
9: $(g, \mathbf{w}) \leftarrow$ `bestEnsemble`$(g^*, \mathbf{w}^*)$ ▷ Get the best ensemble
10: **return** $(g, \mathbf{w})$ ▷ Cell with best generator mixture

Algorithm 2. `Coevolve and train the networks`

Input: τ: Tournament size, X: Input training dataset, β: Mutation probability, n: Cell neighborhood sub-population, M: Loss functions

Return: n: Cell neighborhood sub-population updated

1: $\mathbf{B} \leftarrow$ getMiniBatches(X) ▷ Load minibatches
2: $B \leftarrow$ getRandomMiniBatch($\mathbf{B}$) ▷ Get a minibatch to evaluate GANs
3: **for** $g, d \in \mathbf{g} \times \mathbf{d}$ **do** ▷ Evaluate all GAN pairs
4: $\mathcal{L}_{g,d} \leftarrow$ evaluate(g, d, B) ▷ Evaluate GAN
5: $g, d \leftarrow$ select(n, τ) ▷ Select with min loss($\mathcal{L}$) as fitness
6: **for** $B \in \mathbf{B}$ **do** ▷ Loop over batches
7: $n_\delta \leftarrow$ mutateLearningRate(n_δ, β) ▷ Update learning rate
8: $d' \leftarrow$ getRandomOpponent($\mathbf{d}$) ▷ Get random discriminator to train g
9: $\nabla_g \leftarrow$ computeGradient(g, d') ▷ Compute gradient for g against d'
10: $g \leftarrow$ updateNN(g, ∇_g, B) ▷ Update g with gradient
11: $g' \leftarrow$ getRandomOpponent($\mathbf{g}$) ▷ Get uniform random generator to train g
12: $\nabla_d \leftarrow$ computeGradient(d, g') ▷ Compute gradient for d against g'
13: $d \leftarrow$ updateNN(d, ∇_d, B) ▷ Update d with gradient
14: **for** $g, d \in \mathbf{g} \times \mathbf{d}$ **do** ▷ Evaluate all updated GAN pairs
15: $\mathcal{L}_{g,d} \leftarrow$ evaluate(g, d, B) ▷ Evaluate GAN
16: $\mathcal{L}_g \leftarrow average(\mathcal{L}_{\cdot,d})$ ▷ Fitness of g is the average loss value ($\mathcal{L}$)
17: $\mathcal{L}_d \leftarrow average(\mathcal{L}_{g,\cdot})$ ▷ Fitness of d is the average loss value ($\mathcal{L}$)
18: $n \leftarrow$ replace($n, \mathbf{g}$) ▷ Replace the generator with worst loss
19: $n \leftarrow$ replace($n, \mathbf{d}$) ▷ Replace the discriminator worst loss
20: $n \leftarrow$ setCenterIndividuals(n) ▷ Best g and d are placed in the center
21: **return** n

networks, a *tournament selection operator* is applied to select the parents (a generator and a discriminator) to generate the offspring (lines from 1 to 5).

The selected generator/discriminator is evaluated against a randomly chosen adversary from the sub-population (lines 8 and 11, respectively). The computed losses are used to mutate the parents (i.e., update the networks' parameters) according to the stochastic gradient descent (SGD), which learning rate values n_δ were updated applying Gaussian-based mutation (line 7).

When the training is completed, all models are evaluated again, the least fit generator and discriminator in the sub-populations are replaced with the fittest ones and sets them as the *center* of the cell (lines from 14 to 20).

3.2 Lipizzaner Ablations Analyzed

We conduct an ablation analysis of `Lipizzaner` to evaluate the impact of different degrees of communication/isolation and selection pressure on its COEA GAN training. The ablations are listed in Table 1. They ablate the use of *sub-populations*, the *communication* between cells, and the application of the *selection/replacement* operator. Thus, we define three variations of `Lipizzaner`:

- Spatial Parallel GAN training (`SPaGAN`): It does not apply selection/replacement. After gathering the networks from the neighborhood, it uses them to only train the *center* (i.e., `SPaGAN` does not apply the operations in lines between 2 and 5 and lines between 14 and 20 of `Algorithm` 2).
- Isolated Coevolutionary GAN training (`IsoCoGAN`): It trains sub-populations of GANs without communication between cells. There is no exchange of networks between neighbors after the creation of the initial sub-population (i.e., the operation in line 4 of `Algorithm` 1 is applied only during the first iteration of its main loop).
- Parallel GAN (`PaGAN`): This trains a population of N GANs in parallel. When all the GAN training is finished, it randomly produces N sub-sets of $s \leq N$ generators selected from the entire population of trained generators to define the ensembles and optimizes the mixture weights with ES-(1+1).

Table 1. Key components of the GAN training methods.

Feature	`Lipizzaner`	`SPaGAN`	`IsoCoGAN`	`PaGAN`
Use of sub-populations	✓	✓	✓	–
Communication between sub-populations	✓	✓	–	–
Application of selection/replacement operator	✓	–	✓	–

4 Experimental Setup

This experimental analysis compares `Lipizzaner`, `SPaGAN`, `IsoCoGAN`, and `PaGAN` in creating generative models to produce samples of the MNIST dataset [12]. This dataset is widely used as a benchmark due to its target space and dimensionality.

The communication among the neighborhoods is affected by the grid size [22]. We evaluate SPaGAN and Lipizzaner by using three different grid sizes: 3×3, 4×4, and 5×5, to control for the impact of this parameter.

To evaluate the quality of the generated data we use **FID** score. The **network diversity** of the trained models is evaluated by the L_2 **distance between the parameters** of the generators. The **total variation distance (TVD)** [13], which is a scalar that measures class balance, is used to analyze the diversity of the generated data by the generative models. TVD reports the difference between the proportion of the generated samples of a given digit and the ideal proportion (10% in MNIST). Finally, the **computational cost** is measured in terms of run time. All the analyzed methods apply the same number of training-network steps, i.e., updating the networks' parameters according to SGD. All implementations are publicly available[1] and use the same Python libraries and versions. The distribution of the results is not Gaussian, so a Mann-Whitney U statistical test is applied to evaluate their significance.

The methods evaluated here are configured according to the settings proposed by the authors of Lipizzaner [22]. The experimental analysis is performed on a cloud computing platform that provides 8 Intel Xeon cores 2.2 GHz with 32 GB RAM and an NVIDIA Tesla P100 GPU with 16 GB RAM.

5 Results and Discussion

This section discusses the main results of the experiments carried out to evaluate Lipizzaner, SPaGAN, IsoCoGAN, and PaGAN.

5.1 Generator Quality

Table 2 summarizes the results by showing the best FID scores for the different methods and grid sizes in the 30 independent runs. Lipizzaner obtains the lowest/best **Mean**, **Median**, and **Min** FID scores for all grid sizes. SPaGAN and PaGAN are the second and third best methods, respectively. IsoCoGAN presents the lowest quality by showing the highest (worst) FID scores. The FID values in terms of hundreds indicate the generators are not capable of creating adequate MNIST data samples. The Mann-Whitney U test shows that the methods that exchange individuals among the neighborhoods, i.e., Lipizzaner and SPaGAN, are significantly better than PaGAN and IsoCoGAN ($\alpha << 0.001$). These results are confirmed by posthoc statistical analysis. According to this analysis there is no statistical difference between Lipizzaner 3×3 and SPaGAN for all evaluated grid sizes. In turn, Lipizzaner 4×4 and 5×5 outperform all the other methods and Lipizzaner 5×5 provides the best generators (lowest FID).

IsoCoGAN does not converge since the individuals of one sub-population are evaluated against randomly chosen individuals of the other one. As the accuracy of their fitness evaluation depends subjectively on the quality of the randomly

[1] Lipizzaner and ablations - https://github.com/ALFA-group/lipizzaner-gan.

Table 2. FID results (Low FID indicates more quality)

Grid	Method	Mean ± Std	Median	Iqr	Min	Max
3×3	Lipizzaner	**40.93 ± 8.51**	**39.44**	10.59	**28.04**	62.10
	SPaGAN	43.59 ± 5.53	43.09	5.94	30.65	**51.81**
	IsoCoGAN	881.79 ± 52.67	871.04	66.81	798.03	998.79
	PaGAN	51.15 ± 14.06	46.85	6.40	40.81	112.58
4×4	Lipizzaner	**32.84 ± 6.93**	**32.22**	7.23	**19.15**	**46.53**
	SPaGAN	37.97 ± 8.89	35.98	13.61	23.69	56.36
5×5	Lipizzaner	**28.74 ± 4.91**	**28.14**	7.45	**22.56**	**40.57**
	SPaGAN	39.11 ± 4.00	39.61	5.29	27.39	48.03

chosen opponent, it is likely that the fitness value does not correspond to the real quality of the individual, and therefore, the selection/replacement operator does not promote the objectively best solution in the sub-population.

For all grid sizes, SPaGAN provides higher/worse FIDs than Lipizzaner. SPaGAN has a similar quality for all grid sizes, but Lipizzaner improves when the grid size is increased. Thus, the larger the grid size, the greater the difference between these two methods. We observe the benefits of increasing diversity (population/grid size) when applying the coevolutionary approach with selection and replacement of Lipizzaner. The results of Lipizzaner are mainly due to, first, the larger population sizes' ability to encompass a higher diversity, and second, the selection/replacement process applied by Lipizzaner accelerates the convergence of the population to higher quality generators.

5.2 Quality (FID) Evolution

Figure 2 shows the evolution of the median FID when using PaGAN, SPaGAN, and Lipizzaner in a 3×3 grid. According to this figure, Lipizzaner improves the performance of the generators faster than SPaGAN. After the first 75 to 100 training epochs, the FID does not show such a reduction in Lipizzaner, but SPaGAN is able to keep reducing it until the end of the training process. The faster convergence of Lipizzaner is also illustrated by Fig. 4 that shows the FID scores in the grid of a given independent run at the epoch number 25, 50, 75, and 100. This is mainly due to the capacity of exploitation of this method when using selection/replacement. PaGAN shows a fast FID reduction at the beginning of the training process. But, the FID sharply oscillates during the first 75 iterations and it converges to worse FID scores than Lipizzaner and SPaGAN.

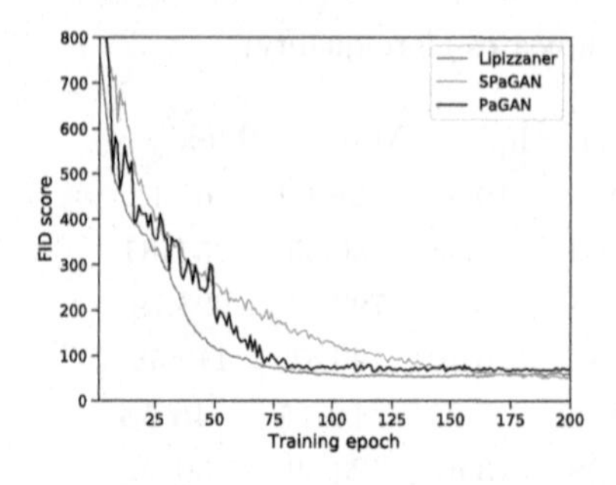

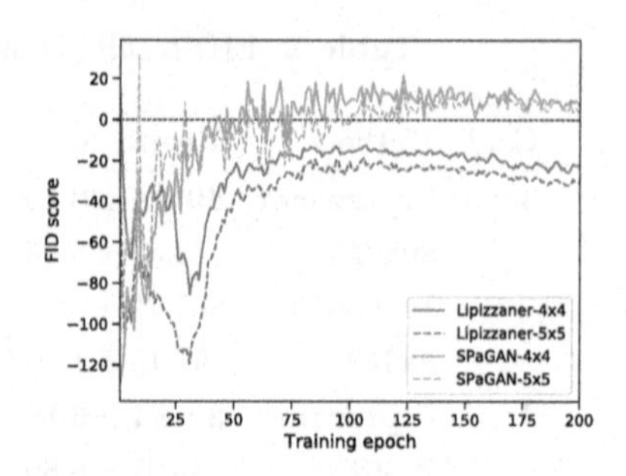

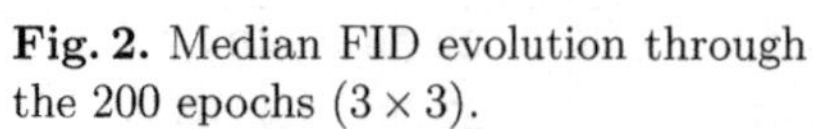

Fig. 2. Median FID evolution through the 200 epochs (3×3).

Fig. 3. FID differences when using differnt grids ($diff\_FID_i^{m \times m}$).

Figure 3 illustrates the differences between the FID score when using 3×3 and 4×4 or 3×3 and 5×5 for the same method. This difference for the grid size m $\times$ m at the i epoch is computed as $diff\_FID_i^{m \times m} = FID_i^{3 \times 3} - FID_i^{m \times m}$. This figure allows evaluating the impact on the FID evolution when using different grid (population) sizes. It shows how Lipizzaner provides lower FIDs and it is able to converge faster as the grid size increases during the first iterations. In contrast, SPaGAN gets better FIDs when increasing the grid size only during the first 50 epochs. Lipizzaner is able to take advantage of the diversity generated when the grid size increase to converge faster to lower FIDs than SPaGAN. Thus, in terms of convergence, selection/replacement helps the convergence of the coevolutionary training method when exchanging solutions with the neighborhoods.

5.3 Generator Output Diversity

The output diversity reports the class distribution of the fake data produced by a generative model. Table 3 summarizes the results in terms of TVD.

For the 3×3 grid experiments, IsoCoGAN shows the worst (highest) TVD values. PaGAN provides good TVDs but it is statistically less competitive than SPaGAN and Lipizzaner according to the Mann-Whitney U test. SPaGAN and Lipizzaner show the best results obtaining the same **Mean**, **Median**, and **Min** values. Therefore, when using communication between the neighborhoods during the training, the generators are able to produce more diverse data samples.

When increasing the grid size, SPaGAN and Lipizzaner improve their results. However, SPaGAN does not show statistical differences between the same algorithm with different grid sizes. Lipizzaner 5×5 provides statistically better results than Lipizzaner 3×3 and than SPaGAN. The coevolutionary approach used in Lipizzaner takes advantage of the divergence generated when increasing the grid size to train generators that create more diverse data samples.

According to these results and the ones in Sect. 5.1, we can answer **RQ1:** *What is the effect on the quality of the generators when training with communication or isolation and the presence or absence of selection pressure?* Communication and selection pressure allowed Lipizzaner to converge to generators

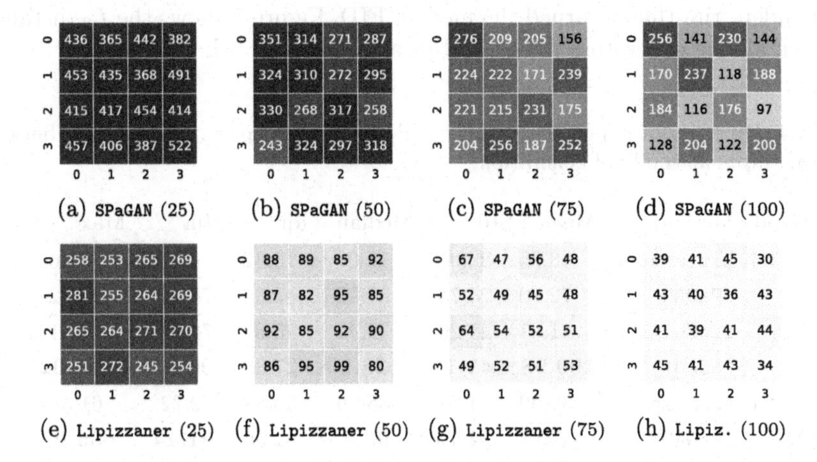

(a) SPaGAN (25) (b) SPaGAN (50) (c) SPaGAN (75) (d) SPaGAN (100)

(e) Lipizzaner (25) (f) Lipizzaner (50) (g) Lipizzaner (75) (h) Lipiz. (100)

Fig. 4. FID score distribution through the grid at different epochs (expresed between parentesis) of an independent run for SPaGAN and Lipizzaner in a 4×4 grid. Lighter blues represent lower (better) FID scores.

Table 3. TVD results (Low TVD indicates more diversity).

Grid	Method	Mean ± Std	Median	Iqr	Min	Max
3×3	Lipizzaner	0.12 ± 0.03	**0.12**	0.04	**0.08**	0.18
	SPaGAN	**0.12 ± 0.02**	**0.12**	0.02	**0.08**	**0.16**
	IsoCoGAN	0.83 ± 0.08	0.82	0.14	0.69	0.91
	PaGAN	0.14 ± 0.02	0.14	0.04	0.10	0.19
4×4	Lipizzaner	**0.11 ± 0.02**	**0.11**	0.02	**0.05**	**0.16**
	SPaGAN	0.12 ± 0.02	0.12	0.03	0.08	**0.16**
5×5	Lipizzaner	**0.10 ± 0.02**	**0.11**	0.02	**0.06**	0.16
	SPaGAN	0.11 ± 0.02	**0.11**	0.03	0.07	**0.15**

with the best quality (FID and TVD). The diversity resulting from communication resulted in the most competitive results, i.e., Lipizzaner and SPaGAN ended with better generators than IsoCoGAN and PaGAN. Isolation in training converged to good solutions when the cell was optimizing only one GAN (PaGAN). However, when isolation is coupled with a sub-population that applies selection/replacement, the algorithm was not able to converge, and therefore, the quality of the generators returned is the worst.

5.4 Genome Space Diversity

We next investigate the diversity of the parameters of the evolved networks. Table 4 summarizes the L_2 distance results for the population at the end of the independent run that returned the median FID. Figure 5 shows the L_2 distances between all the generators in the grid (x and y axes are the cell number).

Table 4. Diversity of population (whole grid) in *genome* space. L_2 distances between the generators at the final generation.

Grid	Method	Mean ± Std	Median	Iqr	Min	Max
3×3	Lipizzaner	13.85 ± 8.29	18.06	15.32	4.09	23.46
	SPaGAN	72.70 ± 25.75	81.72	2.66	78.32	84.82
	IsoCoGAN	71.07 ± 25.29	79.42	5.67	74.00	86.01
	PaGAN	**109.98 ± 41.15**	**128.26**	27.89	**90.82**	**138.02**
4×4	Lipizzaner	30.99 ± 24.28	38.09	55.82	2.72	61.34
	SPaGAN	**87.90 ± 22.71**	**93.75**	1.36	**91.44**	**95.95**
5×5	Lipizzaner	31.98 ± 15.31	36.71	21.32	4.28	52.14
	SPaGAN	**87.96 ± 18.06**	**91.60**	3.30	**87.22**	**96.87**

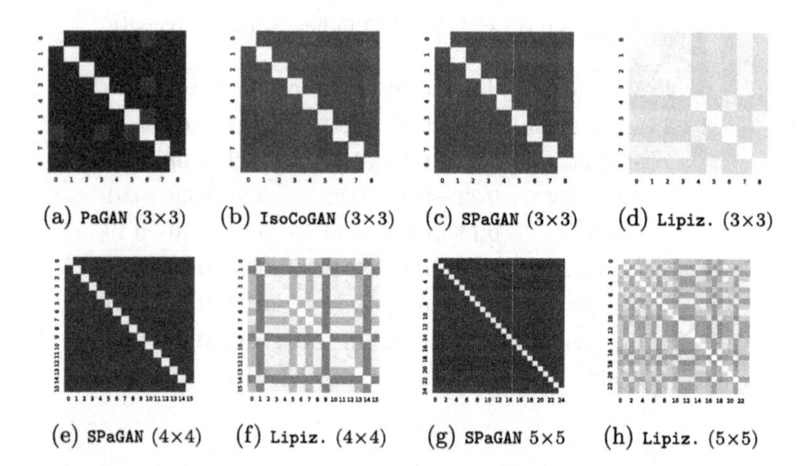

(a) PaGAN (3×3) (b) IsoCoGAN (3×3) (c) SPaGAN (3×3) (d) Lipiz. (3×3)

(e) SPaGAN (4×4) (f) Lipiz. (4×4) (g) SPaGAN 5×5 (h) Lipiz. (5×5)

Fig. 5. Diversity of population in *genome* space. Heatmap of L_2 distance between the generators at the final generation. Dark indicates more diversity.

The highest generator diversity is shown by PaGAN. This is because this method trains a single generator against a single discriminator and it does not use any type of information exchange or communication during the training process. Therefore, each cell converges to different points of the search space.

SPaGAN and IsoCoGAN provide networks with similar diversity in 3 × 3 grid, which is higher than the diversity provided by Lipizzaner. This is mainly due to the combination of both communication and the use of selection/replacement in Lipizzaner provokes the sub-populations to converge to similar accurate individuals, limiting diversity.

The genome diversity in Lipizzaner is increased when increasing the size of the grid (see in Fig. 5 how the heatmap gets darker as the grid size is larger). SPaGAN increases the L_2 distances with the grid size, but in a lower proportion, e.g., from 3×3 to 4×4 Lipizzaner increases 123.75% and SPaGAN 20.90%. The diversity in the genome space helps the creation of ensembles that are able to provide better accuracy. For this reason, Lipizzaner can improve results with an increasing grid size (and diversity).

Answering **RQ2:** *What is the effect on the diversity of the network parameters when training with communication or isolation and the presence or absence of selection pressure?* The combination of communication and selection pressure permits Lipizzaner to converge to similar high-quality generators. Complete isolation and no-population based GAN training (PaGAN) converge to highly different generators, which is expected. SPaGAN illustrates how the absence of selection pressure in the populations keep the individuals diverse in the grid, although there is communication. In 3×3 grid experiments, IsoCoGAN and SPaGAN have similar diversity but highly different quality. This shows how maintaining diversity is not enough to ensure robust GAN training.

5.5 Computational Efficiency

Now, we analyze the computational efficiency of the GAN training methods, taking into account that they use the same number of training epochs but used different components. All these methods apply asynchronous parallelism and the time required by each cell of the grid to perform the same number of training epochs varies. Thus, we report the computational time of a run as the time required by each independent run to finish and return the best ensemble of generators found by all the cells.

Table 5. Computation time in minutes.

Grid	Method	Mean ± Std	Median	Iqr	Min	Max
3×3	Lipizzaner	87.89 ± 1.15	87.73	1.23	85.57	90.23
	SPaGAN	87.20 ± 0.31	87.15	0.34	86.72	88.02
	IsoCoGAN	81.88 ± 4.55	81.34	9.41	74.40	88.19
	PaGAN	**38.07 ± 2.73**	**37.51**	3.49	**33.57**	**44.28**
4×4	Lipizzaner	91.30 ± 0.94	91.07	1.01	90.23	94.26
	SPaGAN	**90.72 ± 0.58**	**90.89**	0.39	**88.97**	**91.27**
5×5	Lipizzaner	105.64 ± 3.25	107.22	4.48	100.25	111.05
	SPaGAN	**101.88 ± 1.64**	**100.91**	2.10	**100.18**	**105.52**

The computational time of PaGAN includes the whole process, i.e., the time of training the GANs plus the time of optimizing the ensemble weights by using the ES-(1+1). This method requires the shortest run time (see Table 5). This is because PaGAN trains a single network in each cell and there is no communication among the different cells. According to the Mann-Whitney U and posthoc statistical tests, Lipizzaner requires the longest computation times. It employs both communication and selection/replacement algorithm components.

When comparing among Lipizzaner, SPaGAN, and IsoCoGAN, the impact on the computation time of applying communication in Lipizzaner and SPaGAN is higher than the use of selection/replacement in Lipizzaner and IsoCoGAN (see Table 5 *3×3-Grid*). This increase in the computation cost may increase in problems that require the use and exchange among cells of bigger models (networks) for the generators and discriminators.

Therefore, answering **RQ3:** *What is the impact on the computational cost of applying migration and selection/replacement?*, the effect of applying selection/replacement is negligible when comparing it with the impact of communicating among the cells, when running the methods on 3×3 grid. However, it increases as the grid size is larger. Moreover, when bigger networks need to be trained to handle harder problems, they will have many more parameters and more need for communications.

6 Conclusions and Future Work

We have empirically shown that the spatially distributed coevolutionary training applied by Lipizzaner is the best choice among options with/without communication and selection/replacement components to train GANs. The combination of selection pressure that promotes convergence in the sub-populations and communication with the overlapped neighborhoods maintains enough diversity to robustly train the networks in the sub-population. Moreover, the use of these two operations does not entail a very significant increase in the computation time (about four minutes in the larger grid size).

SPaGAN illustrates the importance of the communication among the cells (i.e., fostering diversity). It is able to converge to high-quality solutions (generators) although it does not apply selection. This emphasizes the value of exchanging the best individuals even they are just used to train the center of the cells. IsoCoGAN provides the least competitive results. Training the networks with a coevolutionary flavor does not ensure convergence, even when it is done in sub-populations and it uses selection/replacement.

Future work will include further analysis of coevolutionary GAN training in other benchmarks (i.e., problems and data sets). We will evaluate the scalability of this type of training by using larger grids. We will study the effect of training the GANs with different kinds of loss functions. We will assess the impact on the robustness when using different types of neighborhoods. Finally, we will analyze the evolution of the network weights through the generations to better understand the dynamics of this type of GAN training.

Acknowledgments. This research was partially funded by European Union's Horizon 2020 research and innovation program under the Marie Skłodowska-Curie grant agreement No 799078, by the Junta de Andaluía UMA18-FEDERJA-003, European Union H2020-ICT-2019-3, and the Systems that Learn Initiative at MIT CSAIL.

References

1. Al-Dujaili, A., Schmiedlechner, T., Hemberg, E., O'Reilly, U.M.: Towards distributed coevolutionary GANs. In: AAAI 2018 Fall Symposium (2018)
2. Alba, E., Dorronsoro, B.: Cellular Genetic Algorithms. Springer, Heidelberg (2009). https://doi.org/10.1007/978-0-387-77610-1
3. Alba, E., Luque, G., Nesmachnow, S.: Parallel metaheuristics: recent advances and new trends. Int. Trans. Oper. Res. **20**(1), 1–48 (2013)
4. Arjovsky, M., Chintala, S., Bottou, L.: Wasserstein GAN. arXiv preprint arXiv:1701.07875 (2017)
5. Arora, S., Ge, R., Liang, Y., Ma, T., Zhang, Y.: Generalization and equilibrium in generative adversarial nets (GANs). arXiv preprint arXiv:1703.00573 (2017)
6. Chavdarova, T., Fleuret, F.: SGAN: an alternative training of generative adversarial networks. In: Proceedings of the IEEE Conference on Computer Vision and Pattern Recognition, pp. 9407–9415 (2018)
7. Chintala, S., Denton, E., Arjovsky, M., Mathieu, M.: How to train a GAN? tips and tricks to make GANs work(2016). https://github.com/soumith/ganhacks
8. Essaid, M., Idoumghar, L., Lepagnot, J., Brévilliers, M.: GPU parallelization strategies for metaheuristics: a survey. Int. J. Parallel Emergent Distrib. Syst. **34**(5), 497–522 (2019)
9. Goodfellow, I., et al.: Generative adversarial nets. In: Advances in Neural Information Processing Systems, pp. 2672–2680 (2014)
10. Heusel, M., Ramsauer, H., Unterthiner, T., Nessler, B., Hochreiter, S.: GANs trained by a two time-scale update rule converge to a local nash equilibrium, pp. 6626–6637 (2017)
11. Hodjat, B., Hemberg, E., Shahrzad, H., O'Reilly, U.-M.: Maintenance of a long running distributed genetic programming system for solving problems requiring big data. In: Riolo, R., Moore, J. H., Kotanchek, M. (eds.) Genetic Programming Theory and Practice XI. GEC, pp. 65–83. Springer, New York (2014). https://doi.org/10.1007/978-1-4939-0375-7_4
12. LeCun, Y.: The mnist database of handwritten digits (1998). http://yann.lecun.com/exdb/mnist/
13. Li, C., Alvarez-Melis, D., Xu, K., Jegelka, S., Sra, S.: Distributional adversarial networks. arXiv preprint arXiv:1706.09549 (2017)
14. Li, J., Madry, A., Peebles, J., Schmidt, L.: Towards understanding the dynamics of generative adversarial networks. arXiv preprint arXiv:1706.09884 (2017)
15. Mao, X., Li, Q., Xie, H., Lau, R.Y., Wang, Z., Paul Smolley, S.: Least squares generative adversarial networks. In: Proceedings of the IEEE International Conference on Computer Vision, pp. 2794–2802 (2017)
16. Mordido, G., Yang, H., Meinel, C.: Dropout-gan: Learning from a dynamic ensemble of discriminators. arXiv preprint arXiv:1807.11346 (2018)
17. Neyshabur, B., Bhojanapalli, S., Chakrabarti, A.: Stabilizing GAN training with multiple random projections. arXiv preprint arXiv:1705.07831 (2017)
18. Nguyen, T., Le, T., Vu, H., Phung, D.: Dual discriminator generative adversarial nets. In: Advances in Neural Information Processing Systems, pp. 2670–2680 (2017)
19. Popovici, E., Bucci, A., Wiegand, R.P., De Jong, E.D.: Coevolutionary principles. In: Handbook of natural computing, pp. 987–1033. Springer (2012). https://doi.org/10.1007/978-3-540-92910-9_31
20. Radford, A., Metz, L., Chintala, S.: Unsupervised representation learning with deep convolutional generative adversarial networks. arXiv preprint arXiv:1511.06434 (2015)

21. Salimans, T., Ho, J., Chen, X., Sutskever, I.: Evolution strategies as a scalable alternative to reinforcement learning. arXiv:1703.03864 (2017)
22. Schmiedlechner, T., Yong, I.N.Z., Al-Dujaili, A., Hemberg, E., O'Reilly, U.M.: Lipizzaner: a system that scales robust generative adversarial network training. In: the 32nd Conference on Neural Information Processing Systems (NeurIPS 2018) Workshop on Systems for ML and Open Source Software (2018)
23. Stanley, K.O., Clune, J.: Welcoming the era of deep neuroevolution - uber engineering blog, December 2017. https://eng.uber.com/deep-neuroevolution/
24. Talbi, E.G.: Metaheuristics: From Design to Implementation, vol. 74. Wiley, Hoboken (2009)
25. Tolstikhin, I.O., Gelly, S., Bousquet, O., Simon-Gabriel, C.J., Schölkopf, B.: Adagan: boosting generative models. In: Advances in Neural Information Processing Systems, pp. 5430–5439 (2017)
26. Toutouh, J., Hemberg, E., O'Reilly, U.M.: Spatial evolutionary generative adversarial networks. In: Proceedings of the Genetic and Evolutionary Computation Conference. pp. 472–480. GECCO 2019, ACM, New York (2019). https://doi.org/10.1145/3321707.3321860
27. Toutouh, J., Hemberg, E., O'Reilly, U.M.: Data dieting in GAN training. In: Iba, H., Noman, N. (eds.) Deep Neural Evolution: Deep Learning with Evolutionary Computation, pp. 379–400. Springer Singapore, Singapore (2020)
28. Toutouh, J., Hemberg, E., O'Reilly, U.M.: Re-purposing heterogeneous generative ensembles with evolutionary computation. In: Proceedings of the Genetic and Evolutionary Computation Conference. GECCO '2020, Association for Computing Machinery, New York (2020). https://doi.org/10.1145/3377930.3390229, https://doi.org/10.1145/3377930.3390229
29. Wang, C., Xu, C., Yao, X., Tao, D.: Evolutionary generative adversarial networks. IEEE Trans. Evol. Comput. **23**(6), 921–934 (2019)
30. Wang, Y., Zhang, L., van de Weijer, J.: Ensembles of generative adversarial networks. arXiv preprint arXiv:1612.00991 (2016)
31. Wierstra, D., Schaul, T., Peters, J., Schmidhuber, J.: Natural evolution strategies. In: IEEE Congress on Evolutionary Computation, 2008. CEC 2008. (IEEE World Congress on Computational Intelligence), pp. 3381–3387. IEEE (2008)
32. Zhao, J., Mathieu, M., LeCun, Y.: Energy-based generative adversarial network. arXiv preprint arXiv:1609.03126 (2016)

Canonical Correlation Discriminative Learning for Domain Adaptation

Wenjing Wang[1](✉), Yuwu Lu[1,2,3], and Zhihui Lai[1,2,3]

[1] College of Computer Science and Software Engineering, Shenzhen University, Shenzhen, China
wangwenjing2018@email.szu.edu.cn, luyuwu2018@szu.edu.cn, lai_zhi_hui@163.com

[2] Laboratory of Intelligent Information Processing, Shenzhen, China

[3] Guangdong Laboratory of Artificial Intelligence and Digital Economy, Shenzhen, China

Abstract. Domain adaptation aims to diminish the discrepancy between the source and target domains and enhance the classification ability for the target samples by using well-labeled source domain data. However, most existing methods concentrate on learning domain invariant features for cross-domain tasks, but ignore the correlation and discriminative information between different domains. If the learned features from the source and target domains are not correlated, the adaptability of domain adaptation methods will be greatly degraded. To make up for this deficiency, we propose a novel domain adaptation approach, referred to as Canonical Correlation Discriminative Learning (CCDL) for domain adaptation. By introducing a novel correlation representation, CCDL maximizes the correlations of the learned features from the two domains as much as possible. Specifically, CCDL learns a latent feature representation to reduce the difference by jointly adapting the marginal and conditional distributions between the source and target domains, and simultaneously maximizes the inter-class distance and minimizes the intra-class scatter. The experiments certify that CCDL is superior to several state-of-the-art methods on four visual benchmark databases.

Keywords: Domain adaptation · Source and target domains · Canonical correlation discriminative learning

1 Introduction

In realistic research scenarios, well-labeled data commonly play an important role for most machine learning methods. Nevertheless, obtaining large amounts of labeled data is time-consuming and expensive. Therefore, transferring the effective knowledge from the labeled source domain to the related but different unlabeled target domain is a worthwhile goal. In recent years, domain adaptation has been receiving widespread attention from researchers since it is a representative technique in the field of transfer learning, which aims to enhance the performance of a model on the target domain by using information-rich source domain data [1].

T. Bäck et al. (Eds.): PPSN 2020, LNCS 12269, pp. 567–580, 2020.
https://doi.org/10.1007/978-3-030-58112-1_39

In cross-domain problems, the source and target data generally have various distributions. Therefore, the core idea of domain adaptation is to consider how to reduce the distribution difference between the source and target domains. Based on this idea, many methods have been proposed to learn a new feature representation for transfer learning. For example, transfer component analysis (TCA) [5] decreases the marginal distribution between the source and target domains by using maximum mean discrepancy (MMD) [3] distance. However, in reality, the conditional distribution of the source and target domains is usually different. For this reason, Long et al. [4] use MMD distance to jointly adapt the marginal and conditional distributions and integrate them into one formulation for robust transfer learning. These methods can effectively alleviate the problem of domain shift.

In transfer learning or domain adaptation methods, the transferred knowledge from the source domain to the target domain is expected to be as useful as possible. Thus, we hope that the correlations of the learned features in the two domains are maximized as far as possible. However, most existing domain adaptation methods cannot guarantee maximum correlations between the source and target domains in the latent subspace. Canonical correlation analysis (CCA) [6, 7] is a valuable method in multivariate data analysis, which can seek the maximum correlations between two sets of data. If we can find the maximization correlations between the source and target domains, transfer learning performance would be effectively improved. Thus, how to fully use the correlation between the source and target domains is very vital for domain adaptation.

In this paper, to address the low correlation problem in the domain adaptation, we use CCA to learn the maximum correlations between the source and target domains. Meanwhile, we make full use of the label information of the training samples to improve the discriminative ability of the model on the unlabeled target domain. The marginal and conditional distributions of the source and target domains are adopted by MMD in an iterative way so that the difference between them is minimal. We integrated the aforementioned knowledge into a unified formulation and named it canonical correlation discriminative learning (CCDL). In CCDL, we learn a latent low-dimensional feature subspace, that can make the features learned from the source and target domains data most relevant, and the projected samples with the same label are compact and different classes are separated from each other.

The main contributions of our paper are the followings.

(1) By introducing a novel canonical correlation regularization, CCDL learns the features with maximized correlations between the source and target domains.

(2) Two matrices, $\mathbf{K_1}$ and $\mathbf{K_2}$, are introduced in CCDL to overcome the limitation that the number of training samples in the source and target domains must be the same in the CCA regularization term. Thus, CCDL can fully use the correlation between the source and target domains. Specific details are given in Sect. 3.2.

The rest of this paper is organized as follows. Related works are reviewed in Sect. 2. In Sect. 3, we present our model in detail. Experimental results and analysis are provided in Sect. 4. Finally, conclusions are given in Sect. 5

2 Related Works

Domain adaptation is a popular research topic in transfer learning. A great number of domain adaptation methods have been proposed in recent years [8–10, 14, 15], which can be roughly divided into two categories: instance reweighting [8, 10] and feature representation [9, 16, 19, 20].

Most existing instance reweighting methods aim to reweight source data samples to the target domain instances, and learn a set of weights to minimize the distribution difference between them. For example, Li et al. [10] reweighted the predictions of the training classifier (from the source domain) on the testing data (from the target domain) based on the source data signed distance to the domain separator. Huang et al. [11] proposed a method called kernel mean matching (KMM), which reduces the difference between the source and target domains by reweighting the training points, so that the mean value of the two domains in the reproducing kernel Hilbert space (RKHS) are close without estimating the biased densities or selection probabilities.

Feature representation methods reduce the difference between the source and target domains by learning a new feature subspace. For example, Long et al. [2] proposed a transfer joint matching (TJM) method, which aims to jointly match features and reweight source instances in a latent subspace to alleviate the domain shift problem. Li et al. [12] put forward a domain invariant and class discriminative (DICD) method to reduce the distribution difference between the source and target domains by exploiting the discriminative information of training samples and domain-invariant information.

However, all the aforementioned methods ignore the correlations between the source and target domains, which are important for domain adaptation.

Table 1. Notations and descriptions used in this paper.

Notation	Description	Notation	Description
D_s / D_t	source/target domain	$\mathbf{X}_s / \mathbf{X}_t$	source/target data matrix
n_s / n_t	source/target examples	$\mathbf{X}$	original data matrix
C	shared classes	$\mathbf{P}$	adaptation matrix
m	original features	$\mathbf{E}$	embedding matrix
d	projected features	$\mathbf{H}$	centering matrix
N	iterations	$\mathbf{M}$	MMD matrix
β	regularization parameter	$\mathbf{D}_s / \mathbf{D}_d$	intra-class/inter-class distance calculation matrix
λ, ρ	tradeoff parameters	L_s / L_d	intra-class/inter-class distance loss term

3 Proposed Method and Optimization

In this section, we first show the definitions and frequently used notations, and then describe our method in detail; finally, the specific optimization of our method will be given.

3.1 Definitions and Notations

A domain D is comprised of a feature space $\mathcal{X}$ and a marginal probability distribution $P(\mathbf{X})$; that is, $D = \{\mathcal{X}, P(\mathbf{X})\}$, where $\mathbf{X} \in \mathcal{X}$. Given a domain $D = \{\mathbf{X}, P(\mathbf{X})\}$, a task T consists of a label space $\mathcal{Y}$ and a classifier $f(\mathbf{X})$, i.e., $T = \{\mathcal{Y}, f(\mathbf{X})\}$. From the perspective of probability, $f(\mathbf{X}) = Q(y|\mathbf{X})$ can be expressed as a conditional probability distribution, where $y \in \mathcal{Y}$. In this paper, we obtain a labeled source domain D_s and an unlabeled target domain D_t, which have the assumptions that $\mathcal{X}_s = \mathcal{X}_t$, $\mathcal{Y}_s = \mathcal{Y}_t$, $P(\mathbf{X}_s) \neq P(\mathbf{X}_t)$ and $Q(y_s|\mathbf{X}_s) \neq Q(y_t|\mathbf{X}_t)$.

For clarity, the frequently used notations are summarized in Table 1.

3.2 Proposed Approach

It is well known that CCA is a promising technique for finding the correlations between two sets of data. Therefore, we introduce CCA regularization into domain adaptation. That is, we extract the most relevant features between the source and target domains to reduce the difference between them. This can be expressed mathematically as:

$$\left\| \mathbf{P}^T\mathbf{X}_s - \mathbf{P}^T\mathbf{X}_t \right\|_F^2, \tag{1}$$

where $\mathbf{X}_s \in R^{m \times n_s}$ and $\mathbf{X}_t \in R^{m \times n_t}$. Unfortunately, the number of training samples in the source and target domains must be equal; that is, $n_s = n_t$. To overcome this limitation, we introduce two matrices $\mathbf{K_1}$ and $\mathbf{K_2}$, which are respectively defined as:

$$\mathbf{K_1} = \left[\mathbf{I}_{n_s}, \mathbf{0}_{n_s \times n_t}\right], \ \mathbf{K_2} = \left[\mathbf{I}_{n_t}, \mathbf{0}_{n_t \times n_s}\right], \tag{2}$$

where $\mathbf{I}_n$ is an identity matrix of size $n \times n$ and $\mathbf{0}_{m \times n}$ is an m by n matrix the elements of which are all zero.

Therefore, (1) can be rewritten as:

$$\left\| \mathbf{P}^T\mathbf{X}_s\mathbf{K_1} - \mathbf{P}^T\mathbf{X}_t\mathbf{K_2} \right\|_F^2. \tag{3}$$

By introducing the matrices $\mathbf{K_1}$ and $\mathbf{K_2}$, we have solved the constraint that the CCA regularization must have the same training samples for the source and target domains.

To reduce the discrepancy between the source and target domains, we introduce MMD to measure the distribution distance between two domains. The MMD distance between the marginal and conditional distributions of $\mathbf{X}_s \in R^{m \times n_s}$ and $\mathbf{X}_t \in R^{m \times n_t}$ can be expressed as:

$$MMD(\mathbf{X}_s, \mathbf{X}_t) = MMD_0(\mathbf{X}_s, \mathbf{X}_t) + MMD_c(\mathbf{X}_s, \mathbf{X}_t)$$

$$= \left\| \frac{1}{n_s} \sum_{i=1}^{n_s} \mathbf{P}^T x_{si} - \frac{1}{n_t} \sum_{j=1}^{n_t} \mathbf{P}^T x_{tj} \right\|^2 + \left\| \frac{1}{n_s^{(c)}} \sum_{x_{si} \in D_s^{(c)}} \mathbf{P}^T x_{si}^{(c)} - \frac{1}{n_t^{(c)}} \sum_{x_{tj} \in \hat{D}_t^{(c)}} \mathbf{P}^T x_{tj}^{(c)} \right\|^2$$
$$= tr(\mathbf{P}^T \mathbf{X}(\mathbf{M}_0 + \mathbf{M}_c)\mathbf{X}^T \mathbf{P}), \tag{4}$$

where $\mathbf{X} = [\mathbf{X}_s, \mathbf{X}_t]$ and each class $c \in \{1, \cdots, C\}$ in the label set y. $D_s^{(c)} = \{x_{si} : x_{si} \in D_s \wedge y_{si} = c\}$ denotes the set of samples with their true class labels being c in the source data, and $n_s^{(c)}$ is the number of source samples in class c. Correspondingly, $\hat{D}_t^{(c)} = \left\{x_{tj} : x_{tj} \in D_t \wedge \hat{y}_{tj} = c\right\}$ is the set of samples belonging to class c in the target data, $\hat{y}_{tj}$ is the pseudo (predicted) label of x_{tj}, and $n_t^{(c)}$ is the number of target samples in class c. Thus, the MMD matrices $\mathbf{M}_0$ and $\mathbf{M}_c$ are computed as follows:

$$(\mathbf{M}_0)_{ij} = \begin{cases} \frac{1}{n_s n_s}, \ x_i, x_j \in D_s; \\ \frac{1}{n_t n_t}, \ x_i, x_j \in D_t; \\ \frac{-1}{n_s n_t}, \ otherwise. \end{cases}, \quad (\mathbf{M}_c)_{ij} = \begin{cases} \frac{1}{n_s^{(c)} n_s^{(c)}}, & x_i, x_j \in D_s^{(c)}; \\ -\frac{1}{n_s^{(c)} n_t^{(c)}}, & \begin{cases} x_i \in D_s^{(c)}, x_j \in \hat{D}_t^{(c)}; \\ x_j \in D_s^{(c)}, x_i \in \hat{D}_t^{(c)}; \end{cases} \\ \frac{1}{n_t^{(c)} n_t^{(c)}}, & x_i, x_j \in \hat{D}_t^{(c)}; \\ 0, & otherwise. \end{cases} \tag{5}$$

Accordingly, (4) can be rewritten as:

$$MMD(\mathbf{X}_s, \mathbf{X}_t) = \sum\nolimits_{c=0}^{C} tr(\mathbf{P}^T \mathbf{X} \mathbf{M}_c \mathbf{X}^T \mathbf{P}) = tr(\mathbf{P}^T \mathbf{X} \mathbf{M} \mathbf{X}^T \mathbf{P}). \tag{6}$$

We define $\mathbf{M} = \sum_{c=0}^{C} \mathbf{M}_c$, and reduce the difference between the source and target domains by minimizing (6).

To make full use of the discriminative information of the original data, we construct the loss term in two parts. The first part minimizes the distance between samples of the same class and the second part maximizes all the distance between samples of different classes.

(1) Minimizing the discriminative loss term of the source domain: We expect to reduce the intra-class variation of the source domain. It is important note that we can also leverage the different penalty coefficients $n_s / n_s^{(c)}$ to balance the effects of different classes. Specifically, the intra-class distance and inter-class distance of source domain can be formulated as:

$$L_s^{(\mathbf{X}_s)} = \sum_{c=1}^{C} \frac{n_s}{n_s^{(c)}} \sum_{y_{si}, y_{sj} = c} \left\| \mathbf{P}^T x_{si} - \mathbf{P}^T x_{sj} \right\|^2 = tr(\mathbf{P}^T \mathbf{X}_s \mathbf{D}_s^{(\mathbf{X}_s)} \mathbf{X}_s^T \mathbf{P}), \tag{7}$$

$$L_d^{(\mathbf{X}_s)} = \sum_{y_{si} \neq y_{sj}} \left\| \mathbf{P}^T x_{si} - \mathbf{P}^T x_{sj} \right\|^2 = tr(\mathbf{P}^T \mathbf{X}_s \mathbf{D}_d^{(\mathbf{X}_s)} \mathbf{X}_s^T \mathbf{P}), \tag{8}$$

where

$$(\mathbf{D}_s^{(\mathbf{X}_s)})_{ij} = \begin{cases} n_s, & i = j; \\ \frac{-n_s}{n_s^{(c)}}, & i \neq j, y_{si} = y_{sj} = c; \\ 0, & otherwise; \end{cases}, \quad (\mathbf{D}_d^{(\mathbf{X}_s)})_{ij} = \begin{cases} n_s - n_s^{(c)}, & i = j, y_{si} = c; \\ -1, & i \neq j, y_{si} \neq y_{sj}; \\ 0, & otherwise. \end{cases} \tag{9}$$

As a consequence, the discriminative loss term of the source domain can be represented as:

$$L_{dist}^{(\mathbf{X}_s)} = L_s^{(\mathbf{X}_s)} - \rho L_d^{(\mathbf{X}_s)} = tr(\mathbf{P}^T\mathbf{X}_s\mathbf{D}_s^{(\mathbf{X}_s)}\mathbf{X}_s^T\mathbf{P}) - \rho tr(\mathbf{P}^T\mathbf{X}_s\mathbf{D}_d^{(\mathbf{X}_s)}\mathbf{X}_s^T\mathbf{P}), \tag{10}$$

where ρ is a positive parameter used to balance the intra-class and inter-class distances.

(2) *Minimizing the discriminative loss term of the target domain*: The loss term of the target domain is also composed of the intra-class and inter-class distances. The intra-class distance and inter-class distance of the target domain can be represented as:

$$L_s^{(\mathbf{X}_t)} = \sum_{c=1}^{C} \frac{n_t}{n_t^{(c)}} \sum_{\hat{y}_{ti}, \hat{y}_{tj}, = c} \left\| \mathbf{P}^T x_{ti} - \mathbf{P}^T x_{tj} \right\|^2 = tr(\mathbf{P}^T\mathbf{X}_t\mathbf{D}_s^{(\mathbf{X}_t)}\mathbf{X}_t^T\mathbf{P}), \tag{11}$$

$$L_d^{(\mathbf{X}_t)} = \sum_{\hat{y}_{ti} \neq \hat{y}_{tj}} \left\| \mathbf{P}^T x_{ti} - \mathbf{P}^T x_{tj} \right\|^2 = tr(\mathbf{P}^T\mathbf{X}_t\mathbf{D}_d^{(\mathbf{X}_t)}\mathbf{X}_t^T\mathbf{P}), \tag{12}$$

where

$$(\mathbf{D}_s^{(\mathbf{X}_t)})_{ij} = \begin{cases} n_t, & i = j; \\ \frac{-n_t}{n_t^{(c)}}, & i \neq j, \hat{y}_{ti} = \hat{y}_{tj} = c; \\ 0, & otherwise; \end{cases}, \quad (\mathbf{D}_d^{(\mathbf{X}_t)})_{ij} = \begin{cases} n_t - n_t^{(c)}, & i = j, \hat{y}_{ti} = c; \\ -1, & i \neq j, \hat{y}_{ti} \neq \hat{y}_{tj}; \\ 0, & otherwise. \end{cases} \tag{13}$$

Therefore, the discriminative loss term of the target domain is

$$L_{dist}^{(\mathbf{X}_t)} = L_s^{(\mathbf{X}_t)} - \rho L_d^{(\mathbf{X}_t)} = tr(\mathbf{P}^T\mathbf{X}_t\mathbf{D}_s^{(\mathbf{X}_t)}\mathbf{X}_t^T\mathbf{P}) - \rho tr(\mathbf{P}^T\mathbf{X}_t\mathbf{D}_d^{(\mathbf{X}_t)}\mathbf{X}_t^T\mathbf{P}). \tag{14}$$

In CCDL, to extract the class discriminative features in the latent low-dimensional subspace, we must jointly minimize the discriminative loss term of the source and target domains. Therefore, the term L_{dist} can be defined as

$$\begin{aligned} L_{dist} &= L_{dist}^{(\mathbf{X}_s)} + L_{dist}^{(\mathbf{X}_t)} \\ &= tr(\mathbf{P}^T\mathbf{X}_s\mathbf{D}_s^{(\mathbf{X}_s)}\mathbf{X}_s^T\mathbf{P}) - \rho tr(\mathbf{P}^T\mathbf{X}_s\mathbf{D}_d^{(\mathbf{X}_s)}\mathbf{X}_s^T\mathbf{P}) \\ &\quad + tr(\mathbf{P}^T\mathbf{X}_t\mathbf{D}_s^{(\mathbf{X}_t)}\mathbf{X}_t^T\mathbf{P}) - \rho tr(\mathbf{P}^T\mathbf{X}_t\mathbf{D}_d^{(\mathbf{X}_t)}\mathbf{X}_t^T\mathbf{P}). \end{aligned} \tag{15}$$

Supposing that we denote $\mathbf{D}_s = diag(\mathbf{D}_s^{(\mathbf{X}_s)}, \mathbf{D}_s^{(\mathbf{X}_t)})$ and $\mathbf{D}_d = diag(\mathbf{D}_d^{(\mathbf{X}_s)}, \mathbf{D}_d^{(\mathbf{X}_t)})$, (15) can be rewritten as

$$L_{dist} = tr(\mathbf{P}^T\mathbf{X}\mathbf{D}_s\mathbf{X}^T\mathbf{P}) - \rho tr\left(\mathbf{P}^T\mathbf{X}\mathbf{D}_d\mathbf{X}^T\mathbf{P}\right) = tr(\mathbf{P}^T\mathbf{X}(\mathbf{D}_s - \rho\mathbf{D}_d)\mathbf{X}^T\mathbf{P}). \quad (16)$$

By minimizing (16), the projected data of the same class will be close and different classes will be far away from each other in the learned subspace.

3.3 Optimization

The core idea of CCDL is to learn the correlation information, domain-invariant information and discriminative information of the source and target domains. Therefore, by incorporating (3), (6) and (16) into one formulation we can get the objective function of CCDL as follows:

$$\begin{aligned} &\min_{\mathbf{P}} \left\| \mathbf{P}^T\mathbf{X}_s\mathbf{K}_1 - \mathbf{P}^T\mathbf{X}_t\mathbf{K}_2 \right\|_F^2 + \lambda tr(\mathbf{P}^T\mathbf{X}\Theta\mathbf{X}^T\mathbf{P}) + \beta\|\mathbf{P}\|_F^2 \\ &s.t.\ \mathbf{P}^T\mathbf{X}_s\mathbf{X}_s^T\mathbf{P} = \mathbf{I}_d,\ \mathbf{P}^T\mathbf{X}_t\mathbf{X}_t^T\mathbf{P} = \mathbf{I}_d,\ \mathbf{P}^T\mathbf{X}\mathbf{H}\mathbf{X}^T\mathbf{P} = \mathbf{I}_d, \end{aligned} \quad (17)$$

where $\mathbf{\Theta} = \mathbf{M} + \mathbf{D}_s - \rho\mathbf{D}_d$, λ and β are positive parameters, and $\mathbf{H}$ is a centering matrix, which is defined as $\mathbf{H} = \mathbf{I}_{n_s+n_t} - \frac{1}{n_s+n_t}\mathbf{1}$ ($\mathbf{1}$ is the $(n_s+n_t) \times (n_s+n_t)$ matrix of ones). The regularization term $||\mathbf{P}||_F^2$ is usually used to control the complexity of $\mathbf{P}$. The constraint $\mathbf{P}^T\mathbf{X}\mathbf{H}\mathbf{X}^T\mathbf{P} = \mathbf{I}_d$ is derived from principal component analysis (PCA) [17], and it aims to preserve the data properties after projection as much as possible.

It is worth noting that (17) is a nonlinear optimization problem. Therefore, we apply the Lagrangian multiplier method to solve it and obtain the corresponding Lagrange function as follows:

$$\begin{aligned} L(\mathbf{P},\mathbf{\Phi}) = tr(\mathbf{P}^T(&\mathbf{X}_s\mathbf{K}_1\mathbf{K}_1^T\mathbf{X}_s^T - \mathbf{X}_s\mathbf{K}_1\mathbf{K}_2^T\mathbf{X}_t^T - \mathbf{X}_t\mathbf{K}_2\mathbf{K}_1^T\mathbf{X}_s^T + \mathbf{X}_t\mathbf{K}_2\mathbf{K}_2^T\mathbf{X}_t^T + \lambda\mathbf{X}\mathbf{\Theta}\mathbf{X}^T \\ &+\beta\mathbf{I}_m)\mathbf{P}) + tr((\mathbf{I}_d - \mathbf{P}^T\mathbf{X}_s\mathbf{X}_s^T\mathbf{P} + \mathbf{I}_d - \mathbf{P}^T\mathbf{X}_t\mathbf{X}_t^T\mathbf{P} + \mathbf{I}_d - \mathbf{P}^T\mathbf{X}\mathbf{H}\mathbf{X}^T\mathbf{P})\mathbf{\Phi}), \end{aligned} \quad (18)$$

where $\mathbf{\Phi} = diag(\phi_1, \phi_2, \cdots, \phi_d) \in R^{d\times d}$ is the Lagrange multiplier. Setting $\partial L(\mathbf{P},\mathbf{\Phi})/\partial\mathbf{P} = 0$, the generalized eigen-decomposition problem can be obtained as follows:

$$\begin{aligned} (&\mathbf{X}_s\mathbf{K}_1\mathbf{K}_1^T\mathbf{X}_s^T - \mathbf{X}_s\mathbf{K}_1\mathbf{K}_2^T\mathbf{X}_t^T - \mathbf{X}_t\mathbf{K}_2\mathbf{K}_1^T\mathbf{X}_s^T + \mathbf{X}_t\mathbf{K}_2\mathbf{K}_2^T\mathbf{X}_t^T \\ &+\lambda\mathbf{X}\mathbf{\Theta}\mathbf{X}^T + \beta\mathbf{I}_m)\mathbf{P} = (\mathbf{X}_s\mathbf{X}_s^T + \mathbf{X}_t\mathbf{X}_t^T + \mathbf{X}\mathbf{H}\mathbf{X})\mathbf{P}\mathbf{\Phi}. \end{aligned} \quad (19)$$

We can obtain the optimal projection $\mathbf{P}$ by calculating (19) for the d-smallest eigenvectors. The complete CCDL procedure is presented in Algorithm 1.

Algorithm 1 CCDL Algorithm

Input: Labeled source data $\{\mathbf{X}_s, y_s\}$, and Unlabeled target data $\mathbf{X}_t$; Positive parameters: β, λ, ρ; Iterations: N; New subspace dimension: d.

Output: Projection matrix $\mathbf{P}$, embedding matrix $\mathbf{E}$ and adaptive classifier f.

1: Construct matrix $\mathbf{\Theta} = \mathbf{M}_0 + \mathbf{D}_s - \rho\mathbf{D}_d$, where $\mathbf{D}_s = diag(\mathbf{D}_s^{(\mathbf{X}_s)}, \mathbf{0}_{n_t \times n_t})$, and $\mathbf{D}_d = diag(\mathbf{D}_d^{(\mathbf{X}_s)}, \mathbf{0}_{n_t \times n_t})$.

2: **while** $t \leq N$ **do**

3: Obtain the projection matrix $\mathbf{P}$ by calculating the d-smallest eigenvectors of (19)

4: Let $[\mathbf{E}_s, \mathbf{E}_t] = \mathbf{P}^T[\mathbf{X}_s, \mathbf{X}_t]$, and use $\{\mathbf{E}_s, y_s\}$ to train a standard classifier f to predict the target pseudo labels $\hat{y}_t$.

5: Update the matrix $\mathbf{\Theta} = \mathbf{M} + \mathbf{D}_s - \rho\mathbf{D}_d$.

6: $t = t + 1$.

7: **end while**

4 Experiments and Analysis

We conducted experiments on four visual benchmarks to evaluate our method on image classification. Detailed description of databases and analysis of experimental results are introduced in the following subsections.

4.1 Description of Databases

The CMU PIE[1] database includes more than 40,000 face images, and consists of 68 different people. We chose five subsets of CMU PIE: C05 (left pose), C07 (upward pose),

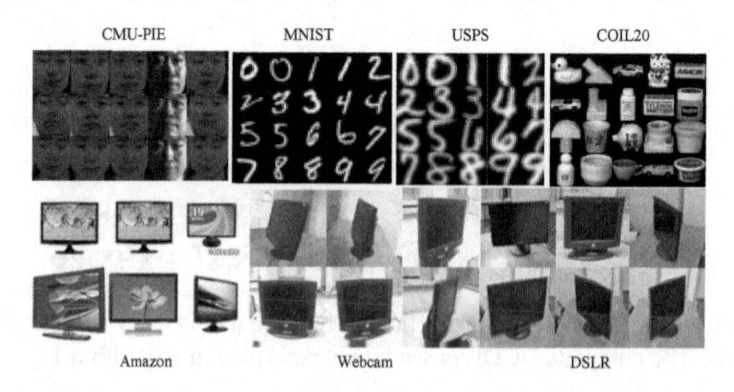

Fig. 1. Image samples from CMU PIE, MNIST, USPS, COIL20 and Office-31 database.

[1] https://github.com/jindongwang/transferlearning.

C09 (downward pose), C27 (frontal pose), and C29 (right pose) for the experiments. Two different subsets are randomly selected as the source and target domains. Thus, 20 cross-domain databases are constructed as: "C05→C07," "C05→C09," ... and "C29→C27."

MNIST (See footnote 1) and USPS (See footnote 1) are two handwritten digits databases, containing ten digits from 0 to 9. We selected 2000 images from the MNIST database and 1800 images from the USPS database to conduct the experiments. All selected images measure 16×16 pixels. Therefore, we constructed two cross-domain databases as follows: "M→U" and "U→M."

The COIL20 (See footnote 1) database comprises 1440 grayscale images of 20 objects with 72 images of each object. In our experiments, we divided the database into two parts: COIL1 (C1) and COIL2 (C2). COIL1 consists of all the images at the directions of $[0°, 85°] \cup [180°, 265°]$ and COIL2 includes all the images taken in the directions of $[90°, 175°] \cup [270°, 355°]$. We randomly selected one subset as the source domain and the rest as the target domain. Thus, two cross-domains are constructed as follows: "C1→C2" and "C2→C1."

The Office-31 (See footnote 1) database contains 31 classes of images from three different domains, i.e., Amazon (A), DSLR (D) and Webcam (W). In our experiments, we adopted 4096-dimensional $DeCAF_7$ features and constructed six cross-domain datasets to evaluate our approach, i.e., "A→D," "A→W," ... "W→D." Figure 1 shows several image examples from the four databases.

4.2 Implementation Details

To confirm the performance of our proposed method, we compared CCDL with several domain adaptation methods: TCA [5], geodesic flow kernel (GFK) [13], joint distribution adaptation (JDA) [4], TJM [2], balanced distribution adaptation (BDA) [18], easy transfer learning (ETL) [21], and DICD [12].

Since the data we obtained are labeled source data and unlabeled target data, we cannot utilize cross-validation to select the optimal parameters. In the experiments, we searched the parameter space by applying an empirical grid search method, and the best results of each method are reported. For the unsupervised domain adaptation methods TCA, GFK, JDA, TJM and DICD, we searched for $d = \{10, 20, 30, \cdots, 100\}$ to determine the optimal dimension of their common subspace. The 1-nearest neighbor (1NN) classifier is used as the final classifier.

The parameter settings in CCDL are the follows: subspace dimension $d = 100$ and number of iterations $T = 10$. The adaptation regularization parameter β and the search range of parameter ρ were selected from 0.10 to 50. We set the value of λ by searching it in the range of [0.1, 500].

4.3 Experimental Results

Extensive experiments were conducted on several cross-domain datasets to evaluate the performance of the proposed method. The experimental results are shown in Table 2 and Table 3.

Table 2. Classification accuracy (%) on CMU PIE database.

Dataset	TCA	GFK	JDA	TJM	BDA	ETL	DICD	CCDL
C05→C07	40.8	26.2	58.8	34.9	54.7	22.6	73.0	**85.6**
C05→C09	41.8	27.3	54.2	44.4	57.4	18.6	72.0	**85.2**
C05→C27	59.6	31.2	84.5	60.7	82.8	20.9	92.2	**99.3**
C05→C29	29.4	17.6	49.8	34.9	40.1	22.1	66.9	**76.4**
C07→C05	41.8	25.2	57.6	38.8	57.6	18.5	69.9	**83.6**
C07→C09	51.5	47.4	62.9	49.8	54.6	19.3	65.9	**98.0**
C07→C27	64.7	54.3	75.8	63.4	76.2	18.8	85.3	**93.7**
C07→C29	33.7	27.1	39.9	35.2	34.4	18.4	48.7	**97.9**
C09→C05	34.7	21.8	51.0	43.0	44.7	20.0	69.4	**87.6**
C09→C07	47.7	43.2	58.0	38.7	55.3	21.0	65.4	**95.6**
C09→C27	56.2	46.4	68.5	65.3	75.0	21.7	83.4	**91.7**
C09→C29	33.2	26.8	40.0	41.1	40.6	21.8	61.4	**99.2**
C27→C05	55.6	34.2	80.6	63.4	81.8	21.8	93.1	**99.7**
C27→C07	67.8	62.9	82.6	60.6	84.3	25.1	90.1	**95.6**
C27→C09	75.9	73.4	87.3	74.4	87.3	26.2	89.0	**92.0**
C27→C29	40.3	37.4	54.7	50.6	56.6	24.6	75.6	**85.9**
C29→C05	27.0	20.4	46.5	37.2	43.6	18.4	62.9	**73.5**
C29→C07	29.9	24.6	42.1	29.3	39.5	18.8	57.0	**96.0**
C29→C09	29.9	28.5	53.3	42.8	36.7	19.1	65.9	**98.5**
C29→C27	33.6	31.3	57.0	46.8	49.7	18.0	74.8	**87.8**
Avg.	44.8	35.4	60.3	47.8	57.6	20.8	73.1	**91.1**

(1) *Results on CMU PIE database*: The experimental results on the CMU PIE database are given in Table 2, and the best classification accuracies are marked in bold. It can be clearly seen from Table 2 that the performance of CCDL is significantly better than that of the other methods. Specifically, the average classification accuracy of CCDL is 91.1%, which overwhelmingly has an 18.0% higher accuracy than the best baseline DICD. It is worth noting that CCDL shows absolute advantages in the 20 cross-domain datasets of CMU PIE and has a large improvement compared with DICD. Compared with TCA and GFK, the classification accuracy of CCDL is much higher, because CCDL jointly reduces the marginal and conditional distributions between the source and target domains. In addition, compared with JDA, TJM, BDA, and DICD, CCDL has a CCA regularization term, which makes the features of the source and target domains more relevant in the latent subspace.

(2) *Results on MNIST + USPS, COIL20, and Office-31 (DeCAF$_7$ features) databases*: The experimental results of CCDL and other methods on the MNSIT+USPS, COIL20, and Office-31 (DeCAF$_7$ features) databases are recorded in Table 3. From Table 3, it can be noted that our proposed method achieves 100.0% classification accuracy on the COIL20 database. This result shows that CCA regularization can help extract the most relevant information from the source and target domains. The difficulty of adaptation varies widely on different tasks. We note that the results of DICD are generally better than

Table 3. Classification accuracy (%) on MNIST + USPS, COIL20, and Office-31 ($DeCAF_7$ features) databases.

Dataset	TCA	GFK	JDA	TJM	BDA	ETL	DICD	CCDL
M→U	56.3	67.2	67.3	63.3	69.8	54.2	77.8	**78.8**
U→M	51.1	46.5	59.7	52.3	59.4	23.3	**65.2**	64.3
C1→C2	88.5	72.5	89.3	91.5	97.2	73.9	95.7	**100.0**
C2→C1	85.8	74.2	88.5	91.8	96.8	74.0	93.3	**100.0**
A→D	61.9	54.0	61.5	62.1	59.0	50.2	64.3	**65.3**
A→W	57.6	45.5	60.1	57.6	54.0	40.4	60.9	**61.0**
D→A	46.3	40.7	49.3	47.1	44.5	41.4	**52.7**	52.6
D→W	93.8	86.0	95.0	93.2	90.2	73.3	**96.2**	96.0
W→A	43.3	40.2	46.2	46.3	45.4	40.8	50.2	**50.6**
W→D	98.6	96.2	98.6	96.8	97.2	85.3	99.2	**99.4**
Avg.	68.3	62.3	71.6	70.2	71.4	55.7	75.6	**76.8**

other comparison methods. The main reason is that DICD leverage the discriminative information of two domains, which makes the samples belonging to the same class as close as possible and the different classes away from each other.

To show the features learned by CCDL more intuitively, we generated the t-SNE [22] visualization plots of the C1→C2 task that the projected representations learned by JDA, DCID, and CCDL, respectively. It can be seen from Fig. 2 that compared with JDA features, the CCDL features have better discrimination, and the specific manifestation is that samples of the same class are more compact, while samples among different classes are more separated. Compared with DICD features, the CCDL features of the same class are tighter and have stronger correlation.

4.4 Parameter Sensitivity and Convergence Analysis

We performed parameter sensitivity analysis to illustrate that CCDL can attain optimal results in a wide range of parameter values. There are three parameters for CCDL: β, λ and ρ. We plotted the results of parameter change on the A→D, C1→C2, M→U, and C05→C27 datasets in Fig. 3.

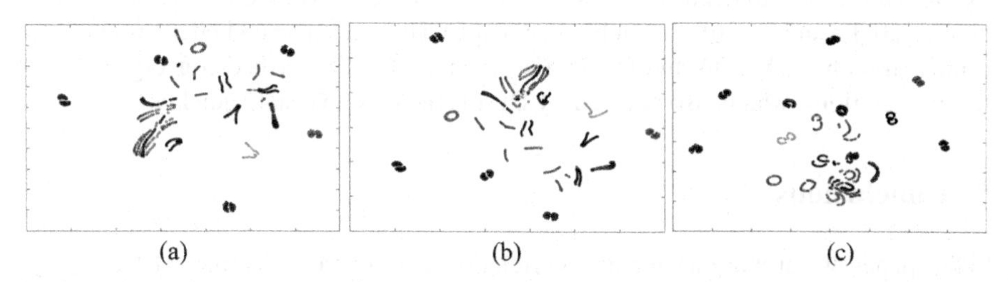

Fig. 2. Feature visualization: data of the COIL20 database embedding by t-SNE of (a) JDA features, (b) DICD features and (c) CCDL features. Samples in different colors represent different classes.

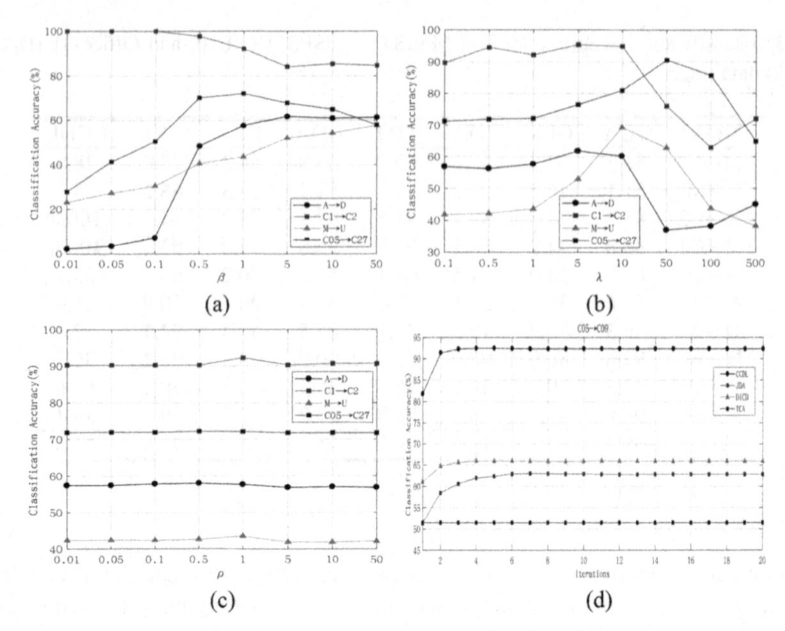

Fig. 3. Parameter sensitivity and convergence study by CCDL. (a), (b), and (c) are the corresponding parameters change graph on A→D, C1→C2, M→U and C05→C27 datasets. (d) The convergence of different methods on C05→C09 dataset.

The value of β varies from 0.01 to 50, with λ and ρ fixed. It can be seen from Fig. 3(a) that the optimal value of β is from 0.01 to 50. In theory, smaller values of β may lead to the optimization problem being ill-defined, while larger values of β could not effectively construct representations for cross-domain datasets.

We fixed β and ρ, and ran CCDL with varying values of λ from 0.1 to 500. We plotted classification accuracy with different values of λ in Fig. 3(b). Figure 3(c) shows the change in classification accuracy with varying values of ρ, where β and λ are fixed. It can be found that the value of ρ has less effect on the classification results. The experimental results illustrate that the values of ρ range from 0.01 to 50; that is, $\rho \in [0.01, 50]$.

We also conducted experiments to check the convergence of CCDL. Figure 3(d) shows the classification accuracy of the learned features from TCA, JDA, DICD, and CCDL with respect to iterations of the C05→C09 datasets. We set the number of iterations as 20. From Fig. 3(d), it can be seen that CCDL can achieved better results faster compared with TCA, JDA and DICD. Furthermore, CCDL reached convergence after several iterations, which also allowed CCDL to reach a stable state quickly.

5 Conclusions

In this paper, we propose a canonical correlation discriminative learning (CCDL) approach for domain adaptation. By solving the limitation that the number of training samples in the source and target domains must be the same in the CCA regularization term, CCDL learns the features with the maximized correlations between the source and target domains. In addition, CCDL matches the marginal and conditional distributions between two domains, and makes full use of the discriminative information of the training samples. The intra-class and inter-class distances in CCDL ensure that samples of the same class are more compact but different classes are separated from each other. Extensive experiments on four visual benchmark databases prove that CCDL is effective and can significantly outperform several state-of-the-art methods in domain adaptation.

Acknowledgements. This work was supported by the National Natural Science Foundation of China (Grant Nos. 61672357, 61732011), the Guangdong Basic and Applied Basic Research Foundation (2019A1515011493), the Natural Science Foundation of Shenzhen University (No. 2019046), and the Science Foundation of Shenzhen (Grant No. JCYJ20160422144110140).

References

1. Pan, S.J., Yang, Q.: A survey on transfer learning. IEEE Trans. Knowl. Data Eng. **22**(10), 1345–1359 (2010)
2. Long, M.S., Wang, J.M., Ding, G.G., Sun, J.G., Yu, P.S.: Transfer joint matching for unsupervised domain adaptation. In: IEEE Conference on Computer Vision and Pattern Recognition (2014)
3. Hu, J.L., Lu, J.W., Tan, Y.P.: Deep transfer metric learning. In: IEEE Conference on Computer Vision and Pattern Recognition, pp. 325–333 (2015)
4. Long, M.S., Wang, J.M., Ding, G.G., Sun, J.G., Yu, P.S.: Transfer feature learning with joint distribution adaptation. In: IEEE Conference on Computer Vision and Pattern Recognition, pp. 2200–2207 (2013)
5. Pan, S.J., Tsang, I.W., Kwok, J.T., Yang, Q.: Domain adaptation via transfer component analysis. IEEE Trans. Neural Netw. **22**(2), 199–210 (2011)
6. Hotelling, H.: Relations between two sets of variates. Biometrika (1936)
7. Chu, D.L., Liao, L.Z., Ng, M.K., Zhang, X.W.: Sparse canonical correlation analysis: new formulation and algorithm. IEEE Trans. Pattern Anal. Mach. Intell. **35**(12), 3050–3065 (2013)
8. Xia, R., Hu, X.L., Lu, J.F., Yang, J., Zong, C.Q.: Instance selection and instance weighting for cross-domain sentiment classification via PU learning. In: International Conference on Artificial Intelligence, pp. 2176–2182 (2013)
9. Wang, J.D., Feng, W.J., Chen, Y.Q., Yu, H., Huang, M.Y., Yu, P.S.: Visual domain adaptation with manifold embedded distribution alignment. In: ACM International Conference on Multimedia (2018)
10. Li, S., Song, S.J., Huang, G.: Prediction reweighting for domain adaptation. IEEE Trans. Neural Netw. Learn. Syst. **28**(7), 1682–1695 (2017)
11. Huang, J.Y., Smola, A.J., Gretton, A., Borgwardt, K.M., Scholkopf, B.: Correcting sample selection bias by unlabeled data. In: Neural Information Processing Systems, pp. 601–608 (2007)
12. Li, S., Song, S.J., Huang, G., Ding, Z.M., Wu, C.: Domain invariant and class discriminative feature learning for visual domain adaptation. IEEE Trans. Image Process. **27**(9), 4260–4272 (2018)
13. Gong, B.Q., Shi, Y., Sha, F., Grauman, K.: Geodesic flow kernel for unsupervised domain adaptation. In: IEEE Conference on Computer Vision and Pattern Recognition (2012)
14. Deng, W.Y., Lendasse, A., Ong, Y.S., Tsang, I.W.H., Chen, L., Zheng, Q.H.: Domain adaptation via feature selection on explicit feature map. IEEE Trans. Neural Netw. Learn. Syst. **30**(4), 1180–1190 (2019)

15. Zhang, L., Wang, P., Wei, W., Lu, H., Shen, C.H.: Unsupervised domain adaptation using robust class-wise matching. IEEE Trans. Circ. Syst. Video Technol. **29**(5), 1339–1349 (2019)
16. Yan, K., Kou, L., Zhang, D.: Learning domain-invariant subspace using domain features and independence maximization. IEEE Trans. Cybern. **48**(1), 288–299 (2018)
17. Wold, S., Esbensen, K., Geladi, P.: Principal component analysis. Chemometr. Intell. Lab. Syst. **2**(1), 37–52 (1987)
18. Wang, J.D., Chen, Y.Q., Hao, S.J., Feng, W.J., Shen, Z.Q.: Balanced distribution adaptation for transfer learning. In: IEEE Conference on Data Mining, pp. 1129–1134 (2017)
19. Zhang, J., Li, W.Q., Ogunbona, P.: Joint geometrical and statistical alignment for visual domain adaptation. In: IEEE Conference on Computer Vision and Pattern Recognition, pp. 5150–5158 (2017)
20. Li, J.J., Jing, M.M., Lu, K., Zhu, L., Shen, H.T.: Locality preserving joint transfer for domain adaptation. IEEE Trans. Image Process. **28**(12), 6103–6115 (2019)
21. Wang, J.D., Chen, Y.Q., Yu, H., Huang, M.Y., Yang, Q.: Easy transfer learning by exploiting intra-domain structures. In: IEEE International Conference on Multimedia and Expo, pp. 1210–1215 (2019)
22. Maaten, L.V.D., Hinton, G.: Visualizing data using t-SNE. J. Mach. Learn. Res. **9**(11), 2579–2605 (2008)

Genetic and Evolutionary Algorithms

Improving Sampling in Evolution Strategies Through Mixture-Based Distributions Built from Past Problem Instances

Stephen Friess[1(✉)], Peter Tiňo[1], Stefan Menzel[2], Bernhard Sendhoff[2], and Xin Yao[1,3]

[1] CERCIA, School of Computer Science, University of Birmingham, Birmingham, UK
{shf814,p.tino,x.yao}@cs.bham.ac.uk
[2] Honda Research Institute Europe, Carl-Legien-Str. 30, 63073 Offenbach a.M., Germany
{stefan.menzel,bernhard.sendhoff}@honda-ri.de
[3] Department of Computer Science and Engineering, Southern University of Science and Technology, Shenzhen, China

Abstract. The notion of learning from different problem instances, although an old and known one, has in recent years regained popularity within the optimization community. Notable endeavors have been drawing inspiration from machine learning methods as a means for algorithm selection and solution transfer. However, surprisingly approaches which are centered around internal sampling models have not been revisited. Even though notable algorithms have been established in the last decades. In this work, we progress along this direction by investigating a method that allows us to learn an evolutionary search strategy reflecting rough characteristics of a fitness landscape. This latter model of a search strategy is represented through a flexible mixture-based distribution, which can subsequently be transferred and adapted for similar problems of interest. We validate this approach in two series of experiments in which we first demonstrate the efficacy of the recovered distributions and subsequently investigate the transfer with a systematic from the literature to generate benchmarking scenarios.

Keywords: Evolution strategies · Model-based optimisation · Continuous optimisation · Algorithm configuration · Transfer learning

1 Introduction

Within recent decades, the field of evolutionary computation has seen a surge of novel algorithms being proposed, frequently with the intent to operate on very specific problem domains. While this reflects on one hand the efficacy of population-based and evolutionary approaches for a wide range of applications, it also reflects deep rooted issues within the current state of the art. Particularly

T. Bäck et al. (Eds.): PPSN 2020, LNCS 12269, pp. 583–596, 2020.
https://doi.org/10.1007/978-3-030-58112-1_40

in regards to: 1) A lack of a prescriptive theory on how to construct efficient algorithms for a given problem and 2) a lack of understanding on what constitutes and characterizes optimisation problems and the similarity thereof in a more generalized way. While the theorists cannot give definite answer to both questions at the moment, one may still legitimately ask whether or not it is possible to approach some of these problems from a pragmatic line of attack. For this reason, two popular trends have emerged within the optimisation community: 1) Research on meta-learning frameworks [13,18,23] and 2) research on transfer learning approaches [4,9,12,14,17]. Both try to boost the efficiency of optimisation algorithms by using prior knowledge from solving problem instances.

In our work, we progress at the intersection of both lines of research by building a model of a search strategy from individual runs which may be then subsequently transferred to similar problem instances. For this reason, we first give in Sect. 2 a brief overview discussing these two existing lines of research and give insight into the state-of-the-art. Section 3 explains the extensions we introduce to consolidate a search strategy. Further, we demonstrate its functionality on an illustrative benchmark function. In Sect. 4.1, we widen the range of considered benchmark problems to a selected variety of multimodal and valley-shaped problems. Subsequently, in Sect. 4.2 we consider the scenario of transferring search strategies across problem instances generated by translations, rotation and various non-linear transformations to the benchmark functions. We conclude our study with a summary in Sect. 5 and give an outlook on future work.

2 Knowledge from Problem Solving Exercises

In principle, within the optimisation community two approaches have been investigated within the recent decades. The first one being related to the construction of meta-learning frameworks for algorithm selection and configuration. The second one relating to instance-based transfer learning through candidate solutions. Both fields, while having gained strong traction with the recent years, can trace their origin back to much earlier roots. With seminal work on algorithm selection being done by Rice et al. [21] in the 1970s and research on transfer learning emerging from the discourse on lifelong machine learning systems in the 1990s [19]. However, their application towards the domain of optimisation has been only considered since recently within the 2000s [17,23].

Meta-learning frameworks attempt to harness high-level knowledge that can be subsequently used in the future to more efficiently solve related tasks. In the classical algorithm selection and algorithm configuration problem, this would equate to predicting the best performing algorithm or configuration for a given problem [13,18]. However, a key problem in optimisation lies in the first place in the extraction and computation of said task specific features. This poses especially an outstanding problem within the domain of continuous optimisation, where unlike in the combinatorial domain, problem features cannot be simply derived from the problem state or definition. Features thus have to be explicitly computed in a cheap and at best informative manner.

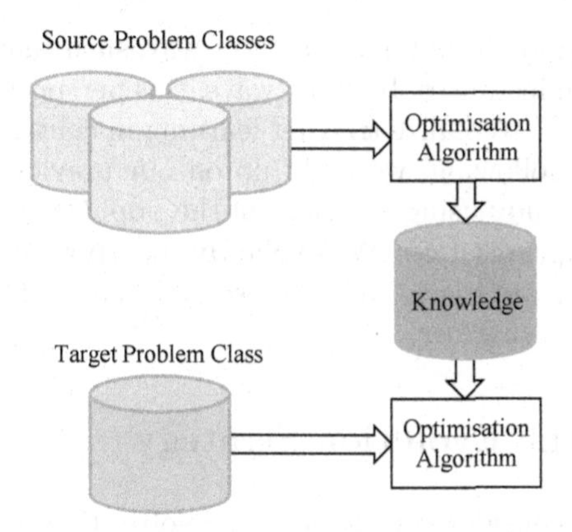

Fig. 1. Diagram of the archetypical pipeline for transfer learning. Roughly adapted from Pan et al. [19]. Similar setups are frequently encountered within literature on population-based optimisation (e.g. [4,5,14]). From previously solved source problem classes knowledge is extracted by the algorithm such that it can subsequently improve the performance on it on a new target problem class.

Transfer learning approaches on the other side may be seen as operating under more relaxed conditions. Essentially, what transfer learning assumes between two problem instances, is that beneficial knowledge which helped solving one problem instance, can be transferred either directly or by means of a transformation to a new problem instance. However, notably it introduces by this further uncertainties. The bulk of transfer learning literature in optimisation draws inspiration from instance-based transfer [19] by means of transferring high performing candidate solutions between tasks (e.g. [5,12,14,22]). Retrieval of the candidate solutions occurs either directly from the previous solved tasks [5,14] or through probabilistic sampling from a continuously built repository (e.g. [4,17]). As a way of determining the probabilistic weights, often times task similarity measures may be used [17]. However, in many scenarios instead simply solution similarity may be used as a proxy of task similarity [4,17]. In general, the lack of satisfying task similarity measures together with being prone to uncontrolled 'negative' knowledge transfer which degrades algorithm performance [5] are known problems of these approaches.

Interestingly, aside from these mentioned works, barely any of the recent literature tries to learn across problem instances explicitly by means of internal sampling models. Although quite notably, many popular algorithms rely upon operators drawing random variables from symmetrical distributions and thus have by default isotropy assumptions built in. However, this assumption becomes broken when given an optimisation problem which does not resemble a flat plane. Quite intuitively, the interplay between algorithm and optimisation problem should enforce characteristic search strategies and behaviors. Modern model-based algorithms [10,15] acknowledge this by adapting a distribution online during the optimisation run. However, they do not attempt to memorize these in a more rough and abstract way, such that these can be transferred across problem instances. In many ways, this perspective might be also the only meaningful notion to realize transfer learning in continuous single-objective optimisation. In the following, we build up on our previous work [7,8] and try tackle the issue in a study using a variant of the popular (μ, λ)-Evolution Strategy for continuous optimisation. We explicitly incorporate strategy parameters through a windowing approach and harness systematics from the literature to build benchmarking scenarios.

3 Extending the Evolution Strategy

In the following, we consider continuous single-objective optimisation problems of the form $f : \chi \subseteq \mathbb{R}^d \rightarrow \mathbb{R}$, where χ denotes the search space and d its associated dimensionality. As a base we use a variant of the Evolution Strategy with (μ, λ) selection mechanism [1]. We keep out explicitly any recombination operators to have the framework reduced to its essentials. Meaning to sample mutations from a multivariate distribution and performing selection in an elitist manner. Note, that from an evolutionary perspective, mutation is the principle source of variation [16]. In many ways, this basic outline may resemble continuous variants of Evolutionary Programming. However, the elitist selection mechanism in Evolution Strategies has been implicated to contribute to performance improvements [2].

In the Evolution Strategy, population members s(j) are represented by tuples $\mathbf{s}(j) = [\mathbf{x}(j), \boldsymbol{\sigma}(j)]$, where $\mathbf{x}(j) = (x_1(j), \cdots, x_d(j))$ is the population member's representation in the solution space and $\boldsymbol{\sigma}(j) = (\sigma_1(j), \cdots, \sigma_d(j))$ are its strategy parameters. The latter can be considered to be a key feature of Evolution Strategy implementations. Strategy parameters essentially control the shape the normal distribution from which mutations

$$\Delta\mathbf{x}(j) \sim \mathcal{N}(\mathbf{0}, \mathrm{diag}[\boldsymbol{\sigma}(j)]) \tag{1}$$

for the individuals j are drawn which shift the individuals $\mathbf{x}'(j) = \mathbf{x}(j) + \Delta\mathbf{x}(j)$ in the solution space. Likewise, variation operators can be defined such that they also vary and recombine the strategy parameters of population members. However, we neglect this extension within our study.

3.1 Quality-Based Filtering of Mutations

In the following, we will further filter performed mutations according to their quality. Thus, we will distinguish between beneficial mutations as defined by

$$f(\mathbf{x}(j)^i_{before}) - f(\mathbf{x}(j)^i_{after}) \geq 0 \tag{2}$$

and detrimental mutations defined by

$$f(\mathbf{x}(j)^i_{before}) - f(\mathbf{x}(j)^i_{after}) < 0. \tag{3}$$

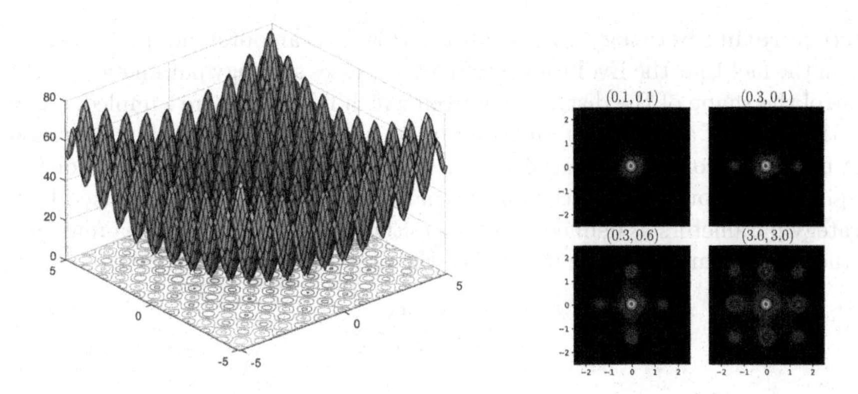

Fig. 2. Left panel: Rastrigin's benchmark function. Right panel: Search distributions for different pairs of strategy parameters $\boldsymbol{\sigma} = (\sigma_x, \sigma_y)$ derived from a 100 component mixture model of the distribution of beneficial mutations from 1000 runs under reweighing according to Eq. (4) & (5).

The idea is, that once we have stored statistics about mutations outside of the algorithm, we can use them to design improved search strategies. Specifically, by means of constructing empirical distributions which serve as basis for model-based mutation operators. These can be seen as reflecting globally averaged characteristics of the fitness landscape. In principle, one would intuitively be interested into enforcing beneficial mutations and suppressing detrimental mutations. However, distributions of detrimental mutations have been implicated to be strongly normal distributed [8]. It is also questionable from the perspective of algorithm design whether suppressing mutations comes at the expense of convergence properties, as every point in the search space should remain reachable by a small finite amount of probability. Thus, we focus in the following only on biasing the algorithm through distributions of beneficial mutations (Fig. 2).

3.2 Constructing Operators from Empirical Distributions

Choosing a Density Estimator. While by default, mutations are sampled in the Evolution Strategy from a multivariate normal distribution as given in Eq. (1), for empirical distributions one explicitly has to use a modeling technique. In principle, many techniques are available for this purpose. However, in the following we will use the Gaussian mixture model as it is a well-studied model which can act as universal density approximator. Mixture models reduce the input data to a small set of descriptive clusters which are parametrized by multivariate normal distributions, such that the full data distribution can then be expressed as $p(\mathbf{x}) = \sum_{k=1}^{K} \pi_k \cdot \mathcal{N}(\mathbf{x}|\boldsymbol{\mu}_k, \boldsymbol{\Sigma}_k)$, with mixture coefficients π_k, which are normalized such that $\Sigma_{k=1}^{K} \pi_k = 1$, and determined together with means μ_k and covariances Σ_k by maximizing the log-likelihood through the expectation-maximization algorithm [3,20].

Incorporating Strategy Parameters. However, an outstanding problem still lies in the fact that the Evolution Strategy possesses strategy parameters $\boldsymbol{\sigma}$ which control the shape of the distribution from which mutations are sampled. Changing the shape of an empirical distribution as basis for improved sampling should not break the contained spatial information. Therefore, we simply window the empirical distribution with the multivariate normal distribution spanned by the strategy parameters as defined by Eq. (1). Effectively, this results in a reweighing of the mixture model where we replace the original mixture coefficients π_k with

$$r_k = \frac{\pi_k c_k}{\sum_{i=1}^{N} \pi_i c_i}, \tag{4}$$

where the coefficients c_k per mixture component quantify the average value of the normal distribution spanned by the strategy parameters over the k-th mixture component. This can be analytically calculated such that

$$\begin{aligned} c_k &:= \int_{\mathbb{R}^n} \mathcal{N}(\boldsymbol{x}|\boldsymbol{\mu}_k, \Sigma_k)\, \mathcal{N}(\boldsymbol{x}|\mathbf{0}, \operatorname{diag}(\boldsymbol{\sigma}))\, \mathrm{d}^n\mathbf{x} \\ &= \int_{\mathbb{R}^n} \frac{\exp\left[-\frac{1}{2}(\mathbf{x}-\boldsymbol{\mu}_k)^T \Sigma_k^{-1} (\mathbf{x}-\boldsymbol{\mu}_k)\right]}{\sqrt{(2\pi)^d |\Sigma_k|}} \times \frac{\exp\left[-\frac{1}{2}\mathbf{x}^T \Sigma_\sigma^{-1} \mathbf{x}\right]}{\sqrt{(2\pi)^d |\Sigma_\sigma|}}\, \mathrm{d}^n\mathbf{x} \\ &= \frac{\exp\left(-\frac{1}{2}\boldsymbol{\mu}_k^T \Sigma_k^{-1} \boldsymbol{\mu}_k + \frac{1}{2}\boldsymbol{\mu}_k^T \left[\Sigma_k^{-1} \Sigma_c\, \Sigma_k^{-1}\right] \boldsymbol{\mu}_k\right)}{\sqrt{(2\pi)^d |\Sigma_k| |\Sigma_c^{-1}| |\Sigma_\sigma|}} \times \int_{\mathbb{R}^n} \mathcal{N}(\mathbf{x}|\boldsymbol{\mu}_c, \Sigma_c)\, \mathrm{d}^n\mathbf{x} \\ &= \frac{\exp\left(-\frac{1}{2}\boldsymbol{\mu}_k^T \left[\Sigma_k^{-1} (\Sigma_k^{-1} + \Sigma_\sigma^{-1})^{-1} \Sigma_\sigma^{-1}\right] \boldsymbol{\mu}_k\right)}{\sqrt{(2\pi)^d |\Sigma_k| |\Sigma_k^{-1} + \Sigma_\sigma^{-1}| |\Sigma_\sigma|}}, \end{aligned} \tag{5}$$

where we further introduced $\Sigma_\sigma := \operatorname{diag}(\boldsymbol{\sigma})$ and $\Sigma_c := (\Sigma_k^{-1} + \Sigma_\sigma^{-1})^{-1}$.

Table 1. Benchmark functions used in this study, grouped from top to bottom according to landscape structure. 1st–3rd row: Unimodal and valley shaped problems. 4th–6th row: Multimodal problems with single global optimum and strong regularity. 7th–9th row: Difficult multimodal problems with single global optimum and high irregularity.

Name	Search Space	Function Definition
Sphere	$[-5.12, 5.12]^d$	$f(\mathbf{x}) = \sum_{i=1}^{d} x_i^2$
Bohachevsky	$[-100, 100]^d$	$f(\mathbf{x}) = \sum_{i=1}^{d-1}[x_i^2 + 2x_{i+1}^2 - 0.3\cos(3\pi x_i) - 0.4\cos(4\pi x_{i+1}) + 0.7]$
Rosenbrock	$[-5, 10]^d$	$f(\mathbf{x}) = \sum_{i=1}^{d-1}[100(x_{i+1} - x_i^2)^2 + (x_i - 1)^2]$
Rastrigin	$[-5.12, 5.12]^d$	$f(\mathbf{x}) = 10\,d + \sum_{i=1}^{d}[x_i^2 - 10\cos(2\pi x_i)]$
Ackley	$[-32.768, 32.768]^d$	$f(\mathbf{x}) = -a\,\mathrm{e}^{0.2(\frac{1}{d}\sum_{i=0}^{d} x_i^2)^{0.5}} + \mathrm{e}^{-\frac{1}{d}\sum_{i=0}^{d}\cos(2\pi x_i)} + a + \mathrm{e}$
Griewank	$[-600, 600]^d$	$f(\mathbf{x}) = 1 + \frac{1}{4000}\sum_{i=1}^{d} x_i^2 + \frac{1}{4000}\prod_{i=1}^{d}\cos(x_i/\sqrt{i})$
Schwefel	$[-500, 500]^d$	$f(\mathbf{x}) = 418.9829d - \sum_{i=1}^{d-1} x_i \sin(\sqrt{\lvert x_i\rvert})$
Eggholder	$[-512, 512]^d$	$f(\mathbf{x}) = -(x_2+47)\sin(\sqrt{\lvert x_2+x_1/2+47\rvert}) - x_1\sin(\sqrt{\lvert x_1-(x_2+47)\rvert})$
Rana	$[-512, 512]^d$	$f(\mathbf{x}) = x_1\sin(\sqrt{\lvert x_2+1-x_1\rvert})\cos(\sqrt{\lvert x_1+x_2+1\rvert}) + (x_2+1)\cos(\sqrt{\lvert x_2+1-x_1\rvert})\sin(\sqrt{\lvert x_1+x_2+1\rvert})$

4 Experimental Study

The following study is based upon the DEAP library for Evolutionary Computation [6] with the extensions as elaborated in Sect. 3. We first investigate in Sect. 4.1 whether distributions of beneficial mutations can be harnessed at all to realize performance improvements on a selected range of different continuous optimisation problems. Subsequently in Sect. 4.2 we investigate different transfer scenarios between problem instances. Particularly, we build these scenarios by harnessing existing systematics from the literature.

4.1 On the Efficacy of Distributions of Beneficial Mutations

In the following we conduct experiments over a range of 9 different optimisation problems listed in Table 1. We group these into unimodal and valley-shaped problems (1st–3rd row), multimodal problems with single global optimum and high regularity (4th–6th row) and multimodal problems with single global optimum and high irregularity (7th–9th row). All experiments are conducted with a population size of $\mu = 10$ and we generate at each generation $\lambda = 30$ offspring members by randomly selecting individuals and either cloning or mutating them with a 30% chance. In all experiments, the population is initialized randomly upon the entire search space, where we use additionally a penalization for the difficult multimodal problems by means of rejecting mutations crossing the search space boundaries. This is necessary, as otherwise in these problems lower optima could be reached in the outer areas. Strategy parameters are initialized such that $\sigma \in [0.1, 4.0]$ for the problems in row 1–6 in Table 1. For the difficult multimodal functions we re-adjust the upper boundaries, where we use for Schwefel's function $\sigma_{\max} = 400$, for Eggholder $\sigma_{\max} = 480$ and for Rana's function $\sigma_{\max} = 150$. We will elaborate further in the succeeding paragraph on the necessity of the re-adjustment. Experiments are conducted over 1000 generations and we accumulate data per experiment from 100 runs. Problem dimension is kept at $d = 2$ in all experiments, as this still allows the interpretation of the retrieved distributions and lifts problems of data sparsity arising with more degrees of freedom. The mixture model is constructed with a total number of $K = 50$ components.

Resulting minimum fitness curves per generation of the optimisation runs are plotted per problem group in Figs. 3 and 4. Where top rows are the runs using default mutation distributions, and the lower rows are runs which use distributions of beneficial mutations with and without considering strategy parameters. Further, median (dark blue), mean (grey) and individual runs (light blue) are plotted. Quite notably, across all considered problems the distribution of beneficial mutations significantly improves the search behavior. Particularly, it reduces late convergences by acting in a regularizing fashion. However, the inclusion of strategy parameters is only helpful when some regularity along the parameter axis can be harnessed. Otherwise, it's effect on the performance is detrimental. The approach can even be shown to work on the difficult multimodal functions of row 7–9 in Table 1. However, we openly admit that further precautions have to be taken for these experiments to work. In particular, for all three we had

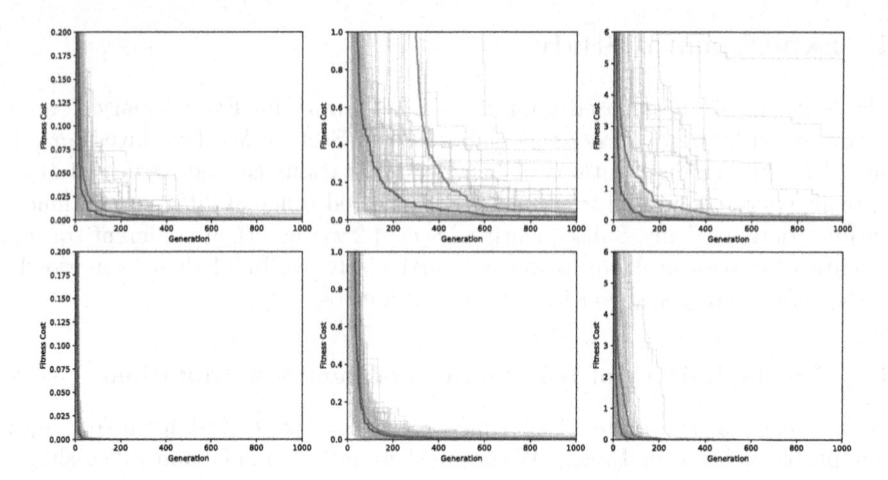

Fig. 3. Column 1–3: Fitness curves (light blue) for the unimodal Sphere, Bohachevsky's and Rosenbrock's function from 100 runs, as well as median (dark blue) and mean (dark grey) curves. Top row: With default sampling. Bottom row: With improved sampling using quality-based mutations. (Color figure online)

to re-adjust the upper bound of the strategy parameter to the previously mentioned values such that we achieved good convergence behavior in the runs with default sampling. Without taking these precautions, we were not able to achieve any improvements using the distribution of beneficial mutations. In fact, for the lower values of the strategy parameters we even found that the distributions of beneficial mutations were detrimental to the optimisation and encouraged premature convergence into local optima. We further list performance values of our experiments, as well as results from a statistical Wilcoxon rank sum test under normal approximation in Table 2. The results indicate that for a significance level of $\alpha = 0.05$, the null hypothesis can be rejected in all experiments.

4.2 Cross-Instance Transfer Scenarios

In the following section we will consider now cross-instance transfer learning scenarios. Meaning we try to transfer a mutation operator learned on a source optimisation problem to a target problem (c.f. Fig. 1) in the hope of realizing performance improvements. To generate variations of the source problem instances we apply in the following a systematic of transformations proposed by Hansen et al. [11].

Transformations of the Fitness Landscape. The following base transformations are designed to explicitly break the well-behavedness of our optimisation problems by acting upon the decision variables $\mathbf{x}$. *Ill-conditioning* introduces fast running components by a means of a linear rescaling

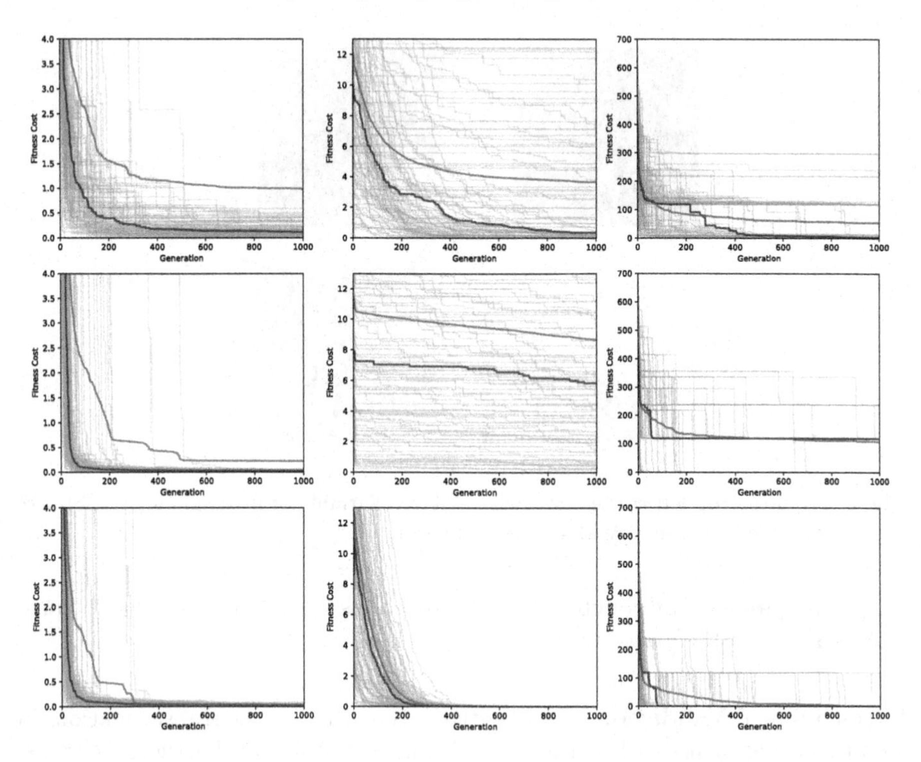

Fig. 4. Column 1–3: Fitness curves (light blue) for the multimodal Rastrigin's, Ackley's and Griewank's function from 100 runs, as well as median (dark blue) and mean (dark grey) curves. Top row: With default sampling. Middle row: With improved sampling considering strategy parameters. Middle row: With improved sampling considering strategy parameters. Bottom row: With improved sampling using quality-based mutations. (Color figure online)

$$T_{ill\text{-}c.} : \mathbb{R}^d \to \mathbb{R}^d,\ \ x_i \longmapsto x_i\ \alpha^{\frac{1}{2}\frac{i-1}{d-1}}, \tag{6}$$

where we choose $\alpha = 10$ in our experiments. The *asymmetrical* transformation breaks the symmetry of components x_i under sign transformations with

$$T_{asy} : \mathbb{R}^d \to \mathbb{R}^d, x_i \longmapsto \begin{cases} x_i^{1+\beta\frac{i-1}{d-1}\sqrt{x_i}} & \text{if } x_i > 0 \\ x_i & \text{otherwise} \end{cases}, \tag{7}$$

such that in the positive quadrant the components scale up exponentially. The *oscillatory* transformation introduces sinusoidal variability of the components by

$$T_{osc} : \mathbb{R}^d \to \mathbb{R}^d, x_i \longmapsto \operatorname{sign}(x_i)\ \exp(\hat{x}_i + 0.049(\sin(c_1\hat{x}_i) + \sin(c_2\hat{x}_i))), \tag{8}$$

$$\hat{x} \longmapsto \begin{cases} \log(|x|) & \text{if } x \neq 0 \\ 0 & \text{otherwise} \end{cases}, \hat{c_1} \longmapsto \begin{cases} 10 & \text{if } x \neq 0 \\ 5.5 & \text{otherwise} \end{cases}, \hat{c_2} \longmapsto \begin{cases} 7.9 & \text{if } x \neq 0 \\ 3.1 & \text{otherwise} \end{cases}$$

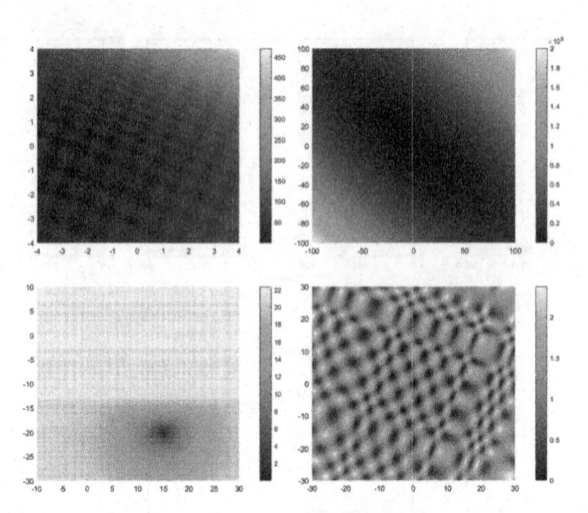

Fig. 5. From the top left corner clockwise: Altered variants of Rastrigin's (R_1), Sphere (S_1), Griewank's (G_1) and Ackley's (A_1) function.

Further, we also use counter-clockwise rotations $T_{rot}(\theta)$ by angle θ and translations T_{trans} of the global optimum.

Experimental Validation. We investigate the utility of the transformations in a set of 9 experiments with 4 transformed standard problems. Further, the transfer from source problem to target problem $P_0 \rightarrow P_1$, and likewise the transfer into the reverse direction $P_0 \rightarrow P_1$. We use in the following the sphere function S_1 with ill-conditioning, 45° rotation and extended search space to $[-100, 100]^2$, the Ackley's function A_1 with a translation of $\mathbf{t} = (-15, 20)$ and subsequently added oscillations and asymmetries, Rastrigin's function R_1 with 22.5° rotation,

Table 2. Medians $\tilde{f}_{\min}$, means $\overline{f}_{\min}$ and standard deviations $\sigma_{\min}$ of the minimum fitness after 1000 generations aggregated from 100 runs for default sampling using a normal distribution ($\mathcal{N}$) and improved sampling using a mixture model of quality-filtered mutations($\mathcal{M}$). Further, normalized ranks z and p-values for a two-tailed Wilcoxon rank sum test have been calculated. For a significance level of $\alpha = 0.05$ the null hypothesis can be considered to be rejected in all experiments.

Benchmark	$\tilde{f}_{\min}$ ($\mathcal{N}$)	$\overline{f}_{\min}$ ($\mathcal{N}$)	$\sigma_{\min}$ ($\mathcal{N}$)	$\tilde{f}_{\min}$ ($\mathcal{M}$)	$\overline{f}_{\min}$ ($\mathcal{M}$)	$\sigma_{\min}$ ($\mathcal{M}$)	$\lvert z \rvert$	p-value
Sphere	5.528e−4	1.303e−3	1.975e−3	2.360e−5	3.600e−5	3.630e−5	9.713e+0	2.670e−22
Bohachevsky	1.668e−2	4.888e−2	8.349e−2	2.754e−3	4.580e−3	5.222e−3	8.107e+0	5.180e−16
Rosenbrock	9.098e−3	1.013e−1	5.641e−1	6.096e−4	9.370e−4	1.081e−3	1.035e+1	4.390e−25
Rastrigin	1.924e−1	5.067e−1	7.974e−1	6.609e−3	9.327e−3	9.066e−3	1.133e+1	9.060e−30
Ackley	1.223e−1	9.722e−1	3.221e+0	2.957e−2	3.297e−2	1.575e−2	9.287e+0	1.580e−20
Griewank	3.686e−1	3.653e+0	5.780e+0	1.011e−4	1.747e−3	3.744e−3	1.172e+0	1.050e−31
Schwefel	3.523e+0	5.413e+1	7.509e+1	2.595e−3	1.189e+0	1.178e+1	1.201e+1	3.270e−33
Eggholder	4.959e+0	6.135e+1	7.786e+1	3.730e−5	1.131e+1	3.494e+1	9.935e+0	2.940e−23
Rana	1.652e+0	1.832e+1	2.939e+1	1.650e+0	4.359e+0	6.102e+0	2.622e+0	8.748e−3

Table 3. Medians $\tilde{f}_{\min}$, means $\overline{f}_{\min}$ and standard deviations $\sigma_{\min}$ of the minimum fitness after 1000 generations aggregated from 100 runs for default sampling (upper table) and transfer scenarios (bottom table). Further, normalized ranks z and p-values for a two-tailed Wilcoxon rank sum test are given. For a significance level of $\alpha = 0.05$ the null hypothesis can be considered to be rejected in all experiments.

Scenarios	$\tilde{f}_{\min}$	$\overline{f}_{\min}$	$\sigma_{\min}$	$\lvert z \rvert$	p-value
S_0^*	8.319e−4	1.443e−3	1.854e−3	–	–
R_0^*	2.173e−1	9.599e+0	4.118e+1	–	–
S_1	2.623e−3	1.859e−1	1.781e+0	–	–
A_1	1.639e−1	3.101e+0	6.812e+0	–	–
R_1	1.813e−1	6.901e−1	1.438e+0	–	–
G_1	3.298e+0	7.139e+0	8.870e+0	–	–
$S_0^* \rightarrow S_1$	5.786e−4	7.114e−4	7.064e−4	7.281e+0	3.306e−13
$S_1 \rightarrow S_0^*$	1.899e−4	2.299e−4	1.751e−4	6.744e+0	1.544e−11
$A_0 \rightarrow A_1$	2.839e−2	1.827e+0	5.716e+0	8.721e+0	2.771e−18
$A_1 \rightarrow A_0$	3.301e−2	3.774e−2	2.379e−2	9.928e+0	3.161e−23
$R_0 \rightarrow R_1$	5.689e−2	8.816e−2	8.894e−2	6.304e+0	2.902e−10
$R_1 \rightarrow R_0$	6.984e−2	1.051e−1	1.111e−1	5.603e+0	2.111e−8
$G_0 \rightarrow G_1$	3.034e−1	1.305e+0	4.749e+0	6.842e+0	7.837e−12
$G_1 \rightarrow G_0$	1.261e−1	1.773e+0	1.793e−1	3.531e+0	4.145e−4
$S_0^* \rightarrow G_0$	3.531e+0	5.110e+0	4.490e+0	3.592e+0	3.284e−4
$G_0 \rightarrow S_0^*$	6.850e−5	1.386e−4	1.488e−4	3.722e+0	1.974e−4

small shift $\mathbf{t} = (3, 2)$, extended search space to $[-100, 100]^2$ and added asymmetry, as well as Griewanks function G_1 with 20° rotation and added oscillations. Further, we denote the Sphere and Rastrigin's function with extended search spaces to $[-100, 100]^2$ as S_0^* and R_0^*. Heightmaps of most altered benchmark problems are plotted in Fig. 5. We find that in most considered transfer scenarios, performance improvements can be realized (Table 3). However, finding difficult and interesting scenarios without making them obvious is a bit of a hurdle. For example, in our experiments the scenario $S_0^* \rightarrow G_0$ features negative transfer, as the transferred distribution is simply adapted for a unimodal fitness landscape with small search space.

5 Conclusions

We have investigated in this paper an approach which allows us to learn an evolutionary search strategy reflecting rough and globally averaged characteristics of a fitness landscape. We represented this search strategy through flexible mixture-based distributions of beneficial mutations as basis for improved operators. Particularly, these distributions can be considered to be improved as they enable us to lift the isotropy assumption usually built into mutation operators, thus ingrain the problem structure and redistribute probability weight radially to more appropriately balance exploration and exploitation on a given problem instance. The distribution can be further adapted through a Gaussian reweighing approach, thus emulating the role strategy parameters have for sampling with a default normal distribution. However, this only seems to be useful on a limited range of scenarios. We showed that unweighted distributions can indeed lead to performance improvements on a large variety problems, however prior good convergence properties of the default sampling approach seems to be an essential prerequisite. Further, we investigated systemically built transfer scenarios and could also realize performance improvements in these. However, we openly acknowledge the difficulty of finding meaningful and difficult transfer scenarios. Part of the problem stems from the fact, as it is unsure to which degree one can alter or change a problem such that it still may be attributed to be an instance of the former. However, introducing and investigating systematic transformations should be one the first key steps towards to resolving the issue. For the future, we plan to investigate the proposed framework in higher dimensions for improved transfer scenarios, as well as look into measures of problem similarity potentially by means of fitness landscape analysis.

Acknowledgements. This research has received funding from the European Union's Horizon 2020 research and innovation programme under grant agreement number 766186. It was also supported by the Program for Guangdong Introducing Innovative and Enterpreneurial Teams (Grant No. 2017ZT07X386), Shenzhen Science and Technology Program (Grant No. KQTD2016112514355531), and the Program for University Key Laboratory of Guangdong Province (Grant No. 2017KSYS008).

References

1. Bäck, T.: Evolutionary Algorithms in Theory and Practice: Evolution Strategies, Evolutionary Programming. Genetic Algorithms. Oxford University Press, Oxford (1996)
2. Bäck, T., Rudolph, G., Schwefel, H.P.: Evolutionary programming and evolution strategies: similarities and differences. In: Proceedings of the Second Annual Conference on Evolutionary Programming. Citeseer (1993)
3. Bishop, C.M.: Pattern Recognition and Machine Learning. Springer, Heidelberg (2006)
4. Da, B., Gupta, A., Ong, Y.: Curbing negative influences online for seamless transfer evolutionary optimization. IEEE Trans. Cybern. **99**, 1–14 (2018). https://doi.org/10.1109/TCYB.2018.2864345
5. Feng, L., Ong, Y.S., Jiang, S., Gupta, A.: Autoencoding evolutionary search with learning across heterogeneous problems. IEEE Trans. Evol. Comput. **21**(5), 760–772 (2017)
6. Fortin, F.A., De Rainville, F.M., Gardner, M.A., Parizeau, M., Gagné, C.: DEAP: evolutionary algorithms made easy. J. Mach. Learn. Res. **13**, 2171–2175 (2012)
7. Friess, S., Tiňo, P., Menzel, S., Sendhoff, B., Yao, X.: Representing experience through problem-tailored search operators. In: 2020 IEEE World Congress on Computational Intelligence (2020)

8. Friess, S., Tiňo, P., Menzel, S., Sendhoff, B., Yao, X.: Learning transferable variation operators in a continuous genetic algorithm. In: 2019 IEEE Symposium Series on Computational Intelligence (SSCI), pp. 2027–2033, December 2019. https://doi.org/10.1109/SSCI44817.2019.9002976
9. Gupta, A., Ong, Y.S., Feng, L.: Insights on transfer optimization: because experience is the best teacher. IEEE Trans. Emerg. Top. Comput. Intell. **2**(1), 51–64 (2018)
10. Hansen, N.: The CMA evolution strategy: a comparing review. In: Lozano, J.A., P. Larrañaga, P., Inza I., Bengoetxea E. (eds.) Towards a New Evolutionary Computation, pp. 75–102. Springer (2006). https://doi.org/10.1007/3-540-32494-1_4
11. Hansen, N., Finck, S., Ros, R., Auger, A.: Real-parameter black-box optimization benchmarking 2009: Noiseless functions definitions (2009)
12. Jiang, M., Huang, Z., Qiu, L., Huang, W., Yen, G.G.: Transfer learning-based dynamic multiobjective optimization algorithms. IEEE Trans. Evol. Comput. **22**(4), 501–514 (2017)
13. Kerschke, P., Hoos, H.H., Neumann, F., Trautmann, H.: Automated algorithm selection: survey and perspectives. Evol. Comput. **27**(1), 3–45 (2019)
14. Koçer, B., Arslan, A.: Genetic transfer learning. Exp. Syst. Appl **37**(10), 6997–7002 (2010)
15. Bengoetxea, E., Larrañaga, P., Bloch, I., Perchant, A.: Estimation of distribution algorithms: a new evolutionary computation approach for graph matching problems. In: Figueiredo, M., Zerubia, J., Jain, A.K. (eds.) EMMCVPR 2001. LNCS, vol. 2134, pp. 454–469. Springer, Heidelberg (2001). https://doi.org/10.1007/3-540-44745-8_30
16. Losos, J.B.: The Princeton Guide to Evolution. Princeton University Press, Princeton (2017)
17. Louis, S.J., McDonnell, J.: Learning with case-injected genetic algorithms. IEEE Trans. Evol. Comput. **8**(4), 316–328 (2004)
18. Muñoz, M.A., Sun, Y., Kirley, M., Halgamuge, S.K.: Algorithm selection for black-box continuous optimization problems: a survey on methods and challenges. Inf. Sci. **317**, 224–245 (2015)
19. Pan, S.J., Yang, Q., et al.: A survey on transfer learning. IEEE Trans. Knowl. Data Eng. **22**(10), 1345–1359 (2010)
20. Pedregosa, F., et al.: Scikit-learn: machine learning in python. J. Mach. Learn. Res. **12**(Oct), 2825–2830 (2011)
21. Rice, J.R., et al.: The algorithm selection problem. Adv. Comput. **15**(65–118), 5 (1976)
22. Ruan, G., Minku, L.L., Menzel, S., Sendhoff, B., Yao, X.: When and how to transfer knowledge in dynamic multi-objective optimization. In: 2019 IEEE Symposium Series on Computational Intelligence (SSCI), pp. 2034–2041. IEEE (2019)
23. Smith-Miles, K.A.: Cross-disciplinary perspectives on meta-learning for algorithm selection. ACM Comput. Surv. (CSUR) **41**(1), 6 (2009)

The Hessian Estimation Evolution Strategy

Tobias Glasmachers[1(✉)] and Oswin Krause[2]

[1] Institute for Neural Computation, Ruhr-University Bochum, Bochum, Germany
tobias.glasmachers@ini.rub.de
[2] Department of Computer Science, University of Copenhagen, Copenhagen, Denmark
oswin.krause@di.ku.dk

Abstract. We present a novel black box optimization algorithm called *Hessian Estimation Evolution Strategy*. The algorithm updates the covariance matrix of its sampling distribution by directly estimating the curvature of the objective function. This algorithm design is targeted at twice continuously differentiable problems. For this, we extend the cumulative step-size adaptation algorithm of the CMA-ES to mirrored sampling. We demonstrate that our approach to covariance matrix adaptation is efficient by evaluating it on the BBOB/COCO testbed. We also show that the algorithm is surprisingly robust when its core assumption of a twice continuously differentiable objective function is violated. The approach yields a new evolution strategy with competitive performance, and at the same time it also offers an interesting alternative to the usual covariance matrix update mechanism.

Keywords: Evolution strategy · Covariance matrix adaptation · Hessian matrix

1 Introduction

We consider minimization of a black-box objective function $f : \mathbb{R}^d \to \mathbb{R}$. Modern evolution strategies (ESs) are highly tuned solvers for such problems [6,7]. The state of the art is marked by the covariance matrix adaptation evolution strategy (CMA-ES) [5,9] and its many variants.

Most modern evolution strategies (ESs) sample offspring from a Gaussian distribution $\mathcal{N}(m, \sigma^2 C)$ around a single mean $m \in \mathbb{R}^d$. Their most crucial mechanism is adaptation of the step size $\sigma > 0$, which enables them to converge at a linear rate on scale-invariant problems [8]. Hence they achieve the fastest possible convergence speed class that can be realized by any comparison-based algorithm [16]. However, for ill-conditioned problems the actual convergence rate can be very slow, i.e., the multiplicative progress per step can be arbitrarily close to one. The main role of CMA is to mitigate this problem: after successful adaptation of the covariance matrix C to a multiple of the inverse of the Hessian H of the problem, the ES makes progress at its optimal rate.

T. Bäck et al. (Eds.): PPSN 2020, LNCS 12269, pp. 597–609, 2020.
https://doi.org/10.1007/978-3-030-58112-1_41

To this end consider a convex quadratic function $f(x) = \frac{1}{2}x^T Hx$. Its Hessian matrix H encodes the curvature of the graph of f. Knowledge of this curvature is valuable for optimization, e.g., turning a simple gradient step $x \leftarrow x - \eta \cdot \nabla f(x)$ into a Newton step $x \leftarrow x - H^{-1}\nabla f(x)$, which jumps straight into the optimum. For an evolution strategy, adapting C to H^{-1} is equivalent to learning a transformation of the input space that turns a convex quadratic function into the sphere function. This way, after successful adaptation, all convex quadratic functions are as easy to minimize as the sphere function $f(x) = \frac{1}{2}\|x\|^2$, i.e., as if the Hessian matrix was $H = I$ (the identity matrix). Due to Taylor's theorem, the advantage naturally extends to local convergence into twice continuously differentiable local optima, covering a large and highly relevant class of problems.

The usual mechanism for adapting the covariance matrix C of the offspring generating distribution is to change it towards a weighted maximum likelihood estimate of successful steps [9]. The update can equally well be understood as following a stochastic natural gradient in parameter space [13].

In this paper we explore a conceptually different and more direct approach for learning the inverse Hessian. It amounts to estimating the curvature of the objective function on random lines through m by means of finite differences. We design a novel CMA mechanism for updating the covariance matrix C based on the estimated curvature information. We call the resulting algorithm *Hessian Estimation Evolution Strategy* (HE-ES).

It is worth pointing out that estimating derivatives destroys an important property of CMA-ES, namely invariance under strictly monotonic transformations of objective values. Our new algorithm is still fully invariant under order-preserving affine transformations of objective values. This is an essentially equally good invariance guarantee only in a local situation, namely if the value transformation is well approximated by its first order Taylor polynomial. We address this potential weakness in our experimental evaluation.
The main goals of this paper are

- to present HE-ES and our novel CMA mechanism, and
- to demonstrate its competitiveness with existing algorithms. To this end we compare HE-ES to CMA-ES as a natural baseline. We also compare with NEWUOA [14], which is based directly on iterative estimates of a quadratic model of the objective function. Furthermore, we include BFGS [12], a "work horse" (gradient-based) non-linear optimization algorithm. It is of interest because being a quasi-Newton method, it implicitly estimates the Hessian matrix.
- Finally, we adapt cumulative step size adaptation to mirrored sampling.

The remainder of the paper is structured as follows. In the next section we describe the new algorithm in detail and briefly discuss its relation to CMA-ES. Our main results are of empirical nature, therefore we present a thorough experimental evaluation of the optimization performance of HE-ES and discuss strengths and limitations. We close with our conclusions.

2 The Hessian Estimation Evolution Strategy

The HE-ES algorithm is designed in as close as possible analogy to CMA-ES. Ideally we would change only the covariance matrix adaptation (CMA) mechanism. However, we end up changing also the offspring generation method to a scheme that is tailored to estimating curvature information. In the following we present the algorithm and motivate and detail all mechanisms that deviate from CMA-ES [9].

Estimating Curvature. A seemingly natural strategy for estimating the Hessian of an unknown black-box function is to estimate single entries H_{ij} of this $d \times d$ matrix by computing finite differences. For a diagonal entry H_{ii} this requires evaluating three points on a line parallel to the i-th coordinate axis, while for an off-diagonal entry H_{ij} we can evaluate four corners of a rectangle with edges parallel to the i-th and j-th coordinate axis. A serious problem with such a procedure is that entries must remain consistent, which makes it difficult to design an online update of a previous estimate of the matrix. Furthermore, the scheme implies that offspring must be sampled along the coordinate axes, which can significantly impair performance, e.g., on ill-conditioned non-separable problems—which are exactly the problems we would like to excel on.

We therefore propose a different solution. To this end we draw mirrored samples $m \pm v$ and evaluate $f(m + \alpha v)$ for $\alpha \in \{-1, 0, +1\}$. Here $v \in \mathbb{R}^d$ is the direction of a line through m. The length $\|v\|$ of that vector controls the scale on which the finite difference estimate is computed. An estimate of the second directional derivative of f in direction $\frac{v}{\|v\|}$ at m is

$$\frac{f(m+v) + f(m-v) - 2f(m)}{\|v\|^2} .$$

For a convex quadratic function $f(x) = \frac{1}{2}(x - x^*)^T H (x - x^*)$ the above quantity coincides with the directional derivative $\frac{v^T H v}{\|v\|^2}$. It can be understood as the "component" of H in direction $\frac{v}{\|v\|}$. The expression simplifies to a diagonal entry H_{ii} if v is a multiple of the i-th standard basis vector $e_i = (0, \dots, 0, 1, 0, \dots, 0)$. For a general twice continuously differentiable function the estimate converges to the above value in the limit $\|v\| \to 0$. Importantly, H is uniquely determined by these components, so given enough directions v there is no need for a sampling procedure that corresponds to estimating off-diagonal entries H_{ij}.

Orthogonal Mirrored Sampling. A single pair of mirrored samples provides information on the curvature in a single direction v. We learn nothing about the curvature in the $d - 1$ dimensional space orthogonal to v. To make best use of the available information we should therefore sample the next pair of mirrored samples orthogonal to v. We apply the following sampling procedure for random orthogonal directions, which was first proposed in [17]. We draw d Gaussian vectors and record their lengths. The vectors are then orthogonalized with the Gram-Schmidt procedure. Placing the vectors into a $d \times d$ matrix yields an orthogonal matrix uniformly distributed in the orthogonal group $O(d, \mathbb{R})$. We then rescale the vectors to their original lengths.

Procedure 1. sampleOrthogonal

1: **input** dimension d
2: $z_1, \ldots, z_d \sim \mathcal{N}(0, I)$
3: $n_1, \ldots, n_d \leftarrow \|z_1\|, \ldots, \|z_d\|$
4: apply the Gram-Schmidt procedure to $z_1, \ldots, z_d$
5: return $y_i = n_i \cdot z_i, \quad i = 1, \ldots, d$

The sampling procedure applied for each block is defined in Procedure 1. It is applicable to up to d pairs of mirrored samples. In general we aim to generate $\tilde{\lambda}$ pairs of mirrored samples, which amounts to $\lambda = 2\tilde{\lambda}$ offspring in total. We therefore split the pairs into $B = \lceil \tilde{\lambda}/d \rceil$ blocks and apply the above procedure B times. The resulting vectors are denoted as b_{ij}, where $i \in \{1, \ldots, B\}$ is the block index and $j \in \{1, \ldots, d\}$ is the index within each block.

Covariance Matrix Update. We aim for an update that modifies an existing covariance matrix in an online fashion. A seemingly straightforward strategy is to adapt the matrix so that after the update it matches the curvature in the sampled directions. This approach is followed in [11], and later in [15]. However, such a strategy disregards the fact that all multiples of the inverse Hessian are optimal covariance matrices. In fact, the update would destroy a perfect covariance matrix simply because it differs from the inverse Hessian by a large factor.

Therefore our goal is to adapt the covariance matrix to the closest *multiple* of the inverse Hessian H^{-1}. To this end we only change curvature values relative to each other: if the measured curvature in direction v_1 is 10 times larger than in direction v_2 then we aim to ensure that the updated matrix represents this relation. Otherwise we modify the matrix as little as possible. In particular, we keep its determinant (encoding the global scale) constant: if an eigenvalue is increased, then another one is decreased accordingly.

The easiest way to achieve the above goals is by means of a multiplicative update [2,4,10]. We decompose the covariance matrix[1] into the form $C = A^T A$. The mirrored samples take the form $x_{ij}^{\pm} = m \pm \sigma \cdot A\, b_{ij}$, resulting in the curvature estimates h_{ij}, which approximate $b_{ij}^T A^T H A\, b_{ij}$. The update takes the form $A' \leftarrow AG$ and hence $C' \leftarrow GCG$, where G is a symmetric positive definite matrix.

In the following we apply the above considerations on curvature estimation to the function $\tilde{f}(x) = f\big(A(x - m)\big)$ using the direction vectors $v = \sigma \cdot b_{ij}$. The actual goal of optimization is to adapt m towards x^*. Turning $\tilde{f}$ into the sphere function greatly facilitates that process. It is achieved by adapting A towards

[1] The decomposition is never computed explicitly in the algorithm. Instead it directly updates the factor A.

(a multiple of) any Cholesky factor of H^{-1}, or in other words, by making all eigenvalues of $A^T HA$ coincide.

In order to understand the update we first review a simplified example. Consider only two vectors b_1 and b_2. For simplicity assume that they fulfill $\sigma\|b_i\| = 1$, and assume that the curvature estimates h_{ii} are exact because the function is convex quadratic. Then the ideal G has an eigenvalue of $\sqrt[4]{h_{22}/h_{11}}$ for eigenvector b_1, an eigenvalue of $\sqrt[4]{h_{11}/h_{22}}$ for eigenvector b_2, and eigenvalue 1 in the space orthogonal to b_1 and b_2. This seemingly very specific choice ensures that $\det(G) = 1$ and it holds

$$
\begin{aligned}
b_1^T (A')^T HA' b_1 &= b_1^T GA^T HAGb_1 \\
&= \sqrt{\frac{h_{22}}{h_{11}}} \cdot b_1^T A^T HA b_1 = \sqrt{\frac{h_{22}}{h_{11}}} \cdot h_{11} = \sqrt{h_{11}h_{22}} \ ,
\end{aligned}
$$

which coincides with $b_2^T (A')^T HA' b_2$ for symmetry reasons. Hence, after the update the curvatures in directions b_1 and b_2 have become equal, while all curvatures orthogonal to the sampling directions remain unchanged. In this sense the resulting problem $\tilde{f}$ has come closer to the sphere function.

Procedure 2. computeG

1: **input** b_{ij}, $f(m)$, $f(x_{ij}^{\pm})$, σ
2: **parameters** κ, η_A
3: $h_{ij} \leftarrow \frac{f(x_{ij}^{+})+f(x_{ij}^{-})-2f(m)}{\sigma^2 \cdot \|b_{ij}\|^2}$ # estimate curvature along b_{ij}
4: **if** $\max(\{h_{ij}\}) \leq 0$ **then return** I
5: $c \leftarrow \max(\{h_{ij}\})/\kappa$
6: $h_{ij} \leftarrow \max(h_{ij}, c)$ # truncate to trust region
7: $q_{ij} \leftarrow \log(h_{ij})$
8: $q_{ij} \leftarrow q_{ij} - \frac{1}{\tilde{\lambda}} \cdot \sum_{ij} q_{ij}$ # subtract mean → ensure unit determinant
9: $q_{ij} \leftarrow q_{ij} \cdot \frac{-\eta_A}{2}$ # learning rate and inverse square root (exponent $-1/2$)
10: $q_{B,j} \leftarrow 0 \quad \forall j \in \{dB - \tilde{\lambda}, \dots, d\}$ # neutral update in the unused directions
11: **return** $\frac{1}{B} \sum_{ij} \frac{\exp(q_{ij})}{\|b_{ij}\|^2} \cdot b_{ij} b_{ij}^T$

A generalization of the above update to an arbitrary number of (unnormalized) update directions $\sigma \cdot b_{ij}$ is implemented by Procedure 2, which computes the matrix G. It forms the algorithmic core of our method.

This core works very well, for example, for smooth convex problems. General non-smooth and non-convex objective functions can exhibit unstable curvature estimates h_{ij}. A noisy objective function can create similar issues. For non-convex problems, the eigenvalues of a Hessian can be zero or even negative, in contrast to the eigenvalues of a covariance matrix. In all of these situations we are better off to smoothen the update. Therefore Procedure 2 takes two measures: first, it bounds the conditioning of the multiplicative update G by a constant κ to limit the effect of outliers (lines 5–6), and second it applies a learning rate $\eta_A \in (0, 1]$ to stabilize estimates through temporal smoothing (line 9). The first of these mechanisms also gracefully handles curvature estimates $h_{ij} \leq 0$ by clipping them to a small positive value. This is a reasonable thing to do since we want to emphasize sampling in these directions without completely destroying the learned information. In our experiments, we use the settings $\kappa = 3$ and $\eta_A = 1/2$. They represent a reasonable compromise between stability and adaptation speed.

Cumulative Step-size Adaptation (CSA) for Orthogonal Frames. The usage of mirrored sampling for CSA was previously explored in [3,17]. It was found that the default algorithm exhibits a strong step size decay on flat or random function surfaces. In the past this issue was alleviated by not considering the mirrored samples from the population when computing the CSA update. This however is inefficient as only half of the samples are used to update the step-size. In this section, we will quantify the step-length bias of CSA with mirrored samples under selection among all offspring. For this, we observe that under mirrored sampling each direction b_{ij} obtains two weights w_{ij}^+ and w_{ij}^- based on the function-values of the mirrored pair $x_{ij}^{\pm}$. Thus, we can write the CSA mean computation [5] as

$$\sum_{i,j} \left(w_{ij}^+ \frac{A^{-1}(x_{ij}^+ - m)}{\sigma} + w_{ij}^- \frac{A^{-1}(x_{ij}^- - m)}{\sigma} \right) = \sum_{i,j} (w_{ij}^+ - w_{ij}^-) b_{ij}, \qquad (1)$$

leading to an update of the evolution path

$$p_s^{(t+1)} \leftarrow (1 - c_s) \cdot p_s^{(t)} + \sqrt{c_s \cdot (2 - c_s) \cdot \mu_{\text{eff}}} \cdot \sum_{ij} (w_{ij}^+ - w_{ij}^-) \cdot b_{ij}. \qquad (2)$$

In the CSA, the evolution path is updated such that its expected length under random selection is the expected value of the $\chi^2(d)$ distribution. To correct for the bias introduced by the weighted mean, the CSA adds the correction $\mu_{\text{eff}} = 1/(\sum_{ij} (w_{ij}^+)^2 + (w_{ij}^-)^2)$. In the mirror-sampling case, the subtraction of the weights on the right hand side of (1) means that the expected length of the vector is smaller than expected under non-mirrored sampling, therefore the step-size update is biased and tends to reduce the step-size prematurely. We fix this problem by computing the correct normalization factor $\mu_{\text{eff}}^{\text{mirrored}} > \mu_{\text{eff}}$.

Under random selection, w_{ij}^+ and w_{ij}^- are randomly picked without replacement from $\boldsymbol{w}$, independently of b_{ij}. Thus, the distribution of the weighted sample-average of (1) is still normal and the expected squared length is:

$$\begin{aligned} E\left\{ \left\| \sum_{i,j} (w_{ij}^+ - w_{ij}^-) b_{ij} \right\|^2 \right\} &= E\left\{ \sum_{i,j} (w_{ij}^+ - w_{ij}^-)^2 \right\} E\left\{ \|y\|^2 \right\} \\ &= E\left\{ \sum_{i,j} (w_{ij}^+ - w_{ij}^-)^2 \right\} d \end{aligned}$$

Note that in the first step, we used that $E\left\{y_i^T y_j\right\} = 0$ for $i \neq j$, while the second step holds because we ensure during sampling that the squared length of the samples is still $\chi^2(d)$-distributed. Next, we will use that the set of all w_{ij}^+ and w_{ij}^- together forms the weight vector $\boldsymbol{w}$ and thus the expectation can be written as permutations τ of the indices of $\boldsymbol{w}$. We can therefore write the expectation in terms of w_i as:

$$
\begin{aligned}
E\left\{\sum_{i,j}^{\tilde{\lambda}}(w_{ij}^+ - w_{ij}^-)^2\right\} &= E\left\{\sum_{i,j}^{\tilde{\lambda}}(w_{ij}^+)^2 + (w_{ij}^-)^2 - 2w_{ij}^+ w_{ij}^-\right\}\\
&= E_\tau\left\{\sum_{k=1}^{\tilde{\lambda}} w_{\tau(k)}^2 + w_{\tau(k+\tilde{\lambda})}^2 - 2w_{\tau(k)}w_{\tau(k+\tilde{\lambda})}\right\}\\
&= \frac{1}{\mu_{\text{eff}}} - 2E_\tau\left\{\sum_{k=1}^{\tilde{\lambda}} w_{\tau(k)}w_{\tau(k+\tilde{\lambda})}\right\}
\end{aligned}
$$

To continue, we expand the expectation and count the number of times each (w_i, w_j)-pair appears. There is a total of $(2\tilde{\lambda})!$ permutations and for each (i,j)-pair with $i \neq j$, there are $(2\tilde{\lambda})!/(2\tilde{\lambda} \cdot (2\tilde{\lambda} - 1))$ permutations such that $\tau(k) = i$ and $\tau(k+\tilde{\lambda}) = j$ for each k, which leads to another factor of $\tilde{\lambda}$. Thus, we obtain:

$$
\begin{aligned}
\frac{1}{\mu_{\text{eff}}} - 2E_\tau\left\{\sum_{k=1}^{\tilde{\lambda}} w_{\tau(k)}w_{\tau(k+\tilde{\lambda})}\right\} &= \frac{1}{\mu_{\text{eff}}} - \frac{2}{(2\tilde{\lambda})!}\sum_{i,j\neq i}^{2\tilde{\lambda}} \frac{\tilde{\lambda}(2\tilde{\lambda})!}{2\tilde{\lambda}\cdot(2\tilde{\lambda}-1)} w_i w_j\\
= \frac{1}{\mu_{\text{eff}}} - \frac{1}{2\tilde{\lambda}-1}\sum_{i,j\neq i}^{2\tilde{\lambda}} w_i w_j &= \frac{1}{\mu_{\text{eff}}} - \frac{1}{2\tilde{\lambda}-1}\sum_{i}^{2\tilde{\lambda}} w_i(1-w_i)\\
= \frac{1}{\mu_{\text{eff}}} - \frac{1}{2\tilde{\lambda}-1}\left(1 - \frac{1}{\mu_{\text{eff}}}\right) &= \frac{1}{\mu_{\text{eff}}}\left(1 - \frac{\mu_{\text{eff}} - 1}{2\tilde{\lambda}-1}\right)
\end{aligned}
$$

In the second step, we use $\sum_{j\neq i}^{2\tilde{\lambda}} w_j = 1 - w_i$. Thus, we remove the step-length bias in CSA by replacing μ_{eff} in Eq. (2) with

$$
\mu_{\text{eff}}^{\text{mirrored}} := \frac{\mu_{\text{eff}}}{1 - \frac{\mu_{\text{eff}}-1}{2\tilde{\lambda}-1}}.
$$

The Algorithm. The resulting HE-ES algorithm is summarized in Algorithm 3. Up to the (significant) changes discussed in the previous sections its design is identical to CMA-ES. In particular, it relies on global intermediate recombination, non-elitist selection, cumulative step-size adaptation (CSA), and it applies the same weights as CMA-ES to the offspring [5]. As default number of mirrored directions, we chose $\tilde{\lambda} = 2 + \lfloor \frac{3}{2}\log(d) \rfloor$. As learning-rates c_s and d_s of the CSA in HE-ES, we chose the same values as the implementation of the CMA in `pycma-2.7.0` with $2\tilde{\lambda}$ offspring. In contrast to CMA-ES, HE-ES needs to evaluate $f(m)$ in each generation. This value is used only for estimating curvatures.

Algorithm 3. Hessian Estimation Evolution Strategy (HE-ES)

1: **input** $m^{(0)} \in \mathbb{R}^d$, $\sigma^{(0)} > 0$, $A^{(0)} \in \mathbb{R}^{d \times d}$
2: **parameters** $\tilde{\lambda} \in \mathbb{N}$, c_s, d_s, $w \in \mathbb{R}^{2\tilde{\lambda}}$
3: $B \leftarrow \lceil \tilde{\lambda}/d \rceil$
4: $p_s^{(0)} \leftarrow 0 \in \mathbb{R}^d$
5: $g_s^{(0)} \leftarrow 0$
6: $t \leftarrow 0$
7: **repeat**
8: **for** $j \in \{1, \dots, B\}$ **do**
9: $b_{1j}, \dots, b_{dj} \leftarrow$ `sampleOrthogonal()`
10: $x_{ij}^- \leftarrow m^{(t)} - \sigma^{(t)} \cdot A^{(t)} b_{ij}$ for $i + (j-1)B \leq \tilde{\lambda}$
11: $x_{ij}^+ \leftarrow m^{(t)} + \sigma^{(t)} \cdot A^{(t)} b_{ij}$ for $i + (j-1)B \leq \tilde{\lambda}$ # mirrored sampling
12: $A^{(t+1)} \leftarrow A^{(t)} \cdot$ `computeG`$(\{b_{ij}\}, f(m), \{f(x_{ij}^\pm)\}, \sigma^{(t)})$ # matrix adaptation
13: $w_{ij}^\pm \leftarrow w_{\text{rank}(f(x_{ij}^\pm))}$
14: $m^{(t+1)} \leftarrow \sum_{ij} w_{ij}^\pm \cdot x_{ij}^\pm$ # mean update
15: $g_s^{(t+1)} \leftarrow (1-c_s)^2 \cdot g_s^{(t)} + c_s \cdot (2 - c_s)$
16: $p_s^{(t+1)} \leftarrow (1-c_s) \cdot p_s^{(t)} + \sqrt{c_s \cdot (2-c_s) \cdot \mu_{\text{eff}}^{\text{mirrored}}} \cdot \sum_{ij} (w_{ij}^+ - w_{ij}^-) \cdot b_{ij}$
17: $\sigma^{(t+1)} \leftarrow \sigma^{(t)} \cdot \exp\left(\frac{c_s}{d_s} \cdot \frac{\|p_s^{(t+1)}\|}{\chi_d} - \sqrt{g_s^{(t+1)}}\right)$ # CSA
18: $t \leftarrow t + 1$
19: **until** *stopping criterion is met*

3 Experimental Evaluation

Our experimental evaluation aims to answer the following research questions:

1. Is Hessian estimation a competitive CMA scheme?
2. What are its strengths and weaknesses compared with CMA-ES?
3. How much does performance change under monotonically increasing but non-affine fitness transformations?

The source code of the algorithm that was used in all experiments is available from the first author's website.[2]

Benchmark Study. Our first experiment is to run the standardized BBOB/COCO procedure, which tests the algorithm on 15 instances of 24 benchmark problems [6]. For handling multi-modal problems we equip HE-ES with an IPOP restart mechanism [1], which restarts the algorithm with doubled population size as soon the standard deviation of the fitness values of a generation falls below 10^{-9}.

The BBOB platform generates a plethora of results. Due to space constraints we show a representative subset thereof. Figures 1 and 2 show ECDF plots on all 24 function for problem dimension 20, with IPOP-CMA-ES, BFGS, and NEWUOA as baselines. Figure 3 shows overall performance in dimensions 2, 5,

[2] https://www.ini.rub.de/the_institute/people/tobias-glasmachers/#software.

10, and 20. The results for IPOP-CMA-ES, BFGS, and NEWUOA were obtained from the BBOB/COCO platform.

Discussion. We observe excellent performance across most problems. On all convex quadratic problems (f_1, f_2, f_{10}, f_{11}, f_{12}) HE-ES performs very well, and on f_{10} (ellipsoid) and f_{11} (discus) it even outperforms the hypothetical "2009-best portfolio algorithm", which picks the best optimizer from the 2009 competition for each problem. Surprisingly, the same holds for problems f_{15} (Rastrigin), f_{16} (Weierstraß, see below), f_{17} (Schaffer F7, condition 10), and f_{19} (Griewank-Rosenbrock). Overall, the performance is much closer to CMA-ES than to NEWUOA and BFGS, which indicates that the character of an ES is preserved, despite the novel mechanism for updating the covariance matrix.

Compared to IPOP-CMA-ES, we observe degraded performance on f_6 (attractive sector), f_{13} (sharp ridge), f_{20} (Schwefel $x \cdot \sin(x)$), f_{21} and f_{22} (Gallagher peaks), f_{23} (Katsuuras), and f_{24} (Lunacek bi-Rastrigin). HE-ES apparently struggles with the asymmetry of the attractive sector problem, which can yield drastically wrong curvature estimates. Similarly, it is conceivable that estimating curvatures on the sharp ridge problem is prone to failure. We believe that these two benchmark functions highlight inherent limitations of the HE-ES update.

Control Experiments. In order to understand the weak performance on the highly multi-modal problems despite IPOP restarts we investigated the behavior of HE-ES on the deceptive Lunacek problem f_{24} in dimension $d = 10$. We ran HE-ES with IPOP restarts 100 times with a reduced budget of $10^4 \cdot d$ function evaluations. This medium-sized budget is 100 times smaller than in the BBOB/COCO experiments. It suffices for 4 to 5 runs, with population sizes ranging from 10 to 160. We found that HE-ES converged to the better of the two funnels in 83 cases, and solved the problem to a high precision of 10^{-10} in 40 out of 100 cases, which corresponds to reaching all BBOB targets. This means that the correct funnel and the best local optimum of the Rastrigin structure were found in 40% of the cases, which is a quite satisfactory behavior. It is noteworthy that CMA mechanisms are not even needed for this problem (indeed, performance is unchanged when disabling CMA), and hence the performance difference to CMA-ES is probably an artifact of a different restart implementation. This is unrelated to the new CMA mechanism and hence of minor relevance for our investigation.

As mentioned above, HE-ES performs surprisingly well on the Weierstraß function f_{16}, which is continuous but nowhere differentiable, and therefore strongly violates the assumption of a twice continuously differentiable objective function. At first glance, this result is surprising. The reason is that HE-ES does not really need correct estimates of the curvature. It is only relevant that the structure (the "global trend") of the objective function at the relevant scale (given by σ) is correctly captured. Of course, HE-ES picks up misleading curvature information. However, since the Weierstraß function does not exhibit a

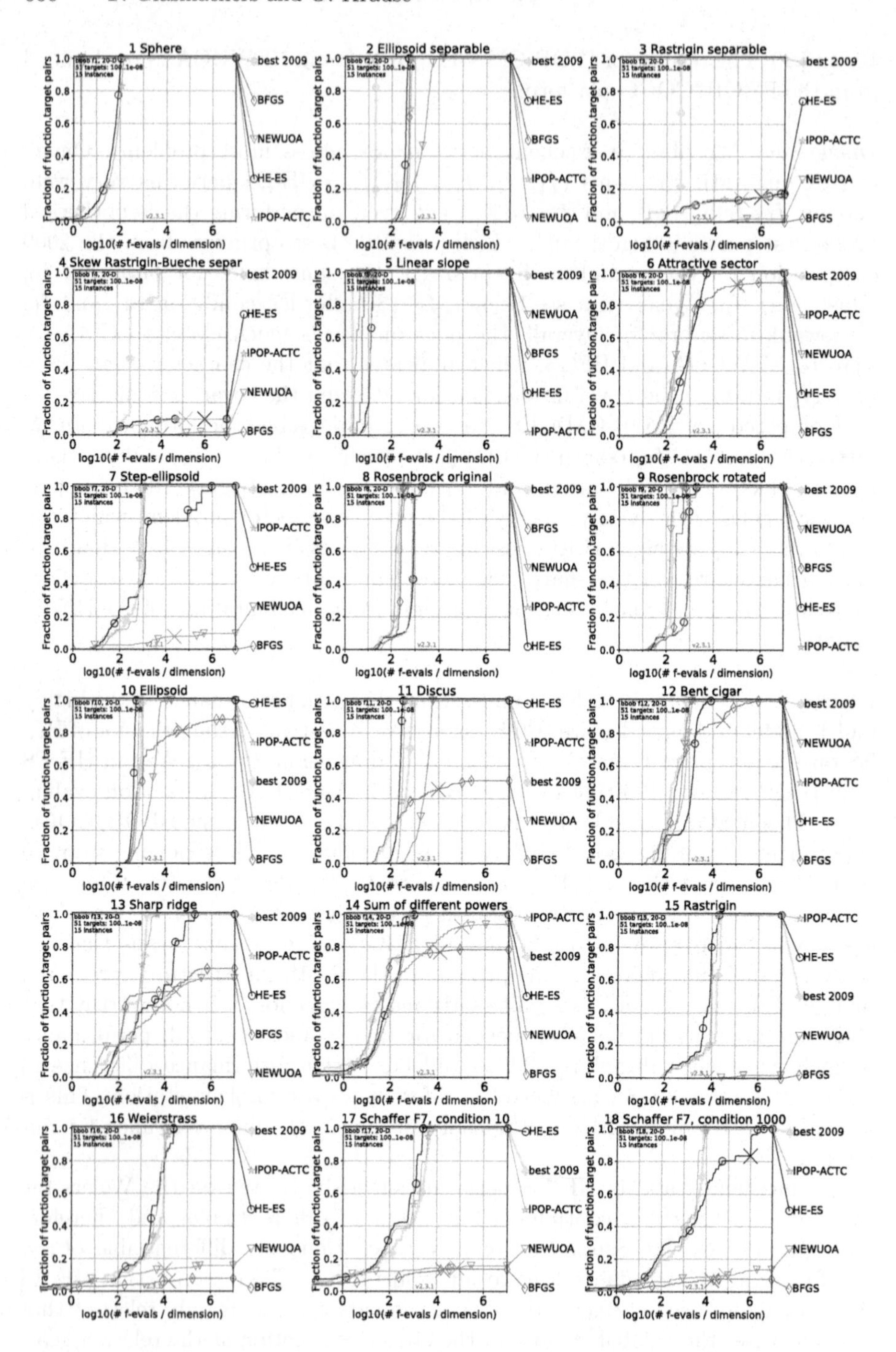

Fig. 1. ECDF plots for the noiseless BBOB problems 1–18 in dimension $d = 20$. We generally observe that HE-ES performs very well on smooth unimodal problems (functions 1, 2, 5, 8, 9, 10, 11, 12, 14).

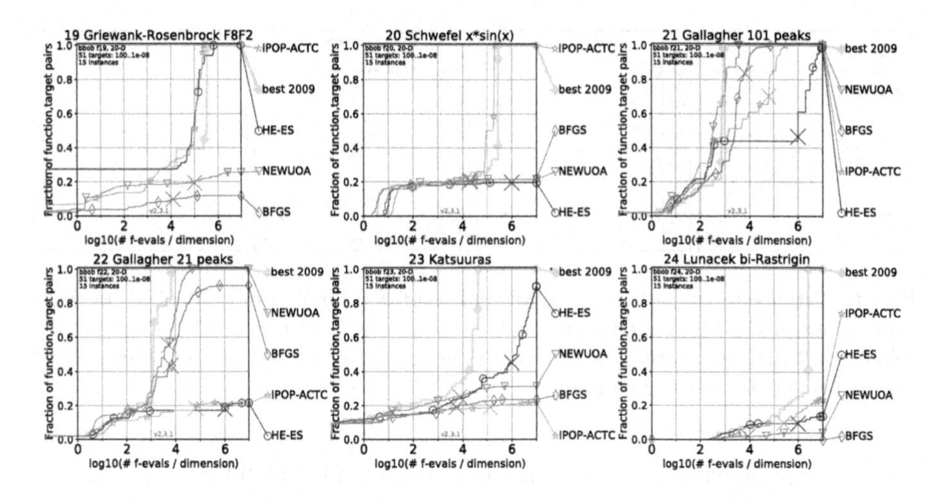

Fig. 2. ECDF plots for the noiseless BBOB problems 19–24 in dimension $d = 20$.

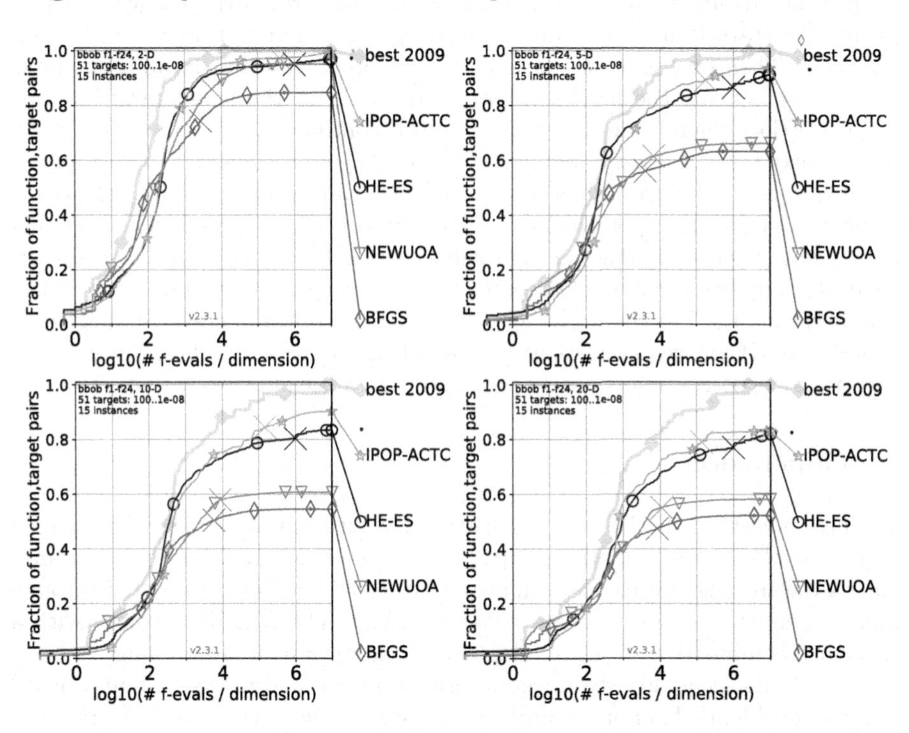

Fig. 3. Aggregated performance on all 24 BBOB functions in dimensions 2, 5, 10 and 20. We observe that HE-ES clearly outperforms NEWUOA and BFGS. IPOP-CMA-ES is more reliable then HE-ES, and this gap slightly increases with increasing dimension. The differences mostly originate from hard multi-modal problems.

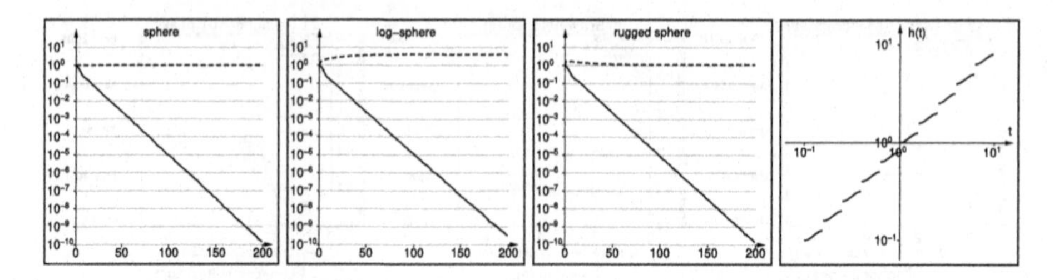

Fig. 4. Distance to the optimum (solid) and condition number of C (dashed) over 200 generations of HE-ES started at $m = (1, 0, \ldots, 0) \in \mathbb{R}^{10}$ with $\sigma = 0.1$ for sphere, log-sphere, and rugged sphere. The curves are medians over 99 independent runs. Right: log-log plot of the transformation h.

systematic preference for a particular direction, such unhelpful information averages out over time and hence does not have a lasting detrimental effect.

In order to investigate this effect closer we performed the following experiment. We start from the 10-dimensional sphere function $f(x) = \frac{1}{2}\|x\|^2$ as a base case. Then we create two variants by monotonically transforming the function values, leaving the level sets intact, resulting in the non-convex function $\log(f(x))$ (log-sphere), and the rugged and discontinuous function $h(f(x))$ with $h(t) = \exp\left(\left[\frac{1}{4} - \frac{1}{2}\cos(\pi(5\log(t) - r(t))) + r(t)\right]/5\right)$, where $r(t) = \lfloor 5\log(t) \rfloor$ (rugged sphere). Figure 4 shows a plot of the transformation h as well as the resulting optimization performance of HE-ES. For sphere the condition number remains at exactly one (the optimal value). Importantly, in the other cases the condition number remains close enough to one so as to not impair optimization performance. For the log-sphere there is a slowdown, however, of negligible magnitude: HE-ES requires about 5% more time. We observe that in this setup, surprisingly, HE-ES suffers nearly not at all from misleading curvature estimates.

4 Conclusion

We have presented the Hessian Estimation Evolution Strategy (HE-ES), an ES with a novel covariance matrix adaptation mechanism. It adapts the covariance matrix towards the inverse Hessian projected to random lines, estimated through finite differences by means of mirrored sampling. The algorithm comes with a specialized cumulative step size adaptation rule for mirrored sampling.

Despite its seemingly strong assumptions the method works well on a broad range of problems. It is particularly well suited for smooth unimodal problems like convex quadratic functions and the Rosenbrock function. Surprisingly, the adaptation mechanism that is based on estimating presumedly positive second derivatives can work well on non-convex and even on discontinuous problems.

We believe that the HE-ES offers an interesting alternative to adapting the covariance matrix of the sampling distribution towards the maximum likelihood estimator of successful steps, corresponding to the natural gradient in parameter space.

References

1. Auger, A., Hansen, N.: A restart CMA evolution strategy with increasing population size. In: IEEE Congress on Evolutionary Computation, vol. 2, pp. 1769–1776 (2005)
2. Beyer, H.G., Sendhoff, B.: Simplify your covariance matrix adaptation evolution strategy. IEEE Trans. Evol. Comput. **21**(5), 746–759 (2017)
3. Brockhoff, D., Auger, A., Hansen, N., Arnold, D.V., Hohm, T.: Mirrored sampling and sequential selection for evolution strategies. In: International Conference on Parallel Problem Solving from Nature. pp. 11–21. Springer (2010)
4. Glasmachers, T., Schaul, T., Sun, Y., Wierstra, D., Schmidhuber, J.: Exponential natural evolution strategies. In: Genetic and Evolutionary Computation Conference (GECCO), pp. 393–400. ACM (2010)
5. Hansen, N., Ostermeier, A.: Completely derandomized self-adaptation in evolution strategies. Evol. Comput. **9**(2), 159–195 (2001)
6. Hansen, N., Auger, A., Mersmann, O., Tušar, T., Brockhoff, D.: COCO: a platform for comparing continuous optimizers in a black-box setting. Technical report arXiv:1603.08785 (2016)
7. Hansen, N., Auger, A., Ros, R., Finck, S., Pošík, P.: Comparing results of 31 algorithms from the black-box optimization benchmarking BBOB-2009. In: Proceedings of the 12th Annual Conference Companion on Genetic and Evolutionary Computation, pp. 1689–1696 (2010)
8. Jebalia, M., Auger, A.: Log-Linear Convergence of the Scale-Invariant and Optimal $(\mu/\mu$ w, $\lambda)$-es and optimal μ for Intermediate Recombination for Large Population Sizes. In: Schaefer, R., Cotta, C., Kołodziej, J., Rudolph, G. (eds.) PPSN 2010. LNCS, vol. 6238, pp. 52–62. Springer, Heidelberg (2010). https://doi.org/10.1007/978-3-642-15844-5_6
9. Kern, S., Müller, S.D., Hansen, N., Büche, D., Ocenasek, J., Koumoutsakos, P.: Learning probability distributions in continuous evolutionary algorithms-a comparative review. Nat. Comput. **3**(1), 77–112 (2004). https://doi.org/10.1023/B:NACO.0000023416.59689.4e
10. Krause, O., Glasmachers, T.: A CMA-ES with multiplicative covariance matrix updates. In: Proceedings of the Genetic and Evolutionary Computation Conference (GECCO) (2015)
11. Leventhal, D., Lewis, A.: Randomized hessian estimation and directional search. Optimization **60**(3), 329–345 (2011)
12. Nocedal, J., Wright, S.: Numerical Optimization. Springer, Heidelberg (2006). https://doi.org/10.1007/978-0-387-40065-5
13. Ollivier, Y., Arnold, L., Auger, A., Hansen, N.: Information-geometric optimization algorithms: a unifying picture via invariance principles. J. Mach. Learn. Res. **18**(1), 564–628 (2017)
14. Powell, M.: The NEWUOA software for unconstrained optimization without derivatives. Technical Rep. DAMTP 2004/NA05, Department of Applied Mathematics and Theoretical Physics, Cambridge University (2004)
15. Stich, S.U., Müller, C.L., Gärtner, B.: Variable metric random pursuit. Math. Program. **156**, 549–579 (2015). https://doi.org/10.1007/s10107-015-0908-z
16. Teytaud, O., Gelly, S.: General lower bounds for evolutionary algorithms. In: Runarsson, T.P., Beyer, H.-G., Burke, E., Merelo-Guervós, J.J., Whitley, L.D., Yao, X. (eds.) PPSN 2006. LNCS, vol. 4193, pp. 21–31. Springer, Heidelberg (2006). https://doi.org/10.1007/11844297_3
17. Wang, H., Emmerich, M., Bäck, T.: Mirrored orthogonal sampling for covariance matrix adaptation evolution strategies. Evol. Comput. **27**(4), 699–725 (2019)

Large Population Sizes and Crossover Help in Dynamic Environments

Johannes Lengler(✉) and Jonas Meier

Department of Computer Science, ETH Zürich, Zürich, Switzerland
johannes.lengler@inf.ethz.ch

Abstract. Dynamic linear functions on the boolean hypercube are functions which assign to each bit a positive weight, but the weights change over time. Throughout optimization, these functions maintain the same global optimum, and never have defecting local optima. Nevertheless, it was recently shown [Lengler, Schaller, FOCI 2019] that the $(1+1)$-Evolutionary Algorithm needs exponential time to find or approximate the optimum for some algorithm configurations. In this experimental paper, we study the effect of larger population sizes for Dynamic BinVal, the extremal form of dynamic linear functions. We find that moderately increased population sizes extend the range of efficient algorithm configurations, and that crossover boosts this positive effect substantially. Remarkably, similar to the static setting of monotone functions in [Lengler, Zou, FOGA 2019], the hardest region of optimization for $(\mu+1)$-EA is not close the optimum, but far away from it. In contrast, for the $(\mu+1)$-GA, the region around the optimum is the hardest region in all studied cases (Extended Abstract. A full version is available on arxiv at [11]).

1 Introduction

The $(\mu+1)$ Evolutionary Algorithm and the $(\mu+1)$ Genetic Algorithm, $(\mu+1)$-EA and $(\mu+1)$-GA for short, are heuristic algorithms that aim to optimize an objective or fitness function $f : \{0,1\}^n \to \mathbb{R}$. Both maintain a population of μ search points, and in each round they create an offspring from the population and discard one of the $\mu+1$ search points, based on their objective values. They differ in how the offspring is created: the $(\mu+1)$-EA chooses a parent from the population and *mutates* it, the $(\mu+1)$-GA uses *crossover* of two parent solutions in addition to mutation.

Two classical theoretical questions for these algorithms have ever been:

- What is the effect of the population size? In which optimization landscapes and regimes are larger (smaller) populations beneficial?
- In which situations does crossover improve performance?

Although these questions have ever been central for studies of the $(\mu+1)$-EA and the $(\mu+1)$-GA, there is still vivid ongoing research on these questions,

T. Bäck et al. (Eds.): PPSN 2020, LNCS 12269, pp. 610–622, 2020.
https://doi.org/10.1007/978-3-030-58112-1_42

see [1,2,4,7,16,19,21,22] for a selection of theoretical works. (Also, the book chapter [20] treats related topics.) More generally, the research question is: which algorithm configurations perform well in which optimization landscapes? Such landscapes are given by a specific benchmark functions or by a class of functions.

Recently, a new type of dynamic landscapes was introduced by Lengler and Schaller [12]. It was called *noisy linear functions* in [12], but we prefer the term *dynamic linear functions*. In this setting, the objective function is of the form $f : \{0,1\}^n \to \mathbb{R}; f(x) = \sum_{i=1}^n W_i x_i$ with positive coefficients $W_i > 0$. However, the twist is that the weights W_i are redrawn for each generation. I.e., we have a distribution $\mathcal{D}$, and for the t-th generation we draw i.i.d. weights $W_i^{(t)}$ from that distribution, which define a function $f^{(t)}$. Then the $\mu + 1$ competing individuals are compared with respect to the fitness function $f^{(t)}$.

To motivate this setting, let us give a grotesquely oversimplified example. Imagine a chess engine has 100 bits as parameters. Each bit switches on/off a database tailored to one specific opening (1 = access, 0 = no access), and this improves performance massively for this opening. E.g., the first bit determines whether the engine plays well in a French Opening, but has no influence whatsoever on the performance of an Italian Opening (the database is ignored since it does not produce matches to that situation). Let us go even further and assume that the engine will always win an opening if the corresponding database is active, and always lose with inactive database. Then we have removed even the slightest ambiguity, and this setting has an obvious optimal solution, which is the all-one string (activate all databases). This situation may seem completely trivial, but crucially, *it is not solved by some standard optimization algorithms.*

To complete the analogy, assume that the engine is trained by playing against different players, where player t has probability $W_i^{(t)}$ to choose the i-th opening. Then the reward is precisely the dynamic linear function introduced above, and it was shown in [12] that *the* $(1+1)$*-EA needs exponential time to approximate the optimum within a constant factor* when configured with bad parameters. These bad parameter settings look quite innocent. With standard bit mutation (i.e., for mutation we flip each bit independently with probability $p = c/n$), any choice $c > c_0 \approx 1.59$ leads asymptotically to an exponential time for finding or approximating the optimum, if the distribution $\mathcal{D}$ is too skewed. On the other hand, for any $c < c_0$ the $(1 + 1)$-EA finds the optimum in time $O(n \log n)$ for any $\mathcal{D}$. This lack of stability motivates our paper: we ask whether larger population sizes and/or crossover can push the threshold c_0 of failure.

Optimization of dynamic functions may occur in various contexts. The chess engine with varying opponents is one such example. Similar examples arise in the context of co-evolution, e.g., a chess-engine trained against itself, or a team of DOTA agents in which some abilities of an agent (good aim, good exploration strategy, good path planning, ...) are always helpful (positive weight), but may be more or less important depending on her current co-agents. A rather different example is planning the timetable of a transportation company: to be efficient in the exploration phase of the optimization algorithm, schedules may be compared only for some partial data, and not for the whole data set. Similarly, consider an optimization process in which the function evaluation involves an offline test, as in drug development or robotic training. Then each test may involve subtly varying outer conditions (e.g., different temperatures or humidity, different lighting), which effectively gives a slightly different fitness function for each test.

Our Results in a Nutshell. Instead of the full range of dynamic linear functions as in [12], we only study the limiting case of these functions, which we call *dynamic binval.* We perform experiments to study the performance of the $(\mu + 1)$-EA and the $(\mu + 1)$-GA for small values of μ. Similarly as for the $(1 + 1)$-EA, we find that for each algorithm there is a threshold c_0 such that the algorithm is efficient for every mutation parameter $c < c_0$, and inefficient for $c > c_0$. This threshold c_0 is our main object of study, and we investigate how it depends on the algorithmic choices. We find that an increased population size helps to push c_0, but that the benefits are much larger when crossover is used. As a baseline, we recover the theoretical result from [12] that for the $(1+1)$-EA the threshold is at $c_0 \approx 1.59$, though experimentally, for $n = 3000$ it seems closer to 1.7. For the $(2+1)$-EA the threshold increases to $c_0 \approx 2.2$, and further to $c_0 \approx 3.1$ for the $(2 + 1)$-GA. If we explicitly forbid that the two parents in crossover are identical then the threshold even shifts to $c_0 \approx 4.2$. We call the resulting algorithm $(2 + 1)$-GA-NoCopy. For larger population sizes we get a threshold of $c_0 \approx 2.6$ for the $(3 + 1)$-EA, $c_0 \approx 3.4$ for the $(5 + 1)$-EA, $c_0 \approx 6.1$ for the $(3 + 1)$-GA, and $c_0 > 20$ for the $(5 + 1)$-GA.

The theoretical results for the $(1 + 1)$-EA predict that the runtime jumps from quasi-linear to exponential. Indeed, we can experimentally confirm huge jumps in the runtime even for slight changes of the mutation parameter c. For example, we obtain a significant p-value for the a posteriori hypothesis that the $(2+1)$-EA with $c = 2.5$ is more than 60 times slower than the $(2+1)$-EA with $c = 2.0$. In fact, this is a highly conservative estimate since we needed to cut off the runs for $c = 2.5$. We systematically list these factors in our result sections.

To get a better understanding of the hardness of the optimization landscape, we compute the *drift of degenerate populations*, inspired by [8]. We call a population *degenerate* if it consists entirely of multiple copies of the same individual. If X_i is the number of zero-bits in the i-th degenerate population, then we estimate the drift $\mathbb{E}[X_i - X_{i+1} \mid X_i = y]$ by Monte-Carlo simulations. Moreover, for y close to 0 we derive precise asymptotic formulas for the degenerate population drift for the $(2 + 1)$-EA and the $(2 + 1)$-GA, which can be found in the full version [11]. In [8] the degenerate population drift was studied theoretically for the $(\mu + 1)$-EA on monotone functions, which is a related, but not identical setup (see below). Still, part of the analysis carries over: if the population drift is negative for some y then the runtime is exponential, while it is $O(n \log n)$ if the population drift is positive everywhere.

Perhaps surprisingly, the $(\mu+1)$-EA and the $(\mu+1)$-GA are not just quantitatively different, but we also find a strong qualitative difference in the hardness landscape. For the $(\mu+1)$-GA, the "hardest" part of the optimization process is close to optimum, in all cases that we have experimentally explored. Formally, we found that if the degenerate population drift is negative *somewhere*, then it is also negative close to the optimum. For the $(\mu+1)$-EA, we found the opposite: the degenerate population drift can be negative for some intermediate ranges, although it is positive around the optimum. This implies that the hard part of optimization (taking exponential time) is getting in the vicinity of the optimum. But once the algorithm is somewhat near the optimum, it will efficiently finish optimization. This behavior is rather counter-intuitive, since common wisdom says that optimization gets harder close to the optimum. Notably, a similar phenomenon has recently been proven for certain monotone functions by Lengler and Zou [8,14], see below.

Related Work. The only previous work on dynamic linear functions is by Lengler and Schaller [12]. As mentioned before, they proved that for every $c > c_0 \approx 1.59$ there is $\varepsilon > 0$ and a distribution $\mathcal{D}$ such that the $(1+1)$-EA with mutation rate c/n needs exponential time to find a search point with at least $(1-\varepsilon)n$ one-bits for dynamic linear functions with weight distribution $\mathcal{D}$. For $c < c_0$ the optimization time is $O(n \log n)$ for all distributions $\mathcal{D}$. Moreover, for any $c > c_0$, they gave a complete characterization of all distributions for which the $(1+1)$-EA with mutation rate c/n is efficient/inefficient.

Other previous work on population-based algorithms in dynamic environments had a slightly different flavor. One goal was to understand how well various algorithms can track a slowly moving optimum in a dynamic environment, e.g. [3,15]. Another was to show that dynamic environments can prevent algorithms from falling into traps [18]. For more details on these topics we refer to the survey by Sudholt [20, Section 8.5], with emphasis on the role of diversity.

An important strand of work that is similar to our setting in spirit, though not in detail, is the study of *monotone functions*. A function is *monotone* if for every bit-string, flipping any zero-bit into a one-bit increases the fitness. Doerr, Jansen, Sudholt, Winzen, and Zarges [5] and Lengler and Steger [13] showed that there are monotone functions for which the $(1+1)$-EA needs exponential time to find or approximate the optimum if the mutation parameter c is too large ($c > c_0 \approx 2.1$), while it is efficient for all monotone functions if $c \leq 1+\varepsilon$ for some small $\varepsilon > 0$ [9]. The construction of hard (static) instances from [13] was named HotTopic in [8], and it resembles dynamic linear functions: the HotTopic function is locally given by linear functions with certain positive weights, but as the algorithm proceeds from one part of the search space ("level") to the next, the weights change. This analogy inspired the introduction of dynamic linear functions in [12].

For HotTopic functions, there is a plethora of results. In [8], the dichotomy between exponential and quasi-linear time from the $(1+1)$-EA was extended to a large number of other algorithms, including the $(1+\lambda)$-EA, the $(\mu+1)$-EA, their so-called "fast" counterparts, and the $(1+(\lambda,\lambda))$-GA. On the other hand, it was shown that the $(\mu+1)$-GA is always efficient for HotTopic functions if the population size is sufficiently large, while for the $(\mu+1)$-EA the population size does not change the threshold c_0 at all. Notably, for the population-based algorithms $(\mu+1)$-EA and $(\mu+1)$-GA, the efficiency result was only obtained for parameterizations of the HotTopic functions in which the weight changes occur close to the optimum. This seemed like a technical detail at first, but in an extremely surprising result, Lengler and Zou [14] showed that this detail was hiding an unexpected core: if the weights are changed far away from the optimum, then increasing the population size has a devastating effect on the performance of the $(\mu + 1)$-EA. For any $c > 0$ (also values much smaller than 1), there is a μ_0 such that the $(\mu + 1)$-EA with $\mu \geq \mu_0$ and mutation rate c/n needs exponential time on some monotone functions. Together with [8], this shows three things for monotone functions:

1. For optimization close to the optimum, the population size has no strong impact on the performance of the $(\mu + 1)$-EA.
2. Close to the optimum, the $(\mu + 1)$-GA outperforms the $(\mu + 1)$-EA massively (quasi-linear instead of exponential) if the population size is large enough. It can cope with any constant mutation parameter c.
3. Far away from the optimum, a larger population size decreases the performance of $(\mu + 1)$-EA massively. There is no safe choice of c if μ is too large.

It would be extremely interesting to understand the $(\mu + 1)$-GA far away from the optimum. Unfortunately, such results are unknown. Theoretical analysis is hard (though perhaps not impossible), and function evaluations of HotTopic are extremely expensive, so experiments are only possible for very small problem sizes. Our paper can be seen as the first work which studies the behavior of the $(\mu + 1)$-GA in a related, though not identical setting.

We conclude this section with a word of caution. HotTopic functions and dynamic linear functions are similar in spirit, but not in actual detail. For example, the analysis of the $(\mu + 1)$-GA in [8] or of the $(\mu + 1)$-EA in [14] rely heavily on the fact that there weights are locally stable in HotTopic functions. Thus it is unclear how far the analogy carries. Some of our experimental findings for the $(\mu + 1)$-EA for dynamic linear functions differ from the theoretical (asymptotic) results for HotTopic in [8,14]. For us, a larger μ is beneficial, as it shifts the theshold c_0 to the right. For HotTopic functions, it does not shift the threshold at all if the algorithm operates close to the optimum, and it shifts the threshold *to the left* (i.e., makes things worse) far away from the optimum. This could either be because the theoretical effects only kick in for very large μ, or because HotTopic and dynamic linear functions are genuinely different. On the other hand, both settings agree in the surprising effect that the hardest part for the algorithm is not close to the optimum, but rather far away from it.

2 Preliminaries

2.1 The Algorithms

All our considered algorithms maintain a population P of search points of size μ. In each round (or *generation*), they create an offspring from the population, and from the $\mu + 1$ search points they remove the one with lowest fitness, breaking ties randomly. They only differ in the offspring creation. The $(\mu + 1)$-EA uses *standard bit mutation*: a random parent is picked from the population, and each bit in the parent is flipped with probability c/n. The genetic algorithms flip a coin in each round whether to use mutation (as above), or whether to use *bitwise uniform crossover*: for the latter, it picks two random parents from the population, and for each bit it randomly chooses the bit of either parent. For the $(\mu + 1)$-GA, the two parents are chosen independently. For the $(\mu + 1)$-GA-NoCopy, they are chosen without repetition. The parameters are thus the mutation parameter $c > 0$, which we will assume to be independent of n, and the population size μ. In our theoretical (asymptotic) results, we will only consider $\mu = 2$. The pseudocode description is given in Algorithm 1.

Algorithm 1: $(\mu + 1)$-EA, $(\mu + 1)$-GA, and $(\mu + 1)$-GA-NoCopy with mutation parameter c for maximizing an unknown function $f : \{0, 1\}^n \to \mathbb{R}$. P is a multiset, i.e., it may contain search points several times.

```
1  Initialize P with μ independent x ∈ {0,1}^n uniformly at random;
2  Optimization: for t = 1, 2, 3, ... do
3      Creation of Offspring: For GAs, flip a fair coin to do either mutation or
       crossover; for EA, always do mutation and no crossover.
4      if mutation then
5          Choose x ∈ P uniformly at random;
6          Create y by flipping each bit in x independently with probability c/n;
7      if crossover then
8          Choose x, x' ∈ X uniformly at random: independently for (μ + 1)-GA;
           without repetition for (μ + 1)-GA-NoCopy;
9          Create y by setting y_i to either x_i or x'_i, each with probability 1/2,
           independently for all bits;
10     Set P ← P ∪ {y};
11     Selection: Select z ∈ arg min{f(x) | x ∈ P} (break ties randomly) and
       update P ← P \ {z};
```

2.2 Dynamic Linear Functions and the Dynamic Binval Function

We have described the algorithms for optimizing a static fitness function f. However, throughout the paper, we will consider dynamic functions that changes in every round. We denote the function in the t-th iteration by $f^{(t)}$. That means that in the selection step (Line 11), we select the worst individual as $z \in \arg\min\{f^{(t)}(x) \mid x \in P^{(t)}\}$, where $P^{(t)}$ is the t-th population. Crucially, we never mix different fitness functions, i.e., we never compare $f^{(t_1)}(x)$ with $f^{(t_2)}(x')$ for different $t_1 \neq t_2$. In other words, the fitness of all individuals changes in each round. Since this requires $\mu + 1$ function evaluations per generation, we define *the runtime as the number of generations* until the algorithm finds the optimum. This deviates from the more standard convention to count the number of function evaluations (essentially by a factor $\mu + 1$), but it makes the performance easier to compare with previous work on static linear functions. Also, note that the runtime equals the number of search points that are sampled, up to an additive $-(\mu - 1)$ from initialization.

We consider two closely related types of dynamic functions, *dynamic linear functions* and *dynamic binval*. A *dynamic linear function* is described by a distribution $\mathcal{D}$ on $\mathbb{R}^+$. For the t-th round, we draw n independent samples $W_1^{(t)}, \dots, W_n^{(t)}$ and set

$$f^{(t)}(x) := \sum_{i=1}^{n} W_i^{(t)} \cdot x_i. \quad (1)$$

So $f^{(t)}(x)$ is a positive value as the weights are positive. Thus all $f^{(t)}$ share the same global optimum $1 \dots 1$, have no other local optima, and they are monotone, i.e., flipping a zero-bit into a one-bit always increases the fitness.

For *dynamic binval*, DYNBV, in the t-th round we draw a permutation $\pi_t : \{1..n\} \to \{1..n\}$ uniformly at random, and define

$$f^{(t)}(x) = \sum_{i=1}^{n} 2^i \cdot x_{\pi_t(i)}. \quad (2)$$

So, we randomly permute the bits of the string, and take the binary value of the permuted string. As for dynamic linear functions, all $f^{(t)}$ share the same global optimum $1 \dots 1$, have no other local optima, and prefer one-bits to zero-bits.

In a certain sense, DYNBV is a limit case of dynamic linear functions, and it was implicitly used in [12, Theorem 7] to construct a hard instance of dynamic linear functions. Due to space constraints, we refer the reader to the full version [11] for a discussion. All simulations are performed for dynamic binval.

2.3 Runtime Simulations

Recall that we count the runtime as the number of generations until the optimum is sampled. We run the different algorithms to observe the distribution of the runtime. A run terminates if either the optimum is found, or an upper limit of generations is reached. Unless otherwise noted, the upper limit is set to be $100e^c/c \cdot n \ln n$, which is 100 times larger than the expected runtime of the (1+1)-EA [22]. The python code of our running time simulation can be found in our GitHub repository [10]. Unless otherwise noted, each data point is obtained by 30 independent runs. We have verified correctness of our implementation by experiments with the (1 + 1)-EA, see the full version [11] for details.

To visualize runtimes, we use plots provided by the IOHprofiler [6]. Note that time (i.e., number of generations) is displayed *on the y-axis*, while the x-axis corresponds to the number of 1-bits. Thus, a steep part of the curve corresponds to slow progress, while a flat part of a curve corresponds to fast progress. Also, mind that the y-axis uses log scale.

Due to the exponential runtimes, we frequently encounter the problem that runs are terminated due to the iteration limit of $100e^c/c \cdot n \ln n$ generations.

In this case, we will often plot two values: the *mean runtime* is a lower bound estimate for the actual runtime, while the *expected runtime* (ERT) as defined by the IOH profiler is an upper bound. For a definition and a discussion of both, see the full version [11] and [6].

Comparison of Runtimes. We want to compare runtimes for different algorithms and values of c. We denote by R_c^{Alg} the random variable describing the runtime of Algorithm Alg for a specific c. Because our sample size is fairly small (10 to 30), we compare runtimes using the Wilcoxon-Mann-Whitney test. The Wilcoxon-Mann-Whitney test is a test of the null hypothesis that with probability at least 1/2 a randomly selected value from one runtime distribution will be at most (at least) a randomly selected value from a second runtime distribution. A small p-value would then indicate that the runtime of one algorithm, treated as random variable, is larger (smaller) than the runtime of the other algorithm in significantly more than half of the cases.

We will also be interested in quantifying *by how much* an algorithm is slower than another algorithm. To this end, we will determine the largest factor $d \geq 1$ by which we can multiply one runtime distribution such that the Wilcoxon-Mann-Whitney test still yields a statistically significant p-value. For example, we will find that even if we multiply the runtime $R_{2.0}^{(2+1)\text{-EA}}$ with $d = 63.15$ then this is still significantly faster than $R_{2.5}^{(2+1)\text{-EA}}$ according to the Wilcoxon-Mann-Whitney test. Note that this is a posteriori hypothesis since the factor d is chosen in hindsight. Therefore, it must not be treated as an actually significant result. Still, it gives useful information, and shows that $R_{2.0}^{(2+1)\text{-EA}}$ is *very much* smaller than $R_{2.5}^{(2+1)\text{-EA}}$. All tests are done with R [17].

2.4 Analysis of Population Drift

As defined in the introduction, X_i is the number of zero-bits in an individual of the i-th degenerated population, where degenerate means that all individuals are copies of the same search point. To be precise, if $P^{(i)}$ is the i-th degenerate population, then $P^{(i+1)}$ is the first degenerate population after $P^{(i)}$ has changed at least once. That does not exclude the possibility $P^{(i)} = P^{(i+1)}$, but we require at least one intermediate step in which an offspring is accepted that is not a copy of the parent(s) in $P^{(i)}$. Then the *degenerate population drift* (population drift for short) is $\mathbf{E}[X_i - X_{i+1} \mid X_i = y]$. We will estimate this drift with Monte-Carlo simulations. Our code is publicly available [10]. Moreover, the full version [11] contains an exact asymptotic formula for the population drift of the $(2+1)$-EA and $(2+1)$-GA in the limit $n \to \infty$ and $y = o(n)$, derived by Markov chain analysis. We omit them for space reasons. It is easy to see [8] that for every constant μ and c, the expected time for degeneration is $O(1)$. Moreover, it was shown in [8] that if the population drift is negative for $y = \alpha n$ for some $\alpha \in (1/2, 1)$ then asymptotically the runtime is exponential in n. On the other hand, if the population drift is positive for all α then the runtime is $O(n \log n)$. Hence, we are trying to identify parameter regimes for which areas of negative population drift occur.

3 Results

3.1 Runtimes

The results of our runtime simulations for different algorithms and values of c can be found in Fig. 1. As expected, we find a threshold behavior, i.e., there is a value c_0 such that the runtime increases dramatically as c crosses this threshold.

We will estimate threshold ranges for the different algorithms by visual inspection. We chose the estimates in a conservative manner, such that the

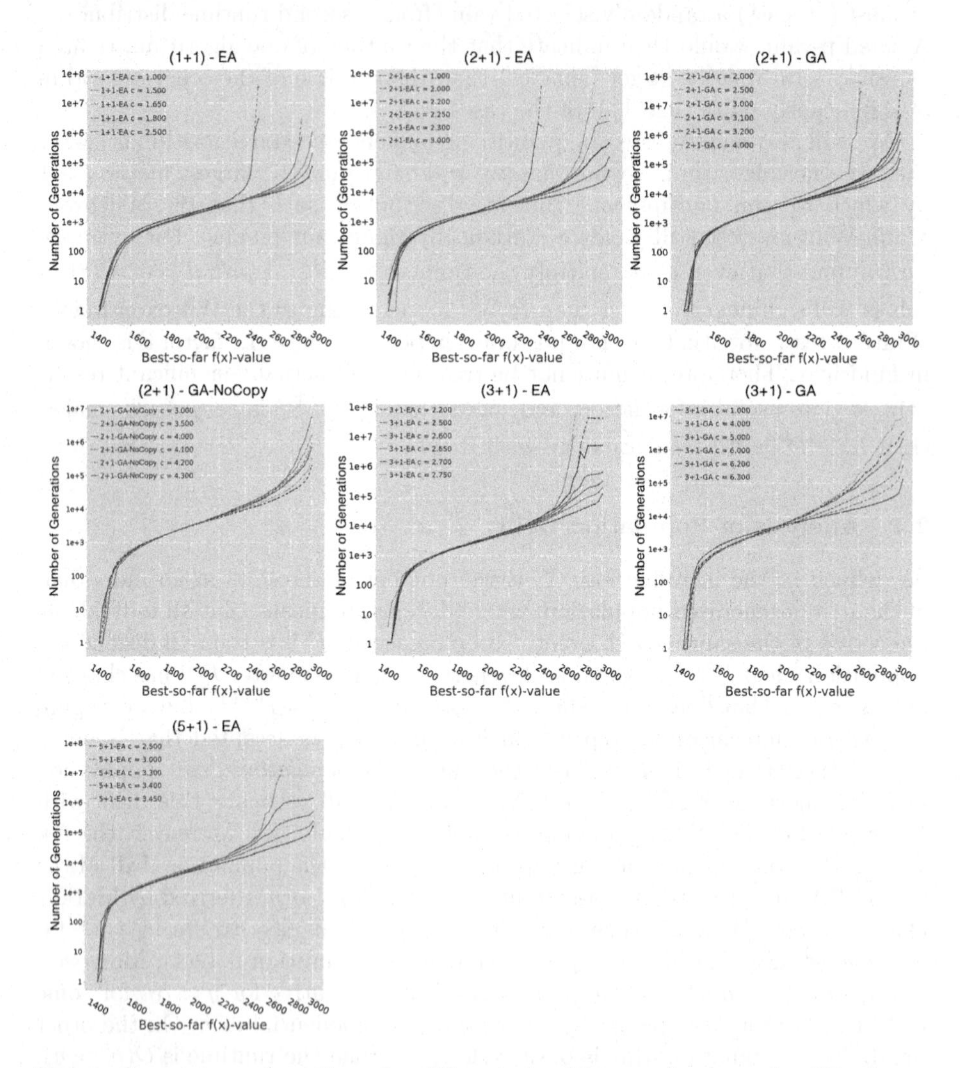

Fig. 1. Comparison of runtimes for different algorithms and values of c. We choose values of c that lie around the threshold for the respective algorithm. We plot both ERT and mean (both in the same color) if they differ significantly (mean is always the lower curve, see Sect. 3.1 for a more detailed explanation). (Color figure online)

ranges will be too big rather than too small. For the (1 + 1)-EA, we observe a threshold in $c_0 \in [1.5, 1.8]$ (in agreement with the theoretically derived threshold $c_0 \approx 1.59$ from [12]). For the (2+1)-EA, it seems to be within $c_0 \in [2.2, 2.3]$, for the (2+1)-GA in [3.0, 3.2] and for the (2+1)-GA-NoCopy in [4.1, 4.3]. Thus we obtain a clear ranking of the algorithms for $\mu \leq 2$, which is (1 + 1)-EA (worst), (2 + 1)-EA, (2 + 1)-GA, and (2 + 1)-GA-NoCopy. For the (3 + 1)-EA, the threshold appears to lie in the interval [2.5, 2.7], for the (3+1)-GA in [6.0, 6.3] and for the (5+1)-EA in [3.3, 3.45]. We were not able to find a threshold behavior for the (5+1)-GA for any $c < 20$. These results further confirm that the GA variants are performing massively better than its EA counterparts. Moreover, both for EAs and GAs, a larger population size shifts the threshold to the right. For GAs, this is analogous to theoretical results for monotone functions, but for EAs the effect goes in the opposite direction than for monotone functions, see the discussion in Sect. 1.

To validate the ranking for the algorithms with $\mu \leq 2$ statistically, we use the following comparisons. If we compare $R_{2.0}^{(1+1)\text{-EA}}$ to $d \cdot R_{2.0}^{(2+1)\text{-EA}}$ the Wilcoxon-Mann-Whitney test yields significant p-values ≤ 0.05 for every $d \leq 57.88$. We conclude that for mutation parameter $c = 2.0$, the (1 + 1)-EA is much slower than the (2 + 1)-EA. In the same manner, $d \cdot R_{2.5}^{(2+1)\text{-GA}}$ is significantly smaller than $R_{2.5}^{(2+1)\text{-EA}}$ for $d \leq 39.09$, and $d \cdot R_{3.5}^{(2+1)\text{-GA-NoCopy}}$ is significantly smaller than $R_{3.5}^{(2+1)\text{-GA}}$ for $d \leq 63.36$. This confirms the aforementioned ranking of the algorithms.

To establish intervals for the critical value c_0 of the algorithms with $\mu \leq 2$, we compare $R_{2.0}^{(1+1)\text{-EA}}$ to $d \cdot R_{1.5}^{(1+1)\text{-EA}}$, and find that the latter is significantly smaller for all $d \leq 38.84$. We interpret this huge drop in performance as strong indication that the threshold lies in the interval $c_0 \in [1.5, 2]$. Likewise, $R_{2.5}^{(2+1)\text{-EA}}$ is larger than $d \cdot R_{2.0}^{(2+1)\text{-EA}}$ for $d \leq 63.15$, $R_{3.5}^{(2+1)\text{-GA}}$ is larger than $d \cdot R_{3.0}^{(2+1)\text{-GA}}$ for $d \leq 29.00$, and $R_{4.5}^{(2+1)\text{-GA-NoCopy}}$ is larger than $d \cdot R_{4.0}^{(2+1)\text{-GA-NoCopy}}$ for $d \leq 29.59$, all with $p < 0.05$.

Degenerate Population Drift. We estimate the degenerate population drift by Monte-Carlo simulation on the (2 + 1)-EA with $c = 2.3$, which is slightly above the threshold. The results are visualized in Fig. 2. We can clearly see that the conditional population drift is negative in the area between 300 and 50 one-bits away from the optimum, but then becomes positive again when being less than 50 one-bits away from the optimum. We conclude that the hardest part for the (2 + 1)-EA is not around the optimum. We obtained similar results for the (3+1)-EA, also visualized in Fig. 2. This surprising result is similar to monotone functions [8,14], see the discussion in Sect. 1.

For the (2 + 1)-GA, the picture looks entirely different, as the drift is now strictly decreasing. We can only observe a negative drift area right at the optimum, starting from about 2900 1-bits. This behavior is similar to the (1+1)-EA and the (μ + 1)-GA-NoCopy, where the most difficult part is also close to the optimum (data not shown). Unfortunately, due to the large value of c_0, we were

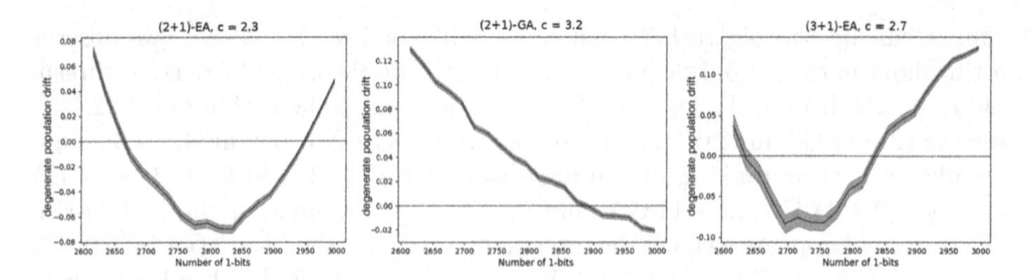

Fig. 2. Degenerate population drift for different algorithms and values of c just above the respective thresholds. The shaded area shows standard deviation.

not able to obtain a conclusive result for the $(3+1)$-GA within reasonable computation time.

For the $(2+1)$-EA and $(2+1)$-GA, we also derive exact asymptotic formulas for the population drift close to the optimum.For space reasons, we refer the reader to the full version [11].Define y to be the number of 0-bits. We compare the formula with the estimates via Monte Carlo simulation for different values of c, using $n = 3000, y = 1$, see Fig. 3. We can see that the curves match closely for small c, and that we get a moderate fit for larger c. We suspect that the deviations for the $(2+1)$-EA come from the expectation of the population drift being influenced by large but rare values of $X_i - X_{i+1}$ for large c. So the Monte Carlo simulations might be missing parts of this heavy tail. This tail is less heavy for the $(2+1)$-GA, since the probability to produce duplicates is always high, and thus degeneration happens quickly even for large c. In both cases, the curves agree perfectly in the *sign* of the population drift, which is our main interest. Negative drift at the optimum occurs at $c > 3.1$ for the $(2+1)$-GA, which matches well the threshold obtained from runtime simulations. However, as expected, there is *no such match* for the $(2+1)$-EA, where the threshold for the population drift at the optimum is 2.5 while the threshold for the runtime

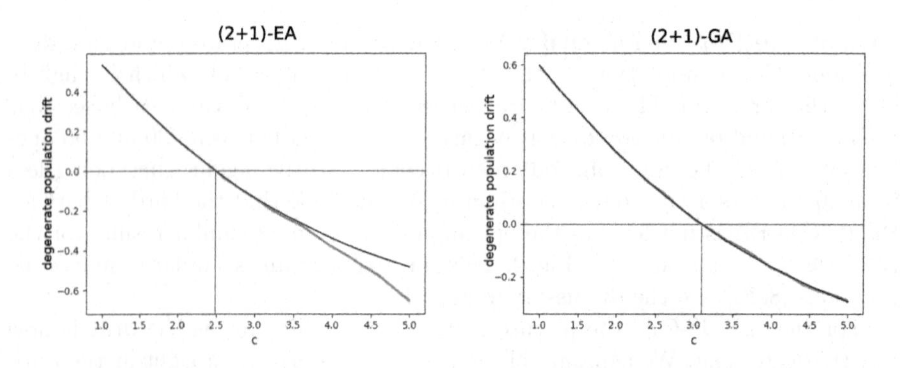

Fig. 3. Degenerate population drift for the $(2+1)$-EA and $(2+1)$-GA at the optimum. Blue: asymptotic formula. Orange: Monte Carlo simulation. (Color figure online)

is below 2.3. I.e., the $(2+1)$-EA already struggles for values of c for which optimization around the optimum is easy. This confirms that the hardest region for the $(2+1)$-EA (but not the $(2+1)$-GA) is not at the optimum, but a bit away from it.

4 Conclusions

We have studied the effect of population size and crossover for the dynamic DynBV benchmark. We have found that the algorithms generally profited from larger population size. Moreover, they profited strongly from crossover, even more so if we forbid crossovers between identical parents.

We have studied the case $\mu \leq 2$ in more depth. Remarkably, there is a strong qualitative difference between the $(2+1)$-EA and the $(2+1)$-GA. While for the latter one, the hardest region for optimization is close to the optimum (as one would expect), the same is not true for the $(2+1)$-EA. We believe that this is an interesting discovery. The only hint at such an effect on OneMax-like functions that we are aware of is for monotone functions [14]. However, the results in [14] predict that large population sizes hurt the $(\mu+1)$-EA, in opposition to our findings. Currently we are lacking any understanding of whether this comes from the small values of μ that we considered here, or whether it is due to the differences between monotone and dynamic linear functions.

For future work, there are many natural questions. We have chosen the $(\mu+1)$-GA to decide randomly between a mutation and a crossover step, but other choices are possible. Even with our formulation, it might be that the probability $1/2$ for choosing crossover has a strong impact. Also, we have exclusively focused on the limiting case DynBV, but dynamic linear functions are also interesting for less extreme case weight distributions. Finally, an interesting variant of dynamic linear functions or DynBV might not change the objective every round, but only every s rounds (our runtime simulation already supports this feature and is publicly available).

References

1. Antipov, D., Doerr, B., Fang, J., Hetet, T.: A tight runtime analysis for the $(\mu+\lambda)$ EA. In: Proceedings of the Genetic and Evolutionary Computation Conference (GECCO), pp. 1459–1466. ACM (2018)
2. Dang, D.C., et al.: Escaping local optima using crossover with emergent diversity. IEEE Trans. Evol. Comput. **22**(3), 484–497 (2017)
3. Dang, D.C., Jansen, T., Lehre, P.K.: Populations can be essential in tracking dynamic optima. Algorithmica **78**(2), 660–680 (2017). https://doi.org/10.1007/s00453-016-0187-y
4. Doerr, B., Happ, E., Klein, C.: Crossover can provably be useful in evolutionary computation. Theoret. Comput. Sci. **425**, 17–33 (2012)
5. Doerr, B., Jansen, T., Sudholt, D., Winzen, C., Zarges, C.: Mutation rate matters even when optimizing monotonic functions. Evol. Comput. **21**(1), 1–27 (2013)

6. Doerr, C., Wang, H., Ye, F., van Rijn, S., Bäck, T.: IOHprofiler: a benchmarking and profiling tool for iterative optimization heuristics. arXiv preprint arXiv:1810.05281 (2018). https://iohprofiler.github.io/
7. Jansen, T., Wegener, I.: The analysis of evolutionary algorithms-a proof that crossover really can help. Algorithmica **34**(1), 47–66 (2002). https://doi.org/10.1007/s00453-002-0940-2
8. Lengler, J.: A general dichotomy of evolutionary algorithms on monotone functions. IEEE Trans. Evol. Comput. (2019)
9. Lengler, J., Martinsson, A., Steger, A.: When does hillclimbing fail on monotone functions: An entropy compression argument. In: Proceedings of the Sixteenth Workshop on Analytic Algorithmics and Combinatorics (ANALCO), pp. 94–102. SIAM (2019)
10. Lengler, J., Meier, J.: Evolutionary algorithms in dynamic environments, source code, March 2020. https://github.com/JomeierFL/BachelorThesis
11. Lengler, J., Meier, J.: Large population sizes and crossover help in dynamic environments, full version. arXiv preprint (2020). http://arxiv.org/abs/2004.09949
12. Lengler, J., Schaller, U.: The (1+1)-EA on noisy linear functions with random positive weights. In: Proceedings of the Symposium Series on Computational Intelligence (SSCI), pp. 712–719. IEEE (2018)
13. Lengler, J., Steger, A.: Drift analysis and evolutionary algorithms revisited. Comb. Probab. Comput. **27**(4), 643–666 (2018)
14. Lengler, J., Zou, X.: Exponential slowdown for larger populations: the (μ+1)-EA on monotone functions. In: Proceedings of the 15th ACM/SIGEVO Conference on Foundations of Genetic Algorithms (FOGA), pp. 87–101. ACM (2019)
15. Lissovoi, A., Witt, C.: MMAS versus population-based EA on a family of dynamic fitness functions. Algorithmica **75**(3), 554–576 (2016)
16. Pinto, E.C., Doerr, C.: A simple proof for the usefulness of crossover in black-box optimization. In: Auger, A., Fonseca, C.M., Lourenço, N., Machado, P., Paquete, L., Whitley, D. (eds.) PPSN 2018. LNCS, vol. 11102, pp. 29–41. Springer, Cham (2018). https://doi.org/10.1007/978-3-319-99259-4_3
17. R Core Team: R: A language and environment for statistical computing. R Foundation for Statistical Computing, Vienna, Austria (2020). https://www.R-project.org/
18. Rohlfshagen, P., Lehre, P.K., Yao, X.: Dynamic evolutionary optimisation: an analysis of frequency and magnitude of change. In: Proceedings of the Genetic and Evolutionary Computation Conference (GECCO), pp. 1713–1720. ACM (2009)
19. Sudholt, D.: How crossover speeds up building block assembly in genetic algorithms. Evol. Comput. **25**(2), 237–274 (2017)
20. Sudholt, D.: The benefits of population diversity in evolutionary algorithms: a survey of rigorous runtime analyses. Theory of Evolutionary Computation. NCS, pp. 359–404. Springer, Cham (2020). https://doi.org/10.1007/978-3-030-29414-4_8
21. Witt, C.: Runtime analysis of the (μ+1)-EA on simple pseudo-Boolean functions. Evol. Comput. **14**(1), 65–86 (2006)
22. Witt, C.: Tight bounds on the optimization time of a randomized search heuristic on linear functions. Comb. Probab. Comput. **22**(2), 294–318 (2013)

Neuromemetic Evolutionary Optimization

Paweł Liskowski[1(✉)], Krzysztof Krawiec[2], and Nihat Engin Toklu[1]

[1] NNAISENSE, Lugano, Switzerland
{pawel,engin}@nnaisense.com
[2] Poznan University of Technology and CAMIL, Poznan, Poland
krawiec@cs.put.poznan.pl

Abstract. Discrete and combinatorial optimization can be notoriously difficult due to complex and rugged characteristics of the objective function. We address this challenge by mapping the search process to a continuous space using recurrent neural networks. Alongside with an evolutionary run, we learn three mappings: from the original search space to a continuous Cartesian latent space, from that latent space back to the search space, and from the latent space to the search objective. We elicit gradient from that last network and use it to perform moves in the latent space, and apply this Neuromemetic Evolutionary Optimization (NEO) to evolutionary synthesis of programs. Evaluation on a range of benchmarks suggests that NEO significantly outperforms conventional genetic programming.

Keywords: Optimization · Neural networks · Program synthesis

1 Motivations

The effectiveness of optimization algorithms is contingent on search operators. Unfortunately, it is often challenging to design a search operator that is both *effective* i.e., offers a reasonably high probability of producing candidate solutions of better quality, and *generic*, i.e. performs well for a range of problem instances or, ideally, for a range of problems. The reason is that the structure of the search space depends on the class of problems, and often on the particular instance of the problem. This is particularly true for combinatorial and discrete optimization, where even minimal modifications of solutions (e.g., swapping the visitation order of just two locations in a traveling-salesperson problem) can lead to very large changes in the value of objective function.

Substantial research effort has been thus invested in designing search operators in evolutionary computation (EC). However, conventional EC search operators like mutation and crossover are not necessarily expected to systematically improve the objective function. The onus of making progress is on the search algorithm, in particular on the selection scheme used to identify the promising parent solutions.

T. Bäck et al. (Eds.): PPSN 2020, LNCS 12269, pp. 623–636, 2020.
https://doi.org/10.1007/978-3-030-58112-1_43

A popular way of making EC search more thorough is to hybridize it with a local search, an approach known as *memetic algorithm*. In this study, we *augment EC operating in discrete/combinatorial search spaces with a gradient-based search.* This has been long considered impossible for two reasons: the discontinuous nature of discrete search spaces and non-differentiablility of the objective function. We tackle these issues by leveraging the capabilities of contemporary neural networks. For the former, we map discrete solutions to a continuous *latent search space* (a Cartesian space of fixed dimensionality). For the latter, we provide a *surrogate model* that maps the images of solutions in the latent space to the search objective. These mappings, by virtue of being continuous and thus end-to-end differentiable, allow us to *define a gradient-based search operator* and use it inside the EC search loop. We thereby address the challenges mentioned earlier: effectiveness, since gradient allows us to explicitly steer towards better candidate solutions, and genericity, since the mappings are trained alongside the EC search and can thus adapt to the problem.

As it will follow from the description of *Neuromemetic Evolutionary Optimization* (NEO), it can be applied to virtually any type of optimization problem (including continuous optimization). Here, we approach program synthesis with genetic programming (GP), a discrete optimization problem that is particularly difficult due to high variability of solution structure (programs being represented as labelled trees of arbitrary shape) and a complex objective function, which requires programs to be evaluated on examples.

2 The Method

To illustrate the universal character of NEO, we present it in abstraction from specific optimization problems, assuming only that a universe of candidate solutions $\mathcal{P}$ is given. The only parameter of the method is the dimensionality n of the *latent space* $\mathcal{Z} \equiv \mathbb{R}^n$. The key components of the system are:

1. The encoder $enc : \mathcal{P} \to \mathcal{Z}$, technically a neural network that embeds candidate solutions in $\mathcal{Z}$,
2. The decoder $dec : \mathcal{Z} \to \mathcal{P}$, a neural network that maps points in $\mathcal{Z}$ to $\mathcal{P}$; when composed with the encoder ($dec \circ enc$) forms an *autoencoder*,
3. The fitness surrogate model $\hat{f} : \mathcal{Z} \to \mathbb{R}$.

Details on these components are provided in Sect. 3, because 1. and 2. are domain-specific by depending on the representation of solutions in $\mathcal{P}$. For the time being we will assume them to be any end-to-end differentiable mappings.

The inputs to the algorithm are:

- a generator $g : () \to \mathcal{P}$ of random candidate solutions,
- a selection method $sel : 2^{\mathcal{P}} \to 2^{\mathcal{P}}$, and
- a domain-specific fitness function $f : \mathcal{P} \to \mathbb{R}$.

The algorithm starts with initializing the working population $P \subset \mathcal{P}$ using g. Each iteration of the search algorithm comprises the following phases.

Phase 1: Selection. We calculate the fitness values $f(p)$ for all $p \in P$ and use *sel* to select well-performing candidate solutions from P, obtaining so a parent/breeding pool $P_s = sel(P)$.

Phase 2: Model Update. In this step, each $p \in P_s$ is first passed through the encoder, which results in its *embedding* $z_p = enc(p)$. Then, the decoder is queried on z_p, resulting in a $\hat{p} = dec(z_p)$. Also the surrogate $\hat{f}$ is queried on z_p, producing $\hat{f}(z_p)$, a fitness estimate for p. We perform then a brief episode of gradient-based training of *enc*, *dec* and $\hat{f}$ using the following loss function:

$$L(p) = L_r(p, \hat{p}) + \alpha L_s(f(p), \hat{f}(p)), \tag{1}$$

where L_r is the reconstruction error, defined as a domain-specific metric representing the divergence of $\hat{p}$ from p, L_s is the surrogate model loss (defined later as mean squared error between its arguments, though in general an arbitrary error measure), and α is a hyperparameter. The details on the divergence measure, training procedure, hyperparameters, etc., are provided in Sect. 5.

Note that L_r and L_s both depend on the encoder, because *dec* and $\hat{f}$ share their input z_p. This may be best illustrated by spelling out Formula 1:

$$L(p) = L_r(p, dec(enc(p))) + \alpha L_s(f(p), \hat{f}(enc(p))). \tag{2}$$

Therefore, by minimizing L, the training algorithm attempts to tune the parameters of *enc* so that the latent representation $\mathcal{Z}$ enables good reconstructions for *dec* on one hand, and enables $\hat{f}$ to produce possibly accurate fitness estimates on the other. Furthermore, L_s acts as a regularizer, reducing the risk of overfitting and providing for better smoothness of the latent representation.

It is important to note that the training episode conducted in any given iteration of the algorithm is relatively short, as its goal is only to 'nudge' the current model so that it provides better predictions for candidate solutions in the current P_s. Performing a full-size training would be questionable, given the relatively small training sample available in P_s and its transient character.

Phase 3: Neural Breeding. In this phase, we produce at most one offspring for each $p \in P_s$ independently. First, we calculate the gradient of the surrogate fitness with respect to the latent representation at z_p:

$$\nabla_z \hat{f}(z_p) \equiv \frac{\partial \hat{f}}{\partial z}(z_p). \tag{3}$$

Note that this is the gradient of $\hat{f}$, i.e. the 'raw' *output* of the network, and not the gradient of $\hat{f}$'s *loss* with respect to the true fitness (i.e. L_s in Formula 1). Also, $\nabla_z \hat{f}$ is calculated with respect to the *inputs* of the surrogate model, and not its parameters (network weights). Therefore, $\nabla_z \hat{f}$ represents the direction of the local increase of $\hat{f}$ in $\mathcal{Z}$ in the vicinity of z_p. We explore z's neighborhood in the direction indicated by ∇_z by appointing a $z'_p \in \mathcal{Z}$ that is meant to represent the latent of the potential offspring:

$$z'_p = z_p + \eta \nabla_z \hat{f}(z_p). \tag{4}$$

Next, we query the decoder on z'_p, obtaining an offspring candidate $p' = dec(z'_p)$. This is essentially one step of gradient ascent in $\mathcal{Z}$ with respect to $\hat{f}$.

The practical difficulty of the above procedure lies in determination of the appropriate step size η. A too small η is likely to cause $dec(z'_p) = p$. A large η may lead to a p' that is completely unrelated to p. Also, a distant p' is not guaranteed to have a better fitness, because $\nabla_z \hat{f}$ describes only the direction of *local* changes of $\hat{f}$, so it is not a reliable basis for making long leaps in the search space. We therefore resort to an iterative procedure: we start with a small $\eta = \eta_0$ and keep incrementing it (in our experiments, we set $\eta_0 = 1$ and increase it by 1 in each iteration). Once η_i results in a $p' \neq p$, we quit the loop and declare p' the offspring of p.

Note that this procedure is entirely deterministic. Nevertheless, as all neural components are trained in each generation, different offspring can be produced from the same parent in different generations.

Discussion. It is interesting to note that, even though $\hat{f} \circ enc$ implements a surrogate model of f, we never assign the obtained fitness estimates to candidate solutions. This would clearly be of no value, as we calculate the exact fitness $f(p)$ for each candidate solution anyway. The sole purpose of $\hat{f}$ to (i) backpropagate the gradient of the error L_s of fitness estimates in Phase 2, in order to guide the training of *enc*, and (ii) to backpropagate the gradient ∇_z of the fitness estimates in Phase 3, to obtain the direction in which the latent representations of the offspring will be sought.

We anticipate the neural models to follow the dynamics of the population, as they are being updated according to the changes in P. However, we do not necessarily expect them to converge to perfect models, i.e., an autoencoder $dec \circ enc$ capable of perfect reconstruction of all possible candidate solutions and a surrogate model $\hat{f} \circ enc$ that accurately predicts program fitness. Such convergence is not essential here – we maintain these components only in order to *learn how to perform potentially improving moves in the latent space.*

3 NEO for Program Synthesis

As mentioned in Sect. 1, NEO is generic: for a given solution space $\mathcal{P}$, it requires only an encoder and decoder that respectively map $\mathcal{P}$ to the latent space and back from it. The remaining requirements listed at the beginning of Sect. 2, i.e., a generator of random candidate solutions, a selection operator, and a fitness function, can be chosen arbitrarily. In the following, we explain how NEO can be applied to program synthesis from examples with genetic programming (GP). We describe the architectures of the encoder and decoder we use in order to connect NEO with the space of programs, and the fitness surrogate model.

Encoder. As is common in GP, we assume that programs are represented as abstract syntax trees (ASTs). However, we transform them into equivalent sequences of tokens using the prefix notation. The encoder's task is thus to map sequences of tokens to a continuous latent vector in $\mathcal{Z}$. The process starts by independently passing each token x_t through an *embedding layer* e, a look-up table that stores vectors of learnable parameters for each token in the programming language. The encoder is a recurrent neural network that 'folds' $e(x_t)$s into the latent representation. At each time-step, the encoder receives both the embedding $e(x_t)$ of the current token x_t and the hidden state from the previous time step h_{t-1}, and outputs a new hidden state h_t:

$$h_t = enc(e(x_t), h_{t-1}) \tag{5}$$

The final hidden state h_T obtained after feeding *enc* with the last token in the sequence becomes the *latent vector* $z = h_T$ that constitutes a continuous representation of the entire sequence. We implement *enc* as a single-layer LSTM network [7].

Decoder. The purpose of a decoder, implemented by a separate recurrent neural network, is to learn the reverse mapping, i.e., to unfold a given latent vector z back into a sequence in autoregressive manner. As with the encoder, this is an iterative process that decodes tokens one at a time. The decoder starts by initializing the hidden state $s_0 = z$ and proceeds to compute the subsequent hidden states by employing a formula analogous to (5):

$$s_t = dec(d(y_t), s_{t-1}), \tag{6}$$

where s_t is decoder's hidden state and $d(y_t)$ is the embedding of decoder's input token y_t. The first token y_1 is always a special start-of-sequence token `<sos>`, while the subsequent ones are obtained in two ways:

- In training, we employ *teacher forcing* [21]. With probability p, $y_{t>1}$ is the actual ground-truth token, i.e., $y_t = x_t$. With probability $1 - p$, we use the token predicted in the previous iteration, i.e. $\hat{y}_{t-1}$ (Eq. 7), even if it is different from x_t. We set $p = 0.5$.
- In querying, y_t is the token produced in the previous iteration, i.e. $\hat{y}_{t-1}$.

To map s_ts to tokens we pass them through a softmax layer

$$\hat{y}_t = \text{softmax}(s_t), \tag{7}$$

producing in this way the predictive distribution over tokens, which we query with arg max whenever we need a categorical output.

The decoding process terminates when the special end-of-sequence token is produced, i.e. $\hat{y}_t$ = `<eos>`, or after a predefined number of time steps. Once decoding is completed, we obtain an output sequence $\hat{Y} = (\hat{y}_1, \hat{y}_2, ..., \hat{y}_{T'})$, that can be compared against the actual input program $X = (x_1, x_2, ..., x_T)$ to calculate loss. If the decoded sequence does not form a syntactically correct program, we fix it by replacing invalid sub-trees with randomly generated ones.

Attention Mechanism. The latent vector z is burdened with carrying all the information about the input sequence that is necessary for correct decoding.

Already the very first step taken by the decoder may cause some of that information to be lost, and this problem aggravates in successive iterations of *dec*. To address it, we allow the decoder to look at the entire input sequence (via its hidden states) at each decoding step using an *attention* mechanism [1].

At each time-step of decoding, we calculate a *context vector* c_t that helps predicting the current token y_t. First we calculate the *alignment weight vector* a_t based on the current target hidden state s_t and all source hidden states h_i:

$$a_t(i) = \frac{\exp(score(s_t, h_i))}{\sum_j \exp(score(s_t, h_j))}, \tag{8}$$

where *score* is a similarity function. Of several scoring function studied in the literature [15], we choose the arguably simplest one based on the dot product:

$$score(s_t, h_i) = s_t^\top h_i. \tag{9}$$

A context vector c_t is then computed as the weighted average, according to a_t, over all the source hidden states h_i. In the subsequent step, the information in the target hidden state s_t and the context vector c_t is fused together by a dense layer *att* that produces an attentional hidden state:

$$\bar{s_t} = \tanh(att([c_t; s_t])) \tag{10}$$

Finally, the vector $\bar{s_t}$ is fed through the softmax layer to make predictions, as in (7). Notice that the attention module we employ here is *global* in being applied to all source tokens (represented by particular hidden states h_t). In such a case the attentional decisions are independent of each other, which is clearly undesirable. To make alignment decision better-informed of the past alignments, we employ an *input-feeding* approach, in which the attentional vectors $\bar{s_t}$ are concatenated with inputs at the next time steps.

Surrogate Fitness Model. The fitness surrogate model $\hat{f}$ is realized as a feed-forward neural network; in this study, we use a network with two dense hidden layers, with details provided in the experimental section. Note that, in contrast to the encoder and decoder, the architecture of $\hat{f}$ is essentially domain-independent, as its signature is $\mathbb{R}^n \to \mathbb{R}$, i.e., it is not applied to the actual candidate solutions, but only to their latent representations. One may however decide to tune the hyperparameters of this network (e.g. the number of layers and their sizes) to the difficulty of the problem.

For program synthesis from examples, the surrogate approximates the conventional GP fitness function based on a set T of *tests cases* of the form (in, out):

$$f(p) = \frac{1}{|T|} \sum_{(in,out) \in T} [p(in) = out], \tag{11}$$

where *in* is an input to the program and *out* is the corresponding desired output, $p(in)$ is the output produced by a program p for *in*, and $[\cdot]$ is the Iverson bracket: $[false] = 0$, $[true] = 1$.

As readily apparent from (11), fitness values are normalized to [0, 1] range, so the output layer of the surrogate model features a single sigmoidal unit that produces predictions in the same range. For p's embedding $enc(p)$ and its fitness $f(p)$, the optimization of $\hat{f}$ aims at minimizing the least-square regression loss

$$L_s(p) = (f(p) - \hat{f}(enc(p))^2. \tag{12}$$

Semantic Surrogate Fitness Model. While the fitness function depicted in Eq. (12) is arguably the simplest characterization of program performance, more fine-grained metrics exist, such as those based on *program semantics*, meant as the vector of outputs produced by a program for all tests in T [11]:

$$sem(p) = (p(in))_{(in,\_) \in T} = (p(in_1), p(in_2), \ldots, p(in_n)). \tag{13}$$

It is easy to see that program semantics is by nature multi-dimensional: each element of $sem(p)$ characterizes p's behavior on a given test. This detailed behavioral characterization opened the door to major advancements in GP, and inspired us to propose a semantically augmented surrogate model that predicts the semantics of a program, rather than predicting just the fitness. Technically, very little changes in the surrogate neural network: the single sigmoidal unit in the output layer is replaced with $|T|$ units, each of them predicting the output of a program for a single test in T. In other words, rather than having one surrogate function, we now have $|T|$ of them, sharing the same hidden layers of the surrogate model network. During training, the losses are computed individually for each output using (12) and then aggregated. When using that network to produce the offspring in the latent space, we perform gradient ascent with respect to all $\hat{y}_i$s simultaneously. The major benefit of the above formulation is that such a semantic surrogate model is much better informed in training, and can be expected to converge better.

4 Related Work

At this point of this study, the justification for the 'neuromemetic' term in NEO should become more clear. Each application of the search operator in NEO engages one step of gradient-based local search. In EC terms, this can be interpreted as an adaptation acquired by an individual in its lifetime, considered as the hallmark of memetic algorithms [16]. Moreover, those adaptations are subsequently inherited by the offspring, which causes NEO to adhere to the Lamarckian-style memetic search (in contrast to the Baldwinian one, in which adaptations are used only to adjust the fitness of an individual, and do not propagate to the offspring). This links NEO to a rich body of metaheuristic literature, some of which involved also GP and semantic GP, e.g., memetic semantic genetic programming proposed in [5].

At its core, NEO aims to aid evolutionary search by eliciting a useful gradient from a surrogate model. Another possibility to improve evolutionary search is to prioritize it directly with neural networks. For instance, in [12], evolutionary program synthesis has been augmented with a network that estimates the likelihoods of instructions based on the samples of input-output behavior.

The other important aspect linking NEO to past work in EC are surrogate models [3]. Also known as *approximate fitness functions* and *response surfaces*, surrogates have been used in EC for a long time, however almost exclusively for the sake of reducing the otherwise high cost of fitness evaluation [8]. That is particularly common and relevant in applications where evaluation involves costly computer simulations. Occasionally, surrogate models have been also used in GP, e.g. in application to job-shop scheduling problems [6], where the proxy fitness value was acquired from the closest neighbor in a database of historical candidate solutions. Another interesting line of work involves applying non-negative matrix factorization to reduces the number of required interactions between programs and tests in GP [13,14].

NEO stands out from those approaches in being motivated not by the evaluation cost, but exclusively by information conveyed by the gradient: recall that the fitness estimates $\hat{f}$ provided by the surrogate are never assigned to candidate solutions, because we already know their true fitness values.

Side-stepping the challenges of combinatorial optimization by using an encoder/decoder network architecture was previously done in the fields of deep supervised learning [20], and reinforcement learning [2,17]. Our study demonstrates that such an architecture operates successfully within the evolutionary memetic optimization framework as well.

5 Experiment

We examine NEO on 16 discrete program synthesis tasks representing two domains: Boolean expressions and the domain of Algebra problems [19].

Benchmark Problems. In Table 1, we summarize the considered 16 benchmark program synthesis tasks, 6 from the Boolean domain and 10 from the Algebra domain: the instruction set used in each domain as well as the number of variables (program inputs) and tests T for every benchmark.

The Boolean benchmarks are defined as follows. For an v-bit comparator Cmp$v$, a program is required to return *true* if the $\frac{v}{2}$ least significant input bits encode a number that is smaller than the number represented by the $\frac{v}{2}$ most significant bits. For the majority Maj$v$ problems, *true* should be returned if more that half of the input variables are *true*. For Mul$v$, the state of the addressed input should be returned (6-bit multiplexer uses two inputs to address the remaining four inputs). In the parity Par$v$ problems, *true* should be returned only for an odd number of *true* inputs.

The Algebra benchmarks originate in Spector *et al.*'s work on evolving algebraic terms [19]. The admissible values of inputs and outputs are $\{0, 1, 2\}$. The programming language comprises a single binary instruction a_i that defines the

Table 1. Considered program synthesis benchmarks.

Domain	Instruction set	Problem	Variables	$\|T\|$
Boolean	and, nand, or, nor	Cmp6	6	64
		Cmp8	8	256
		Maj6	6	64
		Maj8	8	256
		Mux6	6	64
		Par5	5	32
Categorical	a_i	Dsc_i	3	27
		Mal_i	3	15

underlying algebra, and has different semantics in each of the considered benchmarks. For each of the five algebras, we consider two tasks. In the *discriminator term* tasks (*Dsc* in the following), the goal is to synthesize an expression $t^A(x, y, z)$ that is supposed to return x if $x \neq y$, and z otherwise. There are thus $3^3 = 27$ fitness cases. The *Mal* tasks consist in finding the so-called *Mal'cev term* m that satisfies $m(x, x, y) = m(y, x, x) = y$. Thus, the desired outputs for the tests where x, y, and z are all distinct are not determined, and are only 15 fitness cases in each *Mal* task. The motivation for the discriminator and Mal'cev term problems is their interest to mathematicians [4].

Settings. We compare two variants of NEO: the base variant that uses the scalar fitness function (Eq. (11)), and the semantic-aware variant (NEO-S). Our random program generator $g()$ is the ramped-half-and-half method with default parameters, while $sel()$ is the tournament of size 7. When querying the decoder, we set the maximum number of time-steps to 200. We provide two baseline algorithms. The first one is the Koza-style GP [10], also using scalar fitness function from Eq. (11). The second baseline algorithm, NEO-R, is meant to control for the impact of the guidance provided by the gradient back-propagated through the surrogate model. NEO-R diverges from NEO only in one detail: once the gradient vector $\nabla_z \hat{f}(z_p)$ has been calculated in (3), we randomly change its direction while preserving its length (technically, we draw a random unit vector and multiply it by $|\nabla_z \hat{f}(z_p)|$). We then proceed as in NEO. As a result, NEO-R appoints the parent solutions z_p for breeding in the same way as NEO, but produces a different offspring. However, as in NEO, that offspring is guaranteed to be different from its parent, thanks to the iterative scheme described after Formula (4).

We perform 20 independent runs per benchmark and per method. The search process stops when 200 generations elapse or an ideal program is found; the latter outcome is considered a success. All methods maintain a population of 1,000 programs.

Training. We jointly train the encoder, the decoder and the surrogate model by minimizing the combination of reconstruction loss L_r and the surrogate model

Table 2. Success rate of the best-of-run programs (per 20 runs) and average ranks of methods w.r.t. success rate.

PROBLEM	GP	NEO-R	NEO	NEO-S
Cmp6	0.50	0.55	0.80	0.85
Cmp8	0.05	0.00	0.20	0.25
Maj6	0.55	0.60	0.70	0.65
Maj8	0.00	0.00	0.00	0.00
Mux6	0.95	0.90	0.90	0.90
Par5	0.05	0.05	0.10	0.20
Dsc1	0.00	0.05	0.05	0.20
Dsc2	0.00	0.10	0.20	0.15
Dsc3	0.45	0.60	0.65	0.65
Dsc4	0.00	0.00	0.00	0.05
Dsc5	0.10	0.20	0.25	0.35
Mal1	0.85	0.80	0.85	0.85
Mal2	0.00	0.00	0.60	0.55
Mal3	0.65	0.60	0.80	0.80
Mal4	0.00	0.05	0.35	0.35
Mal5	0.85	0.85	0.90	0.95
RANK	3.31	3.22	1.91	1.56

loss L_s, with the trade-off parameters α in (1) set to 0.8. L_r is the cross-entropy loss between the target sequence y (same as the input sequence, i.e. $y \equiv x$) and the sequence $\hat{y}$ returned by the decoder. L_s is the mean squared error between $f(p)$ and $\hat{f}(enc(p))$. We train the complete model for 200 epochs in the first generation of the evolutionary loop in NEO, using the Adam optimizer [9] with a learning rate of 0.001 and 256 programs per batch. In subsequent generations, we decay the number of training epochs until we reach 10 epochs. Both encoder and decoder have 128 hidden units, and their hidden states are normalized to have unit length. The dimensionality of the embeddings is set to 64. We initialize RNN weights using the orthogonal method [18] with gain factor of $\sqrt{2}$. The forget gates biases are initialized to 0.9, and the input gate biases to −0.9. The output gates are initialized to be slightly open by sampling uniformly from the interval [0.1, 0.2]. The weights of the feed-forward surrogate networks are initialized using Xavier initialization.

Results. Table 2 presents the success rates of particular methods on each benchmark and the average ranks of methods (the lower, the better). Table 4 (left) provides statistical analysis of the resulting ranking in terms of Friedman's test. The pattern is clear: NEO and its semantic variant significantly outperform the baselines (GP and NEO-R), while remaining mutually non-dominated.

Table 3. Average and .95-confidence interval of number of nodes in the best-of-run program.

PROBLEM	GP		NEO-R		NEO		NEO-S	
Cmp6	184.6	±35.3	149.1	±22.5	145.2	±28.3	150.3	±20.9
Cmp8	284.5	±33.5	183.7	±20.6	182.3	±26.2	182.0	±19.8
Maj6	260.7	±34.6	166.4	±21.8	159.6	±27.3	157.6	±20.1
Maj8	368.8	±35.0	189.9	±22.2	185.1	±27.7	180.4	±20.6
Mux6	111.9	±26.3	122.2	±13.5	126.7	±19.4	122.7	±11.8
Par5	350.2	±36.1	180.5	±23.3	179.1	±28.8	179.3	±21.6
Dsc1	118.0	±33.2	108.4	±20.3	99.5	±25.8	110.5	±18.7
Dsc2	156.5	±33.7	139.6	±20.8	141.8	±26.4	141.6	±19.2
Dsc3	160.8	±29.5	152.1	±16.6	150.6	±22.2	154.9	±15.9
Dsc4	87.6	±27.1	55.6	±14.2	46.5	±19.8	56.3	±12.6
Dsc5	109.5	±31.7	120.3	±18.8	115.3	±24.4	115.2	±17.2
Mal1	106.1	±27.8	99.8	±15.1	103.3	±20.5	94.1	±13.4
Mal2	151.1	±27.8	145.9	±14.9	144.9	±20.5	136.0	±13.3
Mal3	143.7	±30.1	135.5	±17.3	134.7	±22.8	131.6	±15.7
Mal4	174.3	±31.6	150.2	±18.7	151.8	±24.3	141.5	±17.1
Mal5	78.2	±23.4	55.6	±10.6	54.2	±16.1	59.1	±8.9
RANK	3.62		2.44		2.00		1.94	

In particular, the gradient vector elicited from the surrogate model in NEO and NEO-S indeed tends to steer the evolution towards solutions of better quality: once we replace it with a random direction in NEO-R, the performance deteriorates. Notice that NEO-R explores the search space more thoroughly thanks to randomization. That exploration is however apparently less effective that exploitation of gradient. The slight improvement provided by NEO-S in comparison to NEO suggests that decomposing the loss function according to program's semantics may bring further improvements, though the evidence is not statistically significant there.

Table 4. Post-hoc analysis of Friedman's test conduced on ranks achieved by the configurations from Table 2 (left) and Table 3 (right). Significant values ($\alpha = 0.05$) are marked in bold. Friedman's p-value is 1.0×10^{-04}.

	GP	NEO-R	NEO	NEO-S
GP				
NEO-R	0.996			
NEO	**0.004**	**0.008**		
NEO-S	**0.000**	**0.000**	0.839	

	GP	NEO-R	NEO	NEO-S
GP				
NEO-R	**0.046**			
NEO	**0.002**	0.773		
NEO-S	**0.001**	0.692	0.999	

Tables 3 and 4 (right) presents analogous results for program size, i.e. the number of nodes (instructions) comprising the best-of-run program (whether successful or not). Also on this metric NEO performs better than the control setups, though this time only GP is statistically worse than the other algorithms.

6 Discussion and Future Work

The above results demonstrate that it is possible to elicit a useful gradient from a surrogate model and use it to generate candidate solutions, even if there are no obvious ways of defining a 'natural' coordinate system (and thus gradient) for the candidate solutions. The proposed encoder-decoder architecture learns an embedding of candidate solutions in a latent space that serves as an effective proxy for the original search space. Moreover, this can be achieved on-line, by learning the embedding and the surrogate model alongside with an evolutionary process. Therefore, even though the learned embedding and the surrogate do not capture the entirety of the original search space and the fitness function, NEO can exploit that information to make search effective.

Notice that the performance of NEO is far from obvious in presence of the attention mechanism (Sect. 3). Attention allows the decoder to access *directly* individual hidden states of the encoder, bypassing so the latent representation $\mathcal{Z}$ both in forward propagation (querying) and in backpropagation (training). As a result, $\mathcal{Z}$ is not bound anymore to convey all information about the processed programs. This makes it harder for the surrogate to provide robust fitness estimates. One can thus hypothesize that in absence of attention, the surrogate predictions could be even better.

Given the generic character of NEO, it seems natural to apply it to other domains of black-box optimization. One of them is continuous optimization; there, the constituent components of NEO can be likely implemented using even simpler neural architectures than the RNNs used here for processing the discrete structures of programs. For instance, both encoder and decoder can conveniently be realized with multi-layer dense neural networks, with each of the inputs of the encoder and the outputs of the decoder corresponding directly to one of the dimensions of the original search space $\mathcal{P}$.

Acknowledgments. This research has been partially supported by the statutory funds of Poznan University of Technology.

References

1. Bahdanau, D., Cho, K., Bengio, Y.: Neural machine translation by jointly learning to align and translate. arXiv preprint arXiv:1409.0473 (2014)
2. Bello, I., Pham, H., Le, Q.V., Norouzi, M., Bengio, S.: Neural combinatorial optimization with reinforcement learning. arXiv preprint arXiv:1611.09940 (2016)

3. Brownlee, A.E., Woodward, J.R., Swan, J.: Metaheuristic design pattern: Surrogate fitness functions. In: Proceedings of the Companion Publication of the 2015 Annual Conference on Genetic and Evolutionary Computation, GECCO Companion 2015, pp. 1261–1264. Association for Computing Machinery, New York (2015). https://doi.org/10.1145/2739482.2768499
4. Clark, D.M.: Evolution of algebraic terms 1: term to term operation continuity. Int. J. Algebra Comput. **23**(05), 1175–1205 (2013). https://doi.org/10.1142/S0218196713500227. http://www.worldscientific.com/doi/abs/10.1142/S0218196713500227d
5. Ffrancon, R., Schoenauer, M.: Memetic semantic genetic programming. In: GECCO 2015: Proceedings of the 2015 Annual Conference on Genetic and Evolutionary Computation, pp. 1023–1030. ACM, Madrid, 11–15 July 2015. https://doi.org/10.1145/2739480.2754697
6. Hildebrandt, T., Branke, J.: On using surrogates with genetic programming. Evol. Comput. **23**(3), 343–367 (2015). https://doi.org/10.1162/EVCO_a_00133
7. Hochreiter, S., Schmidhuber, J.: Long short-term memory. Neural Comput. **9**(8), 1735–1780 (1997). https://doi.org/10.1162/neco.1997.9.8.1735
8. Jin, Y., Olhofer, M., Sendhoff, B.: A framework for evolutionary optimization with approximate fitness functions. IEEE Trans. Evol. Comput. **6**, 481–494 (2002)
9. Kingma, D.P., Ba, J.: Adam: a method for stochastic optimization (2014)
10. Koza, J.R.: Genetic Programming: On the Programming of Computers by Means of Natural Selection. MIT Press, Cambridge (1992)
11. Krawiec, K.: Behavioral Program Synthesis with Genetic Programming. Studies in Computational Intelligence, vol. 618. Springer, Heidelberg (2016). https://doi.org/10.1007/978-3-319-27565-9
12. Liskowski, P., Bładek, I., Krawiec, K.: Neuro-guided genetic programming: prioritizing evolutionary search with neural networks. In: Proceedings of the Genetic and Evolutionary Computation Conference, pp. 1143–1150 (2018)
13. Liskowski, P., Krawiec, K.: Non-negative matrix factorization for unsupervised derivation of search objectives in genetic programming. In: Proceedings of the Genetic and Evolutionary Computation Conference, pp. 749–756 (2016)
14. Liskowski, P., Krawiec, K.: Surrogate fitness via factorization of interaction matrix. In: Heywood, M.I., McDermott, J., Castelli, M., Costa, E., Sim, K. (eds.) EuroGP 2016. LNCS, vol. 9594, pp. 68–82. Springer, Cham (2016). https://doi.org/10.1007/978-3-319-30668-1_5
15. Luong, M.T., Pham, H., Manning, C.D.: Effective approaches to attention-based neural machine translation. arXiv preprint arXiv:1508.04025 (2015)
16. Moscato, P.: On evolution, search, optimization, genetic algorithms and martial arts: towards memetic algorithms. Caltech Concurrent Computation Program C3P Report 826 (1989)
17. Nazari, M., Oroojlooy, A., Snyder, L., Takác, M.: Reinforcement learning for solving the vehicle routing problem. In: Advances in Neural Information Processing Systems, pp. 9839–9849 (2018)
18. Saxe, A.M., McClelland, J.L., Ganguli, S.: Exact solutions to the nonlinear dynamics of learning in deep linear neural networks. arXiv preprint arXiv:1312.6120 (2013)
19. Spector, L., Clark, D.M., Lindsay, I., Barr, B., Klein, J.: Genetic programming for finite algebras. In: Proceedings of the 10th Annual Conference on Genetic and Evolutionary Computation, GECCO 2008, pp. 1291–1298. Association for Computing Machinery, New York (2008). https://doi.org/10.1145/1389095.1389343

20. Vinyals, O., Fortunato, M., Jaitly, N.: Pointer networks. In: Advances in Neural Information Processing Systems, pp. 2692–2700 (2015)
21. Williams, R.J., Zipser, D.: A learning algorithm for continually running fully recurrent neural networks. Neural Comput. **1**(2), 270–280 (1989). https://doi.org/10.1162/neco.1989.1.2.270

Evolved Gossip Contracts - A Framework for Designing Multi-agent Systems

Nicola Mc Donnell(✉), Enda Howley, and Jim Duggan

National University of Ireland Galway, Galway, Ireland
nixmcd@gmail.com

Abstract. Multi-agent systems are systems of autonomous interacting agents acting in an environment to achieve a common goal. One of the most interesting aspects of multi-agent systems is when they exhibit emergence; where the whole is considered greater than the sum of the parts. Designing multi-agents systems is challenging, and doing this in an automated way has been described as "one of the holy grails of artificial intelligence and agent-based modelling". In previous research, we presented a novel decentralised cooperation protocol called Gossip Contracts (GC), which is inspired by Contract Net and Gossip Protocol. Here we present Evolved Gossip Contracts (EGC), a new framework which builds on GC and uses evolutionary computing to tailor GC to address a specific problem. We evaluate the EGC framework and the experimental results indicate that it is a promising approach for the automated design of decentralised strategies.

Keywords: Multi-agent systems · Decentralised algorithm · Genetic Programming

1 Introduction

Self-organisation is a phenomenon which is pervasive in nature, such as flocking and shoaling. It refers to a system which is composed of many elements, or agents, which exhibit coordinated behaviour without central control. Serugendo et al. [12] described self-organising within a system as "the mechanism or the process enabling a system to change its organisation without explicit external command during its execution time". One of the most interesting aspects of a self-organising, or multi-agent system, is when they exhibit emergence; where the whole is considered greater than the sum of the parts. More formally, an emergent property of a system is one which does not exist at the lower level of the system but arises from the interactions of elements at the lower level [5]. Perhaps the reason self organising systems are so commonplace in nature is that they are very resilient and effective in a rapidly changing environment as there is no single point of failure.

A Multi-agent system, MAS, is a system of autonomous interacting agents acting in an environment to achieve a common goal [2]. It has been suggested as a

T. Bäck et al. (Eds.): PPSN 2020, LNCS 12269, pp. 637–649, 2020.
https://doi.org/10.1007/978-3-030-58112-1_44

promising technique for problem domains that are distributed, complex and heterogeneous [16]. Davidsson et al. found that agent-based approaches performed better than centralised approaches when the problems are: inherently modular in nature, as opposed to monolithic; are very large in size; are in a rapidly changing environment and where communication cost and communication stability is low [3].

Van Berkel et al. observe that "Automatic algorithm generation for large-scale distributed systems is one of the holy grails of artificial intelligence and agent-based modelling. Inventing such an algorithm usually involves a tedious reasoning process for each individual idea" [15]. Trianni et al. describes the design problem - given the desired macroscopic collective behaviour, it is not trivial to design the corresponding individual behaviours and interaction rules [14].

In previous research, we presented a novel decentralised cooperation protocol called Gossip Contracts (GC) [8]. Here we show a new technique to tailor GC to address a specific problem using evolutionary computing.

This paper is organised as follows. In Sect. 2 we review existing results from the scientific literature. In Sect. 3 we formally define the problem and describe the model. In Sect. 4 we present the results and our analysis of the results. Finally, we report conclusions and outline future research in Sect. 5.

2 Related Work

The first exploration into automatic discovery of a decentralised algorithm was carried out by Mitchell et al. in 1994 [9]. They used a genetic algorithm (GA) to evolve cellular automata that could perform one-dimensional density classification. They noted "epochs of innovation" where new strategies are suddenly discovered and take hold in the GA population, as opposed to a steady gradual improvement.

Van Berkel [15] propose a Meta-Compiler for Automatic Algorithm Discovery (MAAD), a so-called "Global-to-local compiler" to automatically generate local interaction rules between the constituent system elements starting from a global description of the system behaviour. It uses Genetic Programming, specifically Grammatical Evolution, to discover algorithms that fulfil the search goal. They demonstrated success with this technique on five already known decentralised techniques, including Leader election and Churn estimation.

Ferrante et al. developed GESwarm, a new method for the automatic synthesis of collective behaviours for swarm robotics using Grammatical Evolution [4]. They demonstrated that 7 out of 10 outperformed the hand-coded foraging collective behaviour.

Babaoglu et al. developed Anthill, a multi-agent system framework to support the design, implementation and evaluation of Peer-to-peer (P2P) systems [1]. An Anthill P2P system is composed of a network of interconnected nests which can perform computations and host resources. User requests take the form of ants, autonomous agents that travel around the nest network trying to satisfy the request. The ants don't interact directly with one another, but by storing information in visited nests; this is akin to stigmergy in real ants. Anthill uses evolutionary techniques, specifically genetic algorithms, to design the ant algorithms. The parameters which define the behaviour of an ant algorithm are its "genetic code". They used Anthill to implement a document sharing application.

Zhong et al. proposed an evolutionary framework to automatically extract decision rules for agent-based crowd models, so as to reproduce an objective crowd behaviour. The problem of finding optimal decision rules from objective crowd behaviours is formulated as a symbolic regression problem solved using Gene expression programming [17].

3 Methodology

3.1 Gossip Contracts

Gossip Contracts (GC) is inspired by Contract Net (CNET) [13] and Gossip Protocol [6]. In CNET, the agent that tenders a task is called the manager and it sends a task announcement to potential contractors. Contractors can bid for a tender by sending a bid message back to the tendering manager. Tenders have a timeout which allows the contractors to wait until a tender is about to expire before evaluating if it should bid for it or not. This gives time for it to receive other, potentially better, tenders for which it could bid. In Gossip Contracts, managers are also contractors, and so better thought of as agents; thus a GC system is a multi-agent system. In CNET the tender messages are typically sent only to the managers neighbourhood. In GC the tender messages use a counter-based gossip protocol to travel much further across the system; so the agents have a higher number of tenders from which they can choose.

Each agent is connected to k other agents, its neighbourhood, through a random network topology. At the heart of GC are the messages sent between the agents. There are three types of message: a tender message, a bid message and an award message. Every message has a sender, a receiver, an expiration time and a contract id; a unique id for each distinct tender. A tender message also has a task to be completed and a guide price which indicates the relative importance of this tender to the tendering agent. Tender messages have a gossip counter and are forwarded to the receiving agents' neighbours if the counter is greater than zero; it is decremented each time the message is forwarded. This allows tender messages to travel much further across the network than their own neighbourhood. A bid message has a proposal for the completion of a task, and an offer price which indicates the relative importance of this bid to the bidding agent. Award messages are sent from the tendering agent to the contracting agent to whom the contract is being awarded.

Figure 1 shows the flow of messages between two Gossip Contract agents, a tendering agent and a bidding agent.

1. The tendering agent, Agent T, first decides whether or not to tender.
2. Should it decide to tender, it creates a tender message. This message contains the task to be tendered for and its guide price. Agent T sends it to all its neighbours. It also sets a timer which ends at the tender expiration time.

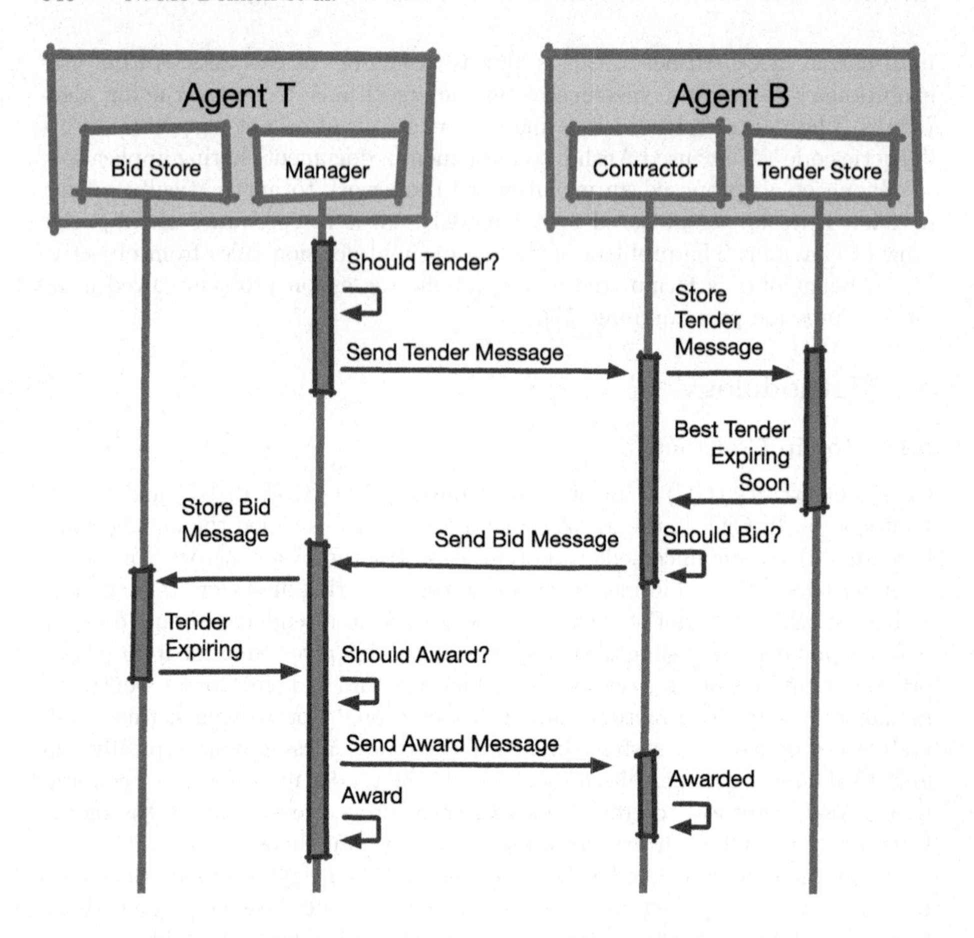

Fig. 1. Shows the flow of messages within Gossip Contracts

3. The bidding agent, Agent B, is a neighbour of Agent T and receives the tender message. Agent B stores the message in it's tender store. Agent B then sets a timer which ends just before the tender message is due to expire. It then checks if the gossip counter of the tender message is greater than zero, and if it is, it decrements the gossip counter by 1, and forwards the tender message to all its neighbours. Agent B receives many tender messages. All are stored, a timer is set for each and they are forwarded to its neighbourhood.
4. As each timer ends Agent B evaluates if the expiring tender message is the best tender message it has. To decide which tender is best, Agent B has a tender comparator which can order tenders. When two or more tenders have the same score as the best tender, one of them will be selected at random to be the best.
5. If the best tender is the tender who's timer is ending, Agent B next decides whether to bid or not.

6. Should it decide to bid, it sends a bid message to Agent T. This message contains its proposal and its offer price.
7. Agent T receives the bid message and stores the bid. Most likely, Agent T will receive bids from other agents which it also stores.
8. When the timer set by Agent T at the start of the tender ends, it evaluates all the bids it has received, and selects the best one. To decide which bid is best, Agent T has a bid comparator which can order bids. When two or more bids have the same score as the best bid, one of them will be selected at random to be the best. Agent T then decides if it should award the tender to this bid, or not.
9. Should Agent T decide to award, it sends an award message to Agent B, and carries out any steps needed to complete the award.
10. Should Agent B receive an award message, it completes any steps needed on its side to complete the award.

In order to get value from the Gossip Contracts paradigm a multi-agent system needs to be specialised to solve the problem at hand. The type of task that the agents tender for must be defined. Similarly, the type of proposal offered by a bidding agent is defined; it is common for the task type, T, and the proposal type, P, to be the same. The agent needs to define when it should tender, bid and award. It must be able to provide the task and guide for the creation of a tender message, and the proposal and offer for the creation of a bid message. It must be able to sort tender and bid messages so that the best ones can be identified. And it should set out the steps to be completed when a tender is awarded, both on the awarding, and the awarded agents.

There are also several parameters which can be used to tune the behaviour of the agent system. These include:

- The *neighbourhood size*, k – the number of agents to which tender messages are sent;
- The *tender gossip counter*, g –the number of times a tender message is gossiped or forwarded;
- The *tender timeout*, t_t – the time before a tender expires;
- The *bid buffer*, b_b – the time window before a tender is due to expire when a bid is sent; and
- The *award timeout*, t_a – the amount of time required to complete the awarding and awarded steps.

3.2 Evolving Gossip Contracts

Here we explain how GC was extended by introducing GP to automate the discovery or design of GC-based decentralised protocols. The extended framework is called Evolved Gossip Contracts (EGC). There are several methods which need to be implemented in order to specialise GC to solve a particular problem. Some of these methods should be human-implemented while others can be evolved. The four human-implemented methods are:

- *getTask()*, which defines the task to be completed;
- *getProposal()*, which defines the proposal to be offered; and
- *award()* and *awarded()*, which define how a task is awarded both on the tendering agent, *award()*, and bidding agent, *awarded()*.

They define broadly how the MAS will coordinate to solve the problem. The seven evolvable methods are:

- *shouldTender()*, to decide if an agent should tender;
- *getGuide()*, to generate a double representing the value of the tender to the tendering agent;
- *tenderToDouble()*, to convert a tender to a double so that all the tenders an agent has received can be ordered to select the best tender;
- *shouldBid()*, to decide if an agent should bid for the best tender;
- *getOffer()*, to generate a double representing the value of the bid to the bidding agent;
- *bidToDouble()*, to convert a bid to a double so that all the bids an agent has received for a tender can be ordered to select the best bid; and
- *shouldAward()*, to decide if a tendering agent should award to the best bid.

Note that the seven evolvable methods are not required to be evolved; they can be human-implemented if preferred.

The Evolved Gossip Contracts (EGC) framework was implemented in Java. Figure 2a shows the architecture for *GCAgents* and *EGCAgents*. The *GCAgent* uses the *PeerSim* network simulator [10] for the multi-agent simulations. The *EGCAgent* extends from the *GCAgent* and uses *ECJ* [11] as the GP evolutionary computation framework.

An alternative way to view the framework is to consider it from the perspective of the computational flow. Figure 2b shows the *ECJ* using the *Agent.params* file to initialise the GP system. It created the first generation of individuals which are evaluated, and so the GP process begins.

Each individual is evaluated as part of the GP process. Figure 2c shows the detail of the individual evaluation using the *PeerSim* simulator. It uses the Agent.properties file to initialise *PeerSim* and runs the simulation.

PeerSim is initiated using the Agent.properties config file. This initialises the *EGCAgents* in the multi-agent system. As the simulation progresses the evolvable methods of the *EGCAgents* are called. When this happens, the method evaluates its response by evaluating the appropriate GP tree as shown in Fig. 2d.

Once the multi-agent system simulation has completed it must be evaluated. Each different system will have a bespoke evaluation. This is done by extending the *EGCProblem* class and implementing an *evaluateSimulation()* method. However, one common evaluation criteria is the convergence speed, which is a measure of how long the multi-agent system ran before it converged. If it does not converge this will be the simulation end time defined by the *PeerSim* configuration property, *simulation.endtime*. To measure this we check for convergence at certain time intervals during the evaluation. This is done by extending the *ConvergenceTest* class and implementing the *isConverged()* method.

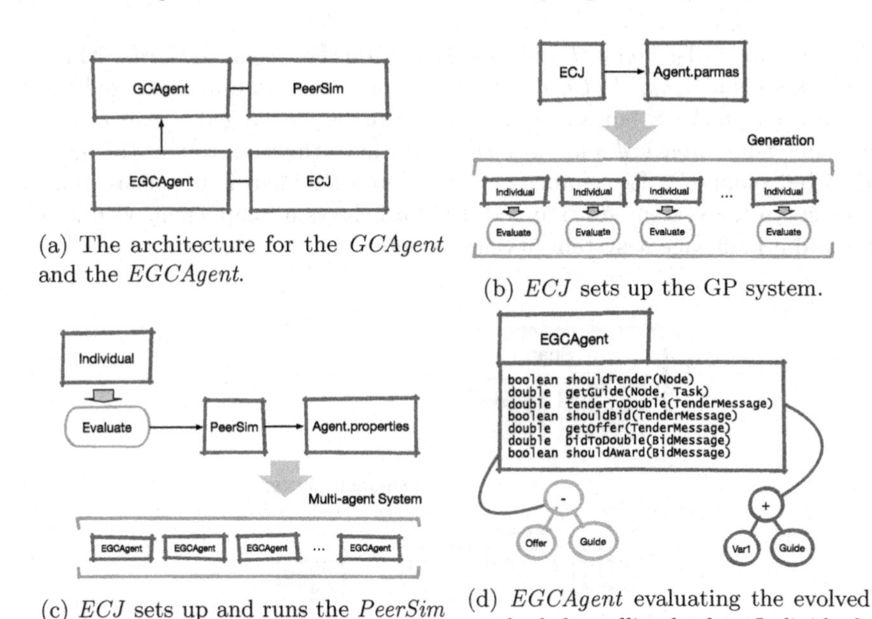

(a) The architecture for the *GCAgent* and the *EGCAgent*.

(b) *ECJ* sets up the GP system.

(c) *ECJ* sets up and runs the *PeerSim* system to evaluate an Individual.

(d) *EGCAgent* evaluating the evolved methods by calling back to Individuals GP trees.

Fig. 2. Computational flow of EGC.

We can assign weights to these two separate criteria: an evaluation weight for how well the multi-agent system solved the problem, and a convergence speed weight for how quickly the multi-agent system converged. These are defined as *eval.problem.evaluation_weight* and *eval.problem.convergence_speed_weight* in the *Agent.params* file; the default values are 1 and 0 respectively. We use a weighted sum model to calculate one fitness value from the two criteria.

In order to make the evolved trees easier to understand we provide a special *ECJ* Statistics class called *OccamStatistics*. It's named after the 14th century logician and Franciscan friar William of Ockham who stated the principle that "Entities should not be multiplied unnecessarily." *OccamStatistics* prints simpler but mathematically equivalent versions of the evolved trees. It calculates the simpler trees using two techniques. The first is to eliminate any redundant nodes. An example of a redundant node would be two not operators in a row, so given a boolean variable called, *maybe*, the subtree !!*maybe* can be simplified to *maybe*. The second is to check if a node evaluates to a constant value, and if so, it is replaced with the constant value. An example of a constant node would be the sum of two constant values, so the subtree $5 + 3$ can be simplified to 8.

3.3 The Snap Problem

To evaluate the EGC framework, we use a simple problem, the Snap Problem, which is loosely inspired by the card game Snap. It consists of a system of agents each with two cards: *myCard* and *swappableCard*. For each *myCard* within the system there is a *swappableCard* with the same value. *myCard* are fixed to the agent it's assigned to, but *swappableCards* can be exchanged between agents. When an agent's *myCard* and *swappableCard* have the same value, the agent is said to be 'snapped'. The challenge in the Snap Problem is to ensure that all the agents in the system are snapped. Figure 3 shows a Snap agent with its two cards and a multi-agent system of Snap agents.

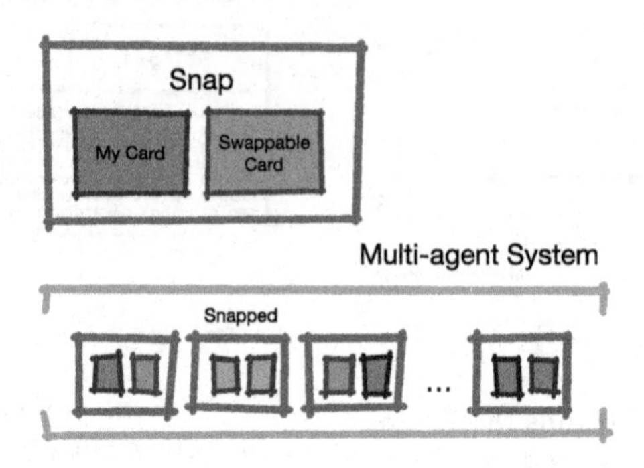

Fig. 3. Shows a Snap agent with its two cards and a multi-agent system of Snap agents.

To implement the Snap Problem with the EGC Framework we define a Snap agent which extends from the *EGCAgent*. The human-implemented methods are defined as follows:

- *getTask()*, returns the *swappableCard* of the tendering agent;
- *getProposal()*, returns the *swappableCard* of the bidding agent; and
- *award()* and *awarded()*, swap the *swappableCards* between the tendering and bidding agents.

To evaluate the evolved *SnapEGCAgent*, we defined a class, *SnapEGCProblem*, which implements the *evaluateSimulation()* method to return the fraction of not snapped agents; if there are a total of 200 *SnapEGCAgents* and 20 are snapped, then the evaluation is 0.9. The evaluation weight was 0.75 and the convergence speed weight was 0.25.

4 Experimental Results and Discussion

For each GP tree there is a unique set of nodes defined which are listed in Table 1. The GP Configuration is based on Koza's classic configuration included with *ECJ* is shown in Table 2. Table 3 shows the EGC configuration. The Snap agents are initialised and connected through a dynamic random network overlay which is rewired every 10 cycles.

Table 1. The functions and terminals and descriptions of the values they return.

<table>
<tr><th>Name</th><th>Description</th></tr>
<tr><td>true</td><td>The constant true</td></tr>
<tr><td>0</td><td>The constant zero</td></tr>
<tr><td>+</td><td>Add two inputs</td></tr>
<tr><td>−</td><td>Subtract two inputs</td></tr>
<tr><td>*</td><td>Multiply two inputs</td></tr>
<tr><td>div</td><td>Protected divide function, divides the first input by the second</td></tr>
<tr><td>==</td><td>Return true if the two inputs are equal</td></tr>
<tr><td><</td><td>Return true if the first input is less than the second input</td></tr>
<tr><td>></td><td>Return true if the first input is greater than the second input</td></tr>
<tr><td>!</td><td>Return true if the input is false, and false if the input is true</td></tr>
<tr><td>& &</td><td>Return true if and only if the two inputs are true</td></tr>
<tr><td>||</td><td>Return true if one of the two inputs are true</td></tr>
<tr><td>min</td><td>Return the smaller of the two inputs</td></tr>
<tr><td>max</td><td>Return the bigger of the two inputs</td></tr>
<tr><td>guide</td><td>The guide price in a tender message</td></tr>
<tr><td>offer</td><td>The offer price in a bid message</td></tr>
<tr><td>myCard</td><td>The agent's myCard</td></tr>
<tr><td>swappableCard</td><td>The agent's swappableCard</td></tr>
<tr><td>task</td><td>The task in the tender, bid or award messages, which is the tendering agent's swappableCard</td></tr>
<tr><td>proposal</td><td>The proposal in the bid or award messages, which is the bidding agent's swappableCard</td></tr>
</table>

Table 2. The GP configuration

Population size	500
Generations	12
Crossover probability	0.9
Mutation probability	0.1
Tree initialisation method	Ramped half-and-half with minimum depth of 2 and maximum depth of 6
Selection method	Tournament selection with parsimony pressure tournament size of 7

Table 3. The EGC configuration

Number of runs	20
Number of cycles	75
Time steps per cycle	300
Number of Agents	200
Neighbourhood size	4
Tender gossip counter, g	4
Tender timeout, t_t	4
Bid buffer, b_b	1
award timeout, t	1

In this section we evaluate the experimental results from two different perspectives. Firstly, we look at how the GP performed, and secondly we explore the best evolved protocol and analyse how it works. To evaluate the GP performance we consider three metrics: the fitness, the convergence cycles and the number of snaps. The fitness is a number between 0 and 1, where 1 is perfect fitness. The average number of cycles is the number of cycles needed to reach convergence averaged across the 20 runs. The number of snaps is the average, across the 20 runs, of the number of snapped agents at the end of the simulations.

Figure 4 shows the results of the GP analysis for the experiment. The top left graph entitled "Fitness by generation" shows how the fitness of the individuals changed for each generation. The fitness of each individual in the population is represented by a grey cross. A box plot of the statistical distribution has been overlaid. The best fitness for each generation is highlighted by a red star. The top right graph entitled "Convergence time by generation" shows the mean convergence time in cycles across the 20 runs for each generation. The individual mean convergence time is represented by a grey cross with a box plot of the statistical distribution overlaid. The convergence time for the fittest individual for each generation is highlighted with a red star. The bottom left graph entitled "Number of snaps by generations" shows the mean number of snaps across the 20 runs for each generation. The individual mean number of snaps are represented by a grey cross with a box plot of the statistical distribution overlaid. The number of snaps for the fittest individual for each generation is highlighted with a red star. From these three graphs we get a sense of how fitness, convergence time and number of snaps are changing for each generation, but we can not ascertain if the individuals with the high number of snaps also have lower convergence times. The bottom right graph entitled "Convergence time by number of snaps" addresses this by showing how for each individual, represented as an x, the convergence time and number of snaps are correlated. There is one sub-graph per generation and the fitness is represented by the colour of the individual cross.

Analysing the graphs we can see the GP making steady progress finding ever fitter individuals. There is a marked improvement in overall fitness of the

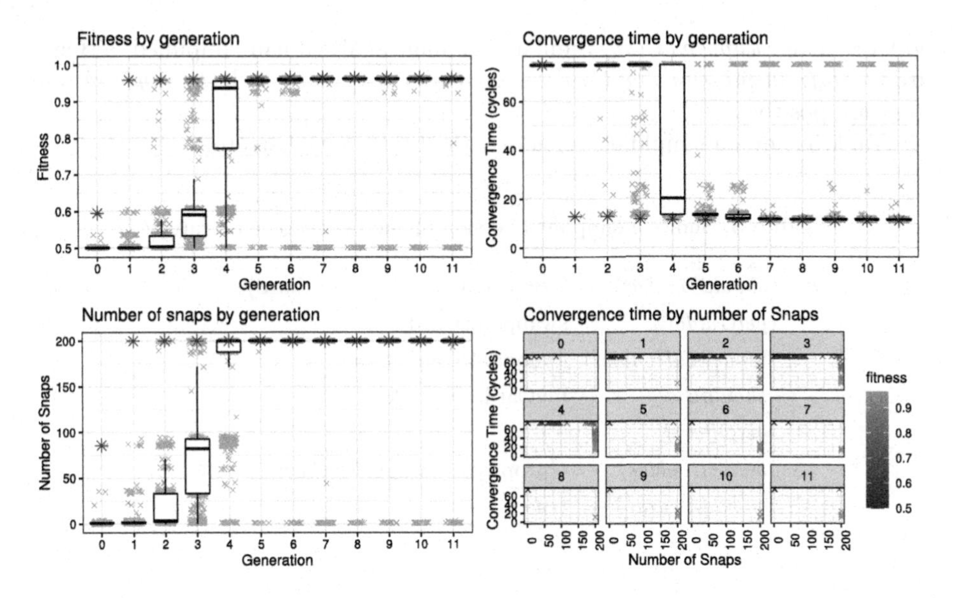

Fig. 4. Shows the results for the experiment.

population from generation 2 to generation 5. In generation 1, an individual achieves 200 snaps. The mean min convergence time drops below 11.5 cycles in generation 6. There is no improvement in fitness after generation 7.

Table 4. The best evolved strategy.

shouldTender()	!(swappableCard == myCard)
getGuide()	swappableCard
tenderToDouble()	(swappableCard + 0.0) max (swappableCard + task)
shouldBid()	myCard == task
getOffer()	(((myCard min task) min task) min 0.0) min (0.0 - 0.0)
bidToDouble()	(proposal - (swappableCard max swappableCard)) - task
shouldAward()	True & & True

Table 4 shows the best evolved strategy. It achieved a fitness of 0.96. The strategy can be simplified using the *OccamStatistics* approach described in Sect. 3.2 and further human analysis of the methods. For example, the *OccamStatistics* approach simplified tenderToDouble from *(swappableCard + 0.0) max (swappableCard + task)* to *max(swappableCard, (swappableCard + task))*, and on further human analysis we simplified it further to *swappableCard + task* because we know *swappableCard* and *task* are positive, so *swappableCard + task* is always greater than *swappableCard*. Similarly, human analysis can simplified getOffer from *min(min(min(min(myCard, task), task), 0.0), 0.0)* to 0, as *myCard* and *task* are positive.

Table 5 shows a simplified version of the best evolved strategy.

Table 5. Shows a simplified version of the best evolved strategy.

shouldTender()	!(swappableCard == myCard)
getGuide()	swappableCard
tenderToDouble()	swappableCard + task
shouldBid()	myCard == task
getOffer()	0
bidToDouble()	((proposal - swappableCard) - task)
shouldAward()	true

The evolved strategy is quite straightforward. It always tenders when the agent is not already snapped, bids if the card offered is the card it needs to snap, and always awards. The *guide* or *offer* values have no impact on the strategy and could simply have been removed. The *tenderToDouble()* returns *swappableCard + task*, but as *swappableCard* is the same for all tenders the lowest *task* is always selected. This is not ideal and a better strategy would be to favour the tender which offers a *task* which matches *myCard* as the *shouldBid()* method invoked later will only return *true* when *myCard == task*. As only one bid is ever received, the *bidToDouble()* function has no impact.

5 Conclusion

Automated development of decentralised algorithms was described by Van Berkel [15] as "one of the holy grails of AI and agent-based modelling". Here we presented the Evolved Gossip Contracts (EGC) framework which builds on Gossip Contracts (GC), a decentralised cooperation protocol, by using Genetic Programming to achieve this very objective. From the experimental results we can see clearly that EGC is a robust and promising approach.

In further work, we aim to evaluate EGC on more complex problems, such as the Bin Packing Problem from the field of Operations Research, which is an NP-Hard combinatorial optimisation problem [7].

References

1. Babaoglu, O., Meling, H., Montresor, A.: Anthill: a framework for the development of agent-based peer-to-peer systems. In: Proceedings 22nd International Conference on Distributed Computing Systems, pp. 15–22. IEEE (2002)

2. Balaji, P., Srinivasan, D.: An introduction to multi-agent systems. In: Srinivasan, D., Jain, L.C. (eds.) Innovations in Multi-Agent Systems and Applications - 1. Studies in Computational Intelligence, vol. 310, pp. 1–27. Springer, Heidelberg (2010). https://doi.org/10.1007/978-3-642-14435-6_1
3. Davidsson, P., Persson, J.A., Holmgren, J.: On the integration of agent-based and mathematical optimization techniques. In: Nguyen, N.T., Grzech, A., Howlett, R.J., Jain, L.C. (eds.) KES-AMSTA 2007. LNCS (LNAI), vol. 4496, pp. 1–10. Springer, Heidelberg (2007). https://doi.org/10.1007/978-3-540-72830-6_1
4. Ferrante, E., Duéñez-Guzmán, E., Turgut, A.E., Wenseleers, T.: Geswarm: Grammatical evolution for the automatic synthesis of collective behaviors in swarm robotics. In: Proceedings of the 15th Annual Conference on Genetic and Evolutionary Computation, pp. 17–24 (2013)
5. Gershenson, C.: Design and control of self-organizing systems. CopIt Arxives (2007)
6. Jelasity, M., Voulgaris, S., Guerraoui, R., Kermarrec, A.M., Van Steen, M.: Gossip-based peer sampling. ACM Trans. Comput. Syst. (TOCS) **25**(3), 8-es (2007)
7. man Jr., E.C., Garey, M., Johnson, D.: Approximation algorithms for bin packing: a survey. In: Approximation Algorithms for NP-Hard Problems, pp. 46–93 (1996)
8. Mc Donnell, N., Howley, E., Duggan, J.: Dynamic virtual machine consolidation using a multi-agent system to optimise energy efficiency in cloud computing. Future Gener. Comput. Syst. **108**, 288–301 (2020)
9. Mitchell, M., Crutchfield, J.P., Hraber, P.T.: Evolving cellular automata to perform computations: mechanisms and impediments. Phys. D **75**(1–3), 361–391 (1994)
10. Montresor, A., Jelasity, M.: PeerSim: a scalable P2P simulator. In: 2009 IEEE Ninth International Conference on Peer-to-Peer Computing, pp. 99–100. IEEE (2009)
11. Scott, E.O., Luke, S.: ECJ at 20: toward a general metaheuristics toolkit. In: Proceedings of the Genetic and Evolutionary Computation Conference Companion, pp. 1391–1398 (2019)
12. Serugendo, G.D.M., Gleizes, M.P., Karageorgos, A.: Self-organization in multi-agent systems. Knowl. Eng. Rev. **20**(2), 165–189 (2005)
13. Smith, R.G.: The contract net protocol: high-level communication and control in a distributed problem solver. IEEE Trans. Comput. **12**, 1104–1113 (1980)
14. Trianni, V., Nolfi, S.: Engineering the evolution of self-organizing behaviors in swarm robotics: a case study. Artif. Life **17**(3), 183–202 (2011)
15. Van Berkel, S., Turi, D., Pruteanu, A., Dulman, S.: Automatic discovery of algorithms for multi-agent systems. In: Proceedings of the 14th Annual Conference Companion on Genetic and Evolutionary Computation, pp. 337–344 (2012)
16. Weiss, G.: Multiagent Systems: A Modern Approach to Distributed Artificial Intelligence. MIT Press (1999)
17. Zhong, J., Luo, L., Cai, W., Lees, M.: Automatic rule identification for agent-based crowd models through gene expression programming. In: Proceedings of the 2014 International Conference on Autonomous Agents and Multi-Agent Systems, pp. 1125–1132 (2014)

A SHADE-Based Algorithm for Large Scale Global Optimization

Oscar Pacheco-Del-Moral(✉) and Carlos A. Coello Coello

Computer Science Department, CINVESTAV-IPN (Evolutionary Computation Group), Av. IPN No. 2508, Col. San Pedro Zacatenco, México D.F. 07300, Mexico
delmoral313@gmail.com, ccoello@cs.cinvestav.mx

Abstract. During the last decade, large-scale global optimization has been a very active research area not only because of its many challenges but also because of its high applicability. It is indeed crucial to develop more effective search strategies to explore large search spaces considering limited computational resources. In this paper, we propose a new hybrid algorithm called *Global and Local search using Success-History Based Parameter Adaptation for Differential Evolution* (GL-SHADE) which was specifically designed for large-scale global optimization. Our proposed approach uses two populations that evolve differently allowing them to complement each other during the search process. One is in charge of exploring the search space while the other is in charge of exploiting it. Our proposed method is evaluated using the CEC'2013 large-scale global optimization (LSGO) test suite with 1000 decision variables. Our experimental results show that the new proposal outperforms one of the best hybrid algorithms available in the state of the art (SHADEILS) in the majority of the test problems adopted while being competitive with respect to several other state-of-the-art algorithms when using the LSGO competition criteria adopted at CEC'2019.

Keywords: Differential Evolution · SHADE · Large scale · Global optimization · Hybrid algorithms

1 Introduction

The general (single-objective) global optimization problem is defined as follows[1]:

$$\begin{array}{ll} \text{Minimize} & f(\boldsymbol{x}) \\ \text{Subject to} & lb_j \leq x_j \leq ub_j, \quad j = 1, 2, ..., D \end{array} \tag{1}$$

[1] Without loss of generality, we will assume minimization.

The first author acknowledges support from CONACyT and CINVESTAV-IPN to pursue graduate studies in Computer Science. The second author gratefully acknowledges support from CONACyT grant no. 2016-01-1920 (*Investigación en Fronteras de la Ciencia 2016*) and from a SEP-Cinvestav grant (application no. 4).

T. Bäck et al. (Eds.): PPSN 2020, LNCS 12269, pp. 650–663, 2020.
https://doi.org/10.1007/978-3-030-58112-1_45

where $\boldsymbol{lb}, \boldsymbol{ub} \in \mathbb{R}^D$ are the lower bound and the upper bound of the decision variables $\boldsymbol{x}$, respectively. $f : \mathbb{R}^D \rightarrow \mathbb{R}$ is the objective function. The feasible solution space is defined as: $\Omega = \{\boldsymbol{x} \in \mathbb{R}^D | lb_j \leq x_j \leq ub_j, \forall j \in \{1, 2, 3, ..., D\}\}$. When $D \geq 1000$, this is called large scale global optimization (LSGO) [11] and because of the limitations of mathematical programming techniques (particularly when dealing with highly nonlinear objective functions [2,17]), the use of metaheuristics (particularly evolutionary algorithms) has become relatively popular [5]. Differential Evolution (DE) [1,14] is a metaheuristic designed for continuous search spaces that has been very successful in solving a variety of complex optimization problems. However, as happens with other meta-heuristics, the performance of DE quickly deteriorates as we increase the number of decision variables of the problem (the so-called "curse of dimensionality") [3]. Additionally, the properties and conditions of the fitness landscape may change (e.g., going from unimodal to multimodal) [2,3].

Many current approaches for large-scale global optimization are based on cooperative coevolution (CC), but several non-CC have also been proposed in recent years [2,4,5,11,12,17]. CC consists in decomposing the original large scale problem into a set of smaller subproblems which are easier to solve (i.e., it is a divide-and-conquer approach) [2,12,17]. On the other hand, the non-CC approaches try to solve the large scale problem as a whole. Most of these non-CC approaches are hybrid schemes that combine several metaheuristics that complement each other in order to overcome their limitations. In fact, several researchers [2,5,11] have combined DE with non-population-based local search methods to boost its overall performance.

In this paper, we propose a new non-CC approach which is based on the so-called *Success-History Based Parameter Adaptation for Differential Evolution* (SHADE) algorithm [15]. Our proposed algorithm consists of three stages: (1) initialization, (2) global search and (3) local search. During the initialization stage a gradient-free non-population-based local search method is applied to one of the best individuals generated in order to make an early enhancement. Afterwards, the global and local search stages are iteratively repeated one after another. Our proposed approach consists of two populations that collaborate with each other since the first population presents a search scheme specialized in exploration (thus carrying out the global search stage) and the second one presents a search engine specialized in exploitation (carrying out the local search stage). The first population evolves according to SHADE's algorithm and the second one according to a new developed SHADE's variant which we've named eSHADE$_{\text{ls}}$. The communication between these two populations is done via a simple migration protocol (happening when switching from the global to the local search stage or vice versa). Our proposal is considered a hybrid and non-CC algorithm since it combines an evolutionary algorithm with a local search method (used just once during the initialization stage) and solves the problem as a whole (decomposition never happens). The performance of our proposed algorithm, referred to as *Global and Local search using SHADE* (GL-SHADE), is evaluated on the CEC'2013 large-scale global optimization (LSGO) test suite.

The remainder of this paper is organized as follows. In Sect. 2, we provide the background required to understand the rest of the paper. Section 3 describes in detail our proposal. Our experimental design and our results are provided in Sect. 4. Finally, our conclusions and some paths for future research work are provided in Sect. 5.

2 Background

2.1 Differential Evolution

Differential Evolution (DE) was originally proposed by Storn and Price in 1995 [14,18]. DE is a stochastic population-based evolutionary algorithm (EA) that has shown to perform well in a variety of complex (including some real-world) optimization problems [1,14]. DE has three control parameters: NP (population size), F (mutation scale factor) and Cr (crossover rate) [3]. It is well known that the performance of DE is very sensitive to these parameters [1]. Like other EAs, an initialization phase is its first task [18]. After that, DE adopts three main operators: mutation, recombination, and selection.

Algorithm 1: Standard Differential Evolution variant $DE/rand/1/bin$ [14].

Data: max_{FEs} , NP , Cr , F
Result: $\boldsymbol{x} \in P$ such that $f(\boldsymbol{x}) \leq f(\boldsymbol{y}), \forall \boldsymbol{y} \in P \setminus \{\boldsymbol{x}\}$

```
1  Create a population P randomly and uniformly distributed over Ω ;
2  current_FEs = 0 + NP;
3  while current_FEs < max_FEs do                               // stopping criterion
4    for i = 0 to i < NP do                      // for every individual in the population
5      Take r1, r2, r3 ∈ [0, NP − 1] randomly ;              // r1 ≠ r2 ≠ r3 ≠ i, rn ∈ ℕ
6      v_i = x_r1 + F * (x_r2 − x_r3) ;              // mutation: F ∈ [0.0, 2.0] and x_rn ∈ P
7      j_rand = randInt(0, D − 1) ;                                      // j_rand ∈ ℕ
8      for j = 0 to j < D do                                  // starting binomial crossover
9        if flip(Cr) || j == j_rand then                              // Cr ∈ [0.0, 1.0]
10         u_i,j = v_i,j ;
11         If u_i,j out of boundary then get it back to the feasible region;
12       else
13         u_i,j = x_i,j ;
14     P_i^next ← fittest between u_i and x_i ;  // selection: take u_i if f(u_i) ≤ f(x_i), x_i otherwise
15     current_FEs + +;
16   P ← P^next ;                                            // advance generation and repeat
17 return (x ∈ P such that f(x) ≤ f(y), ∀y ∈ P \ {x}) ;                          // fittest
```

The initialization phase (see Algorithm 1, line 1) consists in randomly scattering NP guesses (points) over the search space as follows: $x_{i,j} = lb_j + (ub_j - lb_j) * rnd(0,1)$, where $x_{i,j}$ represents the jth gene of the ith individual [18].

The mutation and recombination operators are applied to generate a new trial vector ($\boldsymbol{u_i}$). During mutation, DE creates a new obtained candidate solution called a *donor* solution ($\boldsymbol{v_i}$). The sort of mutation and recombination operator to be adopted is defined based on the DE version that we use. The notation $DE/\alpha/\beta/\gamma$ is adopted to indicate the particular DE variant to be adopted [1]: α specifies the base vector, β is the number of difference vectors used, and γ denotes the type of recombination to be used [3]. Algorithm 1 presents the classical DE variant, called $DE/rand/1/bin$. In this scheme, for the mutation operator, three mutually exclusive individuals ($\boldsymbol{x_{r1}}$, $\boldsymbol{x_{r2}}$ and $\boldsymbol{x_{r3}}$) are randomly selected where

$\boldsymbol{x}_{r1}$ is perturbed using the scaled difference between $\boldsymbol{x}_{r2}$ and $\boldsymbol{x}_{r3}$ in order to obtain $\boldsymbol{v}_i$ (lines 5–6). For the recombination operator, genes are inherited from $\boldsymbol{v}_i$ and from the target vector ($\boldsymbol{x}_i$) and a binomial (*bin*) distribution is adopted in order to obtain $\boldsymbol{u}_i$ (lines 7–13) where at least one gene must be inherited from $\boldsymbol{v}_i$ (see line 9). The last step is the selection operator (line 14) where the fittest between $\boldsymbol{u}_i$ and $\boldsymbol{x}_i$ is chosen to represent the next generation's $\boldsymbol{x}_i$.

Several DE variants exist (see [1,14]) and the selection of any of them will influence the search behavior of DE in different ways. Since there are only two types of recombination (exponential (*exp*) and binomial (*bin*)) the mutation strategy is really the one that better identifies the behavior of a particular DE scheme [1].

2.2 An Enhanced Differential Evolution Algorithm Based on Multiple Mutation Strategies

The *Enhanced Differential Evolution Algorithm Based on Multiple Mutation Strategies* (abbreviated as EDE by its authors) [18] was proposed as an enhancement of the $DE/best/1/bin$ scheme, aiming to overcome the tendency of this scheme to present premature convergence. The core idea of EDE is to take advantage of the direction guidance information of the best individual produced by the $DE/best/1/bin$ scheme, while avoiding being trapped into a local optimum. In the EDE algorithm, an opposition-based learning initialization scheme is combined with a mutation strategy composed of two DE variants ($DE/current/1/bin$ and $DE/pbest/bin/1$) aiming to speed up convergence and to prevent DE from clustering around the global best individual. EDE also incorporates a perturbation scheme for further avoiding premature convergence. Algorithm 2 shows the way in which EDE works.

Algorithm 2: An Enhanced Differential Evolution Algorithm Based on Multiple Mutation Strategies [18].

Data: max_{FEs} , NP , Cr , F, M, r_{max}, r_{min}, w_{max}, w_{min}
Result: $\boldsymbol{x} \in P$ such that $f(\boldsymbol{x}) \leq f(\boldsymbol{y}), \forall \boldsymbol{y} \in P \setminus \{\boldsymbol{x}\}$

1 Create a population P using an opposition-based learning initialization technique;
2 Update $current_{FEs}$ accordingly;
3 **while** $current_{FEs} < max_{FEs}$ **do** // stopping criterion
  // Execute one generation of DE based on multiple mutation strategies
4   **for** $i = 0$ *to* $i < NP$ **do** // for every individual in the population
5   Sort P from best to worst and set $pbest = randInt(0, M-1)$;
6   Take $r_2, r_3 \in [0, NP-1]$ randomly ; // $r_2 \neq r_3 \neq i$, $r_n \in \mathbb{N}$
7     $r_1 = r_{max} - \frac{current_{FEs}}{max_{FEs}} * (r_{max} - r_{min})$; // $r_1 \in \mathbb{R}$
8   **if** $flip(r1)$ **then**
9       $\boldsymbol{v}_i = \boldsymbol{x}_i + F * (\boldsymbol{x}_{r_2} - \boldsymbol{x}_{r_3})$;
10   **else**
11       $\boldsymbol{v}_i = \boldsymbol{x}_{pbest} + F * (\boldsymbol{x}_{r_2} - \boldsymbol{x}_{r_3})$;
12   Apply binomial recombination ; // see Algorithm 1, lines 7-13
13     $P_i^{next} \leftarrow$ fittest between $\boldsymbol{u}_i$ and $\boldsymbol{x}_i$; // selection: take $\boldsymbol{u}_i$ if $f(\boldsymbol{u}_i) \leq f(\boldsymbol{x}_i)$, $\boldsymbol{x}_i$ otherwise
14     $current_{FEs} + +$;
15   $P \leftarrow P^{next}$; // advance generation and repeat
  // Perturb the best (fittest) individual in the population, dimension by dimension
16   **for** $j = 0$ *to* $j < D$ **do**
17     $\boldsymbol{\mu} = \boldsymbol{x}_{best}$; // Copy the best chromosome so far to μ, $\boldsymbol{x}_{best}$
18     $k = rnd(0, NP-1)$ such that $k \neq best$; // $k \in \mathbb{N}$
19     $n = rnd(0, D-1)$ such that $n \neq j$; // $n \in \mathbb{N}$
20     $r_2 = w_{min} + \frac{current_{FEs}}{max_{FEs}} * (w_{max} - w_{min})$; // $r_2 \in \mathbb{R}$
21   **if** $flip(r2)$ **then**
22       $\boldsymbol{\mu}_j = \boldsymbol{x}_{best,n} + (2 * rndreal(0,1) - 1) * (\boldsymbol{x}_{best,n} - \boldsymbol{x}_{k,n})$; // $\boldsymbol{x}_k \in P$
23   **else**
24       $\boldsymbol{\mu}_j = \boldsymbol{x}_{best,j} + (2 * rndreal(0,1) - 1) * (\boldsymbol{x}_{best,n} - \boldsymbol{x}_{k,n})$;
25   If $\boldsymbol{\mu}_j$ out of boundary then get it back to the feasible region ;
26   Evaluate the new chromosome $\boldsymbol{\mu}$ using f and increment $current_{FEs}$;
27   Take the fittest, between $\{\boldsymbol{\mu}, \boldsymbol{x}_{best}\}$ to represent $\boldsymbol{x}_{best}$;
28 **return** ($\boldsymbol{x} \in P$ *such that* $f(\boldsymbol{x}) \leq f(\boldsymbol{y}), \forall \boldsymbol{y} \in P \setminus \{\boldsymbol{x}\}$) ; // fittest

A remarkable feature of EDE is that it first uses a very explorative evolutionary strategy, but as the number of function evaluations increases, it changes to a much less explorative scheme (EDE sets $r_{min} = 0.1$ and $r_{max} = 1$). Another remarkable feature of EDE is the coupled perturbation method that it adopts between generations. Actually, if no mutation strategy switching is incorporated and we decide to use only line 11 (i.e., the mutation operation where $M = 1$), the procedure is transformed into a pure population-based local search method.

2.3 Success-History Based Parameter Adaptation for Differential Evolution

The *Success-History Based Parameter Adaptation for Differential Evolution* (SHADE) [15] is a self-adaptive version of the DE variant called *DE/current-to-pbest/1/bin*. The self-adaptive mechanism is applied to both F and Cr. Therefore, NP is its only control parameter (for further information about the self-adaptive mechanism of SHADE, interested readers are referred to [15]). SHADE also incorporates an external archive (A) built from the defeated parents throughout generations, since such individuals are used during the application of the mutation operator in order to promote a greater diversity. The adopted mutation strategy is the following:

$$\boldsymbol{v_i} = \boldsymbol{x_i} + F_i * (\boldsymbol{x_{pbest}} - \boldsymbol{x_i}) + F_i * (\boldsymbol{x_{r_2}} - \boldsymbol{x_{r_3}}) \quad (2)$$

where F is regenerated for every $x \in P$ (the regeneration procedure can be consulted in [15]). x_{r_2} and x_{r_3} are randomly chosen vectors from P and $P \cup A$, respectively, and x_{pbest} is taken randomly from the p% fittest individuals in the population. The parameter p is regenerated for every $x \in P$ as follows:

$$p_i = rand(\frac{2}{NP}, p_{max}) \quad (3)$$

where $p_{max} = 0.2$. The mutation strategy described in Eq. (2) is explorative as the base vector is practically the target vector. Another important feature is that information about the fittest individuals are taken into account. Finally, binomial recombination is used, but Cr is regenerated in a way analogous to the F regeneration procedure.

3 Our Proposal

Our proposed approach is called *Global and Local Search using SHADE* (GL-SHADE), and its main procedure is illustrated in Algorithm 4. GL-SHADE consists of three main components (lines 1–3):

- **MTS-LS1** (stands for *Multiple Trajectory Search - Local Search 1*) is a gradient-free non-population-based local search method (for a detailed explanation of this method, interested readers should refer to [16]). This method is used at the beginning of the search (line 9) and just once in all the procedure (the idea is to boost the search at the beginning).

- **SHADE** is used for population 1 (line 1). It integrates a global search scheme which takes into account information about the top individuals in the population (this property is remarkable since all enhancements done by other methods to the fittest individual can be used to guide the search) and presents a robust self-adaptive mechanism. Due to the aforementioned features, this scheme is adopted to handle the global search of our proposed approach.
- **eSHADE$_{\mathbf{ls}}$** is used for population 2 (line 2). This is the variant called $SHADE/pbest/1/exp$ coupled with the EDE perturbation method (see Algorithm 2, lines 16–27). This variant uses the following mutation strategy:

$$\boldsymbol{v}_i = \boldsymbol{x}_{pbest} + F_i * (\boldsymbol{x}_{r_2} - \boldsymbol{x}_{r_3}) \tag{4}$$

where p_{max} is set to 0.1 (see Eq. (3)). The strategy described in Eq. (4) is one of the mutation operators that EDE incorporates for scattering new trial vectors close to several top individuals and not just near the fittest one. As can be seen, this strategy is very different to that of Eq. (2), and our proposed approach allows them to complement each other. Additionally, exponential recombination is adopted instead of binomial recombination. The main procedure of eSHADE$_{\text{ls}}$, which was designed to handle the local search, is described in Algorithm 3.

Algorithm 3: eSHADE-ls

```
  Data: L_FEs , NP_2 , w_min, w_max
1 Initialize population and all required parameters ;           // initialization
2 counter_FEs = 0 ;
3 while counter_FEs < L_FEs do                                   // start evolution
4 |  Execute one generation of SHADE/pbest/1/exp;
5 |  Apply the EDE perturbation method to the fittest individual in the population;
6 |  Update counter_FEs accordingly;
7 return (fittest individual so far) ;                           // end evolution
```

The first task in Algorithm 4 is an initialization stage (lines 1–9). During this stage, the GL-SHADE's components are defined (lines 1–3), the corresponding populations are generated (line 5), the fittest individual from population 1 is set as the best solution so far (line 7) and a local search procedure is applied to the best solution recorded so far (line 9); in this case, the MTS-LS1 method is used. When defining the components, it is necessary to provide the stopping criterion (maximum number of evaluations max_{FEs}) since in some cases, it is required to stop the overall search process even when the maximum component's requested number of evaluations hadn't been reached (G_{FEs} or L_{FEs}). For example, MTS-LS1's execution (line 9) stops when $counter_{FEs} \geq L_{FEs}$ or $current_{FEs} \geq max_{FEs}$. In fact, $counter_{FEs}$ and $current_{FEs}$ are updated accordingly as the component's execution goes by.

The second task is to perform a global and local search stage (lines 10–16). During this stage, population 1 receives (line 11) the new best individual (which must be placed at the position where the old best individual is) and then the global search scheme is executed (line 12) while $counter_{FEs} \leq G_{FEs}$ and $current_{FEs} \leq max_{FEs}$. After finishing, population 1 migrates (lines 13–14) its best individual to population 2 (the entry individual is randomly placed) and then the local search scheme is executed (line 15) while $counter_{FEs} \leq L_{FEs}$ and $current_{FEs} \leq max_{FEs}$. Finally, population 2 migrates (lines 16–11) its best individual to population 1 and the procedure is repeated until the maximum number of function evaluations (max_{FEs}) is reached.

It is worth noting that after applying the mutation or perturbation procedures, a variable may fall outside its allowable boundaries. If this happens, we apply the same normalization procedure adopted in [15].

Algorithm 4: GL-SHADE

Data: max_{FEs}, G_{FEs}, L_{FEs}, NP_1, NP_2, $f_{objective}$, w_{min}, w_{max}

// Define components, set maximum number of evaluations and set the objective function

1 $DE_1 \leftarrow SHADE(NP_1, f_{objective}, max_{FEs})$;

2 $DE_2 \leftarrow eSHADE_{ls}(NP_2, f_{objective}, max_{FEs}, w_{min}, w_{max})$;

3 $LS_1 \leftarrow MTS_{LS1}(f_{objective}, max_{FEs})$;

4 $current_{FEs} = 0$; // keep track of the number of evaluations computed so far

// Initialize: populations, parameters and structures needed by the corresponding component

5 $DE_1.initialize()$, $DE_2.initialize()$, $LS_1.initialize()$;

6 $current_{FEs} = current_{FEs} + NP_1 + NP_2$; // update evaluations computed so far

7 $best_{global} \leftarrow DE_1.best$; // update best global

// Early local search: at the end of this stage $current_{FEs} += L_{FEs}$

8 $counter_{FEs} = 0$; // restart counter

9 $LS_1.enhance(best_{global}, L_{FEs}, counter_{FEs}, current_{FEs})$;

// While the number of evaluations is less than the maximum defined, do ...

10 **while** $current_{FEs} < max_{FEs}$ **do**

  // Global search: at the end of this stage $current_{FEs} += G_{FEs}$

11   $DE_1.receive(best_{global})$ and $counter_{FEs} = 0$; // migrate better solution: from pop_2 to pop_1

12   $DE_1.evolve(G_{FEs}, counter_{FEs}, current_{FEs})$; // GS

13   $best_{global} \leftarrow DE_1.best$; // update best global

  // Local search: at the end of this stage $current_{FEs} += L_{FEs}$

14   $DE_2.receive(best_{global})$ and $counter_{FEs} = 0$; // migrate better solution: from pop_1 to pop_2

15   $DE_2.evolve(L_{FEs}, counter_{FEs}, current_{FEs})$; // LS

16   $best_{global} \leftarrow DE_2.best$; // update best global

17 **Report** $best_{global}$

4 Experimental Results

In order to assess the performance of our proposed GL-SHADE, we adopted the test suite used at the large-scale global optimization competition held at the *2013 IEEE Congress on Evolutionary Computation (CEC'2013)* [6], but adopting the experimental conditions and guidelines of the LSGO competition held at the *2019 IEEE Congress on Evolutionary Computation (CEC'2019)* [7].

The previously indicated benchmark consists of 15 test problems, all with 1000 decision variables except for $f13$ and $f14$ which are overlapping functions, where $D = 905$ [2,11]. In general, these test problems can be categorized as shown in Table 1.

For each test problem, 25 independent executions were carried out. Each run is stopped when reaching a pre-defined maximum number of objective function evaluations ($max_{FEs} = 3\times10^6$). Additionally, we also report the results obtained after performing 120,000 and 600,000 objective function evaluations. The parameter values adopted in our experiments are shown in Table 2 and our results are summarized in Table 3 (we report the best, worst, median, mean, and standard deviations calculated over the 25 runs performed for each test problem).

Table 1. Features of the test problems from the CEC'2013 benchmark

Category	Functions
Fully separable	$f1$–$f3$
Partially separable	$f4$–$f11$
Overlapping	$f12$–$f14$
Fully non-separable	$f15$

Table 2. Parameter values used in our experiments

Parameter	Value	Usage
NP_1	100	Size of Population 1
NP_2	100	Size of Population 2
G_{FEs}	25000	Max evaluations requested for global search per iteration
L_{FEs}	25000	Max evaluations requested for local search per iteration
w_{min}	0.0	eSHADE$_{ls}$ algorithm
w_{max}	0.2	eSHADE$_{ls}$ algorithm

Finally, we also provide the convergence curves for $f2$, $f7$, $f11$, $f12$, $f13$ and $f14$ (see Fig. 1). In order to reduce its running time, GL-SHADE and the benchmark set were implemented[2] using C++ and CUDA. All experiments were performed using the Intel(R) Core(TM) i7-3930K CPU @ 3.20 GHz with 8 GB RAM (using the Ubuntu 18.04 operating system), and the GeForce GTX 680 GPU with the CUDA 10.2 version.

4.1 Comparison with Its Components

In order to investigate if GL-SHADE is able to outperform its individual components, we adopted the Wilcoxon signed-rank test. Here, we take $N = 25$ and $\alpha = 0.05$, meaning that we use a sample size of 25 executions and a significance level of 5%. Results are summarized in Table 4. There, we can see the mean as well as its statistical significance next to it. The notation "b/e/w" shown in the last row means that a component is significantly better in "b" test problems, that there was no statistically significant difference in "e" test problems and that it was significantly worse in "w" test problems, with respect to GL-SHADE. These results indicate that our proposed approach presents a better overall performance than any of its components considered separately. This is particularly true for the overlapping and partially separate test problems with a separable subcomponent (i.e., $f4$ to $f7$).

[2] Our source code can be obtained from: https://github.com/delmoral313/gl-shade.

Table 3. Summary of the results obtained by GL-SHADE in the CEC'2013 test problems

1000D		*f1*	*f2*	*f3*	*f4*	*f5*	*f6*	*f7*	*f8*
1.2E+05	Best	1.0922E+05	6.0793E+02	2.0003E+01	2.6578E+10	3.2918E+06	1.0475E+06	4.5536E+08	1.1335E+14
	Median	1.6125E+05	6.8597E+02	2.0003E+01	5.4221E+10	4.5740E+06	1.0547E+06	1.3964E+09	4.4679E+14
	Worst	4.6460E+05	7.8329E+02	2.0004E+01	9.4616E+10	5.6865E+06	1.0597E+06	3.0432E+09	8.5885E+14
	Mean	1.7899E+05	6.8888E+02	2.0003E+01	5.4236E+10	4.5864E+06	1.0546E+06	1.4949E+09	4.4115E+14
	StDev	7.1896E+04	4.1088E+01	3.5612E-04	1.7031E+10	5.8026E+05	3.6927E+03	6.2989E+08	1.9130E+14
6.0E+05	Best	7.2085E-04	1.4647E+01	2.0000E+01	1.0450E+09	1.9819E+06	1.0474E+06	1.2055E+07	1.6256E+12
	Median	3.9288E+01	2.0938E+01	2.0000E+01	2.5779E+09	2.7543E+06	1.0544E+06	2.2112E+07	9.0666E+12
	Worst	1.4477E+02	3.8803E+01	2.0000E+01	5.2019E+09	3.6524E+06	1.0596E+06	8.1126E+07	4.9873E+13
	Mean	4.5649E+01	2.1808E+01	2.0000E+01	2.6494E+09	2.6890E+06	1.0541E+06	2.9485E+07	1.2236E+13
	StDev	4.1617E+01	6.1271E+00	2.4000E-05	1.0419E+09	4.2647E+05	3.5944E+03	1.6787E+07	1.1786E+13
3.0E+06	Best	2.5836E-26	9.9496E-01	2.0000E+01	9.3076E+06	1.3728E+06	1.0107E+06	4.5790E-02	3.9375E+09
	Median	9.4995E-26	4.9748E+00	2.0000E+01	2.4488E+07	2.1891E+06	1.0348E+06	1.0654E+00	3.3649E+10
	Worst	8.7787E-23	2.7859E+01	2.0000E+01	7.8810E+07	2.8653E+06	1.0534E+06	1.0269E+01	3.2976E+11
	Mean	1.0930E-23	6.6067E+00	2.0000E+01	2.7387E+07	2.2180E+06	1.0342E+06	2.1701E+00	8.9404E+10
	StDev	2.3448E-23	6.9108E+00	0.0000E+00	1.5706E+07	3.6639E+05	1.0878E+04	2.5127E+00	1.1417E+11
1000D		*f9*	*f10*	*f11*	*f12*	*f13*	*f14*	*f15*	
1.2E+05	Best	1.2351E+09	9.2531E+07	9.2224E+11	1.9287E+04	1.5994E+10	9.3400E+10	9.0004E+07	
	Median	2.3031E+09	9.4008E+07	9.3948E+11	2.3414E+04	2.6080E+10	3.4485E+11	1.1240E+08	
	Worst	5.7137E+09	9.4342E+07	1.0020E+12	3.0089E+04	4.8688E+10	7.6595E+11	1.4611E+08	
	Mean	2.3688E+09	9.3877E+07	9.4278E+11	2.3987E+04	2.6841E+10	3.7238E+11	1.1368E+08	
	StDev	9.3671E+08	4.4457E+05	1.7790E+10	3.4976E+03	7.3622E+09	1.7564E+11	1.4759E+07	
6.0E+05	Best	1.6273E+09	9.2088E+07	9.1590E+11	1.7099E+02	1.2233E+09	5.3650E+08	9.2158E+06	
	Median	2.1679E+09	9.2980E+07	9.2092E+11	8.9949E+02	2.9356E+09	8.8536E+09	3.2383E+07	
	Worst	4.7715E+09	9.4153E+07	9.4439E+11	1.8122E+03	3.9737E+09	3.0699E+10	6.9700E+07	
	Mean	2.3619E+09	9.3006E+07	9.2554E+11	9.2076E+02	2.7333E+09	1.0825E+10	3.1321E+07	
	StDev	7.3543E+08	4.6799E+05	1.0055E+10	4.3518E+02	8.1207E+08	8.8037E+09	1.7574E+07	
3.0E+06	Best	1.3860E+09	9.0964E+07	9.1543E+11	1.2587E-23	1.3388E+04	4.3995E+06	1.2041E+05	
	Median	2.1147E+09	9.1681E+07	9.2276E+11	2.0667E-23	2.9806E+04	4.7472E+06	1.1409E+06	
	Worst	3.6649E+09	9.2675E+07	9.4943E+11	3.9866E+00	1.2446E+05	5.3146E+06	3.0018E+06	
	Mean	2.1871E+09	9.1750E+07	9.2729E+11	7.9732E-01	3.9831E+04	4.7868E+06	1.2919E+06	
	StDev	6.3160E+08	4.7189E+05	1.0580E+10	1.6275E+00	2.9868E+04	2.1478E+05	1.1058E+06	

4.2 Comparison Between GL-SHADE and SHADEILS

SHADE with iterative local search (SHADEILS) [11] is one of the best hybrid algorithms currently available for large-scale global optimization, according to [9,10]. SHADEILS uses SHADE to handle the global search and it incorporates MTS-LS1 and a gradient-based method to undertake the local search. Additionally, it integrates a re-start mechanism which is launched when stagnation is detected. In order to compare SHADEILS with respect to our proposal, we adopted again the test suite used at the large-scale global optimization competition held at the *2013 IEEE Congress on Evolutionary Computation (CEC'2013)* [6]. We also applied the Wilcoxon signed-rank test with $N = 25$ and $\alpha = 0.05$. Our comparison of results is summarized in Table 5. The results that are better in a statistically significant way are shown in **boldface**. Additionally, we show the convergence curves for both SHADEILS and GL-SHADE in Fig. 1. SHADEILS is significantly better than our proposed approach in 4 out of 15 test problems, while GL-SHADE is significantly better than SHADEILS in 9 out of 15 test problems. Our proposed approach is particularly better than SHADEILS in the overlapping test problems.

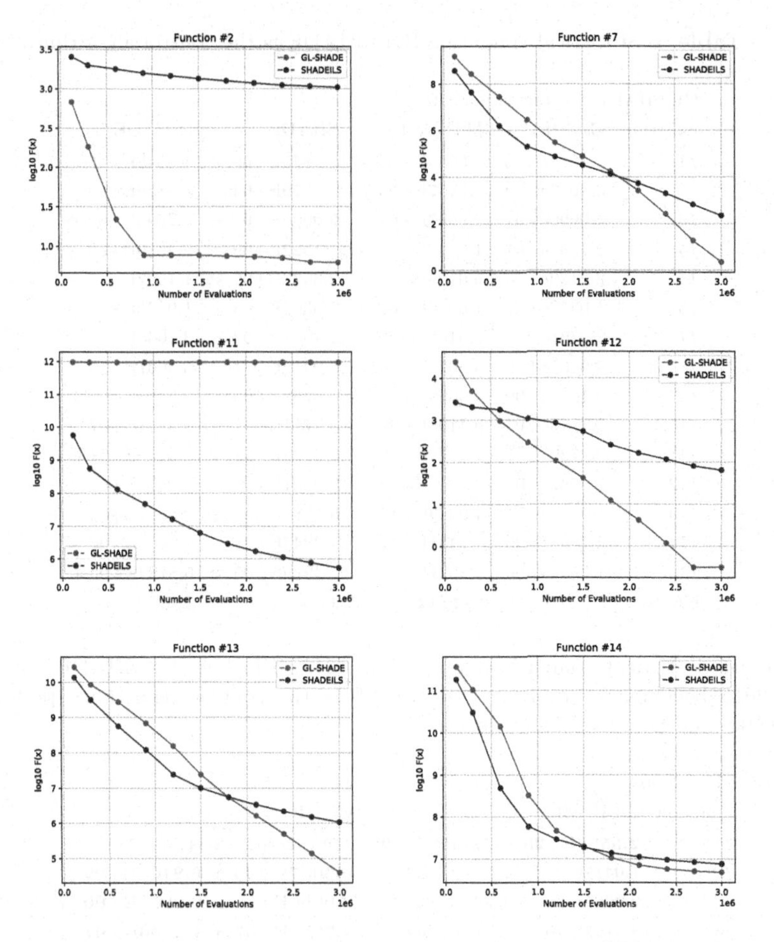

Fig. 1. Convergence curves with logarithmic scale for some CEC'2013 benchmark problems

4.3 Comparison with Respect to State-of-the-art Algorithms

Our proposed approach was also compared with respect to three state-of-the-art algorithms (besides SHADEILS), namely:

- **CC-RDG3** [12]: is a CC-based algorithm. It was the winner of the 2019 LSGO competition [10,13].
- **MOS** [5]: is a hybrid algorithm that was considered as one of the best metaheuristics for LSGO during several years (from 2013 to 2018) [11,13].
- **LSHADE-SPA** [2]: is a hybrid-CC method. This approach together with SHADEILS were the first to outperform MOS [9].

For comparing our results, we adopted the 2019 LSGO competition criteria [10]. Thus, each algorithm is assigned a score (per test problem) using the

Table 4. Statistical validation (GL-SHADE is the control algorithm).

GL-SHADE vs. its components				
Function	GL-SHADE	MTS-LS1	SHADE	eSHADE-ls
$f1$	1.0930e−23	3.0182E−25 ≈	2.1845E+06 −	1.8548E−16 −
$f2$	6.6067e+00	4.0137E+03 −	1.5379E+04 −	3.3829E+00 †
$f3$	2.0000e+01	2.0007E+01 −	2.0060E+01 −	2.0000E+01 ≈
$f4$	2.7387e+07	1.0573E+12 −	1.1522E+09 −	2.3669E+08 −
$f5$	2.2180e+06	5.9109E+07 −	2.3572E+06 ≈	1.1502E+07 −
$f6$	1.0342e+06	1.0506E+06 −	1.0564E+06 −	1.0389E+06 ≈
$f7$	2.1701e+00	7.9165E+09 −	3.1042E+06 −	3.1663E+02 −
$f8$	8.9404e+10	5.5029E+16 −	8.9442E+11 −	5.4067E+13 −
$f9$	2.1871e+09	4.3869E+10 −	1.3792E+09 †	2.3750E+09 ≈
$f10$	9.1750e+07	9.4198E+07 −	9.2711E+07 −	9.2001E+07 ≈
$f11$	9.2729e+11	1.1674E+12 −	9.3339E+11 ≈	9.3258E+11 ≈
$f12$	7.9732e−01	2.1449E+03 −	8.2921E+06 −	3.6372E+02 −
$f13$	3.9831e+04	4.0259E+10 −	5.7585E+07 −	2.9525E+05 −
$f14$	4.7868e+06	1.1431E+12 −	1.2944E+08 −	4.9800E+06 ≈
$f15$	1.2919e+06	2.9521E+08 −	1.4229E+06 ≈	5.8184E+05 †
b/e/w		**0/1/14**	**1/3/11**	**2/6/7**

Table 5. Statistical comparison of results: SHADEILS vs GL-SHADE using the CEC'2013 benchmark problems (SHADEILS is the control algorithm), performing 3,000,000 objective function evaluations

Mean ±Std. Dev.			
Function	SHADEILS	GL-SHADE	Sig.
$f1$	**2.5558E−28 ± 5.3619E−28**	1.0930E−23 ± 2.3448E−23	−
$f2$	1.0415E+03 ± 1.0341E+02	**6.6067E+00 ± 6.9108E+00**	†
$f3$	2.0068E+01 ± 4.7610E-02	**2.0000E+01 ± 0.0000E+00**	†
$f4$	3.0128E+08 ± 1.0458E+08	**2.7387E+07 ± 1.5706E+07**	†
$f5$	**1.3310E+06 ± 2.2657E+05**	2.2180E+06 ± 3.6639E+05	−
$f6$	1.0316E+06 ± 9.8658E+03	1.0342E+06 ± 1.0878E+04	≈
$f7$	2.2356E+02 ± 2.5286E+02	**2.1701E+00 ± 2.5127E+00**	†
$f8$	5.9937E+11 ± 5.4287E+11	**8.9404E+10 ± 1.1417E+11**	†
$f9$	**1.5780E+08 ± 1.4888E+07**	2.1871E+09 ± 6.3160E+08	−
$f10$	9.2556E+07 ± 4.5100E+05	**9.1750E+07 ± 4.7189E+05**	†
$f11$	**5.3888E+05 ± 2.2303E+05**	9.2729E+11 ± 1.0580E+10	−
$f12$	6.4886E+01 ± 2.2987E+02	**7.9732E-01 ± 1.6275E+00**	†
$f13$	1.0706E+06 ± 8.6269E+05	**3.9831E+04 ± 2.9868E+04**	†
$f14$	7.6280E+06 ± 1.2756E+06	**4.7868E+06 ± 2.1478E+05**	†
$f15$	8.6832E+05 ± 6.3444E+05	1.2919E+06 ± 1.1058E+06	≈
b/e/w	**4/2/9**	**9/2/4**	

Formula One Score (FOS) which is based on its position (1st/25pts, 2nd/18pts, 3rd/15pts, 4th/12pts and 5th/10pts). The comparison of results (performing 3,000,000 objective function evaluations) of our proposed GL-SHADE with respect to other state-of-the-art algorithms is shown in Fig. 2. The results from the other algorithms were extracted from the LSGO competition database [7,8,13]. Based on the results summarized in Fig. 2, we obtained the ranking shown in Table 6.

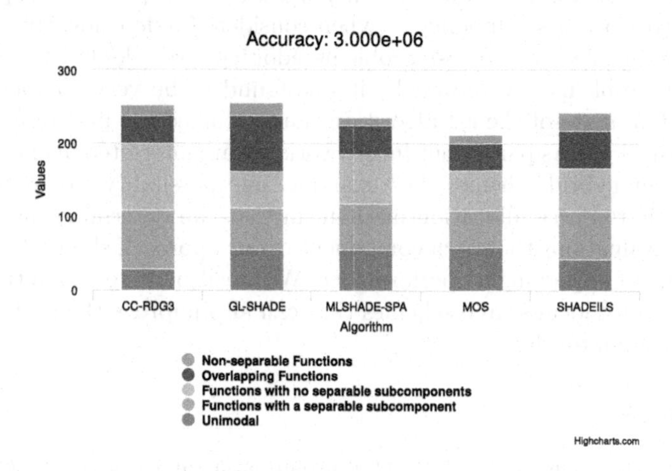

Fig. 2. GL-SHADE vs state-of-the-art algorithms using the CEC'2013 benchmark problems adopting the FOS criterion

Table 6. Ranking according to the FOS criterion

#	Algorithm	Type	FOS
1	*GL-SHADE*	Hybrid	256
2	*CC-RDG3*	CC	253
3	*SHADE-ILS*	Hybrid	243
4	*MLSHADE-SPA*	CC-Hybrid	236
5	*MOS*	Hybrid	212

5 Conclusions and Future Work

In this paper, we proposed a new DE-based algorithm (GL-SHADE) specifically designed to solve LSGO problems. Our proposed approach incorporates a global search engine combined with a local search mechanism. For the global search engine, our proposed approach adopts SHADE. For our local search mechanism, we adopted a population-based self-adaptive algorithm (eSHADE-ls) which is indeed the main difference of our proposal with respect to other state-of-the-art algorithms adopted for LSGO. SHADE and eSHADE-ls collaborate with each other during the evolutionary process. Since our proposed approach adopts very different and complementary mutation strategies, SHADE's strategy scatters a mutated vector around the target vector, while eSHADE-ls' strategy scatters it close to one of the top individuals in the population. Our proposed approach was able to outperform its components (when considered independently), as well as SHADEILS in most of the test problems adopted (we adopted the CEC'2013 LSGO test problems). Additionally, it was found to be very competitive with respect to four state-of-the-art algorithms and obtained the best rank (based on the FOS criterion). As part of our future work, we are interested in experimenting with different hybrid schemes. For example, one possibility would be to have an automatic resource allocation mechanism that can determine the maximum number of evaluations that each component of our approach should be executed, with the aim of improving its performance. We are also interested in trying other (more elaborate) local search schemes that can also improve the performance of our proposed approach.

References

1. Das, S., Suganthan, P.N.: Differential evolution: a survey of the state-of-the-art. IEEE Trans. Evol. Comput. **15**(1), 4–31 (2011)
2. Hadi, A.A., Mohamed, A.W., Jambi, K.M.: LSHADE-SPA memetic framework for solving large-scale optimization problems. Complex Intell. Syst. **5**(1), 25–40 (2018). https://doi.org/10.1007/s40747-018-0086-8
3. Hiba, H., El-Abd, M., Rahnamayan, S.: Improving SHADE with center-based mutation for large-scale optimization. In: 2019 IEEE Congress on Evolutionary Computation (CEC 2019), Wellington, New Zealand, 10–13 June 2019, pp. 1533–1540. IEEE (2019)
4. Jian, J.-R., Zhan, Z.-H., Zhang, J.: Large-scale evolutionary optimization: a survey and experimental comparative study. Int. J. Mach. Learn. Cybern. **11**(3), 729–745 (2019). https://doi.org/10.1007/s13042-019-01030-4
5. LaTorre, A., Muelas, S., Peña, J.M.: Large scale global optimization: experimental results with MOS-based hybrid algorithms. In: 2013 IEEE Congress on Evolutionary Computation (CEC 2013), Cancún, México, 20–23 June 2013, pp. 2742–2749. IEEE (2013)
6. Li, X., Tang, K., Omidvar, M.N., Yang, Z., Qin, K.: Benchmark Functions for the CEC'2013 Special Session and Competition on Large-Scale Global Optimization (2013)
7. Molina, D., LaTorre, A.: Toolkit for the automatic comparison of optimizers: comparing large-scale global optimizers made easy. In: 2018 IEEE Congress on Evolutionary Computation (CEC 2018), Rio de Janeiro, Brazil, 8–13 July 2018 (2018). ISBN 978-1-5090-6018-4
8. Molina, D., LaTorre, A.: Toolkit for the automatic comparison of optimizers (TACO): Herramienta online avanzada para comparar metaheurísticas. In: XIII Congreso Español en Metaheurísticas y Algoritmos Evolutivos y Bioinspirados, pp. 727–732 (2018)
9. Molina, D., LaTorre, A.: WCCI 2018 Large-Scale Global Optimization Competition Results (2018). http://www.tflsgo.org/download/comp2018_slides.pdf. Accessed 29 Feb 2020

10. Molina, D., LaTorre, A.: CEC 2019 Large-Scale Global Optimization Competition Results (2019). http://www.tflsgo.org/assets/cec2019/comp2019_slides.pdf. Accessed 29 Feb 2020
11. Molina, D., LaTorre, A., Herrera, F.: Shade with iterative local search for large-scale global optimization. In: 2018 IEEE Congress on Evolutionary Computation (CEC 2018), Rio de Janeiro, Brazil, 8–13 July 2018. IEEE (2018)
12. Omidvar, M.N., Li, X., Mei, Y., Yao, X.: Cooperative co-evolution with differential grouping for large scale optimization. IEEE Trans. Evol. Comput. **18**(3), 378–393 (2013)
13. Omidvar, M.N., Sun, Y., La Torre, A., Molina, D.: Special Session and Competition on Large-Scale Global Optimization on WCCI 2020 (2020). http://www.tflsgo.org/special_sessions/wcci2020.html. Accessed 22 Feb 2020
14. Storn, R., Price, K.: Differential evolution - a simple and efficient heuristic for global optimization over continuous spaces. J. Glob. Optim. **11**(4), 341–359 (1997)
15. Tanabe, R., Fukunaga, A.: Success-history based parameter adaptation for Differential Evolution. In: 2013 IEEE Congress on Evolutionary Computation (CEC 2013), Cancún, México, 20–23 June 2013, pp. 71–78. IEEE (2013)
16. Tseng, L.Y., Chen, C.: Multiple trajectory search for large scale global optimization. In: 2008 IEEE Congress on Evolutionary Computation (CEC 2008), pp. 3052–3059, Hong Kong, 1–6 June 2008. IEEE (2008)
17. Wu, X., Wang, Y., Liu, J., Fan, N.: A new hybrid algorithm for solving large scale global optimization problems. IEEE Access **7**, 103354–103364 (2019)
18. Xiang, W.L., Meng, X.L., An, M.Q., Li, Y.Z., Gao, M.X.: An enhanced differential evolution algorithm based on multiple mutation strategies. Comput. Intell. Neurosci. **2015** (2015). Article ID 285730

Evolutionary Algorithms with Self-adjusting Asymmetric Mutation

Amirhossein Rajabi(✉) and Carsten Witt

Technical University of Denmark, Kgs. Lyngby, Denmark
{amraj,cawi}@dtu.dk

Abstract. Evolutionary Algorithms (EAs) and other randomized search heuristics are often considered as unbiased algorithms that are invariant with respect to different transformations of the underlying search space. However, if a certain amount of domain knowledge is available the use of biased search operators in EAs becomes viable. We consider a simple (1+1) EA for binary search spaces and analyze an asymmetric mutation operator that can treat zero- and one-bits differently. This operator extends previous work by Jansen and Sudholt (ECJ 18(1), 2010) by allowing the operator asymmetry to vary according to the success rate of the algorithm. Using a self-adjusting scheme that learns an appropriate degree of asymmetry, we show improved runtime results on the class of functions OneMax_a describing the number of matching bits with a fixed target $a \in \{0,1\}^n$.

Keywords: Evolutionary algorithms · Runtime analysis · Self-adjusting algorithms · Parameter control · Asymmetric mutations

1 Introduction

The rigorous runtime analysis of randomized search heuristics, in particular evolutionary algorithms (EAs), is a vivid research area [6,13,20] that has provided proven results on the efficiency of different EAs in various optimization scenarios and has given theoretically guided advice on the choice of algorithms and their parameters. A common viewpoint is that EAs in the absence of problem-specific knowledge should satisfy some *invariance* properties; e. g., for the space $\{0,1\}^n$ of bit strings of length n it is desirable that the stochastic behavior of the algorithm does not change if the bit positions are renamed or the meaning of zeros and ones at certain bit positions is exchanged (formally, these properties can be described by automorphisms on the search space). Black-box complexity theory of randomized search heuristics (e. g., [8]) usually assumes such invariances, also known under the name of unbiasedness [16].

In a given optimization scenario, a certain amount of domain knowledge might be available that invalidates the unbiasedness assumption. For example,

Supported by a grant from the Danish Council for Independent Research (DFF-FNU 8021-00260B).

T. Bäck et al. (Eds.): PPSN 2020, LNCS 12269, pp. 664–677, 2020.
https://doi.org/10.1007/978-3-030-58112-1_46

on bit strings it might be known beforehand that zero-bits have a different interpretation than one-bits and that the total number of one-bits in high-quality solutions should lie within a certain interval. A prominent example is the minimum spanning tree (MST) problem [19], where, when modelled as a problem on bit strings of length m, m being the number of edges, all valid solutions should have exactly $n-1$ one-bits, where n is the number of vertices. In fact, the paper [19] is probably the first to describe a theoretical runtime analysis of an EA with a biased (also called asymmetric) mutation operator: they investigate, in the context of a simple (1+1) EA, an operator that flips each zero-bit of the current bit string x with probability $1/|x|_0$, where $|x|_0$ is the number of zero-bits in x, and each one-bit with probability $1/|x|_1$, where $|x|_1$ is the number of one-bits. As a consequence, the expected number of zero- and one-bits is not changed by this operator. Different biased mutation operators have been studied, both experimentally and theoretically, in, e. g., [18,21,25].

The (1+1) EA with asymmetric mutation operator (for short, asymmetric (1+1) EA) proposed in [19] was revisited in depth by Jansen and Sudholt [14] who investigated its effectiveness on a number of problems on bit strings, including the famous OneMax problem. In particular, they showed the surprising result that the asymmetric (1+1) EA optimizes OneMax in expected time $O(n)$, while still being able to optimize in expected time $O(n \log n)$ all functions from the generalized class $\text{OneMax}_a(x) := n - H(x, a)$, where $a \in \{0,1\}^n$ is an arbitrary target and $H(\cdot,\cdot)$ denotes the Hamming distance. However, the question is open whether this speed-up by a factor $\Theta(\log n)$ for the all-ones target (and the all-zeros target) compared to general targets a is the best possible that can be achieved with an asymmetric mutation. In principle, since the operator knows which bits are ones and which are zeros it is not unreasonable to assume that it could be modified to let the algorithm approach the all-ones string faster than $\Theta(n)$ – however, this must not result in an algorithm that is tailored to the all-ones string and fails badly for other targets.

In this paper, we investigate whether the bias of the asymmetric operator from [14] can be adjusted to put more emphasis on one-bits when it is working on the OneMax_{1^n} function with all-ones string as target (and accordingly for zero-bits with OneMax_{0^n}) while still being competitive with the original operator on general OneMax_a. Our approach is to introduce a *self-adjusting bias*: the probability of flipping a one-bit is allowed to decrease or increase by a certain amount, which is upper bounded by $r/|x|_1$ for a parameter r; the probability of flipping a zero-bit is accordingly adjusted in the opposite direction. A promising setting for this bias is learned through a self-adjusting scheme being similar in style with the 1/5-rule [2] and related techniques: in a certain observation phase of length N two different parameter values are each tried $N/2$ times and the value that is relatively more successful is used in the next phase. Hence, this approach is in line with a recent line of theoretical research of self-adjusting algorithms where the concrete implementation of self-adjustment is an ongoing debate [4,5,7,9–11,15,23,24]. See also the recent survey article [3] for an in-depth coverage of parameter control, self-adjusting algorithms, and theoretical runtime results.

We call our algorithm the *self-adjusting (1+1) EA with asymmetric mutation* (Asym-SA-(1+1) EA) and conduct a rigorous runtime analysis on the ONEMAX_a problem. Since the above-mentioned parameter r also determines the expected number of flipping bits of any type, we allow the bias of zero- resp. one-bits only to change by a small constant. Nevertheless, we can prove a speed-up on the ONEMAX function with target 1^n (and analogously for ZEROMAX with target 0^n) by a factor of at least 2 compared to the previous asymmetric (1+1) EA since the bias is adjusted towards the "right" type of bits. On general ONEMAX_a, where a both contains zeroes and ones, we prove that the bias is not strongly adjusted to one type of bits such that we recover the same asymptotic bound $O(n \log(\min\{|a|_0, |a|_1\}))$ as in [14]. These results represent, to the best of our knowledge, the first runtime analysis of a self-adjusting asymmetric EA and pave the way for the study of more advanced such operators.

This paper is structured as follows: in Sect. 2, we introduce the studied algorithms and fitness function. In Sect. 3, we prove for the function ONEMAX with the all-ones string as target that the Asym-SA-(1+1) EA is by a factor of roughly 2 faster than the original asymmetric (1+1) EA from [14]. For targets a containing both zero- and one-bits, we show in Sect. 4 that our algorithm is asymptotically not slowed down and in fact insensitive to the second parameter α of the algorithm that controls its learning speed. Experiments in Sect. 5 demonstrate that the speedup by a factor of at least 2 already can be observed for small problem sizes while our algorithm is not noticeably slower than the original asymmetric one on the mixed target with half ones and zeros. We finish with some conclusions.

2 Preliminaries

2.1 Algorithms

We consider asymmetric mutation in the context of a (1+1) EA for the maximization of pseudo-boolean functions $f\colon \{0,1\}^n \to \mathbb{R}$, which is arguably the most commonly investigated setting in the runtime analysis of EAs. The framework is given in Algorithm 1, where we consider the following two choices for the mutation operator:

Standard bit mutation Flip each bit in a copy of x independently with probability $\frac{1}{n}$.

Asymmetric mutation Flip each 0-bit in a copy of x with probability $\frac{1}{2|x|_0}$ and each 1-bit with probability $\frac{1}{2|x|_1}$ (independently for all bits).

With standard bit mutation, we call the algorithm the *classical (1+1) EA* and the one with asymmetric mutation the (static) *asymmetric (1+1) EA*. The latter algorithm stems from [14] and differs from the one from [19] by introducing the two factors 1/2 to avoid mutation probabilities above 1/2.

Algorithm 1. (1+1) EA

```
Select x uniformly at random from {0,1}^n
for t ← 1, 2, ... do
    Create y by applying a mutation operator to x.
    if f(y) ≥ f(x) then
        x ← y.
```

The *runtime* (synonymously, *optimization time*), of Algorithm 1, is the smallest t where a search point of maximal fitness (i.e., f-value) has been created; this coincides with the number of f-evaluations until that time. We are mostly interested in the expected value of the runtime and call this the *expected runtime.*

As motivated above, we investigate a biased mutation operator that can vary the individual probabilities of flipping 0- and 1-bits. We propose a rather conservative setting that flips 0-bits of the underlying string x with probability $\frac{r_0}{|x|_0}$ for some value $0 \leq r_0 \leq 1$ and 1-bits with probability $\frac{1-r_0}{|x|_1}$; hence increasing the probability of one type decreases the probability of the other type. Formally, we also allow other ranges $0 \leq r_0 \leq r$ for some $r \geq 1$, which could make sense for problems where more than one bit must flip for an improvement. However, in the context of this paper, we fix $r = 1$. To ease notation, we use $r_1 := r - r_0$, i.e., $r_1 = 1 - r_0$. Since r_i describes the expected number of flipping i-bits, $i \in \{0, 1\}$, we call r_0 the 0-strength and accordingly r_1 the 1-strength.

We now formally define the self-adjusting (1+1) EA with asymmetric mutation (Asym-SA-(1+1) EA), see Algorithm 2. Initially, $r_0 = r_1 = \frac{1}{2}$ so that the operator coincides with the static asymmetric mutation. In an observation phase of length N, which is a parameter to be chosen beforehand, two different pairs of 0- and 1-strengths are tried alternatingly, which differ from each other by an additive term of 2α that controls the speed of change. The pair that leads to relatively more successes (i.e., strict improvements) is used as the ground pair for the next phase. In case of a tie (including a completely unsuccessful phase) a uniform random choice is made. Also, we ensure that the strengths are confined to $[\alpha, 1-\alpha]$. Hereinafter, we say that we *apply asymmetric mutation with probability pair* (p_0, p_1) if we flip every 0-bit of the underlying string x with probability p_0 and every 1-bit with probability p_1 (independently for all bits). Note that the asymmetric (1+1) EA from [14] applies the probability pair $(\frac{1}{2|x|_0}, \frac{1}{2|x|_1})$.

So far, apart from the obvious restriction $\alpha < 1/4$, the choice of the parameter α is open. We mostly set it to small constant values but also allow it to converge to 0 slowly. In experiments, we set the parameter α to 0.1. Note that if the extreme pair of strengths $(\alpha, 1-\alpha)$ is used then 0-bits are only rarely flipped and 1-bits with roughly twice the probability of the static asymmetric operator.

In this paper, we are exclusively concerned with the maximization of the pseudo-Boolean function

$$\textsc{OneMax}_a(x) := n - H(x, a)$$

Algorithm 2. Self-adjusting (1+1) EA with Asymmetric Mutation (Asym-SA-(1+1) EA); parameters: N = length of observation phase, α = learning rate

Select x uniformly at random from $\{0,1\}^n$.
$r \leftarrow 1$. // parameter r fixed to 1 in this paper
$r_0 \leftarrow \frac{1}{2}$, $r_1 \leftarrow r - r_0$.
$b \leftarrow 0$.
for $t \leftarrow 1, 2, \ldots$ **do**
  $p_- \leftarrow (\frac{r_0-\alpha}{|x|_0}, \frac{r_1+\alpha}{|x|_1})$, $p_+ \leftarrow (\frac{r_0+\alpha}{|x|_0}, \frac{r_1-\alpha}{|x|_1})$.
  Create y by applying asymmetric mutation with pair p_- if t is odd and with pair p_+ otherwise.
  if $f(y) \geq f(x)$ **then**
    $x \leftarrow y$.
    if $f(y) > f(x)$ **then**
      if t is odd **then**
        $b = b - 1$.
      else
        $b = b + 1$.
  if $t \equiv 0 \pmod N$ **then**
    if $b < 0$ **then**
      Replace r_0 with $\max\{r_0 - \alpha, 2\alpha\}$ and r_1 with $\min\{r_1 + \alpha, 1 - 2\alpha\}$.
    else if $b > 0$ **then**
      Replace r_0 with $\min\{r_0 + \alpha, 1 - 2\alpha\}$ and r_1 with $\max\{r_1 - \alpha, 2\alpha\}$.
    else
      Perform one of the following two actions with prob. 1/2:
      – Replace r_0 with $\max\{r_0 - \alpha, 2\alpha\}$ and r_1 with $\min\{r_1 + \alpha, 1 - 2\alpha\}$.
      – Replace r_0 with $\min\{r_0 + \alpha, 1 - 2\alpha\}$ and r_1 with $\max\{r_1 - \alpha, 2\alpha\}$.
    $b \leftarrow 0$.

for an unknown target $a \in \{0,1\}^n$, where $H(\cdot,\cdot)$ denotes the Hamming distance. Hence, $\text{OneMax}_a(x)$ returns the number of matching bits in x and a. In unbiased algorithms, the target a is usually and without loss of generality assumed as the all-ones string 1^n, so that we denote $\text{OneMax} = \text{OneMax}_{1^n}$. In the considered asymmetric setting, the choice a makes a difference. Note, however, that only the number $|a|_1$ influences the runtime behavior of the considered asymmetric algorithms and not the absolute positions of these $|a|_1$ 1-bits.

3 Analysis of Self-adjusting Asymmetric Mutation on OneMax

In this section, we show in Theorem 1 that the Asym-SA-(1+1) EA is by a factor of roughly 2 faster on OneMax, i.e., when the target a is the all-ones string, than the static asymmetric (1+1) EA from [14]. The proof shows that the self-adjusting scheme is likely to set the 0-strength to its maximum $1 - 2\alpha$ which makes improvements more likely than with the initial strength of $1/2$. On the way to the proof, we develop two helper results, stated in Lemmas 1 and 2 below.

In the following lemma, we show that the algorithm is likely to observe more improvements with larger 0-strength and smaller 1-strength on OneMax.

Lemma 1. *Consider the Asym-SA-(1+1) EA on* OneMax = OneMax$_{1^n}$ *and let x denote its current search point. For $0 \le \beta \le 1 - r_0$ we have*

$$\Pr\Big(S_{(\frac{r_0+\beta}{|x|_0}, \frac{r_1-\beta}{|x|_1})}\Big) \ge \Pr\Big(S_{(\frac{r_0}{|x|_0}, \frac{r_1}{|x|_1})}\Big) + r_1 r_0 (1 - e^{-\beta}),$$

where $S_{(p_0,p_1)}$ is the event of observing a strict improvement when in each iteration zero-bits and one-bits are flipped with probability p_0 and p_1 respectively.

Proof. Consider the following random experiment which is formally known as a coupling (see [1] for more information on this concept). Assume that we flip bits in two phases: in the first phase, we flip all one-bits with probability $(r_1 - \beta)/|x|_1$ and all zero-bits with probability $r_0/|x|_0$. In the second phase, we only flip zero-bits with probability $\beta/(|x|_0 - r_0)$. We fixed $|x|_0$ and $|x|_1$ before the experiment. At the end of the second phase, the probability of a bit being flipped equals $(r_1 - \beta)/|x|_1$ if the value of the bit is one and $(r_0 + \beta)/|x|_0$ if the value of the bit is zero. The former is trivial (the bit is flipped only once with that probability) but for the latter, the probability of flipping each zero-bit equals $1 - (1 - r_0/|x|_0)(1 - \beta/(|x|_0 - r_0)) = (r_0 + \beta)/|x|_0$. Therefore, the probability of observing an improvement at the end of the second phase is

$$\Pr\Big(S_{(\frac{r_0+\beta}{|x|_0}, \frac{r_1-\beta}{|x|_1})}\Big). \tag{1}$$

Let us now calculate the probability of observing an improvement differently by computing the probability of success in each phase separately. At the end of the first phase, the probability of an improvement is $\Pr(S_{(\frac{r_0}{|x|_0}, \frac{r_1-\beta}{|x|_1})})$. It fails with probability $1 - \Pr(S_{(\frac{r_0}{|x|_0}, \frac{r_1-\beta}{|x|_1})})$ and the probability of success in the second phase is at least $(1 - (r_1 - \beta)/|x|_1)^{|x|_1} \Pr(S_{(\frac{\beta}{|x|_0 - r_0}, 0)})$, where no one-bits are flipped in the first phase and the algorithm finds a better search point when it only flips zero-bits with probability $\beta/(|x|_0 - r_0)$. Altogether we can say that the probability of observing a success is at least

$$\Pr\Big(S_{(\frac{r_0}{|x|_0}, \frac{r_1-\beta}{|x|_1})}\Big) + \Big(1 - \Pr\Big(S_{(\frac{r_0}{|x|_0}, \frac{r_1-\beta}{|x|_1})}\Big)\Big)\Big(1 - \frac{r_1-\beta}{|x|_1}\Big)^{|x|_1} \Pr\Big(S_{(\frac{\beta}{|x|_0 - r_0}, 0)}\Big). \tag{2}$$

By considering (1) and (2), we obtain

$$\begin{aligned}\Pr\Big(S_{(\frac{r_0+\beta}{|x|_0}, \frac{r_1-\beta}{|x|_1})}\Big) &\ge \Pr\Big(S_{(\frac{r_0}{|x|_0}, \frac{r_1-\beta}{|x|_1})}\Big) \\ &\quad + \Big(1 - \Pr\Big(S_{(\frac{r_0}{|x|_0}, \frac{r_1-\beta}{|x|_1})}\Big)\Big)\Big(1 - \frac{r_1-\beta}{|x|_1}\Big)^{|x|_1} \Pr\Big(S_{(\frac{\beta}{|x|_0 - r_0}, 0)}\Big).\end{aligned}$$

Note that $\Pr(S_{(\frac{r_0}{|x|_0},\frac{r_1-\beta}{|x|_1})}) \le (r_0/|x|_0)|x|_0 = r_0$. Also, we have $\Pr(S_{(\frac{r_0}{|x|_0},\frac{r_1-\beta}{|x|_1})}) \ge \Pr(S_{(\frac{r_0}{|x|_0},\frac{r_1}{|x|_1})})$ since in the second setting, the probability of flipping one-bits is larger, with the same probability of flipping zero-bits, which results in flipping more one-bits.

Finally, we have

$$\begin{aligned}
\Pr\left(S_{(\frac{r_0+\beta}{|x|_0},\frac{r_1-\beta}{|x|_1})}\right) &\ge \Pr\left(S_{(\frac{r_0}{|x|_0},\frac{r_1-\beta}{|x|_1})}\right) \\
&\quad + \left(1-\Pr\left(S_{(\frac{r_0}{|x|_0},\frac{r_1-\beta}{|x|_1})}\right)\right)\left(1-\frac{r_1-\beta}{|x|_1}\right)^{|x|_1} \Pr\left(S_{(\frac{\beta}{|x|_0-r_0},0)}\right) \\
&\ge \Pr\left(S_{(\frac{r_0}{|x|_0},\frac{r_1}{|x|_1})}\right) + (1-r_0)(1-r_1+\beta)\left(1-(1-\frac{\beta}{|x|_0})^{|x|_0}\right) \\
&\ge \Pr\left(S_{(\frac{r_0}{|x|_0},\frac{r_1}{|x|_1})}\right) + r_1 r_0(1-e^{-\beta}),
\end{aligned}$$

since $r_1 + r_0 = 1$ and $\beta \ge 0$ as well as for $t \ge 1$ and $0 \le s \le t$, we have the inequality $1 - s \le (1 - s/t)^t \le e^{-s}$. □

We can apply concentration inequalities to show that the larger success probability with the pair p_+ with high probability makes the algorithm to move to the larger 0-strength at the end of an observation phase. This requires a careful analysis since the success probabilities themselves change in the observation phase of length N. Due to space restrictions, the proof of the following lemma has been omitted; it can be found in the preprint [22].

Lemma 2. *Assume that* $\min\{|x|_0, |x|_1\} = \Omega(n^{5/6})$ *for all search points* x *in an observation phase of length* N *of the Asym-SA-(1+1) EA on* OneMax. *With probability at least* $1-\epsilon$, *if* $\alpha = \omega(n^{-1/12}) \cap (0, 1/4)$, *and* $N \ge 8\alpha^{-8}\ln(4/\epsilon)$, *for an arbitrarily small constant* $\epsilon \in (0, 1/2)$, *the algorithm chooses the probability pair* p_+ *on at the end of the phase.*

We can now state and prove our main result from this section.

Theorem 1. *The expected optimization time* $E(T)$ *of the Asym-SA-(1+1) EA with* $\alpha = \omega(n^{-1/12}) \cap (0, 1/4)$, $N \ge \lceil 8\alpha^{-8}\ln(4/\epsilon)\rceil$, *for an arbitrarily small constant* $\epsilon \in (0, 1/2)$, *and* $N = o(n)$ *on* OneMax *and* ZeroMax *satisfies*

$$\frac{n}{2}(1-\alpha)^{-1} \le E(T) \le \frac{n}{2}(1-4\alpha)^{-1} + o(n).$$

This theorem rigorously shows that the presented algorithm outperforms the original asymmetric (1+1) EA from [14] by a constant factor of roughly 2 on OneMax. Indeed, the number of iterations needed for the asymmetric (1+1) EA is at least n since by drift analysis [17], the drift is at most $|x|_0 \cdot 1/(2|x|_0) = 1/2$ (recalling that each 0-bit flips with probability $1/(2|x|_0)$) and $E[H(x^*, 1^n)] = n/2$ for the initial search point x^* while Theorem 1 proves that the Asym-SA-(1+1) EA finds the optimum point within $(1-\Theta(\alpha))^{-1}n/2 + o(n)$ iterations.

We even believe that the speedup of our algorithm is close to $2e^{1/2}$; namely, the original asymmetric (1+1) EA in an improvement usually must preserve existing one-bits, which happens with probability $(1 - 1/(2|x|_1))^{|x|_1} \approx e^{-1/2}$. This probability is close to $1-2\alpha$ in our algorithm when the self-adaptation has increased the 0-strength to $1 - 2\alpha$ and thereby decreased the 1-strength to 2α.

Proof (of Theorem 1). By Chernoff's bound, we have $H(x^*, 1^n) \leq n/2 + n^{3/4}$ with probability $1 - e^{-\Omega(n)}$, where x^* is the initial search point. Having the elitist selection mechanism results in keeping that condition for all future search points.

Now, we divide the run of the algorithm into two epochs according to the number of zero-bits. In the first epoch, we assume that $|x|_0 > n^{5/6}$. Since $|x|_1 > n^{5/6}$ also holds (proved by Chernoff), according to Lemma 2 with $\epsilon < 1/2$, the algorithm sets p_+ as its current probability pair in each observation phase with probability at least $1-\epsilon$. Hence, we have a random walk so the number of phases to reach $p^* := ((1-\alpha)/|x|_0, \alpha/|x|_1)$, by using the Gambler's Ruin Theorem [12], is at most $(1/(2\alpha))/(1-\epsilon-\epsilon)$. Since each phase contains N iterations, the expected number of iterations to reach p^* for the first time is at most $N/(2\alpha(1-2\epsilon))$.

When p_+ is equal to p^*, the algorithm uses the probability pairs

$$p_+ = \left(\frac{1-\alpha}{|x|_0}, \frac{\alpha}{|x|_1}\right) \text{ and } p_- = \left(\frac{1-2\alpha}{|x|_0}, \frac{2\alpha}{|x|_1}\right)$$

in the iterations so the drift is greater than

$$s\frac{1-2\alpha}{s}\left(1 - \frac{2\alpha}{n-s}\right)^{n-s} \geq (1-2\alpha)^2 \geq 1 - 4\alpha.$$

Consequently, via the additive drift theorem [17], we need at most $(n/2)(1-4\alpha)^{-1} + o(n)$ iterations where p_+ equals p^*.

After the first time where p_+ gets equal to p^*, there is a possibility of getting p_+ away from p^*. With the probability of ϵ, the algorithm chooses probability pair p_-. The expected number of phases to reach p^* again is $(1-2\epsilon)^{-1}$ from the Gambler's Ruin Theorem [12], resulting in $N\epsilon(1-2\epsilon)^{-1}$ extra iterations for each phase where p^* is p_+. The expected number of steps needed until we have one step with pair p^* is $\epsilon(1-2\epsilon)^{-1}$ and by linearity of expectation this factor also holds for the expected number of such steps in the drift analysis.

Overall, the expected number of iterations of the first epoch is at most

$$\frac{n}{2}(1-4\alpha)^{-1}\frac{\epsilon}{1-2\epsilon} + \frac{N}{2\alpha(1-2\epsilon)} + o(n) = \frac{n}{2}(1-4\alpha)^{-1}\frac{\epsilon}{1-2\epsilon} + O(\alpha^{-9}) + o(n),$$

where we used $N = o(n)$.

In the second epoch where $|x|_0 \leq n^{5/6}$, the expected number of iterations is at most $O(n^{5/6}/\alpha)$ since in the worst case with the probability pair $p_- = (\alpha/|x|_0, (1-\alpha)/|x|_1)$ the drift is

$$\frac{s\alpha}{s(1-\frac{1-\alpha}{n-s})^{n-s}} \geq 2\alpha.$$

Altogether and by our assumption that $\alpha = \omega(n^{-1/12}) \cap (0, 1/4)$, we have

$$E[T] \leq (1-4\alpha)^{-1} n/2 + o(n).$$

Moreover, in order to compute the lower bound, we have the upper bound $s(1-\alpha)/s = 1-\alpha$ on the drift, so through the additive drift theorem [17], we have

$$E[T] \geq (1-\alpha)^{-1} n/2.$$

□

4 Different Targets Than All-Ones and All-Zeros

After we have seen that the self-adjusting asymmetry is beneficial on the usual OneMax and ZeroMax function, we now demonstrate that this self-adaptation does not considerably harm its optimization time on OneMax_a for targets a different from the all-ones and all-zeros string. A worst case bound, obtained by assuming a pessimistic strength of α for the bits that still have to be flipped, would be $O((n/\alpha)\log z)$ following the ideas from [14]. We show that the factor $1/\alpha$ stemming from this pessimistic assumption does not actually appear since it becomes rare that the strength takes such a worst-case value on OneMax_a if a contains both many zeros and ones. Hence, we obtain the same asymptotic bound as for the original asymmetric (1+1) EA in this case.

Theorem 2. *Assume $z = \min\{|a|_1, |a|_0\} = \omega(\ln n)$, $1/\alpha = O(\log z/\log\log n)$ and $N = O(1)$. Then the expected optimization time of the Asym-SA-(1+1) EA on OneMax_a is $O(n\log z)$, where the implicit constant in the O does not depend on α.*

Proof. W.l.o.g. $|a|_1 \leq n/2$ and a has the form 0^z1^{n-z}, where we refer to the first z bits as the prefix and the final $n-z$ bits as the suffix.

We divide the run of the Asym-SA-(1+1) EA into three epochs according to the Hamming distance $h_t = n - H(x_t, a)$ of the current search point at time t to a, where the division is partially inspired by the analysis in [14]. If $h_t \geq 2z$ there are at least $h_t - z \geq h_t/2$ 0-bits in the suffix that can be flipped to improve and we also have $|x_t|_0 \leq h_t + z \leq (3/2)h_t$. Then the probability of an improvement is at least $((h_t/2)(\alpha/|x|_0))(1-|x|_1/n)^{|x|_1} \geq (\alpha/3)e^{-1-o(1)}$. Hence, the expected length of the first epoch is $O(n/\alpha) = O(n\log z)$ since $1/\alpha = O(\log z/\log\log n)$.

In the second epoch, we have $h_t \leq 2z$. This epoch ends when $h_t \leq n/\log^3 n$ and may therefore be empty. Since nothing is to show otherwise, we assume $z \geq n/\log^3 n$ in this part of the analysis. Since the probability of flipping an incorrect bit and not flipping any other bit is always at least $(\alpha/n)e^{-1+o(1)}$ the expected length of this epoch is at most $\sum_{i=n/\log^3 n}^{z} \frac{e^2 n}{\alpha i} = O((n/\alpha)\ln\ln n)$, which is $O(n\ln z)$ since $1/\alpha = O(\log z/\log\log n)$.

At the start of the third epoch, we have $h_t \leq z^*$, where $z^* = \min\{2z, n/\log^3 n\}$. The epoch ends when the optimum is found and is divided into z^* phases of varying length. Phase i, where $1 \leq i \leq z^*$, starts when the Hamming distance to a has become i and ends before the step where the distance becomes strictly less than i; the phase may be empty. The aim is to show that the expected length of phase i is $O(n/i)$ independently of α. To this end, we concretely consider a phase of length cn/i for a sufficiently large constant c and divide it into subphases of length c_1/α^2 for a sufficiently large constant c_1. The crucial observation we will prove is that such a subphase with probability $\Omega(1)$ contains at least c_2/α^2 steps such that the r_0-strength is in the interval $[1/4, 3/4]$, with c_2 being another sufficiently large constant. During these steps the probability of ending the phase due to an improvement is at least $c_3 i/n$ for some constant $c_3 > 0$. Hence, the probability of an improvement within $O(n\alpha^2/i)$ such subphases is $\Omega(1)$, proving that the expected number of phases of length cn/i is $O(1)$. Altogether, the expected length of phase i is $O(n/i)$. Note that $i \leq z^*$ and $1/\alpha = O(\log n/ \log\log n)$, so that $n\alpha^2/i$ is asymptotically larger than 1. Summing over $i \in \{1, \ldots, z^*\}$, the expected length of the third epoch is $O(n \log z)$.

We are left with the claim that at least $c_2/(2\alpha^2)$ steps in a subphase of length c_1/α^2 have a 0-strength within $[1/4, 3/4]$. To this end, we study the random process of the 0-strength and relate it to an unbiased Markovian random walk on the state space $\alpha, 2\alpha, \ldots, 1-2\alpha, 1-\alpha$, which we rescale to $1, \ldots, \alpha-1$ by multiplying with $1/\alpha$. If the overarching phase of length cn/i leads to an improvement, there is nothing to show. Hence, we pessimistically assume that no improvement has been found yet in the considered sequence of subphases of length c_1/α^2 each. Therefore, since the algorithm makes a uniform choice in the absence of improvements, both the probability of increasing and decreasing the 0-strength equal 1/2. The borders 1 and $\alpha - 1$ of the chain are reflecting with probability 1/2 and looping with the remaining probability.

From any state of the Markov chain, it is sufficient to bridge a distance of $m = 1/(2\alpha)$ to reach 1/2. Pessimistically, we assume strength α (i. e., state 1) at the beginning of the subphase. Using well-known results from the fair Gambler's Ruin Theorem [12], the expected time to reach state m from X_0 conditional on not reaching state 0 (which we here interpret as looping on state 1) is $X_0(m-X_0)$. Also, the probability of reaching m from X_0 before reaching 0 equals X_0/m and the expected number of repetitions until state m is reached is the reciprocal of this. Combining both, the expected time until m is reached from $X_0 = 1$ is at most $m^2 - m$, and by Markov's inequality, the time is at most $2m^2 = \alpha^{-2}$ with probability at least 1/2.

We next show that the time to leave the interval $[m/2, 3m/2]$ (corresponding to strength $[1/4, 3/4]$) is at least c/α^2 with probability at least $\Omega(1)$. To this end, we apply Chernoff bounds to estimate the number of decreasing steps within c/α^2 steps and note that it is less than $1/\alpha$ with probability $\Omega(1)$. Altogether, with probability $\Omega(1)$ a subphase of length c_1/α^2 contains at least c_2/α^2 steps with strength within $[1/4, 3/4]$ as suggested.

The theorem follows by summing up the expected lengths of the epochs. □

5 Experiments

The results in the previous sections are mostly asymptotic. In addition, although the obtained sufficient value for N derived in Lemma 1 is constant for $\alpha = \Omega(1)$, it can be large for a relatively small bit string. Hence, in this section, we present the results of the experiments conducted in order to see how the presented algorithm performs in practice. More information about the experiments is available in [22].

We ran an implementation of Algorithm 2 (Asym-SA-(1+1) EA) on OneMax and OneMax$_a$ with $a = 0^{n/2}1^{n/2}$ and with n varying from 8000 to 20000. The selected parameters of the Algorithm Asym-SA-(1+1) EA are $\alpha = 0.1$ and $N = 50$. We compared our algorithm against the (1+1) EA with standard mutation rate $1/n$ and asymmetric (1+1) EA proposed in [14].

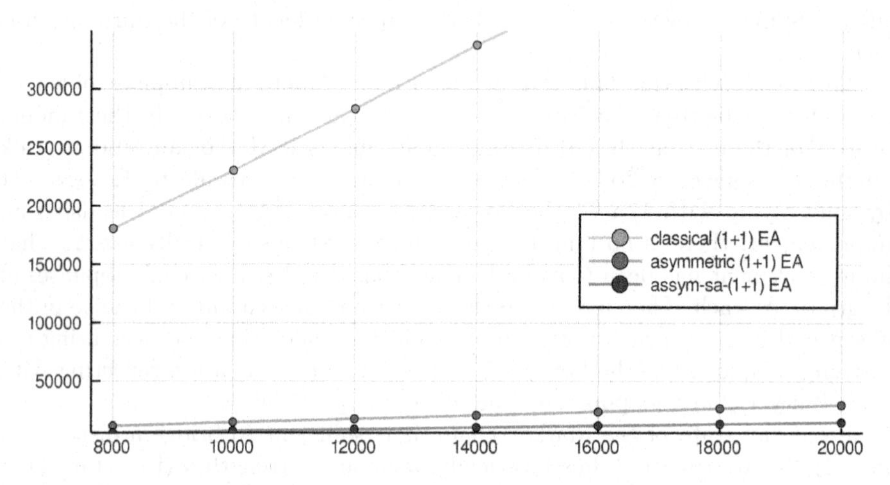

Fig. 1. Average number of fitness calls (over 1000 runs) the mentioned algorithms took to optimize OneMax.

The average optimization time of the experiment carried out on OneMax can be seen in Fig. 1. There is a clear difference between the performance of classical (1+1) EA and the algorithms using asymmetric mutations. It shows that these biased mutations can speed up the optimization time considerably. Also, the Asym-SA-(1+1) EA outperforms asymmetric (1+1) EA although we used relatively small N and large α. In detail, the average number of iterations that the Asym-SA-(1+1) EA and asymmetric (1+1) EA took to optimize for $n = 20000$ are 14864 and 29683 respectively. By considering the standard deviation of 238 and 385 as well, we can argue that our algorithm is faster by a factor of empirically around 2.

In Table 1, the average numbers of iterations which are taken for the algorithms to find the optimum on OneMax$_a$ with $a = 0^{n/2}1^{n/2}$ are available. The

Table 1. Average number of fitness calls (over 1000 runs) the mentioned algorithms took to optimize ONEMAX_a with $a = 0^{n/2}1^{n/2}$.

Algorithm	n						
	8000	10000	12000	14000	16000	18000	20000
Classical (1+1) EA	180019	231404	283965	335668	389859	444544	498244
Asymmetric (1+1) EA	180325	231999	283025	338264	391415	443910	502393
Asym-SA-(1+1) EA	180811	232412	284251	337519	390710	446396	503969

similarity between data for each bit string size suggests that the asymmetric algorithms perform neither worse nor better compared to classical (1+1) EA. More precisely, all p-values obtained from a Mann-Whitney U test between algorithms, with respect to the null hypothesis of identical behavior, are greater than 0.1.

6 Conclusions

We have designed and analyzed a (1+1) EA with self-adjusting asymmetric mutation. The underlying mutation operator chooses 0- and 1-bits of the current search point with different probabilities that can be adjusted based on the number of successes with a given probability profile.

As a proof of concept, we analyzed this algorithm on instances from the function class ONEMAX_a describing the number of matching bits with a target $a \in \{0,1\}^n$. A rigorous runtime analysis shows that on the usual ONEMAX function with target 1^n (and analogously for target 0^n), the asymmetry of the operator is adjusted in a beneficial way, leading to a constant-factor speedup compared to the asymmetric (1+1) EA without self-adjustment from [14]. For different targets a, the asymmetry of our operator does not become considerably pronounced so that the self-adjusting scheme asymptotically does not slow down the algorithm. Experiments confirm that our algorithm is faster than the static asymmetric variant on ONEMAX with target 1^n and not considerably slower for a target with equal number of 0- and 1-bits.

An obvious topic for future research is to consider other ranges for the parameter r that determines the maximum degree of asymmetry chosen by the algorithm. In the present framework, this parameter is linked to the expected number of flipping bits (regardless of their value), so that it is usually not advisable to increase it beyond constant values. Instead, we plan to investigate settings with two parameters where 0- and 1-bits each have their self-adjusted mutation strengths.

References

1. Doerr, B.: Probabilistic tools for the analysis of randomized optimization heuristics. In: Doerr and Neumann [6], pp. 1–87

2. Doerr, B., Doerr, C.: Optimal static and self-adjusting parameter choices for the (1+(λ, λ)) genetic algorithm. Algorithmica **80**(5), 1658–1709 (2018)
3. Doerr, B., Doerr, C.: Theory of parameter control for discrete black-box optimization: provable performance gains through dynamic parameter choices. In: Doerr and Neumann [6], pp. 271–321
4. Doerr, B., Doerr, C., Kötzing, T.: Static and self-adjusting mutation strengths for multi-valued decision variables. Algorithmica **80**(5), 1732–1768 (2018)
5. Doerr, B., Gießen, C., Witt, C., Yang, J.: The $(1 + \lambda)$ evolutionary algorithm with self-adjusting mutation rate. Algorithmica **81**(2), 593–631 (2019)
6. Doerr, B., Neumann, F. (eds.): Theory of Evolutionary Computation - Recent Developments in Discrete Optimization. Springer, Heidelberg (2020). https://doi.org/10.1007/978-3-030-29414-4
7. Doerr, B., Witt, C., Yang, J.: Runtime analysis for self-adaptive mutation rates. In: Proceedings of GECCO 2018, pp. 1475–1482. ACM Press (2018)
8. Doerr, C.: Complexity theory for discrete black-box optimization heuristics. In: Doerr and Neumann [6], pp. 133–212
9. Doerr, C., Wagner, M.: Sensitivity of parameter control mechanisms with respect to their initialization. In: Proceedings of PPSN 2018, pp. 360–372 (2018)
10. Doerr, C., Ye, F., van Rijn, S., Wang, H., Bäck, T.: Towards a theory-guided benchmarking suite for discrete black-box optimization heuristics: profiling (1+λ) EA variants on OneMax and LeadingOnes. In: Proceedings of GECCO 2018, pp. 951–958. ACM Press (2018)
11. Fajardo, M.A.H.: An empirical evaluation of success-based parameter control mechanisms for evolutionary algorithms. In: Proceedings of GECCO 2019, pp. 787–795. ACM Press (2019)
12. Feller, W.: An Introduction to Probability Theory and Its Applications, vol. 1, 3rd edn. Wiley, Hoboken (1968)
13. Jansen, T.: Analyzing Evolutionary Algorithms - The Computer Science Perspective. Springer, Heidelberg (2013). https://doi.org/10.1007/978-3-642-17339-4
14. Jansen, T., Sudholt, D.: Analysis of an asymmetric mutation operator. Evol. Comput. **18**(1), 1–26 (2010)
15. Lässig, J., Sudholt, D.: Adaptive population models for offspring populations and parallel evolutionary algorithms. In: 2011 Proceedings of FOGA 2011, Schwarzenberg, Austria, 5–8 January 2011, Proceedings, pp. 181–192. ACM Press (2011)
16. Lehre, P.K., Witt, C.: Black-box search by unbiased variation. Algorithmica**64**(4), 623–642 (2012)
17. Lengler, J.: Drift analysis. In: Doerr and Neumann [6], pp. 89–131
18. Neumann, A., Alexander, B., Neumann, F.: Evolutionary image transition using random walks. In: Correia, J., Ciesielski, V., Liapis, A. (eds.) EvoMUSART 2017. LNCS, vol. 10198, pp. 230–245. Springer, Cham (2017). https://doi.org/10.1007/978-3-319-55750-2_16
19. Neumann, F., Wegener, I.: Randomized local search, evolutionary algorithms, and the minimum spanning tree problem. Theor. Comput. Sci. **378**(1), 32–40 (2007)
20. Neumann, F., Witt, C.: Bioinspired Computation in Combinatorial Optimization - Algorithms and Their Computational Complexity. Springer, Heidelberg (2010). https://doi.org/10.1007/978-3-642-16544-3
21. Raidl, G.R., Koller, G., Julstrom, B.A.: Biased mutation operators for subgraph-selection problems. IEEE Trans. Evol. Comput. **10**(2), 145–156 (2006)
22. Rajabi, A., Witt, C.: Evolutionary algorithms with self-adjusting asymmetric mutation. CoRR (2020). http://arxiv.org/abs/2006.09126

23. Rajabi, A., Witt, C.: Self-adjusting evolutionary algorithms for multimodal optimization. In: Proceedings of GECCO 2020 (2020, to appear)
24. Rodionova, A., Antonov, K., Buzdalova, A., Doerr, C.: Offspring population size matters when comparing evolutionary algorithms with self-adjusting mutation rates. In: Proceedings of GECCO 2019, pp. 855–863. ACM Press (2019)
25. Sutton, A.M.: Superpolynomial lower bounds for the (1+1) EA on some easy combinatorial problems. Algorithmica **75**(3), 507–528 (2016)

Behavior Optimization in Large Distributed Systems Modeled by Cellular Automata

Franciszek Seredyński and Jakub Gąsior(✉)

Department of Mathematics and Natural Sciences,
Cardinal Stefan Wyszyński University, Warsaw, Poland
{f.seredynski,j.gasior}@uksw.edu.pl

Abstract. We consider a distributed system modeled by the second-order Cellular Automata (CA) and interpreted as a multi-agent system, where interactions between agents are defined by a spatial Prisoner's Dilemma game. The idea of the second-order CA is based on the concept "adapt to the best neighbor. Each agent uses some strategy for the selection of actions used against opponents and can change it during the iterated game. An agent acts in such a way to maximize its income. We intend to study conditions of emerging collective behavior in such systems measured by the average total payoff of agents in the game or by an equivalent measure–the total number of cooperating players. These measures are the external criterion of the game, and players acting selfishly are not aware of them. We show experimentally that collective behavior in such systems can emerge if some conditions related to the game are fulfilled. We propose to introduce an income sharing mechanism to the game, giving a possibility to share incomes locally by agents. We present the results of an experimental study showing that the sharing mechanism is a distributed optimization algorithm that significantly improves the capabilities of emerging collective behavior on a wide range of the game parameters.

Keywords: Collective behavior · Distributed optimization · Income sharing · Multi-agent systems · Spatial Prisoner's Dilemma game · Second–order Cellular Automata

1 Introduction

Recent advances in modern computer-communication technologies resulted in the development of new types of massively distributed systems (e.g., IoT systems, fog/edge/cloud computing systems [17]) oriented on collecting and processing information by a vast number of simple units responsible for collecting information to perform control, analytic and machine-learning tasks. Solving these tasks often can be reduced to solve an optimization problem in a centralized way with a request of full information about the system resources and

T. Bäck et al. (Eds.): PPSN 2020, LNCS 12269, pp. 678–690, 2020.
https://doi.org/10.1007/978-3-030-58112-1_47

users' demands, which becomes intractable for realistic problems. One can relay rather on a distributed problem solving by several independent entities. Therefore we propose a large scale multi-agent system approach, where agents are capable of solving problems in a distributed way. The principle requested by such multi-agent systems is the ability of collective behavior to be a result of local interactions of a considerable number of simple components [1,6,10].

In this paper, we will consider a multi-agent second-order CA-based system working in the framework of CA [16], where an interaction between players is described in terms of non-cooperative game theory [8] with use of the Spatial Prisoner's Dilemma (SPD) game. We will expect a global collective behavior in such a system measured by a total number of cooperating players, i.e., maximizing the average total payoff of agents of the system.

The phenomenon of emerging cooperation in systems described by the SPD game has been a subject of current studies [2,4,9]. They show that it depends on many factors such as payoff function parameters, the type of learning agent, or the way of interaction between agents.

Recently we have shown [13] that the second-order CA [7] based on the concept "adapt to the best neighbor can be successfully applied as simple learning machines in solving problems of collective behavior. In this paper, we introduce a new mechanism of interaction between players, based on the possibility of a local income sharing by agents participating in the game. Our objective is to study the conditions of emerging collective behavior in such a system with local income sharing, i.e., its ability to maximize in a distributed way the average total income. To our knowledge, it is the first attempt to apply this mechanism in the context of CA-based SPD games.

The structure of the paper is the following. In the next section, SPD game is presented and discussed in the context of collective behavior. Section 3 contains a description of the CA-based multi-agent system acting in the SPD game environment. Section 4 presents a basic mechanism of the game, including income sharing. Section 5 presents some results of the experimental study, and the last section concludes the paper.

2 Iterated Spatial Prisoner's Dilemma Game and Collective Behavior

We consider a 2D spatial array of size $n \times m$. We assume that a cell (i, j) will be considered as an agent–player participating in the SPD game [5,7]. We assume that a given player's neighborhood is defined in some way (see, the next Section). Players from this neighborhood will be considered as opponents in the game. At a given discrete moment, each cell can be in one of two states: C or D. The state of a given cell will be considered as an action C (cooperate) or D (defect) of the corresponding player against an opponent from its neighborhood. The payoff function of the game is given in Table 1.

Each player playing a game with an opponent in a single round (iteration) receives a payoff equal to R, T, S or P, where $T > R > P > S$. We assume that

Table 1. Payoff function of a row player participating in SPD game.

Player's action	Opponent's action	
	Cooperate (C)	Defect (D)
Cooperate (C)	$R = 1$	$S = c$
Defect (D)	$T = b$	$P = a$

$R = 1$, $T = b$, $S = c$, and $P = a$, and values of a, b and c can vary depending on the purpose of an experiment.

Let us consider the game with the following values of the payoff table parameters: $R = 1$, $b = 1.4$, $c = 0$, and $a = 0.3$. If a player (i,j) takes action $s_{ij} = C$ and the opponent (i_k, j_k) from a neighborhood also takes action $s_{i_k j_k} = C$, then the player receives payoff $u_{ij}(s_{i,j}, s_{i_k j_k}) = 1$. If the player takes the action D and the opponent player still keeps the action C, the defecting player receives payoff equal to $b = 1.4$. If the player takes the action C while the opponent takes action D, the cooperating player receives payoff equal to $c = 0$. When both players use the action D, then both receive payoff equal to $a = 0.3$.

It is worth noticing that choosing by all players the action D corresponds to the Nash equilibrium point (NE) [8], and it is considered as a solution of the one–shot game. Indeed, if all players select the action D, each of them receives a payoff equal to a, and there is no reason for any of them to change the action to C while the others keep their actions unchanged because it would result in decreasing its payoff to value 0.

We assume that players are rational and act in such a way to maximize their payoff defined by the payoff function. To evaluate the level of collective behavior achieved by the system, we will use an external criterion (not known for players) and ask whether it is possible to expect from players selecting such actions s_{ij} which will maximize the average total payoff (ATP) $\bar{u}()$ of the whole set of players:

$$\bar{u}(s_{11}, s_{12}, ..., s_{nn}) = \sum_{i=1}^{n} \sum_{j=1}^{n} u_{ij}(s_{ij}, s_{i_k j_k})/nm. \tag{1}$$

Game theory predicts the behavior of players oriented towards achieving NE, i.e., choosing the action D by all players. ATP at NE is equal to a, and we will call it the price of NE. In our game example, this value of ATP is low and equal to 0.3. The maximal value of ATP is equal to 1 and corresponds to selecting by all players the action C. We will call this value of ATP the maximal price point. The set of players' actions corresponding to the maximal price point is not NE; therefore, it is challenging to expect to reach global cooperation. The purpose of this study is to recognize conditions of emerging possibly high level of cooperation of players in iterated games.

3 CA–Based Players

Cells (i, j) of the 2D array are considered as CA–based players. At a given discrete moment t, each cell is either in state D or C, which is used by the player as an action with an opponent player. Initial states of CA cells are assigned randomly. For each cell, a local neighborhood is defined. We apply a cyclic boundary condition in order to avoid irregular behavior at the borders. We will assume the Moore neighborhood with eight immediate neighbors. It means that each player has eight opponents in the game.

In discrete moments, CA–based players will use their current states as actions to play games with opponents, they will receive payoffs, and next they will change their states applying assigned to them rules (also called strategies). We will be using some set of rules among which one of them will be initially randomly assigned to a given CA cell, so we deal with a non-uniform CA.

The following set of rules will be used: *all–C* (always cooperate), *all–D* (always defect), and *k–D* (k - level of tolerance). The strategy *k-D* tolerates at most k defections in a local neighborhood. It means that an agent with this strategy will choose the action C when the number of defecting neighbors does not exceed the value of k and will choose the action D in the opposite case. It is worth noticing that the *k–D* strategy is a generalized Tit-For-Tat (TFT) strategy, and when $k = 0$, it is exactly the well known in the literature TFT strategy.

4 Competition and Income Sharing Mechanisms

To study the possibility of the emergence of CA-based players' global collective behavior, we will modify the concept of CA and introduce some local mechanisms of interaction between players.

The first mechanism is a competition that is based on the idea proposed in [7]. It is based on the principle "adapt to the best neighbor and differs from the classical concept of CA. It assumes that each player associated with a given cell plays in a single round a game with each of its neighbors, and this way collects some total score. If the competition mechanism is turned *on*, after q number of rounds (iterations), each agent compares its total payoff with the total payoffs of its neighbors. If a more successful player exists in the neighborhood, this player replaces its own rule by the most successful one. This mechanism converts a classical CA into the *second–order* CA, which can adapt in time.

The second mechanism called an income sharing mechanism (ISM), which we propose provides a possibility of sharing payoffs by players in a neighborhood. Hard local sharing based on mandatory sharing was successfully used [11,12] in the context of 1D evolutionary and learning automata games. Here we propose soft sharing, where a player decides to use it or not on the base of competition for income. It is assumed that each player has a tag indicating whether it wishes (*on*) or not (*off*) to share its payoff with players from the neighborhood. The sharing works in such a way that if two players both wish to share, each of them receives half of the payoff from the sum. Before starting the iterated game, each player turns on its tag with a predefined probability p_{shar}. Due to the competition mechanism, the most successful rules, together with their tags containing information about willing (or not willing) to share incomes, are spreading over the system during its evolution.

5 Experimental Results

The purpose of conducted experiments was to find out under which values of parameters from the payoff table the global cooperation emerges and what is an influence of the ISM mechanism on it. A 2D array of the size 50×50 was used, with an initial state C or D (player actions) of each cell set with probability 0.5. Each strategy was assigned initially to CA cells with probability 0.333. To an agent with the rule k–D, a value k randomly selected from the range $(0 \ldots 7)$ was assigned. Updating rules assigned to agents by the competition mechanism was conducted after each iteration ($q = 1$). Each run lasted 100 iterations. The results presented below were averaged on the base of 20 runs. The experiments have been conducted with a recently developed simulator presented in [14].

5.1 Recognizing Problems

Figure 1 shows results of the first set of experiments for games without ISM for $b = 1.2$ and some settings of parameters a and c from the payoff table. Figure 1 (upper) presents runs of 3 games with different settings. Plots in violet and in red show fractions of cooperating agents/average payoff of players (ATP) in games with $a = 0.3$ and $c = 0.2$, but with different types of CA-based players.

Plots in violet show behavior of players implemented as classical CA, and plots in red show results for the second-order CA-based players. Lower located plots of a given color show change in time of fractions of cooperating agents, while higher plots of a given color show ATP. One can see (in violet) that classical CA does not have any learning capabilities. After an initial assigning of strategies (rules) to CA-cells, these assignments do not change in time, and they define the permanent level of cooperation of players and ATP, which is a result of the initial settings. However, when the second-order CA is used, we can observe (in red) an increasing level of cooperation and ATP. The fraction of cooperating agents reaches the value equal to 0.85, and the corresponding value of ATP equal to 0.92. Figure 1 (middle) shows how distribution of strategies changes in time.

One can see that the strategy *all–C* (in red) becomes the dominating one reaching the value 0.85 while the remaining strategies *all–D* and *k–D* decrease to values 0.08 and 0.07, respectively. How will we change the course of the game if we slightly modify only one setting parameter of the previous game, setting $a = 0.4$ and using the second-order CA? We can notice (see, Fig. 1 (upper) in green) that the cooperation also develops in time but reaches only the values 0.40, a twice lower value than in the previous game. We can also notice that surprisingly the value of ATP equal to 0.64 is higher than we could expect on the base of the

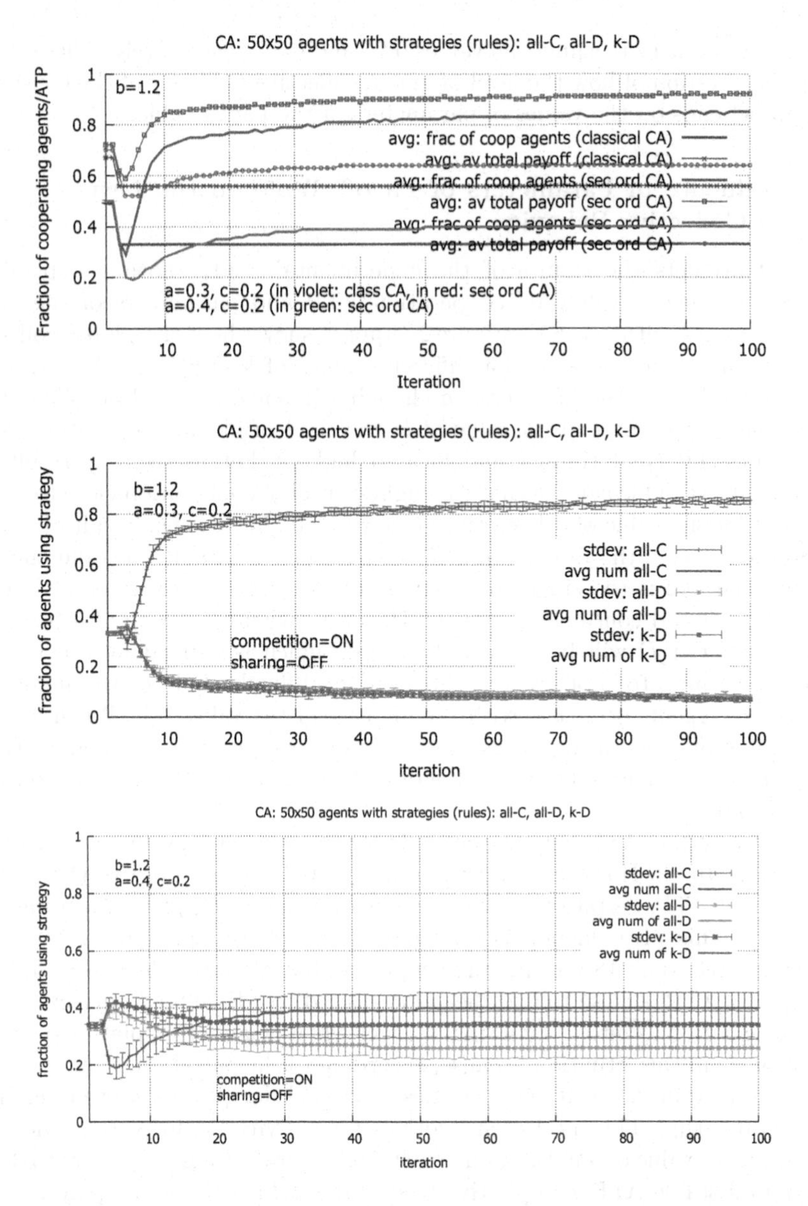

Fig. 1. Behavior of players in games without ISM for $b = 1.2$ and selected values of parameters a and b from payoff table: fraction of cooperating agents/average payoff of players in games with 3 parameters' settings (upper), fractions of agents' strategies in the game with $a = 0.3$ and $c = 0.2$ (middle), and fractions of agents' strategies in the game with $a = 0.4$ and $c = 0.2$ (lower). (Color figure online)

level of cooperation (we return to this issue later). Figure 1 (lower) shows some details of this game. We can see that the strategy *all–C* is still dominant but reaches only the value equal to 0.40, but *all–D* (in blue) and *k–D* (in green) are now much stronger reaching the values 0.26 and 0.34, respectively. These results raise the question of how the level of global collective behavior of the considered system depends on the values of parameters of the payoff table.

5.2 Influence of Parameters of Payoff Table on the Level of Collective Behavior

Figure 2 presents the results of the experimental study showing dependence between the level of players' cooperation/ATP and parameters of the payoff table in games without ISM. Figure 2 (upper) shows how the global collective behavior of players depends, for different values of b ($b = 1.1, 1.2, ..., 1.6$) and $c = 0$, on values of the parameter a changing in some range. Two plots of the same color represent a given value of b: the level of global cooperation (lower plot) and ATP (plot thinner and located higher). Let us see, e.g. results for $b = 1.2$ (in red). One can see that the highest level of cooperation close to 84% is achieved for the value of a from the range $(0 \ldots 0.25)$. It is the result of a small difference between the values b and $R = 1$ (see, payoff table). The difference not greater than 0.2 is too small to be attractive for a player to change its action from C to D and continue to play D. However, when the value of a increases, players selected D will be attracted by NE, which is defined by the value of a, and returning to cooperation will be more difficult. Indeed, the number of cooperating agents decreases with the increase of the value of a. When $a = 0.3$, the number of cooperating players suddenly drops to around 18% and continues to decrease with increasing the value of a. When $a = 0.7$ the cooperation level is close to 0.

Corresponding ATP behaves similarly for the game with values of a from the range $(0 \ldots 0.25)$. However, starting from $a = 0.3$, we can observe the increasing value of ATP, despite that the number of cooperating players continuously decreases. This phenomenon is related to the already mentioned role of NE. When $a = 0.3$, only 18% of players cooperate, but already 82% defect and corresponding ATP is close to 0.4, the value related to NE when all players defect. Changing action D into C by a single player when neighboring players play D is no rationale in this situation because, according to the payoff function, its payoff would drop from 0.4 to 0. NE becomes some lock for players preventing them from cooperating. This mechanism works stronger with the increase of the value of a. A higher value of a in the game results in faster decisions of players to defect and obtain stable ATP, competitive to unstable ATP received by players when they cooperate. Indeed, when $a = 0.4$ already 89% of players defect achieving ATP close to 0.4, when $a = 0.6$ near 92% of players defect with ATP close to 0.6, etc.

For other values of b, the behavior of players is generally similar, but we observe two patterns. The first one is the dependence between the maximal level of cooperation of players and the value of b. For a given value of b, this high level of cooperation can be maintained only over some range of values of a. A smaller value of b means that the higher level of cooperation is possible, and a wider range of values of a exists to support such cooperation. For $b = 1.1$ (in

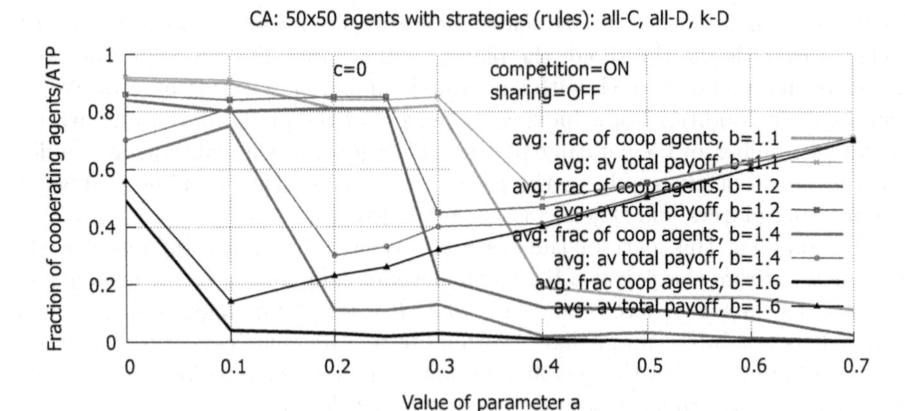

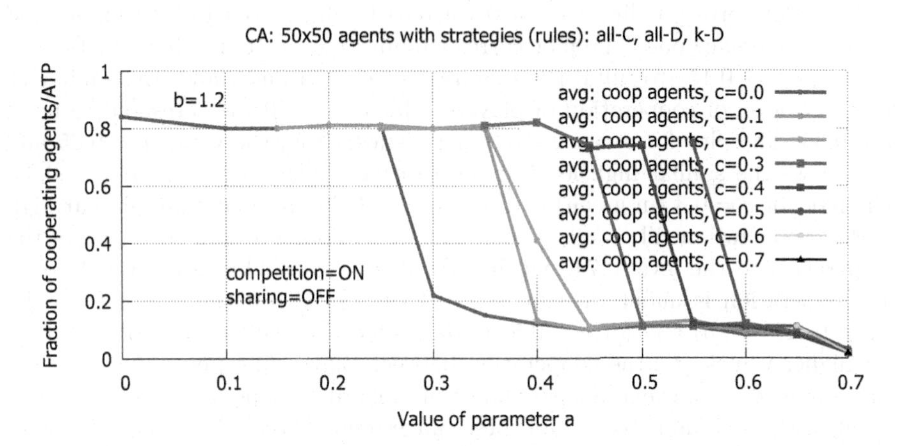

Fig. 2. Behavior of players in games without ISM as a function of payoff table parameters: influence of parameters b and a under $c = 0$ (upper), influence of parameters a and c under $b = 1.2$ (lower). (Color figure online)

orange) the level of cooperation in the range $(0.81 \ldots 0.91)$ can be observed for values of a from the range $(0 \ldots 0.3)$. For $b = 1.4$ (in green) we can observe the cooperation level in the range $(0.63 \ldots 0.72)$ until the threshold value of $a = 0.1$. For $b \geq 1.5$ some limited cooperation is possible but only for values of a close to 0. The second observed feature is that after crossing the threshold value of a, the behavior of players does not depend on the value of b but fully depends on the value of a setting NE. When $a = 0.4$ independently on the value of b, players play the games which are close to NE with ATP close to 0.4, and similarly for increasing values of a.

Looking at these results from the point of view of global optimization, we may say the following. For a given value of parameters b and a, and $c = 0$ players will tend to achieve the value of ATP corresponding to the global maximum (maximal value of ATP corresponding to a maximal number of cooperating players) when the value of a is below some threshold related to the value of b. When a exceeds the threshold, the players will achieve ATP corresponding to a local optimum related to NE with a maximal number of defecting players. We are interested in conditions of achieving by players of the global optimum with the maximal number of cooperating players. Therefore, in the subsequent studies, we will limit ourselves only to the issue of the maximization of the number of cooperating players and skip the issues related to ATP.

To complete this part of the research, we will study the influence of the parameter c from the payoff table on the level of cooperation in the system. We will restrict ourselves here to the case when $b = 1.2$ and a changes in the whole range. It is worth to remind that according to requirements concerning relations between parameters of the payoff function, it is necessary to have $a > c$. The results of the experiments are presented in Fig. 2 (lower).

The plot corresponding to $c = 0$ (in red) is taken from Fig. 2 (upper) and serves as reference one. The plot corresponding to $c = 0.1$ (in blue) starts from the value $a = 0.15$ (to fulfill the requirement $a > c$) and shows the similar as for $a = 0$ level of cooperation of players, which lasts till the threshold value of $a = 0.35$. The plot for $c = 0.2$ (in orange) starts from the value $a = 0.25$ and also shows the similar high level of cooperation which lasts also till $a = 0.35$. The plot for $c = 0.3$ (in green) starts from $a = 0.35$ and maintains similar high level of cooperation till $a = 0.4$, after that we can observe some decrease of the cooperation to the level 0.74, which ends at $a = 0.45$. This new lower level of the cooperation is maintained for $c = 0.4$ (in violet) for values of a from the range $(0.45 \ldots 0.5)$ and for $c = 0.5$ (in dark blue) in a single value of $a = 0.55$. For higher values of c the cooperation between players disappear. One can see that values of $c > 0$ extend possibility of emerging cooperation only in some specific ranges of values of a, and these ranges depend on c. We can understand now behavior of players presented in Fig. 1 (upper) in games with two different settings: ($a = 0.3$, $c = 0.2$) and ($a = 0.4$, $c = 0.2$). Looking at Fig. 2 (lower) we can see (plot in orange) that the first setting is in the range of values of the parameter a providing high level of players cooperation, while the second setting is outside this range.

Summarizing the presented results, we may say that emerging global cooperation is a phenomenon that depends on the values of the payoff function parameters and specific relations between values of them. We can see that it can be observed only in some ranges of values of these parameters. One of such specific settings of parameters is (b = 1.2, c = 0) (see, Fig. 2 (lower). A high degree of cooperation on the level 84% is possible only for the value of a from the range (0..0.25), and after crossing the threshold value of a, cooperation decreases significantly. The question which arises is the following. Is it possible some way of organization of local communities of players providing higher than the observed level of cooperation in the whole range of values of the parameter a? To answer this question, we will consider games with the income sharing mechanism (ISM).

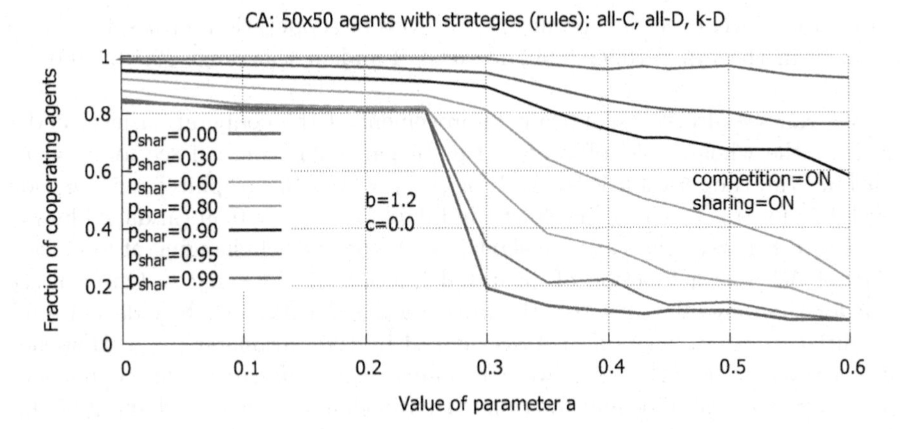

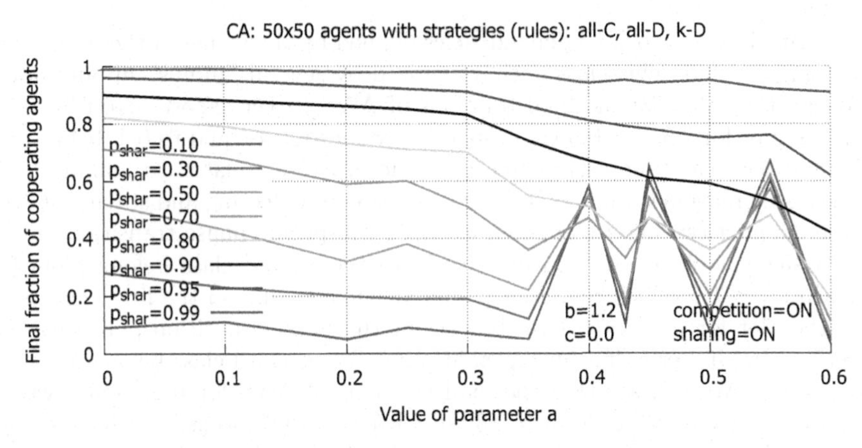

Fig. 3. Behavior of players in games with ISM for payoff table parameters a, $b = 1.2$ and $c = 0$: fraction of cooperating agents (upper), final fraction of players wishing to share income (lower). (Color figure online)

5.3 Games with Income Sharing Mechanism

Figure 3 (upper) shows how the level of cooperation depends on an initial fraction of agents wishing to share (AWS) their incomes with neighbors. This fraction changes from 0 (sharing is *off*, reference plot in red) to 0.99 and is set by the value of parameter p_{shar}. One can see that for $p_{shar} \leq 0.6$ and small values of $a \leq 0.25$, ISM can improve the level of cooperation only slightly (see, curve in orange). However, the situation significantly changes for values of $a \geq 0.3$. We can see that the increase in the level of cooperation under a given value of a depends on the value of p_{shar}.

For $a = 0.3$ the level of cooperation is equal to 0.22 when sharing is *off*. However, when it is *on* the cooperation level increases to 0.34 for $p_{shar} = 0.3$ (in green) and reaches 0.57 for $p_{shar} = 0.8$ (in orange). For $a = 0.5$ these values increase from 0.11 to 0.14 and 0.19, respectively. When $p_{shar} \geq 0.8$ the cooperation level exceeds seen until now levels of cooperation for $a \leq 0.3$ which are now in the range (0.92..0.99) for $a = 0$ and in the range (0.89..0.99) for $a = 0.3$.

We can see also the significant improvement of the cooperation level under ISM for the whole range of values of a. In particular, we can see for $a = 0.6$ reaching the cooperation level 0.58 for $p_{shar} = 0.9$ (in graphite), cooperation level 0.76 for $p_{shar} = 0.95$ (in violet) and 0.92 for $p_{shar} = 0.99$ (in dark blue).

Figure 3 (lower) gives some insight into the process of changing of final fractions of AWS as a function of an initial fraction of AWS and a. One can see that for the range $0 \leq a \leq 0.3$, the final fractions of AWS slightly decrease for all initial values of fraction of AWS defined by the parameter p_{shar}. However, after crossing this value of a, we can observe three classes of the dependence between the initial and final fractions of AWS characterized by changes of this dependency.

To the first class belong dependencies defined by the values of the parameter $0.1 \leq p_{shar} \leq 0.5$. This class is characterized by several "jumps" up and down reducing the values of the final fractions of AWS to very short spectrum (see, e.g., the plot for $p_{shar} = 0.5$ (in orange)) of values for a equal to 0.4, 0.43, 0.45, 0.55 and 0.6. They represent some dynamic processes related to making local decisions by players to choose either cooperation or defecting and staying at NE. The remaining two classes are characterized by specific phase transition points that visually look like a "convex lense." To the second class belong plots for $0.6 \leq p_{shar} \leq 0.8$ which represent a similar behavior like in the first class (see, e.g., the plot for $p_{shar} = 0.7$ (in blue)) but the phase transition points can be described by the term "fuzzy convex lenses." To the third class belong plots for $p_{shar} \geq 0.9$. We can see that the final fractions of AWS for representatives of this class decreases faster for $a > 0.3$ but without specific phase transitions. The speed of decreasing the final fractions of AWS depends on the value of p_{shar} (see, e.g., the plot for $p_{shar} = 0.95$ (in violet)).

The question which arises now is why ISM stimulates developing the cooperation between players. Observing space-time diagrams (not shown here) of changing several AWS, we can see the following. An initial number of AWS are distributed randomly in 2D space as isolated single players wishing to share incomes locally. During the game, the number of AWS changes, increasing or decreasing as a result of the influence of the game parameters, could achieve the final stable fraction of AWS. We can observe the process of creation of local clusters of AWS (players in a neighborhood wishing to share income), which ends, depending on the parameters of the game, reaching some number of isolated different size clusters or a network of clusters. Clusters of AWS represent players supporting local NE around the global optimum. Such local clusters perform actions C and receive the maximal value of the payoff. This value is stable due to converting by ISM this local solution into a local NE, which makes irrational potential players' decisions to defect. An increasing the value of p_{shar} results in the increase of the number of local NE towards converting the global optimum into the strict NE.

6 Conclusions

In this paper, we have studied the conditions of emergence of collective behavior in large second-order CA-based multi-agent systems with agents interacting according to principles of SPD game. We have shown that maximizing the average total payoff can only be observed for some values of the payoff function parameters and specific relations between them. We have proposed a mechanism of income sharing and have shown that it may lead to a significant enlarging the degree of cooperation in such systems. These results can be useful for solving distributed optimization problems in emerging computer-communication technologies. The concepts of collective behavior of automata are currently used in our studies intended to solve problems of scheduling related to cloud computing [3] and wireless sensor networks [15].

References

1. Brambilla, M., Ferrante, E., Birattari, M., Dorigo, M.: Swarm robotics: a review from the swarm engineering perspective. Swarm Intell. **7**(1), 1–41 (2013)
2. Fernández Domingos, E., et al.: Emerging cooperation in N-Person iterated Prisoner's dilemma over dynamic complex networks. Comput. Inform. **36**, 493–516 (2017)
3. Gąsior, J., Seredyński, F.: Security-aware distributed job scheduling in cloud computing systems: a game-theoretic cellular automata-based approach. In: Rodrigues, J.M.F., et al. (eds.) ICCS 2019. LNCS, vol. 11537, pp. 449–462. Springer, Cham (2019). https://doi.org/10.1007/978-3-030-22741-8_32
4. Ishibuchi, H., Namikawa, N.: Evolution of iterated prisoner's dilemma game strategies in structured demes under random pairing in game playing. IEEE Trans. Evol. Comput. **9**(6), 552–561 (2005)
5. Katsumata, Y., Ishida, Y.: On a membrane formation in a spatio-temporally generalized Prisoner's dilemma. In: Umeo, H., Morishita, S., Nishinari, K., Komatsuzaki, T., Bandini, S. (eds.) ACRI 2008. LNCS, vol. 5191, pp. 60–66. Springer, Heidelberg (2008). https://doi.org/10.1007/978-3-540-79992-4_8
6. Khaluf, Y., Ferrante, E., Simoens, P., Huepe, C.: Scale invariance in natural and artificial collective systems: a review. J. R. Soc. Interface **14**, 20170662 (2017)
7. Nowak, M.A., May, R.M.: Evolutionary games and spatial chaos. Nature **359**, 826 (1992)
8. Osborne, M.: An Introduction to Game Theory. Oxford University Press, New York (2009)
9. Peleteiro, A., Burguillo, J.C., Bazzan, A.L.: Emerging cooperation in the spatial IPD with reinforcement learning and coalitions. In: Bouvry, P., González-Vélez, H., Kołodziej, J. (eds.) Intelligent Decision Systems in Large-Scale Distributed Environments. Studies in Computational Intelligence, vol. 362, pp. 187–206. Springer, Heidelberg (2011). https://doi.org/10.1007/978-3-642-21271-0_9
10. Rossi, F., Bandyopadhyay, S., Wolf, M., Pavone, M.: Review of multi-agent algorithms for collective behavior: a structural taxonomy. IFAC-PapersOnLine **51**(12), 112–117 (2018). IFAC Workshop on Networked & Autonomous Air & Space Systems NAASS 2018

11. Seredynski, F.: Loosely coupled distributed genetic algorithms. In: Davidor, Y., Schwefel, H.-P., Männer, R. (eds.) PPSN 1994. LNCS, vol. 866, pp. 514–523. Springer, Heidelberg (1994). https://doi.org/10.1007/3-540-58484-6_294
12. Seredyński, F.: Competitive coevolutionary multi-agent systems: the application to mapping and scheduling problems. J. Parallel Distrib. Comput. **47**(1), 39–57 (1997)
13. Seredyński, F., Gąsior, J.: Collective behavior of large teams of multi-agent systems. In: De La Prieta, F., et al. (eds.) PAAMS 2019. CCIS, vol. 1047, pp. 152–163. Springer, Cham (2019). https://doi.org/10.1007/978-3-030-24299-2_13
14. Seredyński, F., Gąsior, J.: Collective behavior of large teams of multi-agent systems. In: Highlights of Practical Applications of Survivable Agents and Multi-Agent Systems. The PAAMS Collection - International Workshops of PAAMS 2019, Ávila, Spain, 26–28 June 2019, Proceedings, pp. 152–163 (2019)
15. Tretyakova, A., Seredynski, F., Bouvry, P.: Graph cellular automata approach to the maximum lifetime coverage problem in wireless sensor networks. Simulation **92**(2), 153–164 (2016)
16. Wolfram, S.: A New Kind of Science. Wolfram Media (2002)
17. Östberg, P., Byrne, J., et al.: Reliable capacity provisioning for distributed cloud/edge/fog computing applications. In: 2017 European Conference on Networks and Communications (EuCNC), pp. 1–6, June 2017

Learning Step-Size Adaptation in CMA-ES

Gresa Shala[1], André Biedenkapp[1(✉)], Noor Awad[1], Steven Adriaensen[1], Marius Lindauer[2], and Frank Hutter[1,3]

[1] University of Freiburg, Freiburg im Breisgau, Germany
{shalag,biedenka,awad,adriaens,fh}@cs.uni-freiburg.de
[2] Leibniz University Hannover, Hanover, Germany
lindauer@tnt.uni-hannover.de
[3] Bosch Center for Artificial Intelligence, Renningen, Germany

Abstract. An algorithm's parameter setting often affects its ability to solve a given problem, e.g., population-size, mutation-rate or crossover-rate of an evolutionary algorithm. Furthermore, some parameters have to be adjusted dynamically, such as lowering the mutation-strength over time. While hand-crafted heuristics offer a way to fine-tune and dynamically configure these parameters, their design is tedious, time-consuming and typically involves analyzing the algorithm's behavior on simple problems that may not be representative for those that arise in practice. In this paper, we show that formulating dynamic algorithm configuration as a reinforcement learning problem allows us to automatically learn policies that can dynamically configure the mutation step-size parameter of Covariance Matrix Adaptation Evolution Strategy (CMA-ES). We evaluate our approach on a wide range of black-box optimization problems, and show that (i) learning step-size policies has the potential to improve the performance of CMA-ES; (ii) learned step-size policies can outperform the default Cumulative Step-Size Adaptation of CMA-ES; and transferring the policies to (iii) different function classes and to (iv) higher dimensions is also possible.

Keywords: Evolutionary algorithms · Reinforcement learning · Algorithm configuration

1 Introduction

Designing algorithms requires careful design of multiple components. Having the foresight of how these components will interact for all possible applications is an infeasible task. Therefore, instead of hard-wiring algorithms, human developers often expose difficult design decisions as parameters of the algorithm [26]. To make the algorithm usable off-the-shelf, they provide a default configuration that is a myopic compromise for different use-cases and often leads to sub-optimal performance on new applications.

G. Shala and A. Biedenkapp—Equal Contribution.

T. Bäck et al. (Eds.): PPSN 2020, LNCS 12269, pp. 691–706, 2020.
https://doi.org/10.1007/978-3-030-58112-1_48

Automated algorithm configuration can alleviate users from the burden of having to manually configure an algorithm and exceeds human performance in a wide variety of domains [5,7,27,29,42,43]. One shortcoming, however, is that the learned configuration is static. In practice, many algorithms are of an iterative nature and might require different parameter configurations at different stages of their execution. In evolutionary algorithms this kind of "parameter control" is often achieved through so-called self-adaptive mechanisms [2,9,34]. Based on some statistics of the algorithm's behavior, self-adaptation adjusts the parameter on-the-fly and thereby directly influences the algorithm's execution.

Similarly in the well-known CMA-ES [19] the step-size is adapted based on the observed evolution path by a handcrafted heuristic, called CSA [25]. Through this step-size control, CMA-ES is able to avoid premature convergence of the population [21]. However, designing heuristics to adapt not only over a time-horizon but also to the task at hand is more difficult than to simply expose the parameters and configure them at every step.

In this work, we aim to strike a balance between self-adaptive mechanisms and automated algorithm configuration by making use of dynamic algorithm configuration (DAC) [10]. Instead of only learning the optimal *initial* step-size and adapting that by a handcrafted heuristic throughout the run of the algorithm, we learn a DAC policy in a fully automatic and data-driven way that determines how the step-size should be adjusted during the CMA-ES execution.

To learn DAC policies, we make use of guided policy search (GPS) [37], a commonly used reinforcement learning (RL) technique, originating from the robotics community, capable of learning complex non-linear policies from fairly few trials. Our choice for this particular method was motivated by its capability to learn simple first-order optimizers from scratch [39]. An appealing feature of GPS is that it allows us to employ known adaptation schemes as teacher mechanism to warm-start the search. This learning paradigm allows the agent to simply imitate the teacher if it was already optimal for a specific problem, while learning to do better in areas where the teacher struggled to perform well.

We study the potential of this DAC approach to step-size adaptation in CMA-ES for a variety of black-box optimization problems. One important open question so far is how such data-driven approaches can generalize to different settings (e.g., longer optimization runs, higher-dimensional problems or different problem classes) that were not observed during training. More specifically, our contributions are:

1. We address the problem of learning step-size control for CMA-ES from a reinforcement learning perspective;
2. We propose how to model the state space, action space and reward function;
3. To use guided policy search for learning a DAC policy in efficient way, we propose to use a strong teacher guidance.
4. We empirically demonstrate that our learned DAC policies are able to outperform CMA-ES' handcrafted step-size adaptation;

5. We demonstrate the generality of our DAC approach by transferring the learned policies to (i) functions of higher dimensions, (ii) unseen test function and (iii) to a certain degree to longer optimization trajectories.

2 Related Work

Parameter Control using Reinforcement Learning. The potential generality of DAC via RL is widely recognized [1,10,33] and RL has been applied to various specific parameter control settings [8,12,15,18,33,45,46,49,51]. However, RL covers a wide variety of techniques, and our methodology differs from prior-art in the area, in two important ways. First, GPS learns configuration policies *offline*, while most previous research considers the online setting. They attempt to learn how to adapt the parameters of an algorithm "while it is being used", i.e. without separate training phase. While desirable, online learning introduces a challenging exploration-exploitation trade-off. Also, experience is typically not transferred across runs, similar to hand-crafted adaptation mechanisms. That being said, prior-art considering the offline setting does exist, e.g., Battiti et al. [8] for local search SAT solvers and Sharma et al. [51] for EA. Second, GPS belongs to the family of *policy search* methods, which are often able to handle partially observable state spaces and continuous actions spaces better than previously used value-based RL methods.

Black-Box Dynamic Algorithm Configuration. In a sense, our methodology more closely resembles static algorithm configuration (AC) than traditional RL approaches. We represent configuration policies as a neural network; and as in AC, train it offline. Instead of GPS, black-box optimizers, e.g. ES, could also be used to optimize these weights [17,31,50]. In fact, prior-art exists that performs DAC using static AC methods [3,6,32,35]. A limitation of these "black-box" approaches is that they are unaware of the dynamic nature of the problem [1], e.g. which configurations where used at each time step, and how this affected execution. As a consequence, they are not sample-efficient, and practical applications with long trajectories are forced to consider restrictive policy spaces.

Learning to Optimize in Machine Learning. Learning to Learn (L2L) is a form of meta-learning, aiming to use machine learning methods to learn better machine learning systems [53]. Research in the area has recently surged in popularity, resulting in various applications learning better neural network architectures [41,56], hyper-parameters [44], initialization [16], and optimizers [4,11,13,14, 39,55]. As we are learning a component of an optimizer, our work is closely related to *Learning to Optimize* (L2O). Note that most L2O research [4,14,39,55] focuses on learning better gradient-based methods (e.g. Adam [36]), as these are most commonly used to optimize neural networks. Notable exceptions are L2O applications to single-point [13] and multi-point [11] black-box optimization. One L2O approach [4,11,13] models iterative optimizers as a kind of recurrent neural network and trains it in a fully supervised fashion. More general RL methods have also been used in L2O, e.g. REPS [14], PPO [55], and GPS [39]. In this work, we apply GPS in a similar way. The main difference is that, instead of learning a simple first-order optimization method from scratch, we apply this method to dynamically configure a single parameter (step-size) in a state-of-the-art derivative-free method (CMA-ES).

3 Background on CMA-ES

Covariance Matrix Adaptation Evolution Strategy (CMA-ES) [24] is an evolutionary algorithm optimizing a continuous black-box function $f : \mathbb{R}^n \rightarrow \mathbb{R}$ by sampling individuals from a non-stationary multivariate normal search distribution $\mathcal{N}\left(m^{(g)}, {\sigma^{(g)}}^2 C^{(g)}\right)$, with mean $m^{(g)}$ (center), step-size $\sigma^{(g)}$ (scale) and covariance matrix $C^{(g)}$ (shape).

Initially, $C^{(0)} = I$ (identity matrix) and $m^{(0)}$, $\sigma^{(0)}$ are provided by the user. The algorithm then iteratively updates this search distribution to increase the probability of sampling successful individuals. Each generation g, CMA-ES first samples λ individuals $x_1^{(g+1)}, ..., x_\lambda^{(g+1)}$ and chooses the best μ points as the parents of generation *g+1*. Then CMA-ES shifts the mean by a weighted average of μ selected steps:

$$m^{(g+1)} = m^{(g)} + c_m \sum_{i=1}^{\mu} w_i \left(x_{i:\lambda}^{(g+1)} - m^{(g)}\right). \tag{1}$$

where $x_{i:\lambda}$ denotes the i-th best point in terms of the function value and c_m is a learning rate which is usually set to 1. Next, covariance matrix adaptation is performed, which amounts to learning a second order model of the underlying objective function. To control the step-size CMA-ES uses *Cumulative Step Length Adaptation (CSA)* [21]:

$$\sigma^{(g+1)} = \sigma^{(g)} exp\left(\frac{c_\sigma}{d_\sigma}\left(\frac{||p_\sigma^{(g+1)}||}{E||\mathcal{N}(0, I)||} - 1\right)\right), \tag{2}$$

where $c_\sigma < 1$ is the learning rate, $d_\sigma \approx 1$ is the damping parameter, and $p_\sigma^{(g+1)} \in \mathbb{R}^n$ is the conjugate evolution path at generation *g+1*:

$$p_\sigma^{(g+1)} = (1 - c_\sigma)p_\sigma^{(g)} + \sqrt{c_\sigma(2 - c_\sigma)\mu_{eff}}\, C^{(g)-\frac{1}{2}} \frac{m^{(g+1)} - m^{(g)}}{\sigma^{(g)}}. \tag{3}$$

Note that alternatives for CSA have been proposed, e.g. making use of a success rule [30] or facilitate two-point step-size adaptation [20]. More generally, further research has resulted in many variants of CMA-ES suitable for a variety of different problems. A highly modular framework [48] exists that enables easy choice between 4 608 different versions of CMA-ES. This framework has further been used to demonstrate that, theoretically, switching only once during a run between configurations can yield performance improvements [47]. Simply using the switching rules proposed therein did not yield robust results in practice, but could be improved upon to yield better results [54].

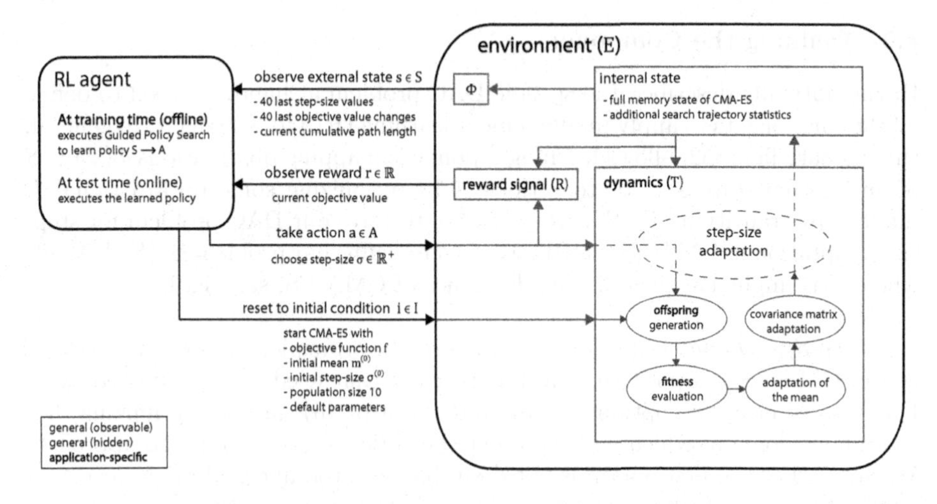

Fig. 1. Interaction of the RL agent with CMA-ES.

4 Learning Step-Size Adaptation

In this section, we will first discuss how we can model the adaptation of the step-size of CMA-ES as a dynamic algorithm configuration problem and propose to use guided policy search to efficiently find well-performing step-size policies.

4.1 The General Objective

The general objective is to adjust step-size $\sigma^{(g+1)}$ for generation $g+1$ based on some state information s_g on how CMA-ES behaved so far. To achieve that, a probabilistic policy π is responsible for the adjustment:

$$\sigma^{(g+1)} \sim \pi(s_g) \tag{4}$$

Along the lines of DAC, we further say that a policy should not only perform well on a single function f, but generalize to many functions $f \in \mathcal{F}$. Note that the policy must not only depend on features of the search trajectory, but could also be enriched by context information about the function at hand. This allows the policy to easily distinguish between different functions and their characteristics.

Dynamic algorithm configuration allows us to minimize an arbitrary cost function $c : \Pi \times \mathcal{F} \rightarrow \mathbb{R}$ that defines how well our algorithm, here CMA-ES, performed by using a policy $\pi \in \Pi$ on a function $f \in \mathcal{F}$. Therefore, our objective is to find a policy π^* that optimally adjusts σ across a set of functions[1]:

$$\pi^* \in \underset{\pi \in \Pi}{\arg\min} \sum_{f \in \mathcal{F}} c(\pi, f) \tag{5}$$

[1] We assume that the cost function is well-defined such that an optimal policy exists.

4.2 Defining the Components

Having formally described the specific DAC problem at hand, we need to define all the components to apply reinforcement learning (RL) for solving it. In general, the RL paradigm [52] allows learning a policy π mapping observations $\Phi(s') \in S$ of an internal state $s' \in S'$ to actions A by optimizing some reward signal R induced by transitions $T : S' \times A \rightarrow S'$. So, to solve our DAC problem for step-size adaptation in CMA-ES via RL, we need to define our problem as $\langle S, A, T, R \rangle$, where T is implicitly given by the dynamics of CMA-ES; see Fig. 1.

The Step-Size Domain and the Action Space. In principle, the step-size parameter of CMA-ES is a positive, continuous scalar that needs to be adjusted by π. For this, we have two options: (i) discretizing it or (ii) directly optimizing in a continuous domain, which will represent the action space for our RL approach. We argue that the first option is not desirable, as a too fine grid might lead to a large action space with many potentially irrelevant choices; whereas a too coarse grid might not contain all relevant choices. Hence, we model A as the continuous domain of the step size parameter.

State Representation of CMA-ES. To enable DAC for the step-size, it is crucial that S encodes sufficient information about the CMA-ES run. Given that our aim is to learn from, and possibly improve over, the performance of CSA for step-size control, we encode the information CSA uses in the state. Additionally, we include information on the optimization process by keeping a history of a fixed number h of past step-size values (in our experiments, $h = 40$) and past objective values. Specifically, our chosen state description contains:

1. the current step-size value $\sigma^{(g)}$ (see Eq. 2)
2. the current cumulative path length $p_\sigma^{(g)}$ (see Eq. 2)
3. the history of changes in objective value (i.e. the differences between successive objective values from h previous iterations)
4. the step-size history from h previous iterations[2]

The Cost Function and the Reward. The overall objective of CMA-ES is to find the minimizer of a function f at hand. So, we can say that the cost function should directly reflect the function value found by CMA-ES. Because the optimization budget (e.g., the number of allowed function evaluations) is not always known beforehand, we argue that it is desired to optimize for any-time performance. Since RL maximizes a cumulative reward over time, we can simply define the reward function per step (i.e. generation) as the negative function value of the current incumbent. By doing that, we optimize for any-time performance.

[2] When such a long history is not available yet, the missing values are filled with zeros.

4.3 Using Guided Policy Search for Efficient Learning of the Policy

Prior work showed [49,51] that value-based RL can be used for learning a DAC policy. However, this approach is typically not very sample-efficient, making it a very expensive approach in general. For example, Biedenkapp et al. [10] needed more than 10 000 algorithm runs to learn a simple sigmoid function.

A key insight of our work is that RL does not need to learn a well-performing policy from scratch but can use existing self-adaptive heuristics as a teacher. Here, for example, we propose to use CSA as a teacher to learn a policy that either imitates or improves upon it. In addition to better learning stability of the policy, we will show in our experiments that learning a policy for step-size adaptation is comparably cheap by using less than 1 000 runs of CMA-ES.

Similar to Li and Malik [39] in learning to optimize, we propose to use guided policy search (GPS) under unknown dynamics [37] to learn step-size policies. In essence GPS learns arbitrary parameterized policies through supervised learning by fitting a policy to guiding trajectories [37,38]. From teaching trajectories, a teacher distribution is computed such that it maximizes the reward and the agreement with the current policy. The policy parameters are then updated in a supervised fashion such that new sample trajectories produced by the policy do not deviate too far from the teacher. For a detailed explanation we refer to [37].

4.4 Extending GPS-Based DAC by a Stronger Teacher

For our purposes, we initialize the teacher to fit trajectories generated by CMA-ES with CSA. This initial teacher thus closely resembles CSA. As GPS updates the teacher over time to improve the reward, the teacher is likely to move away from CSA over time as only student and teacher are constrained to stay close to each other. If both teacher and student stray too far from CSA, the learned policy might not be able to recover CSAs behaviour in cases where it is beneficial. Thus we would like to encourage the student policy to also continually learn from CSA, to gain a more diverse teaching experience.

Instead of restricting the student policies through hard divergence criterion to not go too far away from CSA, we propose to add additional sample trajectories from running CMA-ES with CSA and not only the teacher to train the student policy. Thereby CSA acts as an additional fixed teacher. We extend GPS by introducing a *sampling rate* to determine the fraction of sample trajectories obtained from CSA when training the policy. Finally, in order to ensure exploration during learning, the initial step-size values and values for the mean of the initial distribution for the functions in the training set are randomly sampled from a uniform distribution.

5 Experiments

In this section we empirically evaluate the effectiveness of our proposed approach. We demonstrate the ability of our policies to generalize to new settings.

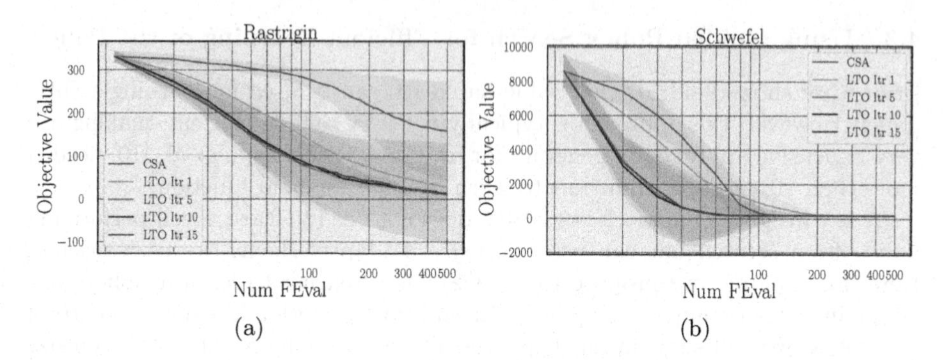

Fig. 2. Performance comparison of CMA-ES default step-size adaptation (CSA) to that of our methods incumbent policy after 1, 5, 10 and 15 training iterations on the Rastrigin function (a) and the Schwefel function (b).

5.1 Setup

For guided policy search we used the implementation provided by Li and Malik [39]. We incorporated our policy[3] in the python version of CMA-ES (pycma) in version 2.7.0 [22]. We only optimized the step-size adaptation with our approach and left all other pycma parameters as specified by the default, except we used a fixed population size of 10. As functions, we used a representative set with different characteristics as introduced by Hansen et al. in the BBOB-2009 [23].

We used 10 runs of SMAC [28, 40] to tune the initial step-size of CSA for each of the 10 considered BBOB functions individually, giving us a strong baseline. On the unseen functions we used an initial step-size of 0.5. In all experiments we used the same initial step-size for both our method and the baseline.

We trained our step-size policy for 50 steps (i.e. generations) of CMA-ES. We model the policy as a neural network consisting of two hidden layers with 50 hidden units each and ReLU activations. During training, the trajectory samples are obtained from the teaching policy with a probability of 0.7, whereas with a probability of 0.3 we sample trajectories from running CMA-ES with CSA. We obtain the final policy after training for 15 iterations of GPS.

We show performance comparisons of CMA-ES with the learned policy for step-size adaptation and CMA-ES with CSA from 25 independent runs of each method. The tables show an estimate of how likely it is for our learned policy to outperform CSA, based on pairwise comparisons of final objective values from the 25 runs for each method. The online appendix[4] describes this metric in detail, including its relation to statistical significance[5].

---

[3] Code and trained policies available at https://github.com/automl/LTO-CMA.

[4] https://ml.informatik.uni-freiburg.de/papers/20-PPSN-LTO-CMA.pdf.

[5] Estimates $\geq 0.64 \implies$ our learned policy significantly outperformed CSA ($\alpha = 0.05$).

Table 1. Probability of our method to outperform the baseline when training with different sampling rates. 1.0 indicates that we always outperform the baseline and 0.0 indicates we are always outperformed. The best sampling rate per function are marked in bold.

	Sampling Rate									
	0.0	0.1	0.2	0.3	0.4	0.5	0.6	0.7	0.8	0.9
BentCigar	0.53	0.84	0.46	**0.96**	0.38	0.33	0.14	0.26	0.25	0.08
Discus	0.00	0.66	0.23	**0.74**	0.34	0.35	0.30	0.37	0.29	0.32
Ellipsoid	0.59	**0.97**	0.51	**0.97**	0.51	0.48	0.35	0.48	0.56	0.44
Katsuura	0.64	0.91	0.66	**0.96**	0.64	0.63	0.63	0.64	0.64	0.61
Rastrigin	0.81	0.94	0.83	**1.00**	0.97	0.87	0.79	0.85	0.79	0.80
Rosenbrock	0.67	0.28	0.43	**0.89**	0.61	0.17	0.12	0.51	0.57	0.22
Schaffers	0.75	0.68	0.87	0.78	0.92	**0.98**	0.45	0.57	0.90	0.94
Schwefel	0.93	**1.00**	**1.00**	**1.00**	**1.00**	**1.00**	**1.00**	**1.00**	**1.00**	**1.00**
Sphere	0.77	0.92	0.48	0.78	0.58	0.25	0.94	**0.99**	0.93	0.94
Weierstrass	0.35	**1.00**	0.54	**1.00**	0.32	0.52	0.58	0.52	0.49	0.42
Average	0.60	0.82	0.60	0.91	0.63	0.56	0.53	0.62	0.64	0.58

5.2 Function-Specific Policy

Comparison against our Teacher CSA. We begin by exploring our method's ability to learn step-size policies when trained on a single 10D function for which we sampled 18 different starting points. In each training iteration of GPS, we evaluated CMA-ES 5 times on all starting conditions. In most cases, we already learn a well performing policy after 10 training iterations, which amounts only to $18 \times 5 \times 10 = 900$ runs of CMA-ES. Figure 2 depicts the training performance of our learned step-size policy after 1, 5, 10 and 15 training iterations of GPS on the Rastrigin and Rosenbrock functions. From Fig. 2a we can see that even though our policy starts out with samples from the default step-size adaptation of CMA-ES, already after one iteration, the learned policy can outperform the hand-crafted baseline. After four more training steps, our learned policy continues improving and still outperforms CSA. Finally when having trained for 15 iterations, our learned policy readily outperforms CSA, leading not only to a much better final performance, but also to a much better anytime performance on the Rastrigin function. We observe a similar behaviour when training on the Schwefel function, but the learned policy does not drastically outperform CSA.

Studying the Sampling Rate. We further used this setting to determine the influence of training length and sampling rate on the final performance of our policies, see Table 1. The sampling rate is crucial for our method as it determines how similar the learned policy's behavior is to CSA.

The performance of the learned policy improved by introducing sample trajectories from CSA compared to only sampling from the time-varying linear

Table 2. Probability of our method to outperform the baseline (a) for varying trajectory lengths, when having only trained with trajectories of length 50, and (b) for different dimensions when training them on functions of dimension $5-30$ and applying the learned policies to functions of dimensionality > 30.

	Trajectory Length							Dimensions					
	50	100	150	200	250	500	1000	35	40	45	50	55	60
BentCigar	**0.89**	0.00	0.00	0.00	0.00	0.05	0.04	0.87	0.98	0.56	0.49	0.76	**1.00**
Discus	0.90	**0.95**	0.76	0.40	0.00	0.00	0.00	0.89	0.86	0.93	0.94	0.94	**0.97**
Ellipsoid	**0.94**	0.92	0.90	0.86	0.61	0.00	0.00	**1.00**	**1.00**	**1.00**	**1.00**	**1.00**	**1.00**
Katsuura	**1.00**	**1.00**	**1.00**	**1.00**	**1.00**	**1.00**	**1.00**	0.92	0.92	0.96	**1.00**	0.96	0.87
Rastrigin	**1.00**	0.81	0.80	0.83	0.92	0.73	0.74	**1.00**	**1.00**	**1.00**	**1.00**	**1.00**	**1.00**
Rosenbrock	**0.93**	0.77	0.78	0.90	0.62	0.24	0.04	**1.00**	**1.00**	**1.00**	**1.00**	**1.00**	**1.00**
Schaffers	**0.60**	0.55	0.40	0.39	0.48	0.39	0.57	0.31	0.58	0.78	**0.87**	0.76	0.74
Schwefel	**0.99**	0.52	0.76	0.79	0.87	0.84	0.65	**1.00**	0.96	0.96	**1.00**	**1.00**	0.98
Sphere	**0.89**	0.00	0.00	0.00	0.00	0.00	0.00	0.41	0.38	0.56	0.65	0.64	**0.72**
Weierstrass	0.97	0.97	0.89	0.92	**1.00**	**1.00**	**1.00**	0.97	**1.00**	0.95	**1.00**	**1.00**	0.93
Average	0.91	0.65	0.63	0.61	0.55	0.43	0.40	0.84	0.87	0.87	0.89	0.91	0.92

(a) Different Trajectory Lengths (b) Different # Dimensions

Gaussian teacher. Results on some functions are more strongly affected by this change, e.g. BentCigar, than others, such as Schwefel. The final row shows the average performance of the sampling rate over all 10 considered training functions. Further, it becomes apparent that a sampling rate of 0.3 results in the strongest performance of our method, indicating that sampling also from our default teacher can improve performance. As a conclusion of this meta-parameter study, we will use 0.3 for our following experiments on generalization.

Generalization to Longer Trajectory Length. Finally, in this setting we explore the capability of the agent to transfer to longer trajectories. During training we opted to limit the training samples to be of maximal length 50, which corresponds to 500 function evaluations, to keep the cost for training low. Naturally the question thus arises if it is possible to further make use of such policies on longer optimization trajectories. From Table 2a we can observe that, even if a policy is trained with trajectories of at most 500 function evaluations, the policies are generally capable of generalizing to optimization trajectories that are 5 times longer while struggling to generalize to even longer trajectories.[6] On functions where the learned policies perform very well, only a small performance decrease is noticeable over a longer trajectory. On other functions the final performance lacks behind that of the handcrafted baseline over longer optimization trajectories, whereas on Weierstrass, the opposite is the case. On average, we can observe a decline in final performance of our learned policy, the further the

[6] The learned policies outperform CSA on anytime performance as shown in the Appendix, but CSA is better in terms of end objective values.

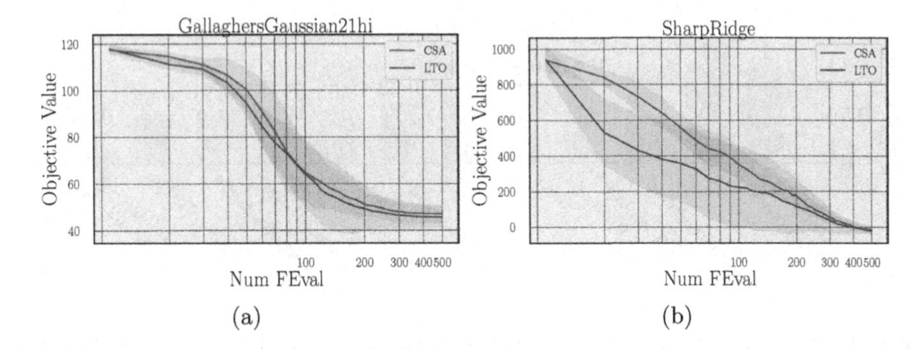

Fig. 3. Optimization trajectories of CMA-ES using CSA (blue) and our learned policy on two prior unseen test functions. The solid lines depict the mean performance and the shaded area the standard deviation over 25 repetitions.

optimization trajectory length is from the one used originally for training. A limiting factor as of yet is scaling the training to much longer trajectories. With increased trajectory length more training iterations will be needed to learn well performing policies.

5.3 Function-Class Specific Policy

We are generally not interested in policies that are only of use for one specific function; a more desirable policy would be capable of handling a broader range of functions. As a first step, we are interested in generalizing to similar functions of a specific function class. A very interesting, yet challenging task is hereby to generalize to higher dimensions. For this purpose we trained our policies on functions of dimension $5-30$ and evaluated them on dimensions $35-60$.

From Table 2b we can see that with increasing dimensionality, the probability that our policies outperform the handcrafted baseline actually *increases*. Upon inspection of the results, we see that with increasing dimensionality, the baseline optimization trajectories need more and more generations before reaching a good performance. Similarly, with increase in dimensionality, optimization trajectories guided by our policies require more generations to reach good final performances, however they are less affected by the dimensionality than the baseline. Especially on functions like Rosenbrock or Ellipsoid this effect seems to be very strongly pronounced. We can observe this trend for both training and testing our policies (see appendix for results on training).

5.4 Generalization to New Functions

Policies scaling to higher dimensions already promise great generalization capability. However, in practice, the problems, to which a solver is applied, could be fairly heterogeneous. To look into a more realistic scenario, we trained our agent on the 10 black-box functions we have mentioned before and assess its generalization capability on 12 black-box functions unseen during training.

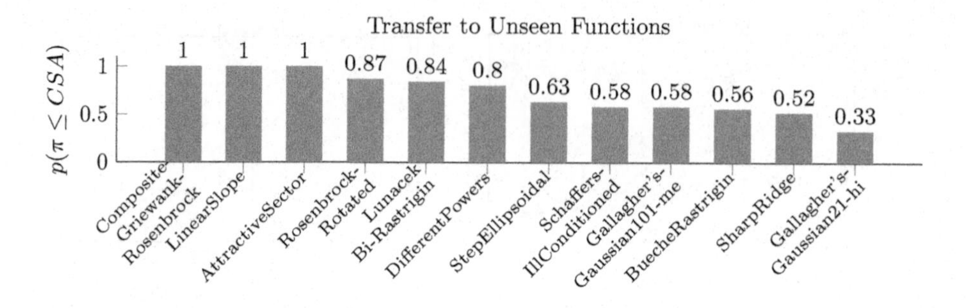

Fig. 4. Probability of our learned policies outperforming the default baseline on prior unseen test functions when training on all 10 BBOB functions.

Figure 3 shows two exemplary optimization trajectories that are achievable with our learned policies, compared to that of the default CSA. On Gallhager's Gaussian 21-hi we see that the optimization trajectory of CMA-ES with our learned policy closely resembles that of the handcrafted baseline as the step-sizes follow the same trend, see Fig. 3a. On SharpRidge (Fig. 3b) the learned policy is able to find well performing regions quicker; however in the end the baseline catches up.

Figure 4 summarizes the result for all 12 test functions. On 6 out of the 12 test functions, the learned policy significantly (≥ 0.64, $\alpha = 0.05$) outperformed the baseline, while being significantly outperformed (≤ 0.36) on one.

6 Conclusion

We demonstrated that we can automatically learn policies to dynamically configure the mutation step-size parameter of CMA-ES using reinforcement learning. To the best of our knowledge, we are the first to use policy search for the dynamic configuration of evolutionary algorithms, rather than value-based reinforcement learning. In particular, we described how *guided* policy search can be used to learn configuration policies, starting from a known handcrafted default policy. We conducted a comprehensive empirical investigation, and observed that (i) the learned policies are capable of outperforming the default policy on a wide range of black-box optimization problems; (ii) using a fixed teacher can further improve the performance; (iii) our learned policies can generalize to higher dimensions as well as to unseen functions.

These results open the door for promising future research in which policy search is used to learn policies that jointly configure multiple parameters (e.g. population *and* step-size) of CMA-ES. Another line of future research could improve the employed policy search mechanism, e.g. by learning from a variety of teachers at the same time. A more diverse set of teachers, might facilitate even better generalization as the learned policies could make use of strengths of individual teachers on varying problem domains. Finally, the development of a benchmark platform for dynamic algorithm configuration would facilitate apple-to-apple comparisons of different reinforcement learning techniques, driving future research.

Acknowledgements. The authors acknowledge funding by the Robert Bosch GmbH.

References

1. Adriaensen, S., Nowé, A.: Towards a white box approach to automated algorithm design. In: Kambhampati, S. (ed.) Proceedings of the 26th International Joint Conference on Artificial Intelligence (IJCAI 2016), pp. 554–560 (2016)
2. Aleti, A., Moser, I.: A systematic literature review of adaptive parameter control methods for evolutionary algorithms. ACM Comput. Surv. **49**(3), 56:1–56:35 (2016)
3. Andersson, M., Bandaru, S., Ng, A.H.: Tuning of multiple parameter sets in evolutionary algorithms. In: 2016 Proceedings of the Genetic and Evolutionary Computation Conference, pp. 533–540 (2016)
4. Andrychowicz, M., et al.: Learning to learn by gradient descent by gradient descent. In: Lee, D., Sugiyama, M., von Luxburg, U., Guyon, I., Garnett, R. (eds.) Proceedings of the 30th International Conference on Advances in Neural Information Processing Systems (NeurIPS 2016), pp. 3981–3989 (2016)
5. Ansótegui, C., Malitsky, Y., Sellmann, M.: MaxSAT by improved instance-specific algorithm configuration. In: Brodley, C., Stone, P. (eds.) Proceedings of the Twenty-Eighth National Conference on Artificial Intelligence (AAAI 2014), pp. 2594–2600. AAAI Press (2014)
6. Ansótegui, C., Pon, J., Sellmann, M., Tierney, K.: Reactive dialectic search portfolios for MaxSAT. In: Singh, S., Markovitch, S. (eds.) Proceedings of the Conference on Artificial Intelligence (AAAI 2017). AAAI Press (2017)
7. Ansótegui, C., Sellmann, M., Tierney, K.: A gender-based genetic algorithm for the automatic configuration of algorithms. In: Gent, I.P. (ed.) CP 2009. LNCS, vol. 5732, pp. 142–157. Springer, Heidelberg (2009). https://doi.org/10.1007/978-3-642-04244-7_14
8. Battiti, R., Campigotto, P.: An investigation of reinforcement learning for reactive search optimization. In: Hamadi, Y., Monfroy, E., Saubion, F. (eds.) Autonomous Search, pp. 131–160. Springer, Heidelberg (2011). https://doi.org/10.1007/978-3-642-21434-9_6
9. Battiti, R., Brunato, M., Mascia, F.: Reactive Search and Intelligent Optimization, vol. 45. Springer, Heidelberg (2008). https://doi.org/10.1007/978-0-387-09624-7
10. Biedenkapp, A., Bozkurt, H.F., Eimer, T., Hutter, F., Lindauer, M.: Dynamic algorithm configuration: foundation of a new meta-algorithmic framework. In: Lang, J., Giacomo, G.D., Dilkina, B., Milano, M. (eds.) Proceedings of the Twenty-fourth European Conference on Artificial Intelligence (ECAI 2020), June 2020
11. Cao, Y., Chen, T., Wang, Z., Shen, Y.: Learning to optimize in swarms. In: Wallach, H., Larochelle, H., Beygelzimer, A., d'Alché-Buc, F., Fox, E., Garnett, R. (eds.) Advances in Neural Information Processing Systems 32: Annual Conference on Neural Information Processing Systems 2019, (NeurIPS 2019), pp. 15018–15028 (2019)
12. Chen, F., Gao, Y., Chen, Z., Chen, S.: SCGA: controlling genetic algorithms with sarsa (0). In: International Conference on Computational Intelligence for Modelling, Control and Automation and International Conference on Intelligent Agents, Web Technologies and Internet Commerce (CIMCA-IAWTIC 2006), vol. 1, pp. 1177–1183. IEEE (2005)

13. Chen, Y., et al.: Learning to learn without gradient descent by gradient descent. In: Precup, D., Teh, Y. (eds.) Proceedings of the 34th International Conference on Machine Learning (ICML 2017), vol. 70, pp. 748–756. Proceedings of Machine Learning Research (2017)
14. Daniel, C., Taylor, J., Nowozin, S.: Learning step size controllers for robust neural network training. In: Schuurmans, D., Wellman, M. (eds.) Proceedings of the Thirtieth National Conference on Artificial Intelligence (AAAI 2016). AAAI Press (2016)
15. Eiben, A.E., Horvath, M., Kowalczyk, W., Schut, M.C.: Reinforcement learning for online control of evolutionary algorithms. In: Brueckner, S.A., Hassas, S., Jelasity, M., Yamins, D. (eds.) ESOA 2006. LNCS (LNAI), vol. 4335, pp. 151–160. Springer, Heidelberg (2007). https://doi.org/10.1007/978-3-540-69868-5_10
16. Finn, C., Abbeel, P., Levine, S.: Model-agnostic meta-learning for fast adaptation of deep networks. In: Precup, D., Teh, Y. (eds.) Proceedings of the 34th International Conference on Machine Learning (ICML 2017), vol. 70, pp. 1126–1135. Proceedings of Machine Learning Research (2017)
17. Fuks, L., Awad, N., Hutter, F., Lindauer, M.: An evolution strategy with progressive episode lengths for playing games. In: Kraus, S. (ed.) Proceedings of the Twenty-Eighth International Joint Conference on Artificial Intelligence (IJCAI), pp. 1234–1240. ijcai.org (2019)
18. Di Gaspero, L., Urli, T.: Evaluation of a family of reinforcement learning cross-domain optimization heuristics. In: Hamadi, Y., Schoenauer, M. (eds.) LION 2012. LNCS, pp. 384–389. Springer, Heidelberg (2012). https://doi.org/10.1007/978-3-642-34413-8_32
19. Hansen, N.: The CMA evolution strategy: a comparing review. In: Lozano, J., Larranaga, P., Inza, I., Bengoetxea, E. (eds.) Towards a New Evolutionary Computation. Studies in Fuzziness and Soft Computing, vol. 192, pp. 75–102. Springer, Heidelberg (2006). https://doi.org/10.1007/3-540-32494-1_4
20. Hansen, N.: CMA-ES with two-point step-size adaptation. arXiv:0805.0231 [cs.NE] (2008)
21. Hansen, N.: The CMA evolution strategy: a tutorial. arXiv:1604.00772v1 [cs.LG] (2016)
22. Hansen, N., Akimoto, Y., Baudis, P.: CMA-ES/pycma on GitHub. Zenodo, February 2019. https://doi.org/10.5281/zenodo.2559634
23. Hansen, N., Finck, S., Ros, R., Auger, A.: Real-parameter black-box optimization benchmarking 2009: noiseless functions definitions. Research report RR-6829, INRIA (2009)
24. Hansen, N., Ostermeier, A.: Convergence properties of evolution strategies with the derandomized covariance matrix adaptation: the $(\mu/\mu_I,\lambda)$-CMA-ES. In: Proceedings of the 5th European Congress on Intelligent Techniques and Soft Computing, pp. 650–654 (1997)
25. Hansen, N., Ostermeier, A.: Completely derandomized self-adaptation in evolution strategies. Evol. Comput. **9**, 159–195 (2001)
26. Hoos, H.: Programming by optimization. Commun. ACM **55**(2), 70–80 (2012)
27. Hutter, F., Hoos, H.H., Leyton-Brown, K.: Automated configuration of mixed integer programming solvers. In: Lodi, A., Milano, M., Toth, P. (eds.) CPAIOR 2010. LNCS, vol. 6140, pp. 186–202. Springer, Heidelberg (2010). https://doi.org/10.1007/978-3-642-13520-0_23

28. Hutter, F., Hoos, H.H., Leyton-Brown, K.: Sequential model-based optimization for general algorithm configuration. In: Coello, C.A.C. (ed.) LION 2011. LNCS, vol. 6683, pp. 507–523. Springer, Heidelberg (2011). https://doi.org/10.1007/978-3-642-25566-3_40
29. Hutter, F., Lindauer, M., Balint, A., Bayless, S., Hoos, H., Leyton-Brown, K.: The configurable SAT solver challenge (CSSC). Artif. Intell. **243**, 1–25 (2017)
30. Igel, C., Hansen, N., Roth, S.: Covariance matrix adaptation for multi-objective optimization. Evol. Comput. **15**, 1–28 (2001)
31. Igel, C.: Neuroevolution for reinforcement learning using evolution strategies. In: The 2003 Congress on Evolutionary Computation. CEC 2003, vol. 4, pp. 2588–2595. IEEE (2003)
32. Kadioglu, S., Sellmann, M., Wagner, M.: Learning a reactive restart strategy to improve stochastic search. In: Battiti, R., Kvasov, D.E., Sergeyev, Y.D. (eds.) LION 2017. LNCS, vol. 10556, pp. 109–123. Springer, Cham (2017). https://doi.org/10.1007/978-3-319-69404-7_8
33. Karafotias, G., Eiben, A., Hoogendoorn, M.: Generic parameter control with reinforcement learning. In: Proceedings of the 2014 Annual Conference on Genetic and Evolutionary Computation, pp. 1319–1326 (2014)
34. Karafotias, G., Hoogendoorn, M., Eiben, Á.: Parameter control in evolutionary algorithms: trends and challenges. IEEE Trans. Evol. Comput. **19**(2), 167–187 (2015)
35. Karafotias, G., Smit, S.K., Eiben, A.E.: A generic approach to parameter control. In: Di Chio, C., et al. (eds.) EvoApplications 2012. LNCS, vol. 7248, pp. 366–375. Springer, Heidelberg (2012). https://doi.org/10.1007/978-3-642-29178-4_37
36. Kingma, D., Ba, J.: Adam: a method for stochastic optimization. In: Proceedings of the International Conference on Learning Representations (ICLR 2015) (2015). Published online: iclr.cc
37. Levine, S., Abbeel, P.: Learning neural network policies with guided policy search under unknown dynamics. In: Ghahramani, Z., Welling, M., Cortes, C., Lawrence, N., Weinberger, K. (eds.) Proceedings of the 28th International Conference on Advances in Neural Information Processing Systems (NeurIPS 2014), pp. 1071–1079 (2014)
38. Levine, S., Koltun, V.: Guided policy search. In: Dasgupta, S., McAllester, D. (eds.) Proceedings of the 30th International Conference on Machine Learning (ICML 2013), pp. 1–9. Omnipress (2013)
39. Li, K., Malik, J.: Learning to optimize. In: Proceedings of the International Conference on Learning Representations (ICLR 2017) (2017). Published online: iclr.cc
40. Lindauer, M., Eggensperger, K., Feurer, M., Falkner, S., Biedenkapp, A., Hutter, F.: SMAC v3: Algorithm configuration in Python (2017). https://github.com/automl/SMAC3
41. Liu, H., Simonyan, K., Yang, Y.: DARTS: differentiable architecture search. In: Proceedings of the International Conference on Learning Representations (ICLR 2019) (2019). Published online: iclr.cc
42. López-Ibáñez, M., Dubois-Lacoste, J., Stützle, T., Birattari, M.: The irace package, iterated race for automatic algorithm configuration. Technical report, IRIDIA, Université Libre de Bruxelles, Belgium (2011). http://iridia.ulb.ac.be/IridiaTrSeries/IridiaTr2011-004.pdf
43. López-Ibáñez, M., Stützle, T.: Automatic configuration of multi-objective ACO algorithms. ANTS 2010. LNCS, vol. 6234, pp. 95–106. Springer, Heidelberg (2010). https://doi.org/10.1007/978-3-642-15461-4_9

44. Maclaurin, D., Duvenaud, D., Adams, R.: Gradient-based hyperparameter optimization through reversible learning. In: Bach, F., Blei, D. (eds.) Proceedings of the 32nd International Conference on Machine Learning (ICML 2015), vol. 37, pp. 2113–2122. Omnipress (2015)
45. Muller, S., Schraudolph, N., Koumoutsakos, P.: Step size adaptation in evolution strategies using reinforcement learning. In: Proceedings of the 2002 Congress on Evolutionary Computation. CEC 2002 (Cat. No. 02TH8600), vol. 1, pp. 151–156. IEEE (2002)
46. Pettinger, J., Everson, R.: Controlling genetic algorithms with reinforcement learning. In: Proceedings of the 4th Annual Conference on Genetic and Evolutionary Computation, pp. 692–692 (2002)
47. van Rijn, S., Doerr, C., Bäck, T.: Towards an adaptive CMA-ES configurator. In: Auger, A., Fonseca, C.M., Lourenço, N., Machado, P., Paquete, L., Whitley, D. (eds.) PPSN 2018. LNCS, vol. 11101, pp. 54–65. Springer, Cham (2018). https://doi.org/10.1007/978-3-319-99253-2_5
48. van Rijn, S., Wang, H., van Leeuwen, M., Bäck, T.: Evolving the structure of evolution strategies. In: 2016 IEEE Symposium Series on Computational Intelligence (SSCI), pp. 1–8. IEEE (2016)
49. Sakurai, Y., Takada, K., Kawabe, T., Tsuruta, S.: A method to control parameters of evolutionary algorithms by using reinforcement learning. In: Proceedings of the Sixth International Conference on Signal-Image Technology and Internet Based Systems, pp. 74–79. IEEE (2010)
50. Salimans, T., Ho, J., Chen, X., Sutskever, I.: Evolution strategies as a scalable alternative to reinforcement learning. arXiv:1703.03864 [stat.ML] (2017)
51. Sharma, M., Komninos, A., López-Ibáñez, M., Kazakov, D.: Deep reinforcement learning based parameter control in differential evolution. In: Proceedings of the Genetic and Evolutionary Computation Conference, pp. 709–717 (2019)
52. Sutton, R.S., Barto, A.G.: Reinforcement Learning: An Introduction. MIT Press, Cambridge (2018)
53. Thrun, S., Pratt, L.: Learning to Learn. Springer, Heidelberg (2012)
54. Vermetten, D., van Rijn, S., Bäck, T., Doerr, C.: Online selection of CMA-ES variants. In: Auger, A., Stützle, T. (eds.) Proceedings of the Genetic and Evolutionary Computation Conference (GECCO 2019), pp. 951–959. ACM (2019)
55. Xu, Z., Dai, A.M., Kemp, J., Metz, L.: Learning an adaptive learning rate schedule. arXiv:1909.09712 [cs.LG] (2019)
56. Zoph, B., Le, Q.V.: Neural architecture search with reinforcement learning. In: Proceedings of the International Conference on Learning Representations (ICLR 2017) (2017). Published online: iclr.cc

Sparse Inverse Covariance Learning for CMA-ES with Graphical Lasso

Konstantinos Varelas[1,2(✉)], Anne Auger[1], and Nikolaus Hansen[1]

[1] Inria, RandOpt Team, CMAP, École Polytechnique, Palaiseau, France
{konstantinos.varelas,anne.auger,nikolaus.hansen}@inria.fr
[2] Thales LAS France SAS, Limours, France

Abstract. This paper introduces a variant of the Covariance Matrix Adaptation Evolution Strategy (CMA-ES), denoted as gl-CMA-ES, that utilizes the Graphical Lasso regularization. Our goal is to efficiently solve partially separable optimization problems of a certain class by performing stochastic search with a search model parameterized by a sparse precision, i.e. inverse covariance matrix. We illustrate the effect of the global weight of the l_1 regularizer and investigate how Graphical Lasso with non equal weights can be combined with CMA-ES, allowing to learn the conditional dependency structure of problems with sparse Hessian matrices. For non-separable sparse problems, the proposed method with appropriately selected weights, outperforms CMA-ES and improves its scaling, while for dense problems it maintains the same performance.

1 Introduction

The challenge of estimating covariance or precision, i.e. inverse covariance matrices when the dimension is large, due to the need of large numbers of samples and to the cumulation of significant amounts of estimation errors, leads to the need of discovering efficient high dimensional estimation methods and developing corresponding tools. For this purpose, several approaches have been proposed, often with the assumption of sparsity properties of the matrix to be estimated, which include soft/hard thresholding, l_1 penalization, column-by-column estimation and others [6].

The Covariance Matrix Adaptation Evolution Strategy [9] is a gradient-free continuous optimization algorithm that addresses problems of the form

$$\underset{\mathbf{x}\in\mathcal{X}\subset\mathbb{R}^n}{\text{minimize}}\, f(\mathbf{x}) \tag{1}$$

by performing stochastic search in the parameter space $\mathcal{X}$ with an adaptive normal distribution $\mathcal{N}(\mathbf{m},\sigma^2\mathbf{C})$. The adaptation of the covariance matrix is performed with a learning rate of the order of $\mathcal{O}(1/n^2)$, n being the dimension. Large scale variants which impose different restrictions to the search model have been proposed, in order to increase the adaptation speed [1–3,16].

In this paper, we apply one of the high dimensional tools mentioned above, namely the Graphical Lasso [4,7] within CMA-ES, to obtain an algorithm which,

T. Bäck et al. (Eds.): PPSN 2020, LNCS 12269, pp. 707–718, 2020.
https://doi.org/10.1007/978-3-030-58112-1_49

in cases where the Hessian matrix of the objective function has a sparse structure, exploits this property. Eventually, the degrees of freedom of the search model, and thus the adaptation time, can be reduced without deteriorating the performance of CMA-ES.

The paper is organised as follows: in Sect. 2 we describe the fundamentals of CMA-ES as well as of estimation using the Graphical Lasso. In Sect. 3, we introduce the novel algorithm and in Sect. 4 we test the method in selected examples with known sparse structure. Finally, in Sect. 5 we summarize and discuss future steps for further improvement of our approach.

2 Graphical Lasso and CMA-ES

The adaptation of the search distribution in CMA-ES is based on principles of invariance. Through sampling candidate solutions from $\mathcal{N}(\mathbf{m}, \sigma^2\mathbf{C})$ and ranking the corresponding f values, CMA-ES estimates a better distribution via weighted recombination, which eventually generates solutions with improved f-values.

In cases where the objective function f in (1) is convex and quadratic, strong empirical evidence indicate that the search covariance matrix approximates the inverse Hessian of f up to a scalar. This property allows, after adapting $\mathbf{C}$, to optimize any convex quadratic function with the same convergence rate as the Sphere function $f(\mathbf{x}) = \|\mathbf{x}\|_2^2$, and efficiently address ill-conditioned problems.

2.1 Partial Separability, Hessian Sparsity and Conditionally Independent Models

In large-scale optimization, parsimony becomes essential, either for representing a Hessian matrix, or for developing update techniques and learning the matrix economically. The notion of partial separability [8,15] becomes useful in this case: f is called partially separable if it can be decomposed in a form:

$$f(\mathbf{x}) = \sum_{i=1}^{p} f_i(\mathbf{x}) \tag{2}$$

where each f_i has a large invariant subspace. We recall that the invariant subspace $\mathcal{N}_h$ of a function $h : \mathbb{R}^n \to \mathbb{R}$ is defined as the largest subspace of $\mathbb{R}^n$ such that for all vectors $\mathbf{y} \in \mathcal{N}_h$ and for all $\mathbf{x} \in \mathbb{R}^n$, the relation $h(\mathbf{x} + \mathbf{y}) = h(\mathbf{x})$ holds. In cases where each term f_i is a function of few search coordinates, this property is obviously satisfied (even though partial separability is in general a weaker assumption). If in addition f_i are twice differentiable, the Hessian matrix of f will be sparse.

Motivated by the above observations, we propose a method which is developed based on the CMA-ES principles and performs stochastic search with a distribution with conditionally independent coordinates. Our goal is to reduce if possible the number of free parameters of the search model, thus increase the adaptation speed, without deteriorating the performance of CMA-ES. In the case of a smooth objective function f, this would mean to exploit sparsity of the Hessian. However, no assumption is imposed on the regularity of f.

2.2 Graphical Lasso

The Graphical Lasso was introduced to estimate distributions with a sparse precision, i.e. inverse covariance, matrix. With this property, one introduces parametric models with conditionally independent search coordinates, a procedure also known as covariance selection [5]. In particular, if Σ is the sample estimation of a covariance matrix, the solution of

$$\underset{X \in \mathbb{S}^n_{++}}{\text{minimize}} \operatorname{tr}(\Sigma X) - \log\det X + \alpha \|X\|_1 \tag{3}$$

provides the sparse model estimation, where X represents the precision matrix to be estimated, $\mathbb{S}^n_{++}$ is the set of symmetric positive definite $n \times n$ matrices and the penalty factor α controls the tradeoff between the log-likelihood and the penalization term $\|X\|_1 = \sum_{i,j=1}^n |X_{ij}|$. For $\alpha = 0$, the solution X^* of (3) is $X^* = \Sigma^{-1}$, since the Kullback-Leibler divergence of the distributions parameterized by X and Σ is decomposed as:

$$\begin{aligned} D_{\mathrm{KL}}(\mathcal{N}(\mathbf{0},\Sigma)||\mathcal{N}(\mathbf{0},X^{-1})) &= \frac{1}{2}\left(\operatorname{tr}(\Sigma X) - n + \log\left(\frac{\det X^{-1}}{\det \Sigma}\right)\right) \\ &= \frac{1}{2}\left(\operatorname{tr}(\Sigma X) - \log\det X\right) - \frac{1}{2}\left(n + \log\det \Sigma\right) \end{aligned} \tag{4}$$

The l_1 penalization in (3) is employed to force sparsity on the precision matrix X, or equivalently on the partial correlation matrix $\operatorname{diag}X^{-1/2}X\operatorname{diag}X^{-1/2}$, and can be viewed as a convex relaxation of the number of non zero entries of X, $\mathbf{Card}(X)$ (which makes (3) a NP-hard problem [4]).

In a black-box scenario where the sparsity structure is unknown, estimating the precision matrix with the Graphical Lasso serves exactly the purpose of discovering this structure. In the context of CMA-ES, in order to learn sparse search models, the candidate solutions are generated from the regularized distribution that solves (3) and the original update rules are used. In practice, the Graphical Lasso is applied to standardized variables, thus when solving (3) Σ is the correlation matrix provided by the CMA-ES update.

2.3 Equal Weights and Effect on Conditioning

Problem (3) imposes the same penalization factor α on all precision entries, and the alternative regularization term $\alpha\|X^-\|_1 = \alpha\sum_{i\neq j}|X_{ij}|$, which penalizes only the off-diagonal entries, has been proposed e.g. in [14]. This kind of penalization leads to a consistent reduction of the axes length ratios learned by CMA-ES after the regularization step, as illustrated in Fig. 1 for a 4 dimensional block diagonal case.

Recently, tools for an extension of (3) with non-equal penalization factors, i.e. for solving:

$$\underset{X \in \mathbb{S}^n_{++}}{\text{minimize}} \operatorname{tr}(\Sigma X) - \log\det X + \sum_{i,j=1}^{n} \alpha_{ij}|X_{ij}| \tag{5}$$

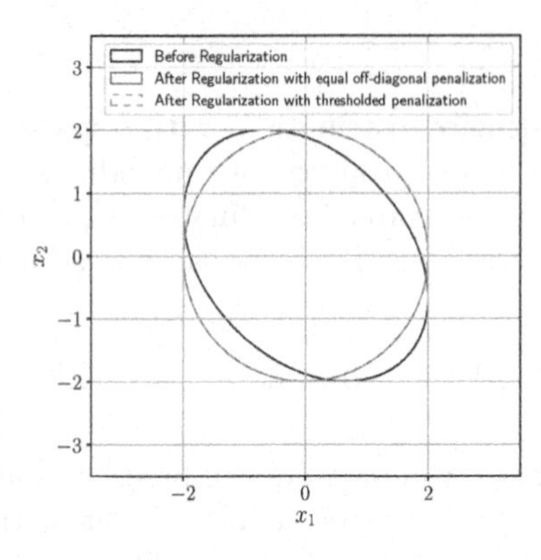

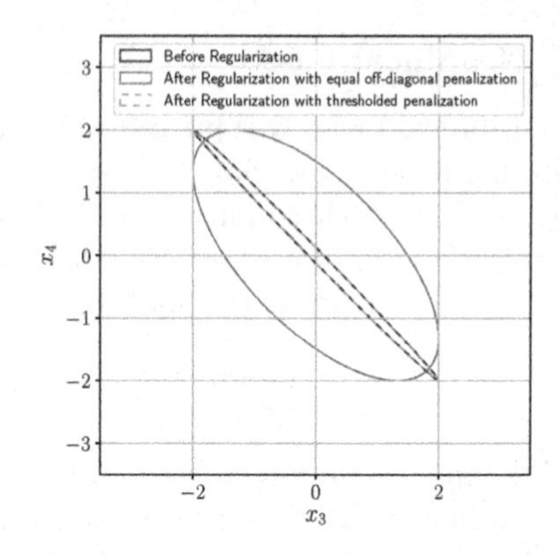

Fig. 1. Equal-density ellipses of the marginal distributions of (x_1, x_2) and (x_3, x_4) before and after regularization with off-diagonal only and with thresholded penalty factors. The random vector $\mathbf{x} = (x_1, \dots, x_4)$ is distributed as $\mathbf{x} \sim \mathcal{N}(\mathbf{0}, \mathbf{C})$ where $\mathbf{C}$ is a block-diagonal covariance matrix of the form $\mathbf{C} = \left(\begin{smallmatrix} \mathbf{C}_1 & \mathbf{0} \\ \mathbf{0} & \mathbf{C}_2 \end{smallmatrix}\right)$, $\mathbf{C}_1, \mathbf{C}_2$ being of size 2×2. The solid magenta line illustrates the effect of regularization when only off-diagonal elements are penalized with the same factor α. The factor value is chosen as the minimal value that achieves an isotropic distribution for the pair of weakly dependent variables (x_1, x_2) (left, with an axis ratio $\sqrt{2}$), i.e. $\alpha = 1/3$ in this particular example. The search distribution of the strongly dependent pair (x_3, x_4) (right, with an axis ratio $\sqrt{1000}$) is drastically affected. The dashed line (green) corresponds to thresholded regularization according to (5). In this case, only the precision matrix entry corresponding to the pair (x_1, x_2) is penalized, i.e. the chosen factors are: $\alpha_{ij} = 1/3$ if $(i, j) \in \{(1, 2), (2, 1)\}$ else $\alpha_{ij} = 0$. (Color figure online)

with selected $\alpha_{ij} \geq 0$ have been developed [12]. In the following, this formulation is used along with a simple rule for selecting the penalty factors in order to surpass the above effect: precision entries are penalized only if the corresponding partial correlations, i.e. the entries of $\mathrm{diag} X^{-1/2} X \mathrm{diag} X^{-1/2}$, are below a threshold τ.

3 Algorithm

In this section we introduce the proposed algorithm, denoted as gl-CMA-ES. It only uses recombination with positive weights for the update of the covariance matrix, in order to ensure its positive definiteness. The differences with respect to the original CMA-ES setting with positive recombination weights [1] are highlighted in Algorithm 1, while Algorithm 2 describes the regularization step. The minimization step in line 6 of Algorithm 2 is solved using [12].

For reasons of stability, and since the number of degrees of freedom for the covariance matrix is $n(n+1)/2$, the corresponding learning rate in CMA-ES

[1] An extension of CMA-ES, called Active CMA-ES [10], that performs recombination with both positive and negative weights using all sampled solutions has been proposed.

is of the order of $\mathcal{O}(1/n^2)$. In other large scale variants of CMA-ES, e.g. the Separable CMA-ES [16], the degrees of freedom of the search model are reduced and the adaptation is performed faster. Similarly, in our approach the learning rates depend on the number of non zero entries of the Lasso estimated precision matrix, ranging from $\mathcal{O}(1/n)$ for sparse to $\mathcal{O}(1/n^2)$ for dense matrices. Furthermore, limited memory methods have been proposed [11,13], aiming at reducing the internal space and time complexity of CMA-ES. Such methods, though, do not exploit properties such as separability in order to accelerate the convergence [11,17].

The algorithm coincides with CMA-ES if the threshold is chosen as $\tau = 0$, that is if the l_1 penalization is not applied. If this holds, the sampling matrix $\mathbf{C}_{t+1}^{\text{reg}}$ is equal to $\mathbf{C}_t$ in line 4 of Algorithm 1, thus the candidate solutions are sampled from $\mathcal{N}(\mathbf{m}_t, \sigma_t^2 \mathbf{C}_t)$, see lines 9 and 10. Additionally, the evolution path for the adaptation of the step size follows the same update rule as in CMA-ES, see line 14 of Algorithm 1. The learning rates c_1, c_μ are defined in a compatible way with CMA-ES in line 6, when the precision matrix is fully dense, i.e. when $n_z = n^2$.

Note that the invariance property to strictly monotonic transformations of the objective function f that CMA-ES possesses is maintained in the algorithm. However, invariance to affine transformations of the search space breaks when regularization is applied, i.e. when setting $\tau > 0$.

4 Results

We present experimental results on representative problems included in Table 1, in order to verify whether the proposed approach is able to identify the correct sparse structure of the objective function's Hessian matrix. All experiments were performed with an initial step size $\sigma_0 = 1$ and with a starting point $\mathbf{x}_0$ defined in Table 1.

The first test function f_{ellisub} is constructed by composing the Ellipsoid function f_{elli} with a rotational search space transformation as defined in Table 1. This results in a non-separable ill-conditioned problem with maximal sparsity, since the upper triangular part of the Hessian of f_{ellisub} has exactly one non zero entry. Figure 2 presents the gain in convergence speed in terms of number of function evaluations of gl-CMA-ES compared to the default CMA-ES (with positive recombination weights). It also shows the performance scaling with dimension, compared to other large scale variants of CMA-ES, namely the Separable CMA-ES [16], the VkD-CMA-ES [1] and the dd-CMA-ES [3], as well as with the Active CMA-ES [10], i.e. the algorithm that uses the entire sample population additionally with negative recombination weights.

The second test case is the non-convex Rosenbrock function $f_{\text{rosen}}(\mathbf{x}) = \sum_{i=1}^{n-1} 100(x_{i+1} - x_i^2)^2 + (1 - x_i)^2$, for which the Hessian matrix is (globally) tridiagonal. Figure 3 presents the speed-up obtained for different values of the threshold parameter τ and the scaling with dimension. In dimension $n = 5$ the convergence speed is almost the same with the speed of CMA-ES, while in

Algorithm 1. gl-CMA-ES

1: **Set parameters:** $\lambda = 4 + \lfloor 3 \ln n \rfloor$, $\mu = \lfloor \lambda/2 \rfloor$, $w_i = \frac{\ln(\mu+\frac{1}{2})-\ln i}{\sum_{j=1}^{\mu} \ln(\mu+\frac{1}{2})-\ln j}$ for $i = 1 \ldots \mu$, $\mu_w = \frac{1}{\sum_{i=1}^{\mu} w_i^2}$, $c_\sigma = \frac{\mu_w+2}{n+\mu_w+3}$, $d_\sigma = 1 + 2\max\{0, \sqrt{\frac{\mu_w-1}{n+1}} - 1\} + c_\sigma$, $c_c = \frac{4+\mu_w/n}{n+4+2\mu_w/n}$,
2: **Initialize:** $\mathbf{p}_t^c \leftarrow \mathbf{0}$, $\mathbf{p}_t^\sigma \leftarrow \mathbf{0}$, $\mathbf{C}_t \leftarrow \mathbf{I}$, $t \leftarrow 0$, τ,
3: **while** termination criteria not met **do**
4: $\quad \mathbf{C}_{t+1}^{\text{reg}} \leftarrow \text{REGULARIZE}(\mathbf{C}_t, \tau)$,
5: $\quad n_z \leftarrow \# {\mathbf{C}_{t+1}^{\text{reg}}}^{-1} > 0$,
6: $\quad c_1 \leftarrow \frac{2}{(n_z/n+1.3)(n+1.3)+\mu_w}$, $c_\mu \leftarrow \min\{1 - c_1, 2\frac{\mu_w+1/\mu_w-1.75}{(n_z/n+2)(n+2)+\mu_w}\}$,
7:
8: $\quad$ **for** $k \leftarrow 1, \ldots, \lambda$ **do**
9: $\quad\quad \mathbf{z}_k \sim \mathcal{N}(\mathbf{0}, \mathbf{C}_{t+1}^{\text{reg}})$
10: $\quad\quad \mathbf{x}_k \leftarrow \mathbf{m}_t + \sigma_t \mathbf{z}_k$
11: $\quad\quad f_k \leftarrow f(\mathbf{x}_k)$
12: $\quad$ **end for**
13: $\quad \mathbf{m}_{t+1} \leftarrow \sum_{k=1}^{\mu} w_k \mathbf{x}_{k:\mu}$
14: $\quad \mathbf{p}_{t+1}^\sigma \leftarrow (1 - c_\sigma)\,\mathbf{p}_t^\sigma + \sqrt{c_\sigma(2 - c_\sigma)\mu_w}\,{\mathbf{C}_{t+1}^{\text{reg}}}^{-\frac{1}{2}} \frac{\mathbf{m}_{t+1}-\mathbf{m}_t}{\sigma_t}$,
15: $\quad h_\sigma \leftarrow \mathbb{1}_{\|\mathbf{p}_{t+1}^\sigma\| < (1.4+\frac{2}{n+1})\sqrt{1-(1-c_\sigma)^{2(t+1)}}\mathbb{E}\|\mathcal{N}(\mathbf{0},\mathbf{I})\|}$,
16: $\quad \delta(h_\sigma) \leftarrow (1 - h_\sigma)c_c(2 - c_c)$,
17: $\quad \mathbf{p}_{t+1}^c \leftarrow (1 - c_c)\,\mathbf{p}_t^c + h_\sigma \sqrt{c_c(2 - c_c)\mu_w} \frac{\mathbf{m}_{t+1}-\mathbf{m}_t}{\sigma_t}$,
18: $\quad \mathbf{C}_{t+1}^\mu \leftarrow \sum_{k=1}^{\mu} w_k \mathbf{z}_{k:\mu} {\mathbf{z}_{k:\mu}}^T$,
19: $\quad \mathbf{C}_{t+1} \leftarrow (1 + c_1\delta(h_\sigma) - c_1 - c_\mu)\mathbf{C}_t + c_1 \mathbf{p}_{t+1}^c {\mathbf{p}_{t+1}^c}^T + c_\mu \mathbf{C}_{t+1}^\mu$,
20: $\quad \sigma_{t+1} \leftarrow \sigma_t \exp\left(\frac{c_\sigma}{d_\sigma}\left(\frac{\|\mathbf{p}_{t+1}^\sigma\|}{\mathbb{E}\|\mathcal{N}(\mathbf{0},\mathbf{I})\|} - 1\right)\right)$,
21: $\quad t \leftarrow t + 1$
22: **end while**

Algorithm 2. Regularization

1: **function** REGULARIZE($\mathbf{C}$, τ) $\quad \triangleright$ $\mathbf{C}$ is a covariance matrix
2: $\quad \tilde{\mathbf{C}} \leftarrow \text{diag}\mathbf{C}^{-1/2}\mathbf{C}\text{diag}\mathbf{C}^{-1/2}$,
3: $\quad \mathbf{P} \leftarrow \tilde{\mathbf{C}}^{-1}$
4: $\quad \tilde{\mathbf{P}} \leftarrow \text{diag}\mathbf{P}^{-1/2}\mathbf{P}\text{diag}\mathbf{P}^{-1/2}$,
5: $\quad \mathbf{W}_{ij} \leftarrow 1$ if $\tilde{\mathbf{P}}_{ij} < \tau$ else 0
6: $\quad \mathbf{P}_{\text{reg}} \leftarrow \arg\min_{\boldsymbol{\Theta} \in \mathbb{S}_{++}^n} \text{tr}(\tilde{\mathbf{C}}\boldsymbol{\Theta}) - \log\det\boldsymbol{\Theta} + \sum_{i,j=1}^{n} \mathbf{W}_{ij}|\theta_{ij}|$ $\quad \triangleright$ Initialized at $\mathbf{P}$
7: $\quad \tilde{\mathbf{C}}_{\text{reg}} \leftarrow {\mathbf{P}_{\text{reg}}}^{-1}$
8: $\quad$ **return** $\text{diag}\mathbf{C}^{1/2}\tilde{\mathbf{C}}_{\text{reg}}\text{diag}\mathbf{C}^{1/2}$
9: **end function**

dimension $n = 80$, the method becomes more than 3 times faster. The conditional dependency graphs learned by the proposed approach for 2 different values of τ are shown in Fig. 4.

Table 1. Benchmark functions. The matrix $\mathbf{R}$ ($\mathbf{R}_k$) is a random 2×2 ($k \times k$) rotation matrix drawn from the Haar distribution in SO(2) (SO(k) respectively). The block diagonal matrix $\mathbf{B}$ has the form $\mathbf{B} = \begin{pmatrix} \mathbf{B}_1 & \mathbf{0} \\ \mathbf{0} & \mathbf{B}_2 \end{pmatrix}$, where $\mathbf{B}_1$ and $\mathbf{B}_2$ are random rotation matrices of size $\frac{n}{2} \times \frac{n}{2}$, $\mathbf{Q}$ is a random rotation matrix of size $n \times n$ and $\mathbf{P}_1, \mathbf{P}_2$ are random permutation matrices.

Name	Definition	$\mathbf{x}_0$
Sphere	$f_{\text{sphere}}(\mathbf{x}) = \sum_{i=1}^{n} x_i^2$	$3 \cdot \mathbf{1}$
Ellipsoid	$f_{\text{elli}}(\mathbf{x}) = \sum_{i=1}^{n} 10^{6\frac{i-1}{n-1}} x_i^2$	$3 \cdot \mathbf{1}$
Cigar	$f_{\text{cig}}(\mathbf{x}) = x_1^2 + \sum_{i=2}^{n} 10^{6\frac{i-1}{n-1}} x_i^2$	$3 \cdot \mathbf{1}$
Tablet	$f_{\text{tab}}(\mathbf{x}) = 10^6 x_1^2 + \sum_{i=2}^{n} x_i^2$	$3 \cdot \mathbf{1}$
Twoaxes	$f_{\text{twoax}}(\mathbf{x}) = \sum_{i=1}^{\lfloor n/2 \rfloor} 10^6 x_i^2 + \sum_{i=\lfloor n/2 \rfloor+1}^{n} x_i^2$	$3 \cdot \mathbf{1}$
Subspace Rotated Ellipsoid	$f_{\text{ellisub}}(\mathbf{x}) = \sum_{i=2}^{n-1} 10^{6\frac{i-1}{n-1}} x_i^2 + \begin{pmatrix} x_1 & x_n \end{pmatrix} \mathbf{R}^T \begin{pmatrix} 1 & 0 \\ 0 & 10^6 \end{pmatrix} \mathbf{R} \begin{pmatrix} x_1 \\ x_n \end{pmatrix}$	$3 \cdot \mathbf{1}$
Rosenbrock	$f_{\text{rosen}}(\mathbf{x}) = \sum_{i=1}^{n-1} 100(x_{i+1} - x_i^2)^2 + (1 - x_i)^2$	$\mathbf{0}$
2-Blocks Ellipsoid	$f_{\text{2-blocks elli}}(\mathbf{x}) = f_{\text{elli}}(\mathbf{Bx})$	$3 \cdot \mathbf{1}$
2-Blocks Cigar	$f_{\text{2-blocks cig}}(\mathbf{x}) = f_{\text{cig}}(\mathbf{Bx})$	$3 \cdot \mathbf{1}$
2-Blocks Tablet	$f_{\text{2-blocks tab}}(\mathbf{x}) = f_{\text{tab}}(\mathbf{Bx})$	$3 \cdot \mathbf{1}$
Permuted 2-Block Rotated Ellipsoid	$f_{\text{perm ellisub}}(\mathbf{x}) = f_{\text{elli}}(\mathbf{P}_2\mathbf{B}\mathbf{P}_1\mathbf{x})$	$3 \cdot \mathbf{1}$
Rotated Ellipsoid	$f_{\text{ellirot}}(\mathbf{x}) = f_{\text{elli}}(\mathbf{Qx})$	$3 \cdot \mathbf{1}$
k-Rotated Quadratic	$f_{k\text{-rot}}(\mathbf{x}) = \sum_{i=1}^{n-k} x_i^2 + \begin{pmatrix} x_{n-k+1} & \cdots & x_n \end{pmatrix} \mathbf{R}_k^T \begin{pmatrix} \mathbf{I}_{k-1} & \mathbf{0} \\ \mathbf{0} & 10^6 \end{pmatrix} \mathbf{R}_k \begin{pmatrix} x_{n-k+1} \\ \vdots \\ x_n \end{pmatrix}$	$3 \cdot \mathbf{1}$

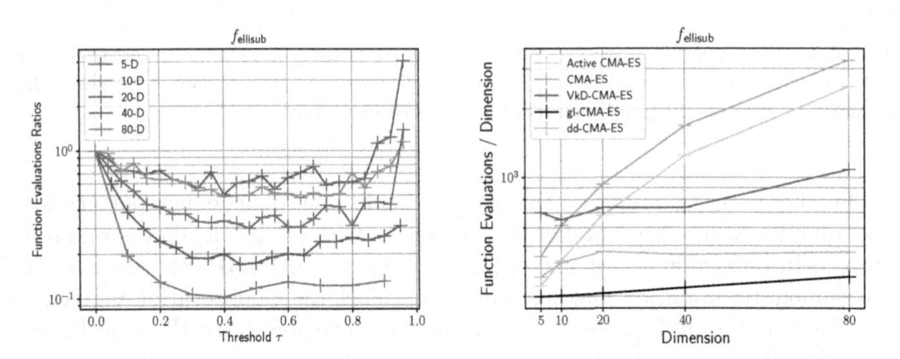

Fig. 2. Left: Ratios of number of function evaluations performed in single runs to reach a precision 10^{-10} close to the global optimal f value of the proposed approach over CMA-ES depending on the regularization threshold τ. One run has been performed per each value of τ. Right: Scaling with dimension of the average number of function evaluations to reach a 10^{-10} precision. The average is taken over 10 independent runs of each algorithm. The chosen threshold value is $\tau = 0.4$.

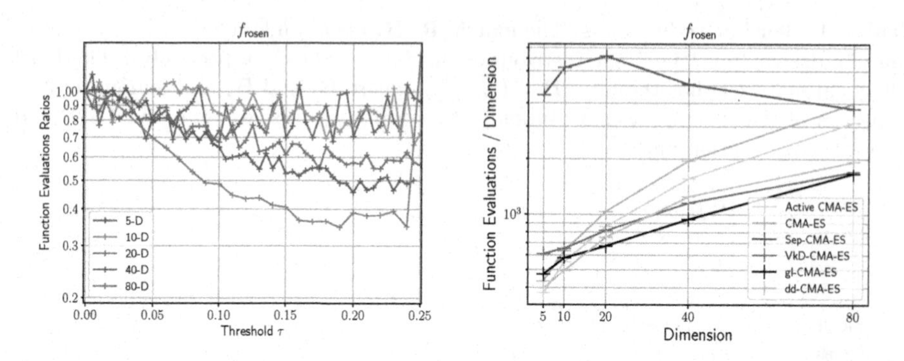

Fig. 3. Left: Ratios of number of function calls performed in single runs to reach a precision of 10^{-10} of the proposed approach over CMA-ES. Right: Performance scaling for $\tau = 0.24$ with averaging over 10 independent runs.

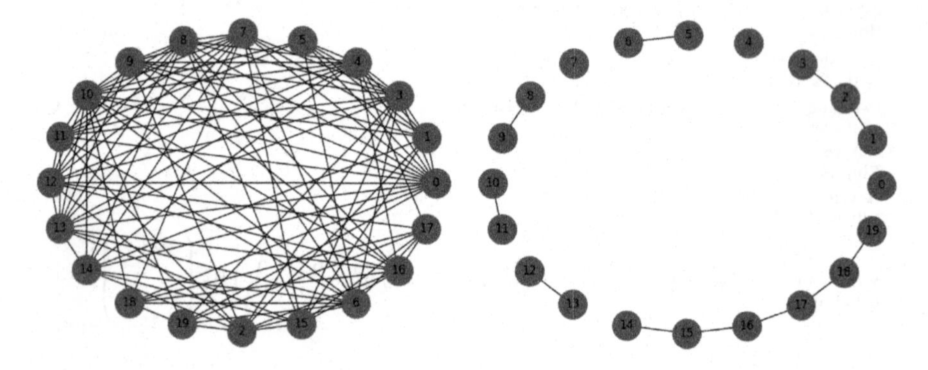

Fig. 4. Conditional dependency graphs in the last iteration of a single run for thresholds $\tau = 0.05$ (left) and $\tau = 0.24$ (right) on the 20-dimensional Rosenbrock function. Edges depict non zero off-diagonal entries of the precision matrix.

Furthermore, we illustrate the learned conditional dependency pattern for a test function where the number of non zero entries of the Hessian is quadratic with n. In particular, we define $f_{\text{perm ellisub}}(\mathbf{x}) = f_{\text{elli}}(\mathbf{P}_2\mathbf{B}\mathbf{P}_1\mathbf{x})$, where $\mathbf{P}_1, \mathbf{P}_2$ are random permutation matrices and $\mathbf{B}$ a 2-block diagonal rotation matrix, see also Table 1. Figure 5 presents the graph that corresponds to the true Hessian sparsity pattern and the final conditional dependency graph resulting from gl-CMA-ES.

The next example is the function $f_{k-\text{rot}}$, defined in Table 1, which results from an ill-conditioned separable function after performing a (random) rotation in a $k-$dimensional subspace of $\mathbb{R}^n$. This forms a group of k strongly dependent search coordinates (with high probability) and the Hessian's sparsity decreases with increasing k. Figure 6 illustrates the convergence speed for different threshold values and for varying values of k. Threshold values between 0.3 and 0.5 reveal similar and close to optimal performance.

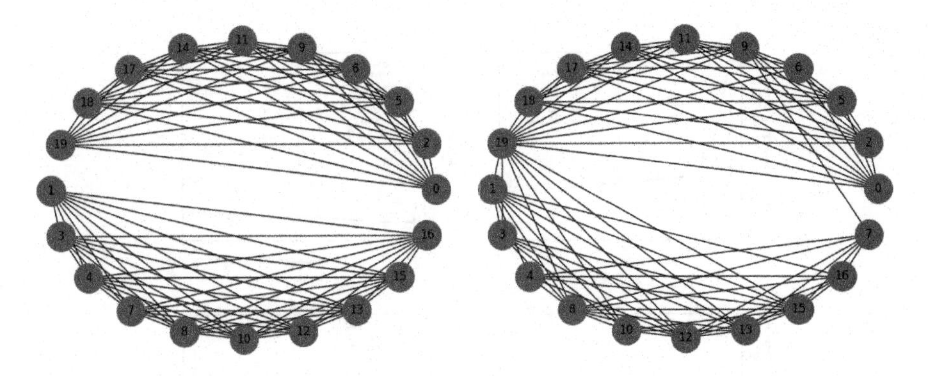

Fig. 5. Adjacency graph of the true Hessian matrix (left) and conditional dependency graph in the last iteration of a single run of gl-CMA-ES with $\tau = 0.1$ on $f_{\text{perm ellisub}}$.

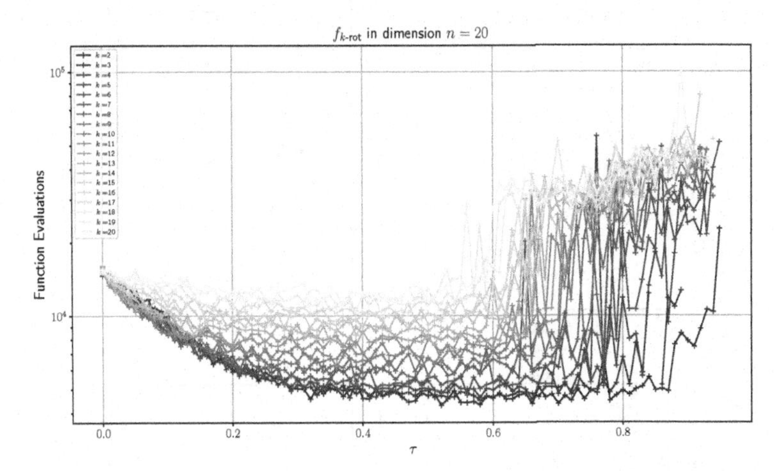

Fig. 6. Number of function evaluations performed by single runs of gl-CMA-ES to reach the global optimum of $f_{k-\text{rot}}$, for different values of k versus the threshold τ. The sparsity of the objective function's Hessian is determined by the block size k. Missing values of the number of function evaluations (typically for large threshold values, which lead to an axis-parallel search model) correspond to single runs where gl-CMA-ES does not reach the optimum within a precision of 10^{-10}. The maximal gain in convergence speed is observed when the sparsity is maximal, i.e. for $k = 2$.

Finally, the performance scaling on the rest of the benchmark functions of Table 1 is shown in Fig. 7 for selected threshold parameter values, chosen after preliminary experimentation in a way that the estimated precision's sparsity pattern is not less rich than the Hessian's true pattern. Separable problems allow a choice of a large value for τ and we obtain a behaviour very similar to Separable CMA-ES. Also, the scaling of the method on sparse non-separable problems such as the $f_{\text{2-blocks tablet}}$ and $f_{\text{2-blocks elli}}$ functions is advantageous over all other methods, with the exception of dd-CMA-ES, which shows better scaling

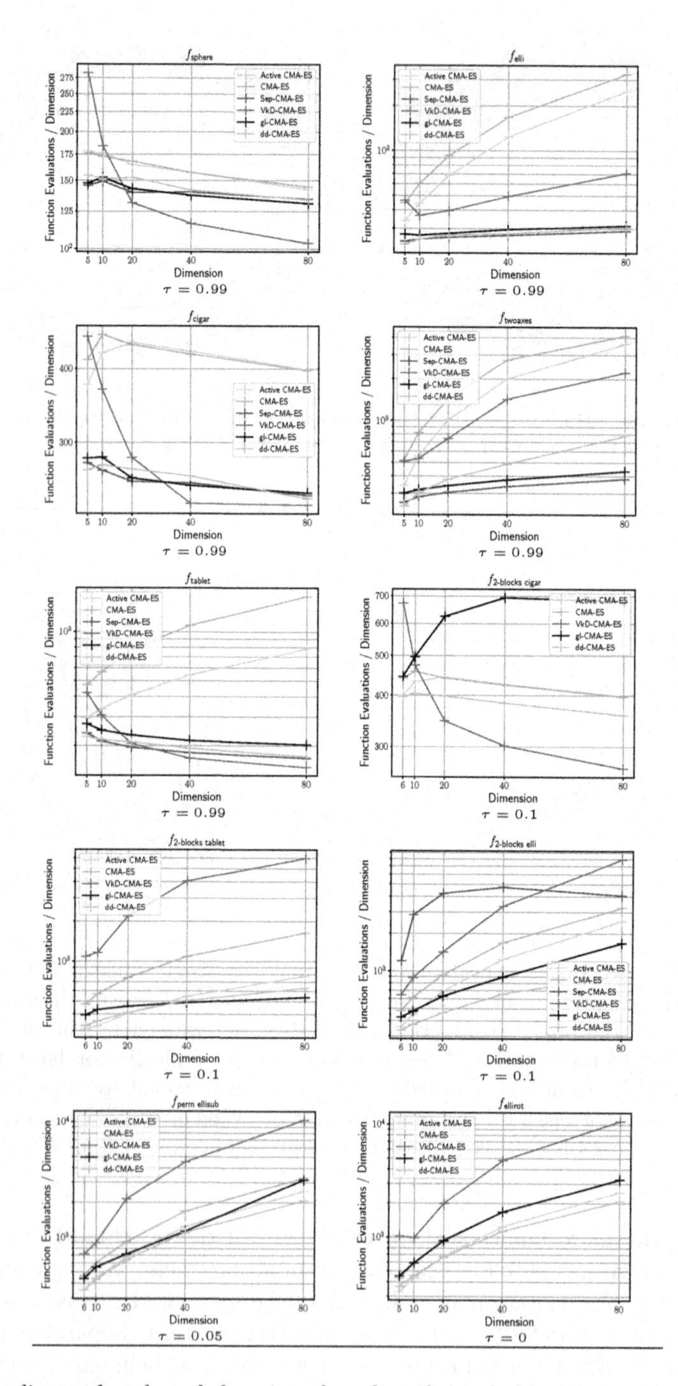

Fig. 7. Scaling on benchmark functions for selected thresholds. The performance measure is the number of function evaluations to reach a target precision 10^{-10} close to the global optimal f value. Average is taken over 10 independent runs of each method.

on the latter function, due to less conservative learning rate values compared to gl-CMA-ES. For both functions, in dimension $n = 6$, gl-CMA-ES and CMA-ES differ by a factor smaller than 1.3 and in dimension $n = 80$, gl-CMA-ES is more than twice as fast as CMA-ES. One exception is the $f_{\text{2-blocks cigar}}$ function, where all other methods outperform gl-CMA-ES, for dimensions $n \geq 10$. This is also the only case of worse performance compared to CMA-ES indicating that the choice of τ is too large. On fully dense non-separable problems such as the rotated Ellipsoid function f_{ellirot}, the value $\tau = 0$ reduces to the default setting of CMA-ES and the performance is identical.

5 Discussion

We proposed a method of l_1 regularization within CMA-ES, attempting to increase the adaptation speed of the search distribution. We investigated its behaviour and showed the gain in convergence speed in representative sparse problems and for selected values of the threshold parameter.

The setting of the threshold is crucial for the richness of the search model and thus for the performance of gl-CMA-ES, and good choices depend on properties of the function to be optimized, for example separability. As a result, a future step for improving this approach is the inclusion of an adaptive mechanism for this parameter, rather than being predefined and static.

Furthermore, in this study we only investigate the performance gain in terms of convergence speed through accelerating the covariance matrix adaptation. No focus has been given on the internal cost of the regularization step using the Graphical Lasso. A possible strategy would be to perform this step in the updated search distribution once every a certain number of iterations while sampling with the same regularized search model in between, in cases where the dimension is large and the computational cost becomes a burden.

Finally, in order to guarantee the positive definiteness of the covariance matrix, only positive recombination weights are used, as mentioned in the algorithm's description. Therefore, another interesting aspect and future step for improvement is to employ negative recombination weights.

Acknowledgement. The PhD thesis of Konstantinos Varelas is funded by the French MoD DGA/MRIS and Thales Land & Air Systems.

References

1. Akimoto, Y., Hansen, N.: Online model selection for restricted covariance matrix adaptation. In: Handl, J., Hart, E., Lewis, P.R., López-Ibáñez, M., Ochoa, G., Paechter, B. (eds.) PPSN 2016. LNCS, vol. 9921, pp. 3–13. Springer, Cham (2016). https://doi.org/10.1007/978-3-319-45823-6_1
2. Akimoto, Y., Hansen, N.: Projection-based restricted covariance matrix adaptation for high dimension. In: Genetic and Evolutionary Computation Conference (GECCO 2016), Denver, United States, pp. 197–204, July 2016

3. Akimoto, Y., Hansen, N.: Diagonal acceleration for covariance matrix adaptation evolution strategies. Evol. Comput., 1–31 (2019)
4. d'Aspremont, A., Banerjee, O., El Ghaoui, L.: First-order methods for sparse covariance selection. SIAM J. Matrix Anal. Appl. **30**(1), 56–66 (2008)
5. Dempster, A.P.: Covariance selection. Biometrics **28**, 157–175 (1972)
6. Fan, J., Liao, Y., Liu, H.: An overview of the estimation of large covariance and precision matrices. Econom. J. **19**(1), C1–C32 (2016)
7. Friedman, J., Hastie, T., Tibshirani, R.: Sparse inverse covariance estimation with the graphical lasso. Biostatistics **9**(3), 432–441 (2008)
8. Griewank, A., Toint, P.: On the unconstrained optimization of partially separable functions. In: Nonlinear Optimization 1981, pp. 301–312. Academic Press (1982)
9. Hansen, N., Ostermeier, A.: Completely derandomized self-adaptation in evolution strategies. Evol. Comput. **9**(2), 159–195 (2001)
10. Jastrebski, G.A., Arnold, D.V.: Improving evolution strategies through active covariance matrix adaptation. In: 2006 IEEE International Conference on Evolutionary Computation, pp. 2814–2821. IEEE (2006)
11. Knight, J.N., Lunacek, M.: Reducing the space-time complexity of the CMA-ES. In: Proceedings of the 9th Annual Conference on Genetic and Evolutionary Computation. GECCO 2007, pp. 658–665. Association for Computing Machinery, New York (2007). https://doi.org/10.1145/1276958.1277097
12. Laska, J., Narayan, M.: skggm 0.2.7: A scikit-learn compatible package for Gaussian and related Graphical Models, July 2017. https://doi.org/10.5281/zenodo.830033
13. Loshchilov, I.: LM-CMA: an alternative to L-BFGS for large scale black-box optimization. Evol. Comput. **25**, 143–171 (2017)
14. Mazumder, R., Hastie, T.: Exact covariance thresholding into connected components for large-scale graphical lasso. J. Mach. Learn. Res. **13**(1), 781–794 (2012)
15. Nocedal, J., Wright, S.: Numerical Optimization. Springer, New York (2006). https://doi.org/10.1007/978-0-387-40065-5
16. Ros, R., Hansen, N.: A simple modification in CMA-ES achieving linear time and space complexity. In: Rudolph, G., Jansen, T., Beume, N., Lucas, S., Poloni, C. (eds.) PPSN 2008. LNCS, vol. 5199, pp. 296–305. Springer, Heidelberg (2008). https://doi.org/10.1007/978-3-540-87700-4_30
17. Varelas, K., et al.: A comparative study of large-scale variants of CMA-ES. In: Auger, A., Fonseca, C.M., Lourenço, N., Machado, P., Paquete, L., Whitley, D. (eds.) PPSN 2018. LNCS, vol. 11101, pp. 3–15. Springer, Cham (2018). https://doi.org/10.1007/978-3-319-99253-2_1

Adaptive Stochastic Natural Gradient Method for Optimizing Functions with Low Effective Dimensionality

Teppei Yamaguchi(✉), Kento Uchida, and Shinichi Shirakawa

Graduate School of Environment and Information Sciences, Yokohama National University, Yokohama, Japan
{yamaguchi-teppei-yc,uchida-kento-nc}@ynu.jp, shirakawa-shinichi-bg@ynu.ac.jp

Abstract. Black-box optimization algorithms, such as evolutionary algorithms, have been recognized as useful tools for real-world applications. Several efficient probabilistic model-based evolutionary algorithms, such as the compact genetic algorithm (cGA) and the covariance matrix adaptation evolution strategy (CMA-ES), can be regarded as a stochastic natural gradient ascent on statistical manifolds. Our baseline algorithm is the adaptive stochastic natural gradient (ASNG) method which automatically adapts the learning rate based on the signal-to-noise ratio (SNR) of the approximated natural gradient. ASNG has shown effectiveness in a practical application, but the convergence speed of ASNG deteriorates on objective functions with low effective dimensionality (LED), where LED means that part of the design variables is ineffective or does not affect the objective value significantly. In this paper, we propose an element-wise adjustment method for the approximated natural gradient based on the element-wise SNR and introduce the proposed adjustment method into ASNG. The proposed method suppresses the natural gradient elements with the low SNRs, helping to accelerate the learning rate adaptation in ASNG. We incorporate the proposed method into the cGA and demonstrate the effectiveness of the proposed method on the benchmark functions of binary optimization.

Keywords: Probabilistic model-based black-box optimization · Natural gradient · Low effective dimensionality · Learning rate adaptation

1 Introduction

A lot of problems in real-world applications, such as the simulation-based optimization in engineering and the hyperparameter optimization in machine learning, are formulated as black-box optimization problems, that is, the gradient of the objective function cannot be accessed. The population-based black-box optimization methods, including evolutionary algorithms, have succeeded in a wide

T. Bäck et al. (Eds.): PPSN 2020, LNCS 12269, pp. 719–731, 2020.
https://doi.org/10.1007/978-3-030-58112-1_50

range of applications. In general, the performance of evolutionary algorithms depends on the choice of the hyperparameter, for example, the population size and learning rate. Tuning such hyperparameters in real-world applications is not realistic because the number of function evaluations is usually limited, and the computational cost for function evaluations is expensive. Therefore, a robust parameter adaptation mechanism is required in practical situations.

Probabilistic model-based evolutionary algorithms [3,5,6] are promising black-box optimization methods that define a parametric probability distribution on the search space and iteratively update the parameters of the distribution to improve the objective function value of the samples generated from the distribution. Several efficient probabilistic model-based evolutionary algorithms, such as the compact genetic algorithm (cGA) [6] or the covariance matrix adaptation evolution strategy (CMA-ES) [5], can be regarded as a stochastic natural gradient ascent on statistical manifolds [9]. In this paper, we focus on the adaptive stochastic natural gradient (ASNG) method [1] that adapts the learning rate in the stochastic natural gradient methods such as the cGA. In the black-box optimization, the natural gradient has to be estimated by the Monte Carlo approximation. ASNG measures the reliability of the estimated natural gradient direction by means of the signal-to-noise ratio (SNR) and tries to keep the SNR around a constant value by controlling the learning rate. The literature [1] shows that ASNG can achieve an efficient and robust optimization in the problem of a one-shot neural architecture search for deep learning. However, the convergence speed of ASNG becomes slow on objective functions with low effective dimensionality (LED), in which part of the variables is ineffective or does not affect the objective value significantly. Particularly, when redundant variables exist, the learning rate becomes smaller than necessary because of the low SNR due to the random walk in the redundant dimensions.

Because LED often appears in high-dimensional problems, including real-world applications [4,7,8], several works tackling LED exist. REMBO [11] projects a high dimensional search space into a low dimensional subspace using a random embedding to solve efficiently high-dimensional problems with LED by Bayesian optimization. REMEDA [10] applies the same idea to the estimation of distribution algorithm. However, the random projection does not reflect the landscape information, and the number of subspace dimensions still remains as a hyperparameter, which should be carefully chosen depending on the target problems.

In this paper, we propose a method for improving ASNG for objective functions with LED. When optimizing the objective functions with LED by ASNG, the reliabilities of the elements of the estimated natural gradient differ. Precisely, the SNR of the estimated natural gradient corresponding to the nonsensitive dimensions becomes small. Our idea is to adjust the update direction of the distribution parameters by exploiting the element-wise SNR of the natural gradient. The proposed method can be regarded as using the element-wise learning rate in ASNG, that is, the larger learning rate is adopted for the large SNR elements.

We incorporate the proposed method into ASNG without breaking its theoretical principal. The experimental results demonstrate that the proposed method, termed ASNG-LED, works efficiently on binary objective functions with LED.

2 Preliminaries

2.1 Stochastic Natural Gradient for Black-Box Optimization

We consider a black-box objective function $f : \mathcal{X} \to \mathbb{R}$ to be maximized on an arbitrary search space $\mathcal{X}$. To realize the black-box optimization, we introduce the technique called *stochastic relaxation* to transform the original problem into a differentiable objective function. Note that the following transformation is the same as the one considered in the information geometric optimization (IGO) framework [9] and natural evolution strategies [12].

Let us consider a parametric family of probability distributions $\mathcal{P} = \{P_\theta : \theta \in \Theta \subseteq \mathbb{R}^{D_\theta}\}$ on $\mathcal{X}$. The transformed objective function is the expectation of f under P_θ, that is,

$$J(\theta) := \int_{x \in \mathcal{X}} f(x) p_\theta(x) \mathrm{d}x = \mathbb{E}_{P_\theta}[f(x)], \tag{1}$$

where p_θ is the density function of P_θ with respect to (w.r.t.) the reference measure $\mathrm{d}x$ on $\mathcal{X}$. For any x, we assume that the log-likelihood $\ln p_\theta(x)$ is differentiable w.r.t. θ, and there exists a sequence of the distributions approaching the Dirac Delta distribution $\boldsymbol{\delta}_x$ around x. Then, the maximization of J has the same meaning as the original problem in the sense that $\sup_{\theta \in \Theta} J(\theta) = \mathbb{E}_{\boldsymbol{\delta}_{x^*}}[f(x)] = f(x^*) = \sup_{x \in \mathcal{X}} f(x)$, where x^* is the global optimal solution.

In this paper, we focus on the special case, as also assumed in [1], that $\mathcal{P}$ is represented by an exponential family of probability distributions whose density function is given as $p_\theta(x) = h(x) \exp(\eta(\theta)^{\mathrm{T}} T(x) - A(\theta))$, where $\eta : \Theta \to \mathbb{R}^{D_\theta}$ is the normal (canonical) parameter, $T : \mathcal{X} \to \mathbb{R}^{D_\theta}$ is the sufficient statistics, and $A(\theta)$ is the normalization factor. To make it simple, we assume $h(x) = 1$, and the parameter of the distribution is represented by $\theta = \mathbb{E}_{p_\theta}[T(x)]$, which is called the expectation parameter. Then, the natural gradient of the log-likelihood is given by $\tilde{\nabla} \ln p_\theta(x) = T(x) - \theta$, and the inverse of the Fisher information matrix is $\mathrm{F}(\theta)^{-1} = \mathbb{E}_{p_\theta}[(T(x) - \theta)(T(x) - \theta)^{\mathrm{T}}]$. We note that several known families of probability distributions, such as the Gaussian distribution and the Bernoulli distribution, are included in the exponential family and can be represented by the expectation parameter.

To maximize J, we consider the update of θ in the metric of the Kullback-Leibler (KL) divergence. Then, the steepest direction is given by the natural gradient direction [2] w.r.t. the Fisher metric defined by $\mathrm{F}(\theta)$. Because the natural gradient cannot be obtained analytically in the black-box optimization scenario, we approximate it by Monte Carlo with independent and identically distributed (i.i.d.) samples $x_1^{(t)} \cdots x_N^{(t)}$ from $P_{\theta^{(t)}}$. Moreover, we apply the utility transformation $u(x_i^{(t)})$ based on the ranking of $x_i^{(t)}$ w.r.t. f-value. As a result, the estimated natural gradient is obtained as

$$G(\theta^{(t)}) = \frac{1}{N}\sum_{i=1}^{N} u(x_i^{(t)})(T(x_i^{(t)}) - \theta^{(t)}). \tag{2}$$

Introducing the learning rate $\epsilon_\theta > 0$, the update rule of $\theta^{(t)}$ reads

$$\theta^{(t+1)} = \theta^{(t)} + \epsilon_\theta G(\theta^{(t)}). \tag{3}$$

2.2 Adaptive Stochastic Natural Gradient (ASNG)

Akimoto et al. [1] developed the adaptive stochastic natural gradient (ASNG) method by introducing a learning rate adaptation mechanism into the stochastic natural gradient method using an exponential family of probability distributions with the expectation parameters. Although the joint optimization of the differentiable and non-differentiable variables is considered in [1], ASNG can apply to a naive black-box optimization without any modification. Here, we explain the outline of ASNG in the black-box optimization scenario.

In ASNG, the learning rate is represented by

$$\epsilon_\theta = \delta_\theta / \|G(\theta^{(t)})\|_{\mathrm{F}(\theta^{(t)})} \tag{4}$$

The update rule (3) with the learning rate in (4) is similar to the trust region method under the KL divergence with the trust region radius δ_θ. The adaptation of δ_θ in ASNG is based on the theoretical insight of the stochastic natural gradient method. According to [1, Theorem 4], the monotonic improvement of J is ensured when it holds

$$D_{\mathrm{KL}}(\theta^{(t+1)}, \theta^{(t)} + \epsilon_\theta \tilde{\nabla} J(\theta^{(t)})) \leq \zeta D_{\mathrm{KL}}(\theta^{(t)} + \epsilon_\theta \tilde{\nabla} J(\theta^{(t)}), \theta^{(t)}) \tag{5}$$

for some $\zeta > 0$ and if $\epsilon_\theta < (\zeta f(x^*) + J(\theta))^{-1}$ with some other mild assumptions, where $D_{\mathrm{KL}}(\theta, \theta')$ is the KL divergence between P_θ and $P_{\theta'}$. ASNG relaxes the condition for monotonic improvement to the improvement over $\tau \propto \delta_\theta^{-1}$ iterations. If ϵ_θ is small enough, it allows the approximation where $\theta^{(t)} \approx \theta^{(t+k)}$ and $G(\theta^{(t+k)})$ is i.i.d. for $k = 0, \cdots, \tau - 1$. Then, approximation of the KL divergence with the Fisher metric allows the transformation of the condition in (5) as

$$\left\| \sum_{k=0}^{\tau-1} \frac{\epsilon_\theta G(\theta^{(t+k)}) - \epsilon_\theta \mathbb{E}[G(\theta^{(t)})]}{\sqrt{\tau}} \right\|_{\mathrm{F}(\theta^{(t)})}^2 \leq \zeta \tau \epsilon_\theta^2 \|\mathbb{E}[G(\theta^{(t)})]\|_{\mathrm{F}(\theta^{(t)})}^2. \tag{6}$$

Then, the limitation of $\tau \to \infty$ in the LHS leads

$$\frac{\|\mathbb{E}[G(\theta^{(t)})]\|_{\mathrm{F}(\theta^{(t)})}^2}{\mathrm{Tr}(\mathrm{Cov}[G(\theta^{(t)})]\mathrm{F}(\theta^{(t)}))} \geq \frac{1}{\zeta\tau} \in \Omega(\delta_\theta). \tag{7}$$

The LHS in (7) is the SNR of the estimated natural gradient measured w.r.t Fisher metric. The above discussion indicates that the SNR should be greater than a constant value proportional to δ_θ.

To estimate the lower bound of the SNR, the following accumulations were proposed in [1]:

$$s^{(t+1)} = (1-\beta)s^{(t)} + \sqrt{\beta(2-\beta)}\mathrm{F}(\theta^{(t)})^{\frac{1}{2}}G(\theta^{(t)}) \tag{8}$$

$$\gamma^{(t+1)} = (1-\beta)^2\gamma^{(t)} + \beta(2-\beta)\|G(\theta^{(t)})\|^2_{\mathrm{F}(\theta^{(t)})}, \tag{9}$$

where $s^{(0)} = \mathbf{0}$ and $\gamma^{(0)} = 0$. Finally, introducing hyperparameters $\alpha > 1$ and $\beta \propto \delta_\theta$, the adaptation of δ_θ is written as

$$\delta_\theta \leftarrow \delta_\theta \exp\left(\beta\left(\frac{\|s^{(t+1)}\|^2}{\alpha} - \gamma^{(t+1)}\right)\right). \tag{10}$$

This adaptation tries to maintain $\|s^{(t+1)}\|^2/\gamma^{(t+1)}$ around α, which makes the lower bound of the SNR proportional to δ_θ. Note that the condition in (7) is satisfied under this design principle.

3 ASNG for Low Effective Dimensionality

ASNG increases the learning rate when the estimated SNR becomes large and decreases it when the SNR becomes small. We can intuitively regard the SNR as a measurement of the reliability of the natural gradient direction. If there are low-effective or redundant variables in the objective function, the variance of the natural gradient elements corresponding to such variables becomes large, resulting in the estimated SNR of $G(\theta^{(t)})$ and the learning rate becoming smaller than necessary. Therefore, the performance of ASNG deteriorates on objective functions with LED. To prevent this, we propose a natural gradient adjustment method for ASNG by using the element-wise SNR.

3.1 Estimation of Element-Wise SNR

Let us denote i-th element of $G(\theta^{(t)})$ as $G_i^{(t)}$ for short. By using a one-hot vector $h_{(i)}$ whose i-th element is one and other elements are zero, we define the SNR of $G_i^{(t)}$ on the Fisher metric as follows:

$$\frac{\|\mathbb{E}[h_{(i)} \circ G(\theta^{(t)})]\|^2_{\mathrm{F}(\theta^{(t)})}}{\mathrm{Tr}(\mathrm{Cov}[h_{(i)} \circ G(\theta^{(t)})]\mathrm{F}(\theta^{(t)}))} = \frac{(\mathbb{E}[G_i^{(t)}])^2}{\mathrm{Var}[G_i^{(t)}]}. \tag{11}$$

To estimate the element-wise SNR, we accumulate $\hat{s}^{(t)}$ and $\hat{\gamma}^{(t)}$ similar to (8) and (9) as

$$\hat{s}^{(t+1)} = \left(1-\hat{\beta}\right)\hat{s}^{(t)} + \sqrt{\hat{\beta}\left(2-\hat{\beta}\right)G\left(\theta^{(t)}\right)} \tag{12}$$

$$\hat{\gamma}^{(t+1)} = (1-\hat{\beta})^2\hat{\gamma}^{(t)} + \hat{\beta}(2-\hat{\beta})G(\theta^{(t)}) \circ G(\theta^{(t)}), \tag{13}$$

where $\hat{\beta} \in (0, 1)$ is a smoothing factor and $\circ$ means the element-wise product. When the learning rate ϵ_θ is small enough, we can consider $\theta^{(t+k)}$ stays around the $\theta^{(t)}$ for $\tau \propto \epsilon_\theta^{-1}$ updates after t iterations. Therefore, the expectations of i-th elements of $\hat{s}^{(t+1)}$ and $\hat{\gamma}^{(t+1)}$ are approximated under mild condition as

$$\mathbb{E}[\hat{s}_i^{(t+1)}] \approx \left(\sum_{k=0}^{\tau-1}(1-\hat{\beta})^k\right)\sqrt{\hat{\beta}(2-\hat{\beta})}\mathbb{E}[G_i^{(t)}] \overset{\tau\to\infty}{\longrightarrow} \sqrt{\frac{2-\hat{\beta}}{\hat{\beta}}}\mathbb{E}[G_i^{(t)}], \tag{14}$$

$$\mathbb{E}[\hat{\gamma}_i^{(t+1)}] \approx \left(\sum_{k=0}^{\tau-1}(1-\hat{\beta})^{2k}\right)\hat{\beta}(2-\hat{\beta})\mathbb{E}[(G_i^{(t)})^2] \overset{\tau\to\infty}{\longrightarrow} \mathbb{E}[(G_i^{(t)})^2]. \tag{15}$$

Moreover, the variance of $\hat{s}_i^{(t+1)}$ is approximated under mild condition as

$$\mathrm{Var}[\hat{s}_i^{(t+1)}] \approx \left(\sum_{k=0}^{\tau-1}(1-\hat{\beta})^{2k}\right)\hat{\beta}(2-\hat{\beta})\mathrm{Var}[G_i^{(t)}] \overset{\tau\to\infty}{\longrightarrow} \mathrm{Var}[G_i^{(t)}]. \tag{16}$$

Therefore, the SNR of the i-th element of $G(\theta^{(t)})$ can be approximated as

$$\frac{(\mathbb{E}[G_i^{(t)}])^2}{\mathrm{Var}[G_i^{(t)}]} \approx \frac{\mathbb{E}[(\hat{s}_i^{(t+1)})^2]/\mathbb{E}[\hat{\gamma}_i^{(t+1)}] - 1}{2\hat{\beta}^{-1} - 1 - \mathbb{E}[(\hat{s}_i^{(t+1)})^2]/\mathbb{E}[\hat{\gamma}_i^{(t+1)}]} \approx \frac{\xi_i^{(t+1)}}{2\hat{\beta}^{-1} - 2 - \xi_i^{(t+1)}} \tag{17}$$

where $\xi_i^{(t+1)} := (\hat{s}_i^{(t+1)})^2/\hat{\gamma}_i^{(t+1)} - 1$.

3.2 Natural Gradient Adjustment Using the Element-Wise SNR

The small value of the element-wise SNR estimated by (17) indicates that the corresponding natural gradient element is unreliable and has a large variance. We control the element-wise strength of $G(\theta^{(t)})$ using the element-wise estimation of the SNR; that is, we suppresses the strength of the natural gradient elements with the low SNRs. Because the denominator in (17) can be approximated by $2\hat{\beta}^{-1}$ when $\hat{\beta}$ is significantly small, the estimation of the SNR is approximately proportional to $\xi_i^{(t)}$. Therefore, we employ a D_θ-dimensional adjustment vector $\sigma^{(t)}$ using $\xi_i^{(t)}$ and define $\sigma^{(t)}$ by the following sigmoid function as

$$\sigma_i^{(t)} = \frac{1}{1 + \exp\left(-\omega \xi_i^{(t)}\right)}, \tag{18}$$

where ω (> 0) is the gain parameter. Then, we adjust the natural gradient direction by

$$H(\theta^{(t)}) = \sigma^{(t)} \circ G(\theta^{(t)}). \tag{19}$$

When we set $\omega = \infty$, the sigmoid function coincides with the step function, and the natural gradient update with $H(\theta^{(t)})$ behaves as a dimensionality reduction method. We use the adjusted natural gradient $H(\theta^{(t)})$ in the proposed method instead of $G(\theta^{(t)})$, that is, the natural gradient $G(\theta^{(t)})$ in (3) and (4) is replaced by $H(\theta^{(t)})$.

3.3 Combination with ASNG

Let us assume $\mathrm{F}(\theta^{(t)})$ is given by a diagonal matrix; for example, the Bernoulli distribution satisfies this assumption. In such a case, we show the proposed method can be combined with ASNG without breaking the theoretical principle described in Sect. 2.2.

From the condition of monotonic improvement in (5), the search efficiency of the proposed method with ASNG is guaranteed in a similar way with that of ASNG. Let us assume $\hat{\beta}$ and ϵ_θ are sufficiently small so that we can consider $\sigma^{(t)}$ and $\theta^{(t)}$ stay at the same point for $\tau \propto \delta_\theta^{-1}$ iterations. Then, we consider the relaxation of (5) in the same way as described in Sect. 2.2, which is given by

$$\left\| \sum_{k=0}^{\tau-1} \frac{\epsilon_\theta H(\theta^{(t+k)}) - \epsilon_\theta \mathbb{E}[G(\theta^{(t)})]}{\sqrt{\tau}} \right\|_{\mathrm{F}(\theta^{(t)})}^2 \leq \zeta \tau \epsilon_\theta^2 \|\mathbb{E}[G(\theta^{(t)})]\|_{\mathrm{F}(\theta^{(t)})}^2 . \tag{20}$$

When taking the limit of τ as τ goes to infinity, the LHS in (20) can be transformed and bounded from upper as

$$\epsilon_\theta^2 \sum_{i=1}^{D_\theta} \mathrm{F}_{ii}(\theta^{(t)}) \left(\mathrm{Var}[H_i(\theta^{(t)})] + \left(\frac{1-\sigma_i^{(t)}}{\sigma_i^{(t)}} \right)^2 \mathbb{E}[H_i(\theta^{(t)})]^2 \right) \tag{21}$$

$$\leq \epsilon_\theta^2 \mathrm{Tr}(\mathrm{Cov}[H(\theta^{(t)})]\mathrm{F}(\theta^{(t)})) + \epsilon_\theta^2 \|\mathbb{E}[H(\theta^{(t)})]\|_{\mathrm{F}(\theta^{(t)})}^2 K^{(t)}, \tag{22}$$

where $K^{(t)} = \sum_{i=1}^{D_\theta} ((1-\sigma_i^{(t)})/\sigma_i^{(t)})^2$. Moreover, because $\sigma_i^{(t)} \leq 1$, we get

$$\|\mathbb{E}[\epsilon_\theta H(\theta^{(t)})]\|_{\mathrm{F}(\theta^{(t)})}^2 \leq \|\mathbb{E}[\epsilon_\theta G(\theta^{(t)})]\|_{\mathrm{F}(\theta^{(t)})}^2 . \tag{23}$$

As a result, we get the sufficient condition of (20) under the limitation of τ as

$$\frac{\|\mathbb{E}[H(\theta^{(t)})]\|_{\mathrm{F}(\theta^{(t)})}^2}{\mathrm{Tr}(\mathrm{Cov}[H(\theta^{(t)})]\mathrm{F}(\theta^{(t)}))} \geq \frac{1}{\zeta\tau - K^{(t)}} \geq \frac{1}{\zeta\tau} \in \Omega(\delta_\theta), \tag{24}$$

where we assume τ satisfies $\tau > K^{(t)}/\zeta$. We note $K^{(t)}$ can be transformed as

$$K^{(t)} = \sum_{i=1}^{D_\theta} \left(\frac{1-\sigma_i^{(t)}}{\sigma_i^{(t)}} \right)^2 = \sum_{i=1}^{D_\theta} \exp\left(-2\omega \left(\frac{(\hat{s}_i^{(t)})^2}{\hat{\gamma}_i^{(t)}} - 1 \right) \right). \tag{25}$$

Here, we consider the case that the accumulation of $\hat{s}_i^{(t)}$ and $\hat{\gamma}_i^{(t)}$ works ideally. Then, replacing $(\hat{s}_i^{(t)})^2$ and $\hat{\gamma}_i^{(t)}$ with their expectations given by (14), (15) and (16) approximates $K^{(t)}$ as

$$K^{(t)} \approx \sum_{i=1}^{D_\theta} \exp\left(-\frac{4\omega(1-\hat{\beta})}{\hat{\beta}} \frac{(\mathbb{E}[G_i^{(t-1)}])^2}{(\mathbb{E}[G_i^{(t-1)}])^2 + \mathrm{Var}[G_i^{(t-1)}]} \right) \leq D_\theta . \tag{26}$$

Algorithm 1: ASNG-LED

Require: $\theta^{(0)}$ {initial distribution parameter}
Require: $\alpha = 1.5$, $\delta_\theta^{\text{init}} = 1$, $N = 2$, $\hat{\beta} = D_\theta^{-1}$
1: $t = 0$, $\Delta = 1$, $\gamma = 0$, $s = \mathbf{0}$, $\hat{\gamma}_i = 0$, $\hat{s}_i = 0$
2: **repeat**
3: set $\delta_\theta = \delta_\theta^{\text{init}}/\Delta$ and $\beta = \delta_\theta / D_\theta^{1/2}$
4: compute $G_\theta(\theta^t)$ and $\sigma^{(t)}$ using (2) and (18)
5: set $H(\theta^{(t)}) = \sigma^{(t)} \circ G(\theta^{(t)})$
6: **for** $i = 1$ to D_θ **do**
7: **if** $G_i^{(t)} \neq 0$ **then**
8: $\hat{s}_i \leftarrow (1-\hat{\beta})\hat{s}_i + \sqrt{\hat{\beta}(2-\hat{\beta})} G_i^{(t)}/|G_i^{(t)}|$
9: $\hat{\gamma}_i \leftarrow (1-\hat{\beta})^2 \hat{\gamma}_i + \hat{\beta}(2-\hat{\beta})$
10: **end if**
11: **end for**
12: compute ϵ_θ and $\theta^{(t+1)}$ by (4) and (3) replacing $G(\theta^{(t)})$ with $H(\theta^{(t)})$
13: $s \leftarrow (1-\beta)s + \sqrt{\beta(2-\beta)} \mathrm{F}(\theta^{(t)})^{\frac{1}{2}} H(\theta^{(t)}) / \|H(\theta^{(t)})\|_{\mathrm{F}(\theta^{(t)})}$
14: $\gamma \leftarrow (1-\beta)^2 \gamma + \beta(2-\beta)$
15: $\Delta \leftarrow \Delta \exp(\beta(\gamma - \|s\|^2/\alpha))$
16: $t \leftarrow t+1$
17: **until** termination conditions are met

Thus, we can expect that $K^{(t)}$ becomes not so large and (24) is established when considering $\tau > D_\theta/\zeta$. Moreover, we can expect that $K^{(t)}$ becomes small when $\hat{\beta}$ is set as the small value.

Motivated from the above discussion, we modify the accumulation rule in (8) and (9) by replacing $G(\theta^{(t)})$ with $H(\theta^{(t)})$. Namely, the learning rate adaptation tries to make the lower bound of the SNR of $H(\theta^{(t)})$ proportional to δ_θ.

3.4 Implementation of ASNG-LED

Referring to [1], we replace $\mathrm{F}(\theta^{(t)})^{\frac{1}{2}} H(\theta^{(t)})$ and $\|H(\theta^{(t)})\|^2_{\mathrm{F}(\theta^{(t)})}$ in the accumulations of $s^{(t+1)}$ and $\gamma^{(t+1)}$ with $\mathrm{F}(\theta^{(t)})^{\frac{1}{2}} H(\theta^{(t)}) / \|H(\theta^{(t)})\|_{\mathrm{F}(\theta^{(t)})}$ and 1 for the stable updates. In the same manner, $\hat{s}_i^{(t+1)}$ and $\hat{\gamma}_i^{(t+1)}$ accumulate $G_i^{(t)}/|G_i^{(t)}|$ and 1 instead of $G_i^{(t)}$ and $|G_i^{(t)}|^2$. Because of this modification, $\hat{s}_i^{(t+1)}$ and $\hat{\gamma}_i^{(t+1)}$ are not updated when $G_i^{(t)} = 0$. We set the sample size N as two and apply the ranking-based utility transformation introduced in [1], where $u(x_1^{(t)}) = 1$ and $u(x_2^{(t)}) = -1$ when $f(x_1^{(t)}) > f(x_2^{(t)})$ (and vice versa), and $u(x_1^{(t)}) = u(x_2^{(t)}) = 0$ when $f(x_1^{(t)}) = f(x_2^{(t)})$. This utility transformation can be generalized for an arbitrary sample size as follows: $u(x_i^{(t)}) = 1$ for best $\lceil N/4 \rceil$ samples, $u(x_i^{(t)}) = -1$ for worst $\lceil N/4 \rceil$ samples and $u(x_i^{(t)}) = 0$ otherwise. The proposed method implemented with ASNG, termed ASNG-LED, is summarized in Algorithm 1.

4 Experiment

4.1 Experimental Setting

We evaluate ASNG-LED on several benchmark functions on D-dimensional binary domain. To simply demonstrate that ASNG-LED works well on functions with LED, we construct the benchmark functions with LED by injecting the redundant variables that do not affect the function value at all. The modified benchmark functions listed in Table 1 have d ($\leq D$) effective dimensions and $D - d$ redundant (ineffective) dimensions. The optimal solutions on these functions are given by vectors whose first d elements are given by 1. These functions become the same as the widely used binary benchmark functions when $D = d$. In LeadingOnes, the effective variables change during the optimization. All the variables of the OneMax function with $D = d$ have an equal contribution to the objective value, whereas the effects of the variables in BinVal significantly differ in each dimension. When $D > d$, these functions have redundant dimensions, and we particularly expect that ASNG-LED works well in such a situation.

Table 1. The benchmark functions used in the experiment. They are D-dimensional functions of which only d dimensions affect the function value.

Name	LeadingOnes	OneMax	BinVal
Definition	$\sum_{i=1}^{d} \prod_{j=1}^{i} x_j$	$\sum_{i=1}^{d} x_i$	$\sum_{i=1}^{d} 2^{i-1} x_i$

In ASNG-LED, we set the strategy parameters as $\alpha = 1.5$, $\hat{\beta} = D_\theta^{-1}$, and $N = 2$. The gain parameter in (18) is set to $\omega = 1$. In our preliminary experiment, we observed that the impact of the setting of ω is not so significant if it is set to around one. We use the multivariate Bernoulli distribution, whose probability mass function is defined as $p_\theta(x) = \prod_{i=1}^{D_\theta} \theta_i^{x_i}(1 - \theta_i)^{1-x_i}$, where $D_\theta = D$, to apply ASNG to binary optimization problems. The natural gradient of the log-likelihood and the Fisher information matrix are given by $\tilde{\nabla} \ln p_\theta(x) = x - \theta$ and a diagonal matrix with diagonal elements equal to $\mathrm{F}_{ii}(\theta) = \theta_i^{-1}(1 - \theta_i)^{-1}$, respectively. We compare the proposed method to ASNG and the cGA. We use the default parameter setting proposed in [1] and set a sample size of two. The cGA can be regarded as a stochastic natural gradient with a sample size $N = 2$, which is derived by applying the family of Bernoulli distributions to the IGO framework. In other words, the cGA is an algorithm without the learning rate adaptation in ASNG. In our experiments, the learning rate of the cGA is fixed as $\epsilon_\theta = D^{-1}$. Moreover, we incorporate the margin of D^{-1} into all methods as done in [1], i.e., the range of θ_i is restricted to $[D^{-1}, 1 - D^{-1}]$, to leave the probability of generating arbitrary binary vectors. We ran 50 independent trials for each method, where all algorithm settings succeeded in finding an optimal solution in all trials.

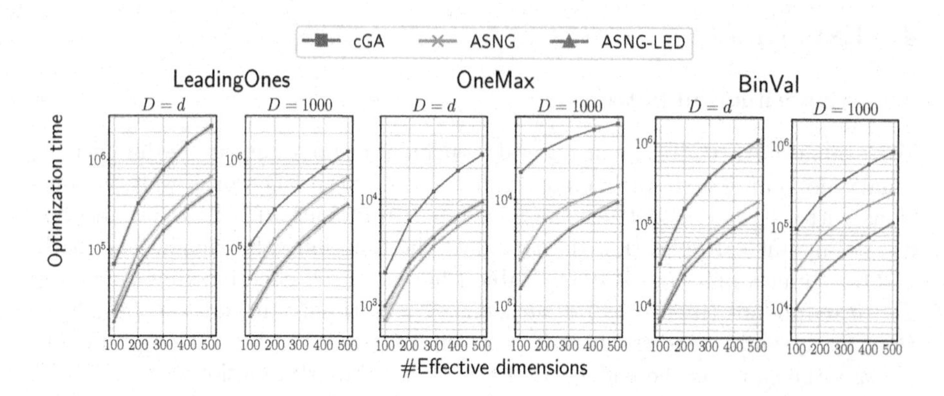

Fig. 1. Comparison of the optimization time (the number of function evaluations) on the benchmark functions with redundant dimensions ($D = 1000$) and without redundant dimensions ($D = d$). The median values and inter-quartile ranges over 50 trials are displayed.

4.2 Experimental Result and Discussion

Optimization Time: We compare the search efficiency of ASNG-LED and baseline algorithms on the functions with and without redundant dimensions. We vary the number of effective dimensions as $d = 100, 200, 300, 400$, and 500, while the numbers of dimensions are set as $D = d$ and $D = 1000$ for each benchmark function. We call the former the benchmark functions without redundant dimensions and the latter the benchmark functions with redundant dimensions.

Figure 1 shows the relation between the number of effective dimensions and the optimization time, which is the number of iterations needed to find one of the optimal solutions. We observe that ASNG-LED can reduce the optimization time on the functions with redundant dimensions compared with ASNG. Also, the optimization time of ASNG-LED does not increase significantly on the functions with redundant dimensions compared with on the ones without redundant dimensions. The performance improvement against ASNG is highlighted in the case of the functions with redundant dimensions. We believe the reason is that ASNG-LED suppressed the strength of the estimated natural gradient corresponding to the redundant dimensions, allowing the adaptation mechanism for the learning rate to work effectively.

Focusing on LeadingOnes and BinVal, ASNG-LED outperforms ASNG on functions both with and without redundant dimensions. This is because part of the dimensions on these functions does not contribute to the function value significantly or at all during the optimization. More precisely, in LeadingOnes, any element following the first 0 in a binary string does not affect the function value at all. On BinVal, the weights for each bit are greatly different, and the lower parts of dimensions do not change the function value significantly. Because of such properties, ASNG-LED successfully adjusts the natural gradient estimates

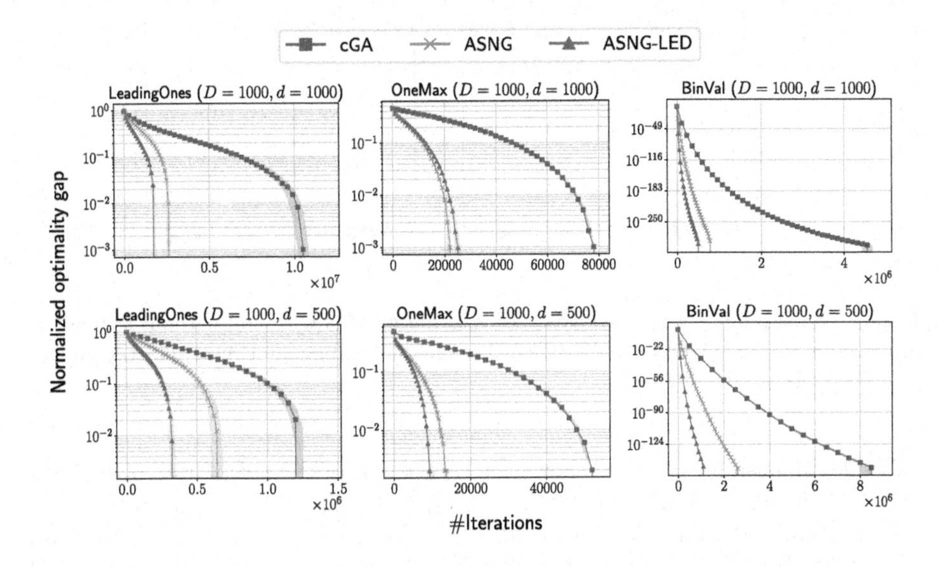

Fig. 2. Transitions of the optimality gap on the benchmark functions without redundant dimensions ($D = d = 1000$) and with redundant dimensions ($d = 500$). The median values and inter-quartile ranges over 50 trials are displayed.

and reduces the undesirable effect of the nonsensitive dimensions for the learning rate adaptation.

For the results of OneMax without redundant dimensions, we observe that ASNG-LED requires more optimization time than ASNG. OneMax is a linear function that has the same weight for each dimension. Therefore, adjusting the element-wise strength of the update direction does not provide a positive effect on the learning rate adaptation on OneMax. We note, however, the optimization time on OneMax is greatly shorter than the one on LeadingOnes and BinVal. Meanwhile, ASNG-LED performs better on OneMax when there is a large number of redundant dimensions. This is because the increased number of redundant dimensions makes the SNR value smaller in the original ASNG, and the performance of ASNG is degraded. In contrast, the accumulated value $\|s\|^2$ for learning rate adaptation in the proposed method does not become so small because of the modification of the accumulation rule.

We note that the performances of the cGA differ between the functions with and without redundant dimensions. This performance difference is caused by the different learning rate and margin settings of the distribution parameters described in Sect. 4.1.

Optimization Process on Functions with and Without Redundant Dimensions: We apply each algorithm to both $D = 1000$ dimensional benchmark functions without redundant dimensions ($d = D$) and ones with $d = 500$

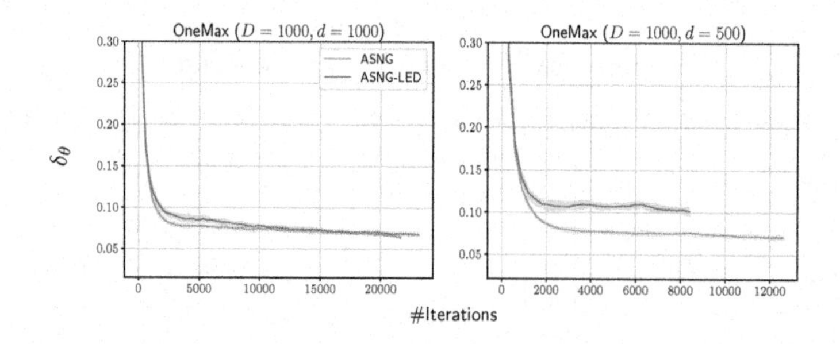

Fig. 3. Transition of the trust-region radius δ_θ in OneMax with and without redundant dimensions in ASNG and ASNG-LED. The median values and inter-quartile ranges over 50 trials are displayed.

effective dimensions. Figure 2 shows the transitions of the normalized optimality gap $\phi(t)$, which is defined as $\phi(t) := 1 - f_{\text{best}}(t)/f_{\text{opt}}$, where f_{opt} and $f_{\text{best}}(t)$ are the function values for an optimal solution and for the best so far solution at t-th iteration, respectively. From Fig. 2, we can observe that ASNG-LED converges faster than ASNG except for OneMax without redundant dimensions ($D = d = 1000$). This implies that the adaptation mechanism of the proposed method works more efficiently than that of ASNG on the functions with redundant or insignificant dimensions.

To demonstrate how the modification in the proposed method accelerates the learning rate adaptation, we show the transitions of the trust-region radius δ_θ on both OneMax without and with redundant dimensions in Fig. 3. We observe that the transitions of the trust-region radius δ_θ of ASNG and ASNG-LED are almost the same in OneMax without redundant dimensions, while δ_θ of ASNG-LED is larger than δ_θ of ASNG in OneMax with redundant dimensions. Because the exact natural gradient elements corresponding to the redundant dimensions are zero, the estimates of such elements behave like a random walk. Therefore, $\|s\|^2$ becomes small against γ if the redundant dimensions exist, resulting in an unnecessarily small learning rate in ASNG. On the other hand, ASNG-LED tries to suppress the influence of redundant dimensions on the SNR value to avoid unnecessarily small learning rate on the adaptation mechanism.

5 Conclusion

We have proposed a method to improve the performance of ASNG on objective functions with LED. The proposed ASNG-LED adjusts the estimated natural gradient based on the elemental-wise SNR estimation. We confirmed the proposed adjustment can be combined with ASNG without breaking the theoretical aspect. We implemented ASNG-LED by applying the Bernoulli distribution and evaluated the performance on several benchmark functions on binary domain. The experimental results showed that ASNG-LED could accelerate the learning rate adaptation in ASNG and outperform the original ASNG on the functions with LED. In future work, ASNG-LED with the Gaussian distribution for continuous optimization should be implemented and evaluated. Also, the effectiveness of ASNG-LED should be verified on more realistic problems such as hyperparameter optimization and feature selection. The limitation of ASNG-LED is that it assumes the irrelevant directions are aligned with the axes, i.e., the performance of ASNG-LED will degrade by the rotation of the coordinate system. Making ASNG-LED rotational invariant is another important future work.

Acknowledgments. This work is partially supported by the SECOM Science and Technology Foundation and JSPS KAKENHI Grant Number JP20H04240.

References

1. Akimoto, Y., Shirakawa, S., Yoshinari, N., Uchida, K., Saito, S., Nishida, K.: Adaptive stochastic natural gradient method for one-shot neural architecture search. In: Proceedings of the 36th International Conference on Machine Learning (ICML), pp. 171–180 (2019)
2. Amari, S.: Natural gradient works efficiently in learning. Neural Comput. **10**, 251–276 (1998)
3. Baluja, S., Caruana, R.: Removing the genetics from the standard genetic algorithm. In: Proceedings of the 12th International Conference on Machine Learning (ICML), pp. 38–46 (1995)
4. Bergstra, J., Bengio, Y.: Random search for hyper-parameter optimization. J. Mach. Learn. Res. **13**, 281–305 (2012)
5. Hansen, N., Müller, S.D., Koumoutsakos, P.: Reducing the time complexity of the derandomized evolution strategy with covariance matrix adaptation (CMA-ES). Evol. Comput. **11**(1), 1–18 (2003)
6. Harik, G.R., Lobo, F.G., Goldberg, D.E.: The compact genetic algorithm. IEEE Trans. Evol. Comput. **3**, 287–297 (1999)
7. Hutter, F., Hoos, H., Leyton-Brown, K.: An efficient approach for assessing hyperparameter importance. In: Proceedings of the 31st International Conference on Machine Learning (ICML), pp. 754–762 (2014)
8. Lukaczyk, T., Constantine, P., Palacios, F., Alonso, J.: Active subspaces for shape optimization. In: Proceedings of the 10th AIAA Multidisciplinary Design Optimization Conference, pp. 1–18 (2014)
9. Ollivier, Y., Arnold, L., Auger, A., Hansen, N.: Information-geometric optimization algorithms: a unifying picture via invariance principles. J. Mach. Learn. Res. **18**, 1–65 (2017)
10. Sanyang, M.L., Kabán, A.: REMEDA: random embedding EDA for optimising functions with intrinsic dimension. In: Handl, J., Hart, E., Lewis, P.R., López-Ibáñez, M., Ochoa, G., Paechter, B. (eds.) PPSN 2016. LNCS, vol. 9921, pp. 859–868. Springer, Cham (2016). https://doi.org/10.1007/978-3-319-45823-6_80
11. Wang, Z., Hutter, F., Zoghi, M., Matheson, D., de Freitas, N.: Bayesian optimization in a billion dimensions via random embeddings. J. Artif. Intell. Res. **55**, 361–387 (2016)
12. Wierstra, D., Schaul, T., Glasmachers, T., Sun, Y., Peters, J., Schmidhuber, J.: Natural evolution strategies. J. Mach. Learn. Res. **15**, 949–980 (2014)

Author Index

Printed in the United States
By Bookmasters